Vector Optimization

Series editor
Johannes Jahn
Erlangen, Germany

The series in Vector Optimization contains publications in various fields of optimization with vector-valued objective functions, such as multiobjective optimization, multi criteria decision making, set optimization, vector-valued game theory and border areas to financial mathematics, biosystems, semidefinite programming and multiobjective control theory. Studies of continuous, discrete, combinatorial and stochastic multiobjective models in interesting fields of operations research are also included. The series covers mathematical theory, methods and applications in economics and engineering. These publications being written in English are primarily monographs and multiple author works containing current advances in these fields.

More information about this series at http://www.springer.com/series/8175

Qamrul Hasan Ansari • Elisabeth Köbis •
Jen-Chih Yao

Vector Variational Inequalities and Vector Optimization

Theory and Applications

Springer

Qamrul Hasan Ansari
Department of Mathematics
Aligarh Muslim University
Aligarh, India

Elisabeth Köbis
Institute of Mathematics
Martin Luther University Halle-Wittenberg
Halle, Germany

Jen-Chih Yao
Center for General Education
China Medical University
Taichung, Taiwan

ISSN 1867-8971 ISSN 1867-898X (electronic)
Vector Optimization
ISBN 978-3-319-63048-9 ISBN 978-3-319-63049-6 (eBook)
DOI 10.1007/978-3-319-63049-6

Library of Congress Control Number: 2017951114

Printed on acid-free paper

This Springer imprint is published by Springer Nature
The registered company is Springer International Publishing AG
The registered company address is: Gewerbestrasse 11, 6330 Cham, Switzerland

Preface

Going back to the groundbreaking works by Edgeworth (1881) and Pareto (1906), the notion of optimality in multiobjective optimization is an efficient tool for describing optimal solutions of real-world problems with conflicting criteria. This branch of optimization has formally started with the pioneering work by Kuhn and Tucker (1951). The concept of multiobjective optimization is further generalized from finite-dimensional spaces to vector spaces leading to the field of vector optimization. This theory has bourgeoned tremendously due to rich application fields in economics, management science, engineering design, etc.

A powerful tool to study vector optimization problems is the theory of vector variational inequalities which was started with the fundamental work of F. Giannessi in 1980, where he extended classical scalar variational inequalities to the vector setting. Later, he has shown equivalence between optimal solutions of vector optimization problems with differentiable convex objective function and solutions of vector variational inequalities of Minty type.

It is well known that many practical equilibrium problems with vector payoff can be formulated as vector variational inequalities. In the last two decades, extensive research has been devoted to the existence theory of their solutions.

The objective of this book is to present a mathematical theory of vector optimization, vector variational inequalities, and vector equilibrium problems. The well-posedness and sensitivity analysis of vector equilibrium problems are also studied. The reader is expected to be familiar with the basic facts of linear algebra, functional analysis, optimization, and convex analysis.

The outline of the book is as follows. Chapter 1 collects basic notations and results from convex analysis, functional analysis, set-valued analysis, and fixed point theory for set-valued maps. A brief introduction to variational inequalities and equilibrium problems is also presented. Chapter 2 gives an overview on analysis over cones, including continuity and convexity of vector-valued functions. Several notions for solutions of vector optimization problems are presented in Chap. 3. Classical linear and nonlinear scalarization methods for solving vector optimization problems are studied in Chap. 4. Chapter 5 is devoted to the vector variational inequalities and existence theory for their solutions. The relationship between a vector variational inequality and a vector optimization problem with smooth objective function is given. Chapter 6 deals with scalarization methods for vector variational inequalities. Such scalarization methods are used to study several existence results for solutions of vector variational inequalities. In Chap. 7, we consider nonsmooth vector variational inequalities defined by means of a bifunction and present several existence results for their solutions. The relationship between nonsmooth vector variational inequalities and vector optimization problems in which the objective function is not necessarily differentiable but has some kind of generalized directional derivative is discussed. Chapter 8 presents vector variational inequalities for set-valued maps, known as generalized vector variational inequalities, and gives several existence results for their solutions. It is shown that the generalized vector variational inequalities provide the optimal solutions of nonsmooth vector optimization problems. Chapter 9 is devoted to the detailed study of vector equilibrium problems, e.g., existence results, duality, and sensitivity analysis. It is worth mentioning that the vector equilibrium problems include vector variational inequalities, nonsmooth vector variational inequalities, and vector optimization problems as special cases. Chapter 10 deals with vector equilibrium problems defined by means of a set-valued bifunction, known as generalized vector equilibrium problems. The generalized vector equilibrium problems include generalized vector variational inequalities and vector optimization problems with nonsmooth objective function as special cases. The existence of solutions, duality, and sensitivity analysis of generalized vector equilibrium problems are studied in detail.

We would like to take this opportunity to express our most sincere thanks to Kathrin Klamroth, Anita Schöbel, and Christiane Tammer for their support and collaboration. The second author is truly grateful to her husband Markus Köbis and her parents for patience and encouragement.

Moreover, we are thankful to Johannes Jahn for encouraging and supporting our plan to write this monograph. We are grateful to Christian Rauscher, Senior Editor, Springer, for taking a keen interest in publishing this monograph.

This book is dedicated to our families. We are grateful to them for their support and understanding.

Finally, we thank our coauthors for their support, understanding, and hard work for this fruitful collaboration. We are also grateful to all researchers whose work is cited in this monograph.

Any comment on this book will be accepted with sincere thanks.

Aligarh, India — Qamrul Hasan Ansari
Halle, Germany — Elisabeth Köbis
Taichung, Taiwan — Jen-Chih Yao

Contents

List of Notations and Symbols

$[A]$	Linear hull of A
A^c	Complement of A
$\text{aff}(A)$	Affine hull of A
$\text{bd}(A)$	Boundary of A
$B_r(x)$	Open ball with centered at x and radius r
$B_r[x]$	Closed ball with centered at x and radius r
$\mathbb{C}$	Set of complex numbers
C_∞	Asymptotic cone or recession cone
C°	Polar cone of C
C^*	Dual cone of C
$C^\#$	Quasi-interior of a dual cone C^* of C
C_+^*	Strict dual cone of C
$\text{cl}(A)$ or $\overline{A}$	Closure of A
$\text{cl}_A B$	Closure of B in A
$\text{co}(A)$	Convex hull of A
$\text{cone}(A)$	Conic hull of A
$\text{cor}(A)$	Core of A (or algebraic interior of A)
$d_A(x)$	Distance function from a point x to set A
$\mathcal{D}(A)$	Image of set A under $\mathcal{D}$
$\langle Df(\bar{x}); v\rangle$	Gâteaux derivative of f at $\bar{x}$ in the direction v
$D_+f(\bar{x}; v)$	Lower Dini directional derivative of f at $\bar{x}$ in direction v
$D^+f(\bar{x}; v)$	Upper Dini directional derivative of f at $\bar{x}$ in direction v
$\text{dom}\, f$	Domain of f
$\mathbb{E}(A, C)$	Set of efficient elements of the set A with respect to C
$\text{epi}(f)$	Epigraph of f
$f^\circ(\bar{x}; v)$	Clarke generalized derivative of f at $\bar{x}$ in direction v
f^*	Conjugate of f
$\langle f'(\bar{x}); v\rangle$	Directional derivative of f at $\bar{x}$ in the direction v
$F(A)$	Image set of A under F
$\mathscr{F}(A)$	Family of all nonempty finite subsets of A

$\mathcal{F}(X)$	Family of all closed subsets of a topological space X
$\mathcal{G}(\mathcal{F})$	Graph of F
$\mathrm{int}(A)$	Interior of A
$\mathrm{int}_A(B)$	Interior of B in A
K_{eff}	Efficient solutions set
$K_{\text{s-eff}}$	Strongly efficient solution set
$K_{\text{w-eff}}$	Weakly efficient solution set
$\mathcal{L}(X,Y)$	Space of continuous linear maps from X to Y
$\mathbb{N}$	Set of natural numbers
$N_\varepsilon(x)$	Neighborhood of x
$\mathcal{P}(Y)$ or 2^Y	Power set of Y
$\mathbb{PE}(A,C)$	Set of properly efficient elements of the set A with respect to C
$\mathbb{Q}$	Set of rational numbers
$\mathbb{R}$	Set of real numbers
$\overline{\mathbb{R}}$	Extended real line $\mathbb{R} \cup \{\pm\infty\}$
$\mathbb{R}_+$	$[0,\infty)$
$\mathbb{R}^n$	n-dimensional Euclidean space
$\mathbb{R}^n_+$	Nonnegative orthant in $\mathbb{R}^n$
$\mathrm{relb}(A)$ or $\mathrm{rb}(A)$	Relative boundary of a set A
$\mathrm{relint}(A)$	Relative interior of a set A
$\mathbb{SE}(A,C)$	Set of strictly efficient elements of the set A with respect to C
$\mathcal{S}^n_+$	Cone of real symmetric positive semidefinite matrices
$\top$	Transpose
$T(A,\bar{x})$	Contingent cone of A at $\bar{x}$
$\xrightarrow{w^*}$	Weak* convergence
$\mathbb{WE}(A,C)$	Set of weakly efficient elements of the set A with respect to C
$\|x\|$	Norm of vector x
Y^*	Topological dual space of Y
$\mathcal{Y}$	Feasible objective region $f(K)$ of a vector optimization problem
$\mathcal{Y}_{\mathrm{eff}}$	Set of efficient elements
$\mathcal{Y}_{\text{s-eff}}$	Set of strongly efficient elements
$\mathcal{Y}_{\text{w-eff}}$	Set of weakly efficient elements $\mathbb{WE}(f(K),C)$
∇f	Gradient of f
$\emptyset$	Empty set
$\langle \cdot,\cdot \rangle$	Duality pairing between Y and its topological dual Y^*
$\leq_C$	Partial ordering induced by the cone C
$\|\cdot\|_*$	Norm in the dual space Y^*
$\|\cdot\|_p$	l_p norm
$\partial f(\bar{x})$	Subdifferential of f at $\bar{x}$
$\partial^c f(\bar{x})$	Clarke generalized subdifferential
$\Pi(A)$	Family of all nonempty subsets of A

Acronyms

DIWVVIP	Dual Implicit Weak Vector Variational Problem
EP	Equilibrium Problem
EPs	Equilibrium Problems
IVP	Implicit Variational Problem
IWVVP	Implicit Weak Vector Variational Problem
IWVVPs	Implicit Weak Vector Variational Problems
MVEP	Minty Vector Equilibrium Problem
MSVEP	Minty Strong Vector Equilibrium Problem
MWVEP	Minty Weak Vector Equilibrium Problem
SNVVIP	Strongly Nonlinear Vector Variational Inequality Problem
SVEP	Strong Vector Equilibrium Problem
VEP	Vector Equilibrium Problem
VEPs	Vector Equilibrium Problems
VO	Vector Optimization
VOP	Vector Optimization Problem
VVI	Vector Variational Inequality
VVIs	Vector Variational Inequalities
VVIP	Vector Variational Inequality Problem
VVIPs	Vector Variational Inequality Problems
WVEP	Weak Vector Equilibrium Problem

Chapter 1
Preliminaries

This chapter deals with basic definitions from convex analysis and nonlinear analysis, such as convex sets and cones, convex functions and their properties, generalized derivatives, and continuity for set-valued maps. We also gather some known results from fixed point theory for set-valued maps, namely, Nadler's fixed point theorem, Fan-KKM lemma and its generalizations, Fan section lemma and its generalizations, Browder fixed point theorem and its generalizations, maximal element theorems and Kakutani fixed point theorem. A brief introduction of scalar variational inequalities, nonsmooth variational inequalities, generalized variational inequalities and equilibrium problems is given.

1.1 Convex Sets and Cones

Throughout the book, all vector spaces are assumed to be defined over the field of real numbers, and we adopt the following notations.

We denote by $\mathbb{R}$, $\mathbb{Q}$ and $\mathbb{N}$ the set of all real numbers, rational numbers and natural numbers, respectively. The interval $[0, \infty)$ is denoted by $\mathbb{R}_+$. We denote by $\mathbb{R}^n$ the n-dimensional Euclidean space and by $\mathbb{R}^n_+$ the nonnegative orthant in $\mathbb{R}^n$. The zero element in a vector space will be denoted by $\mathbf{0}$. Let A be a nonempty set. We denote by 2^A (respectively, $\Pi(A)$) the family of all subsets (respectively, nonempty subsets) of A and by $\mathscr{F}(A)$ the family of all nonempty finite subsets of A. If A and B are nonempty subsets of a topological space X such that $B \subseteq A$, we denote by $\text{int}_A(B)$ (respectively, $\text{cl}_A(B)$) the interior (respectively, closure) of B in A. We also denote by $\text{int}(A)$, $\text{cl}(A)$ (or $\overline{A}$), and $\text{bd}(A)$ the interior of A in X, the closure of A in X, and the boundary of A, respectively. Also, we denote by A^c the complement of the set A. If X and Y are topological vector spaces, then we denote by $\mathcal{L}(X, Y)$ the space of all continuous linear functions from X to Y.

Q.H. Ansari et al., *Vector Variational Inequalities and Vector Optimization*,
Vector Optimization, DOI 10.1007/978-3-319-63049-6_1

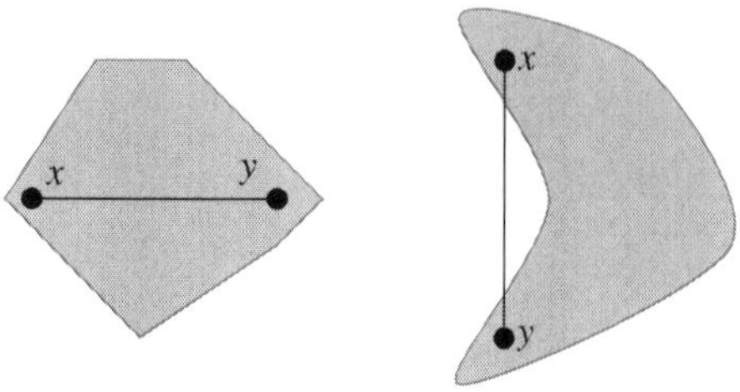

Fig. 1.1 Illustration of a convex set and of a nonconvex set, respectively

Definition 1.1 Let X be a vector space, and x and y be distinct points in X. The set $L = \{z : z = \lambda x + (1 - \lambda)y \text{ for all } \lambda \in \mathbb{R}\}$ is called the *line through x and y*.

The set $[x, y] = \{z : z = \lambda x + (1 - \lambda)y \text{ for } 0 \leq \lambda \leq 1\}$ is called a *line segment* with the endpoints x and y.

Definition 1.2 A subset W of a vector space X is said to be a *subspace* if for all $x, y \in W$ and $\lambda, \mu \in \mathbb{R}$, we have $\lambda x + \mu y \in W$.

Geometrically speaking, a subset W of X is a subspace of X if for all $x, y \in W$, the plane through the origin, x and y lies in W.

Definition 1.3 A subset M of a vector space X is said to be an *affine set* if for all $x, y \in M$ and $\lambda, \mu \in \mathbb{R}$ such that $\lambda + \mu = 1$ imply that $\lambda x + \mu y \in M$, that is, for all $x, y \in M$ and $\lambda \in \mathbb{R}$, we have $\lambda x + (1 - \lambda)y \in M$.

Geometrically speaking, a subset M of X is an affine set if it contains the whole line through any two of its points.

Definition 1.4 A subset K of a vector space X is said to be a *convex set* if for all $x, y \in K$ and $\lambda, \mu \geq 0$ such that $\lambda + \mu = 1$ imply that $\lambda x + \mu y \in K$, that is, for all $x, y \in K$ and $\lambda \in [0, 1]$, we have $\lambda x + (1 - \lambda)y \in K$.

Geometrically speaking, a subset K of X is *convex* if it contains the whole line segment with endpoints through any two of its points (see Fig. 1.1).

Definition 1.5 A subset C of a vector space X is said to be a *cone* if for all $x \in C$ and $\lambda \geq 0$, we have $\lambda x \in C$.

A subset C of X is said to be a *convex cone* if it is convex and a cone; that is, for all $x, y \in K$ and $\lambda, \mu \geq 0$ imply that $\lambda x + \mu y \in C$ (see Fig. 1.2 and 1.3).

Remark 1.1 If C is a cone, then $\mathbf{0} \in C$. In the literature, it is mostly assumed that the cone has its apex at the origin. This is the reason why $\lambda \geq 0$ is chosen in the definition of a cone. However, some references define a set $C \subset X$ to be a cone if $\lambda x \in C$ for all $x \in C$ and $\lambda > 0$. In this case, the apex of the "shifted" cone may not be at the origin, or $\mathbf{0}$ may not belong to C.

Remark 1.2 It is clear from the above definitions that every subspace is an affine set as well as a convex cone, and every affine set and every convex cone are convex. But the converse of these statements may not be true in general.

Evidently, the empty set, each singleton set $\{x\}$ and the whole space X are all both affine and convex. In $\mathbb{R}^n$, straight lines, circular discs, ellipses and interior of

Fig. 1.2 A convex cone

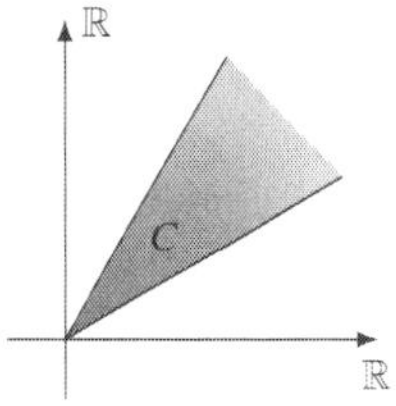

Fig. 1.3 A convex cone in $\mathbb{R}^3$

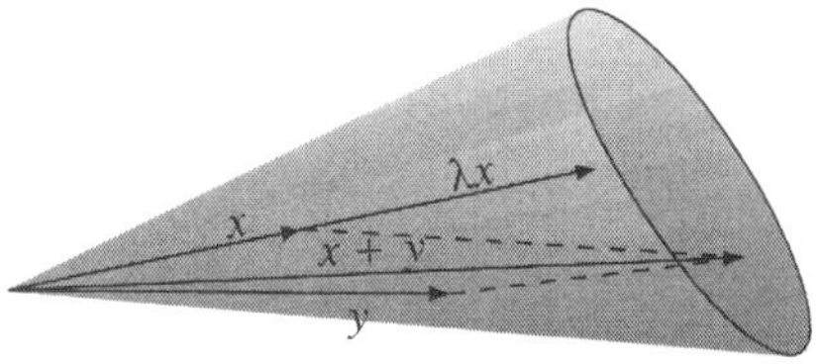

Fig. 1.4 A cone which is not convex

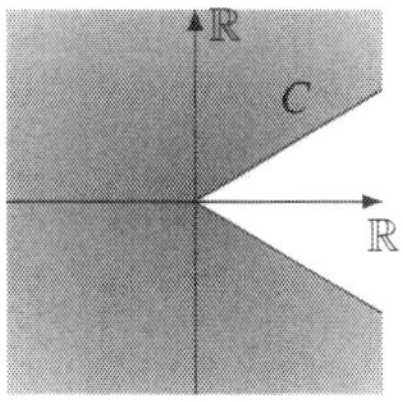

triangles are all convex. A ray, which has the form $\{x_0 + \lambda v : \lambda \geq 0\}$, where $v \neq \mathbf{0}$, is convex, but not affine.

Remark

(a) A cone C may or may not be convex (see Figs. 1.2 - 1.4).
(b) A cone C may be open, closed or neither open nor closed.
(c) A set C is a convex cone if it is both convex as well as a cone.
(d) If C_1 and C_2 are convex cones, then $C_1 \cap C_2$ and $C_1 + C_2$ are also convex cones.

Definition 1.6 Let X be a vector space. Given $x_1, x_2, \ldots, x_m \in X$, a vector $x = \lambda_1 x_1 + \lambda_2 x_2 + \cdots + \lambda_m x_m$ is called

(a) a *linear combination* of $x_1, x_2, \ldots, x_m$ if $\lambda_i \in \mathbb{R}$ for all $i = 1, 2, \ldots, m$;
(b) an *affine combination* of $x_1, x_2, \ldots, x_m$ if $\lambda_i \in \mathbb{R}$ for all $i = 1, 2, \ldots, m$ with $\sum_{i=1}^{m} \lambda_i = 1$;
(c) a *convex combination* of $x_1, x_2, \ldots, x_m$ if $\lambda_i \geq 0$ for all $i = 1, 2, \ldots, m$ with $\sum_{i=1}^{m} \lambda_i = 1$;
(d) a *cone combination* of $x_1, x_2, \ldots, x_m$ if $\lambda_i \geq 0$ for all $i = 1, 2, \ldots, m$.

A set K is a subspace, affine, convex or a cone if it is closed under linear, affine, convex or cone combination, respectively, of points of K.

Theorem 1.1 *A subset K of a vector space X is convex (respectively, subspace, affine, convex cone) if and only if every convex (respectively, linear, affine, cone) combination of points of K belongs to the set K.*

Proof Since a set that contains all convex combinations of its points is obviously convex, we only consider K is convex and prove that it contains any convex combination of its points, that is, if K is convex and $x_i \in K$, $\lambda_i \geq 0$ for all $i = 1, 2, \ldots, m$ with $\sum_{i=1}^{m} \lambda_i = 1$, then we have to show that $\sum_{i=1}^{m} \lambda_i x_i \in K$. We prove this by induction on the number m of points of K occurring in a convex combination. If $m = 1$, the assertion is simply $x_1 \in K$ implies $x_1 \in K$, evidently true. If $m = 2$, then $\lambda_1 x_1 + \lambda_2 x_2 \in K$ for $\lambda_i \geq 0$, $i = 1, 2$, $\sum_{i=1}^{2} \lambda_i = 1$, holds because K is convex. Now suppose that the result is true for m. Then (for $\lambda_{m+1} \neq 1$)

$$\begin{aligned}
\sum_{i=1}^{m+1} \lambda_i x_i &= \sum_{i=1}^{m} \lambda_i x_i + \lambda_{m+1} x_{m+1} \\
&= \sum_{i=1}^{m} (1 - \lambda_{m+1}) \frac{\lambda_i x_i}{1 - \lambda_{m+1}} + \lambda_{m+1} x_{m+1} \\
&= (1 - \lambda_{m+1}) \sum_{i=1}^{m} \frac{\lambda_i}{1 - \lambda_{m+1}} x_i + \lambda_{m+1} x_{m+1} \\
&= (1 - \lambda_{m+1}) \sum_{i=1}^{m} \mu_i x_i + \lambda_{m+1} x_{m+1},
\end{aligned}$$

where $\mu_i = \dfrac{\lambda_i}{(1 - \lambda_{m+1})}$, $i = 1, 2, \ldots, m$. But then $\mu_i \geq 0$ for $i = 1, 2, \ldots, m$ and

$$\sum_{i=1}^{m} \mu_i = \frac{\sum_{i=1}^{m} \lambda_i}{1 - \lambda_{m+1}} = \frac{1 - \lambda_{m+1}}{1 - \lambda_{m+1}} = 1,$$

so by the result for m, $y = \sum_{i=1}^{m} \mu_i x_i \in K$. Immediately, by convexity of K, we have

$$\sum_{i=1}^{m+1} \lambda_i x_i = (1 - \lambda_{m+1}) y + \lambda_{m+1} x_{m+1} \in K.$$

The proof for subspace, affine and convex cone cases follows exactly the same pattern. □

Remark 1.3

(a) The intersection of any number of convex sets (respectively, subspaces, affine sets, convex cones) is a convex set (respectively, subspace, affine set, convex cone).
(b) The union of any number of convex sets need not be convex.
(c) For $i \in \mathbb{N}$, let K_i be convex. If $K_i \subseteq K_{i+1}$, $i \in \mathbb{N}$, then $\bigcup_{i=1}^{\infty} K_i$ is convex.

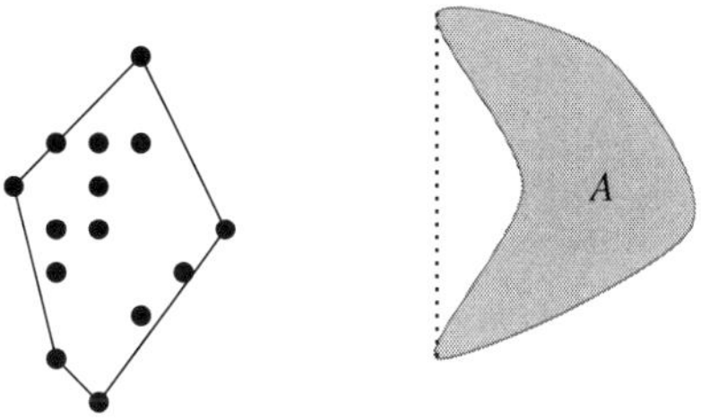

Fig. 1.5 Illustration of the convex hull of 14 points and a convex hull of a set A

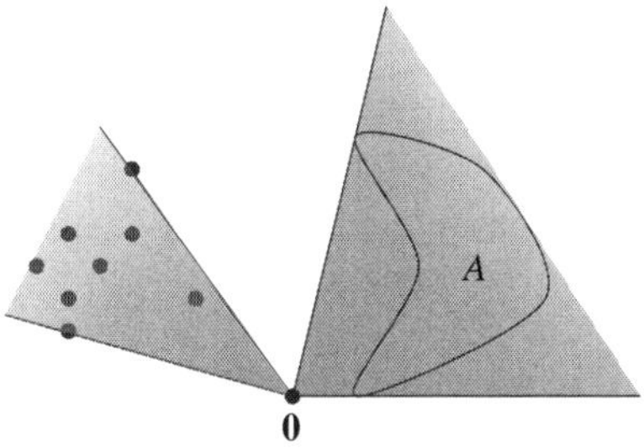

Fig. 1.6 Conic hull of 8 points and the cone generated by the set A

(d) If K_1 and K_2 are convex subsets of a vector space X and $\alpha \in \mathbb{R}$, then $K_1 + K_2 = \{x + y : x \in K_1,\ y \in K_2\}$ and $\alpha K_1 = \{\alpha x : x \in K_1\}$ are convex sets.
(e) A subset K of a vector space X is convex if and only if $(\lambda + \mu)K = \lambda K + \mu K$ for all $\lambda \geq 0$, $\mu \geq 0$.

Definition 1.7 Let A be a nonempty subset of a vector space X. The intersection of all convex sets (respectively, subspaces, affine sets) containing A is called a *convex hull* (respectively, *linear hull, affine hull*) of A, and it is denoted by $\text{co}(A)$ (respectively, $[A]$, $\text{aff}(A)$) (see Fig. 1.5). Similarly, the intersection of all convex cones containing A is called a *conic hull* of A, and it is denoted by $\text{cone}(A)$ (see Fig. 1.6).

By Remark 1.3 (a), the convex (respectively, affine, conic) hull is a convex set (respectively, affine set, convex cone). In fact, $\text{co}(A)$ (respectively, $\text{aff}(A)$, $\text{cone}(A)$) is the smallest convex set (respectively, affine set, convex cone) containing A.

The cone $\text{cone}(A)$ can also be written as

$$\text{cone}(A) = \{x \in X : x = \lambda y \text{ for some } \lambda \geq 0 \text{ and some } y \in A\}.$$

It is also called a *cone generated* by A (see Fig. 1.6).

Theorem 1.2 *Let A be a nonempty subset of a vector space X. Then $x \in \text{co}(A)$ if and only if there exist x_i in A and $\lambda_i \geq 0$, for $i = 1, 2, \ldots, m$, for some positive integer m, where $\sum_{i=1}^{m} \lambda_i = 1$ such that $x = \sum_{i=1}^{m} \lambda_i x_i$.*

Proof Since $\text{co}(A)$ is a convex set containing A, therefore, from Theorem 1.1, every convex combination of its points lies in it, that is, $x \in \text{co}(A)$.

Conversely, let $K(A)$ be the set of all convex combinations of elements of A. We claim that the set

$$K(A) = \left\{ \sum_{i=1}^{m} \lambda_i x_i : x_i \in A,\ \lambda_i \geq 0, i = 1, 2, \ldots, m,\ \sum_{i=1}^{m} \lambda_i = 1,\ m \geq 1 \right\}$$

is convex. Indeed, consider $y = \sum_{i=1}^{m} \lambda_i y_i$ and $z = \sum_{j=1}^{\ell} \mu_j z_j$ where $y_i \in A$, $\lambda_i \geq 0$, $i = 1, 2, \ldots, m$, $\sum_{i=1}^{m} \lambda_i = 1$ and $z_j \in A$, $\mu_j \geq 0, j = 1, 2, \ldots, \ell$, $\sum_{j=1}^{\ell} \mu_j = 1$, and let $0 \leq \lambda \leq 1$. Then

$$\lambda y + (1-\lambda) z = \sum_{i=1}^{m} \lambda \lambda_i y_i + \sum_{j=1}^{\ell} (1-\lambda) \mu_j z_j,$$

where $\lambda \lambda_i \geq 0, i = 1, 2, \ldots, m$, $(1-\lambda)\mu_j \geq 0, j = 1, 2, \ldots, \ell$ and

$$\sum_{i=1}^{m} \lambda \lambda_i + \sum_{j=1}^{\ell} (1-\lambda)\mu_j = \lambda \sum_{i=1}^{m} \lambda_i + (1-\lambda) \sum_{j=1}^{\ell} \mu_j = \lambda + (1-\lambda) = 1.$$

Also, the set $K(A)$ of convex combinations contains A (each x in A can be written as $x = 1 \cdot x$). By the definition of co(A) as the intersection of all convex supersets of A, we deduce that co(A) is contained in $K(A)$.

Thus the convex hull of A is the set of all (finite) convex combinations from within A. □

The above result also holds for an affine set and a convex cone.

Corollary 1.1

(a) *The set A is convex if and only if* $A = \mathrm{co}(A)$.
(b) *The set A is affine if and only if* $A = \mathrm{aff}(A)$.
(c) *The set A is a convex cone if and only if* $A = \mathrm{cone}(A)$.
(d) *The set A is a subspace if and only if* $A = [A]$.

Definition 1.8 The *relative interior* of a set C in a topological vector space X, denoted by relint(C), is defined as

$$\mathrm{relint}(C) = \{x \in C : N_\varepsilon(x) \cap \mathrm{aff}(C) \subseteq C \text{ for some } \varepsilon > 0\},$$

where $N_\varepsilon(x)$ denotes the neighborhood of x.

Remark 1.4

(a) We have $\mathrm{relint}(C) \subseteq \mathrm{aff}(C)$.
(b) $\mathrm{relint}(C) = \mathrm{aff}(C)$ if and only if $\mathrm{aff}(C) = X$.

Example 1.1

(a) Consider the set $C = \{(x, y, z) \in \mathbb{R}^3 : x^2 + y^2 \leq 1, z = 0\}$. Then $\text{int}(C) = \emptyset$, but $\text{relint}(C) = \{(x, y, z) \in \mathbb{R}^3 : x^2 + y^2 < 1, z = 0\}$.
(b) For the set $C = \{(x, y, z) \in \mathbb{R}^3 : x^2+y^2+z^2 \leq 1\}$, we have $\text{int}(C) = \text{relint}(C) = \{(x, y, z) \in \mathbb{R}^3 : x^2 + y^2 + z^2 < 1\}$.

Definition 1.9 The *relative boundary* of a set C in a topological vector space X, denoted by $\text{relb}(C)$ or $\text{rb}(C)$, defined as

$$\text{relb}(C) = \text{cl}(C) \setminus \text{relint}(C).$$

Example 1.2 Consider a square in the (x_1, x_2)-plane in $\mathbb{R}^3$ defined as

$$C = \left\{x = (x_1, x_2, x_3) \in \mathbb{R}^3 : -1 \leq x_1 \leq 1,\ -1 \leq x_2 \leq 1,\ x_3 = 0\right\}.$$

Its affine hull is the (x_1, x_2)-plane, that is,

$$\text{aff}(C) = \left\{x = (x_1, x_2, x_3) \in \mathbb{R}^3 : x_3 = 0\right\}.$$

The interior of C is empty, but the relative interior is

$$\text{relint}(C) = \left\{x = (x_1, x_2, x_3) \in \mathbb{R}^3 : -1 < x_1 < 1,\ -1 < x_2 < 1,\ x_3 = 0\right\}.$$

Its boundary (in $\mathbb{R}^3$) is itself; its relative boundary is the wire-frame outline,

$$\text{relb}(C) = \left\{x = (x_1, x_2, x_3) \in \mathbb{R}^3 : \max\{|x_1|, |x_2|\} = 1,\ x_3 = 0\right\}.$$

Definition 1.10 A subset C of a topological vector space X is called *relatively open* if $\text{relint}(C) = C$.

Remark 1.5 If $C_1 \subseteq C_2$, then

(a) $\text{cl}(C_1) \subseteq \text{cl}(C_2)$ and
(b) $\text{int}(C_1) \subseteq \text{int}(C_2)$.

Note that property (b) does not hold for relative interior, that is, $\text{relint}(C_1) \subseteq \text{relint}(C_2)$ is not true in general. For example, if C_2 is a cube in $\mathbb{R}^3$ and C_1 is one of the faces of C_2. Then $\text{relint}(C_2)$ and $\text{relint}(C_1)$ are both nonempty but disjoint.

Remark 1.6 Let C be a subset of a topological vector space X.

(a) Every affine set is relatively open by definition and at the same time closed.
(b) $\text{cl}(C) \subset \text{cl}(\text{aff}(C)) = \text{aff}(C)$ for every $C \subseteq X$.
(c) Any line through two different points of $\text{cl}(C)$ lies entirely in $\text{aff}(C)$.

If $C \subseteq X$ is convex, then we have the following assertions:

(d) $\text{int}(C)$ and $\text{relint}(C)$ are convex.

(e) $\text{cl}(C)$ is also convex.
(f) If $C \subseteq X$ is a convex set with nonempty interior, then $\text{cl}(\text{int}(C)) = \text{cl}(C)$.
(g) If $C \subseteq X$ is a convex set with nonempty interior, then $\text{int}(\text{cl}(C)) = \text{int}(C)$.
(h) $\text{relint}(C) = \text{relint}(\text{cl}(C))$. Moreover, it holds $\text{int}(C) = \text{int}(\text{cl}(C))$.
(i) $\text{cl}(C) = \text{cl}(\text{relint}(C))$ as well as $\text{cl}(C) = \text{cl}(\text{int}(C))$ if $\text{int}(C) \neq \emptyset$.

Remark 1.7 ([127, Corollary 6.3.2]) If C is a convex set in $\mathbb{R}^n$, then every open set which meets $\text{cl}(C)$ also meets $\text{relint}(C)$.

Proposition 1.1 ([86]) *Let Y be a topological vector space with a cone C, $c_0 \in \text{int}(C)$ and $V := \text{int}(C) - c_0$. Then $Y = \{\lambda V : \lambda \geq 0\}$.*

Definition 1.11 A cone C in a vector space X is said to be

(a) *nontrivial* or *proper* if $C \neq \{\mathbf{0}\}$ and $C \neq X$;
(b) *reproducing* if $C - C = X$;
(c) *pointed* if for $x \in C$, $x \neq \mathbf{0}$, the negative $-x \notin C$, that is, $C \cap (-C) = \{\mathbf{0}\}$.

Definition 1.12 A cone C in a topological vector space X is said to be a

(a) *closed cone* if it is also closed;
(b) *solid cone* if it has nonempty interior.

Below we give some properties of a cone.

Remark 1.8

(a) If C is a cone, then the convex hull of C, $\text{co}(C)$ is a convex cone.
(b) If C_1 and C_2 are convex cones, then $C_1 + C_2 = \text{co}(C_1 \cup C_2)$.

Example 1.3 Let

$$\mathbb{R}^n_+ = \{x = (x_1, x_2, \ldots, x_n) \in \mathbb{R}^n : x_i \geq 0 \text{ for all } i = 1, 2, \ldots, n\}.$$

Then $\mathbb{R}^n_+$ is a proper, closed, pointed, reproducing convex cone in the vector space $\mathbb{R}^n$.

Example 1.4 Let $C[0, 1]$ be the vector space of all real-valued continuous linear functionals defined on the interval $[0, 1]$. Then

$$C_+[0, 1] = \{f \in C[0, 1] : f(t) \geq 0 \text{ for all } t \in [0, 1]\}$$

is a proper, reproducing, pointed, convex cone in $C[0, 1]$. Note that the set

$$C^+ = \{f \in C_+[0, 1] : f \text{ is nondecreasing}\}$$

is also a proper, pointed, convex cone in the space $C[0, 1]$ but it is not reproducing as $C^+ - C^+$ is the proper subspace of all functions with bounded variation of $C[0, 1]$.

Example 1.5 Let

$$C = \{x = (x_1, x_2, \ldots, x_n) \in \mathbb{R}^n : x_1 > 0, \text{ or} \\ x_1 = 0,\ x_2 > 0, \text{ or} \\ \ldots \\ x_1 = \cdots = x_{n-1} = 0,\ x_n > 0, \text{ or} \\ x = \mathbf{0}\},$$

where $\mathbf{0}$ is the zero vector in $\mathbb{R}^n$. Then C is a proper, closed, pointed, reproducing convex cone in the vector space $\mathbb{R}^n$.

Let C be a subset of a vector space X. We denote by $\ell(C) = C \cap (-C)$.

Definition 1.13 Let X be a topological vector space. A convex cone C in X is said to be

(a) *acute* if its closure $\mathrm{cl}(C)$ is pointed;
(b) *correct* if $\mathrm{cl}(C) + C \setminus \ell(C) \subseteq C$.

Example 1.6

(a) The nonnegative orthant $\mathbb{R}^n_+$ of all vectors of $\mathbb{R}^n$ with nonnegative coordinates is a convex, closed, acute and correct cone. The set $\{\mathbf{0}\}$ is also such a cone, but it is a trivial cone. The set composed of zero and of the vectors with the first coordinates being positive, is a pointed, correct cone, but it is not acute.
(b) Let

$$C = \{(x, y, z) \in \mathbb{R}^3 : x > 0,\ y > 0,\ z > 0\} \\ \cup \{(x, y, z) \in \mathbb{R}^3 : x \geq y \geq 0,\ z = 0\}.$$

Then C is a convex, acute cone but not correct.
(c) Let Ω be the vector space of all sequences $x = \{x_m\}$ of real numbers. Let

$$C = \{x = \{x_m\} \in \Omega : x_m \geq 0 \text{ for all } m\}.$$

Then C is a convex pointed cone. We cannot say whether it is correct or acute because no topology has been given on the space.

Proposition 1.2 *A cone C is correct if and only if* $\mathrm{cl}(C) + C \setminus \ell(C) \subseteq C \setminus \ell(C)$.

Proof If $\mathrm{cl}(C) + C \setminus \ell(C) \subseteq C \setminus \ell(C)$, then the cone is obviously correct because $C \setminus \ell(C) \subseteq C$.

Conversely, assume that C is a correct convex cone. Since $\ell(C)$ is a subspace and C is convex, for all $a, b \in C$, $a + b \in \ell(C)$ implies $a, b \in \ell(C)$. Therefore,

$$C \setminus \ell(C) + C \setminus \ell(C) = C \setminus \ell(C),$$

and

$$C + C \setminus \ell(C) \subseteq C \setminus \ell(C).$$

Thus,

$$\begin{aligned}\mathrm{cl}(C) + C \setminus \ell(C) &= \mathrm{cl}(C) + C \setminus \ell(C) + C \setminus \ell(C) \\ &\subseteq C + C \setminus \ell(C) \subseteq C \setminus \ell(C).\end{aligned}$$

This completes the proof. □

The cone $\mathbb{R}^n_+ \subset \mathbb{R}^n$ has the following interesting property: Consider the set

$$B = \left\{ x = (x_1, x_2, \ldots, x_n) \in \mathbb{R}^n_+ : \sum_{i=1}^{n} x_i = 1 \right\}.$$

Then for every $x \in \mathbb{R}^n_+ \setminus \{\mathbf{0}\}$, there exist a unique $b \in B$ and $\lambda > 0$ such that $x = \lambda b$. Indeed, consider $\lambda = x_1 + x_2 + \cdots + x_n$ (> 0) and $b = \lambda^{-1} x$. In view of this property, we have the following definition.

Definition 1.14 Let X be a vector space and C be a proper cone in X. A nonempty subset $B \subset C$ is called a *base* for C if each nonzero element $x \in C$ has a unique representation of the form $x = \lambda b$ for some $\lambda > 0$ and some $b \in B$ (Figs. 1.7 and 1.8).

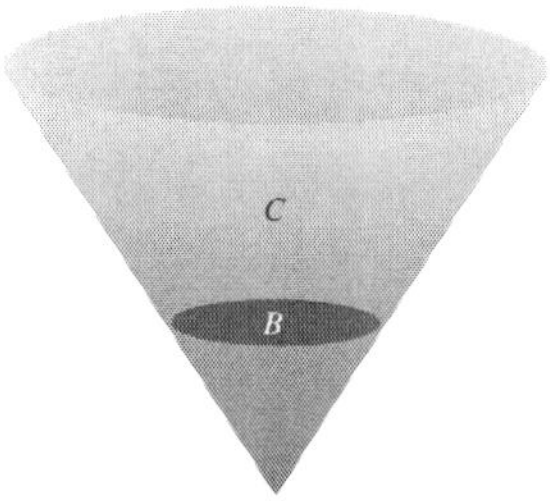

Fig. 1.7 B is a base for the cone C

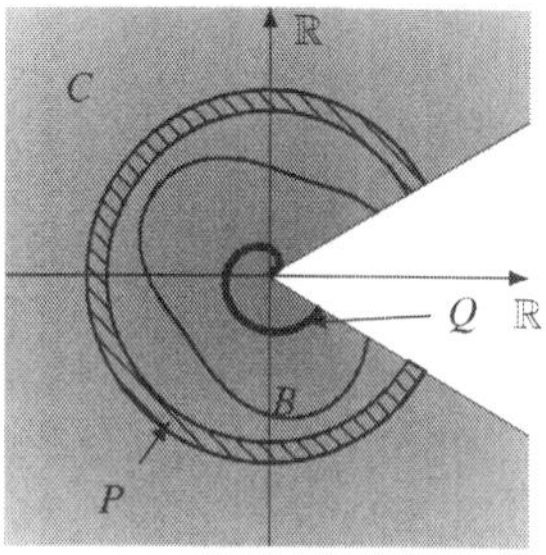

Fig. 1.8 B is base for C, but Q and P are not a base for C

Remark 1.9 Note that if B is a convex base of a proper convex cone C, then $\mathbf{0} \notin B$. Indeed, suppose that $\mathbf{0} \in B$. Since B is convex, for every element $b \in B$, the convex combination of $\mathbf{0}$ and b also belongs to B. Then we also have $b = 2 \cdot \frac{1}{2}b \in B$, contradicting the uniqueness of the representation of $b \in C \setminus \{\mathbf{0}\}$.

Theorem 1.3 *Let C be a proper convex cone in a vector space X and $B \subset X$ be a convex set. Then the following assertions are equivalent:*

(a) *B is a base for C;*
(b) *$C = \mathbb{R}_+B$ and $\mathbf{0} \notin \mathrm{aff}(B)$;*
(c) *There exists a linear functional $\phi : X \to \mathbb{R}$ such that $\phi(x) > 0$ for every $x \in C \setminus \{\mathbf{0}\}$ and $B = \{x \in C : \phi(x) = 1\}$.*

Proof (a) $\Rightarrow$ (b) Let B be a base for C. Then by Definition 1.14, $C = \mathbb{R}_+B$. Because B is convex, $\mathrm{aff}(B) = \{\mu b + (1-\mu)b' : b, b' \in B,\ \mu \in \mathbb{R}\}$. Assume that $\mathbf{0} \in \mathrm{aff}(B)$, then $\mathbf{0} = \mu b + (a-\mu)b'$ for some $b, b' \in B$ and $\mu \in \mathbb{R}$. Since $\mathbf{0} \notin B$, $\mu \notin [0,1]$. Thus, there exists some $\mu_0 > 1$, $b_0, b_0' \in B$ such that $\mu_0 b_0 = (\mu_0 - 1)b_0' \in C$, in contradiction to the definition of the base. Therefore, $\mathbf{0} \notin \mathrm{aff}(B)$.

(b) $\Rightarrow$ (c) Assume that $C = \mathbb{R}_+B$ and $\mathbf{0} \notin \mathrm{aff}(B)$. Consider $b_0 \in B$ and $X_0 := \mathrm{aff}(B) - b_0$. Then X_0 is a linear subspace of X and $b_0 \notin X_0$. Let $L_0 \subset X_0$ be a base of X_0. Then $L_0 \cup \{b_0\}$ is linearly independent, so, we can complete $L_0 \cup \{b_0\}$ to a base L of X. There exists a unique linear function $\phi : X \to \mathbb{R}$ such that $\phi(x) = 0$ for all $x \in L \setminus \{b_0\}$ and $\phi(b_0) = 1$. Since $\mathrm{aff}(B) = b_0 + X_0$, it holds $\phi(x) = 1$ for all $x \in \mathrm{aff}(B)$, thus, $B \subset \{x \in C : \phi(x) = 1\}$. Conversely, let $x \in C$ be such that $\phi(x) = 1$. Then $x = tb$ for some $t > 0$ and $b \in B$. It follows that $1 = \phi(x) = t\phi(b) = t$, thus, $x \in B$.

(c) $\Rightarrow$ (a) Assume that $\phi : X \to \mathbb{R}$ is linear, $\varphi(x) > 0$ for every $x \in C \setminus \{\mathbf{0}\}$, and $B = \{x \in C : \phi(x) = 1\}$. Consider $x \in C \setminus \{\mathbf{0}\}$ and take $t := \phi(x) > 0$ and $b := t^{-1}x$. Then $x = tb$. Since $b \in C$ and $\phi(b) = 1$, we have $b \in B$. Suppose that $x = t'b'$ for some $t' > 0$ and $b' \in B$. Then $t = \phi(x) = t'\phi(b') = t'$, whence $b = b'$. So, every nonzero element x of C has a unique representation tb with $t > 0$ and $b \in B$. This means that B is a base of C. □

Lemma 1.1 *Each proper convex cone with a convex base in a vector space is pointed.*

Proof Let C be a proper convex cone with a convex base B. Take any $x \in C \cap (-C)$ and assume that $x \neq \mathbf{0}$. Then there are $b_1, b_2 \in B$ and $\lambda_1, \lambda_2 > 0$ with $x = \lambda_1 b_1 = -\lambda_2 b_2$. Since B is convex, we have

$$\frac{\lambda_1}{\lambda_1+\lambda_2}b_1 + \frac{\lambda_2}{\lambda_1+\lambda_2}b_2 = \mathbf{0} \in B,$$

a contradiction to Remark 1.9. □

Example 1.7 The cone $C = \{x : x = \lambda \cdot (1,2), \lambda \geq 0\} \cup \{x : x = \lambda \cdot (2,1), \lambda \geq 0\}$ is pointed, proper and has a base $B = \{(1,2), (2,1)\}$, but C is not convex.

Remark 1.10 If B is a base of a cone C, then $\text{cone}(B) = C$. If $\mathbf{0} \in \text{cor}(C)$, the core of C, for a nonempty subset C of a vector space X, then $\text{cone}(C) = X$.

The following result can be found in Jameson [82, p. 80] and known as Jameson lemma.

Proposition 1.3 (Jameson Lemma) *Let X be a Hausdorff topological vector space with its zero vector being denoted by $\mathbf{0}$. Then a cone $C \subset X$ with a closed convex bounded base B is closed and pointed.*

Proof We show that C is closed. Let $\{c_\alpha\} \subseteq C$ be a net converging to c. Since B is a base, there exist a net $\{b_\alpha\} \subseteq B$ and a net $\{t_\alpha\}$ of positive numbers such that $c_\alpha = t_\alpha b_\alpha$. We claim that t_α is bounded. Suppose, contrary, that $\lim_\alpha t_\alpha = \infty$. Then the net $\left\{b_\alpha = \frac{c_\alpha}{t_\alpha}\right\}$ converges to $\mathbf{0}$ as X is Hausdorff. Since B is closed, $\mathbf{0} = \lim_\alpha b_\alpha \in B$ which contradicts to the fact that B does not contain the zero element. So, we may assume that $\{t_\alpha\}$ converges to some $t_0 \geq 0$. If $t_0 = 0$, then by the boundedness of B, $\lim_\alpha t_\alpha b_\alpha = \mathbf{0}$. Hence, $c = \mathbf{0}$ and, of course, $c = \mathbf{0} \in C$. If $t_0 > 0$, we may assume that $t_\alpha > \varepsilon$ for all α and some positive ε. Now, $b_\alpha = \frac{c_\alpha}{t_\alpha}$ converges to $\frac{c}{t_0}$ and again by the closedness of B, $\frac{c}{t_0} \in B$. Hence, $c \in C$ and so C is closed. The pointedness of C can be easily seen. □

Definition 1.15 Let Y be a topological vector space with its topological dual Y^*, and C be a convex cone in Y. The *dual cone* C^* of C is defined as

$$C^* = \{\xi \in Y^* : \langle \xi, y \rangle \geq 0 \text{ for all } y \in C\},$$

where $\langle \xi, y \rangle$ denotes the evaluation of ξ at y. The *strict dual cone* C^*_+ of C is defined as

$$C^*_+ = \{\xi \in Y^* : \langle \xi, y \rangle > 0 \text{ for all } y \in C\}.$$

The *quasi-interior* of C^* is defined as

$$C^\# := \{\xi \in Y^* : \langle \xi, y \rangle > 0 \text{ for all } y \in C \setminus \{\mathbf{0}\}\}.$$

If C is empty, then C^* interprets as the whole space Y^*.

For example, in $\mathbb{R}^2$ the dual of a convex cone C consists of all vectors making a non-acute angle with all vectors of the cone C (see Fig. 1.9). For an example of a dual cone in $\mathbb{R}^3$, see Fig. 1.10.

The following proposition can be proved easily by using the definition. Therefore, we omit the proof.

Proposition 1.4 *Let Y be a topological vector space with its topological dual Y^*. Let C, C_1 and C_2 be convex cones in Y.*

(a) *The dual cone C^* is a closed convex cone.*
(b) *The strict dual cone C^* is a convex cone.*

Fig. 1.9 A dual cone in $\mathbb{R}^2$

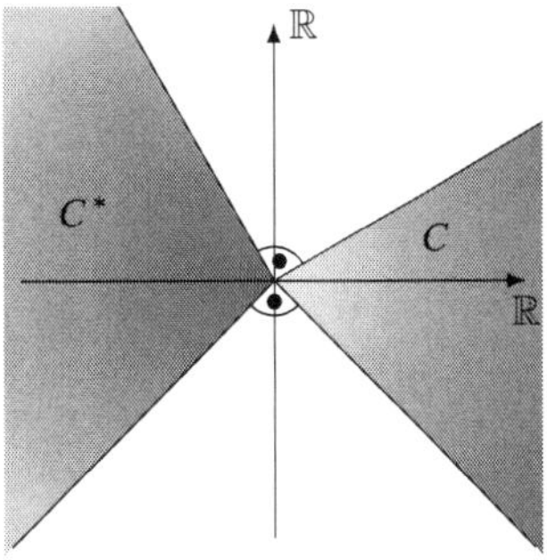

Fig. 1.10 A dual cone in $\mathbb{R}^3$

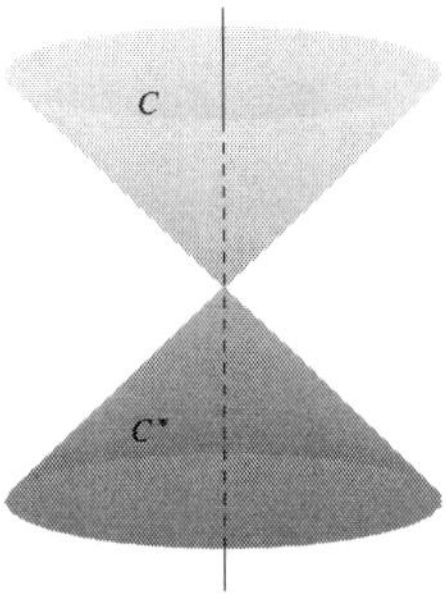

(c) $C^* = (\mathrm{cl}(C))^*$.
(d) *If* $C_1 \subset C_2$, *then* $C_2^* \subset C_1^*$ *and* $C_{2+}^* \subset C_{1+}^*$.
(e) $(C^*)^* = C^{**} = \mathrm{cl}(C)$.
(f) $(C_1 + C_2)^* = C_1^* \cap C_2^* = (C_1 \cap C_2)^*$.
(g) $(C_1 \cap C_2)^* \supset C_1^* + C_2^* = \mathrm{co}\,(C_1 \cup C_2)^*$.
(h) *If* C_1 *and* C_2 *are closed convex cones with nonempty intersection, then*

$$(C_1 \cap C_2)^* = \mathrm{cl}\left(C_1^* + C_2^*\right) = \mathrm{cl}\left(\mathrm{co}\,(C_1 \cup C_2)^*\right).$$

We now define the recession cone and asymptotic cone and discuss their properties.

Definition 1.16 Let C be a nonempty subset of a vector space Y. A vector $d \in Y$ is said to be a *direction of recession* if for any $x \in C$, the ray $\{x + \lambda d : \lambda \geq 0\}$ (starting from x and going indefinitely along d) lies in C (or never crosses the relative boundary of C).

Definition 1.17 Let C be a nonempty subset of a vector space Y. The set of all directions of recession is called *recession cone* and it is denoted by C_∞ (see Fig. 1.11). That is, for any $x \in C$,

$$C_\infty = \{d \in Y : x + \lambda d \in C \text{ for all } \lambda \geq 0\}.$$

Below we collect some properties of a recession cone.

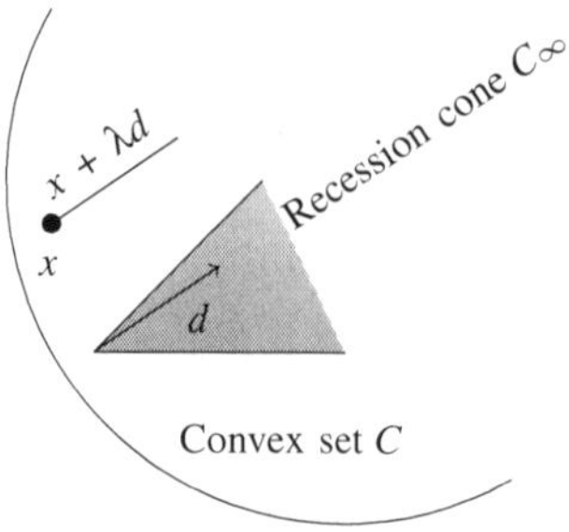

Fig. 1.11 A recession cone (see [21])

Remark 1.11

(a) C_∞ depends only on the behavior of C at infinity. In fact, $x + \lambda d \in C$ implies $x + \alpha d \in C$ for all $\alpha \in [0, \lambda]$. Thus, C_∞ is just the set of all directions from which one can go straight from x to infinity, while staying in C.
(b) If C is closed and convex, then for all $x \in C$, we have

$$C_\infty = \bigcap_{\lambda>0} \frac{C-x}{\lambda}.$$

(c) C_∞ does not depend on $x \in C$.

Definition 1.18 Let Y be a topological vector space. A recession cone of a nonempty closed convex set $C \subset Y$ is called *asymptotic cone*.

In other words, if C is a nonempty closed convex subset of Y, then the asymptotic cone of C is defined as

$$C_\infty = \left\{ d \in Y : \exists\, \lambda_m \to +\infty,\ \exists\, x_m \in C \text{ with } \lim_{m\to\infty} \frac{x_m}{\lambda_m} = d \right\}.$$

If Y is a reflexive Banach space and C is a weakly closed convex set C in Y, then the asymptotic cone C_∞ of C is defined as

$$C_\infty = \{x \in X :\ \exists\, \lambda_m \downarrow 0 \text{ and } \exists\, x_m \in C \text{ such that } \lambda_m x_m \rightharpoonup x\},$$

where "$\rightharpoonup$" means convergence in the weak topology.

We set $\emptyset_\infty = \emptyset$.

Example 1.8

(a) If $C = \left\{(x_1, x_2) \in \mathbb{R}^2 : |x_1| \leq x_2\right\}$, then C is unbounded and $C_\infty = C$.
(b) If $C = \left\{(x_1, x_2) \in \mathbb{R}^2 : |x_1| < x_2\right\}$, then C is unbounded and $C_\infty = \mathrm{cl}(C)$.
(c) If $C = \left\{(x_1, x_2) \in \mathbb{R}^2 : |x_1|^k \leq x_2,\ k > 1\right\}$, then C is unbounded and

$$C_\infty = \{(0, x_2) : x_2 \geq 0\}.$$

(d) If $C = \left\{(x_1, x_2) \in \mathbb{R}^2 : x_1 > 0,\ x_2 \geq \frac{1}{x_1}\right\}$, then C is unbounded and

$$C_\infty = \left\{(x_1, x_2) \in \mathbb{R}^2 : x_1 \geq 0,\ x_2 \geq 0\right\}.$$

(e) If $C = \left\{(x_1, x_2) \in \mathbb{R}^2 : x_2 \geq x_1^2\right\}$, then C is unbounded and

$$C_\infty = \left\{(x_1, x_2) \in \mathbb{R}^2 : x_1 = 0,\ x_2 \geq 0\right\}.$$

(f) If $C = \left\{(x_1, x_2) \in \mathbb{R}^2 : x_1^2 + x_2^2 \leq 1\right\}$, then C is bounded and

$$C_\infty = \left\{(x_1, x_2) \in \mathbb{R}^2 : x_1 = x_2 = 0\right\} = \left\{(0, 0)\right\}.$$

(g) If $C = \left\{(x_1, x_2) \in \mathbb{R}^2 : x_1 > 0,\ x_2 > 0\right\} \cup \{(0, 0)\}$, then C is unbounded and $C_\infty = C$.
(h) The recession cone of a nonempty affine set M is the subspace L parallel to M.

Theorem 1.4 *Let Y be a topological vector space and C be a nonempty closed convex subset of Y.*

(a) *The recession cone C_∞ is a closed convex cone containing the origin, that is, $C_\infty = \{d : C + d \subset C\}$.*
(b) *Furthermore, let $(Y, \|\cdot\|)$ be a normed vector space. Then $d \in C_\infty$ if and only if there exists a vector $x \in C$ such that $x + \lambda d \in C$ for all $\lambda \geq 0$, that is,*

$$C_\infty = \{d : \text{there exists } x \in C,\ x + \lambda d \in C \text{ for all } \lambda \geq 0\}.$$

(c) *If $(Y, \|\cdot\|)$ is a normed vector space, then C is bounded if and only if $C_\infty = \{\mathbf{0}\}$.*

Proof

(a) Let $d \in C_\infty$, then $x + d \in C$ for any $x \in C$, that is, $C + d \subset C$.

On the other hand, if $C + d \subset C$, then

$$C + 2d = (C + d) + d \subset C + d \subset C,$$

and so forth, implying $x + md \in C$ for any $x \in C$ and for any positive integer m. The line segments joining the points $x, x + d, x + 2d, \ldots$, are then all contained in C by convexity, so that $x + \lambda d \in C$ for every $\lambda \geq 0$. Thus, $d \in C_\infty$. Since positive scalar multiplication does not change directions, C_∞ is truly a cone.

It remains to show that C_∞ is convex. Let $d_1, d_2 \in C_\infty$ and $0 \leq \lambda \leq 1$, then we have

$$\begin{aligned}(1 - \lambda)d_1 + \lambda d_2 + C &= (1 - \lambda)(d_1 + C) + \lambda(d_2 + C)\\ &\subset (1 - \lambda)C + \lambda C = C.\end{aligned}$$

Hence, $(1 - \lambda)d_1 + \lambda d_2 \in C_\infty$.

(b) If $d \in C_\infty$, then $x + \lambda d \in C$ for all $\lambda \geq 0$ for all $x \in C$ by the definition of C_∞.

Conversely, let $d \neq \mathbf{0}$ be such that there exists a vector $x \in C$ such that $x + \lambda d \in C$ for all $\lambda \geq 0$. We fix $\bar{x} \in C$ and $\lambda > 0$, and we show that $\bar{x}+\lambda d \in C$. It is sufficient to show that $\bar{x}+d \in C$, that is, to assume that $\lambda = 1$, since the general case where $\lambda > 0$ can be reduced to the case where $\lambda = 1$ by replacing d with d/λ.

Let $z_m = x+md$ for $m = 1, 2, \ldots$ and note that $z_m \in C$ for all m, since $x \in C$ and $d \in C_\infty$. If $\bar{x} = z_m$ for some m, then $\bar{x} + d = x + (m+1)d$, which belongs to C and we are done. We thus assume that $\bar{x} \neq z_m$ for all m, and we define

$$d_m = \frac{(z_m - \bar{x})}{\|z_m - \bar{x}\|}\|d\|, \quad m = 1, 2, \ldots.$$

so that $\bar{x} + d_m$ lies on the line that starts at $\bar{x}$ and passes through z_m. We have

$$\frac{d_m}{\|d\|} = \frac{\|z_m - x\|}{\|z_m - \bar{x}\|}\frac{z_m - x}{\|z_m - x\|} + \frac{x - \bar{x}}{\|z_m - \bar{x}\|} = \frac{\|z_m - x\|}{\|z_m - \bar{x}\|}\frac{d}{\|d\|} + \frac{x - \bar{x}}{\|z_m - \bar{x}\|}.$$

Since $\{z_m\}$ is an unbounded sequence,

$$\frac{\|z_m - x\|}{\|z_m - \bar{x}\|} \to 1, \qquad \frac{x - \bar{x}}{\|z_m - \bar{x}\|} \to 0,$$

so by combining the preceding relations, we have $d_m \to d$. The vector $\bar{x} + d_m$ lies between $\bar{x}$ and z_m in the segment connecting $\bar{x}$ and z_m for all m such that $\|z_m - \bar{x}\| \geq \|d\|$, so by the convexity of C, we have $\bar{x}+d_m \in C$ for all sufficiently large m. Since $\bar{x} + dm \to \bar{x} + d$ and C is closed, it follows that $\bar{x} + d \in C$.

(c) If C is bounded, then it is clear that C_∞ can not contain any nonzero direction.

Conversely, let $\{x_m\} \subseteq C$ be such that $\|x_m\| \to +\infty$ (we assume $x_m \neq \mathbf{0}$). The sequence $\left\{d_m : \dfrac{x_m}{\|x_m\|}\right\}$ is bounded, so we can extract a convergent subsequence, namely, $\{d_k\}$ such that $\lim_{k\to\infty} d_k = d$ with $k \in \mathbb{K} \subseteq \mathbb{N}$ ($\|d\| = 1$). Now, given $x \in C$ and $\lambda > 0$, take k so large that $\|x_k\| \geq \lambda$. Then we see that

$$x + \lambda d = \lim_{k\to\infty}\left[\left(1 - \frac{\lambda}{\|x_k\|}\right)x + \frac{\lambda}{\|x_k\|}x_k\right]$$

is in the closed convex set C and hence $d \in C_\infty$. □

Below we present some properties of recession cones and asymptotic cones.

Remark 1.12 Let X be a topological vector space.

(a) For a nonempty convex set $C \subseteq X$, we have $(\mathrm{cl}(C))_\infty = (\mathrm{relint}(C))_\infty$, where $\mathrm{relint}(C)$ denotes the relative interior of C; Furthermore, for any $x \in \mathrm{relint}(C)$, one has $d \in (\mathrm{cl}(C))_\infty$ if and only if $x + \lambda d \in \mathrm{relint}(C)$ for all $\lambda > 0$.

(b) Moreover, for a nonempty convex set $C \subseteq X$, it holds $(C + x)_\infty = C_\infty$ for all $x \in X$.
(c) For two nonempty closed convex sets $\tilde{C}, \hat{C} \subset Y$, $\tilde{C} \subseteq \hat{C}$ implies $\tilde{C}_\infty \subseteq \hat{C}_\infty$.
(d) Let $\{C_\alpha\}_{\alpha \in \Lambda}$ be any family of nonempty sets in X, then

$$\left(\bigcap_{\alpha \in \Lambda} C_\alpha\right)_\infty \subset \bigcap_{\alpha \in \Lambda} (C_\alpha)_\infty.$$

If, in addition, $\bigcap_{\alpha \in \Lambda} C_\alpha \neq \emptyset$ and each set C_α is closed and convex, then we obtain an equality in the previous inclusion.

Moreover, If $C_1 \subseteq X_1, C_2 \subseteq X_2, \ldots, C_m \subseteq X_m$ are closed convex sets, where X_i, $i = 1, 2, \ldots, m$ are topological vector spaces, then

$$(C_1 \times C_2 \times \cdots \times C_m)_\infty = (C_1)_\infty \times (C_2)_\infty \times \cdots \times (C_m)_\infty.$$

We present the definition of a contingent cone and its properties.

Definition 1.19 Let C be a nonempty subset of a normed space X.

(a) Let $\bar{x} \in \mathrm{cl}(C)$ be given. An element $u \in X$ is said to be a *tangent* to C at $\bar{x}$ if there exist a sequence $\{x_m\}$ of elements $x_m \in C$ and a sequence $\{\lambda_m\}$ of positive real numbers λ_m such that $\lim_{m \to \infty} x_m \to \bar{x}$ and $\lim_{m \to \infty} \lambda_m (x_m - \bar{x}) = u$.
(b) The set $T(C, \bar{x})$ of all tangents to C at $\bar{x}$ is called the *contingent cone* (or the *Bouligand tangent cone*) to C at $\bar{x}$.

In other words, a contingent cone $T(C, \bar{x})$ to C at $\bar{x}$ is defined as

$$T(C, \bar{x}) = \big\{u \in X : \exists \{x_m\} \subset C \text{ and } \{\lambda_m\} \subset (0, \infty) \\ \text{such that } x_m \to \bar{x} \text{ and } \lambda_m (x_m - \bar{x}) \to u\big\}.$$

Figure 1.12 visualizes two contingent cones.

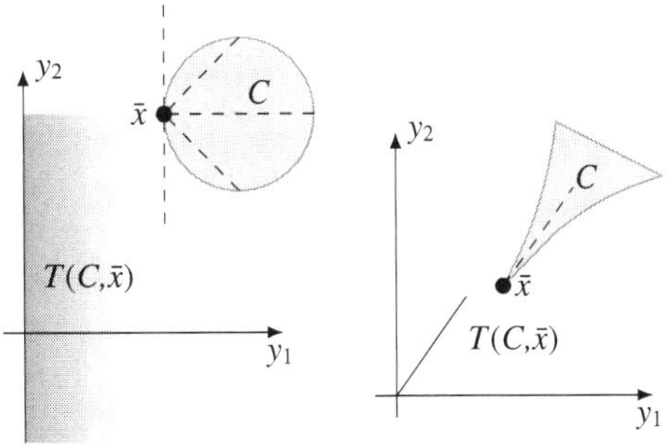

Fig. 1.12 Tangent to C at $\bar{x}$ and contingent cone $T(C, \bar{x})$

It is easy to see that the above definition of contingent cone can be written as

$$T(C,\bar{x}) = \Big\{ u \in X : \exists \{x_m\} \subset C \text{ and } \{\lambda_m\}$$

$$\text{such that } x_m \to \bar{x},\ \lambda_m \to 0^+ \text{ and } \frac{x_m - \bar{x}}{\lambda_m} \to u \Big\}.$$

If $\bar{x} \in \operatorname{int}(C)$, then $T(C,\bar{x})$ is clearly the whole space. That is why we considered $\bar{x} \in \operatorname{cl}(C)$.

If $u_m = \dfrac{x_m - \bar{x}}{\lambda_m}$ $(\to u)$, that is, $x_m = \bar{x} + \lambda_m u_m$ $(\in C)$, then we have $d \in T(C,\bar{x})$ if and only if there exist sequences $\{u_m\} \to u$ and $\{\lambda_m\} \to 0^+$ such that $\bar{x} + \lambda_m u_m \in C$ for all $m \in \mathbb{N}$.

It is equivalent to saying that $u \in T(C,\bar{x})$ if and only if there exist sequences $\{u_m\} \to u$ and $\{\lambda_m\} \subset \mathbb{R}_+$ such that

$$\bar{x} + \lambda_m u_m \in C, \quad \text{for all } m \in \mathbb{N} \quad \text{and} \quad \{\lambda_m x_m\} \to 0.$$

Remark 1.13

(a) A contingent cone to a set C at a point $\bar{x} \in \operatorname{cl}(C)$ describes a local approximation of the set $C - \{\bar{x}\}$. This concept is very helpful for the investigation of optimality conditions.
(b) From the definition of $T(C,\bar{x})$, we see that $\bar{x}$ belongs to the closure of the set C. It is evident that the contingent cone is really a cone.

Lemma 1.2 *Let C and D be nonempty subsets of a normed space X.*

(a) *If $\bar{x} \in \operatorname{cl}(C) \subset \operatorname{cl}(D)$, then $T(C,\bar{x}) \subset T(D,\bar{x})$.*
(b) *If $\bar{x} \in \operatorname{cl}(C \cap D)$, then $T(C \cap D,\bar{x}) \subset T(C,\bar{x}) \cap T(D,\bar{x})$.*

Definition 1.20 Let C be a subset of a vector space X is called *starshaped* at $\bar{x} \in C$ if for all $x \in C$ and for every $\lambda \in [0, 1]$,

$$\lambda x + (1 - \lambda)\bar{x} \in C.$$

An example for a starshaped set $C \subset \mathbb{R}^2$ is given in Fig. 1.13.

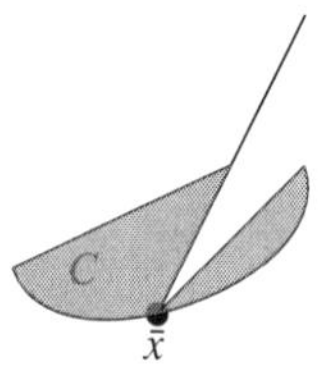

Fig. 1.13 A starshaped set C

Theorem 1.5 *Let C be a nonempty subset of a normed space X. If C is starshaped at some $\bar{x} \in C$, then* $\text{cone}\,(C - \{\bar{x}\}) \subset T(C, \bar{x})$.

Proof Take any $x \in C$. Then we have

$$x_m = \bar{x} + \frac{1}{m}(x - \bar{x}) = \frac{1}{m}x + \left(1 - \frac{1}{m}\right)\bar{x} \in C, \quad \text{for all } m \in \mathbb{N}.$$

Hence, we get $\bar{x} = \lim_{m\to\infty} x_m$ and $x - \bar{x} = \lim_{m\to\infty} m(x_m - \bar{x})$. But this implies that $x - \bar{x} \in T(C, \bar{x})$ and therefore, we obtain $C - \{\bar{x}\} \subset T(C, \bar{x})$.

Since $T(C, \bar{x})$ is a cone, it follows further that $\text{cone}\,(C - \{\bar{x}\}) \subset T(C, \bar{x})$. □

Theorem 1.6 *Let C be a nonempty subset of a normed space X. For every $\bar{x} \in \text{cl}(C)$, we have $T(C, \bar{x}) \subset \text{cl}\,(\text{cone}(C - \{\bar{x}\}))$.*

Proof Take an arbitrary tangent u to C at $\bar{x}$. Then there exist a sequence $\{x_m\}$ of elements in X and a sequence $\{\lambda_m\}$ of positive real numbers such that

$$\bar{x} = \lim_{m\to\infty} x_m \quad \text{and} \quad u = \lim_{m\to\infty} \lambda_m(x_m - \bar{x}).$$

The last equality implies $u \in \text{cl}\,(\text{cone}(C - \{\bar{x}\}))$. □

Theorem 1.7 *Let C be a nonempty subset of a normed space X. The contingent cone $T(C, \bar{x})$ is closed for every $\bar{x} \in \text{cl}(C)$.*

Proof Let $\{u_m\}$ be an arbitrary sequence in $T(C, \bar{x})$ with $\lim_{m\to\infty} u_m = u \in X$. For every tangent u_m, there exist a sequence $\{x_{m_i}\}$ of elements in C and a sequence $\{\lambda_{m_i}\}$ of positive real numbers such that

$$\bar{x} = \lim_{i\to\infty} x_{m_i} \quad \text{and} \quad u_m = \lim_{i\to\infty} \lambda_{m_i}(x_{m_i} - \bar{x}).$$

Consequently, for every $m \in \mathbb{N}$, there exists an $i(m) \in \mathbb{N}$ such that

$$\|x_{m_i} - \bar{x}\| \leq \frac{1}{m}, \quad \text{for all } i \geq i(m),$$

and

$$\|\lambda_{m_i}(x_{m_i} - \bar{x}) - u_m\| \leq \frac{1}{m}, \quad \text{for all } i \geq i(m).$$

If we define $y_m = x_{m_{i(m)}} \in C$ and $\mu_m = \lambda_{m_{i(m)}} > 0$ for all $m \in \mathbb{N}$, then we get $\bar{x} = \lim_{m\to\infty} y_m$ and

$$\begin{aligned}\|\mu_m(y_m - \bar{x}) - u\| &\leq \|\mu_m(y_m - \bar{x}) - u_m\| + \|u_m - u\| \\ &\leq \frac{1}{m} + \|u_m - u\|, \quad \text{for all } m \in \mathbb{N}.\end{aligned}$$

This implies that $u = \lim_{m\to\infty} \mu_m(y_m - \bar{x})$. Hence, $u \in T(C, \bar{x})$ and so $T(C, \bar{x})$ is closed. □

Corollary 1.2 *Let C be a nonempty subset of a normed space X. If C is starshaped at some $\bar{x} \in C$, then $T(C, \bar{x}) = \text{cl}\,(\text{cone}(C - \{\bar{x}\}))$.*

Theorem 1.8 *Let C be a nonempty convex subset of a normed space X. The contingent cone $T(C, \bar{x})$ is convex for every $\bar{x} \in \text{cl}(C)$.*

Proof Since C is convex, $C - \{\bar{x}\}$ and $\text{cone}(C - \{\bar{x}\})$ are convex as well. Since the closure of a convex set is convex, we have $\text{cl}\,(\text{cone}(C - \{\bar{x}\}))$ is also convex. Finally, from above corollary, we have $T(C, \bar{x}) = \text{cl}\,(\text{cone}(C - \{\bar{x}\}))$. □

1.2 Convex Functions and Their Properties

Definition 1.21 Let X be a vector space. A function $f : X \to \mathbb{R}$ is said to be

(a) *positively homogeneous* if for all $x \in X$ and all $r > 0$, we have $f(rx) = rf(x)$;
(b) *subodd* if for all $x \in X \setminus \{\mathbf{0}\}$, we have $f(x) \geq -f(-x)$.

Example 1.9

(a) Every linear function is positively homogeneous.
(b) The function $f(x) = |x|$ is positively homogeneous.
(c) Every norm is positively homogeneous.
(d) The function $f(x) = \begin{cases} x, & \text{if } x > 0, \\ -\frac{1}{2}x, & \text{if } x \leq 0 \end{cases}$ is positively homogeneous.
(e) $f(x) = x^2$ is subodd.

Definition 1.22 Let K be a subspace of a vector space X. A function $f : K \to \mathbb{R}$ is said to be *linear* if for all $x, y \in K$ and all $\lambda, \mu \in \mathbb{R}$,

$$f(\lambda x + \mu y) = \lambda f(x) + \mu f(y). \tag{1.1}$$

Definition 1.23 Let K be a nonempty affine subset of a vector space X. A function $f : K \to \mathbb{R}$ is said to be *affine* if (1.1) holds for all $x, y \in K$ and all $\lambda, \mu \in \mathbb{R}$ such that $\lambda + \mu = 1$.

In other words, f is affine if and only if

$$f(\lambda x + (1 - \lambda)y) = \lambda f(x) + (1 - \lambda)f(y), \tag{1.2}$$

for all $x, y \in K$ and all $\lambda \in \mathbb{R}$.

Definition 1.24 Let K be a nonempty convex subset of a vector space X. A function $f : K \to \mathbb{R}$ is said to be *convex* if for all $x, y \in K$ and all $\lambda, \mu \geq 0$ with $\lambda + \mu = 1$, we have

$$f(\lambda x + \mu y) \leq \lambda f(x) + \mu f(y). \tag{1.3}$$

In other words, f is convex if and only if

$$f(\lambda x + (1-\lambda)y) \leq \lambda f(x) + (1-\lambda)f(y), \tag{1.4}$$

for all $x, y \in K$ and all $\lambda \in [0, 1]$.

The functional f is said to be *strictly convex* if inequality (1.4) is strict for all $x \neq y$.

A function f is said to be *concave* if $-f$ is convex.

Example 1.10

(a) Let $K = X = \mathbb{R}$ and $f(x) = x^2$ for all $x \in K$. Then f is a convex function.
(b) Let $K = [0, \pi]$ and $f(x) = \sin x$ for all $x \in K$. Then f is a convex function.
(c) Let $K = X = \mathbb{R}$ and $f(x) = |x|$ for all $x \in K$. Then f is a convex function. In fact, the functions in (a) and (b) are strictly convex but the function in (c) is not.
(d) The functions $f(x) = \ln |x|$ for $x > 0$, and $g(x) = +\sqrt{1-x^2}$ for $x \in [-1, 1]$ are concave.

Remark 1.14 An affine function is both convex and concave.

Definition 1.25 Let K be a nonempty subset of a vector space X and $f : K \to \mathbb{R}$ be a function. The set

$$\text{epi}(f) = \{(x, \alpha) \in K \times \mathbb{R} : f(x) \leq \alpha\}$$

is called *epigraph* of f.

Theorem 1.9 *Let K be a nonempty convex subset of a vector space X and $f : K \to \mathbb{R}$ be a function. Then f is convex if and only if its epigraph is a convex set.*

Proof Let f be a convex function. Then for any (x, α) and $(y, \beta) \in \text{epi}(f)$, we have $f(x) \leq \alpha$ and $f(y) \leq \beta$. Also, for all $\lambda \in [0, 1]$, we have

$$f(\lambda x + (1-\lambda)y) \leq \lambda f(x) + (1-\lambda)f(y) \leq \lambda\alpha + (1-\lambda)\beta.$$

Thus,

$$((\lambda x + (1-\lambda)y), \lambda\alpha + (1-\lambda)\beta) = \lambda(x, \alpha) + (1-\lambda)(y, \beta) \in \text{epi}(f).$$

Hence, epi(f) is convex.

Conversely, let epi(f) be a convex set, and $(x, f(x)) \in \text{epi}(f)$ and $(y, f(y)) \in \text{epi}(f)$. Then for all $x, y \in K$ and all $\lambda \in [0, 1]$, we have

$$\lambda\,(x, f(x)) + (1-\lambda)\,(y, f(y)) \in \text{epi}(f).$$

This implies that

$$(\lambda x + (1-\lambda)y, \lambda f(x) + (1-\lambda)f(y)) \in \text{epi}(f)$$

and thus,

$$f(\lambda x + (1-\lambda)y) \leq f(x) + (1-\lambda)f(y).$$

Hence, f is convex. □

Theorem 1.10 *Let K be a nonempty convex subset of a vector space X and $f : K \to \mathbb{R}$ be a convex function. Then the lower level set $L_\alpha = \{x \in K : f(x) \leq \alpha\}$ is convex for every $\alpha \in \mathbb{R}$.*

Proof Let $x, y \in L_\alpha$. Then $(x, \alpha) \in \text{epi}(f)$ and $(y, \alpha) \in \text{epi}(f)$. Therefore, for all $\lambda \in [0, 1]$,

$$\lambda(x, \alpha) + (1-\lambda)(y, \alpha) \in \text{epi}(f),$$

equivalently,

$$(\lambda x + (1-\lambda)y, \lambda\alpha + (1-\lambda)\alpha) \in \text{epi}(f)$$

and thus

$$f(\lambda x + (1-\lambda)y) \leq \lambda\alpha + (1-\lambda)\alpha = \alpha.$$

Hence, $\lambda x + (1-\lambda)y \in L_\alpha$ and so L_α is convex. □

Remark 1.15 The converse of above theorem may not hold. For example, the function $f(x) = x^3$ defined on $\mathbb{R}$ is not convex but its lower level set $L_\alpha = \{x \in \mathbb{R} : x \leq \alpha^{1/3}\}$ is convex for every $\alpha \in \mathbb{R}$.

Theorem 1.11 *Let K be a nonempty convex subset of a vector space X. A function $f : K \to \mathbb{R}$ is convex if and only if for all $x_1, x_2, \ldots, x_m \in K$ and $\lambda_i \in [0, 1]$, $i = 1, 2, \ldots, m$ with $\sum_{i=1}^m \lambda_i = 1$,*

$$f\left(\sum_{i=1}^m \lambda_i x_i\right) \leq \sum_{i=1}^m \lambda_i f(x_i). \tag{1.5}$$

The inequality (1.5) is called *Jensen's inequality*.

Proof Suppose that the Jensen's inequality (1.5) holds. Then trivially, f is convex.

Conversely, we assume that the function f is convex. Then we show that the Jensen's inequality (1.5) holds. We prove it by induction on m. For $m = 1$ and $m = 2$, the inequality (1.5) trivially holds. Assume that the inequality (1.5) holds for m. We shall prove the result for $m + 1$. If $\lambda_{m+1} = 1$, the result holds because

$\lambda_i = 0$, for $i = 1, 2, \ldots, m$ and the result is true for $m = 1$. If $\lambda_{m+1} \neq 1$, we have

$$\begin{aligned} f\left(\sum_{i=1}^{m+1} \lambda_i x_i\right) &= f\left(\sum_{i=1}^{m} \lambda_i x_i + \lambda_{m+1} x_{m+1}\right) \\ &= f\left(\sum_{i=1}^{m} (1-\lambda_{m+1}) \frac{\lambda_i x_i}{1-\lambda_{m+1}} + \lambda_{m+1} x_{m+1}\right) \\ &= f\left((1-\lambda_{m+1}) \sum_{i=1}^{m} \frac{\lambda_i}{1-\lambda_{m+1}} x_i + \lambda_{m+1} x_{m+1}\right) \\ &= f\left((1-\lambda_{m+1}) \sum_{i=1}^{m} \mu_i x_i + \lambda_{m+1} x_{m+1}\right) \\ &\leq (1-\lambda_{m+1}) f\left(\sum_{i=1}^{m} \mu_i x_i\right) + \lambda_{m+1} f(x_{m+1}) \\ &\leq (1-\lambda_{m+1}) \sum_{i=1}^{m} \mu_i f(x_i) + \lambda_{m+1} f(x_{m+1}), \end{aligned}$$

where $\mu_i = \dfrac{\lambda_i}{(1-\lambda_{m+1})}$, $i = 1, 2, \ldots, m$ with $\mu_i \geq 0$ for $i = 1, 2, \ldots, m$ and

$$\sum_{i=1}^{m} \mu_i = \frac{\sum_{i=1}^{m} \lambda_i}{1-\lambda_{m+1}} = \frac{1-\lambda_{m+1}}{1-\lambda_{m+1}} = 1.$$

This completes the proof. □

The following theorems provide some properties of convex functions. The proof of these theorems is quite trivial, and hence, omitted.

Theorem 1.12 *Let K be a nonempty convex subset of a vector space X.*

(a) *If $f_1, f_2 : K \to \mathbb{R}$ are two convex functions, then $f_1 + f_2$ is a convex function on K.*
(b) *If $f : K \to \mathbb{R}$ is a convex function and $\alpha \geq 0$, then αf is a convex function on K.*
(c) *For each $i = 1, 2, \ldots, m$, if $f_i : K \to \mathbb{R}$ is a convex function and $\alpha_i \geq 0$, then $\sum_{i=1}^{m} \alpha_i f_i$ is a convex function. Further, if at least one of the functions f_i is strictly convex with the corresponding $\alpha_i > 0$, then $\sum_{i=1}^{m} \alpha_i f_i$ is strictly convex on K.*

Theorem 1.13 *Let K be a nonempty convex subset of a vector space X. For each $i = 1, 2, \ldots, m$, if $f_i : K \to \mathbb{R}$ is a convex function, then $\max\{f_1, f_2, \ldots, f_m\}$ is also a convex function on K.*

Next we provide characterizations of a differentiable convex function.

Theorem 1.14 ([10, 110]) *Let K be a nonempty open convex subset of $\mathbb{R}^n$ and $f : K \to \mathbb{R}$ be a differentiable function. Then*

(a) *f is convex if and only if for all $x, y \in K$,*

$$\langle \nabla f(x), y - x \rangle \leq f(y) - f(x). \tag{1.6}$$

(b) *f is strictly convex if and only if the inequality (1.6) is strict for $x \neq y$.*

Proof

(a) If f is a convex function, then for all $\lambda \in [0, 1]$

$$f((1-\lambda)x + \lambda y) \leq (1-\lambda)f(x) + \lambda f(y).$$

For $\lambda > 0$, we have

$$\frac{f((1-\lambda)x + \lambda y) - f(x)}{\lambda} \leq f(y) - f(x),$$

which on taking limit $\lambda \to 0^+$ leads to (1.6) as f is a differentiable function.

Conversely, let $\lambda \in [0, 1]$ and $u, v \in K$. On taking $x = (1-\lambda)u + \lambda v$ and $y = u$ in (1.6), we have

$$\lambda \langle \nabla f((1-\lambda)u + \lambda v), u - v \rangle \leq f(u) - f((1-\lambda)u + \lambda v). \tag{1.7}$$

Similarly, on taking $x = (1-\lambda)u + \lambda v$ and $y = v$ in (1.6), we have

$$-(1-\lambda)\langle \nabla f((1-\lambda)u + \lambda v), u - v \rangle \leq f(v) - f((1-\lambda)u + \lambda v). \tag{1.8}$$

Multiplying inequality (1.7) by $(1-\lambda)$ and inequality (1.8) by λ, and then adding the resultants, we obtain

$$f((1-\lambda)u + \lambda v) \leq (1-\lambda)f(u) + \lambda f(v).$$

(b) Suppose that f is strictly convex and $x, y \in K$ be such that $x \neq y$. Since f is convex, the inequality (1.6) holds. We need to show the inequality is strict. Suppose on the contrary that

$$\langle \nabla f(x), y - x \rangle = f(y) - f(x).$$

Then for $\lambda \in]0, 1[$, we have

$$f((1-\lambda)x + \lambda y) < (1-\lambda)f(x) + \lambda f(y) = f(x) + \lambda \langle \nabla f(x), y - x \rangle.$$

Let $z = (1 - \lambda)x + \lambda y$, then $z \in K$ and the above inequality can be written as

$$f(z) < f(x) + \langle \nabla f(x), z - x \rangle,$$

which contradicts the inequality (1.6). Proof of the converse part follows as given for the convex case. □

Theorem 1.15 ([63, 110]) *Let K be a nonempty open convex subset of $\mathbb{R}^n$ and $f : K \to \mathbb{R}$ be a differentiable function. Then f is convex if and only if for all $x, y \in K$,*

$$\langle \nabla f(y) - \nabla f(x), y - x \rangle \geq 0.$$

Proof Let f be a differentiable convex function. Then by Theorem 1.14 (a), we have

$$\langle \nabla f(x), y - x \rangle \leq f(y) - f(x), \quad \text{for all } x, y \in K.$$

By interchanging the roles of x and y, we have

$$\langle \nabla f(y), x - y \rangle \leq f(x) - f(y), \quad \text{for all } x, y \in K.$$

On adding the above inequalities we get the conclusion.

Conversely, by mean value theorem, for all $x, y \in K$, there exists $z = (1-\lambda)x+\lambda y$ for some $\lambda \in]0, 1[$ such that

$$\begin{aligned} f(y) - f(x) &= \langle \nabla f(z), y - x \rangle = (1/\lambda)\langle \nabla f(z), z - x \rangle \\ &\geq (1/\lambda)\langle \nabla f(x), z - x \rangle = \langle \nabla f(x), y - x \rangle, \end{aligned}$$

where the above inequality is obtained on using the given hypothesis. Hence, by Theorem 1.14 (a), f is a convex function. □

The following example illustrates the above theorem.

Example 1.11 The function $f(x) = x_1^2 + x_2^2$, where $x = (x_1, x_2) \in \mathbb{R}^2$, is a convex function on $\mathbb{R}^2$ and $\nabla f(x) = 2(x_1, x_2)$. For $x, y \in \mathbb{R}^2$,

$$\begin{aligned} \langle \nabla f(y) - \nabla f(x), y - x \rangle &= \langle 2(y_1 - x_1, y_2 - x_2), (y_1 - x_1, y_2 - x_2) \rangle \\ &= 2(y_1 - x_1)^2 + 2(y_2 - x_2)^2 \geq 0. \end{aligned}$$

Definition 1.26 Let K be a nonempty subset of a normed space X and $x \in K$ be a given point. A function $f : K \to \mathbb{R}$ is said to be *locally Lipschitz around* x if for some $k > 0$

$$|f(y) - f(z)| \leq k\|y - z\|, \quad \text{for all } y, z \in N(x) \cap K, \tag{1.9}$$

where $N(x)$ is a neighborhood of x. The constant k is called *Lipschitz constant* and it varies as the point x varies.

The function f is said to be *Lipschitz continuous* on K if the inequality (1.9) holds for all $y, z \in K$.

A continuously differentiable function always satisfies the Lipschitz condition (1.9). However, a locally Lipschitz function at a given point need not be differentiable at that point. For example, the function $f : \mathbb{R} \to \mathbb{R}$, defined by $f(x) = |x|$, satisfies the Lipschitz condition on $\mathbb{R}$. But f is not differentiable at 0.

The class of Lipschitz continuous functions is quite large. It is invariant under usual operations of sum, product and quotient.

It is clear that every Lipschitz continuous function is continuous. Also, every convex function is not only continuous but also locally Lipschitz in the interior of its domain.

Theorem 1.16 ((See [6, Theorem 1.14])) *Let K be a nonempty convex subset of $\mathbb{R}^n$, $f : K \to \mathbb{R}$ be a convex function and x be an interior point of K. Then f is locally Lipschitz at x.*

As we have seen, the convex functions cannot be characterized by lower level sets. However, if the function is convex then lower level sets are convex but the converse is not true. Now we define a class of such functions, called quasiconvex functions, which are characterized by convexity of their level sets.

Definition 1.27 Let K be a nonempty convex subset of a vector space X. A function $f : K \to \mathbb{R}$ is said to be

(a) *quasiconvex* if for all $x, y \in K$ and all $\lambda \in]0, 1[$,

$$f(x + \lambda(y - x)) \leq \max\{f(x), f(y)\};$$

(b) *strictly quasiconvex* if for all $x, y \in K$, $x \neq y$ and all $\lambda \in]0, 1[$,

$$f(x + \lambda(y - x)) < \max\{f(x), f(y)\};$$

(c) *semistrictly quasiconvex* if for all $x, y \in K$ with $f(x) \neq f(y)$,

$$f(x + \lambda(y - x)) < f(x), \quad \text{for all } \lambda \in]0, 1[.$$

A function $f : K \to \mathbb{R}$ is said to be (*strictly, semistrictly*) *quasiconcave* if $-f$ is (strictly, semistrictly) quasiconvex.

Note that in Definition 1.27 (b), the premise excludes the case $f(x) = f(y)$. Therefore, the formulation of the definition of semistrictly quasiconvexity differs from (a). Also, note that a strictly quasiconvex function was referred to as a strongly quasiconvex function in [17] and a semistrictly quasiconvex function was referred to as a strictly quasiconvex function in [17, 19, 110].

Example 1.12

(a) Every convex function is quasiconvex.
(b) The function $f : \mathbb{R} \to \mathbb{R}$, defined by $f(x) = \sqrt{|x|}$, is quasiconvex, but not convex.
(c) Every strictly convex function is semistrictly quasiconvex.
(d) The function $f : \mathbb{R} \to \mathbb{R}$, defined by $f(x) = x$, is semistrictly quasiconvex, but f is not strictly convex.

Obviously, every (strictly) convex function is (strictly) quasiconvex but the converse is not necessarily true. The function $f : \mathbb{R} \to \mathbb{R}$ defined by $f(x) = x^3$ is a quasiconvex function but not a convex function. Also, a convex function is semistrictly quasiconvex but the converse may not be true. Again, we see that $f(x) = x^3$ is semistrictly quasiconvex but not convex. We note that the strict quasiconvexity is not a generalization of convexity as a constant function is convex but not strictly quasiconvex. Obviously, a strictly quasiconvex function is quasiconvex but the converse is not true, For example, the greatest integer function $f(x) = [x]$ is quasiconvex but not strictly quasiconvex on $\mathbb{R}$.

We now give the characterization of a quasiconvex function in terms of convexity of its lower level sets.

Theorem 1.17 *Let K be a nonempty convex subset of a vector space X. A function $f : K \to \mathbb{R}$ is quasiconvex if and only if the lower level sets $L(f, \alpha)$ are convex for all $\alpha \in \mathbb{R}$.*

Proof Let f be a quasiconvex function and for $\alpha \in \mathbb{R}$, let $x, y \in L(f, \alpha)$. Then $f(x) \leq \alpha$ and $f(y) \leq \alpha$. Since f is a quasiconvex function, for all $\lambda \in [0, 1]$, we have

$$f((1 - \lambda)x + \lambda y) \leq \max\{f(x), f(y)\} \leq \alpha,$$

that is, $(1 - \lambda)x + \lambda y \in L(f, \alpha)$ for all $\lambda \in [0, 1]$. Hence, $L(f, \alpha)$ is convex.

Conversely, let $x, y \in K$ and $\bar{\alpha} = \max\{f(x), f(y)\}$. Then $x, y \in L(f, \bar{\alpha})$, and by convexity of $L(f, \bar{\alpha})$, we have $(1 - \lambda)x + \lambda y \in L(f, \bar{\alpha})$ for all $\lambda \in [0, 1]$. Thus for all $\lambda \in [0, 1]$,

$$f((1 - \lambda)x + \lambda y) \leq \bar{\alpha} = \max\{f(x), f(y)\}.$$

This completes the proof. □

Next result gives the characterization of a quasiconvex function in terms of its gradient.

Theorem 1.18 ([6, 10, 110]) *Let K be a nonempty open convex subset of $\mathbb{R}^n$ and $f : K \to \mathbb{R}$ be a differentiable function. Then f is quasiconvex if and only if for all $x, y \in K$,*

$$f(y) \leq f(x) \quad \Rightarrow \quad \langle \nabla f(x), y - x \rangle \leq 0. \tag{1.10}$$

It can be easily seen that if $f_i : K \to \mathbb{R}$, $i = 1, 2, \ldots, m$, is a quasiconvex function on a nonempty convex subset K of a vector space X, then $\max\{f_1, f_2, \ldots, f_m\}$ is also a quasiconvex function on K.

Definition 1.28 Let X be a topological space. A function $f : X \to \mathbb{R} \cup \{\pm\infty\}$ is said to be *lower semicontinuous* (respectively, *upper semicontinuous*) at a point $x \in X$ if for every $\varepsilon > 0$, there exists a neighborhood U of x such that $f(y) \leq f(x) + \varepsilon$ (respectively, $f(y) \geq f(x) - \varepsilon$) for all $y \in U$ when $f(x) > -\infty$, and $f(y) \to -\infty$ as $y \to x$ when $f(x) = -\infty$ (respectively, $f(x) < +\infty$, and $f(y) \to +\infty$ as $y \to x$ when $f(x) = +\infty$).

A function f is lower semicontinuous (respectively, upper semicontinuous) on X if it is lower semicontinuous (respectively, upper semicontinuous) at every point of X.

If X is a metric space then it can be expressed as

$$\limsup_{y \to x} f(y) \leq f(x) \quad \text{(respectively, } \liminf_{y \to x} f(y) \geq f(x)\text{)}.$$

For non-metric spaces, an equivalent definition using nets may be stated.

It can be easily seen that a function $f : X \to \mathbb{R} \cup \{\pm\infty\}$ is lower semicontinuous (respectively, upper semicontinuous) on X if and only if the set $\{x \in X : f(x) \leq r\}$ (respectively, $\{x \in X : f(x) \geq r\}$) is closed for all $r \in \mathbb{R}$.

Theorem 1.19 *Let X be a topological space and $f : X \to \mathbb{R} \cup \{\pm\infty\}$ be a function. Then f is lower semicontinuous if and only if* $\operatorname{epi}(f) := \{(x, r) \in X \times \mathbb{R} : f(x) \leq r\}$ *is closed.*

The following theorem gives a sufficient condition for a semistrictly quasiconvex function to be quasiconvex.

Theorem 1.20 *Every lower semicontinuous semistrictly quasiconvex function on a convex set is quasiconvex.*

Proof Let f be a semistrictly quasiconvex function defined on a convex subset K of a vector space X. Then for all $x, y \in K$, $f(x) \neq f(y)$ and $\lambda \in]0, 1[$, we have

$$f((1 - \lambda)x + \lambda y) < \max\{f(x), f(y)\}.$$

It remains to show that if $f(x) = f(y)$ and $\lambda \in]0, 1[$, then

$$f((1 - \lambda)x + \lambda y) \leq \max\{f(x), f(y)\}.$$

Assume contrary that $f(z) > f(x)$ for some $z \in]x, y[$. Then $z \in \Omega := \{z \in]x, y[: f(z) > f(x)\}$. Since f is a lower semicontinuous function, the set Ω is open. Therefore, there exists $z_0 \in]x, z[$ such that $z_0 \in \Omega$. Since $z, z_0 \in \Omega$, by semistrict quasiconvexity of f, we have

$$f(x) < f(z) \quad \Rightarrow \quad f(z_0) < f(z),$$

and

$$f(y) < f(z_0) \quad \Rightarrow \quad f(z) < f(z_0),$$

which is a contradiction. □

Definition 1.29 Let K be a nonempty open subset of $\mathbb{R}^n$. A differentiable function $f : K \to \mathbb{R}$ is said to be

(a) *pseudoconvex* if for all $x, y \in K$,

$$\langle \nabla f(x), y - x \rangle \geq 0 \quad \Rightarrow \quad f(y) \geq f(x);$$

(b) *strictly pseudoconvex* if for all $x, y \in K$, $x \neq y$,

$$\langle \nabla f(x), y - x \rangle \geq 0 \quad \Rightarrow \quad f(y) > f(x).$$

A function f is (strictly) pseudoconcave if $-f$ is (strictly) pseudoconvex.

Clearly, every differentiable convex function is pseudoconvex, but the converse is not true. For example, the function $f : \mathbb{R} \to \mathbb{R}$, defined by $f(x) = x + x^3$, is pseudoconvex, but not convex.

Theorem 1.21 ([6, 19]) *Let $K \subseteq \mathbb{R}^n$ be a nonempty, open and convex set and $f : K \to \mathbb{R}$ be a differentiable and pseudoconvex function. Then f is both semistrictly quasiconvex and quasiconvex.*

The converse of above theorem does not hold. For example, the function $f : \mathbb{R} \to \mathbb{R}$, defined by $f(x) = x^3$, is quasiconvex, but not pseudoconvex because for $x = 0$ and $y = -1$, $\langle \nabla f(x), y - x \rangle = 0$ and $f(y) < f(x)$.

Definition 1.30 Let K be a nonempty open convex subset of $\mathbb{R}^n$. A differentiable function $f : K \to \mathbb{R}$ is said to be *pseudolinear* if it is both pseudoconvex and pseudoconcave.

Some of the examples of pseudolinear function defined on $\mathbb{R}$ are e^x, $x + x^3$ and $\tan^{-1} x$.

We present certain characterizations of a pseudolinear function given by Chew and Choo [44].

Theorem 1.22 *Let K be a nonempty open convex subset of $\mathbb{R}^n$ and $f : K \to \mathbb{R}$ be a differentiable function. Then the following statements are equivalent:*

(a) *f is a pseudolinear function;*
(b) *For any $x, y \in K$, $\langle \nabla f(x), y - x \rangle = 0$ if and only if $f(x) = f(y)$;*
(c) *There exists a real-valued function p defined on $K \times K$ such that for any $x, y \in K$,*

$$p(x, y) > 0 \quad \textit{and} \quad f(y) = f(x) + p(x, y)\langle \nabla f(x), y - x \rangle.$$

The function p obtained in the above theorem is called the *proportional function* of f.

Theorem 1.23 ([6, Thoerem 1.39]) *Let K be a nonempty open convex subset of $\mathbb{R}^n$ and $f : K \to \mathbb{R}$ be a continuously differentiable function. Then f is pseudolinear if and only if for any $x, y \in K$,*

$$\langle \nabla f(x), y-x \rangle = 0 \quad \Rightarrow \quad f(x) = f((1-\lambda)x + \lambda y), \quad \textit{for all } \lambda \in [0, 1]. \tag{1.11}$$

Now we give a brief introduction of the concept of monotonicity and give some characterizations of convex and generalized convex functions in terms of monotonicity of their gradient function.

Definition 1.31 Let K be a nonempty subset of $\mathbb{R}^n$. A map $F : K \to \mathbb{R}^n$ is said to be

(a) *monotone* if for all $x, y \in K$, $x \neq y$, we have

$$\langle F(y) - F(x), y - x \rangle \geq 0;$$

(b) *strictly monotone* if for all $x, y \in K$, $x \neq y$, we have

$$\langle F(y) - F(x), y - x \rangle > 0;$$

(c) *strongly monotone* with modulus σ if there exists a real number $\sigma > 0$ such that for all $x, y \in K$, $x \neq y$, we have

$$\langle F(y) - F(x), y - x \rangle \geq \sigma \|y - x\|^2.$$

It is clear that a strictly monotone map is monotone but the converse is not true. For example, the map $F : \mathbb{R}^2 \to \mathbb{R}^2$ defined by $F(x_1, x_2) = (2x_1, 0)$, is monotone but not strictly monotone as the definition fails at $x = (0, 1)$, $y = (0, 2)$.

Also, every strongly monotone map is strictly monotone but the converse is not true. For instance, the map $F : \mathbb{R} \to \mathbb{R}$ defined by

$$F(x) = \begin{cases} 1 + x^2, & \text{if } x \geq 0, \\ 1 - x^2, & \text{if } x < 0, \end{cases}$$

is strictly monotone but it is not strongly monotone. We observe that if we restrict the domain of the function F defined above to $[1, \infty)$, then it is strongly monotone with modulus $\sigma = 2$.

In view of Theorem 1.15, we have the following result.

Theorem 1.24 *Let K be a nonempty open convex subset of $\mathbb{R}^n$. A differentiable function $f : K \to \mathbb{R}$ is*

(a) *convex if and only if its gradient ∇f is monotone;*
(b) *strictly convex if and only if its gradient ∇f is strictly monotone.*

Analogous to Theorem 1.24, we have the following theorem.

Theorem 1.25 ([6, Theorem 4.2]) *Let K be a nonempty open convex subset of $\mathbb{R}^n$. A differentiable function $f : K \to \mathbb{R}$ is strongly convex with modulus $\rho > 0$ if and only if its gradient ∇f is strongly monotone with modulus $\sigma = 2\rho$.*

Next we define generalized monotone maps and relate generalized convexity with generalized monotonicity of its gradient function. Karamardian [88] introduced the concept of pseudomonotone maps whereas the notions of strict pseudomonotonicity and quasimonotonicity were introduced by Hassouni [73] and independently by Karamardian and Schaible [89].

Definition 1.32 Let K be a nonempty subset of $\mathbb{R}^n$. A map $F : K \to \mathbb{R}^n$ is said to be

(a) *quasimonotone* if for all $x, y \in K$, $x \neq y$, we have

$$\langle F(x), y - x \rangle > 0 \quad \Rightarrow \quad \langle F(y), y - x \rangle \geq 0;$$

(b) *pseudomonotone* if for all $x, y \in K$, $x \neq y$, we have

$$\langle F(x), y - x \rangle \geq 0 \quad \Rightarrow \quad \langle F(y), y - x \rangle \geq 0;$$

(c) *strictly pseudomonotone* if for all $x, y \in K$, $x \neq y$, we have

$$\langle F(x), y - x \rangle \geq 0 \quad \Rightarrow \quad \langle F(y), y - x \rangle > 0.$$

It is clear that a (strictly) monotone map is (strictly) pseudomonotone but the converse is not true. For example, the map $F : \mathbb{R} \to \mathbb{R}$ defined by $F(x) = xe^{-x^2}$ is pseudomonotone but the definition of monotonicity fails at $x = 1$, $y = 2$.

A strictly pseudomonotone map is pseudomonotone but the converse is not true. For instance, the map $F : \mathbb{R} \to \mathbb{R}$ defined as $F(x) = \max\{x, 0\}$ is pseudomonotone but it is not strictly pseudomonotone.

Also, every pseudomonotone map is quasimonotone but the converse is not true as the map $F(x) = x^2$ is quasimonotone on $\mathbb{R}$ but it is not pseudomonotone on $\mathbb{R}$.

The following result gives a characterization of quasiconvex functions.

Theorem 1.26 ([6, Theorem 4.3]) *Let K be a nonempty open convex subset of $\mathbb{R}^n$. A differentiable function $f : K \to \mathbb{R}$ is quasiconvex if and only if its gradient ∇f is quasimonotone.*

As expected we have a similar characterization for (strict) pseudoconvexity of a function in terms of the (strict) pseudomonotonicity of the gradient map.

Theorem 1.27 ([6, Theorem 4.4]) *Let K be a nonempty open convex subset of $\mathbb{R}^n$. A differentiable function $f : K \to \mathbb{R}$ is pseudoconvex (respectively, strictly pseudoconvex) if and only if its gradient ∇f is pseudomonotone (respectively, strictly pseudomonotone).*

Proof Assume that f is pseudoconvex but ∇f is not pseudomonotone. Then there exist $x, y \in K$, $x \neq y$, such that

$$\langle \nabla f(x), y - x \rangle \geq 0 \quad \text{and} \quad \langle \nabla f(y), x - y \rangle > 0.$$

Since f is pseudoconvex, the first inequality leads to $f(y) \geq f(x)$, and the second one leads to $f(x) > f(y)$ as every pseudoconvex function is quasiconvex. We thus arrive at a contradiction as the two conclusions are contradictory to each other.

Conversely, assume on the contrary that ∇f is pseudomonotone but f is not pseudoconvex. Then there exist $x, y \in K$ such that

$$\langle \nabla f(x), y - x \rangle \geq 0 \quad \text{and} \quad f(y) < f(x).$$

By the mean value theorem, there exists $z = (1 - \lambda)x + \lambda y$ for some $\lambda \in]0, 1[$ such that

$$f(y) - f(x) = \langle \nabla f(z), y - x \rangle = (1/\lambda)\langle \nabla f(z), z - x \rangle.$$

Since $f(y) < f(x)$, it follows that $\langle \nabla f(z), z - x \rangle < 0$. Now by pseudomonotonicity of ∇f, we have

$$\langle \nabla f(x), z - x \rangle < 0, \text{ that is, } \langle \nabla f(x), y - x \rangle < 0,$$

which leads to a contradiction. □

The following concepts of strict and semistrict quasimonotonicity were introduced by Blum and Oettli [28].

Definition 1.33 Let K be a nonempty convex subset of $\mathbb{R}^n$. A map $F : K \to \mathbb{R}^n$ is said to be

(a) *strictly quasimonotone* if F is quasimonotone and for all $x, y \in K$, $x \neq y$, there exists $z \in]x, y[$ such that $\langle F(z), y - x \rangle \neq 0$;
(b) *semistrictly quasimonotone* if F is quasimonotone and for $x, y \in K$, $x \neq y$,

$$\langle F(x), y - x \rangle > 0 \Rightarrow \text{ there exists } z \in](x + y)/2, y[\quad \text{such that } \langle F(z), y - x \rangle > 0.$$

Obviously, a pseudomonotone map is semistrictly quasimonotone, and a strictly pseudomonotone map is strictly quasimonotone.

The following diagram gives the relationship among different classes of monotone maps defined above.

Strong monotonicity
⇓
Strict monotonicity ⇒ Monotonicity
⇓ ⇓
Strict pseudomonotonicity ⇒ Pseudomonotonicity ⇒ Semistrict Quasimonotonicity
⇓ ⇓
Strict quasimonotonicity ⇒ Quasimonotonicity

Theorem 1.28 *Let K be a nonempty convex subset of $\mathbb{R}^n$. If $F : K \to \mathbb{R}^n$ is strictly quasimonotone, then it is semistrictly quasimonotone.*

Proof If $\langle F(x), y-x\rangle > 0$ for all $x, y \in K$, $x \neq y$, then

$$\langle F(x), z-x\rangle > 0, \quad \text{for all } z \in]x, y[.$$

Since F is quasimonotone, we have $\langle F(z), z-x\rangle \geq 0$ which implies that

$$\langle F(z), y-x\rangle \geq 0, \quad \text{for all } z \in]x, y[.$$

Since F is strictly quasimonotone, there exists $\hat{z} \in](x+y)/2, y[$ such that $\langle F(\hat{z}), y-x\rangle \neq 0$. Thus, we have $\langle F(\hat{z}), y-x\rangle > 0$, that is, F is semistrictly quasimonotone.

□

We now link strict quasiconvexity of a function with strict quasimonotonicity of its gradient.

Theorem 1.29 ([6, Theorem 4.7]) *Let K be a nonempty open convex subset of $\mathbb{R}^n$. A differentiable function $f : K \to \mathbb{R}$ is strictly quasiconvex if and only if its gradient ∇f is strictly quasimonotone.*

The following theorem relates semistrict quasiconvexity of a function with semistrict quasimonotonicity of its gradient.

Theorem 1.30 ([6, Theorem 4.8]) *Let K be a nonempty open convex subset of $\mathbb{R}^n$. A differentiable function $f : K \to \mathbb{R}$ is semistrictly quasiconvex if and only if its gradient ∇f is semistrictly quasimonotone.*

1.3 Generalized Derivatives

In order to deal with the optimality conditions for optimization problems of functions whose ordinary derivative does not exist but they have some kind of generalized derivatives, we give the concept of some generalized derivatives.

Definition 1.34 Let X and Y be locally convex topological vector spaces, K be a nonempty convex subset of X, and $f : X \to Y$ be a given mapping.

(a) If for some $x \in K$ and some $d \in X$, the limit

$$\langle f'(x), d\rangle := \lim_{t\to 0} \frac{1}{t}\left[f(x+td) - f(x)\right]$$

exists, then $\langle f'(x), d\rangle$ is called the *directional derivative* of f at x in the direction d. If this limit exists for all $d \in X$, then f is called directionally differentiable at x.

(b) If for some $x \in K$ and all $d \in X$, the limit

$$\langle Df(x), d\rangle := \lim_{t\to 0} \frac{1}{t}\left[f(x+td) - f(x)\right]$$

exists and $Df(x)$ is a continuous linear map from X to Y, then $Df(x)$ is called the *Gâteaux derivative* of f at x, and f is called Gâteaux differentiable at x. If f is Gâteaux differentiable at every $x \in K$, then we say that f is Gâteaux differentiable on K.

Example 1.13 It is well known that the function $f : \mathbb{R} \to \mathbb{R}$, defined by $f(x) = |x|$, is not Gâteaux differentiable, but it is directionally differentiable at 0.

Definition 1.35 Let $(X, \|\cdot\|_X)$ and $(Y, \|\cdot\|_Y)$ be normed spaces, K be a nonempty open subset of X, and $f : X \to Y$ be a mapping. Let $x \in K$ be given. If there is a continuous linear map $f'(x) : X \to Y$ with

$$\lim_{\|h\|_X \to 0} \frac{\|f(x+h) - f(x) - \langle f'(x), h\rangle\|_Y}{\|h\|_X} = 0,$$

then $f'(x)$ is called the *Fréchet derivative* of f at x and f is called *Fréchet differentiable* at x.

Lemma 1.3 ([81, Lemma 2.17]) *Let $(X, \|\cdot\|_X)$ and $(Y, \|\cdot\|_Y)$ be normed spaces, K be a nonempty open subset in X, and $f : X \to Y$ be a mapping. If the Fréchet derivative of f at $x \in K$ exists, then the Gâteaux derivative of f at x exists and both are equal.*

Definition 1.36 Let $K \subseteq \mathbb{R}^n$ be an open convex set and $f : \mathbb{R}^n \supseteq K \to \mathbb{R}$ be a real-valued function. The *upper* and *lower Dini directional derivatives of f at $x \in K$* in the direction $d \in \mathbb{R}^n$ are defined as

$$D^+f(x; d) = \limsup_{t \downarrow 0} \frac{f(x+td) - f(x)}{t},$$

and

$$D_+f(x; d) = \liminf_{t \downarrow 0} \frac{f(x+td) - f(x)}{t},$$

respectively.

Remark 1.16 It is easy to see that $D_+f(x; d) \leq D^+f(x; d)$. If the function f is convex, then the upper and lower Dini directional derivatives are equal to the directional derivative.

Definition 1.37 Let K be a nonempty convex subset of $\mathbb{R}^n$. The function $f : K \to \mathbb{R}$ is called *radially upper (lower) semicontinuous* (also known as *upper (lower) hemicontinuous* on K) if for every pair of distinct points $x, y \in K$, the function f is upper (lower) semicontinuous along the line segment $[x, y]$.

Theorem 1.31 (Diewert Mean Value Theorem) [50] *Let K be a nonempty convex subset of $\mathbb{R}^n$ and $f : K \to \mathbb{R}$ be radially upper semicontinuous on K. Then for any* *pair x, y of distinct points of K, there exists $w \in [x, y[$ such that*

$$f(y) - f(x) \geq D^+f(w; y - x),$$

where $[x, y[$ denotes the line segment joining x and y including the endpoint x. In other words, there exists $\lambda \in [0, 1[$ such that

$$f(y) - f(x) \geq D^+f(w; y - x), \quad \text{where } w = x + \lambda(y - x).$$

If f is radially lower semicontinuous on K, then for any pair x, y of distinct points of K, there exists $v \in [x, y[$ such that

$$f(y) - f(x) \leq D_+f(v; y - x),$$

Definition 1.38 [97] Let $K \subseteq \mathbb{R}^n$ be a nonempty set and $q : K \times \mathbb{R}^n \to \mathbb{R}$ be a bifunction. A function $f : K \to \mathbb{R}$ is said to be

(a) *q-quasiconvex* if for all $x, y \in K$,

$$f(x) \leq f(y) \quad \Rightarrow \quad q(y; x - y) \leq 0;$$

(b) *q-quasiconcave* if $-g$ is q-quasiconvex;
(c) *q-pseudoconvex* if for all $x, y \in K$, $x \neq y$,

$$f(x) < f(y) \quad \Rightarrow \quad q(y; x - y) < 0;$$

(d) *strictly q-pseudoconvex* if for all $x, y \in K$, $x \neq y$,

$$f(x) \leq f(y) \quad \Rightarrow \quad q(y; x - y) < 0;$$

(e) *q-pseudoconcave* if $-f$ is q-pseudoconvex;
(f) *q-pseudolinear* if it is both q-pseudoconvex as well as q-pseudoconcave.

If $q(x; d) = D^+f(x; d)$ ($q(x; d) = D_+f(x; d)$), then the above definitions are called D^+-quasiconvex, D^+-quasiconcave, D^+-pseudoconvex, strictly D^+-pseudoconvex, D^+-pseudoconcave, and D^+-pseudolinear (D_+-quasiconvex, D_+-quasiconcave, D_+-pseudoconvex, strictly D_+-pseudoconvex, D_+-pseudoconcave, and D_+-pseudolinear), respectively.

Remark 1.17 It is clear that strict q-pseudoconvexity implies q-pseudoconvexity and also q-quasiconvexity. But, as pointed out in [97], neither q-quasiconvexity implies q-pseudoconvexity nor the reverse implication holds.

Example 1.14 Let $K = [-1, 1]$ and

$$f(x) = \begin{cases} x, & \text{if } x \geq 0 \\ \frac{1}{2}x, & \text{if } x < 0. \end{cases}$$

Then f is D^+-pseudolinear over K.

Remark 1.18 Let K be a nonempty convex subset of $\mathbb{R}^n$ and $f : \mathbb{R}^n \to \mathbb{R}$ be a function.

(a) If f is D^+-pseudoconvex over K, then it is pseudoconvex over K in the sense of Diewert [50], that is, for all $x, y \in K, f(x) < f(y)$ implies $D_+f(y; x - y) < 0$.
(b) If f is D^+-pseudoconvex (D_+-pseudoconcave) over K and lower semicontinuous (upper semicontinuous), then it is quasiconvex (quasiconcave) over K (see Corollary 15 in [50]).
(c) If f is quasiconvex over K, then for all $x, y \in K$,

$$f(x) \le f(y) \quad \Rightarrow \quad D^+f(y; x - y) \le 0.$$

(d) If f is quasiconcave over K, then for all $x, y \in K$,

$$f(x) \ge f(y) \quad \Rightarrow \quad D^+f(y; x - y) \ge 0.$$

(e) Any linear fractional function whose denominator is positive over K is D^+-pseudolinear.

Lemma 1.4 ([50]) *Let $K \subseteq \mathbb{R}^n$ be nonempty set and $f : K \to \mathbb{R}$ be upper semicontinuous and D^+-pseudoconvex, that is, for all $x, y \in K$, $f(x) < f(y) \Rightarrow D^+f(y; x - y) < 0$. Then f is quasiconvex and semistrictly quasiconvex.*

Lemma 1.5 ([97]) *Let $K \subseteq \mathbb{R}^n$ be a nonempty set, $f : K \to \mathbb{R}$ be a function, and $p, q : K \times \mathbb{R}^n \to \mathbb{R}$ be bifunctions such that for all $x \in K$ and all $d \in \mathbb{R}^n$, $p(x; d) \le q(x; d)$. Then q-quasiconvexity, q-pseudoconvexity, and strict q-pseudoconvexity imply p-quasiconvexity, p-pseudoconvexity, and strict p-pseudoconvexity, respectively.*

Proof Let f be q-quasiconvex. Then we have for all $x, y \in K$,

$$f(x) \le f(y) \quad \Rightarrow \quad q(y; x - y) \le 0.$$

Because $p(y; x - y) \le q(y; x - y)$ for all $x, y \in K$, the implication

$$f(x) \le f(y) \quad \Rightarrow \quad p(y; x - y) \le 0$$

also holds, and thus, f is p-quasiconvex. The remaining assertions can be proven in a similar way. □

Definition 1.39 [68, 97] Let $K \subseteq \mathbb{R}^n$ be a nonempty set. A bifunction $q : K \times \mathbb{R}^n \to \mathbb{R}$ is said to be *pseudomonotone* if for every pair of distinct points $x, y \in K$, we have

$$q(x; y - x) \ge 0 \quad \Rightarrow \quad q(y; x - y) \le 0. \tag{1.12}$$

Remark 1.19 The above implication (1.12) is equivalent to the following implication:

$$q(y; x - y) > 0 \quad \Rightarrow \quad q(x; y - x) < 0. \tag{1.13}$$

Lemma 1.6 *A bifunction $q : K \times \mathbb{R}^n \to \mathbb{R}$ is pseudomonotone if and only if for every pair of distinct points $x, y \in K \subseteq \mathbb{R}^n$, we have*

$$q(x; y-x) > 0 \quad \Rightarrow \quad q(y; x-y) < 0. \tag{1.14}$$

Proof The implication (1.14) is equivalent to the following implication:

$$q(y; x-y) \geq 0 \quad \Rightarrow \quad q(x; y-x) \leq 0.$$

Interchanging x and y, we get (1.12). □

Lemma 1.7 ([129]) *Let $f : K \to \mathbb{R}$ be radially upper semicontinuous on $K \subseteq \mathbb{R}^n$ and $q : K \times \mathbb{R}^n \to \mathbb{R}$ be subodd and positively homogeneous in the second argument such that for all $x \in K$, $q(x; \cdot) \leq D^+f(x; \cdot)$. Then*

(a) *f is quasiconvex over K if and only if it is q-quasiconvex;*
(b) *f is q-pseudoconvex if and only if q is pseudomonotone.*

Definition 1.40 Let K be a nonempty subset of a Banach space, $f : K \to \mathbb{R}$ be locally Lipschitz at a given point $x \in K$. The *Clarke directional derivative* of f at $x \in K$ in the direction of a vector $v \in K$, denoted by $f^\circ(x; v)$, is defined by

$$f^\circ(x; v) = \limsup_{\substack{y \to x \\ t \downarrow 0}} \frac{f(y+tv) - f(y)}{t}.$$

Clearly, for all $x, v \in K$, we have $D^+f(x; v) \leq f^\circ(x; v)$.

Definition 1.41 Let K be a nonempty subset of a Banach space with its dual space X^*, $f : K \to \mathbb{R}$ be locally Lipschitz at a given point $x \in K$. The *Clarke generalized subdifferential* of f at $x \in K$, denoted by $\partial^c f(x)$, is defined by

$$\partial^c f(x) = \{\xi \in X^* : f^\circ(x; v) \geq \langle \xi, v \rangle \text{ for all } v \in K\}.$$

Remark 1.20 It follows from the definition that for every $v \in K$,

$$f^\circ(x; v) = \max\{\langle \xi, v \rangle : \xi \in \partial^c f(x)\}.$$

If f is convex, then the Clarke generalized subdifferential coincides with the subdifferential of f in the sense of convex analysis [127].

Proposition 1.5 ([48, Proposition 2.1.1]) *Let K be a nonempty subset of a normed space X and $f : K \to \mathbb{R}$ be a locally Lipschitz at a point $x \in K$.*

(a) *The function $v \mapsto f^\circ(x; v)$ is finite, positively homogeneous, and subadditive, and satisfies $|f^\circ(x; v)| \leq k\|v\|$.*
(b) *$f^\circ(x; v)$ is upper semicontinuous as a function of $(x; v)$ and, satisfies the Lipschitz condition as a function of v alone.*
(c) *$f^\circ(x; -v) = (-f)^\circ(x; v)$.*

Lemma 1.8 ([48]) *Let K be a nonempty subset of a Banach space and $f : K \to \mathbb{R}$ be locally Lipschitz. Then the set-valued map $\partial^c f$ is upper semicontinuous (see, Sect. 1.4 for upper semicontinuity of a set-valued map).*

Theorem 1.32 (Lebourg's Mean Value Theorem) [48] *Let x and y be points in a Banach space X, and suppose that f is Lipschitz on an open set containing the line segment $[x, y]$. Then there exists a point u in $]x, y[$ such that*

$$f(y) - f(x) \in \langle \partial^c f(u), y - x \rangle.$$

Since when f is convex, the Clarke subdifferential coincides with the subdifferential of f in the sense of convex analysis, Theorem 1.32 also holds for subdifferential in the sense of convex analysis.

1.4 Tools from Nonlinear Analysis

In this section, we recall some concepts and results from nonlinear analysis which will be used in the sequel.

1.4.1 Continuity for Set-Valued Maps

Definition 1.42 ([15, 20]) Let X and Y be topological spaces. A set-valued map $T : X \to 2^Y$ is said to be

(a) *upper semicontinuous at* $x_0 \in X$ if for any open set V in Y containing $T(x_0)$, there exists an open neighborhood U of x_0 in X such that $T(x) \subseteq V$ for all $x \in U$;
(b) *lower semicontinuous at* $x_0 \in X$ if for any open set V in Y such that $V \cap T(x_0) \neq \emptyset$, there exists an open neighborhood U of x_0 in X such that $T(x) \cap V \neq \emptyset$ for all $x \in U$;
(c) *upper semicontinuous* (respectively, *lower semicontinuous*) on X if it upper semicontinuous (respectively, lower semicontinuous) at every point $x \in X$;
(d) *continuous* on X if it is upper semicontinuous as well as lower semicontinuous on X;
(e) *compact* if there exists a compact subset $\mathcal{K} \subseteq Y$ such that $T(X) \subseteq \mathcal{K}$;
(f) *closed* if its graph $\mathcal{G}(T) := \{(x, y) : x \in X,\ y \in T(x)\}$ is closed in $X \times Y$.

Remark 1.21 If $T(x)$ is a singleton in a neighborhood of x, then the upper semicontinuous and the lower semicontinuous of T at x are equivalent.

Example 1.15 Let $X = \mathbb{R}$, $Y = \mathbb{R}^2$, and consider the set-valued mapping $T : X \to 2^Y$ given by

$$T(x) := [(1 - x, x), (1, 1)],$$

Fig. 1.14 An illustration of the set-valued mapping T defined in Example 1.15

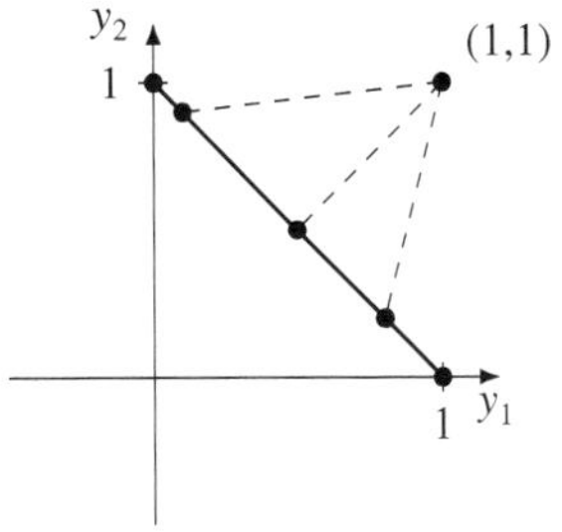

where $[(a,b),(c,d)]$ is the line segment between (a,b) and (c,d) (see Fig. 1.14). Then T is upper and lower semicontinuous, and therefore continuous. If the set-valued map is changed slightly to

$$T_1(x) := \begin{cases} [(1-x,x),(1,1)], & \text{if } x \in [0,1], \\ \emptyset, & \text{else,} \end{cases}$$

then T_1 is upper semicontinuous, but T_1 is not lower semicontinuous. If we choose

$$T_2(x) := \begin{cases} [(1-x,x),(1,1)], & \text{if } x \in\,]0,1[, \\ \emptyset, & \text{else,} \end{cases}$$

then T_2 is lower semicontinuous, but not upper semicontinuous, and therefore not continuous.

Several other examples of upper semicontinuous and lower semicontinuous set-valued maps can be found in [3, 20].

Lemma 1.9 ([15, 20]) *Let X and Y be topological spaces. A set-valued map $T : X \to 2^Y$ is lower semicontinuous at $x \in X$ if and only if for any net $\{x_\alpha\} \subset X$, $x_\alpha \to x$ and for any $y \in T(x)$, there is a net $\{y_\alpha\}$ such that $y_\alpha \in T(x_\alpha)$ and $y_\alpha \to y$.*

Lemma 1.10 *Let X be a topological space, Y be a topological vector space and $T : X \to 2^Y$ be a set-valued map such that $T(x)$ is nonempty and compact for all $x \in X$. Then T is upper semicontinuous at $x \in X$ if and only if for any nets $\{x_\mu\} \subset X$ with $x_\mu \to x$ and $\{y_\mu\} \subset Y$ with $y_\mu \in T(x_\mu)$, there exists a subnet $\{y_\nu\} \subset \{y_\mu\}$ such that $y_\nu \to y$ for some $y \in T(x)$.*

Proof Let T be upper semicontinuous at $x \in X$. Assume that $\{x_\mu\} \subseteq X$ with $x_\mu \to x$ and $\{y_\mu\} \in Y$ with $y_\mu \in T(x_\mu)$. Let $\mathcal{V} \in \mathfrak{V}$, where $\mathfrak{V}$ stands for a basis of neighborhoods of $\mathbf{0}$. Then $\bigcup_{u \in T(x)} \{u + \mathcal{V}\}$ is an open covering of $T(x)$. Since $T(x)$ is compact, there exists a finite subset $\{u_1, u_2, \ldots, u_m\} \subseteq T(x)$ such that $\bigcup_{i=1}^{m} \{u_i + \mathcal{V}\} \supseteq T(x)$. Since T is upper semicontinuous at $x \in X$, there exists a neighborhood $\mathcal{U}$ of x such that

$$T(x') \subseteq \bigcup_{i=1}^{n} \{u_i + \mathcal{V}\}, \quad \text{for all } x' \in \mathcal{U}.$$

Since $x_\mu \to x$, there exists μ' such that $\{x_\mu\} \subset \mathcal{U}$ for all $\mu \geq \mu'$. Hence for each $\mu \geq \mu'$, $y_\mu \in u_i + \mathcal{V}$ for some $u_i \in \{u_1, u_2, \ldots, u_m\}$. Therefore, there exist a subnet $\{y_\nu\} \subseteq \{y_\mu\}$ and $u_i \in \{u_1, u_2, \ldots, u_m\}$ such that $\{y_\nu\} \subset u_i + \mathcal{V}$. Let this $u_i =: u^{\mathcal{V}}$. Corresponding to $\mathcal{V} \in \mathfrak{V}$, that is, for each $\mathcal{V} \in \mathfrak{V}$, there exist a subnet $\{y_\nu\} \subset \{y_\mu\}$ and $u^{\mathcal{V}} \in T(x)$ such that

$$\{y_\nu\} \subseteq u^{\mathcal{V}} + \mathcal{V}. \tag{1.15}$$

Let $\mathcal{V}_2 \geq \mathcal{V}_1$ for each $\mathcal{V}_1, \mathcal{V}_2 \in \mathfrak{V}$, if $\mathcal{V}_2 \subset \mathcal{V}_1$. Then $\mathfrak{V}$ is a directed set and $\{u^{\mathcal{V}}\}$ is a net of $T(x)$. Since $T(x)$ is compact, there exist $\mathfrak{V}' \subset \mathfrak{V}$ and $y \in T(x)$ such that $u^{\mathcal{V}'} \to y$, where $\mathcal{V}' \in \mathfrak{V}'$. Let V be a neighborhood of y. Since Y is a topological vector space, there exists $\bar{\mathcal{V}} \in \mathfrak{V}$ such that $\bar{\mathcal{V}} + \bar{\mathcal{V}} \in V - y$. Since $u^{\mathcal{V}'} \to y$, there exists $\mathcal{V}'' \in \mathfrak{V}'$ such that

$$u^{\mathcal{V}'} \in y + \bar{\mathcal{V}}, \quad \text{for all } \mathcal{V}' \geq \mathcal{V}''.$$

Hence for any $\Upsilon \in \mathfrak{V}'$ with $\Upsilon \geq \bar{\mathcal{V}} \cap \mathcal{V}''$,

$$\{u^{\Upsilon}\} + \mathcal{V}' \subset (y + \bar{\mathcal{V}}) + \bar{\mathcal{V}} \subset V.$$

Therefore by (1.15), there exists a subnet $\{y_\nu\} \subset \{y_\mu\}$ such that $\{y_\nu\} \subset V$. Thus there exists a subnet $\{y_\nu\} \subset \{y_\mu\}$ such that $y_\nu \to y$ for some $y \in T(x)$.

Suppose that T is not upper semicontinuous at $x \in X$. Then there exists an open set $\mathcal{V}$ containing $T(x)$ such that for any neighborhood $\mathcal{U}_\mu$ of x, there is a point $x_\mu \in \mathcal{U}_\mu$ with $T(x_\mu) \cap \mathcal{V}^c \neq \emptyset$. Hence, there exist $\{x_\nu\} \subset X$ converging to x and $y_\nu \in T(x_\nu) \cap \mathcal{V}^c$. Since $y_\nu \notin T(x)$ for all ν, $\{y_\nu\}$ does not have subnet converging to some point of $T(x)$. □

The following results provide the characterization of upper semicontinuity and lower semicontinuity, respectively.

Proposition 1.6 *Let X and Y be topological spaces and $T : X \to 2^Y$ be a set-valued map such that $T(x)$ is compact for each $x \in X$. Then T is upper semicontinuous if and only if for each open subset G of Y, the set*

$$T_+^{-1}(G) = \{x \in X : T(x) \subseteq G\}$$

is open.

Proposition 1.7 *Let X and Y be topological spaces. A set-valued map $T : X \to 2^Y$ is lower semicontinuous if and only if $T^{-1}(G) = \{x \in X : T(x) \cap G \neq \emptyset\}$ is open for every open subset G of Y.*

Proposition 1.8 ([20, p.112, Theorem 6]) *Let X and Y be topological spaces and $T : X \to 2^Y$ be a set-valued map. If T is upper semicontinuous, then T is closed.*

Proposition 1.9 ([20, p.112, Theorem 7]) *Let X and Y be topological spaces and $T_1, T_2 : X \to 2^Y$ be set-valued maps. If T_2 is upper semicontinuous, then the mapping $T = T_1 \cap T_2$ is upper semicontinuous.*

The following proposition shows that under the compactness assumption on the space Y, a set-valued map is closed if and only if it is upper semicontinuous. Notice that this proposition cannot be applied to Example 1.15, as $\mathbb{R}^2$ is not compact.

Proposition 1.10 ([20, p.112, Corollary]) *Let X and Y be topological spaces such that Y is compact and $T : X \to 2^Y$ be a set-valued map. Then T is closed if and only if it is upper semicontinuous.*

Proof Assume that T is closed. Let $\widetilde{T}$ be a set-valued map such that $\widetilde{T}(x) = Y$ for each $x \in X$. Then, by Proposition 1.9, $T = T \cap \widetilde{T}$ is upper semicontinuous, because $\widetilde{T}$ is upper semicontinuous.

The reverse implication follows from Proposition 1.8. □

Lemma 1.11 ([20, Theorem 3]) *Let X and Y be topological spaces. If $T : X \to 2^Y$ is upper semicontinuous on X and D is a compact subset of X, then $T(D)$ is compact.*

Example 1.16 The function T_1, defined in Example 1.15, is upper semicontinuous, and for $D_1 = [0, 1]$, $T_1(D_1)$ is compact. However, for the compact set $D_2 = \left[\frac{1}{4}, \frac{1}{2}\right]$, the set $T_2(D_2)$, where T_2 is also defined in Example 1.15, is compact, but T_2 is not upper semicontinuous.

The following lemma has been applied to the study of game theory (see [141]).

Lemma 1.12 ([141]) *Let X and Y be Hausdorff topological vector spaces such that Y is compact. Let $f : X \times Y \to \mathbb{R}$ be a lower semicontinuous function and for each fixed $y \in Y$, the function $x \mapsto f(x, y)$ be upper semicontinuous on X. Then the function $\Psi : X \to \mathbb{R}$ defined by*

$$\Psi(x) = \min_{y \in Y} f(x, y), \quad \textit{for all } x \in X$$

is continuous on X.

Let X be a metric space with metric d. We use the following notations:

$$2^X_q = \text{ set of all nonempty and compact subsets of } X;$$

$$2^X_{cl} = \text{ set of all nonempty, closed and bounded subsets of } X;$$

For any nonempty subset M and N of X and for any $x \in M$, we define the distance from x to N by

$$d(x, N) = \inf_{y \in N} d(x, y).$$

We define the number $d(M, N)$ as

$$d(M.N) = \sup_{x \subset M} d(x, N) = \sup_{x \in M} \inf_{y \in N} d(x, y).$$

The *Hausdorff metric* $\mathscr{H}(M,N)$ on 2^X_{cl} is defined as

$$\mathscr{H}(M,N) = \max\{d(M,N), d(N,M)\}, \quad \text{for all } M,N \in 2^X_{cl}.$$

Then $\mathscr{H}$ is metric on 2^X_{cl}. If (X,d) is complete metric space with metric d, then $(2^X_{cl}, \mathscr{H})$ is a complete metric space.

Lemma 1.13 (Nadler's Theorem) [115] *Let (X,d) be a metric space and $\mathscr{H}$ be a Hausdorff metric on 2^X_{cl}. If M and N are compact sets in X, then for each $x \in M$, there exists $y \in N$ such that*

$$d(x,y) \leq \mathscr{H}(M,N).$$

Now we define the continuity of a set-valued map in terms of ε and δ.

Definition 1.43 Let (X,d) and (Y,ρ) be metric spaces. A set-valued map $T : X \to 2^Y_q$ is said to be *$\mathscr{H}$-continuous* on X if for every $\varepsilon > 0$, there exists a $\delta > 0$ such that for all $x, y \in X$

$$\mathscr{H}(T(x), T(y)) < \varepsilon \quad \text{whenever} \quad d(x,y) < \delta.$$

Remark 1.22 The notions of continuity in the sense of Definitions 1.42 and 1.43 are equivalent if T is compact-valued.

Definition 1.44 ($\mathcal{H}$-Hemicontinuity) [145] Let K be a nonempty convex subset of a normed space X and Y be a normed vector space. A nonempty compact-valued map $T : K \to 2^{\mathcal{L}(X,Y)}$ is said to be *$\mathscr{H}$-hemicontinuous* if for any $x, y \in K$, the mapping $\alpha \mapsto \mathscr{H}(T(x + \alpha(y - x)), T(x))$ is continuous at 0^+, where $\mathscr{H}$ is the Hausdorff metric defined on $2^{\mathcal{L}(X,Y)}_{cl}$.

Definition 1.45 (u-Hemicontinuity) Let X and Y be topological vector spaces. A set-valued map $T : X \supseteq K \to 2^Y \setminus \{\emptyset\}$ is said to be *u-hemicontinuous* if for any $x, y \in K$ and $\alpha \in [0,1]$, the set-valued map $\alpha \mapsto T(\alpha x + (1 - \alpha)y)$ is upper semicontinuous at 0^+.

1.4.2 Fixed Point Theory for Set-Valued Maps

In 1929, Knaster, Kuratowski and Mazurkiewicz [96] formulated the so-called KKM principle in the finite dimensional Euclidean space. Later, in 1961, it has been generalized to infinite dimensional Hausdorff topological vector spaces by Ky Fan [59]. Fan also established an elementary but very basic geometric lemma for set-valued maps which is called Fan's geometric lemma. In 1968, Browder gave a fixed point version of Fan's geometric lemma and this result is known as Browder fixed point theorem. Since then there have been numerous generalizations of Browder fixed point theorem their applications to coincidence and fixed point theory, minimax inequalities, variational inequalities, convex analysis, game theory, mathematical economics, social sciences, and so on.

It is well known that the famous Browder fixed point theorem [33] is equivalent to a maximal element theorem (see [138]). Such kind of maximal element theorems are useful to establish the existence of solutions of vector variational inequalities, vector equilibrium problems and their generalizations.

In this section, we recall some basic definitions from nonlinear analysis and present Fan-KKM lemma and its generalizations and some famous fixed point theorems for set-valued maps, namely, Nadler's fixed point theorem, Browder fixed point theorem and its generalizations, Kakutani fixed point theorem, etc.

Definition 1.46 Let X be a metric space and $T : X \to 2^X$ be a set-valued map with nonempty values. A point $x \in X$ is said to be a *fixed point* of T if $x \in T(x)$.

Definition 1.47 Let (X, d) and (Y, ρ) be metric spaces and $\mathscr{H}$ be a Hausdorff metric on 2^Y_{cl}. A set-valued map $T : X \to 2^Y_{cl}$ is said to be *set-valued Lipschitz map* if there exists a constant $\alpha > 0$ such that

$$\mathscr{H}(T(x), T(y)) \leq \alpha d(x, y), \quad \text{for all } x, y \in X.$$

The constant α is called a *Lipschitz constant* for T. If $\alpha < 1$, then T is called a set-valued contraction map. If $\alpha = 1$, then T is called *nonexpansive*.

In 1969, Nadler [115] extended the well-known Banach contraction principle for set-valued maps and established the following fixed point theorem.

Theorem 1.33 (Nadler's Fixed Point Theorem) [115] *Let (X, d) be a complete metric space. If $T : X \to 2^X_{cl}$ is a set-valued contraction map, then T has a fixed point.*

Definition 1.48 Let X be a topological vector space and K be a nonempty subset of X. A set-valued map $T : K \to 2^X$ is said to be a *KKM-map* if

$$\text{co}\,(\{x_1, x_2, \dots, x_m\}) \subseteq \bigcup_{i=1}^{m} T(x_i)$$

for every finite subset $\{x_1, x_2, \dots, x_m\}$ of X.

Obviously, if T is a KKM-map, then $x \in T(x)$ for every $x \in K$.

Example 1.17 Let $X = K = \mathbb{R}$ and the set-valued map $T : X \to 2^X$ be defined by $T(x) = [0, x]$, where $[0, x]$ is the line segment between 0 and x. Then T is a KKM-map.

Lemma 1.14 (Fan-KKM Lemma) [59] *Let X be a Hausdorff topological vector space and K a nonempty subset of X. Let $T : K \to 2^X$ be a KKM-map such that $T(x)$ is a closed subset of X for all $x \in K$ and compact for at least one $x \in K$. Then $\bigcap_{x \in K} T(x) \neq \emptyset$.*

Chang and Zhang [41] introduced the following concept of generalized KKM mapping.

Definition 1.49 Let K be a nonempty subset of a Hausdorff topological vector space X. A set-valued map $T : K \to 2^X$ is called a *generalized KKM map* if for any finite set $\{x_1, x_2, \ldots, x_m\} \subset K$, there is a finite subset $\{y_1, y_2, \ldots, y_m\} \subset X$ such that for any subset $\{y_{i_1}, y_{i_2}, \ldots, y_{i_k}\} \subset \{y_1, y_2, \ldots, y_m\}$, $1 \leq k \leq m$, we have

$$\text{co}\left(\{y_{i_1}, y_{i_2}, \ldots, y_{i_k}\}\right) \subset \bigcup_{j=1}^{k} T(x_{i_j}).$$

Clearly, if $T : K \to 2^X$ is a KKM map, then it is generalized KKM map. Indeed, for any finite set $\{x_1, x_2, \ldots, x_m\} \subset K$, taking $y_i = x_i$, $i = 1, 2, \ldots, m$, then since T is a KKM map, we have

$$\text{co}\left(\{y_{i_1}, y_{i_2}, \ldots, y_{i_k}\}\right) \subset \bigcup_{j=1}^{k} T(x_{i_j}).$$

However, if T is a generalized KKM map, then it may not be a KKM map.

Example 1.18 [41] Let $X = \mathbb{R}$, $K = [-2, 2]$ and $T : K \to 2^X$ be defined by

$$T(x) = \left[-\left(1 + \frac{x^2}{5}\right), 1 + \frac{x^2}{5}\right], \quad \text{for all } x \in K.$$

Then $\bigcup_{x \in K} T(x) = \left[-\frac{9}{5}, \frac{9}{5}\right]$ and $x \notin T(x)$ for all $x \in [-2, -9/5) \cup (9/5, 1]$. It follows that T is not a KKM map. Next we prove that T is a generalized KKM map. If for any finite subset $\{x_1, x_2, \ldots, x_m\}$ of K, we take $\{y_1, y_2, \ldots, y_m\} \subset [-1, 1]$, then for any finite subset $\{y_{i_1}, y_{i_2}, \ldots, y_{i_k}\} \subset \{y_1, y_2, \ldots, y_m\}$, we have

$$\text{co}\left(\{y_{i_1}, y_{i_2}, \ldots, y_{i_k}\}\right) \subset [-1, 1] = \bigcap_{x \in K} T(x) \subset \bigcup_{j=1}^{k} T(x_{i_j}),$$

that is, T is a generalized KKM map.

The following lemma is proved in [41] where convexity on K is assumed. However, Ansari et al. [7] pointed out that this lemma is true without convexity assumption on K.

Lemma 1.15 *Let K be a nonempty subset of a Hausdorff topological vector space X. If $T : K \to 2^X$ is a set-valued map and for each $x \in K$, the set $T(x)$ is finitely closed (i.e., for every finite-dimensional subspace L in X, $T(x) \cap L$ is closed in the Euclidean topology in L). Then the family of sets $\{T(x) : x \in K\}$ has the finite intersection property if and only if $T : K \to 2^X$ is a generalized KKM mapping.*

Definition 1.50 [135] Let X and Y be topological spaces. A set-valued map $T : X \to 2^Y$ is said to be *transfer open-valued* (respectively, *transfer closed-valued*) if for every $x \in X$, $y \in T(x)$ (respectively, $y \notin T(x)$), there exists a point $z \in X$ such that $y \in \text{int}(T(z))$ (respectively, $y \notin \text{cl}(T(z))$).

It is easy to see that a open-valued (respectively, closed-valued) set-valued map is a transfer open-valued (respectively, transfer closed-valued) set-valued map. But the converse is not true.

Lemma 1.16 ([42]) *Let X be a nonempty set, Y be a topological space and $T : X \to 2^Y$ be a set-valued map.*

(a) *T is transfer closed-valued if and only if $\bigcap_{x\in X} T(x) = \bigcap_{x\in X} \text{cl}(T(x))$.*
(b) *T is transfer open-valued if and only if $\bigcup_{x\in X} T(x) = \bigcup_{x\in X} \text{int}(T(x))$.*
(c) *X is a topological space, $T(x)$ is nonempty for each $x \in X$ and T^{-1} is transfer open-valued, then $X = \bigcup_{y\in Y} \text{int}(T^{-1}(y))$.*

Ansari et al. [7] established the following generalized form of Fan-KKM lemma.

Theorem 1.34 *Let K be a nonempty convex subset of a Hausdorff topological vector space X. Let $T : K \to 2^X$ be a transfer closed-valued map such that $\text{cl}(T(x_0))$ is compact for at least one $x_0 \in K$, and let $\text{cl}\, T : K \to 2^X$ be a generalized KKM map. Then $\bigcap_{x\in K} T(x) \neq \emptyset$.*

Proof Since $\text{cl}\, T : K \to 2^X$ is defined by $(\text{cl}\, T)(x) = \text{cl}(T(x))$ for all $x \in K$, we have that $\text{cl}\, T$ is a generalized KKM map with closed values. By Lemma 1.15, the family of sets $\{T(x) : x \in K\}$ has the finite intersection property. Since $\text{cl}(T(x_0))$ is compact, we have $\bigcap_{x\in K} \text{cl}(T(x)) \neq \emptyset$. Since T is transfer closed-valued,

$$\bigcap_{x\in K} T(x) = \bigcap_{x\in X} \text{cl}(T(x)) \neq \emptyset.$$

This completes the proof. □

The following section lemma, due to Xiang and Debnath [137], is a generalization of Fan section lemma [61] which can be derived by using Fan-KKM Lemma 1.14.

Lemma 1.17 (Fan Section Lemma) *Let K be a nonempty subset of a Hausdorff topological vector space X. Let A be a subset of $K \times K$ such that the following conditions hold.*

(i) *$(x, x) \in A$ for all $x \in K$;*
(ii) *For all $y \in K$, the set $A_y = \{x \in K : (x, y) \in A\}$ is closed in K;*
(iii) *For all $x \in K$, the set $A_x = \{y \in K : (x, y) \notin A\}$ is convex or empty;*
(iv) *For a nonempty compact convex subset $D \subset K$ with each $x \in K$, there exists $y \in D$ such that $(x, y) \notin A$.*

Then there exists $\bar{x} \in K$ such that $\{\bar{x}\} \times K \subset A$.

The following lemma is a generalization of Lemma 1.14.

Lemma 1.18 ([46]) *Let K be a nonempty convex subset of a topological vector space X. Let $T : K \to 2^K$ be a KKM-map such that the following conditions hold.*

(i) $\mathrm{cl}_K(T(\tilde{x}))$ *is compact for some* $\tilde{x} \in K$*;*
(ii) *For each* $A \in \mathscr{F}(K)$ *with* $\tilde{x} \in A$ *and each* $x \in \mathrm{co}(A)$, $T(y) \cap \mathrm{co}(A)$ *is closed in* $\mathrm{co}(A)$.
(iii) *For each* $A \in \mathscr{F}(K)$ *with* $\tilde{x} \in A$,

$$\left(\mathrm{cl}_K\left(\bigcap_{x\in\mathrm{co}(A)} T(x)\right)\right) \cap \mathrm{co}(A) = \left(\bigcap_{x\in\mathrm{co}(A)} T(x)\right) \cap \mathrm{co}(A).$$

Then $\bigcap_{x\in K} T(x) \neq \emptyset$.

Definition 1.51 Let X be a topological space and Y be a nonempty set. A set-valued map $T : X \to 2^Y$ is said to have *open lower section* if the set $T^{-1}(y) = \{x \in X : y \in T(x)\}$ is open in X for every $y \in Y$.

Lemma 1.19 ([136]) *Let X be a topological space and Y be a convex subset of a topological vector space. Let $S, T : X \to 2^Y$ be set-valued maps with open lower sections. Then*

(a) *the set-valued map* $H : X \to 2^Y$*, defined by* $H(x) = \mathrm{co}(S(x))$ *for all* $x \in X$*, has open lower sections;*
(b) *the set-valued map* $J : X \to 2^Y$*, defined by* $J(x) = S(x) \cap T(x)$ *for all* $x \in X$*, has open lower sections.*

The following fixed-point theorem has been proven by Browder [33].

Lemma 1.20 (Browder Fixed Point Theorem) *Let K be a nonempty compact convex subset of a Hausdorff topological vector space X. Suppose that $T : K \to 2^K$ is a set-valued map with nonempty convex values and has open lower sections. Then T has a fixed point.*

We present a Browder type fixed point theorem for set-valued maps under noncompact setting.

Theorem 1.35 ([9]) *Let K be a nonempty convex subset of a Hausdorff topological vector space X. Let $S, T : K \to 2^K$ be set-valued maps such that the following conditions hold.*

(i) *For all* $x \in K$, $\mathrm{co}(S(x)) \subseteq T(x)$ *and* $S(x) \neq \emptyset$*;*
(ii) $K = \bigcup\{\mathrm{int}_K(S^{-1}(x)) : x \in K\}$*;*
(iii) *If K is not compact, assume that there exist a nonempty compact convex subset B of K and a nonempty compact subset D of K such that for each $x \in K \setminus D$ there exists $\tilde{y} \in B$ such that $x \in \mathrm{int}_K(S^{-1}(\tilde{y}))$.*

Then there exists $\bar{x} \in K$ such that $\bar{x} \in T(\bar{x})$.

If S has open lower sections, then condition (ii) in Theorem 1.35 holds, and hence, we have the following result.

Corollary 1.3 [9] *Let K be a nonempty convex subset of a Hausdorff topological vector space X. Let $S, T : K \to 2^K$ be set-valued maps such that the following conditions hold.*

(i) *For all $x \in K$, $\mathrm{co}(S(x)) \subseteq T(x)$ and $S(x) \neq \emptyset$;*
(ii) *The set $S^{-1}(y) = \{x \in K : y \in S(x)\}$ is open;*
(iii) *If K is not compact, assume that there exist a nonempty compact convex subset B of K and a nonempty compact subset D of K such that for each $x \in K \setminus D$ there exists $\tilde{y} \in B$ such that $x \in S^{-1}(\tilde{y})$.*

Then there exists $\bar{x} \in K$ such that $\bar{x} \in T(\bar{x})$.

Chowdhury and Tan [47] establish the following version of Browder type fixed point theorem for non-Hausdorff spaces.

Theorem 1.36 *Let K be a nonempty convex subset of a topological vector space Y and $S, T : K \to 2^K$ be set-valued maps. Assume that the following conditions hold:*

(a) *For all $x \in K$, $S(x) \subseteq T(x)$.*
(b) *For all $x \in K$, $T(x)$ is convex and $S(x)$ is nonempty.*
(c) *For all $y \in K$, $S^{-1}(y) = \{x \in K : y \in S(x)\}$ is compactly open.*
(d) *There exists a nonempty closed compact (not necessarily convex) subset D of K and an element $\tilde{y} \in D$ such that $K \setminus D \subset T^{-1}(\tilde{y})$.*

Then there exists $\hat{x} \in K$ such that $\hat{x} \in T(\hat{x})$.

The following maximal element theorem for a set-valued map is equivalent to Corollary 1.3.

Theorem 1.37 ([49, 105]) *Let K be a nonempty convex subset of a Hausdorff topological vector space X. Let $S, T : K \to 2^K$ be set-valued maps satisfying the following conditions:*

(i) *For all $x \in K$, $\mathrm{co}(S(x)) \subseteq T(x)$;*
(ii) *For all $x \in K$, $x \notin T(x)$;*
(iii) *For all $y \in K$, $S^{-1}(y) = \{x \in K : y \in S(x)\}$ is open in K;*
(iv) *There exist a nonempty compact subset D of K and a nonempty compact convex subset B of K such that for all $x \in K \setminus D$, $S(x) \cap B \neq \emptyset$.*

Then there exists $\bar{x} \in K$ such that $S(\bar{x}) = \emptyset$.

Definition 1.52 (Φ-Condensing Map) [125, 126] Let X be a Hausdorff topological vector space, L be a lattice with a minimal element, and let $\Phi : 2^X \to L$ be a measure of noncompactness on X and $D \subseteq X$. A set-valued map $T : D \to 2^X$ is called *Φ-condensing* if $M \subseteq D$ with $\Phi(T(M)) \geq \Phi(M)$ implies that M is precompact.

Remark 1.23 Note that every set-valued map defined on a compact set is necessarily Φ-condensing. If X is locally convex, then a compact set-valued map (that is, $T(D)$ is precompact) is Φ-condensing for any measure of noncompactness Φ. Obviously, if $T : D \to 2^X$ is Φ-condensing and if $S : D \to 2^X$ satisfies $S(x) \subseteq T(x)$ for all $x \in D$, then S is also Φ-condensing.

Remark 1.24 If K is a nonempty, closed and convex subset of a locally convex Hausdorff topological vector space X, then condition (iii) of Theorem 1.35 and condition (iv) of Theorem 1.37 can be replaced by the following condition:

(iv)$'$ The set-valued map $S : K \to 2^K$ is Φ-condensing,

see Corollary 4 in [43].

Theorem 1.38 (Kakutani Fixed Point Theorem) [85] *Let K be a nonempty compact convex subset of a locally convex topological vector space X and Y be a topological vector space. Let $T : K \to 2^Y$ be a set-valued map such that for each $x \in K$, $T(x)$ is nonempty, compact and convex. Then T has a fixed point, that is, there exists $\bar{x} \in K$ such that $\bar{x} \in T(\bar{x})$.*

The Kakutani fixed point theorem is a set-valued version of the following Brouwer fixed point theorem.

Theorem 1.39 (Brouwer's Fixed Point Theorem) *Let K be a nonempty, compact and convex subset of a finite dimensional space $\mathbb{R}^n$ and $f : K \to K$ be a continuous map. Then there exists $x \in K$ such that $f(x) = x$.*

1.5 Variational Inequalities

Theory of variational inequalities is one of the powerful tools of current mathematical technology, introduced separately by G. Fichera and G. Stampacchia in early sixties. The ideas and techniques of variational inequalities are being applied in various fields of mathematics, engineering, management and social sciences including fluid flow through porous media, contact problems in elasticity, optimal control, nonlinear optimization, transportation and economic equilibria, etc. During the last three decades, variational inequalities are used as tools to solve optimization problems; See for example [2, 6, 7, 11–16, 18, 30–33, 45, 55, 56, 64, 67, 69, 94, 98, 116, 121, 124, 132, 134, 139, 140, 142] and the references therein. In this section, we give a brief introduction to the theory of variational inequalities.

Let X be a topological vector space with its topological dual X^*, and K be a nonempty convex subset of X. The value of $l \in X^*$ at x is denoted by $\langle l, x \rangle$. Let $F : K \to X^*$ be a mapping. The *variational inequality problem* (in short, VIP) is to find $\bar{x} \in K$ such that

$$\langle F(\bar{x}), y - \bar{x} \rangle \geq 0, \quad \text{for all } y \in K. \tag{1.16}$$

The inequality (1.16) is called *variational inequality*.

Roughly speaking, the variational inequality (1.16) states that the vector $F(\bar{x})$ must be at a non-obtuse angle with all the feasible vectors emanating from $\bar{x}$. In other words, the vector $\bar{x}$ is a solution of VIP if and only if $F(\bar{x})$ forms a non-obtuse angle with every vector of the form $y - \bar{x}$ for all $y \in K$.

First let us consider an application of varianal inequalities in Partial Differential Equations.

Example 1.19 (Inverse Problems in Partial Differential Equations) There is a large number of examples in applied sciences that can be modeled by means of partial differential equations (PDEs). The corresponding PDEs often involve certain unknown variable parameters when a measurement of a solution of the PDE is available. This leads to so-called *inverse problems*. The direct problem, on the other hand, is to solve the PDE. As an example, let us consider the following elliptic boundary value problem (BVP)

$$-\nabla \cdot (q\nabla u) = f \text{ in } \Omega, \quad u = 0 \text{ on } \partial\Omega, \tag{1.17}$$

where Ω denotes a domain in $\mathbb{R}^2$ or $\mathbb{R}^3$ and $\partial\Omega$ is its boundary. Problems of the form (1.17) have been studied in great detail in the literature due to their wide real-world applications. For example, $u = u(x)$ may be the steady-state temperature at a fixed point x of a body. Then q would represent a variable thermal conductivity coefficient and f would constitute the external heat source. When the problem (1.17) needs to be solved, one can choose from a large number of concepts proposed in the literature. Most approaches either regard problem (1.17) as a hyperbolic PDE in q or pose an optimization problem whose solution is an estimate of q. There exist two approaches involving the reformulation of (1.17) as an optimization problem: The problem (1.17) can either be formulated as an unconstrained optimization problem, or it can be handled as a constrained optimization problem which involves the PDE in as a constraint. Since the solution of equations corresponds to minimization problems and therefore to variational inequalities as optimality conditions, the results to be presented in this chapter are directly applicable to (1.17). For further applications and in-depth analysis of inverse problems, we refer to [76].

The simplest example of a variational inequality problem is the problem of solving a system of nonlinear equations.

Proposition 1.11 *Let* $F : \mathbb{R}^n \to \mathbb{R}^n$ *be a mapping. A vector* $\bar{x} \in \mathbb{R}^n$ *is a solution of* VIP *if and only if* $F(\bar{x}) = \mathbf{0}$.

Proof Let $F(\bar{x}) = \mathbf{0}$. Then, obviously, inequality (1.16) holds with equality.

Conversely, suppose that $\bar{x}$ satisfies the inequality (1.16). Then, by taking $y = \bar{x} - F(\bar{x})$ in (1.16), we get

$$\langle F(\bar{x}), \bar{x} - F(\bar{x}) - \bar{x}\rangle = \langle F(\bar{x}), -F(\bar{x})\rangle \geq 0,$$

that is, $-\|F(\bar{x})\|^2 \geq 0$, which implies that $F(\bar{x}) = \mathbf{0}$. □

If $F(x)$ is the gradient of a differentiable convex function $f : \mathbb{R}^n \to \mathbb{R}$, then the VIP provides the necessary and sufficient condition for a solution of an optimization problem.

Proposition 1.12 *Let K be a nonempty convex subset of $\mathbb{R}^n$ and $f : K \to \mathbb{R}$ be a differentiable function. If $\bar{x}$ is a solution of the following optimization problem:*

$$\text{minimize } f(x), \quad \text{subject to } x \in K, \tag{1.18}$$

then $\bar{x}$ is a solution of VIP *with $F \equiv \nabla f$.*

Proof For any $y \in K$, define a function $\varphi : [0, 1] \to \mathbb{R}$ by

$$\varphi(\lambda) = f(\bar{x} + \lambda(y - \bar{x})), \quad \text{for all } \lambda \in [0, 1].$$

Since $\varphi(\lambda)$ attains its minimum at $\lambda = 0$, therefore, $\varphi'(0) \geq 0$, that is,

$$\langle \nabla f(\bar{x}), y - \bar{x} \rangle \geq 0, \quad \text{for all } y \in K. \tag{1.19}$$

Hence, $\bar{x}$ is a solution of VIP with $F \equiv \nabla f$. □

Proposition 1.13 *Let K be a nonempty convex subset of $\mathbb{R}^n$ and $f : K \to \mathbb{R}$ be a pseudoconvex function. If $\bar{x}$ is a solution of* VIP *with $F(\bar{x}) = \nabla f(\bar{x})$, then it is a solution of the optimization problem* (1.18).

Proof Suppose that $\bar{x}$ is a solution of VIP, but not an optimal solution of the optimization problem (1.18). Then there exists a vector $y \in K$ such that $f(y) < f(\bar{x})$. By pseudoconvexity of f, we have $\langle \nabla f(\bar{x}), y - \bar{x} \rangle < 0$, which is a contradiction to the fact that $\bar{x}$ is a solution of VIP. □

Let K be a closed convex cone in a topological vector space X and $F : K \to X^*$ be a mapping. The *nonlinear complementarity problem* (NCP) is to find a vector $\bar{x} \in K$ such that

$$F(\bar{x}) \in K^* \quad \text{and} \quad \langle F(\bar{x}), \bar{x} \rangle = 0, \tag{1.20}$$

where K^* is the dual cone of K.

For further details and applications of complementarity problems, we refer to [56, 64, 74, 75, 87–89, 131] and the references therein.

The next result provides the equivalence between a nonlinear complementarity problem and a variational inequality problem.

Proposition 1.14 *If K is a closed convex pointed cone in a topological vector space X, then* VIP *and* NCP *have precisely the same solution sets.*

Proof Let $\bar{x} \in K$ be a solution of VIP. Then

$$\langle F(\bar{x}), y - \bar{x} \rangle \geq 0, \quad \text{for all } y \in K. \tag{1.21}$$

In particular, taking $y = x + \bar{x}$ in the above inequality, we get

$$\langle F(\bar{x}), x \rangle \geq 0, \quad \text{for all } x \in K,$$

which implies that $F(\bar{x}) \in K^*$.

By substituting $y = 2\bar{x}$ in inequality (1.21), we obtain

$$\langle F(\bar{x}), \bar{x} \rangle \geq 0, \tag{1.22}$$

and again taking $y = \mathbf{0}$ in inequality (1.21), we get

$$\langle F(\bar{x}), -\bar{x} \rangle \geq 0. \tag{1.23}$$

Inequalities (1.22) and (1.23) together imply that $\langle F(\bar{x}), \bar{x} \rangle = 0$. Hence, $\bar{x}$ is a solution of NCP.

Conversely, suppose that $\bar{x} \in K$ is a solution of NCP, then we have

$$\langle F(\bar{x}), \bar{x} \rangle = 0 \text{ and } \langle F(\bar{x}), y \rangle \geq 0, \quad \text{for all } y \in K.$$

Thus,

$$\langle F(\bar{x}), y - \bar{x} \rangle \geq 0, \quad \text{for all } y \in K.$$

Hence, $\bar{x}$ is a solution of VIP. □

Let K be a nonempty subset of a normed space X and $T : K \to K$ be a mapping. The *fixed point problem* (FPP) is to find $\bar{x} \in K$ such that

$$T(\bar{x}) = \bar{x}. \tag{1.24}$$

Now we give a relationship between a VIP and a FPP.

Proposition 1.15 *Let K be a nonempty subset of a normed space X and $T : K \to K$ be a mapping. If the mapping $F : K \to K$ is defined by*

$$F(x) = x - T(x), \tag{1.25}$$

then VIP (1.16) *coincide with* FPP (1.24).

Proof Let $\bar{x} \in K$ be a fixed point of the problem (1.24). Then, $F(\bar{x}) = \mathbf{0}$, and thus, $\bar{x}$ solves (1.16).

Conversely, suppose that $\bar{x}$ solves (1.16) with $F(\bar{x}) = \bar{x} - T(\bar{x})$. Then $T(\bar{x}) \in K$ and letting $y = T(\bar{x})$ in (1.16) gives $-\|\bar{x} - T(\bar{x})\|^2 \geq 0$, that is, $\bar{x} = T(\bar{x})$. □

A problem closely related to the VIP is the following problem, known as *Minty variational inequality problem* (MVIP): Find $\bar{x} \in K$ such that

$$\langle F(y), y - \bar{x} \rangle \geq 0, \quad \text{for all } y \in K. \tag{1.26}$$

The inequality (1.26) is known as *Minty variational inequality* (MVI). Minty [113] gave a complete characterization of the solutions of VIP in terms of the solutions of MVIP. Since the origin of VIP, most of the existence results for a solution of a VIP are established by showing the equivalence between VIP and MVIP.

To distinguish between a variational inequality and Minty variational inequality, we sometimes write Stampacchia variational inequality (SVI) instead of a variational inequality.

Contrary to the Stampacchia variational inequality problem (SVIP), Minty variational inequality problem (MVIP) is a sufficient optimality condition for the optimization problem (1.18) which becomes necessary if the objective function f is pseudoconvex and differentiable.

Theorem 1.40 (Giannessi 1998) *Let K be a nonempty convex subset of $\mathbb{R}^n$ and $f : K \to \mathbb{R}$ be a differentiable function. The following statements hold:*

(a) *If $\bar{x} \in K$ is a solution of* MVIP *with $F \equiv \nabla f$, then $\bar{x}$ is a solution of optimization problem* (1.18).
(b) *If f is pseudoconvex and $\bar{x} \in K$ is a solution of the optimization problem* (1.18), *then it is a solution of* MVIP *with $F \equiv \nabla f$.*

Proof

(a) Let $y \in K$ be arbitrary. Consider the function $\varphi(\lambda) = f(\bar{x} + \lambda(y - \bar{x}))$ for all $\lambda \in [0, 1]$. Since $\varphi'(\lambda) = \langle \nabla f(\bar{x} + \lambda(y - \bar{x})), y - \bar{x} \rangle$ and $\bar{x}$ is a solution of MVIP with $F \equiv \nabla f$, it follows that

$$\varphi'(\lambda) = \langle \nabla f(\bar{x} + \lambda(y - \bar{x})), y - \bar{x} \rangle \geq 0, \quad \text{for all } \lambda \in [0, 1].$$

This implies that φ is a nondecreasing function on $[0, 1]$, and therefore,

$$f(y) = \varphi(1) \geq \varphi(0) = f(\bar{x}).$$

Thus, $\bar{x}$ is a solution of the optimization problem (1.18).

(b) Let $\bar{x}$ be an optimal solution of the optimization problem (1.18). Then for all $y \in K$, $f(\bar{x}) \leq f(y)$. Since f is a pseudoconvex differentiable function, by Theorem 1.21, f is quasiconvex. Then by Theorem 1.18, we have

$$\langle \nabla f(y), y - \bar{x} \rangle \geq 0, \quad \text{for all } y \in K.$$

Thus, $\bar{x}$ is a solution of MVIP. □

Definition 1.53 Let K be a nonempty convex subset of topological vector space X. A mapping $F : K \to X^*$ is said to be

(a) *lower hemicontinuous* or *radially lower semicontinuous* if for any fixed $x, y \in K$, the function $\lambda \mapsto F(x + \lambda(y - x))$ defined on $[0, 1]$ is lower semicontinuous;
(b) *upper hemicontinuous* or *radially upper semicontinuous* if for any fixed $x, y \in K$, the function $\lambda \mapsto F(x + \lambda(y - x))$ defined on $[0, 1]$ is upper semicontinuous;

(c) *hemicontinuous* or *radially semicontinuous* if for any fixed $x, y \in K$, the mapping $\lambda \mapsto F(x + \lambda(y - x))$ defined on $[0, 1]$ is continuous, that is, if F is continuous along the line segments in K.

The following Minty lemma is an important tool in the theory of variational inequalities when the mapping is monotone and the domain is convex.

Lemma 1.21 (Minty Lemma) *Let K be a nonempty subset of a topological vector space X and $F : K \to X^*$ be a mapping. The following assertions hold.*

(a) *If K is convex and F is hemicontinuous, then every solution of* MVIP *is a solution of* VIP.
(b) *If F is pseudomonotone, then every solution of* VIP *is a solution of* MVIP.

Proof

(a) Let $\bar{x} \in K$ be a solution of MVIP. Then for any $y \in K$ and $\lambda \in]0, 1]$, $z = \bar{x} + \lambda(y - \bar{x}) \in K$, and hence,

$$\langle F(z), z - \bar{x} \rangle \geq 0, \quad \text{for all } \lambda \in]0, 1],$$

which implies that

$$\langle F(y + \lambda(\bar{x} - y)), y - \bar{x} \rangle \geq 0, \quad \text{for all } \lambda \in]0, 1].$$

By the hemicontinuity of F, we have

$$\langle F(\bar{x}), y - \bar{x} \rangle \geq 0, \quad \text{for all } y \in K.$$

Hence, $\bar{x}$ is a solution of VIP.
(b) Obvious, by pseudomonotonicity of F. □

It can be easily seen that if K is a nonempty closed convex subset of X and $F : K \to X^*$ be hemicontinuous and pseudomonotone, then the solution set of VIP is closed and convex. Moreover, if the F is strictly monotone, then the solution of VIP is unique, provided it exists. Finally, we present a result on the existence of a solution of VIP (1.16).

Theorem 1.41 ([131, Theorem 3.1]) *Let X be a reflexive Banach space, K be a nonempty bounded closed convex subset of X and $T : K \to X^*$ be a mapping. Suppose that T is pseudomonotone and hemicontinuous. Then there exists a solution $x \in K$ of* VIP (1.16). *Furthermore, if in addition T is strictly pseudomonotone, then the solution is unique.*

1.5.1 Nonsmooth Variational Inequalities

Motivated by the optimality conditions in terms of the generalized directional derivatives, we associate an optimization problem with the variational inequality problem defined by means of a bifunction h.

Let K be a nonempty subset of $\mathbb{R}^n$ and $h : K \times \mathbb{R}^n \to \mathbb{R} \cup \{\pm\infty\}$ be a bifunction. The *variational inequality problem in terms of a bifunction h* is defined as follows:

$$\text{Find } \bar{x} \in K \text{ such that } h(\bar{x}; y - \bar{x}) \geq 0, \quad \text{for all } y \in K. \tag{(VIP)$_h$}$$

When $h(x; y - x) = \langle F(x), y - x \rangle$, where $F : \mathbb{R}^n \to \mathbb{R}^n$, then (VIP)$_h$ reduces to the VIP studied in the previous section.

As we have seen in the previous section that the Minty variational inequality problem is closely related to VIP, and also provides a necessary and sufficient optimality condition for a differentiable optimization problem under convexity or pseudoconvexity assumption. Therefore, the study of Minty variational inequality defined by means of a bifunction h is also very important in the theory of nonsmooth variational inequalities. The *Minty variational inequality problem in terms of a bifunction h* is defined as follows:

$$\text{Find } \bar{x} \in K \text{ such that } h(y; \bar{x} - y) \leq 0, \quad \text{for all } y \in K. \tag{(MVIP)$_h$}$$

To prove the equivalence between (VIP)$_h$ and (MVIP)$_h$, we introduce the following concept of upper sign continuity.

Definition 1.54 Let K be a nonempty convex subset of $\mathbb{R}^n$. A bifunction $h : K \times \mathbb{R}^n \to \mathbb{R} \cup \{\pm\infty\}$ is said to be *upper sign continuous* if for all $x, y \in K$ and $\lambda \in]0, 1[$,

$$h(x + \lambda(y - x); x - y) \leq 0 \quad \text{implies} \quad h(x; y - x) \geq 0.$$

This notion of upper sign continuity for a bifunction extends the concept of upper sign continuity introduced in [70].

Clearly, every subodd radially upper semicontinuous bifunction is upper sign continuous.

The following lemma is a generalization of Minty Lemma 1.21.

Lemma 1.22 *Let K be a nonempty convex subset of $\mathbb{R}^n$ and $h : K \times \mathbb{R}^n \to \mathbb{R} \cup \{\pm\infty\}$ be a pseudomonotone and upper sign continuous bifunction such that h is positively homogeneous in the second argument. Then $\bar{x} \in K$ is a solution of* (VIP)$_h$ *if and only if it is a solution of* (MVIP)$_h$.

Proof The pseudomonotonicity of h implies that every solution of (VIP)$_h$ is a solution of (MVIP)$_h$.

Conversely, let $\bar{x} \in K$ be a solution of (MVIP)$_h$. Then

$$h(y; \bar{x} - y) \leq 0, \quad \text{for all } y \in K. \tag{1.27}$$

Since K is convex, we have $y_\lambda = \bar{x} + \lambda(y - \bar{x}) \in K$ for all $\lambda \in]0,1[$. Therefore, inequality (1.27) becomes

$$h(y_\lambda; \bar{x} - y_\lambda) \leq 0.$$

As $\bar{x} - y_\lambda = \lambda(\bar{x} - y)$ and h is positively homogeneous in the second argument, we have

$$h(y_\lambda; \bar{x} - y) \leq 0.$$

Thus, the upper sign continuity of h implies that $\bar{x} \in K$ is a solution of $(\mathrm{VIP})_h$. □

Let us recall the optimization problem:

$$\text{minimize } f(x), \quad \text{subject to } x \in K, \tag{P}$$

where K is a nonempty convex subset of $\mathbb{R}^n$ and $f : K \to \mathbb{R}$ is a function.

In the subsequent theorems, we relate the solutions of the problem (P) and $(\mathrm{VIP})_h$.

Theorem 1.42 *Let K be a nonempty convex subset of $\mathbb{R}^n$, $f : K \to \mathbb{R}$ be a function and $h : K \times \mathbb{R}^n \to \mathbb{R} \cup \{\pm\infty\}$ be a bifunction. If f is h-convex and $\bar{x} \in K$ is a solution of* $(\mathrm{VIP})_h$*, then $\bar{x}$ solves the problem* (P).

Proof By h-convexity of f, we have

$$f(y) - f(\bar{x}) \geq h(\bar{x}; y - \bar{x}), \quad \text{for all } y \in K.$$

Since $\bar{x}$ is a solution of $(\mathrm{VIP})_h$, we have

$$h(\bar{x}; y - \bar{x}) \geq 0, \quad \text{for all } y \in K.$$

The last two inequalities together imply that

$$f(y) - f(\bar{x}) \geq 0, \quad \text{for all } y \in K,$$

that is, $\bar{x}$ is a solution of problem (P). □

The h-convexity assumption in the above theorem can be weakened to h-pseudoconvexity.

For the converse of Theorem 1.42 to hold, we do not require the function f to be h-convex. However, we assume that the function f and the bifunction h satisfy the following condition:

$$\forall\, x \in K,\ d \in \mathbb{R}^n : \ D_+f(x; d) \leq h(x; d). \tag{1.28}$$

Theorem 1.43 *Let K be a nonempty convex subset of $\mathbb{R}^n$, $f : K \to \mathbb{R}$ be a function and $h : K \times \mathbb{R}^n \to \mathbb{R} \cup \{\pm\infty\}$ satisfy the condition* (1.28)*. If $\bar{x}$ is an optimal solution of the problem* (P)*, then $\bar{x} \in K$ is a solution of* $(\mathrm{VIP})_h$.

Proof Since K is convex and $\bar{x}$ is an optimal solution of problem (P), for any $y \in K$, we have

$$f(\bar{x}) \leq f(\bar{x} + \lambda(y - \bar{x})), \quad \text{for all } \lambda \in]0, 1].$$

This implies that

$$\frac{f(\bar{x} + \lambda(y - \bar{x})) - f(\bar{x})}{\lambda} \geq 0, \quad \text{for all } \lambda \in]0, 1].$$

Taking lim inf as $\lambda \to 0^+$, we obtain

$$D_+ f(\bar{x}; y - \bar{x}) \geq 0, \quad \text{for all } y \in K,$$

which on using (1.28) implies that

$$h(\bar{x}; y - \bar{x}) \geq 0, \quad \text{for all } y \in K.$$

Hence, $\bar{x}$ is a solution of $(\text{VIP})_h$. □

Thus, it is possible to identify the solutions of the optimization problem (P) with those of the $(\text{VIP})_h$ provided the objective function is h-convex or h-pseudoconvex.

Theorem 1.44 *Let K be a nonempty convex subset of $\mathbb{R}^n$ and $f : K \to \mathbb{R}$ be a function such that*

$$h(x; y - x) > f(y) - f(x), \quad \textit{for all } x, y \in K \textit{ and } x \neq y. \tag{1.29}$$

Then every solution of the problem (P) *is a solution of* $(\text{VIP})_h$.

Proof Assume that $\bar{x}$ is a solution of the problem (P) but not a solution of $(\text{VIP})_h$. Then there exists $y \in K$ such that

$$h(\bar{x}; y - \bar{x}) < 0. \tag{1.30}$$

From (1.29), we reach to a contradiction to our assumption that $\bar{x}$ is a solution of the problem (P). Hence, $\bar{x}$ is a solution of $(\text{VIP})_h$. □

Next we establish that a solution of the Minty variational inequality problem $(\text{MVIP})_h$ is an optimal solution of the problem (P) under specific assumptions.

Theorem 1.45 *Let K be a nonempty convex subset of $\mathbb{R}^n$, $f : K \to \mathbb{R}$ be a radially lower semicontinuous function and $h : K \times \mathbb{R}^n \to \mathbb{R} \cup \{\pm\infty\}$ satisfy condition* (1.28) *and be positively homogeneous in the second argument. If $\bar{x} \in K$ is a solution of* $(\text{MVIP})_h$, *then it is a solution of the problem* (P).

Proof Let $\bar{x} \in K$ be a solution of $(\text{MVIP})_h$. Then

$$h(y; \bar{x} - y) \leq 0, \quad \text{for all } y \in K. \tag{1.31}$$

Let $y \in K$, $y \neq \bar{x}$ be arbitrary. Since f is radially lower semicontinuous, by Theorem 1.31, there exists $\theta \in [0, 1[$ such that for $w = y + \theta(\bar{x} - y)$, we have

$$D_{+}f(w; \bar{x} - y) \geq f(\bar{x}) - f(y). \tag{1.32}$$

As $\theta < 1$, by the positive homogeneity of h in the second argument, we have from relation (1.32) and the condition (1.28) that

$$(1 - \theta)^{-1} h(w; \bar{x} - w) \geq f(\bar{x}) - f(y).$$

From (1.31), we have

$$0 \geq h(w; \bar{x} - w) \geq (1 - \theta)(f(\bar{x}) - f(y)),$$

and as $\theta < 1$, it follows that $f(\bar{x}) - f(y) \leq 0$. Since $y \in K$ was arbitrary, it follows that $\bar{x}$ is a solution of problem (P). □

As in the differentiable case, the problem $(\text{MVIP})_h$ is a necessary optimality condition under the assumption of the convexity (or pseudoconvexity) of f.

Theorem 1.46 *Let K be a nonempty convex subset of $\mathbb{R}^n$, $h : K \times \mathbb{R}^n \to \mathbb{R} \cup \{\pm\infty\}$ be a bifunction and $f : K \to \mathbb{R}$ be a h-convex function. If $\bar{x} \in K$ is solution of problem* (P)*, then it solves* $(\text{MVIP})_h$.

Proof Since f is h-convex, we have

$$f(\bar{x}) - f(y) - h(y; \bar{x} - y) \geq 0, \quad \text{for all } y \in K.$$

Since $\bar{x}$ is a solution of problem (P), we obtain

$$0 \geq f(\bar{x}) - f(y) \geq h(y; \bar{x} - y), \quad \text{for all } y \in K,$$

thus $\bar{x}$ solves $(\text{MVIP})_h$. □

In the following theorem, we relax the h-convexity assumption but we add some other assumptions.

Theorem 1.47 *Let K be a nonempty convex subset of $\mathbb{R}^n$, $h : K \times \mathbb{R}^n \to \mathbb{R} \cup \{\pm\infty\}$ satisfy the condition*

$$\forall\, x \in K,\ d \in \mathbb{R}^n :\ h(x; d) \leq D^{+}f(x; d) \tag{1.33}$$

and be positively homogeneous and subodd in the second argument and $f : K \to \mathbb{R}$ be a h-pseudoconvex function. If $\bar{x} \in K$ is solution of problem (P)*, then it solves* $(\text{MVIP})_h$.

Proof Since $\bar{x}$ is a solution of problem (P), we have

$$f(\bar{x}) \leq f(y), \quad \text{for all } y \in K.$$

By Lemma 1.7(a), f is h-quasiconvex and hence,

$$h(y;\bar{x}-y) \leq 0, \quad \text{for all } y \in K,$$

thus $\bar{x}$ solves $(\text{MVIP})_h$. □

We close this section by giving the following existence result for a solution of $(\text{VIP})_h$.

Theorem 1.48 ([6, Theorem 6.8]) *Let K be a nonempty compact convex subset of $\mathbb{R}^n$ and $h : K \times \mathbb{R}^n \to \mathbb{R} \cup \{\pm\infty\}$ be a pseudomonotone bifunction such that h is proper subodd in the second argument and the function $x \mapsto h(y; x-y)$ is lower semicontinuous. Then $(\text{MVIP})_h$ has a solution $\bar{x} \in K$. Furthermore, if h is upper sign continuous and positively homogeneous as well as subodd in the second argument, then $\bar{x} \in K$ is a solution of $(\text{VIP})_h$.*

For a thorough study on nonsmooth variational inequalities, we refer to [6].

1.5.2 Generalized Variational Inequalities

Let X be a topological vector space with its dual X^*, K be a nonempty subset of X, $F : K \to 2^{X^*}$ be a set-valued map with nonempty values. The *generalized variational inequality problem* (GVIP) is to find $\bar{x} \in K$ and $\bar{u} \in F(\bar{x})$ such that

$$\langle \bar{u}, y-\bar{x} \rangle \geq 0, \quad \text{for all } y \in K. \tag{1.34}$$

An element $\bar{x} \in K$ is said to be a *strong solution* of GVIP if there exists $\bar{u} \in F(\bar{x})$ such that the inequality (1.34) holds.

The weak form of the GVIP is the problem of finding $\bar{x} \in K$ such that for each $y \in K$, there exists $\bar{u} \in F(\bar{x})$ satisfies

$$\langle \bar{u}, y-\bar{x} \rangle \geq 0. \tag{1.35}$$

It is called a *weak generalized variational inequality problem* (WGVIP). An element $\bar{x} \in K$ is said to be a *weak solution* of GVIP if for each $y \in K$, there exists $\bar{u} \in F(\bar{x})$ such that the inequality (1.35) holds. It should be noted that $\bar{u}$ in WGVIP depends on y. Of course, if F is a single-valued map, then both the problems mentioned above reduce to the variational inequality problem (1.16).

Clearly, every strong solution of GVIP is a weak solution. However, the converse is not true in general, see, for example, Example 8.1 in [6].

For the next result, we need the following theorem.

Theorem 1.49 (Kneser Minimax Theorem) [95] *Let K be a nonempty convex subset of a vector space X and D be a nonempty compact convex subset of a topological vector space Y. Suppose that $f : K \times D \to \mathbb{R}$ is lower semicontinuous* *and convex in the second argument and concave in the first argument. Then*

$$\min_{y\in D}\sup_{x\in K} f(x,y) = \sup_{x\in K}\min_{y\in D} f(x,y).$$

The following lemma says that every weak solution of GVIP is a strong solution if the set-valued map F is nonempty, compact and convex valued.

Lemma 1.23 *Let K be a nonempty convex subset of X and $F : K \to 2^{X^*}$ be a set-valued map such that for each $x \in K$, $F(x)$ is nonempty, compact and convex. Then every weak solution of* GVIP *is a strong solution.*

Proof Let $\bar{x} \in K$ be a weak solution of GVIP. Then for each $y \in K$, there exists $\bar{u} \in F(\bar{x})$ such that

$$\langle \bar{u}, \bar{x} - y\rangle \leq 0,$$

that is,

$$\inf_{u\in F(\bar{x})} \langle u, \bar{x} - y\rangle \leq 0, \quad \text{for all } y \in K.$$

Define a functional $f : K \times F(\bar{x}) \to \mathbb{R}$ by

$$f(y,u) := \langle u, \bar{x} - y\rangle.$$

Then for each $y \in K$, the real-valued functional $u \mapsto f(y,u)$ is lower semicontinuous and convex, and for each $u \in F(\bar{x})$, the functional $y \mapsto f(y,u)$ is concave. Since $F(\bar{x})$ is compact and convex, by Theorem 1.49, we have

$$\inf_{u\in F(\bar{x})}\sup_{y\in K}\langle u, \bar{x} - y\rangle = \sup_{y\in K}\inf_{u\in F(\bar{x})}\langle u, \bar{x} - y\rangle \leq 0. \tag{1.36}$$

Since $F(\bar{x})$ is compact, there exists $\bar{u} \in F(\bar{x})$ such that

$$\sup_{y\in K}\langle \bar{u}, \bar{x} - y\rangle \leq 0,$$

and hence

$$\langle \bar{u}, y - \bar{x}\rangle \geq 0, \quad \text{for all } y \in K.$$

□

If $K = X$, then clearly, WGVIP reduces to the following *set-valued inclusion problem* : Find $\bar{x} \in X$ such that

$$\mathbf{0} \in F(\bar{x}). \tag{1.37}$$

We consider the generalized complementarity problem which is one of the most important problems from operations research. For details on complementarity problems and their generalizations, we refer [2, 56, 64, 74, 75, 87–89, 130, 131, 135] and the references therein.

Let K be a convex cone in X with its dual cone $K^* = \{u \in X^* : \langle u, x \rangle \geq 0$ for all $x \in K\}$. The *generalized complementarity problem* (GCP) is to find $\bar{x} \in K$ and $\bar{u} \in F(\bar{x})$ such that

$$\bar{u} \in K^* \quad \text{and} \quad \langle \bar{u}, \bar{x} \rangle = 0. \tag{1.38}$$

Proposition 1.16 ([6, Proposition 8.1]) *$(\bar{x}, \bar{u})$ is a solution of* GVIP *if and only if it is a solution of* GCP.

Let K be a nonempty subset of a normed space X and $T : K \to 2^K$ be a set-valued map with nonempty values. The *set-valued fixed point problem* (in short, SVFPP) associated with T is to find $\bar{x} \in K$ such that

$$\bar{x} \in T(\bar{x}). \tag{1.39}$$

The point $\bar{x} \in K$ is called a *fixed point* of T if the relation (1.39) holds. This problem can be converted into a generalized variational inequality formulation as shown below in the set-valued version of Proposition 1.15.

Proposition 1.17 ([6, Proposition 8.2]) *Let K be a nonempty subset of a normed space X and $T : K \to 2^K$ be a set-valued map with nonempty values. If the set-valued map $F : K \to 2^X$ is defined by*

$$F(x) = x - T(x), \tag{1.40}$$

then an element $\bar{x} \in K$ is a strong solution of GVIP (1.34) *if and only if it is a fixed point of T.*

Let K be a nonempty convex subset of a Banach space X and $f : K \to \mathbb{R}$ be a function. Consider the following optimization problem:

$$\text{minimize} f(x), \quad \text{subject to } x \in K. \tag{1.41}$$

The following result shows that the GVIP with $F(x) = \partial f(x)$, the subdifferential of a convex function f, is a necessary and sufficient optimality condition for the optimization problem (1.41).

Proposition 1.18 *Let K be a nonempty convex subset of a Banach space X and $f : K \to \mathbb{R}$ be a convex function. If $\bar{x} \in K$ is a solution of the minimization problem* (1.41), *then it is a strong solution of* GVIP *with $F(x) = \partial f(x)$ for all $x \in K$. Conversely, if $(\bar{x}, \bar{u})$ is a solution of* GVIP *with $\bar{u} \in \partial f(\bar{x})$, then $\bar{x}$ solves the optimization problem* (1.41).

Proof Let $\bar{x} \in K$ be a solution of the minimization problem (1.41). Then, $f(\bar{x}) \leq f(y)$ for all $y \in K$. By the definition of subdifferential of a convex function, $\mathbf{0} \in \partial f(\bar{x})$. Hence, $(\bar{x}, \mathbf{0})$ is a solution of GVIP, that is, $\bar{x}$ is a strong solution of GVIP with $F(x) = \partial f(x)$ for all $x \in K$.

Conversely, assume that $(\bar{x}, \bar{u})$ is a solution of GVIP with $F(x) = \partial f(x)$ for all $x \in K$. Then $\bar{u} \in \partial f(\bar{x})$ and

$$\langle \bar{u}, y - \bar{x} \rangle \geq 0, \quad \text{for all } y \in K. \tag{1.42}$$

Since $\bar{u} \in \partial f(\bar{x})$, we have

$$\langle \bar{u}, y - \bar{x} \rangle \leq f(y) - f(\bar{x}), \quad \text{for all } y \in X. \tag{1.43}$$

By combining inequalities (1.42) and (1.43), we obtain

$$f(\bar{x}) \leq f(y), \quad \text{for all } y \in K.$$

Hence, $\bar{x}$ is a solution of the minimization problem (1.41). □

It can be easily seen that if $\bar{x}$ is a weak solution of GVIP, even then it is a solution of the minimization problem (1.41).

Theorem 1.50 ([130]) *Let K be a nonempty compact convex subset of $\mathbb{R}^n$ and $F : K \to 2^{\mathbb{R}^n}$ be an upper semicontinuous set-valued map such that for each $x \in K$, $F(x)$ is nonempty, compact and convex. Then there exists a solution $(\bar{x}, \bar{u})$ of* GVIP.

If K is not necessarily bounded, then we have the following result.

Theorem 1.51 ([6, Theorem 8.2]) *Let K be a nonempty closed convex subset of $\mathbb{R}^n$ and $F : K \to 2^{\mathbb{R}^n}$ be an upper semicontinuous set-valued map such that for each $x \in K$, $F(x)$ is nonempty, compact and convex. If there exist an element $\tilde{y} \in K$ and a constant $r > \|\tilde{y}\|$ such that*

$$\max_{u \in F(x)} \langle u, \tilde{y} - x \rangle \leq 0, \tag{1.44}$$

for all $x \in K$ with $\|x\| = r$, then there exists a solution $(\bar{x}, \bar{u})$ of GVIP.

Theorems 1.50 and 1.51 also hold in the setting of Banach spaces. Some existence results for a solution of GVIP under the assumption that the underlying set K is convex but neither bounded nor closed, are derived in [62].

The following problem is the set-valued version of the Minty variational inequality problem, known as *generalized Minty variational inequality problem* (in short, GMVIP): Find $\bar{x} \in K$ such that for all $y \in K$ and all $v \in F(y)$, we have

$$\langle v, y - \bar{x} \rangle \geq 0. \tag{1.45}$$

A weak form of the generalized Minty variational inequality problem is the following problem which is called *weak generalized Minty variational inequality problem* (WGMVIP): Find $\bar{x} \in K$ such that for all $y \in K$, there exists $v \in F(y)$ satisfying the inequality (1.45).

A solution of WGMVIP is called a *weak solution* of GMVIP. It is clear that every solution of GMVIP is a weak solution of GMVIP.

The following result provides a necessary and sufficient condition for a solution of the minimization problem (1.41).

Proposition 1.19 *Let K be a nonempty convex subset of a Banach space X and $f : K \to \mathbb{R}$ be a convex function. Then $\bar{x} \in K$ is a solution of the minimization problem* (1.41) *if and only if it is a solution of* GMVIP (1.45) *with $F(x) = \partial f(x)$.*

Proof Let $\bar{x} \in K$ be a solution of GMVIP (1.45) but not a solution of minimization problem (1.41). Then there exists $z \in K$ such that

$$f(z) < f(\bar{x}). \tag{1.46}$$

By Theorem 1.32, there exist $\lambda \in]0, 1[$ and $v \in \partial f(z(\lambda))$, where $z(\lambda) = \lambda z + (1 - \lambda)\bar{x}$, such that

$$\langle v, z - \bar{x}\rangle = f(z) - f(\bar{x}). \tag{1.47}$$

By combining (1.46) and (1.47), we obtain

$$\langle v, z - \bar{x}\rangle < 0.$$

Since $\lambda(z - \bar{x}) = z(\lambda) - \bar{x}$, we have

$$\langle v, z(\lambda) - \bar{x}\rangle < 0,$$

a contradiction to our supposition that $\bar{x}$ is a solution of GMVIP (1.45).

Conversely, suppose that $\bar{x} \in K$ is a solution of the minimization problem (1.41). Then we have

$$f(y) - f(\bar{x}) \geq 0, \quad \text{for all } y \in K. \tag{1.48}$$

Since f is convex, we deduce that

$$\langle v, y - \bar{x}\rangle \geq f(y) - f(\bar{x}), \quad \text{for all } y \in K \text{ and all } y \in \partial f(y). \tag{1.49}$$

From inequalities (1.48) and (1.49), it follows that $\bar{x}$ is a solution of GMVIP (1.45). □

Definition 1.55 Let K be a nonempty convex subset of a topological vector space X. A set-valued map $F : K \to 2^X$ is said to be *generalized hemicontinuous* if for any $x, y \in K$ and for all $\lambda \in [0, 1]$, the set-valued map

$$\lambda \mapsto \langle F(x + \lambda(y - x)), y - x\rangle = \bigcup_{w \in F(x+\lambda(y-x))} \langle w, y - x\rangle$$

is upper semicontinuous at $\mathbf{0}$.

Definition 1.56 Let K be a nonempty convex subset of a topological vector space X. A set-valued map $F : K \to 2^{X^*}$ is said to be *generalized pseudomonotone* if for every pair of distinct points $x, y \in K$ and for any $u \in F(x)$ and $v \in F(y)$, we have

$$\langle u, y - x \rangle \geq 0 \quad \Rightarrow \quad \langle v, y - x \rangle \geq 0.$$

F is called *generalized weakly pseudomonotone* if for every pair of distinct points $x, y \in K$ and for any $u \in F(x)$, we have

$$\langle u, y - x \rangle \geq 0 \quad \Rightarrow \quad \langle v, y - x \rangle \geq 0 \text{ for some } v \in F(y).$$

Now we present some existence results for solutions of GVIP under different kinds of generalized monotonicities.

The following result which was established by Konnov and Yao [103], is a set-valued version of the Minty lemma.

Lemma 1.24 (Generalized Linearization Lemma) [6, Lemma 8.2] *Let K be a nonempty convex subset of a topological vector space X and $F : K \to 2^{X^*}$ be a set-valued map with nonempty values. The following assertions hold.*

(a) *If F is generalized hemicontinuous, then every solution of* WGMVIP *is a solution of* WGVIP.
(b) *If F is generalized pseudomonotone, then every solution of* WGVIP *is a solution of* GMVIP.
(c) *If F is generalized weakly pseudomonotone, then every solution of* WGVIP *is a solution of* WGMVIP.

Theorem 1.52 *Let K be a nonempty compact convex subset of a Banach space X and $F : K \to 2^{X^*}$ be a generalized pseudomonotone and generalized hemicontinuous set-valued map such that for each $x \in K$, $F(x)$ is nonempty. Then there exists a solution $\bar{x} \in K$ of* WGVIP. *If, in addition, the set $F(\bar{x})$ is also compact and convex, then $\bar{x} \in K$ is a strong solution of* GVIP.

Proof For each $y \in K$, define two set-valued maps $S, T : K \to 2^K$ by

$$S(y) = \{x \in K : \exists u \in F(x),\ \langle u, y - x \rangle \geq 0\},$$

and

$$T(y) = \{x \in K : \forall v \in F(y),\ \langle v, y - x \rangle \geq 0\},$$

respectively. We divide the proof into five steps.

(i) We claim that S is a KKM map, that is, the convex hull $\text{co}(\{y_1, y_2, \ldots, y_m\})$ of every finite subset $\{y_1, y_2, \ldots, y_m\}$ of K is contained in the corresponding union $\bigcup_{i=1}^m S(y_i)$.

Let $\hat{x} \in \text{co}(\{y_1, y_2, \ldots, y_m\})$. Then

$$\hat{x} = \sum_{i=1}^{m} \lambda_i y_i, \quad \text{for some } \lambda_i \geq 0 \text{ with } \sum_{i=1}^{m} \lambda_i = 1.$$

If $\hat{x} \notin \bigcup_{i=1}^{m} S(y_i)$, then for all $w \in F(\hat{x})$,

$$\langle w, y_i - \hat{x} \rangle < 0, \quad \text{for all } i = 1, 2, \ldots, m.$$

For all $w \in F(\hat{x})$, it follows that

$$\begin{aligned} 0 &= \langle w, \hat{x} - \hat{x} \rangle \\ &= \left\langle w, \sum_{i=1}^{m} \lambda_i y_i - \sum_{i=1}^{m} \lambda_i \hat{x} \right\rangle \\ &= \left\langle w, \sum_{i=1}^{m} \lambda_i (y_i - \hat{x}) \right\rangle \\ &= \sum_{i=1}^{m} \lambda_i \langle w, y_i - \hat{x} \rangle < 0, \end{aligned}$$

which is a contradiction. Therefore, we must have

$$\text{co}(\{y_1, y_2, \ldots, y_m\}) \subseteq \bigcup_{i=1}^{m} S(y_i),$$

and hence, S is a KKM map.

(ii) We show that $S(y) \subseteq T(y)$ for all $y \in K$, and hence T is a KKM map.

By generalized pseudomonotonicity of F, we have that $S(y) \subseteq T(y)$ for all $y \in K$. Since S is a KKM map, so is T.

(iii) We assert that $\bigcap_{y \in K} S(y) = \bigcap_{y \in K} T(y)$.

From step (ii), we have

$$\bigcap_{y \in K} S(y) \subseteq \bigcap_{y \in K} T(y),$$

and from Lemma 1.24, we have

$$\bigcap_{y \in K} S(y) \supseteq \bigcap_{y \in K} T(y).$$

Therefore, the conclusion follows.

(iv) We prove that for each $y \in K$, $T(y)$ is a closed subset of K.

For any fixed $y \in K$, let $\{x_m\}$ be a sequence in $T(y)$ such that $x_m \to \tilde{x} \in K$. Since $x_m \in T(y)$, for all $v \in F(y)$, we have $\langle v, y - x_m \rangle \geq 0$ for all m. As $\langle v, y - x_m \rangle$ converges to $\langle v, y - \tilde{x} \rangle$, therefore $\langle v, y - \tilde{x} \rangle \geq 0$, and hence, $\tilde{x} \in T(y)$. Consequently, $T(y)$ is closed.

(v) Finally, we show that the WGVIP is solvable.

From step (iv), $T(y)$ is a closed subset of the compact set K, and hence, it is compact. By step (ii) and Lemma 1.14, we have $\bigcap_{y \in K} T(y) \neq \emptyset$. Consequently, by step (iii), we also have $\bigcap_{y \in K} S(y) \neq \emptyset$. Hence, there exists $\bar{x} \in K$ such that

$$\forall y \in K, \ \exists \bar{u} \in F(\bar{x}) \ : \ \langle \bar{u}, y - \bar{x} \rangle \geq 0. \tag{1.50}$$

Thus, $\bar{x}$ is a solution of WGVIP.

If, in addition, the set $F(\bar{x})$ is also compact and convex, then by Lemma 1.23, $\bar{x} \in K$ is a strong solution of GVIP. □

1.6 Equilibrium Problems

Investigations of equilibrium states of a system play a central role in such diverse fields as economics, mechanics, biology and social sciences. There are many general mathematical problems which were suggested for modeling and studying various kinds of equilibria. Many researchers were / are considering these problems in order to obtain existence and uniqueness results and to propose solution methods. The Ky Fan [59–61] type inequality is one of such problems, which plays an important role in the theory of nonlinear analysis and optimization. It was W. Oettli who coined the name "Equilibrium Problem" to the Ky Fan type inequality, perhaps, because it is equivalent to find the equilibrium point of an optimization problem under certain conditions. The mathematical formulation of an equilibrium problem (in short, EP) is to find an element $\bar{x}$ of a set K such that

$$f(\bar{x}, y) \geq 0, \quad \text{for all } y \in K, \tag{1.51}$$

where $f : K \times K \to \mathbb{R}$ is a bifunction such that $f(x, x) \geq 0$ for all $x \in K$. It seems the most general problem and includes other equilibrium type ones such as optimization problem, saddle point problem, fixed point problem, complementarity problems, variational inequality problems, Nash equilibrium problem, etc. In this general form, EP was first considered by H. Nikaido and K. Isoda [119] as an auxiliary problem to establish existence results for Nash equilibrium points in non-cooperative games [117, 118]. This transformation allows one to extend various iterative methods, which were proposed for saddle point problems, for the case of EP. In the theory of EPs, the key contribution was made by Ky Fan [59–61], whose new existence results contained the original technique which became a basis for most further existence theorems in topological spaces. The work of Ky Fan perhaps motivated by the min-max problems appearing in economic equilibrium. Within the context of calculus of variations, motivated mainly by the work of Stampacchia [132], there arises the work of Brézis, Niremberg and Stampacchia [31] establishing a more general result than that in [61]. After the work of Blum and Oettli [29], it emerged as a new direction of research in nonlinear analysis, optimization, optimal control, game theory, mathematical economics, etc.

Example 1.20

(a) **Minimization Problem.** Let K be a nonempty set and $\varphi : K \to \mathbb{R}$ be a real-valued function. The *minimization problem* (in short, MP) is to find $\bar{x} \in K$ such that

$$\varphi(\bar{x}) \leq \varphi(y), \quad \text{for all } y \in K. \tag{1.52}$$

If we set $f(x, y) = \varphi(y) - \varphi(x)$ for all $x, y \in K$, then MP is equivalent to EP.

(b) **Saddle Point Problem.** Let K_1 and K_2 be nonempty sets and $\ell : K_1 \times K_2 \to \mathbb{R}$ be a real-valued bifunction. The *saddle point problem* (in short, SPP) is to find $(\bar{x}_1, \bar{x}_2) \in K_1 \times K_2$ such that

$$\ell(\bar{x}_1, y_2) \leq \ell(y_1, \bar{x}_2), \quad \text{for all } (y_1, y_2) \in K_1 \times K_2. \tag{1.53}$$

Set $K := K_1 \times K_2$ and define $f : K \times K \to \mathbb{R}$ by

$$f((x_1, x_2), (y_1, y_2)) = \ell(y_1, x_2) - \ell(x_1, y_2) \tag{1.54}$$

for all $(x_1, x_2), (y_1, y_2) \in K_1 \times K_2$. Then SPP coincides with EP.

(c) **Nash Equilibrium Problem.** Let $I = \{1, 2, \ldots, m\}$ be the set of players. For each player $i \in I$, let K_i be the strategy set of the ith player. Let $K = \prod_{i=1}^{m} K_i$. For every player $i \in I$, let $\varphi_i : K \to \mathbb{R}$ be the loss function of the ith player, depending on the strategies of all players. For $x = (x_1, x_2, \ldots, x_m) \in K$, we define $x^i = (x_1, \ldots, x_{i-1}, x_{i+1}, \ldots, x_m)$. Then $\bar{x} = (\bar{x}_1, \bar{x}_2, \ldots, \bar{x}_m) \in K$ is called a *Nash equilibrium point* if for all $i \in I$,

$$\varphi_i(\bar{x}) \leq \varphi_i(\bar{x}^i, y_i), \quad \text{for all } y_i \in K_i. \tag{1.55}$$

This means that no player can reduce his loss by varying his strategy alone. We now define

$$f(x, y) = \sum_{i=1}^{m} \left(\varphi_i(x^i, y_i) - \varphi_i(x) \right).$$

For such f, EP coincides with *Nash equilibrium problem* (in short, NEP) of finding $\bar{x} = (\bar{x}_1, \bar{x}_2, \ldots, \bar{x}_m) \in K$ such that (1.55) holds. Indeed, if (1.55) holds for all $i \in I$, obviously (1.51) is fulfilled. If, for some $i \in I$, we choose $y \in K$

such that $y^i = \bar{x}^i$, then

$$f(\bar{x}, y) = \varphi_i\left(\bar{x}^i, y_i\right) - \varphi_i(\bar{x}).$$

Thus EP implies NEP.

(d) **Fixed Point Problem.** Let X be an inner product space whose inner product is denoted by $\langle ., . \rangle$. Let K be a nonempty subset of X and $\varphi : K \to K$ be a given mapping. The *fixed point problem* (in short, FPP) is to find $\bar{x} \in K$ such that $\varphi(\bar{x}) = \bar{x}$.

Setting $f(x, y) = \langle x - \varphi(x), y - x \rangle$. Then $\bar{x}$ is a solution of FPP if and only if it is a solution of EP.

Indeed, FPP implies EP is obvious. If EP is satisfied, then by choosing $y = \varphi(\bar{x})$, we obtain

$$0 \leq f(\bar{x}, \varphi(\bar{x})) = -\|\bar{x} - \varphi(\bar{x})\|^2.$$

(e) **Variational Inequality Problem.** Let X, X^*, K and F be the same as in the formulation of variational inequality problem defined by (1.16). We set $f(x, y) = \langle F(x), y - x \rangle$ for all $x, y \in K$. Then VIP is equivalent to EP.

Let K and h be the same as defined in the formulation of nonsmooth variational inequality problem $(\text{VIP})_h$. If we define $f(x, y) = h(x; y - x)$, then $(\text{VIP})_h$ is equivalent to EP.

For further details on different special cases of EP, we refer to [3, 29, 65, 66, 80, 91–93] and the references therein.

Most of the results on the existence of solutions for equilibrium problems are derived in the setting of topological vector spaces either by using Browder type or Kakutani type fixed point theorems or by using Fan-KKM type theorems. Blum, Oettli and Théra [29, 120] have studied the existence of solutions of equilibrium problems in the setting of complete metric spaces inspired by the well-known Ekeland's variational principle [53, 54]. They extended Ekelend's variational principle for bifunctions and established several equivalent formulations, namely, Takahashi's minimization theorem [133] and Caristi-Kirk's fixed point theorem [34]. After the work of Blum, Oettli and Théra, several people have started working in this direction and established existence results for solutions of equilibrium problems in different settings or under different assumptions, see, for example, [1, 3–5, 26, 90, 106, 122, 123] and the references therein.

For solution methods for equilibrium problems, we refer to [51, 79, 99–101, 104, 109, 144] and the references therein.

Let X be a topological vector space, K be a nonempty convex subset of X and $f : K \times K \to \mathbb{R}$ be a bifunction such that $f(x, x) = 0$ for all $x \in K$. A problem closely related to EP is the following problem, called *dual equilibrium problem* (in short, DEP) or *Minty equilibrium problem* (in short, MEP): find $\bar{x} \in K$ such that

$$f(y, \bar{x}) \leq 0, \quad \text{for all } y \in K. \tag{1.56}$$

Konnov and Schaible [102] defined the duality for equilibrium problems by using the rule that the dual of the dual is the primal, and used dual equilibrium problem. They proposed various duals of EP. The duality of equilibrium problems is also studied by Martínez-Legaz and Sosa [111] and Bigi et al. [27] but by using different approaches. However, Mastroeni [112] studied gap functions for equilibrium problems which convert an equilibrium problem to an optimization problem.

When $f(x, y) = g(x, y) + h(x, y)$ for all $x, y \in K$ with $g, h : K \times K \to \mathbb{R}$ such that $g(x, x) \geq 0$ and $h(x, x) = 0$ for all $x \in K$, then EP reduces to find $\bar{x} \in K$ such that

$$g(\bar{x}, y) + h(\bar{x}, y) \geq 0, \quad \text{for all } y \in K. \tag{1.57}$$

It was first proposed by Blum and Oettli [29] and further studied by Chadli et al. [38, 39], Kalmoun [83] and Chadli et al. [40] with applications to eigenvalue problems, hemivariational inequalities and anti-periodic solutions for nonlinear evolution equations, see also [35, 37] and the references therein.

Let $l : K \times K \to \mathbb{R}$ be a function. The *implicit variational problem* (for short, IVP) is to find $\bar{x} \in K$ such that

$$l(\bar{x}, \bar{x}) + g(\bar{x}, \bar{x}) \geq l(\bar{x}, y) + g(\bar{x}, y), \quad \text{for all } y \in K. \tag{1.58}$$

It is considered and studied by Mosco [114] and it contains EP (1.51) and (1.57) as special cases. It also includes variational and quasi-variational inequalities [18], fixed point problem and saddle point problem, Nash equilibrium problem of non-cooperative games as special cases. The existence of solutions of IVP was studied by Mosco [114], while Dolcetta and Matzeu [52] discussed its duality and applications.

Let $F, G : K \to \mathcal{L}(X, Y)$ be nonlinear operators. Set

$$l(x, y) = \langle F(x), y - x \rangle \quad \text{and} \quad g(x, y) = \langle G(x), y - x \rangle, \quad \text{for all } x, y \in K.$$

Then IVP reduces to the problem of finding $\bar{x} \in K$ such that

$$\langle F(\bar{x}) + G(\bar{x}), y - \bar{x} \rangle \geq 0, \quad \text{for all } y \in K. \tag{1.59}$$

It is known as *strongly nonlinear variational inequality problem* (in short, SNVIP).

Now we present some basic results on the existence of solutions for EP (1.51).

Theorem 1.53 *Let K be a nonempty convex subset of a Hausdorff topological vector space X and $f : K \times K \to \mathbb{R}$ be a bifunction vanishing on the diagonal, i.e. $f(x, x) = 0$ for all $x \in K$ such that the following conditions hold.*

(i) *f is quasiconvex in the second variable;*
(ii) $\liminf_{x \to x^*} f(x, y) \leq f(x^*, y)$ *for all $y \in K$ whenever $x \to x^* \in K$;*
(iii) *There exist a nonempty compact convex subset B of K and a nonempty compact subset D of K such that for each $x \in K \setminus D$, there exists $\tilde{y} \in B$ such that $f(x, \tilde{y}) < 0$.*

Then EP (1.51) *has a solution in K.*

Proof For each $y \in K$, define

$$P(y) = \{x \in K : f(x, y) \geq 0\}.$$

Then the solution set of EP (1.51) is $\mathbb{S} = \bigcap_{y \in K} P(y)$. By condition (ii), for each $y \in K$, $P(y)$ is closed.

Now we prove that the solution set $\mathbb{S}$ is nonempty. Assume contrary that $\mathbb{S} = \emptyset$. Then for each $x \in K$, the set

$$S(x) := \{y \in K : x \notin P(y)\} = \{y \in K : f(x, y) < 0\}\} \neq \emptyset.$$

By quasiconvexity of f in the second variable, we have that $S(x)$ is convex for each $x \in K$. Thus, $S : K \to 2^K$ defines a set-valued map such that for each $x \in K$, $S(x)$ is nonempty and convex. Now for each $y \in K$, the set

$$\begin{aligned} S^{-1}(y) &= \{x \in K : y \in S(x)\} = \{x \in K : f(x, y) < 0\} \\ &= \{x \in K : f(x, y) \geq 0\}^c = [P(y)]^c \end{aligned}$$

is open in K. Then the set-valued map $S : K \to 2^K$ satisfies all the conditions of Corollary 1.3 (with $S = T$), and therefore, there exists a point $\hat{x} \in K$ such that $\hat{x} \in S(\hat{x})$, that is, $0 = f(\hat{x}, \hat{x}) < 0$, which is a contradiction. Hence the solution set $\mathbb{S}$ of EP (1.51) is nonempty. □

Allen [2] also proved a similar result with different coercivity condition (iii) but by using Fan-KKM Lemma. If K is compact, then the condition (iii) in the above theorem is satisfied. Therefore, if K is compact and f is upper semicontinuous in the first argument, then Theorem 1.53 reduces to the well-known Ky Fan theorem [61].

The following result is a slight generalization of a particular form of Theorem 10 in [37].

Theorem 1.54 *Let X be a Hausdorff topological vector space, K be a closed convex subset of X and $f : K \times K \to \mathbb{R}$ be a bifunction such that $f(x, x) = 0$ for all $x \in K$. Suppose that*

(i) *for each finite subset E of K, $\min_{x \in \text{co}(E)} \max_{y \in E} f(x, y) \geq 0$;*
(ii) *for each fixed $y \in K$, the function $x \mapsto f(x, y)$ is upper semicontinuous;*
(iii) *there exist a nonempty compact convex subset B of K and a nonempty compact subset D of K such that for each $x \in K \setminus D$, there exists $\tilde{y} \in B$ such that $f(x, \tilde{y}) < 0$.*

Then EP (1.51) *has a solution.*

Now we present a theorem which will be used in the sequel.

Theorem 1.55 ([29, Lemma 1]) *Let X be a Hausdorff topological vector space, K be a nonempty compact convex subset of X, D be a convex subset of X and $f : K \times D \to \mathbb{R}$ be concave and upper semicontinuous in the first argument, and convex* *in the second argument. Assume that*

$$\max_{x \in K} f(x,y) \geq 0, \quad \textit{for all } y \in D.$$

Then there exists $\bar{x} \in K$ *such that* $f(\bar{x}, y) \geq 0$ *for all* $y \in D$.

Proof Assume contrary that the conclusion does not hold. Then for every $x \in K$, there exist $y \in D$ and $\varepsilon > 0$ such that $f(x,y) < -\varepsilon$. Therefore, the open sets

$$S_\varepsilon(y) := \{x \in K : f(x,y) < -\varepsilon\}, \quad \text{for } y \in D,\ \varepsilon > 0,$$

cover the compact set K. Hence there exists a finite subcover $\{S_{\varepsilon_i}(y_i)\}_{i=1}^m$ of K. Let $\varepsilon := \min_{1 \leq i \leq m} \varepsilon_i$. Then from the fact that $K \subseteq \bigcup_{i=1}^m S_{\varepsilon_i}(y_i)$, we have

$$\min_{1 \leq i \leq m} f(x, y_i) \leq -\varepsilon, \quad \text{for all } x \in K.$$

Since the function $x \mapsto f(x, y_i)$ is concave, it follows from [127, Theorem 21.1] that there exist real numbers $\mu_i \geq 0$ for $i = 1, 2, \ldots, m$ with $\sum_{i=1}^m \mu_i = 1$ such that $\sum_{i=1}^m f(x, y_i) \leq -\varepsilon$ for all $x \in K$. The convexity of $y \mapsto f(x,y)$ implies with $\tilde{y} := \sum_{i=1}^m \mu_i y_i \in D$ that $f(x, \tilde{y}) \leq -\varepsilon$ for all $x \in K$. Hence $\max_{x \in K} f(x, \tilde{y}) < 0$, a contradiction of our hypothesis. □

Definition 1.57 ([146]) Let K be a nonempty convex subset of a topological vector space X. A bifunction $f : K \times K \to \mathbb{R}$ is said to be *diagonally quasiconvex* in y if for any finite set $\{y_1, y_2, \ldots, y_m\} \subset K$ and any $x_0 \in \text{co}(\{y_1, y_2, \ldots, y_m\})$, we have

$$f(x_0, x_0) \leq \max_{1 \leq i \leq m} f(x_0, y_i).$$

f is said to be *diagonally quasiconcave* in y if $-f$ is diagonally quasiconvex in y.

A bifunction $f : K \times K \to \mathbb{R}$ is said to be γ*-diagonally quasiconvex* in y for some $\gamma \in \mathbb{R}$ if for any finite set $\{y_1, y_2, \ldots, y_m\} \subset K$ and any $x_0 \in \text{co}(\{y_1, y_2, \ldots, y_m\})$, we have

$$\gamma \leq \max_{1 \leq i \leq m} f(x_0, y_i).$$

f is said to be γ*-diagonally quasiconcave* in y for some $\gamma \in \mathbb{R}$ if $-f$ is $-\gamma$-diagonally quasiconvex in y.

Definition 1.58 ([41]) Let K be a nonempty convex subset of a topological vector space X. A bifunction $f : K \times K \to \mathbb{R}$ is said to be γ*-generalized diagonally quasiconvex* in y if for any finite set $\{y_1, y_2, \ldots, y_m\} \subset K$, there is a finite set $\{x_1, x_2, \ldots, x_m\} \subset K$ such that for any set $\{x_{i_1}, x_{i_2}, \ldots, x_{i_k}\} \subset \{x_1, x_2, \ldots, x_m\}$ and any $x_0 \in \operatorname{co}(\{x_{i_1}, x_{i_2}, \ldots, x_{i_k}\})$, we have

$$\gamma \leq \max_{1 \leq j \leq k} f(x_0, y_{i_j}).$$

f is said to be *γ-generalized diagonally quasiconcave* in y if $-f$ is $-\gamma$-generalized diagonally quasiconvex in y.

Chang and Zhang [41] gave the relation between generalized KKM maps and γ-generalized diagonally quasiconvexity (quasiconcavity).

Proposition 1.20 *Let K be a nonempty convex subset of a topological vector space X, $f : K \times K \to \mathbb{R}$ be a bifunction and $\gamma \in \mathbb{R}$. Then the following statements are equivalent:*

(a) *The set-valued map $T : K \to 2^K$ defined by*

$$T(y) = \{x \in K : f(x, y) \leq \gamma\} \text{ (respectively, } T(y) = \{x \in K : f(x, y) \geq \gamma\})$$

is a generalized KKM map.

(b) *$f(x, y)$ is γ-generalized diagonally quasiconcave (respectively, γ-generalized diagonally quasiconvex) in y.*

Tian [136] introduced the following definition of γ-transfer lower semicontinuous functions.

Definition 1.59 Let X and Y be topological spaces. A bifunction $f : X \times Y \to \mathbb{R}$ is said to be *γ-transfer lower semicontinuous* (respectively, *γ-transfer lower semicontinuous*) function in the first argument for some $\gamma \in \mathbb{R}$ if for all $x \in X$ and $y \in Y$ with $f(x, y) > \gamma$ (respectively, $f(x, y) < \gamma$), there exist a point $z \in Y$ and a neighborhood $N(x)$ of x such that $f(u, z) > \gamma$ (respectively, $f(u, z) < \gamma$) for all $u \in N(x)$.

The bifunction f is said to be to *γ-transfer lower semicontinuous (respectively,* γ-transfer lower semicontinuous*) in the first argument if it is γ-transfer lower semicontinuous (respectively, γ-transfer lower semicontinuous) for every $\gamma \in \mathbb{R}$.*

Ansari et al. [7] established the following minimax inequality theorem.

Theorem 1.56 *Let K be a nonempty closed convex subset of a Hausdorff topological vector space X and $f, g : K \times K \to \mathbb{R}$ be bifunctions such that the following conditions hold.*

(i) *For any fixed $y \in K$, the function $x \mapsto f(x, y)$ is 0-transfer upper semicontinuous.*
(ii) *For any fixed $x \in K$, the function $y \mapsto g(x, y)$ is 0-generalized diagonally quasiconvex.*
(iii) *$f(x, y) \geq g(x, y)$ for all $(x, y) \in K \times K$.*
(iv) *The set $\{x \in K : f(x, y_0) \geq \gamma\}$ is precompact (that is, its closure is compact) for at least one $y_0 \in K$.*

Then there exists a solution $\bar{x} \in K$ of EP (1.51).

Proof Define set-valued maps $S, T : K \to 2^K$ by

$$S(y) = \{x \in K : f(x, y) \geq \gamma\} \quad \text{and} \quad T(y) = \{x \in K : g(x, y) \geq \gamma\},$$

for all $y \in K$. Condition (i) implies that S is a transfer closed-valued map. Indeed, if $x \notin S(y)$, then $f(x, y) < 0$. Since $f(x, y)$ is 0-transfer lower semicontinuous in x, there is a $z \in K$ and a neighborhood $N(x)$ of x such that $f(u, z) < 0$ for all $u \in N(x)$. Then $S(z) \subset K \setminus N(x)$. Hence, $x \in \mathrm{cl}\ (S(z))$. Thus, S is transfer closed-valued.

From condition (ii) and Proposition 1.20, T is a generalized KKM map. From (iii), we have that $T(y) \subset S(y)$ for all $y \in K$, and hence S is also a generalized KKM map. So, cl S is also a KKM map. Condition (iv) implies that $S(y_0)$ is precompact. Hence, cl $S(y_0)$ is compact. By Theorem 1.34,

$$\bigcap_{y \in K} S(y) \neq \emptyset.$$

As a result, there exists $\bar{x} \in K$ such that $f(\bar{x}, y) \geq \gamma$ for all $y \in K$. □

Remark 1.25

(a) If for every fixed $y \in K$, the function $x \mapsto f(x, y)$ is upper semicontinuous in x, then condition (i) of Theorem 1.56 is satisfied immediately.
(b) The following condition implies condition (iv) in Theorem 1.56.
(iv)′ There exist a compact subset D of K and $y_0 \in K$ such that for all $x \in K \setminus D, f(x, y_0) < 0$.

For further details, existence results and applications of equilibrium problems, we refer [1, 3–5, 7, 8, 22–29, 31, 35–41, 46, 47, 49, 51, 52, 57, 58, 65, 66, 71, 72, 77, 78, 80, 83, 84, 90–93, 99–102, 104, 107–109, 111, 112, 120, 128, 135, 142, 143] and the references therein.

References

1. S. Al-Homidan, Q.H. Ansari, J.-C. Yao, Some generalizations of Ekeland-type variational principle with applications to equilibrium problems and fixed point theory. Nonlinear Anal. **69**(1), 126–139 (2008)
2. G. Allen, Variational inequalities, complementarity problems, and duality theorems. J. Math. Anal. Appl. **58**, 1–10 (1977)
3. Q.H. Ansari, *Metric Spaces—Including Fixed Point Theory and Set-Valued Maps* (Narosa Publishing House, New Delhi, 2010). Also Published by Alpha Science International Ltd. Oxford, U.K. (2010)
4. Q.H. Ansari, Ekeland's variational principle and its extensions with applications, in *Topics in Fixed Point Theory*, ed. by S. Almezel, Q.H. Ansari, M.A. Khamsi (Springer, Cham, 2014), pp. 65–100

5. Q.H. Ansari, L.-J. Lin, Ekeland type variational principle and equilibrium problems, in *Topics in Nonconvex Optimization, Theory and Applications*, ed. by S.K. Mishra (Springer, New York, 2011), pp. 147–174
6. Q.H. Ansari, C.S. Lalitha, M. Mehta, *Generalized Convexity, Nonsmooth Variational Inequalities and Nonsmooth Optimization* (CRC Press/Taylor & Francis Group, Boca Raton, 2014)
7. Q.H. Ansari, Y.C. Lin, J.C. Yao, General KKM theorem with applications to minimax and variational inequalities. J. Optim. Theory Appl. **104**(1), 41–57 (2000)
8. Q.H. Ansari, N.-C. Wong, J.C. Yao, The existence of nonlinear inequalities. Appl. Math. Lett. **12**, 89–92 (1999)
9. Q.H. Ansari, J.C. Yao, A fixed point theorem and its applications to a system of variational inequalities. Bull. Aust. Math. Soc. **59**, 433–442 (1999)
10. K.J. Arrow, A.C. Enthoven, Quasi-concave programming. Econometrica **29**, 779–800 (1961)
11. J.-P. Aubin, *Mathematical Methods of Game and Economic Theory* (North-Holland Publishing Company, Amsterdam, 1979)
12. J.-P. Aubin, *Optima and Equilibria* (Springer, Berlin, 1993)
13. J.-P. Aubin, A. Cellina, *Differential Inclusions* (Springer, Berlin, 1994)
14. J.-P. Aubin, I. Ekeland, *Applied Nonlinear Analysis* (John Wiley & Sons, New York, 1984)
15. J.-P. Aubin, H. Frankowska, *Set-Valued Analysis* (Birkhäuser, Boston, 1990)
16. A. Auslender, *Optimisation: Méthodes Numériques* (Masson, Paris, 1976)
17. M. Avriel, *Nonlinear Programming: Analysis and Methods* (Prentice-Hall, Inc., Englewood Cliffs, 1976)
18. C. Baiocchi, A. Capelo, *Variational and Quasivariational Inequalities, Applications to Free Boundary Problems* (John Wiley & Sons, New York, 1984)
19. M.S. Bazara, H.D. Sherali, C.M. Shetty, *Nonlinear Programming: Theory and Applications* (John Wiley & Sons, New York, 1993)
20. C. Berge, *Topological Space* (Oliver & Boyd, Edinburgh, 1963)
21. D.P. Bertsekas, A. Nedic, A.E. Ozdaglar, *Convex Analysis and Optimization* (Athena Scientific, Belmont, 2003)
22. M. Bianchi, R. Pini, A note on equilibrium problems for properly quasimonotone bifunctions. J. Global Optim. **20**, 67–76 (2001)
23. M. Bianchi, R. Pini, Coercivity conditions for equilibrium problems. J. Optim. Theory Appl. **124**, 79–92 (2005)
24. M. Bianchi, S. Schaible, Generalized monotone bifunctions and equilibrium problems. J. Optim. Theory Appl. **90**, 31–43 (1996)
25. M. Bianchi, S. Schaible, Equilibrium problems under generalized convexity and generalized monotonicity. J. Global Optim. **30**, 121–134 (2004)
26. M. Bianchi, G. Kassay and R. Pini, Existence of equilibria via Ekeland's principle. J. Math. Anal. Appl. **305**, 502–512 (2005)
27. G. Bigi, M. Castellani, G. Kassay, A dual view of equilibrium problems. J. Math. Anal. Appl. **342**, 17–26 (2008)
28. E. Blum, W. Oettli, Variational principles for equilibrium problems, in *Parametric Optimization and Related Topics III*, ed. by J. Guddat et al. (Peter Lang, Frankfurt am Main, 1993), pp. 79–88
29. E. Blum, W. Oettli, From optimization and variational inequalities to equilibrium problems. Math. Student **63**, 123–145 (1994)
30. H. Brézis, *Analyse Fonctionnelle* (Masson, Paris, 1983)
31. H. Brézis, L. Nirenberg, G. Stampacchia, A remark on Ky Fan's minimax principle. Boll. Unione Mat. Ital. **6**(4) (1972), 293–300.
32. F.E. Browder, Nonlinear monotone operators and convex sets in Banach spaces. Bull. Am. Math. Soc. **71**, 780–785 (1965)
33. F.E. Browder, The fixed point theory of multivalued mappings in topological vector spaces. Math. Ann. **177**, 283–301 (1968)

34. J. Caristi, W.A. Kirk, Geometric fixed point theory and inwardness conditions, in *The Geometry of Metric and Linear Spaces*. Lecture Notes in Mathematics, vol. 490 (Springer, Berlin, 1975), pp. 74–83
35. O. Chadli, Z. Chbani, H. Riahi, Recession methods for equilibrium problems and applications to variational and hemivariational inequalities. Discret. Cont. Dynamic. Syst. **5**, 185–195 (1999)
36. O. Chadli, Z. Chbani, H. Riahi, Equilibrium problems and noncoercive variational inequalities. Optimization **49**, 1–12 (1999)
37. O. Chadli, Z. Chbani, H. Riahi, Equilibrium problems with generalized monotone bifunctions and applications to variational inequalities. J. Optim. Theory Appl. **105**, 299–323 (2000)
38. O. Chadli, N.C. Wong, J.C. Yao, Equilibrium problems with application to eigenvalue problems. J. Optim. Theory Appl. **117**(2), 245–266 (2003)
39. O. Chadli, S. Schaible, J.C. Yao, Regularized equilibrium problems with application to noncoercive hemivariational inequalities. J. Optim. Theory Appl. **121**, 571–596 (2004)
40. O. Chadli, Q.H. Ansari, J.-C. Yao, Mixed equilibrium problems and anti-periodic solutions for nonlinear evolution equations. J. Optim. Theory Appl. **168**(2), 410–440 (2016)
41. S.S. Chang, Y. Zhang, Generalized KKM theorem and variational inequalities. J. Math. Anal. Appl. **159**, 208–223 (1991)
42. S.S. Chang, B.S. Lee, X. Wu, J. Cho, G.M. Lee, On the generalized quasi-variational inequality problems. J. Math. Anal. Appl. **203**, 686–711 (1996)
43. S. Chebbi, M. Florenzano, Maximal elements and equilibria for condensing correspondences. Nonlinear Anal. **38**, 995–1002 (1999)
44. K.L. Chew, E.U. Choo: pseudolinearity and efficiency. Math. Program. **28**, 226–239 (1984)
45. M. Chipot, *Variational Inequalities and Flow in Porous Media* (Springer, New York, 1984)
46. M.S.R. Chowdhury, K.K. Tan, Generalization of Ky Fan's minimax inequality with applications to generalized variational inequalities for pseudo-monotone operators and fixed point theorems. J. Math. Anal. Appl. **204**, 910–929 (1996)
47. M.S.R. Chowdhury, K.K. Tan, Generalized variational inequalities for quasi-monotone operators and applications. Bull. Polish Acad. Sci. Math. **45**, 25–54 (1997)
48. F.H. Clarke, *Optimization and Nonsmooth Analysis* (SIAM, Philadelphia, 1990)
49. P. Deguire, K.K. Tan, G.X.-Z. Yuan, The study of maximal element, fixed point for L_S-majorized mappings and their applications to minimax and variational inequalities in the product topological spaces. Nonlinear Anal. **37**, 933–951 (1999)
50. W.E. Diewert, Alternative characterizations of six kinds of quasiconcavity in the nondifferentiable case with applications to nonsmooth programming, in *Generalized Concavity in Optimization and Economics*, ed. by S. Schaible, W.T. Ziemba (Academic Press, New York, 1981), pp. 51–93
51. B.V. Dinh, L.D. Muu, A projection algorithm for solving pseudomontone equilibrium problems and its applications to a class of bilevel equilibria. Optimization **64**(3), 559–575 (2015)
52. I.C. Dolcetta, M. Matzeu, Duality for implicit variational problems and numerical applications. Numer. Funct. Anal. Optim. **2**(4), 231–265 (1980)
53. I. Ekeland, Sur les problèmes variationnels. C.R. Acad. Sci. Paris **275**, 1057–1059 (1972)
54. I. Ekeland, On the variational principle. J. Math. Anal. Appl. **47**, 324–353 (1974)
55. I. Ekeland, R. Temam, *Convex Analysis and Variational Problems* (North-Holland Publishing Company, Amsterdam, 1976)
56. F. Facchinei, J.-S. Pang, *Finite Dimensional Variational Inequalities and Complementarity Problems*, vol. I & II (Springer, New York, 2003)
57. M. Fakhar, J. Zafarani, Generalized equilibrium problems for quasimonotone and pseudomonotone bifunctions. J. Optim. Theory Appl. **123**(2), 349–364 (2004)
58. M. Fakhar, J. Zafarani, Generalized vector equilibrium problems for pseudomonotone multivalued bifunctions. J. Optim. Theory Appl. **126**, 109–124 (2005)
59. K. Fan, A generalization of Tychonoff's fixed point theorem. Math. Ann. **142**, 305–310 (1961)
60. K. Fan, Extension of two fixed point theorems of F.E. Browder. Math. Z. **112**, 234–240 (1969)

61. K. Fan, A minimax inequality and application, in *Inequalities III*, ed. by O. Shisha (Academic Press, New York, 1972), pp. 103–113
62. S.C. Fang, E.L. Peterson, Generalized variational inequalities. J. Optim. Theory Appl. **38**, 363–383 (1982)
63. W. Fenchel, *Convex Cones, Sets and Functions*, Mimeographed Lecture Notes (Princeton University, Princeton, 1953)
64. M. Ferris, J.S. Pang, Engineering and economic applications of complementarity problems. SIAM Rev. **39**, 669–713 (1997)
65. F. Flores-Bazán, Existence theorems for generalized noncoercive equilibrium problems: the quasi-convex case. SIAM J. Optim. **11**(3), 675–690 (2000)
66. F. Flores-Bazán, Existence theory for finite-dimensional pseudomonotone equilibrium problems. Acta Appl. Math. **77**, 249–297 (2003)
67. A. Friedman, *Variational Principles and Free Boundary Problems* (R. E. Krieger Publishing Company, Malabar, 1988)
68. G. Giorgi, S. Komlósi, Dini derivatives in optimization—Part III. Rivista Mat. Sci. Econom. Soc. **18**(1), 47–63 (1995)
69. R. Glowinski, J.L. Lions, R. Tremoliéres, *Numerical Analysis of Variational Inequalities* (North-Holland, Amsterdam, 1981)
70. N. Hadjisavvas, Continuity amd maximality properties of pseudomonotone operators. J. Convex Anal. **10**(2), 459–469 (2003)
71. N. Hadjisavvas, S. Schaible, From scalar to vector equilibrium problems in the quasimonotone case. J. Optim. Theory Appl. **96**, 297–305 (1998)
72. N. Hadjisavvas, S. Schaible, Quasimonotonicity and pseudomonotonicity in variational inequalities and equilibrium problems, in *Generalized Convexity, Generalized Monotonicity: Recent Results*, ed. by J.P. Crouzeix, J.E. Martinez-Legaz, M. Volle (Kluwer Academic Publications, Dordrecht, 1998), pp. 257–275
73. A. Hassouni, Sous-differentiels des fonctions quasiconvexes, Doctorat de 3ème cycle de l'Université Paul Sabatier (Toulouse 3), 1983
74. G. Isac, *Complementarity Problems*. Lecture Notes in Mathematics, vol. 1528 (Springer, Berlin, 1992)
75. G. Isac, *Topological Matheods in Complementarity Theory* (Kluwer Academic Publishers, Dordrecht, 2000)
76. V. Isakov, *Inverse Problems for Partial Differential Equations* (Springer, New York, 2006)
77. A.N. Iusem, G. Kassay, W. Sosa, An result for equilibrium problem with surjectivity consequences. J. Convex Anal. **16**, 807–826 (2009)
78. A.N. Iusem, G. Kassay, W. Sosa, On certain conditions for the existence of solutions of equilibrium problems. Math. Program. **116**, 259–273 (2009)
79. A.N. Iusem, W. Sosa, Iterative algorithms for equilibrium problems. Optimization **52**(3), 301–316 (2003)
80. A.N. Iusem, W. Sosa, New existence results for equilibrium problems. Nonlinear Anal. **52**, 621–635 (2003)
81. J. Jahn, *Vector Optimization: Theory, Applications, and Extensions* (Springer, Berlin, 2004)
82. G. Jameson, *Ordered Linear Spaces* (Springer, Berlin, 1970)
83. E.M. Kalmoun, On Ky Fan's minimax inequalities, mixed equilibrium problems and hemivariational inequalities. J. Inequal. Pure Appl. Math. 12, **2**(1), (2001)
84. E.M. Kalmoun, H. Riahi, Topological KKM theorems and generalized vector equilibria on G-convex spaces with applications. Proc. Am. Math. Soc. **129**, 1335–1348 (2001)
85. S. Kakutani, A generalization of Brouwer's fixed point theorem. Duke Math. J. **8**, 457–459 (1941)
86. L.W. Kantorowitsch, G.P. Akilow, *Funktionalanalysis in Normierten Räumen* (Akademie-Verlag, Berlin, 1978)
87. S. Karamardian, Generalized complementarity problem. J. Optim. Theory Appl. **8**, 161–168 (1971)

88. S. Karamardian, Complementarity over cones with monotone and pseudomonotone maps. *J. Optim. Theory Appl.* **18**, 445-454 (1976)
89. S. Karamardian, S. Schaible, Seven kinds of monotone maps. J. Optim. Theory Appl. **66**, 37–46 (1990)
90. P. Kas, G. Kassay, Z.B. Sensoy, On generalized equilibrium points. J. Math. Anal. Appl. **296**, 619–633 (2004)
91. G. Kassay, *The Equilibrium Problem and Related Topics* (Cluj-Napoca Risoprint, Cluj-Napoca, 2000)
92. G. Kassay, On equilibrium problems, in *Optimization and Optimal Control, Theory and Applications*, ed. by A. Chinchuluun, P.M. Pardalos, R. Enkhbat, I. Tseveendorj (Springer, New York, 2010), pp. 55–83
93. G. Kassay, The equilibrium problem and its applications to optimization, minimax problems and Nash equilibria, in *Topics in Nonlinear Analysis and Optimization*, ed. by Q.H. Ansari (World Education, Delhi, 2012), pp. 203–226
94. D. Kinderlehrer, G. Stampacchia, *An Introduction to Variational Inequalities and Their Applications* (Academic Press, New York, 1980)
95. H. Kneser, Sur le théorème fondamentale de la thérie des jeux. C. R. Acad. Sci. Paris **234**, 2418–2420 (1952)
96. H. Knaster, C. Kuratowski, S. Mazurkiewicz, Ein Beweis des Fixpunktsatzes für n-dimensionale Simplexe. Fund. Math. **14**, 132–137 (1929)
97. S. Komlòsi, Generalized monotonicity and generalized convexity. J. Optim. Theory Appl. **84**(2), 361–376 (1995)
98. I.V. Konnov, *Combined Relaxation Methods for Variational Inequalities*. Lecture Notes in Mathematical Economics (Springer, Berlin, 2000)
99. I.V. Konnov, Combined relaxation method for monotone equilibrium problems. J. Optim. Theory Appl. **111**(2), 327–340 (2001)
100. I.V. Konnov, Application of the proximal point method to nonmonotone equilibrium problems. J. Optim. Theory Appl. **119**(2), 317–333 (2003)
101. I.V. Konnov, M.S.S. Ali, Descent methods for monotone equilibrium problems in Banach spaces. J. Comput. Appl. Math. **188**, 156–179 (2006)
102. I.V. Konnov, S. Schaible, Duality for equilibrium problems under generalized monotonicity. J. Optim. Theory Appl. **104**, 395–408 (2000)
103. I.V. Konnov, J.-C. Yao, On the generalized vector variational inequality problem. J. Math. Anal. Appl. **206**, 42–58 (1997)
104. I.V. Konnov, S. Schaible, J.C. Yao, Combined relaxation method for mixed equilibrium problems. J. Optim. Theory Appl. **126**(2), 309–322 (2005)
105. L.-J. Lin, Q.H. Ansari, Collective fixed points and maximal elements with applications to abstract economies. J. Math. Anal. Appl. **296**, 455–472 (2004)
106. L.-J. Lin, W.-S. Du, On maximal element theorems, variants of Ekeland's variational principle and their applications. Nonlinear Anal. **68**, 1246–1262 (2008)
107. L.-J. Lin, S. Park, Z.-T. Yu, Remarks on fixed points, maximal elements, and equilibria of generalized games. J. Math. Anal. Appl. **233**, 581–596 (1999)
108. L.J. Lin, Z.T. Yu, G. Kassay, Existence of equilibria for multivalued mappings and its applications to vectorial equilibria. J. Optim. Theory Appl. **114**, 189–208 (2002)
109. R. López, Approximations of equilibrium problems. SIAM J. Control Optim. **50**(2), 1038–1070 (2012)
110. O. Mangasarian, *Nonlinear Programming* (McGraw Hill, New York, 1969)
111. J.E. Martínez-Legaz, W. Sosa, Duality for equilibrium problems. J. Global Optim. **35**, 311–319 (2006)
112. G. Mastroeni, Gap functions for equilibrium problems. J. Global Optim. **27**, 411–426 (2003)
113. G.J. Minty, Monotone (nonlinear) operators in Hilbert space. Duke Math. J. **29**, 341–346 (1962)
114. U. Mosco, Implicit variational problems and quasi-variational inequalities, in Lecture Notes in Mathematics, vol. 543 (Springer, Berlin, 1976)

115. S.B. Nadler, Jr., Multi-valued contraction mappings. Pacific J. Math. **30**, 475–488 (1969)
116. A. Nagurney, *Network Economics: A Variational Inequality Approach* (Kluwer Academic Publishers, Dordrecht, 1993)
117. J.F. Nash, Non-cooperative games. Ann. Math. **54**, 286–295 (1951)
118. J.F. Nash, Two-persons cooperative games. Econometrica **21**, 128–140 (1953)
119. H. Nikaido, K. Isoda, Note on noncooperative convex games. Pacific J. Math. **5**, 807–815 (1955)
120. W. Oettli, M. Théra, Equivalents of Ekeland's principle. Bull. Aust. Math. Soc. **48**, 385–392 (1993)
121. P.D. Panagiotopoulus, *Inequality Problems in Mechanics and Applications* (Birkhauser, Boston, 1985)
122. S. Park, Another generalizations of the Ekeland type variational principles. Math. Sci. Res. Hot-Line **1**(10), 1–6 (1997)
123. S. Park, On generalizations of the Ekeland-type variational principles. Nonlinear Anal. **39**, 881–889 (2000)
124. M. Patriksson, *Nonlinear Programming and Variational Inequality Problems* (Kluwer Academic Publishers, Dordrecht, 1999)
125. W.V. Petryshyn, P.M. Fitzpatrick, Fixed point theorems of multivalued noncompact inward maps. J. Math. Anal. Appl. **46**, 756–767 (1974)
126. W.V. Petryshyn, P.M. Fitzpatrick, Fixed point theorems of multivalued noncompact acyclic mappings. Pac. J. Math. **54**, 17–23 (1974)
127. R.T. Rockafellar, *Convex Analysis* (Princeton University Press, Princeton, 1970)
128. B.D. Rouhani, E. Tarafdar, P.J. Watson, Existence of solutions to some equilibrium problems. J. Optim. Theory Appl. **126**(1), 97–107 (2005)
129. P.H. Sach, J.-P. Penot, Charaterizations of generalized convexities via generalized directional derivative. Numer. Funct. Anal. Optim. **19**(5–6), 615–634 (1998)
130. R. Saigal, Extension of generalized complementarity problem. Math. Oper. Res. **1**, 260–266 (1976)
131. S. Schaible, J.C. Yao, On the equivalence of nonlinear complementarity problems. Math. Program. **70**, 191–200 (1995)
132. G. Stampacchia, Formes bilinéaires coercitives sur les ensembles convexes. C. R. Acad. Sci. Paris **258**, 4413–4416 (1964)
133. W. Takahashi, Existence theorems generalizing fixed point theorems for multivalued mappings, in *Fixed Point Theory and Applications*, ed. by J.-B. Baillon. Théra, Pitman Research Notes in Mathematics, vol. 252 (Longman, Harlow, 1991), pp. 397–406
134. E. Tarafdar, G.X.-Z. Yuan, Generalized variational inequalities and its applications. Nonlinear Anal. Theory Methods Appl. **30**, 4171–4181 (1997)
135. G. Tian, Generalization of the FKKM theorem and the Ky Fan minimax inequality, with applications to maximal elements, price equilibrium, and complementarity. J. Math. Anal. Appl. **170**, 457–471 (1992)
136. G. Tian, Generalized quasi-variational-like inequality problem. Math. Oper. Res. **18**, 213–225 (1993)
137. F.-N. Xiang, L. Debnath, Fixed point theorems and variational inequalities with non-compact sets. Int. J. Math. Math. Sci. **19**, 111–114 (1996)
138. N.C. Yannelis, N.D. Prabhakar, Existence of maximal elements and equilibria in linear topological spaces. J. Math. Econ. **12**, 233–245 (1983)
139. J.C. Yao, Variational inequalities with generalized monotone operators. Math. Oper. Res. **19**, 691–705 (1994)
140. J.C. Yao, Multi-valued variational inequalities with K-pseudomonotone operators. J. Optim. Theory Appl. **83**, 391–403 (1994)
141. J. Yu, Essential equilibria of n-person noncooperative games. J. Math. Econ. **31**, 361–372 (1999)
142. G.X.-Z. Yuan, *KKM Theory and Applications in Nonlinear Analysis* (Marcel Dekker, Inc., New York, 1999)

143. L.C. Zeng, S.Y. Wu, J.C. Yao, Generalized KKM theorem with applications to generalized minimax inequalities and generalized equlibrium problems. Taiwan. J. Math. **10**(6), 1497–1514 (2006)
144. L.C. Zeng, J.C. Yao, Generalized Minty's lemma for generalized vector equilibrium problems. Appl. Math. Lett. **19**, 1320–1326 (2006)
145. L.C. Zeng, J.C. Yao, Existence of solutions of generalized vector variational inequalities in reflexive Banach spaces. J. Global Optim. **36**, 483–497 (2006)
146. J.X. Zhou, G. Chen, Diagonally convexity conditions for problems in convex analysis and quasi-variational inequalities. J. Math. Anal. Appl. **132**, 213–225 (1988)

Chapter 2
Analysis over Cones

An optimization problem is called a *vector optimization problem* if the objective function is vector-valued. In general, a vector optimization problem is stated as follows:

$$\begin{aligned} &\text{minimize } g(x) \\ &\text{subject to } x \in K, \end{aligned} \tag{2.1}$$

where $g : K \to Y$ is a vector-valued function, X is a vector space, and Y is a topological vector space and $K \subseteq X$ is a nonempty set. Specifically, we consider the special case of problem (2.1) with $X = \mathbb{R}^n$ and $Y = \mathbb{R}^\ell$, namely, we study the problem

$$\begin{aligned} &\text{minimize } f(x) = \big(f_1(x), f_2(x), \ldots, f_\ell(x)\big) \\ &\text{subject to } x \in K, \end{aligned} \tag{2.2}$$

where $f = \big(f_1, f_2, \ldots, f_\ell\big) : \mathbb{R}^n \to \mathbb{R}^\ell$ is called an *objective function*, $x = (x_1, x_2, \ldots, x_n)$ is called a *decision (variable) vector*, $K \subseteq \mathbb{R}^n$ is called the *feasible region*, $\mathbb{R}^n$ is called the *decision variable space*, and $f(K)$ is called the *feasible objective region* and it is a subset of the *objective space* $\mathbb{R}^\ell$. The problem (2.2) is also called *multiobjective optimization problem* or *multicriteria optimization problem*.

Now, the question is, what does the word 'minimize' mean? Do we want to minimize all the objective functions simultaneously? If yes, then it can be achieved if there is no conflict between the objective functions. In this case, a solution can be found, without requiring any special method, where every objective function attains its minimum. Almost every real-world application of mathematics has conflictive multiple criteria. For example, we consider a formal mathematical problem with feasible set $K = \{x \in \mathbb{R} : -5 \leq x \leq 5\}$ and objective functions

$$f_1(x) = x^2, \quad f_2(x) = x$$

Q.H. Ansari et al., *Vector Variational Inequalities and Vector Optimization*, Vector Optimization, DOI 10.1007/978-3-319-63049-6_2

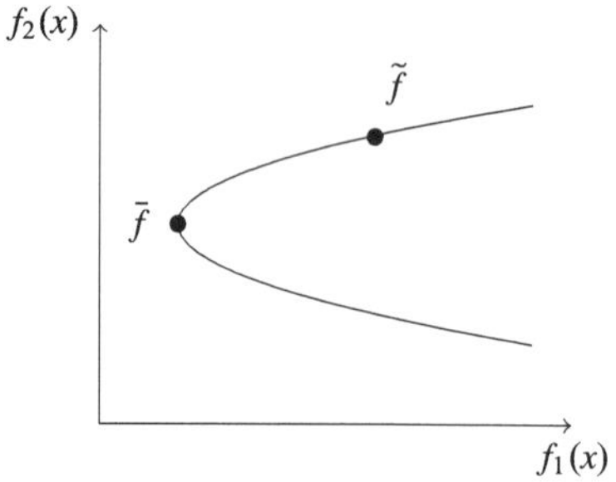

Fig. 2.1 Visualization of the functions $f_1(x) = x^2$, $f_2(x) = x$ for $x \in [-5, 5]$

and we want to minimize both objective functions over K:

$$\text{minimize } \big(f_1(x), f_2(x)\big)$$
$$\text{subject to } x \in K.$$

In this problem, we have two criteria and one decision variable. Note that for each function individually the corresponding optimization problem is easy, and $x_1 = 0$ and $x_2 = -5$ are the (unique) minimizers for f_1 and f_2 on K, respectively. Now the questions are: What are the 'minima' and the 'minimizer' in this situation? Does $x_1 = 0$ or $x_2 = -5$ or both simultaneously minimize f_1 and f_2? Of course, the answer to the latter question is 'no'. Both functions f_1 and f_2 are plotted against each other at the right hand side in Fig. 2.1. We can see that there does not exist a solution that minimizes f_1 and f_2 at the same time. In order to determine minimal solutions of this problem, consider, for instance, the vector $\tilde{f}$ in Fig. 2.1. We can move $\tilde{f}$ along the parabola, minimizing both objective functions at the same time, until the vertex $\bar{f}$ of the parabola. If we want to minimize f_2 further, we can only do so by maximizing f_1. The lower bound of the parabola is called *Pareto frontier*, and it symbolizes that one objective function cannot be minimized without the other being worsened. In this example, both objectives f_1 and f_2 are clearly conflicting, as minimizing f_1 results in a maximization of f_2 on the Pareto frontier.

The following example describes a well-known application of vector optimization in financial mathematics.

Example 2.1 (Portfolio Optimization) A shareholder would like to invest in a portfolio consisting of n shares that maximizes his wins and minimizes the risk associated with the shares at the same time. Let $x = (x_1, x_2, \dots, x_n)$ denote the vector of shares and $r = (r_1, r_2, \dots, r_n)$ be the vector of returns of the respective shares. The return of the whole portfolio is then $r_p := \langle r, x\rangle$. Of course, the return vector r is subject to uncertainties, and r is a vector of random variables, such that we write $\mathbb{E}(r) =: \mu$ (here $\mathbb{E}(\cdot)$ denotes the expected values of r) and $\mathbb{E}(r_i) =: \mu_i$, $i = 1, 2, \dots, n$. The covariance matrix that represents the risk is denoted by

$C = \begin{pmatrix} c_{11} & \cdots & c_{1n} \\ \vdots & \ddots & \vdots \\ c_{1n} & \cdots & c_{nn} \end{pmatrix}$, which is assumed to be positive definite. The entries in the covariance matrix C can be computed by means of $c_{ij} = \mathbb{E}[(r_i - \mathbb{E}(r_i)) \cdot (r_j - \mathbb{E}(r_j))]$ for $i, j = 1, 2, \ldots, n$. The values in the main diagonal of C are the variances of the respective shares. The risk of a portfolio x can then be described by $\langle x, Cx \rangle$. Moreover, it is assumed that no short sales are allowed, i.e., $x_i \geq 0, i = 1, 2, \ldots, n$, and all available capital shall be used and normed to one, i.e., $\sum_{i=1}^{n} x_i = 1$. The problem of finding a portfolio of shares then reads

$$\begin{aligned} &\text{minimize } \big(f_1(x), f_2(x)\big) \\ &\text{subject to } x \in K, \end{aligned}$$

where $f_1 := \langle -\mu, x \rangle$, $f_2(x) := \langle x, Cx \rangle$, and $K := \{x \in \mathbb{R}^n : x_i \geq 0, i = 1, 2, \ldots, n, \ \sum_{i=1}^{n} x_i = 1\}$. These objective functions are contradictive, because higher returns are usually accompanied by higher risk.

The following example describes an application of vector optimization in location theory.

Example 2.2 (Location Theory) Let m locations a_i, $i = 1, 2, \ldots, m$, in the plane $\mathbb{R}^2$ be given. Consider the problem of choosing a location for a new facility $x \in \mathbb{R}^2$ with minimal distance to all given facilities. The problem then reads

$$\begin{aligned} &\text{minimize } \big(||a_1 - x||_p \,, ||a_2 - x||_p \,, \ \ldots \ , \ ||a_m - x||_p\big) \\ &\text{subject to } x \in \mathbb{R}^2, \end{aligned}$$

where $|| \cdot ||_p$ is an arbitrary norm in $\mathbb{R}^2$. Figure 2.2 shows a visualization of this problem with three existing locations a_i, $i = 1, 2, 3$. One searches for one new location x with minimal Euclidian distance to all three existing locations. Now, when moving x towards $\bar{x}$, we can see that the distance to a_2 declines, whereas the distances towards a_1 and a_2 increase. Thus, here we have the typical situation of conflicting goals in vector optimization.

As we have seen above, because of the contradiction and possible incommensurability of the objective functions, it is not possible to find a single solution that would be optimal for all the objectives simultaneously. Vector optimization problems are in a sense ill-defined. There is no natural ordering in the objective space because

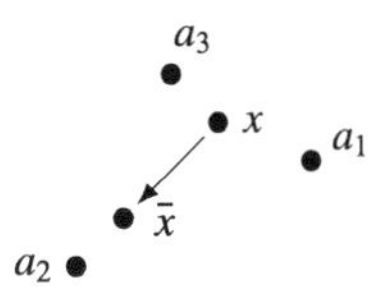

Fig. 2.2 Visualization of Example 2.2 with three existing locations

it is only partially ordered, meaning that, for example, $(1, 1)$ can be said to be less than $(3, 3)$, but how to compare $(1, 3)$ and $(3, 1)$? This means that the objective functions are at least partially conflicting. They may also be incommensurable (that is, in different units). Before defining the solution concepts of a vector optimization problem, we study (preference) orders.

2.1 Orders

A preference order represents a preference attitude of the decision maker in the objective space. It is a binary relation on a set $f(K) := \bigcup_{x \in K} f(x)$, where f is a vector-valued function and K is a feasible decision set. Below we give the definition of a binary relation.

Definition 2.1 (Binary Relation) Let A be an arbitrary nonempty set and let $A \times A$ represent the set of ordered pairs $\{(a, b) : a, b \in A\}$. Then a *binary relation* (or an *order relation*) $\mathcal{R}$ on A is a subset of $A \times A$.

Definition 2.2 A binary relation $\mathcal{R}$ on an arbitrary set A is called

(a) *reflexive* if for all $a \in A$, $(a, a) \in \mathcal{R}$;
(b) *irreflexive* if for all $a \in A$, $(a, a) \notin \mathcal{R}$;
(c) *symmetric* if for all $a, b \in A$, $(a, b) \in \mathcal{R}$ implies $(b, a) \in \mathcal{R}$;
(d) *asymmetric* if for all $a, b \in A$, $(a, b) \in \mathcal{R}$ implies $(b, a) \notin \mathcal{R}$;
(e) *antisymmetric* if for all $a, b \in A$, $(a, b) \in \mathcal{R}$ and $(b, a) \in \mathcal{R}$ imply $a = b$;
(f) *transitive* if for all $a, b, c \in A$, $(a, b) \in \mathcal{R}$ and $(b, c) \in \mathcal{R}$ imply $(a, c) \in \mathcal{R}$;
(g) *negatively transitive* if for all $a, b, c \in A$, $(a, b) \notin \mathcal{R}$ and $(b, c) \notin \mathcal{R}$ imply $(a, c) \notin \mathcal{R}$;
(h) *connected* or *complete* if for all $a, b \in A$, $(a, b) \in \mathcal{R}$ or $(b, a) \in \mathcal{R}$ (possibly both);
(i) *weakly connected* if for all $a, b \in A$, $a \neq b$ implies $(a, b) \in \mathcal{R}$ or $(b, a) \in \mathcal{R}$.

As a first example, consider for a nonempty set A the binary relation

$$\mathcal{R}_A := \{(x, x) : x \in A\}.$$

Then the relation $\mathcal{R}_A$ is reflexive, transitive and antisymmetric. Furthermore, an illustrative example is given by the binary relation

$$\mathcal{R}_{\leq} := \{(x, y) \in \mathbb{R}^2 : x \leq y\},$$

which is also reflexive, transitive and antisymmetric.

The following proposition gives some interrelations between binary relations.

Proposition 2.1

(a) *An asymmetric binary relation is irreflexive.*
(b) *A transitive and irreflexive binary relation is asymmetric.*
(c) *A negatively transitive and asymmetric relation is transitive.*

Proof

(a) It is obvious.
(b) Suppose that $\mathcal{R}$ is a transitive and irreflexive binary relation, but assume that $\mathcal{R}$ is not asymmetric. Then there exist two elements $a, b \in A$ with $(a, b) \in \mathcal{R}$ and $(b, a) \in \mathcal{R}$. Because $\mathcal{R}$ is transitive, we conduct $(a, a) \in \mathcal{R}$, in contradiction to $\mathcal{R}$ being irreflexive.
(c) Let $\mathcal{R}$ be a negatively transitive and asymmetric relation, but assume that $\mathcal{R}$ is not transitive. Then there exist $a, b, c \in A$ such that $(a, b) \in \mathcal{R}$ and $(b, c) \in \mathcal{R}$, but $(a, c) \notin \mathcal{R}$. Since $\mathcal{R}$ is asymmetric, it follows that $(b, a) \notin \mathcal{R}$. Because $\mathcal{R}$ is negatively transitive, $(b, a) \notin \mathcal{R}$ and $(a, c) \notin \mathcal{R}$ imply $(b, c) \notin \mathcal{R}$, contradicting $(b, c) \in \mathcal{R}$. □

Let $\mathcal{R}$ be a binary relation on an arbitrary set A. For any $a, b \in A$, we also write $a\mathcal{R}b$ instead of $(a, b) \in \mathcal{R}$.

Definition 2.3 (Orders) A binary relation $\mathcal{R}$ on an arbitrary set A is called

(a) a *preorder* or *quasi-order* if it is reflexive and transitive;
(b) a *weak order* if it is asymmetric and negatively transitive;
(c) a *partial order* if it is reflexive, antisymmetric and transitive;
(d) a *strict partial order* if it is irreflexive and transitive (or equivalently, if it is asymmetric and transitive);
(e) a *total order* or *linear order* if it is reflexive, antisymmetric, transitive (that is, $\mathcal{R}$ is a partial order) and $\mathcal{R}$ weakly connected;
(f) a *strict* or *strong order* if it is transitive, irreflexive and weakly connected;
(g) an *equivalence relation* if it is reflexive, symmetric and transitive.

The following result can be easily proved and therefore we omit the proof.

Proposition 2.2 *A total order is a weak order, and a weak order is a strict partial order.*

Definition 2.4 The pair $(A, \mathcal{R})$ is called an *ordered structure* if A is an arbitrary set and $\mathcal{R}$ is a binary relation (or an order relation). An ordered structure $(A, \mathcal{R})$ is called *well-ordered* if each nonempty subset B of A has first element b, meaning that $b \in B$ and $(b, a) \in \mathcal{R}$ for all $a \in A$.

At this point it is interesting to recall Zermelo's Theorem: For every nonempty set A there exists a partial order $\mathcal{R}$ on A such that the ordered structure $(A, \mathcal{R})$ is well-ordered.

In the context of orders, the relation $\mathcal{R}$ is usually written as $\preccurlyeq$. We use the following convention for an arbitrary set A and for $a, b \in A$:

$$a \preccurlyeq b \;\Leftrightarrow\; (a, b) \in \mathcal{R}, \quad \text{and} \quad a \not\preccurlyeq b \;\Leftrightarrow\; (a, b) \notin \mathcal{R}.$$

Given any preorder $\preccurlyeq$, two other relations closely associated with $\preccurlyeq$ are defined for $a, b \in A$ as follows:

$$a \prec b \quad \text{if and only if } a \preccurlyeq b \text{ and } b \not\preccurlyeq a, \tag{2.3}$$

$$a \sim b \quad \text{if and only if } a \preccurlyeq b \text{ and } b \preccurlyeq a. \tag{2.4}$$

The relation $\prec$ is called *strict preference relation* associated with the preference given by the preorder $\preccurlyeq$. The relation $\sim$ is called *equivalence* (or *indifference*) relation. The binary relation $\succ$ means: $a \succ b$ implies that a is preferred to b.

Proposition 2.3 *Let $\preccurlyeq$ be a preorder on any arbitrary set A. Then the relation $\prec$ defined by* (2.3) *is irreflexive and transitive and the relation $\sim$ defined by* (2.4) *is an equivalence relation.*

Proof Since $\preccurlyeq$ is a preorder, it is reflexive and transitive. Therefore, $\sim$ is reflexive and it is symmetric by definition. Now, we prove that $\sim$ is transitive. Let $a, b, c \in A$ such that $a \sim b$ and $b \sim c$. Then by using transitivity of $\preccurlyeq$, we have

$$\left.\begin{array}{l} a \preccurlyeq b \preccurlyeq c \;\Rightarrow\; a \preccurlyeq c \\ c \preccurlyeq b \preccurlyeq a \;\Rightarrow\; c \preccurlyeq a \end{array}\right\} \;\Rightarrow\; a \sim c. \tag{2.5}$$

Hence, $\preccurlyeq$ is transitive and thus it is an equivalence relation. By definition, $\prec$ is irreflexive. We have only to show that $\prec$ is transitive. Let $a, b, c \in A$ such that $a \prec b$ and $b \prec c$. Then $a \preccurlyeq b \preccurlyeq c$ and from the transitivity of $\preccurlyeq$, we have $a \preccurlyeq c$. To show that $a \prec c$, we assume that $c \preccurlyeq a$. Since $a \preccurlyeq b$, by transitivity of $\preccurlyeq$, we have $c \preccurlyeq b$. This contradiction implies that $c \not\preccurlyeq b$, that is, $a \prec c$. □

Definition 2.5 (Total Preorder) A binary relation $\preccurlyeq$ on an arbitrary set A is called a *total preorder* if it is reflexive, transitive and weakly connected.

In the light of Definition 2.5, we can say that a binary relation is a total order if it is antisymmetric and a total preorder. In other words, a binary relation $\preccurlyeq$ on an arbitrary set A is a total order if $\preccurlyeq$ is a partial order and for any $a, b \in A$, either $a \preccurlyeq b$ or $b \preccurlyeq a$.

Proposition 2.4 *If $\preccurlyeq$ is a total preorder on A, then the associated relation $\prec$ is a weak order. If $\prec$ is a weak order on A, then $\preccurlyeq$ defined by*

$$a \preccurlyeq b \;\Leftrightarrow\; \textit{either } a \prec b \textit{ or } (a \not\prec b \textit{ and } b \not\prec a) \tag{2.6}$$

is a total preorder.

For the proof of above proposition, we refer to Proposition 1.3 in [5, p. 9].

Example 2.3 [5, Table 1.2.] Here we give some orders on the Euclidean space $\mathbb{R}^n$. For all $x = (x_1, x_2, \ldots, x_n) \in \mathbb{R}^n$ and $y = (y_1, y_2, \ldots, y_n) \in \mathbb{R}^n$, we have

Name	Definition
Weak componentwise order	if $x_i \leq y_i, \quad i = 1, 2, \ldots, n$
Componentwise order	if $x_i \leq y_i, \quad i = 1, 2, \ldots, n; x \neq y$
Strict componentwise order	if $x_i < y_i, \quad i = 1, 2, \ldots, n$
Max order	if $\max_{i=1,2,\ldots,n} x_i \leq \max_{i=1,2,\ldots,n} y_i$

Example 2.4 [15, Example 2.1.5 (1)] Let X be a nonempty set and $\mathcal{P}(X)$ be the class of all subsets of X. The binary relation $\mathcal{R} = \{(A, B) \in \mathcal{P}(X) \times \mathcal{P}(X) : A \subset B\}$ is a partial order on $\mathcal{P}(X)$. However, if X contains at least two elements, then $\mathcal{R}$ is not weakly connected, and therefore $\mathcal{R}$ is not a total order. Moreover, $\mathcal{R}$ is not connected.

Another example for a binary relation that is not a total order is given below.

Example 2.5 Let $X := \{a, b, c\}$. The binary relation $\mathcal{R} = \{(a, a), (b, b), (c, c)\}$ is a partial order on X, but $\mathcal{R}$ is not connected, not weakly connected and therefore not a total order.

Example 2.6 [15, Example 2.1.5 (2)] Let $\mathbb{N}$ be the set of nonnegative integers and

$$\mathcal{R}_{\mathbb{N}} = \{(n, m) \in \mathbb{N} \times \mathbb{N} : \exists p \in \mathbb{N} \text{ such that } m = n + p\}.$$

Then $\mathbb{N}$ is well-ordered by $\mathcal{R}_{\mathbb{N}}$, because $(0, m) \in \mathcal{R}_{\mathbb{N}}$ for all $m \in \mathbb{N}$. Clearly, $\mathcal{R}_{\mathbb{N}}$ defines the usual order relation on $\mathbb{N}$, and $(n, m) \in \mathcal{R}_{\mathbb{N}}$ is denoted by $n \leq m$ or, equivalently, $m \geq n$.

Below we give an example of a set which is not well-ordered by a total order relation.

Example 2.7 Let $\mathbb{Z}$ be the set of integers and

$$\mathcal{R}_{\mathbb{Z}} = \{(n, m) \in \mathbb{Z} \times \mathbb{Z} : \exists p \in \mathbb{N} \text{ such that } m = n + p\}.$$

Apparently, $\mathbb{Z}$ is not well-ordered by $\mathcal{R}_{\mathbb{Z}}$, although $\mathcal{R}_{\mathbb{Z}}$ is a total order.

Example 2.8 [15, Example 2.1.5 (4)] Let $n \in \mathbb{N}$ and $n \geq 2$ be two given integers. We define a binary relation $\mathcal{R}_n$ on $\mathbb{R}^n$ by

$$\mathcal{R}_n = \{(x, y) \in \mathbb{R}^n \times \mathbb{R}^n : \text{ for all } i = 1, 2, \ldots, n, \ \ x_i \leq y_i\},$$

where $x = (x_1, x_2, \ldots, x_n)$ and $y = (y_1, y_2, \ldots, y_n)$. Then $\mathcal{R}_n$ is a partial order on $\mathbb{R}^n$, but $\mathcal{R}_n$ is not a total order. For instance, the vectors $e_1 = (1, 0, 0, \ldots, 0)$ and $e_2 = (0, 1, 0, \ldots, 0)$ cannot be compared by means of $\mathcal{R}_n$.

Example 2.9 Consider the binary relation

$$\mathcal{R}_2 = \{(n,m) \in \mathbb{N} \times \mathbb{N} : m = n + 1\}.$$

Then the binary relation $\mathcal{R}_2$ is irreflexive and asymmetric.

Example 2.10 Let $M := \{1,2,3,6\}$, and consider the binary relation

$$\mathcal{R}_3 = \{(n,m) \in M \times M : n \mid m\},$$

where $n \mid m$ denotes that $\frac{n}{m} \in \mathbb{N}$, i.e., n is divisible by m. Then the binary relation $\mathcal{R}_3$ is a partial order on M.

Since the preference orders (and, more generally binary relations) on a set A are subsets of the product space $A \times A$, we can treat them as a graph of a set-valued map T from A to 2^A. Namely, we identify the preference order $\prec$ with

$$\text{Graph}(T) = \{(a,b) \in A \times A : b \in T(a)\} = \{(a,b) \in A \times A : a \prec b\}$$

where Graph(T) denotes the graph of a set-valued map $T : A \to 2^A$ defined by

$$T(a) = \{b \in A : a \prec b\}.$$

Another way of representing preference order by a set-valued map is the concept of dominated structures.

Let X be a vector space, and $A \subset X$. For each $a \in A$, the *domination factor* is defined as

$$D(a) = \{b \in X : a \succ a + b\} \cup \{\mathbf{0}\}.$$

Then, clearly, the set-valued map $D : A \to 2^A$ represents the given preference order and it is called *domination structure*.

Definition 2.6 The domination structure $D(\cdot)$ is said to be

(a) *asymmetric* if for all $a \in A \subset X$, $b \in D(a)$ and $b \neq \mathbf{0}$ implies $-b \notin D(a+b)$, that is, if $a \in c + D(c)$ and $c \in a + D(a)$ then $a = c$;
(b) *transitive* if for all $a \in A \subset X$, $b \in D(a)$ and $c \in D(a+b)$ imply $b + c \in D(a)$, that is, if $a \in c + D(c)$ and $c \in d + D(d)$ then $a \in d + D(d)$;
(c) *negatively transitive* if for all $a \in A \subset X$, $b \notin D(a)$ and $c \notin D(a+b)$ imply $b + c \notin D(a)$.

For further detail on domination structure and its properties, we refer to [20, 28, 29, 35] and the references therein.

Definition 2.7 Let A be a partially ordered set with partial order $\preccurlyeq$ on A, and let B be a nonempty subset of A.

(a) An element $a \in A$ is called a *lower bound* of B if $a \preccurlyeq b$ for all $b \in B$.
(b) An element $a \in A$ is called an *upper bound* of B if $b \preccurlyeq a$ for all $b \in B$.
(c) An element $a \in A$ is called a *minimal element* of A if for all $c \in A$ such that $c \preccurlyeq a$ implies $a \preccurlyeq c$.
(d) An element $a \in A$ is called a *maximal element* of A if for all $c \in A$ such that $a \preccurlyeq c$ implies $c \preccurlyeq a$.

Remark 2.1 Since in Definition 2.7 the binary relation $\preccurlyeq$ is a partial order and hence it is antisymmetric, $a \in A$ is a minimal element of the set A if and only if

$$\text{for } c \in A : \; c \preccurlyeq a \;\Rightarrow\; c = a$$

and $a \in A$ is a maximal element of A if and only if

$$\text{for } c \in A : \; a \preccurlyeq c \;\Rightarrow\; a = c.$$

The existence of minimal and maximal elements of sets is an important issue in vector optimization that is addressed in the following *Zorn's Lemma* (or *Zorn's Axiom*).

Lemma 2.1 (Zorn's Lemma) *Let $(A, \preccurlyeq)$ be a preordered set. If every nonempty totally ordered subset B of A has an upper bound (lower bound), then A has at least one maximal element (minimal element).*

Next we show how the set of nonnegative elements of a set (here $\mathbb{R}^n$, and $\mathbb{R}^2$ for purposes of illustration) can be used to derive a geometric interpretation of properties of orders.

We use a cone to define an ordering which is compatible with scalar multiplication.

Proposition 2.5 *Let C be a cone in a vector space X. Then $\leq_C$ defined as*

$$x \leq_C y \;\Leftrightarrow\; y - x \in C \tag{2.7}$$

is compatible with scalar multiplication and addition in X, that is,

$$\textit{for all } x, y \in X \textit{ and } \lambda \geq 0 \quad : \quad x \leq_C y \Rightarrow \lambda x \leq_C \lambda y \tag{2.8}$$

and

$$\textit{for all } x, y, z \in X \quad : \quad x \leq_C y \Rightarrow (x + z) \leq_C (y + z). \tag{2.9}$$

Furthermore,

(a) *$\leq_C$ is reflexive;*
(b) *C is convex if and only if $\leq_C$ is transitive;*
(c) *C is pointed if and only if $\leq_C$ is antisymmetric.*

Conversely, if $\preccurlyeq$ is a reflexive relation on X such that

$$\text{for all } x, y \in X \text{ and } \lambda \geq 0 \quad : \quad x \preccurlyeq y \;\Rightarrow\; \lambda x \preccurlyeq \lambda y \tag{2.10}$$

and

$$\text{for all } x, y, z \in X \quad : \quad x \preccurlyeq y \;\Rightarrow\; (x+z) \preccurlyeq (y+z), \tag{2.11}$$

then $C = \{x \in X : \mathbf{0} \preccurlyeq x\}$ is a cone and $\preccurlyeq$ and $\leq_C$ are equivalent.

Proof Let $x, y, z \in X$ and $\lambda \geq 0$ be arbitrarily chosen. If $x \leq_C y$, then by (2.7) we have $y - x \in C$ and $\lambda(y - x) \in C$ because C is a cone. Again, by (2.7), $\lambda x \leq_C \lambda y$. Also, $x \leq_C y$ implies $y - x = (y+z) - (x+z) \in C$, and therefore $(x+z) \leq_C (y+z)$.

(a) Let $x \in C$. Because C is a cone, $\mathbf{0} \cdot x = x - x = \mathbf{0} \in C$, i.e., $x \leq_C x$, and hence, $\leq_C$ is reflexive.

(b) We show that $\leq_C$ is transitive if C is convex.

Let $x, y, z \in X$ such that $x \leq_C y$ and $y \leq_C z$. Then by (2.7), $y - x \in C$ and $z - y \in C$. Since the cone C is convex, we have $(y - x) + (z - y) = z - x \in C$ and so $x \leq_C z$. Hence, $\leq_C$ is transitive.

Conversely, assume that $\leq_C$ is transitive. Let $x, y \in C$. Then by (2.7), $\mathbf{0} \leq_C x$ and $\mathbf{0} \leq_C y$ and so by (2.9), we have $x \leq_C x + y$. The transitivity of $\leq_C$ implies that $\mathbf{0} \leq_C x + y$ and so $x + y \in C$. Hence, the cone C is convex.

(c) We show that $\leq_C$ is antisymmetric if C is pointed. Let $x, y \in X$ such that $x \leq_C y$ and $y \leq_C x$. These yield $y - x \in C$ and $x - y \in C$, that is, $y - x \in C \cap (-C) = \{\mathbf{0}\}$. Therefore, $y = x$.

Conversely, assume that $\leq_C$ is antisymmetric. If $x \in C \cap (-C)$, then by (2.7) $\mathbf{0} \leq_C x$ and $x \leq_C \mathbf{0}$. The antisymmetry of $\leq_C$ implies that $x = \mathbf{0}$.

Finally, let $\preccurlyeq$ be a reflexive relation on X that satisfies (2.10) and (2.11). Consider the set

$$C = \{x \in X : \mathbf{0} \preccurlyeq x\}. \tag{2.12}$$

Then C is a cone. Indeed, let $\lambda \geq 0$ and $x \in C$. Then, $\mathbf{0} \preccurlyeq x$. Since $\preccurlyeq$ satisfies (2.10), we obtain that $\lambda \mathbf{0} \preccurlyeq \lambda x$, and so $\lambda x \in C$. Moreover,

$$x \preccurlyeq y \underset{\text{by (2.11)}}{\Leftrightarrow} (x - x) \preccurlyeq (y - x) \underset{\text{by (2.12)}}{\Leftrightarrow} y - x \in C \underset{\text{by (2.7)}}{\Leftrightarrow} x \leq_C y.$$

Therefore, $\preccurlyeq$ and $\leq_C$ are equivalent. □

Remark 2.2 The preceding proposition shows that when $\{\mathbf{0}\} \neq C \subset X$, then the relation $\leq_C$ defined by (2.7) is a preorder if and only if C is a convex cone, and $\leq_C$ is a partial order if and only if C is a pointed convex cone.

The preceding proposition and remark say that with the help of a pointed convex cone, we can always define a partial ordering $\leq_C$ on a vector space X. We also write

$x \geq_C y$ for $y \leq_C x$. Similarly, if C is a solid closed convex cone in a topological vector space X, we define an associated strict partial ordering by

$$x <_C y \quad \Leftrightarrow \quad y - x \in \text{int}(C), \tag{2.13}$$

and write $x >_C y$ for $y <_C x$. Note that the term *strict partial ordering* is adopted here, although $<_C$ is not a partial order, as $<_C$ is not reflexive.

Proposition 2.6 *Let C be a solid closed convex cone in a topological vector space X. Then the strict partial ordering $<_C$ given by* (2.13) *has the following properties:*

(a) $x \not<_C x$ *(i.e., $<_C$ is irreflexive).*
(b) *If $x <_C y$, then $x \leq_C y$.*
(c) *If $x <_C y$ and $y <_C z$, then $x <_C z$ (transitivity).*
(d) *If $x <_C y$ and $u <_C v$, then $x + u <_C y + v$.*
(e) *If $x <_C y$ and $\lambda > 0$, then $\lambda x <_C \lambda y$ (compatibility with positive scalar multiplication).*
(f) *If $x <_C y$, then for u and v small enough, $x + u <_C y + v$ (compatibility with addition).*

When $C = \mathbb{R}_+$, the partial ordering $\leq_C$ is the usual ordering $\leq$ on $\mathbb{R}$, and strict partial ordering $<_C$ is the same as the usual strict ordering $<$ on $\mathbb{R}$.

Example 2.11 The nonnegative orthant $C = \mathbb{R}^n_+$ is a closed, solid, pointed and convex cone. The associated partial ordering $\leq_C$ corresponds to componentwise inequality between vectors, such that $x \leq_C y$ means that $x_i \leq y_i$, $i = 1, 2, \ldots, n$. The associated strict partial ordering $<_C$ corresponding to componentwise inequality between vectors is then introduced as $x <_C y$, which means that $x_i < y_i$, $i = 1, 2, \ldots, n$.

Example 2.12 The positive semidefinite cone $\mathbf{S}^n_+ = \{X \in \mathbf{S}^n : X \succeq 0\}$ is a closed, solid, pointed and convex cone in the set of symmetric $n \times n$ matrices, $\mathbf{S}^n = \{X \in \mathbb{R}^{n \times n} : X = X^\top\}$ which is a vector space of dimension $n(n+1)/2$. The set of all symmetric positive definite matrices is denoted by $\mathbf{S}^n_{++} = \{X \in \mathbf{S}^n : X \succ 0\}$. The interior of $\mathbf{S}^n_+$ in $\mathbf{S}^n$ is $\mathbf{S}^n_{++}$. The partial ordering $\leq_C$ on $\mathbf{S}^n$ associated with $C = \mathbf{S}^n_+$ is defined as $X \leq_C Y$, means $Y - X$ is positive semidefinite. The strict partial ordering $<_C$ on $\mathbf{S}^n$ associated with $C = \mathbf{S}^n_{++}$ is defined as $X <_C Y$, means $Y - X$ is positive definite.

Example 2.13 [33, p. 6] Let K be a compact set in $\mathbb{R}$ and let $\mathcal{C}[K]$ denote the space of continuous functionals defined on K. Then the set

$$\mathcal{C}[K]_+ := \{x \in \mathcal{C}[K] : x(t) \geq 0, \ \forall\, t \in K\}$$

is a convex pointed cone. The associated partial ordering $\leq_{\mathcal{C}[K]}$ corresponds to

$$x \leq_{\mathcal{C}[K]} y \ \Leftrightarrow \ (y - x)(t) \geq 0, \quad \text{for all } t \in K.$$

Furthermore, we have

$$\text{int}(\mathcal{C}[K]_+) := \{x \in \mathcal{C}[K] : x(t) > 0, \ \forall\, t \in K\} \neq \emptyset,$$

and the strict partial order then means that

$$x <_{\mathcal{C}[K]} y \ \Leftrightarrow \ (y - x)(t) > 0, \quad \text{for all } t \in K.$$

Example 2.14 [33, p. 7] Let $K := [a, b]$ be an interval for $a, b \in \mathbb{R}$, and let $\mathcal{C}[K]$ be the space of all continuous functionals defined on K. Then the set

$$\mathcal{C}[K]^m_+ := \{x \in \mathcal{C}[K] : x(t) \geq 0, \ \forall t \in K, \ x \text{ is monotone increasing}\}$$

is a convex pointed cone. Notice that $\text{int}(\mathcal{C}[K]^m_+) = \emptyset$, because for every $\bar{x} \in \mathcal{C}[K]^m_+$ and for every $\varepsilon > 0$, we can find a functional $x_\varepsilon \in \mathcal{C}[K]$ with the property $\|\bar{x} - x_\varepsilon\| \leq \varepsilon$ (where $||x|| := \sup\limits_{t \in K} |x(t)|$) such that x_ε is not monotone increasing on an arbitrary value $t \in K$. Thus no strict partial ordering can be defined here using $\mathcal{C}[K]^m_+$.

Throughout the book, we adopt the following notations.

Let Y be a topological vector space, C be a solid (that is, $\text{int}(C) \neq \emptyset$), closed and pointed convex cone in Y, and $\mathbf{0}$ denote the zero element in the considered space. We denote by $C_0 = C \setminus \{\mathbf{0}\}$. With respect to the cones C, C_0 and $\text{int}(C)$, we define the following partial ordering and strict partial ordering relationships: For all $x, y \in Y$,

$$x \leq_C y \ \Leftrightarrow \ y - x \in C; \qquad x \nleq_C y \ \Leftrightarrow \ y - x \notin C;$$

$$x \geq_C y \ \Leftrightarrow \ x - y \in C; \qquad x \ngeq_C y \ \Leftrightarrow \ x - y \notin C;$$

$$x \leq_{C_0} y \ \Leftrightarrow \ y - x \in C \setminus \{\mathbf{0}\}; \qquad x \nleq_{C_0} y \ \Leftrightarrow \ y - x \notin C \setminus \{\mathbf{0}\};$$

$$x <_C y \ \Leftrightarrow \ y - x \in \text{int}(C); \qquad x \not<_C y \ \Leftrightarrow \ y - x \notin \text{int}(C);$$

$$x >_C y \ \Leftrightarrow \ x - y \in \text{int}(C); \qquad x \not>_C y \ \Leftrightarrow \ x - y \notin \text{int}(C).$$

For two given subsets A and B of Y, the following partial ordering and strict partial ordering relationships are defined as follows:

$$A \leq_C B \ \Leftrightarrow \ x \leq_C y, \quad \text{for all } x \in A, \ y \in B;$$

$$A \nleq_C B \ \Leftrightarrow \ x \nleq_C y, \quad \text{for all } x \in A, \ y \in B;$$

$$A <_C B \ \Leftrightarrow \ x <_C y, \quad \text{for all } x \in A, \ y \in B;$$

$$A \not<_C B \ \Leftrightarrow \ x \not<_C y, \quad \text{for all } x \in A, \ y \in B.$$

A topological vector space Y with a pointed convex cone C which induces the partial ordering is called an *ordered topological vector space* and it is denoted by (Y, C).

Let Y be a topological vector space with partial ordering generated by a pointed convex cone and $x, y \in Y$, then $y \geq_C \mathbf{0}$ implies $y \nless_C \mathbf{0}$. Furthermore,

$$x \geq_C \mathbf{0} \text{ and } y \geq_C \mathbf{0} \text{ imply } (x+y) \geq_C \mathbf{0},$$
$$x \ngtr_C y \text{ and } x \geq_C \mathbf{0} \text{ imply } y \nless_C \mathbf{0},$$
$$x >_C \mathbf{0} \text{ and } y \geq_C \mathbf{0} \text{ imply } (x+y) >_C \mathbf{0},$$

since $C + \text{int}(C) \subseteq \text{int}(C)$.

In view of Proposition 2.5, we have the following lemmas.

Lemma 2.2 *Let C be a solid pointed convex cone in a topological vector space Y. Then for all $x, y, z \in Y$, we have the following implications:*

(a) $x \geq_C y$ *implies* $x + z \geq_C y + z$;
(b) $x >_C y$ *implies* $x + z >_C y + z$;
(c) $x \geq_{C_0} y$ *implies* $x + z \geq_{C_0} y + z$;
(d) $x \ngeq_C y$ *implies* $x + z \ngeq_C y + z$;
(e) $x \ngtr_C y$ *implies* $x + z \ngtr_C y + z$;
(f) $x \ngeq_{C_0} y$ *implies* $x + z \ngeq_{C_0} y + z$.

The same is true for $\leq_C$, $<_C$, $\leq_{C_0}$, $\nleq_C$, $\nless_C$, and $\nleq_{C_0}$, respectively.

Proof We prove this lemma for the binary relations $\geq_C$, $>_C$ and $\geq_{C_0}$. The proof for the relations $\leq_C$, $<_C$, $\leq_{C_0}$, $\nleq_C$, $\nless_C$, and $\nleq_{C_0}$ follows analogously.

(a) The compatibility with addition is proven in Proposition 2.5.
(b) If $x >_C y$, then $x - y \in \text{int}(C)$ and $x - y = (x + z) - (y + z) \in \text{int}(C)$, and therefore $(x + z) >_C (y + z)$.
(c) Let $x \geq_{C_0} y$, and thus $x - y \in C \setminus \{\mathbf{0}\}$ and $x - y = (x + z) - (y + z) \in C \setminus \{\mathbf{0}\}$, and therefore $(x + z) \geq_{C_0} (y + z)$.
(d) Let $x \ngeq_C y$, which means $x - y \notin C$. Suppose that for arbitrary $z \in Y$, $x + z \geq_C y + z$. But this means that $(x + z) - (y + z) = x - y \in C$, a contradiction.
(e) and (f) can be proven analogously. □

Lemma 2.3 *Let C be a solid pointed convex cone in a topological vector space Y. Then for all $x, y, z \in Y$, we have the following implications:*

(a) $x \leq_C y \leq_C z$ *implies* $x \leq_C z$;
(b) $x \leq_C y \leq_{C_0} z$ *implies* $x \leq_{C_0} z$;
(c) $x \leq_C y <_C z$ *implies* $x <_C z$;
(d) $x \nleq_C y >_C z$ *implies* $x \nless_C z$;
(e) $x \nless_C y \geq_C z$ *implies* $x \nless_C z$;
(f) $x \ngtr_C y <_C z$ *implies* $x \nless_C z$;
(g) $x \ngtr_C y \leq_C z$ *implies* $x \nless_C z$.

The same is true for $\leq_C$, $<_C$, $\leq_{C_0}$, $\nleq_C$, $\nless_C$, and $\nleq_{C_0}$, respectively.

Lemma 2.4 ([1]) *Let C be a solid pointed convex cone in a topological vector space Y, and $x, y \in Y$ with $x <_C \mathbf{0}$ and $y <_C \mathbf{0}$. Then the set of upper bounds of x and y is nonempty and intersects with $(-\operatorname{int}(C))$.*

Proof We have to show that there exists $c <_C \mathbf{0}$ such that $x \leq_C c$ and $y \leq_C c$. It is sufficient to choose $c = \alpha y$ with $\alpha > 0$ close to zero. □

Lemma 2.5 ([1]) *Let C be a solid pointed convex cone in a topological vector space Y, and $x, y \in Y$ with $x <_C \mathbf{0}$ and $y \not\geq_C \mathbf{0}$. Then the set of upper bounds of x and y is nonempty and intersects with $Y \setminus C$.*

Proof We have to show that there exists $c \not\geq_C \mathbf{0}$ such that $x \leq_C c$ and $y \leq_C c$. Since $\operatorname{int}(C) \neq \emptyset$, there exists $d \in \operatorname{int}(C)$ such that $d - y \in C$ (see [18]). For $t \in [0, 1]$, set $d_t := td + (1 - t)y$. Since C is closed convex, there exists $t_0 \in]0, 1[$ such that $d_t \in C$ for all $t \in [t_0, 1]$, and $d_t \notin C$ for all $t \in [0, t_0[$. In particular, we have that $d_{t_0} \geq_C \mathbf{0} >_C x$ implying $d_{t_0} - x \in \operatorname{int}(C)$. Hence, for $t_1 < t_0$ sufficiently close to t_0, we still have $d_{t_1} - x \in \operatorname{int}(C)$. Set $x = d_{t_1}$. Then $c \notin C$ and thus $c \not\geq_C \mathbf{0}$. Furthermore, we have $c \geq_C x$ and $c - y = t_1(d - b) \geq_C \mathbf{0}$. □

2.2 Some Basic Properties

Proposition 2.7 *Let Y be a topological vector space, and let K and E be two nonempty subsets of Y such that K is open, E is convex and* $\operatorname{int}(E) \neq \emptyset$. *Then*

$$K + \operatorname{cl}(E) = K + E = K + \operatorname{int}(E).$$

Proof Clearly, $K + \operatorname{int}(E) \subseteq K + E \subseteq K + \operatorname{cl}(E)$. We next show that $K + \operatorname{int}(E) \supseteq K + \operatorname{cl}(E)$. Let $y \in K + \operatorname{cl}(E)$. Then there exist $k \in K$ and $e \in \operatorname{cl}(E)$ such that $y = k + e$. Since Y is a topological vector space and K is an open subset of Y, there is a balanced neighborhood V of $\mathbf{0}$ such that $V + k \subset K$. On the other hand, since E is convex and $e \in \operatorname{cl}(E)$, for any neighborhood U of $\mathbf{0}$, $(U + e) \cap \operatorname{int}(E) \neq \emptyset$. Hence, there exists $v \in V$ such that $v + e \in \operatorname{int}(E)$. Since V is balanced, $-v \in V$. Therefore, $y = k + e = (k - v) + (e + v)$ with $(k - v) \in K$ and $(e + v) \in \operatorname{int}(E)$. Hence $y \in K + \operatorname{int}(E)$. □

Proposition 2.8 *Let Y be a topological vector space and C be a solid pointed convex cone in Y. Then*

$$\operatorname{cl}(C) + \operatorname{int}(C) = \operatorname{int}(C).$$

Proof By Proposition 2.7, $\operatorname{cl}(C) + \operatorname{int}(C) = \operatorname{int}(C) + \operatorname{int}(C)$. Since C is convex cone, $\operatorname{int}(C) + \operatorname{int}(C) = \operatorname{int}(C)$. □

Proposition 2.9 *Let Y be a topological vector space, A be a subset of Y and C a solid pointed convex cone in Y. If $A \cap (-\operatorname{int}(C)) = \emptyset$, then*

$$(A + \operatorname{cl}(C)) \cap (-\operatorname{int}(C)) = \emptyset.$$

Proof Suppose to the contrary that there exists $y \in (A + \operatorname{cl}(C)) \cap (-\operatorname{int}(C))$. Then there exist $a \in A$, $c' \in \operatorname{cl}(C)$ and $c \in \operatorname{int}(C)$ such that $y = a + c' = -c$. Hence, $a = -(c' + c) \in -\operatorname{int}(C)$ by Proposition 2.8. This contradicts to the hypothesis that $A \cap (-\operatorname{int}(C)) = \emptyset$. □

Proposition 2.10 *Let Y be a topological vector space and C be a solid pointed convex cone in Y. Then*

$$(\operatorname{cl}(C))^{c} - \operatorname{cl}(C) = (\operatorname{int}(C))^{c} - \operatorname{int}(C) = (\operatorname{cl}(C))^{c}.$$

Proof Since $(\operatorname{cl}(C))^{c}$ is open and $\operatorname{cl}(C)$ is convex with $\operatorname{int}(C) \neq \emptyset$, by Proposition 2.7, we have

$$(\operatorname{cl}(C))^{c} - \operatorname{cl} C = (\operatorname{cl} C)^{c} - \operatorname{int}(C), \tag{2.14}$$

because $\operatorname{int}(\operatorname{cl}(C)) = \operatorname{int}(C)$. Since $(\operatorname{cl}(C))^{c} \subset (\operatorname{int}(C))^{c}$, we obtain

$$(\operatorname{cl}(C))^{c} - \operatorname{int}(C) \subset (\operatorname{int}(C))^{c} - \operatorname{int}(C). \tag{2.15}$$

By combining (2.14) and (2.15), we get

$$(\operatorname{cl}(C))^{c} - \operatorname{cl}(C) \subset (\operatorname{int}(C))^{c} - \operatorname{int}(C). \tag{2.16}$$

Now, we claim that

$$(\operatorname{int}(C))^{c} - \operatorname{int}(C) \subset (\operatorname{cl}(C))^{c}. \tag{2.17}$$

Indeed, suppose that there exist $z \in (\operatorname{int}(C))^{c}$, $c' \in \operatorname{int}(C)$ and $\tilde{c} \in \operatorname{cl}(C)$ such that $z - c' = \tilde{c}$. Then by Proposition 2.8, $z = \tilde{c} + c' \in \operatorname{int}(C)$, a contradiction.

Finally,

$$(\operatorname{cl}(C))^{c} = (\operatorname{cl}(C))^{c} - \mathbf{0} \subset (\operatorname{cl}(C))^{c} - \operatorname{cl}(C). \tag{2.18}$$

Then, from (2.16) – (2.18), we have

$$(\operatorname{cl}(C))^{c} - \operatorname{cl}(C) \subset (\operatorname{int}(C))^{c} - \operatorname{int}(C) \subset (\operatorname{cl}(C))^{c} \subset (\operatorname{cl}(C))^{c} - \operatorname{cl}(C),$$

that is,

$$(\mathrm{cl}(C))^{\mathrm{c}} - \mathrm{cl}(C) = (\mathrm{int}(C))^{\mathrm{c}} - \mathrm{int}(C) = (\mathrm{cl}(C))^{\mathrm{c}}. \qquad \square$$

Proposition 2.11 *Let Y be a topological vector space and C be a pointed convex cone in Y with $k \in \mathrm{int}(C)$. Then the following statements hold:*

(a) *For every $y \in Y$, there exists $\lambda \in \mathbb{R}$ such that $y \in \lambda \cdot k + \mathrm{int}(C)$;*
(b) *For every $y \in \mathrm{int}(C)$, there exists $\lambda > 0$ such that $y - \lambda \cdot k \in \mathrm{int}(C)$.*

Proof

(a) Let $y \in Y$. Since $k \in \mathrm{int}(C)$, $-k + \mathrm{int}(C)$ is a neighborhood of $\mathbf{0}$. Since Y is a topological vector space, each neighborhood of $\mathbf{0}$ is absorbing. Hence, there exists $\lambda > 0$ such that $y \in \lambda(-k + \mathrm{int}(C))$, that is, $y \in (-\lambda \cdot k + \mathrm{int}(C))$.
(b) Let $y \in \mathrm{int}(C)$. Then there exists a neighborhood U of $\mathbf{0}$ such that $y - U \subseteq \mathrm{int}(C)$. Since Y is a topological vector space, there exists $\lambda > 0$ such that $k \in \lambda \cdot U$. Hence $y - \dfrac{1}{\lambda} \cdot k \in y - U \subset \mathrm{int}(C)$. $\square$

Lemma 2.6 ([12]) *Let Y be a topological vector space and C be a cone in Y and $e \in \mathrm{int}(C)$. Then*

$$Y = \bigcup \{\lambda e - \mathrm{int}(C) : \lambda > 0\}.$$

Proof Let $U := e - \mathrm{int}(C)$. Then U is an open set in Y. Since $e \in \mathrm{int}(C)$, $\mathbf{0} \in U$. Since C is a cone, we get

$$\lambda U = \lambda(e - \mathrm{int}(C)) \subset \lambda e - \mathrm{int}(C), \quad \text{for all } \lambda > 0.$$

Thus,

$$\bigcup \{\lambda U : \lambda \geq 0\} \subset \bigcup \{\lambda e - \mathrm{int}(C) : \lambda > 0\},$$

because $\lambda e \in \mathrm{int}(C)$. From $Y = \bigcup \{\lambda U : \lambda \geq 0\}$ (see Proposition 1.1), we conclude that

$$Y = \bigcup \{\lambda e - \mathrm{int}(C) : \lambda > 0\}. \qquad \square$$

Lemma 2.7 ([3, Lemma 1.51]) *Let Y be a topological vector space, C be a proper, closed and convex cone in Y and $e \in \mathrm{int}(C)$. For $\lambda \in \mathbb{R}$, we set $C_\lambda = \lambda e - C$.*

(a) *If* $z \in C_\lambda$ *for some* $\lambda \in \mathbb{R}$, *then*

$$z \in \mu e - \operatorname{int}(C), \quad \text{for each } \mu > \lambda;$$

moreover,

$$z \in \mu e - C, \quad \text{for each } \mu > \lambda.$$

(b) *For each* $z \in Y$, *there exists a real number* $\lambda \in \mathbb{R}$ *such that* $z \notin C_\lambda$.
(c) *Let* $z \in Y$. *If* $z \notin C_\lambda$ *for some* $\lambda \in \mathbb{R}$, *then* $z \notin C_\mu$ *for each* $\mu < \lambda$.

Proof

(a) Let $\mu > \lambda$ and $z \in C_\lambda$. Then we have

$$\mu e - z = (\mu - \lambda)e + \lambda e - z \in \operatorname{int}(C) + C \subset \operatorname{int}(C).$$

Thus,

$$z \in \mu e - \operatorname{int}(C) \subset \mu e - C.$$

(b) Suppose contrary that there exists $z_0 \in Y$ such that for all $\lambda \in \mathbb{R}$, $z_0 \in C_\lambda$. From (a), we have

$$z_0 \in \lambda e - \operatorname{int}(C), \quad \text{for all } \lambda \in \mathbb{R}.$$

Thus,

$$\{\lambda e - z_0 : \lambda \in \mathbb{R}\} \subset \operatorname{int}(C), \quad \text{equivalently,} \quad \{-\lambda e - z_0 : \lambda \in \mathbb{R}\} \subset \operatorname{int}(C).$$

From Lemma 2.6, we have

$$Y = \{\lambda e - \operatorname{int}(C) : \lambda \in \mathbb{R}_+ \setminus \{0\}\}.$$

Therefore, for each $y \in Y$, there exist $c \in \operatorname{int}(C)$ and $\alpha > 0$ such that $-y = \alpha e - c$, because $-y \in Y$ as Y is a vector space. Then,

$$\begin{aligned} y = -\alpha e + c &= (-\alpha e - z_0) + c + z_0 \\ &\in \operatorname{int}(C) + \operatorname{int}(C) + z_0 = z_0 + \operatorname{int}(C). \end{aligned}$$

Thus, $Y \subset z_0 + \operatorname{int}(C)$ which contradicts $C \neq Y$.
(c) Let $z \notin C_\lambda$ for some $\lambda \in \mathbb{R}$. Suppose contrary that for some $\mu < \lambda$, $z \in C_\mu$. From (a), we have that $z \in C_\lambda$ which is a contradiction of our supposition. □

2.3 Cone Topological Concepts

In scalar optimization theory, compactness assumptions on the feasible solution set plays an important role in existence results for a solution of an optimization problem. It is well-known by the Weierstrass theorem that if a functional $f : \mathbb{R}^n \supseteq K \to \mathbb{R}$ is continuous and the set K is compact, then f attains its extremum on K. A generalization of the Weierstrass theorem states that f has a minimal point if f is lower semicontinuous and K is a compact set. When we consider the image of K, we observe that $f(K) := \bigcup_{x \in K} f(x)$ attains its minimum if $f(K) + \mathbb{R}_+$ is closed and bounded from below. In this section, we focus on corresponding results for vector-valued functions.

We start with a definition of C-closed and C-bounded sets.

Definition 2.8 Let Y be a Hausdorff topological vector space, and C be a convex cone in Y. A set $K \subseteq Y$ is said to be

(a) *C-closed* if $K + \mathrm{cl}(C)$ is closed;
(b) *C-bounded* if $K_\infty \cap (-\mathrm{cl}(C)) = \{\mathbf{0}\}$, where K_∞ is the recession cone of the set K.

Example 2.15 Consider the set $K = \{(x_1, x_2) \in \mathbb{R}^2 : x_1 \geq 0,\ x_2 \geq 0\}$ and $C = \{(x_1, x_2) \in \mathbb{R}^2 : x_1 = x_2 \geq 0\}$. Then the set K is C-closed as well as C-bounded.

We note that there is no implication between the closedness of K and its C-closedness. Let $C_1 := \{x \in \mathbb{R}^2 : x = r \cdot (0, 2),\ r \in \mathbb{R}_+\}$ be a given cone. The set $K_1 := \{(x_1, x_2) \in \mathbb{R}^2 : x_1 x_2 = -1,\ x_1 > 0\}$ is closed but not C_1-closed. For a visualization of K_1 and $K_1 + C_1$, see the left illustration in Fig. 2.3. Furthermore, the set $K_1 := \{(x_1, x_2) \in \mathbb{R}^2 : x_1 x_2 = -1,\ x_1 > 0\}$ is closed but not $\mathbb{R}^2_+$-closed, see the second illustration in Fig. 2.3. We can see that $K_1 + \mathbb{R}^2_+$ yields the open halfplane $\{(x_1, x_2) \in \mathbb{R}^2 : x_1 > 0\}$. Also, the set $K_2 := \{(x_1, x_2) \in \mathbb{R}^2 : x_1^2 + x_2^2 < 1\} \cup \{(x_1, x_2) \in \mathbb{R}^2 : x_1^2 + x_2^2 = 1, x_1 \leq 0,\ x_2 \leq 0\}$ is not closed but $\mathbb{R}^2_+$-closed (see the third illustration in Fig. 2.3). Moreover, the set $K_2 + \mathbb{R}^2_+$ is $\mathbb{R}^2_+$-bounded.

Remark 2.3 A bounded set is also C-bounded.

Remark 2.4 The following definition of C-boundedness is given by Luc [25]: $K \subseteq Y$ is C-bounded if for each neighborhood U of $\mathbf{0}$, there exists a positive number λ such that $K \subseteq \lambda U + C$. We note that these two definitions of C-boundedness are not comparable. For example, consider the set $K = C = \{(x_1, x_2) \in \mathbb{R}^2 : x_1 = x_2\}$. Then K is C-bounded in the sense of Luc. But there exist some points of K_∞, namely $(1, 1)$, that belong to $(-\mathrm{cl}(C)) \setminus \{\mathbf{0}\}$. Hence, K is not C-bounded.

Lemma 2.8 *Let Y be a Hausdorff topological vector space, K a nonempty subset of Y and C a pointed, closed and convex cone in Y. Then the set K is C-bounded if and only if $K + C$ is C-bounded.*

Proof Since $K_\infty \subset (K+C)_\infty$, we obtain that K is C-bounded if $(K+C)_\infty \cap (-C) = \{\mathbf{0}\}$. Conversely, assume contrary that $(K + C)_\infty \cap (-C) \neq \{\mathbf{0}\}$. Then there exist

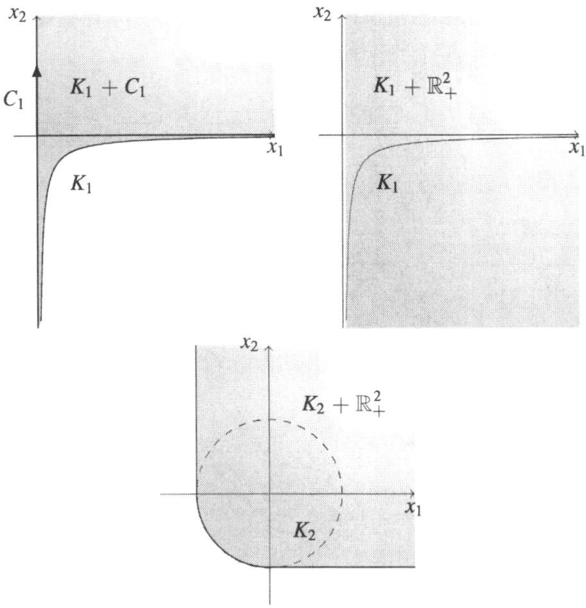

Fig. 2.3 *Left*: The set K_1 is closed but not C_1-closed. *Right*: The set K_1 is closed but not $\mathbb{R}^2_+$-closed. *Below*: Here, the set K_2 is not closed but $\mathbb{R}^2_+$-closed. Moreover, K_2 is $\mathbb{R}^2_+$-bounded

sequences $\{\lambda_n\} \subset \mathbb{R}_+$ with $\lim_{n\to+\infty} \lambda_n = 0$ and $\{y_n + c_n\} \subset Y + C$ with $y_n \in Y$ and $c_n \in C$ such that $\lim_{n\to+\infty} \lambda_n(y_n + c_n) = -c \in (-C) \setminus \{\mathbf{0}\}$.

Let us suppose that $\{\lambda_n c_n\}$ has a convergent subsequence. We may assume that $\lim_{n\to+\infty} \lambda_n c_n = \tilde{c} \in C$. Therefore, $\{\lambda_n y_n\}$ converges to $-c - \tilde{c}$ which is a nonzero vector in $K_\infty \cap (-C)$ since C is a pointed cone. So that we have $K_\infty \cap (-C) \neq \{\mathbf{0}\}$.

If we suppose that $\{\lambda_n c_n\}$ has no convergent subsequence, then $\{\lambda_n c_n\}$ is unbounded. Then from the fact that K is bounded if and only if $K_\infty = \{\mathbf{0}\}$, we have $(\{\lambda_n c_n\})_\infty \neq \{\mathbf{0}\}$. Namely, we may assume by taking a subsequence of $\{\lambda_n c_n\}$ that there exists another sequence $\alpha_n \subset \mathbb{R}_+$ with $\lim_{n\to+\infty} \alpha_n = 0$ and $\lim_{n\to+\infty} \alpha_n(\lambda_n c_n) = \bar{c} \neq 0$. Naturally, $\bar{c} \in C$. Since

$$\begin{aligned} \|\alpha_n\lambda_n c_n + \bar{c}\| &= \|\alpha_n[\lambda_n(y_n + c_n) + c] - (\alpha_n\lambda_n c_n - \bar{c}) - \alpha_n c\| \\ &\le \alpha_n\|\lambda_n(y_n + c_n) + c\| + \|\alpha_n\lambda_n c_n - \bar{c})\| + \alpha_n\|c\|, \end{aligned}$$

it follows that $\lim_{n\to+\infty} \alpha_n\lambda_n y_n = -\bar{c}$ or $K_\infty \cap (-C) \neq \{\mathbf{0}\}$. □

Definition 2.9 (*C*-**Compact Set**) Let C be a convex cone in a Hausdorff topological vector space Y. A nonempty set $K \subseteq Y$ is called

(a) *C-compact* if for all $y \in K$, the set $(y - \text{cl}(C)) \cap K$ is compact;
(b) *C-semicompact* if every open cover of K of the form $\{(y_\alpha - \text{cl}(C))^c : y_\alpha \in K,\ \alpha \in \Lambda\}$ has a finite subcover. In other words, whenever $K \subseteq \bigcup_{\alpha\in\Lambda} (y_\alpha - \text{cl}(C))^c$, there is $m \in \mathbb{N}$ and $\{\alpha_1, \alpha_2, \ldots, \alpha_m\} \subset \Lambda$ such that

$$K \subseteq \bigcup_{i=1}^{m} (y_{\alpha_i} - \text{cl}(C))^c,$$

where $(y_\alpha - \text{cl}(C))^c$ denotes the complement of $(y_\alpha - \text{cl}(C))$ in Y.

Note that the complement $Y \setminus (y_\alpha - \text{cl}(C))$ of $(y_\alpha - \text{cl}(C))$ in Y is always open.

Remark 2.5 Notice that a compact set is always C-compact, while the reverse statement is generally not true. Consider, for instance, the set $K_1 := \{(x_1, x_2) \in \mathbb{R}^2 : x_1 + x_2 \geq 0\}$ and $C = \mathbb{R}^2_+$. Then the set K_1 is $\mathbb{R}^2_+$-compact, while K_1 is not a compact set. Alternatively, the set $K_2 := \mathbb{R}^2_+$ is not compact but $\mathbb{R}^2_+$-compact. For a visualization, see Fig. 2.4.

Example 2.16

(a) The set

$$K = \{(x_1, x_2) \in \mathbb{R}^2 : x_1^2 + x_2^2 \leq 0,\ x_1 < 0,\ x_2 < 0\}$$

is $\mathbb{R}^2_+$-compact, $\mathbb{R}^2_+$-bounded, but not $\mathbb{R}^2_+$-closed.
(b) The set

$$K = \left\{(x_1, x_2) \in \mathbb{R}^2 : x_1 x_2 = -1,\ x_1 > 0\right\}$$

is $\mathbb{R}^2_+$-closed, $\mathbb{R}^2_+$-bounded, but not $\mathbb{R}^2_+$-compact.

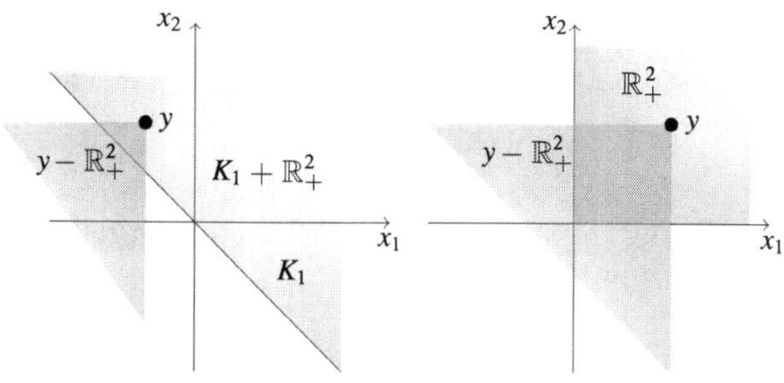

Fig. 2.4 *Left*: The set $K_1 := \{(x_1, x_2) \in \mathbb{R}^2 : x_1 + x_2 \geq 0\}$ is $\mathbb{R}^2_+$-compact, while K_1 is not a compact set. *Right*: $\mathbb{R}^2_+$ is not compact but $\mathbb{R}^2_+$-compact (see Remark 2.5)

(c) The set

$$K = \{(x_1, x_2) \in \mathbb{R}^2 : x_1 x_2 = 1,\ x_1 < 0\}$$

is $\mathbb{R}^2_+$-compact, $\mathbb{R}^2_+$-closed, but not $\mathbb{R}^2_+$-bounded.

Proposition 2.12 *Every C-compact set is C-semicompact.*

Proof Let K be a C-compact set and $\{(y_\alpha - \text{cl}(C))^c : y_\alpha \in K,\ \alpha \in \Lambda\}$ be an open covering of K. For arbitrary $y_{\alpha'} \in K$, $(y_{\alpha'} - C) \cap K$ is a compact set because K is C-compact. Consider an open covering

$$\mathcal{O} = \left\{(y_\alpha - \text{cl}(C))^c : y_\alpha \in K,\ \alpha \in \Lambda,\ \alpha \neq \alpha'\right\}$$

of $(y_{\alpha'} - \text{cl}(C)) \cap K$. Since $(y_{\alpha'} - \text{cl}(C)) \cap K$ is a compact set, the open cover $\mathcal{O}$ has a finite subcover of $(y_{\alpha'} - \text{cl}(C)) \cap K$. This subcover together with $(y_{\alpha'} - \text{cl}(C))^c$ yields a finite cover of K, and so K is C-semicompact. □

As we have seen in Proposition 2.12 that every C-compact set is C-semicompact. But converse assertion may not be true in general.

Example 2.17 Consider $C = \mathbb{R}^2_+$ and the set $K = \{(y_1, y_2) \in \mathbb{R}^2 : y_1^2 + y_2^2 \leq 1,\ y_1 > 0,\ y_2 > 0\} \cup \{(0, 0)\}$. Then K is C-semicompact but not C-compact.

Lemma 2.9 *Let Y be a Hausdorff topological vector space and C be a closed convex cone in Y. If M + C is C-semicompact, then M is C-semicompact.*

Proof Let $\{(y_\alpha - C)^c : y_\alpha \in K,\ \alpha \in \Lambda\}$ be an open cover of K. For any $y \in K$, let $y \in (y_{\alpha_0} - C)^c$ with $y_{\alpha_0} \in K$. Since C is a convex cone, we have $y + C \subset (y_{\alpha_0} - C)^c$. In fact, if $y + c \in (y_{\alpha_0} - C)$ for some $c \in C$, then $y \in y_{\alpha_0} - C$ which is a contradiction. Hence, $\{(y_\alpha - C)^c : y_\alpha \in K,\ \alpha \in \Lambda\}$ is also an open cover of $K + C$. Since this last set is C-semicompact, this cover has a finite subcover which is a subcover of K. □

Proposition 2.13 *Let Y be a Hausdorff topological vector space and C be a closed convex cone in Y. If K is C-closed and C-bounded, then K + C is C-compact, and K is C-semicompact.*

Proof For any $y \in K + C$, let us consider the set $(y - C) \cap (K + C)$. Since $(K_1 \cap K_2)_\infty \subseteq (K_1)_\infty \cap (K_2)_\infty$ for any $K_1, K_2 \subseteq Y$ and by Lemma 2.8, we have

$$((y - C) \cap (K + C))_\infty \subset (y - C)_\infty \cap (K + C)_\infty = (-C) \cap (K + C)_\infty = \{\mathbf{0}\}$$

as K is C-bounded. Therefore, $(y - C) \cap (K + C)$ is a closed and bounded set, that is, a compact set for all $y \in Y + C$. By the definition of C-compactness, we have $K + C$ is C-compact set and, by Proposition 2.12, also C-semicompact. Hence by Lemma 2.9, K is C-semicompact. □

Definition 2.10 Let Y be a Hausdorff topological vector space, and let C be a closed cone in Y. A net $\{y_\alpha\}_{\alpha \in \Lambda} \subseteq Y$ is said to be *decreasing* (with respect to C) if $y_\alpha - y_\beta \in C \setminus \{\mathbf{0}\}$ for every $\alpha, \beta \in \Lambda$, $\beta > \alpha$.

Definition 2.11 Let Y be a Hausdorff topological vector space with its ordering cone C.

(a) The convex cone C is called *Daniell* if every decreasing net which has a lower bound converges to its infimum.
(b) Y is called *boundedly order complete* if every bounded decreasing net has an infimum.

Below we give an example of a Daniell cone, and we present an example of a convex cone which is not Daniell in a space that is not boundedly order complete.

Example 2.18

(a) A convex cone with a weakly compact base is a Daniell cone (see [20, Example 2.2.8, 3]).
(b) Let $Y := \mathcal{C}[K]$ be the space of all continuous functionals defined on the compact set $K \subset \mathbb{R}^n$. The natural ordering cone in $\mathcal{C}[K]$ is $\mathcal{C}[K]_+ := \{y \in \mathcal{C}[K] : y(t) \geq 0,\ \forall\ t \in K\}$. Let us consider the special case for $K := [0, 1]$. We choose the net $y_k : [0, 1] \to \mathbb{R}$, $y_k := -t^{1/k}$ (illustrated in Fig. 2.5), which is decreasing, because for $l > k$, we have

$$y_k - y_l \in \mathcal{C}[K]_+ \setminus \{\mathbf{0}\},$$

as $-t^{1/k} - (-t^{1/l}) = t^{1/l} - t^{1/k} > 0$ for all $t \in [0, 1]$. One lower bound for this net is given by the constant function $\bar{y} :\equiv -1$, which is also the infimum of the net. But this net does not possess a limit in $\mathcal{C}[0, 1]$, and is therefore not convergent. The (pointwise) limit function of the net is the function

$$\tilde{y} := \begin{cases} 0, & \text{if } t = 0, \\ -1, & \text{otherwise.} \end{cases}$$

But $\tilde{y}$ is not continuous, thus $\tilde{y} \notin \mathcal{C}[0, 1]$. Furthermore, the space $\mathcal{C}[0, 1]$ is not boundedly order complete. This is shown by considering the net

$$z_k := \begin{cases} 0, & \text{if } t \leq \frac{1}{2}, \\ -(2 \cdot (x - \frac{1}{2}))^{(1/k)}, & \text{if } t > \frac{1}{2}, \end{cases} \tag{2.19}$$

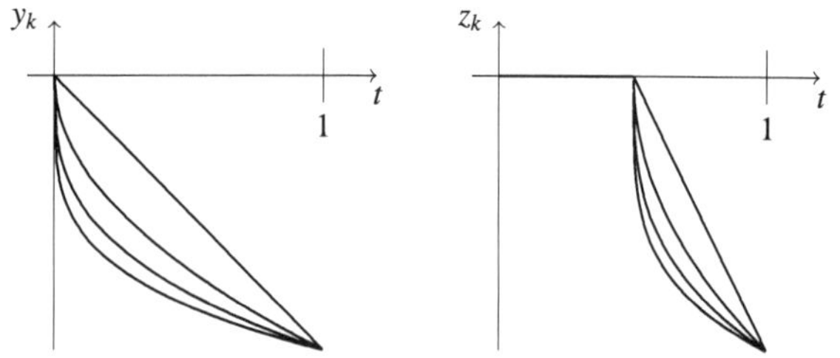

Fig. 2.5 *Left*: The net $y_k := -t^{1/k}$ for $k = 1, 2, 3, 4$, which is considered in Example 2.18. *Right*: The net z_k given in (2.19) for $k = 1, 2, 3, 4$

see the right image in Fig. 2.5. The net z_k does not possess an infimum, and thus $C[0,1]$ is not boundedly order complete.

Definition 2.12 Let C be a convex cone in a Hausdorff topological vector space Y. A set $K \subseteq Y$ is said to be *C-complete* (respectively, *strongly C-complete*) if it has no covers of the form $\{(y_\alpha - \mathrm{cl}(C))^c : \alpha \in \Lambda\}$ (respectively, $\{(y_\alpha - C)^c : \alpha \in \Lambda\}$) for every decreasing net $\{y_\alpha\}$ in K.

Example 2.19 The set $K_1 := \{(y_1, y_2) \in \mathbb{R}^2 : y_1^2 + y_2^1 < 1\}$ is not $\mathbb{R}^2_+$-complete, whereas $K_2 := K_1 \cup \{(y_1, y_2) \in \mathbb{R}^2 : y_1^2 + y_2^1 = 1, y_1, y_2 \leq 0\}$ is $\mathbb{R}^2_+$-complete.

Remark 2.6 It is obvious that whenever C is closed, C-completeness and strong C-completeness coincide. Also note that if K is an open set, K cannot be C-complete.

We present some criteria for a set to be C-complete.

Lemma 2.10 *A set $K \subseteq Y$ is C-complete if any one of the following conditions holds.*

(a) *K is C-semicompact, in particular, K is C-compact or compact;*
(b) *K is weakly compact and Y is a locally convex space;*
(c) *K is closed bounded and C is Daniell, Y is boundedly order complete;*
(d) *K is closed minorized (that is, there is $x \in X$ such that $K \subseteq x + C$) and C is Daniell.*

Proof

(a) Assume, to the contrary, that there is a cover as required in the definition of C-completeness. By the C-semicompactness, there is a finite number of indexes, say, $1, 2, \ldots, m$ from Λ such that $\{(y_i - \mathrm{cl}(C))^c : i = 1, 2, \ldots, m\}$ covers K, where $y_1 >_C y_2 >_C \cdots >_C y_m$. This is a contradiction because

$$y_m \in y_i - C \subseteq y_i - \mathrm{cl}(C), \quad \text{for all } i = 1, 2, \ldots, m,$$

consequently, no element of that cover contains $y_m \in K$.
(b) Consider K and C in the weak topology and taking into account the fact that a closed convex set in a locally convex space is also weak closed. Then by (a), we obtain the result.
(c) and (d) It suffices to observe the following fact: If a net $\{y_\alpha\}$ is a decreasing net in K, then it has an infimum to which it converges. Moreover, this infimum must be in K and therefore it belongs to $y_\alpha - \mathrm{cl}(C)$ for every α. Hence, the net cannot provide a cover of the form as in the definition of C-completeness. □

Theorem 2.1 *Let C be a closed convex cone in a Hausdorff topological vector space Y. A set $K \subseteq Y$ is C-complete if it is C-semicompact.*

Proof Assume to the contrary that K has a cover of the form $\{(y_\alpha - C)^c\}_{\alpha \in \Lambda}$, where $\{y_\alpha\}$ is a decreasing net in K. Since K is C-semicompact, there is $m \in \mathbb{N}$ and

$\{\alpha_1, \alpha_2, \ldots, \alpha_m\} \subseteq \Lambda$ such that

$$K \subseteq \bigcup_{i=1}^{m} (y_{\alpha_i} - C)^c \quad \text{or} \quad K \subseteq (y_{\alpha_m} - C)^c$$

which contradicts the fact that $y_{\alpha_m} \notin (y_{\alpha_m} - C)$. □

The following example shows that a C-complete set need not be a C-semicompact set.

Example 2.20 Let

$$K = \{(y_1, y_2) \in \mathbb{R}^2 : y_2 < 0,\ y_1 y_2 = 1\} \cup \{(y_1, y_2) \in \mathbb{R}^2 : y_2 \leq 0,\ y_2 = -y_1\}.$$

Then K is $\mathbb{R}^2_+$-complete, but it is not $\mathbb{R}^2_+$-semicompact, as its cover

$$\{(y_m - \mathbb{R}^2_+)^c\},$$

with $y_m = (m, -m)$, has no finite subcovers.

2.4 Cone Convexity

It is well-known that convex functions and their generalizations play an important role in scalar optimization theory as a local minimum of a convex function is also a global minimum. Corresponding to scalar convex functions and their generalizations, we have the concept of vector-valued functions, so-called cone convexity. In this section, we present such cone convexity and its generalizations.

Definition 2.13 (***C*-Convexity**) Let Y be a vector space with a partial ordering defined by a solid pointed convex cone C, and K be a nonempty convex subset of a vector space X. A vector-valued function $\varphi : K \to Y$ is said to be

(a) *C-convex* if for all $x, y \in K$ and $t \in [0, 1]$,

$$\varphi(tx + (1-t)y) \leq_C t\varphi(x) + (1-t)\varphi(y),$$

that is,

$$\varphi(tx + (1-t)y) \in t\varphi(x) + (1-t)\varphi(y) - C;$$

(b) *strictly C-convex* if for all $x, y \in K$, $x \neq y$ and $t \in]0, 1[$,

$$\varphi(tx + (1-t)y) <_C t\varphi(x) + (1-t)\varphi(y),$$

that is,

$$\varphi(tx + (1-t)y) \in t\varphi(x) + (1-t)\varphi(y) - \operatorname{int}(C);$$

(c) *C-quasiconvex* if for each $\alpha \in Y$, and for all $x, y \in K$ and $t \in [0, 1]$,

$$\varphi(x),\ \varphi(y) \in \alpha - C \quad \text{imply} \quad \varphi(tx + (1-t)y) \in \alpha - C,$$

equivalently, for all $x, y \in K$ and $t \in [0, 1]$,

$$\varphi(tx + (1-t)y) \in \alpha - C, \quad \text{for all } \alpha \in C(\varphi(x), \varphi(y)),$$

where $C(\varphi(x), \varphi(y))$ is the set of upper bounds of $\varphi(x)$ and $\varphi(y)$, that is,

$$C(\varphi(x), \varphi(y)) := \{\alpha \in Y : \alpha \in \varphi(x) + C \text{ and } \alpha \in \varphi(y) + C\};$$

(d) *strictly C-quasiconvex* if for each $\alpha \in Y$ and for all $x, y \in K$, $x \neq y$ and $t \in]0, 1[$,

$$\varphi(x),\ \varphi(y) \in \alpha - C \quad \text{imply} \quad \varphi(tx + (1-t)y) \in \alpha - \operatorname{int}(C);$$

(e) *properly C-quasiconvex* if for all $x, y \in K$ and $t \in [0, 1]$, either

$$\varphi(tx + (1-t)y) \leq_C \varphi(x),$$

or

$$\varphi(tx + (1-t)y) \leq_C \varphi(y);$$

(f) *strict properly C-quasiconvex* if for all $x, y \in K$ and $t \in]0, 1[$, either

$$\varphi(tx + (1-t)y) <_C \varphi(x),$$

or

$$\varphi(tx + (1-t)y) <_C \varphi(y);$$

(g) *naturally C-quasiconvex* if for all $x, y \in K$ and $t \in [0, 1]$, there exists $s \in [0, 1]$ such that

$$\varphi(tx + (1-t)y) \leq_C s\varphi(x) + (1-s)\varphi(y),$$

equivalently, for all $x, y \in K$ and $t \in [0, 1]$,

$$\varphi(tx + (1-t)y) \in \operatorname{co}\{\varphi(x), \varphi(y)\} - C,$$

where co$A$ denotes the convex hull of a set A;

(h) *explicitly C-quasiconvex* if it is C-quasiconvex and for all $x, y \in K$ such that $\varphi(x) <_C \varphi(y)$, we have

$$\varphi(tx + (1-t)y) <_C \varphi(y), \quad \text{for all } t \in]0, 1[;$$

(i) *C-convex-like* if for all $x, y \in K$ and $t \in [0, 1]$, there exists $z \in K$ such that

$$\varphi(z) \leq_C t\varphi(x) + (1-t)\varphi(y);$$

(j) *C-subconvex-like* if for all $x, y \in K$, $t \in]0, 1[$ and $\varepsilon > 0$, there exist $z \in K$ and $c \in \text{int}(C)$ such that

$$\varphi(z) \leq_C t\varphi(x) + (1-t)\varphi(y) + \varepsilon c.$$

The function φ is said to be *C-concave* (respectively, *strictly C-concave*, *C-quasiconcave*, *strictly C-quasiconcave*, *properly C-quasiconcave*, *strict properly C-quasiconcave*, *naturally C-quasiconcave*, *explicitly C-quasiconcave*, *C-concave-like*, *C-subconcave-like*) if $-\varphi$ is C-convex (respectively, strictly C-convex, C-quasiconvex, strictly C-quasiconvex, properly C-quasiconvex, strict properly C-quasiconvex, naturally C-quasiconvex, explicitly C-quasiconvex, C-convex-like, C-subconvex-like).

When $Y = \mathbb{R}$ and $C = \mathbb{R}_+$, then the above definitions of different types of C-convexity reduce to the ordinary definition of different types of corresponding convexity.

Remark 2.7

(a) It can be easily seen that φ is C-convex if and only if for any $x_i \in K$ and $t_i \in [0, 1]$, $i = 1, 2, \ldots, m$ with $\sum_{i=1}^m t_i = 1$,

$$\varphi\left(\sum_{i=1}^m t_i x_i\right) \leq_C \sum_{i=1}^m t_i \varphi(x_i),$$

that is,

$$\varphi\left(\sum_{i=1}^m t_i x_i\right) \in \sum_{i=1}^m t_i \varphi(x_i) - C.$$

(b) Let $K \subseteq \mathbb{R}^n$, $Y = \mathbb{R}^\ell$ and $C = \mathbb{R}^\ell_+$. The vector-valued function $\varphi = (\varphi_1, \varphi_2, \ldots, \varphi_\ell) : K \to \mathbb{R}^\ell$ is (strictly) $\mathbb{R}^\ell_+$-convex if and only if every component $\varphi_i : K \to \mathbb{R}$ of φ is (strictly) convex. Similarly, φ is (strictly) $\mathbb{R}^\ell_+$-quasiconvex if and only if every component $\varphi_i : K \to \mathbb{R}$ of φ is (strictly) quasiconvex. Moreover, φ is naturally $\mathbb{R}^\ell_+$-quasiconvex if and only if each φ_i is naturally $\mathbb{R}^\ell_+$-quasiconvex, where $i = 1, 2, \ldots, \ell$. A corresponding relationship also holds for explicitly C-quasiconvexity, C-convex-like functions as well as C-subconvex-like functions. These assertions can be easily proved by contradiction. However, this relationship is not true for properly $\mathbb{R}^\ell_+$-quasiconvexity. The reason for this is the lack of a total order in $\mathbb{R}^\ell$. Consider, for instance, the functions $\varphi_1(x) = x$ and $\varphi_2(x) = -x$ and $K = [0, 1]$. These functions are properly $\mathbb{R}_+$-quasiconvex. However, the function $\varphi := (\varphi_1, \varphi_2) = (x, -x)$ is not properly $\mathbb{R}^2_+$-quasiconvex. To see this, consider, for example, $x = 0$, $y = 1$, $t = 0.5$. Then, $\varphi(tx + (1-t)y) = \varphi(0.5) = (0.5, -0.5)$. But we have

$$(0.5, -0.5) \notin \varphi(x) - \mathbb{R}^2_+$$

as well as

$$(0.5, -0.5) \notin \varphi(y) - \mathbb{R}^2_+.$$

Therefore, φ is not properly $\mathbb{R}^2_+$-quasiconvex.

(c) If $\varphi : K \to Y$ is explicitly C-quasiconvex, then it can be easily proved that

$$\varphi(x) <_C \varphi(y) \text{ and } \varphi(y) \not>_C \mathbf{0} \quad \Rightarrow \varphi(y + t(x - y) <_C \varphi(y), \text{ for all } t \in]0, 1[.$$

For a first idea on C-convexity, consider the following simple example.

Example 2.21 Let $K = [-5, 5]$, $Y = \mathbb{R}^2$, $C = \mathbb{R}^2_+ = \{(y_1, y_2) \in \mathbb{R}^2 : y_1 \geq 0,\ y_2 \geq 0\}$ and define a function $\varphi : K \to Y$ by $\varphi(x) = (x^2, x)$. Then we have for all $x, y \in K$ and for every $t \in [0, 1]$,

$$\left((tx + (1-t)y)^2, tx + (1-t)y\right) \in t\left(x^2, x\right) + (1-t)\left(y^2, y\right) - C,$$

and thus, φ is a C-convex function. Notice that φ is not strictly C-convex, because

$$tx + (1-t)y \not<_C tx + (1-t)y,$$

and thus,

$$\varphi(tx + (1-t)y) \notin t\varphi(x) + (1-t)\varphi(y) - \operatorname{int}(C),$$

for arbitrary $x, y \in K$ and $t \in [0, 1]$.

Remark 2.8 From Definition 2.13, it is clear that C-convexity implies natural C-quasiconvexity. Moreover, proper C-quasiconvexity implies natural C-quasiconvexity. Furthermore, from Definition 2.13, it is evident that C-convexity implies the C-convex-like condition, and the C-convex-like condition implies the C-subconvex-like condition.

In the following proposition (see Fig. 2.6), we present an equivalent condition for C-convex-like vector-valued functions.

Proposition 2.14 ([31, Lemma 1]) *Let Y be a vector space with a partial ordering defined by a pointed convex cone C, and let K be a nonempty convex subset of a vector space X. Let a vector-valued function $\varphi : K \to Y$ be given, and define $\varphi(K) := \bigcup_{x \in K} \varphi(x)$. Then φ is C-convex-like if and only if the set $\varphi(K) + C$ is convex.*

Proof Let φ be C-convex-like. Choose $z_1, z_2 \in \varphi(K) + C$. Then there exist $x, y \in K$ and $c_1, c_2 \in C$ with $z_1 = \varphi(x) + c_1$, $z_2 = \varphi(y) + c_2$. Because φ is C-convex-like, we have for all $t \in [0, 1]$, $t\varphi(x) + (1 - t)\varphi(x) \in \varphi(K) + C$. This yields

$$tz_1 - tc_1 + (1 - t)z_2 - (1 - t)c_2 \in \varphi(K) + C,$$

which due to the cone property of C results in

$$tz_1 + (1 - t)z_2 \in \varphi(K) + C.$$

Thus, $\varphi(K) + C$ is a convex set.

Conversely, let $\varphi(K) + C$ be a convex set. Now choose $z_1, z_2 \in \varphi(K) + C$. Then there exist $x, y \in K$ and $c_1, c_2 \in C$ such that $z_1 = \varphi(x) + c_1$ and $z_2 = \varphi(y) + c_2$. Since $\varphi(K) + C$ is convex, we have $tz_1 + (1 - t)z_2 \in \varphi(K) + C$, and thus for every $t \in [0, 1]$, $t\varphi(x) + (1 - t)\varphi(y) \in \varphi(K) + C$. This means that φ is C-convex-like. □

In the following proposition, we show that the condition defining C-quasiconvexity can be replaced by a convexity condition on a specifically defined set.

Proposition 2.15 ([31, Lemma 3]) *Let Y be a vector space with a partial ordering defined by a pointed convex cone C, and let K be a nonempty convex subset of a vector space X. A vector-valued function $\varphi : K \to Y$ is C-quasiconvex if and only if the set*

$$A(\alpha) := \{x \in K : \varphi(x) \leq_C \alpha\}, \quad \textit{for all } \alpha \in Y$$

is convex or empty.

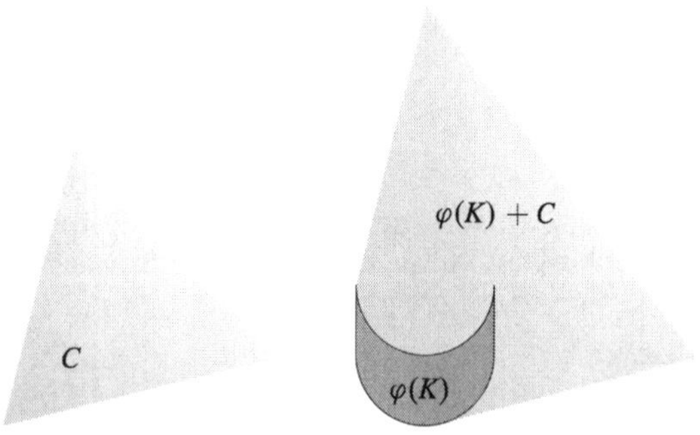

Fig. 2.6 *Left*: The cone C. *Right*: The function φ is C-convex-like, because the set $\varphi(K) + C$ is convex (see Proposition 2.14)

Proof Let $x, y \in K$ and let $t \in [0, 1]$ be arbitrarily chosen. For all $\alpha \in C(\varphi(x), \varphi(y)) := \{\alpha \in Y : \alpha \in \varphi(x) + C \text{ and } \alpha \in \varphi(y) + C\}$, we have $x, y \in A(\alpha)$. Now let $A(\alpha)$ be a convex and nonempty set. Then for all $t \in [0, 1]$, it holds $tx + (1 - t)y \in A(\alpha)$, and thus $\varphi(tx + (1 - t)y) \leq_C \alpha$. This means that φ is C-quasiconvex.

Conversely, let φ be C-quasiconvex. Now choose $x, y \in A(\alpha)$ and $t \in [0, 1]$. Then we have $tx + (1 - t)y \in K$, $\varphi(x) \leq_C \alpha$ and $\varphi(y) \leq_C \alpha$. Hence, $\alpha \in C(\varphi(x), \varphi(y))$. Because φ is C-quasiconvex, we conclude with $\varphi(tx + (1 - t)y) \leq_C \alpha$, and thus $tx + (1 - t)y \in A(\alpha)$. This means that $A(\alpha)$ is a convex set. □

Remark 2.9 If φ is C-quasiconvex, then the set $\{x \in K : \varphi(x) <_C \alpha\}$ is convex.

The following theorem describes a relationship between natural C-quasiconvexity and C-quasiconvexity of a vector-valued function.

Theorem 2.2 ([31, Theorem 2]) *Let Y be a vector space with a partial ordering defined by a pointed convex cone C, and let K be a nonempty convex subset of a vector space X. If a vector-valued function $\varphi : K \to Y$ is naturally C-quasiconvex, then it is C-quasiconvex.*

Proof Let φ be naturally C-quasiconvex. Then for all $x, y \in K$ and $t \in [0, 1]$, there exists $s \in [0, 1]$ such that

$$\varphi(tx + (1 - t)y) \leq_C s\varphi(x) + (1 - s)\varphi(y). \tag{2.20}$$

For all $\alpha \in C(\varphi(x), \varphi(y)) := \{\alpha \in Y : \alpha \in \varphi(x) + C \text{ and } \alpha \in \varphi(y) + C\}$, it holds

$$s\varphi(x) + (1 - s)\varphi(y) \in s\alpha + (1 - s)\alpha - C \subseteq \alpha - C.$$

Thus, together with (2.20), this yields for all $x, y \in K$ and $t \in [0, 1]$, $\varphi(tx + (1 - t)y) \leq_C \alpha$ for every $\alpha \in C(\varphi(x), \varphi(y))$. □

The following Theorem 2.3 verifies that a naturally C-quasiconvex vector-valued function φ is C-convex-like if φ is continuous and C is closed.

Theorem 2.3 ([31, Theorem 3]) *Let Y be a vector space with a partial ordering defined by a solid pointed closed convex cone C, and let K be a nonempty convex subset of a vector space X. Furthermore, let a vector-valued function $\varphi : K \to Y$ be continuous. If φ is naturally C-quasiconvex, then φ is C-convex-like.*

Example 2.22 Let $K = [0, 1]$, $Y = \mathbb{R}^2$, $C = \mathbb{R}^2_+ = \{(y_1, y_2) \in \mathbb{R}^2 : y_1 \geq 0,\ y_2 \geq 0\}$ and define a function $\varphi : K \to Y$ by $\varphi(x) = \left(x^2, 1 - x^2\right)$. Then the function φ is continuous and naturally C-quasiconvex (and thus C-quasiconvex), but neither C-convex nor properly C-quasiconvex.

The following figure comprises the relationships among some of the convexity notions (Fig. 2.7).

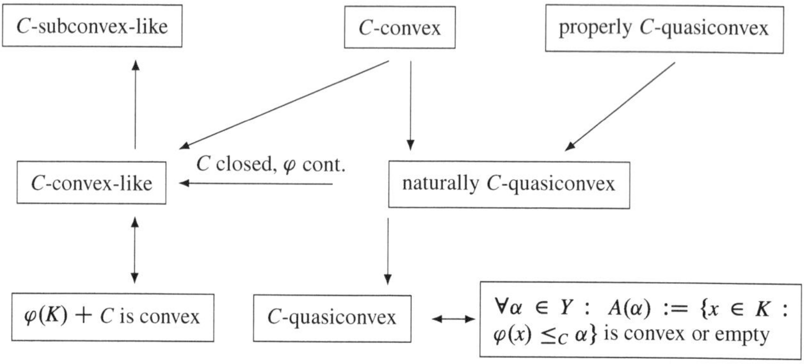

Fig. 2.7 Relationships between some of the convexity notions

Example 2.23 Let X, Y, C be the same as in Example 2.22. We define the function $\xi : K \to Y$ by

$$\xi(x) = \left(\cos\left(\frac{\pi x}{2}\right), \sin\left(\frac{\pi x}{2}\right)\right)$$

and the function $\tau : K \to Y$ by

$$\tau(x) = (\cos(2\pi x), \sin(2\pi x)).$$

Then the function ξ is continuous and C-quasiconvex, but not naturally C-quasiconvex, and the function τ is continuous and C-convex-like, but not naturally C-quasiconvex and hence, not C-convex.

Example 2.24 Let $\phi : \mathbb{R}^2 \to \mathbb{R}$ be a real-valued function defined by

$$\phi(x, y) = x^2 + y^2, \quad \text{for all } (x, y) \in \mathbb{R}^2.$$

Let $\varphi : (x, y) \mapsto (x, y, \varphi(x, y))$ and $C = \{(z_1, z_2, z_3) : z_3 > 0\} \cup \{\mathbf{0}\}$ be the ordering cone in $\mathbb{R}^3$. Then φ is strict properly C-quasiconvex on $\mathbb{R}^2$.

Indeed, let $(x_1, x_2), (y_1, y_2) \in \mathbb{R}^2$ with $(x_1, x_2) \neq (y_1, y_2)$ and $\phi(x_1, x_2) \geq \phi(y_1, y_2)$. Then by the strict convexity of ϕ, we have

$$\phi(t(x_1, x_2) + (1 - t)(y_1, y_2)) <_C \phi(x_1, x_2), \quad \text{for all } t \in]0, 1[.$$

Therefore,

$$\varphi(t(x_1, x_2) + (1 - t)(y_1, y_2)) \in \varphi(x_1, x_2) - \operatorname{int}(C), \quad \text{for all } t \in]0, 1[.$$

Thus, φ is strict properly C-quasiconvex on $\mathbb{R}^2$.

The following proposition follows directly from the definitions of (strictly) C-(quasi)convexity.

Proposition 2.16 *Let K be a nonempty convex subset of a vector space X, Y a vector space, and C a pointed convex cone in Y. Let $\varphi, \phi : K \to Y$ be vector-valued functions. Then the following statements hold.*

(a) *If φ is C-convex (respectively, strictly C-convex), then $t\varphi$ is also C-convex (respectively, strictly C-convex) for all $t > 0$.*
(b) *If φ and ϕ are C-convex (respectively, C-quasiconvex), then so is $\varphi + \phi$ (respectively, C-quasiconvex).*
(c) *If φ and ϕ are C-convex and at least one of them is strictly C-convex, then $\varphi + \phi$ is strictly C-convex.*

Definition 2.14 Let X and Y be vector spaces, K a nonempty subset of X and C a pointed convex cone in Y. Let $\varphi : K \to Y$ be a vector-valued function. The set

$$\text{epi}(\varphi) = \{(x, y) \in X \times Y : x \in K,\ y \in \{\varphi(x)\} + C\} \tag{2.21}$$

is called *epigraph* of φ.

We note that (2.21) can also be written as

$$\text{epi}(\varphi) = \{(x, y) \in X \times Y : x \in K,\ \varphi(x) \leq_C y\}.$$

The following theorem shows that a C-convex vector-valued function can be characterized by its epigraph.

Theorem 2.4 *Let X and Y be vector spaces, K be a nonempty convex subset of X and C be a pointed convex cone in Y. A vector-valued function $\varphi : K \to Y$ is C-convex if and only if $\text{epi}(\varphi)$ is a convex set.*

Proof Let φ be a C-convex vector-valued function and let $z_1 = (x_1, y_1)$, $z_2 = (x_2, y_2) \in \text{epi}(\varphi)$ and $t \in [0, 1]$. Then we have $tx_1 + (1-t)x_2 \in K$ and

$$\begin{aligned} ty_1 + (1-t)y_2 &\in t\left(\{\varphi(x_1)\} + C\right) + (1-t)\left(\varphi(\{x_2)\} + C\right) \\ &= \{t\varphi(x_1) + (1-t)\varphi(x_2)\} + C \\ &\subseteq \{\varphi(tx_1 + (1-t)x_2)\} + C. \end{aligned}$$

Consequently, we have $tz_1 + (1-t)z_2 \in \text{epi}(\varphi)$. Thus, $\text{epi}(\varphi)$ is a convex set.

Conversely, let $\text{epi}(\varphi)$ be a convex set, $x_1, x_2 \in K$ and $t \in [0, 1]$. Then we have $t(x_1, \varphi(x_1)) + (1-t)(x_2, \varphi(x_2)) \in \text{epi}(\varphi)$ and

$$\varphi(tx_1 + (1-t)x_2) \leq_C t\varphi(x_1) + (1-t)\varphi(x_2).$$

Hence, φ is C-convex. □

Proposition 2.17 *Let K be a nonempty convex subset of a topological vector space X and Y be a Hausdorff topological vector space ordered by a closed convex pointed cone C. If $\varphi : X \to Y$ is Gâteaux differentiable and C-convex on K, then for every $x, y \in K$,*

$$\varphi(y) \geq_C \varphi(x) + \langle D\varphi(x), y - x \rangle, \tag{2.22}$$

where $D\varphi(x)$ is the Gâteaux derivative of φ at x.

Proof Since φ is C-convex on K, for every $x, y \in K$ and $t \in]0, 1[$, we have

$$\varphi(x + t(y - x)) = \varphi(ty + (1 - t)x) \in t\varphi(y) + (1 - t)\varphi(x) - C,$$

that is,

$$\varphi(y) \in \varphi(x) + \frac{\varphi(x + t(y - x)) - \varphi(x)}{t} + C.$$

Taking $t \to 0^+$ and using the definition of the Gâteaux derivative, we obtain $\varphi(y) \geq_C \varphi(x) + \langle D\varphi(x), y - x \rangle$. □

In the same way, we obtain the following result.

Proposition 2.18 *Let K be a nonempty convex subset of a topological vector space X and Y be a Hausdorff topological vector space ordered by a closed convex pointed cone C. If $\varphi :\to Y$ is Gâteaux differentiable and C-concave on K, then for every $x, y \in K$,*

$$\varphi(y) \leq_C \varphi(x) + \langle D\varphi(x), y - x \rangle,$$

where $D\varphi(x)$ is the Gâteaux derivative of φ at x.

Below we present another notion of generalized convexity.

Definition 2.15 (C-Pseudoconvexity) Let K be a nonempty convex subset of a vector space X, Y be a topological vector space with a convex cone C and $\varphi : K \to Y$ be a Gâteaux differentiable vector-valued function. Then φ is called *C-pseudoconvex* at x if for all $y \in K$,

$$\langle D\varphi(x), y - x \rangle \in C \quad \text{implies} \quad \varphi(y) - \varphi(x) \in C.$$

If φ is C-pseudoconvex at all $x \in K$, then we say that φ is *C-pseudoconvex*.

Remark 2.10 Let $K \subseteq \mathbb{R}^n$, $Y = \mathbb{R}^\ell$ and $C = \mathbb{R}^\ell_+$. The vector-valued function $\varphi = (\varphi_1, \varphi_2, \ldots, \varphi_\ell) : K \to \mathbb{R}^\ell$ is $\mathbb{R}^\ell_+$-pseudoconvex if and only if every component $\varphi_i : K \to \mathbb{R}$ of φ is pseudoconvex.

The following theorem shows that the class of C-pseudoconvex functions is broader than the class of C-convex functions.

Theorem 2.5 *Let K be a nonempty convex subset of a vector space X, Y be a topological vector space with a closed convex pointed cone C and $\varphi : K \to Y$ be a Gâteaux differentiable vector-valued function. If φ is C-convex, then φ is C-pseudoconvex.*

Proof According to (2.22), we have

$$\varphi(y) \geq_C \varphi(x) + \langle D\varphi(x), y - x\rangle,$$

being equivalent to

$$\varphi(y) - \varphi(x) \in \langle D\varphi(x), y - x\rangle + C.$$

If $\langle D\varphi(x), y - x\rangle \in C$, then due to the convexity of the cone C, we obtain $\varphi(y) - \varphi(x) \in C$. Thus, φ is C-pseudoconvex. □

In the following example we present a function which is $\mathbb{R}^2_+$-pseudoconvex, but not $\mathbb{R}^2_+$-convex.

Example 2.25 The function $\varphi : \mathbb{R} \to \mathbb{R}^2$ with

$$\varphi(x) = (\varphi_1(x), \varphi_2(x)) := \left(-\frac{1}{1+x^2}, x^2\right)$$

is $\mathbb{R}^2_+$-pseudoconvex, because

$$\langle D\varphi_1(x), y - x\rangle = \frac{2x}{(1+x)^2}(y - x) \geq 0 \quad \Leftrightarrow \quad x(y - x) \geq 0.$$

So, necessarily $\varphi_1(y) - \varphi_1(x) \geq 0$ is satisfied. For the convex function $\varphi_2(x) = x^2$, the implication is fulfilled accordingly. Note that $\varphi(x)$ is not $\mathbb{R}^2_+$-convex, because $\varphi_2(x)$ is not convex.

Let us give the brief description of the scalarization procedure due to Jayakumar et al. [19]. Let $\mathcal{Z}^*$ be the dual of a locally convex topological vector space Z, P a convex cone in Z, and $P^* \subseteq \mathcal{Z}^*$ the dual cone of P, that is, $P^* = \{z^* \in \mathcal{Z}^* : \langle z^*, z\rangle \geq 0, \text{ for all } z \in P\}$. We assume that P^* has a weak* compact convex base B^*. This means that $B^* \subseteq P^*$ is a weak* compact convex set such that $\mathbf{0} \notin B^*$ and $P^* = \bigcup_{\lambda \geq 0} \lambda B^*$; See, for example, [19].

For each $z \in Z$, consider the scalar function $\psi : Z \to \mathbb{R}$ defined by

$$\psi(z) = \max_{\lambda \in B^*} \langle \lambda, z\rangle.$$

Then the function ψ is sublinear, hence convex, and lower semicontinuous. For all $z \in Z$, there holds for $P = \text{int}(C) \cup \{\mathbf{0}\}$ and $C \subset Z$ a convex cone (see, for example, [19])

$$z \in P \quad \Leftrightarrow \quad z \in (-\text{int}(C))^c \quad \Leftrightarrow \quad \psi(z) \geq 0. \tag{2.23}$$

Proposition 2.19 *Let X be a Hausdorff topological vector space, $P = \text{int}(C) \cup \{\mathbf{0}\}$ and $h : X \to Z$ be a vector-valued function. If h is P-convex (respectively, P-concave), then the real-valued function $g : X \to \mathbb{R}$ defined by $g(x) = \psi(h(x))$ for all $x \in X$, is convex (respectively, concave).*

Proof Let $x, y \in X$ and $t \in [0, 1]$. Suppose that h is P-convex, then

$$th(x) + (1 - t)h(y) - h(tx + (1 - t)y) \in P.$$

It follows from (2.23) that

$$\max_{\lambda \in B^*} \langle \lambda, h(tx + (1 - t)y) - th(x) - (1 - t)h(y) \rangle \leq 0,$$

and hence,

$$\max_{\lambda \in B^*} \langle \lambda, h(tx + (1 - t)y) \rangle \leq t \max_{\lambda \in B^*} \langle \lambda, h(x) \rangle + (1 - t) \max_{\lambda \in B^*} \langle \lambda, h(y) \rangle.$$

Therefore,

$$g(tx + (1 - t)y) \leq tg(tx) + (1 - t)g(y)$$

and thus, g is convex. The case that h is P-concave can be treated similarly and hence the details will be omitted. □

Usually, we define cone concavity of f by the fact that $-f$ is cone convex. However the following definition for cone concavity is also natural.

Definition 2.16 (**C-Quasiconcave-Like Function**) Let Y be a vector space with a partial ordering defined by a pointed solid convex cone C and K be a nonempty convex subset of a vector space X. A vector-valued function $\varphi : K \to Y$ is said to be *C-quasiconcave-like* on K if for every $\alpha \in Y$, the set

$$\{x \in K : \varphi(x) \notin \alpha - \text{int}(C)\}$$

is convex or empty. The function φ is said to be *strictly C-quasiconcave-like* on K if for every $\alpha \in Y$, the set

$$\{x \in K : \varphi(x) \notin \alpha - \text{cl}(C)\}$$

is convex or empty.

The following example illustrates the notion of a C-quasiconcave-like function.

Example 2.26 Consider the function $\varphi : [-1, 1] \to \mathbb{R}$ defined as

$$\varphi(x) = \begin{cases} -x^2 + 1, & \text{if } -1 \leq x \leq 0, \\ 0, & \text{if } 0 < x \leq 1. \end{cases}$$

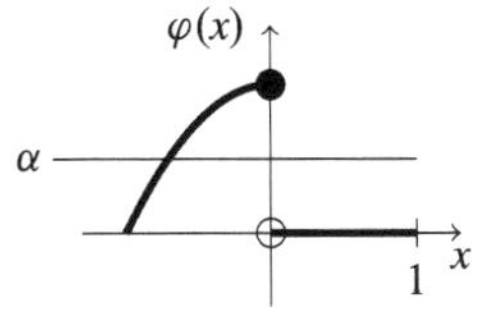

Fig. 2.8 The function φ is $\mathbb{R}_+$-quasiconcave-like. For example, for $\alpha = 0.5$, we obtain the convex set $\{x \in K : \varphi(x) \notin \alpha - \text{int}(C)\} = \left[-\frac{1}{\sqrt{2}}, 0\right]$

Apparently, φ is $\mathbb{R}_+$-quasiconcave-like, but φ is not concave (see Fig. 2.8).

Proposition 2.20 *Let K be a nonempty convex subset of a vector space X and Y be a topological vector space with a proper solid pointed convex cone C. If $\varphi : K \to Y$ is properly $(-C)$-quasiconvex on K, then it is C-quasiconcave-like on K.*

Proof Let $\alpha \in Y$ and $x_1, x_2 \in \{x \in K : \varphi(x) \notin \alpha - \text{int}(C)\}$. Since φ is properly $(-C)$-quasiconvex on K, for every $t \in [0, 1]$

$$\varphi(tx_1 + (1-t)x_2) \in \{\varphi(x_1), \varphi(x_2)\} + \text{int}(C).$$

Hence, $\varphi(tx_1 + (1-t)x_2) \notin \alpha - \text{int}(C)$. Therefore, $\{x \in K : \varphi(x) \notin \alpha - \text{int}(C)\}$ is convex on X, that is, φ is C-quasiconcave-like on K. □

Definition 2.17 (Finite Concave-Like Function) Let Y be a vector space with a partial ordering defined by a pointed convex cone C, and K be a nonempty convex subset of a vector space X. A vector-valued function $\varphi : K \to Y$ is said to be *finite concave-like* if for any finite subset K_0 of K, and for any $x, y \in K$ and $t \in [0, 1]$, there exists $z \in K_0$ such that

$$t\varphi(x) + (1-t)\varphi(y) \leq_C \varphi(z).$$

Definition 2.18 (Downward Directed Set) Let Y be a vector space with partially ordering defined by a pointed convex cone C. A nonempty subset D of Y is said to be a *downward directed set* if for all $d_1, d_2 \in D$, there exists $d \in D$ such that $d \leq_C d_1$ and $d \leq_C d_2$.

Definition 2.19 (Downward Directed Function) Let Y be a vector space with partially ordering defined by a pointed and convex cone C, and K be a nonempty subset of a vector space X. A vector-valued function $\varphi : K \to Y$ is said to be *downward directed* if for any $x, y \in K$, there exists $z \in K$ such that

$$\varphi(z) \leq_C \varphi(x) \quad \text{and} \quad \varphi(z) \leq_C \varphi(y).$$

Remark 2.11 If K is a nonempty convex subset of X and $\varphi : K \to Y$ is C-lower semicontinuous (see Definition 2.25) and finite concave-like, then φ is downward directed.

Definition 2.20 (Arc-Concave-Like Function) Let Y be a topological vector space with a partially ordering defined by a pointed closed and convex cone C, and K be a nonempty convex subset of a topological vector space X. A vector-valued function $\varphi : K \to Y$ is said to be *arc-concave-like* if for any $x, y \in K$, there exists an arc $I_{x,y}(t) \subset K$ such that for each $t \in [0, 1]$,

$$t\varphi(x) + (1-t)\varphi(y) \leq_C \varphi(I_{x,y}(t)),$$

where $I_{x,y} : [0, 1] \to K$ is a continuous map with $I_{x,y}(0) = x$ and $I_{x,y}(1) = y$.

Remark 2.12 It is easy to see that every arc-concave-like function is finite concave-like.

Definition 2.21 Let X and Y be vector spaces and K be a nonempty convex set of X. A set-valued map $T : X \to 2^Y$ with nonempty values is said to be

(a) *convex* on K if for each $x, y \in K$ and $t \in [0, 1]$

$$T(tx + (1-t)y) \supseteq tT(x) + (1-t)T(y);$$

(b) *concave* on K if for each $x, y \in K$ and $t \in [0, 1]$

$$T(tx + (1-t)y) \subseteq tT(x) + (1-t)T(y);$$

(c) *affine* on K if T is convex as well as concave on K.

The proof of the following proposition is straightforward and hence is omitted.

Proposition 2.21 *Let K and D be nonempty convex subsets of two vector spaces, respectively, Y be a topological vector space with a solid pointed convex cone C and $g : K \times D \times D \to Y$ be a function. Suppose that for each $y \in D$, $g(\cdot, \cdot, y)$ is properly $(-C)$-quasiconvex and that $S : K \to 2^D$ is defined by*

$$S(p) = \{x \in D : g(p, x, y) \notin -\operatorname{int}(C) \text{ for all } y \in D\}$$

is nonempty for each $p \in K$. Then S is a convex set-valued mapping.

Definition 2.22 (C-Quasiconcavity) Let K be a nonempty convex subset of a topological vector space X and C be a proper closed convex solid cone in a topological vector space Y. A set-valued map $T : K \to 2^Y \setminus \{\emptyset\}$ is said to be *C-quasiconcave* on K if for each $x_1, x_2 \in K$ and $y \in Y$ such that $T(x_1) \nsubseteq y - \operatorname{int}(C)$ and $T(x_2) \nsubseteq y - \operatorname{int}(C)$, we have

$$T(x_\mu) \nsubseteq y - \operatorname{int}(C), \quad \text{for all } x_\mu \in]x_1, x_2[.$$

We also say that T is *strictly C-quasiconcave* on K if for each $x_1, x_2 \in K$ and $y \in Y$ such that $T(x_1) \not\subset y - \operatorname{int}(C)$ and $T(x_2) \not\subset y - \operatorname{int}(C)$, we have

$$T(x_\mu) \not\subset y - \operatorname{cl}(\mathrm{C}), \quad \text{for all } x_\mu \in]x_1, x_2[.$$

Definition 2.23 (**C-Proper Quasiconcavity**) Let K be a nonempty convex subset of a topological vector space X and C be a proper closed convex solid cone in a topological vector space Y. A set-valued map $T : K \to 2^Y \setminus \{\emptyset\}$ is said to be *C-proper quasiconcave* on K if for each $x_1, x_2 \in K$ and $y \in Y$ such that $T(x_1) \cap \big(y - \text{int}(C)\big) = \emptyset$ and $T(x_2) \cap \big(y - \text{int}(C)\big) = \emptyset$, we have

$$T(x_\mu) \cap \big(y - \text{int}(C)\big) = \emptyset, \quad \text{for all } x_\mu \in]x_1, x_2[.$$

We also say that T is *strictly C-properly quasiconcave* on K if for each $x_1, x_2 \in K$ and $y \in Y$ such that $T(x_1) \cap \big(y - \text{int}(C)\big) = \emptyset$ and $T(x_2) \cap \big(y - \text{int}(C)\big) = \emptyset$ imply

$$T(x_\mu) \cap \big(y - \text{cl}\, C\big) = \emptyset, \quad \text{for all } x_\mu \in]x_1, x_2[.$$

Remark 2.13 If T is single-valued, Definitions 2.22 and 2.23 reduce to the definition of $(-C)$-proper quasiconvexity.

Definition 2.24 Let K be a nonempty convex subset of X and C be a proper closed convex cone in Y. A set-valued map $S : K \to 2^Y \setminus \{\emptyset\}$ is called:

(a) *C-convex* if

$$\alpha S(x) + (1-\alpha)S(y) \subseteq S(\alpha x + (1-\alpha)y) + C, \quad \text{for all } x, y \in K \text{ and all } \alpha \in [0, 1];$$

(b) *C-quasiconvex* if for all $y_1, y_2 \in K$ and $\alpha \in [0, 1]$, we have either

$$S(y_1) \subseteq S(\alpha y_1 + (1-\alpha)y_2) + C$$

or

$$S(y_2) \subseteq S(\alpha y_1 + (1-\alpha)y_2) + C;$$

(c) *C-quasiconvex-like* if for all $y_1, y_2 \in K$ and $\alpha \in [0, 1]$, we have either

$$S(\alpha y_1 + (1-\alpha)y_2) \subseteq S(y_1) - C$$

or

$$S(\alpha y_1 + (1-\alpha)y_2) \subseteq S(y_2) - C;$$

(d) *properly C-quasiconvex* if for all $x, y \in K$ and all $\alpha \in]0, 1[$,

$$S(x) \subseteq S(\alpha x + (1-\alpha)y) + C \quad \text{or} \quad S(y) \subseteq S(\alpha x + (1-\alpha)y) + C;$$

(e) *explicitly δ-C-quasiconvex* [21] if for all $x, y \in K$ and all $\alpha \in]0, 1]$, we have either

$$S(x) \subseteq S(\alpha x + (1-\alpha)y) + C \text{ or } S(y) \subseteq S(\alpha x + (1-\alpha)y) + C$$

and if $(S(y) - S(x)) \cap (-\operatorname{int}(C)) \neq \emptyset$, we have

$$S(x) \subseteq S(\alpha x + (1-\alpha)y) + \operatorname{int}(C), \quad \text{for all } \alpha \in]0,1].$$

(f) *explicitly C-quasiconvex-like* if it is C-quasiconvex-like and, in case

$$S(y_2) - S(y_1) \nsubseteq -\operatorname{int}(C), \text{ for all } y_1, y_2 \in K \text{ and } \alpha \in]0,1[,$$

we have

$$S(\alpha y_1 + (1-\alpha)y_2) \subseteq S(y_2) - \operatorname{int}(C).$$

Clearly, every C-quasiconvex set-valued map is explicitly δ-C-quasiconvex.

To show that the class of C-quasiconvex-like set-valued maps is nonempty, we give the following example.

Example 2.27 Let $K = [0,1]$ and $C = [0,+\infty)$. We define $S : K \to 2^{\mathbb{R}}$ by

$$S(x) = [0, x+1], \quad \text{for all } x \in K.$$

For all $x, x_1, x_2 \in K$ and $0 \leq \alpha \leq 1$, we note that

$$\text{if} \quad x_1 \leq x_2, \quad \text{then} \quad \alpha x_1 + (1-\alpha)x_2 \leq x_2$$

and

$$\text{if} \quad x_1 > x_2, \quad \text{then} \quad \alpha x_1 + (1-\alpha)x_2 \leq x_1.$$

Therefore, we have for each $t \in S(\alpha x_1 + (1-\alpha)x_2)$,

$$t = \begin{cases} (x_2+1) - [(x_2+1) - t], & x_1 \leq x_2 \\ (x_1+1) - [(x_1+1) - t], & x_1 > x_2. \end{cases}$$

Hence we have either

$$S(\alpha x_1 + (1-\alpha)x_2) \subseteq S(x_1) - C$$

or

$$S(\alpha x_1 + (1-\alpha)x_2) \subseteq S(x_2) - C.$$

Thus S is C-quasiconvex-like.

2.5 Cone Continuity

Definition 2.25 (**C-Continuity**) Let K be a nonempty subset of a topological vector space X and Y be a topological vector space with a partial ordering defined by a pointed closed convex cone C. A function $\varphi : K \to Y$ is said to be *C-lower semicontinuous* (respectively, *C-upper semicontinuous*) *at* $x_0 \in K$ if for any open neighborhood $V \subseteq Y$ of $\varphi(x_0) \in Y$, there exists an open neighborhood $U \subseteq X$ of x_0 such that

$$\varphi(x) \in V + C \quad \text{for all } x \in U \cap K \tag{2.24}$$

$$(\text{respectively, } \varphi(x) \in V - C \quad \text{for all } x \in U \cap K). \tag{2.25}$$

Furthermore, φ is *C-lower semicontinuous* (respectively, *C-upper semicontinuous*) on K if it is C-lower semicontinuous (respectively, C-upper semicontinuous) at every point of K.

φ is called *C-continuous* on K if it is C-lower semicontinuous as well as C-upper semicontinuous on K.

Definition 2.26 (**C-Pseudocontinuity**) [13] Let X be a topological space, and Y a vector space with a pointed convex cone C. A vector-valued function $f : X \to Y$ is said to be *C-pseudocontinuous at* $x \in X$ if for each $k \in C \setminus \{\mathbf{0}\}$, there exists a neighbourhood $U_x \subset X$ of x such that $f(u) \in f(x) - k + C$ for all $u \in U_x$. Moreover, f is said to be *C-psudocontinuous on* X if it is C-pseudocontinuous at every of X.

Example 2.28 Consider the mapping $\varphi: \mathbb{R} \to \mathbb{R}^2$ defined by

$$\varphi(x) := (\varphi_1(x), \varphi_2(x)) = \begin{cases} \left(\frac{1}{x}, x\right), & \text{for } x > 0, \\ (-1, -1), & \text{for } x \leq 0. \end{cases}$$

The function values are depicted in Fig. 2.9. One can see that φ is $\mathbb{R}^2_+$-lower semicontinuous: For $x_0 \neq 0$, there is nothing to show. As long as the neighborhood

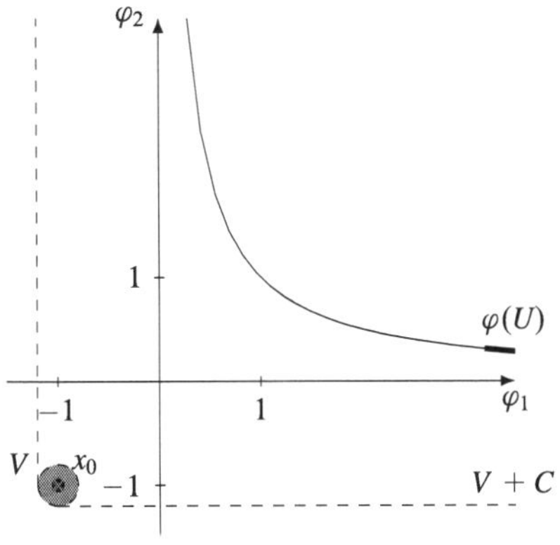

Fig. 2.9 The function φ is $\mathbb{R}^2_+$-lower semicontinuous

U of x_0 is chosen small enough, the inclusion follows from the continuity of φ_1 and φ_2, respectively. For $x_0 = 0$, the neighborhood U also contains $x > 0$ but since for those arguments $\varphi_2(x) > 0$ holds, the inclusion is still fulfilled. Note that φ is not $\mathbb{R}^2_+$-upper semicontinuous, and thus φ is not $\mathbb{R}^2_+$-continuous.

Remark 2.14 Whenever $Y = \mathbb{R}$ and $C = \mathbb{R}_+$, C-lower semicontinuity and C-upper semicontinuity are the same as ordinary lower and upper semicontinuity, respectively. In [25, Definition 5.1, p. 22], a C-lower semicontinuous function (respectively, C-upper semicontinuous function) is called C-continuous function (respectively, ($-C$)-continuous function). If the function φ is $\{\mathbf{0}\}$-continuous, then it is continuous in ordinary sense.

The following lemma can be easily proved, and therefore we omit the proof.

Lemma 2.11 *A vector-valued function $\varphi : \mathbb{R}^n \to \mathbb{R}^\ell$ is C-upper semicontinuous (respectively, C-lower semicontinuous) if and only if each $\varphi_i : \mathbb{R}^n \to \mathbb{R}$, $i = 1, 2, \ldots, \ell$, is upper semicontinuous (respectively, lower semicontinuous), where C is a pointed closed convex cone in $\mathbb{R}^\ell$.*

Proposition 2.22 *Let X be a topological space and Y be a topological vector space with a partial ordering defined by a solid pointed convex cone C. A vector-valued function $\varphi : X \to Y$ is C-lower semicontinuous on X if and only if for each $\alpha \in Y$, $\varphi^{-1}(\alpha + \text{int}(C))$ is an open subset of X.*

Proof Suppose that $\varphi^{-1}(\alpha + \text{int}(C))$ is open for all $\alpha \in Y$. For each $x_0 \in X$ and any $d \in \text{int}(C)$, we have $x_0 \in \varphi^{-1}(\text{int}(C) + (\varphi(x_0) - d))$, which is open. Hence, there exists an open neighborhood $U \subset X$ of x_0 such that $\varphi(x) \in \varphi(x_0) - d + \text{int}(C)$ for all $x \in U$.

Conversely, let $x \in \varphi^{-1}(\alpha + \text{int}(C))$ and $d := \varphi(x_0) - \alpha$. Since $d \in \text{int}(C)$, there exists an open neighborhood $U \subset X$ of x_0 such that $\varphi(x) \in \varphi(x_0) - d + \text{int}(C)$ for all $x \in U$, and hence $\varphi(x) \in \alpha + \text{int}(C)$. This implies that $\varphi^{-1}(\alpha + \text{int}(C))$ is open. □

Similarly, we can prove the following propositions.

Proposition 2.23 *Let X be a topological space and Y be a topological vector space with a partial ordering defined by a solid pointed convex cone C. A vector-valued function $\varphi : X \to Y$ is C-upper semicontinuous on X if and only if for each $\alpha \in Y$, $\varphi^{-1}(\alpha - \text{int}(C))$ is an open subset of X.*

Proposition 2.24 *Let X be a topological space and Y be a topological vector space with a partial ordering defined by a pointed closed convex cone C. A vector-valued function $\varphi : X \to Y$ is C-upper semicontinuous on X if and only if for each $\alpha \in Y$, $\varphi^{-1}(\alpha - C) = \{x \in X : \varphi(x) \leq_C \alpha\}$ is a closed subset of X.*

Proposition 2.25 *Let K be a nonempty subset of a topological vector space X, Y be a topological vector space with a partial ordering defined by a pointed closed convex cone C and $\varphi : K \to Y$ be a function. The following statements are equivalent:*

(a) *$\varphi : K \to Y$ is C-lower semicontinuous on K.*

(b) *For each* $\alpha \in Y$, *the (lower level) set*

$$L(\alpha) = \{x \in K : \varphi(x) \not>_C \alpha\}$$

is closed in K.

(c) *For each* $x_0 \in K$ *and any* $\lambda \in \text{int}(C)$, *there exists an open neighborhood* $U \subset X$ *of* x_0 *such that* $\varphi(x) \in \varphi(x_0) - \lambda + \text{int}(C)$, *for all* $x \in U \cap K$.

Proof (a) $\Rightarrow$ (b): Since

$$L(\alpha) = \{x \in K : \varphi(x) - \alpha \notin \text{int}(C)\} = \varphi^{-1}[(\alpha + \text{int}(C))^c].$$

In view of Proposition 2.22, it is sufficient to prove that for each $\alpha \in Y$,

$$[L(\alpha)]^c = \{x \in K : \varphi(x) - \alpha \in \text{int}(C)\} = \varphi^{-1}[(\alpha + \text{int}(C))]$$

is open in K.

Suppose that φ is C-lower semicontinuous on K. Let $x^* \in [L(\alpha)]^c$. Then $\alpha + \text{int}(C)$ is an open neighborhood of $\varphi(x^*)$. Hence, there exists an open neighborhood U of x^* such that

$$\varphi(x) \in (\alpha + \text{int}(C)) + C = \alpha + \text{int}(C), \quad \text{for all } x \in U \cap K,$$

that is, for all $x \in U \cap K$, we have $x \in [L(\alpha)]^c$. Then, $[L(\alpha)]^c$ is open in K.

(b) $\Rightarrow$ (c): It is sufficient to show that for each $\alpha \in Y$,

$$[L(\alpha)]^c = \{x \in K : \varphi(x) - \alpha \in \text{int}(C)\} = \varphi^{-1}[(\alpha + \text{int}(C))]$$

is open in K implies (c).

Suppose that for each $\alpha \in Y$, $[L(\alpha)]^c$ is open in K. For each $x_0 \in K$ and $z \in \text{int}(C)$, we have $x_0 \in \varphi^{-1}(\text{int}(C) + (\varphi(x_0) - z))$ which is open in K. Hence, there exists an open neighborhood U of x_0 such that $\varphi(x) \in \varphi(x_0) - z + \text{int}(C)$ for all $x \in U \cap K$

(c) $\Rightarrow$ (a): Let $x_0 \in K$ and V an open neighborhood of $\varphi(x_0)$. There is $\lambda \in \text{int}(C)$ such that $\varphi(x_0) - \lambda \in V$. Then there exists an open neighborhood U of x_0 such that $\varphi(x) \in \varphi(x_0) - \lambda + \text{int}(C)$ for all $x \in U \cap K$. Since V is open and so by Proposition 2.7, we have $V + \text{int}(C) = V + C$, and hence $\varphi(x) \in V + C$ for all $x \in U \cap K$. Thus, φ is C upper semicontinuous. □

Similarly, we can prove the following proposition.

Proposition 2.26 *Let* K *be a nonempty subset of a topological vector space* X, Y *be a topological vector space with a partial ordering defined by a pointed closed convex cone* C *and* $\varphi : K \to Y$ *be a function. The following statements are equivalent:*

(a) $\varphi : K \to Y$ *is* C*-upper semicontinuous on* K.

(b) *For each* $\alpha \in Y$, *the (upper level) set* $L(\alpha) = \{x \in K : \varphi(x) \nless_C \alpha\}$ *is closed in* K.
(c) *For each* $x_0 \in K$ *and any* $\lambda \in \text{int}(C)$, *there exists an open neighborhood* U *of* x_0 *such that* $\varphi(x) \in \varphi(x_0) - \lambda - \text{int}(C)$, *for all* $x \in U \cap K$.

Proposition 2.27 ([2]) *Let* K *be a nonempty subset of a topological vector space* X *and* Y *be a topological vector space with a partial ordering defined by a pointed closed convex cone* C. *A vector-valued function* $\varphi : K \to Y$ *is* C*-upper semicontinuous on* K *if and only if for each* $x \in K$, $d \in \text{int}(C)$ *and for any net* $\{x_\alpha\}_{\alpha\in\Lambda}$ *converging to* x, *there exists* α_0 *in the index set* Λ *such that*

$$\text{cl}(\{\varphi(x_\beta) : \beta \geq \alpha\}) \subset \varphi(x) + d - \text{int}(C), \quad \textit{for all } \alpha \geq \alpha_0.$$

Proof For $\alpha \in \Lambda$, let $A_\alpha = \{\varphi(x_\beta) : \beta \geq \alpha\}$. Suppose φ is C-upper semicontinuous. Then there exists an open neighborhood U of x such that $\varphi(y) \in \varphi(x) + \frac{v}{2} - \text{int}(C)$ for all $y \in U$. Hence, there is α_0 in the index set Λ such that

$$\alpha \geq \alpha_0 \quad \Rightarrow \quad x_\alpha \in U \text{ and } \varphi(x_\alpha) \in \varphi(x) + \frac{1}{2}v - \text{int}(C).$$

This implies that $A_\alpha \subseteq \varphi(x) + \frac{1}{2}v - \text{int}(C)$ and $\text{cl}(A_\alpha) \subseteq \varphi(x) + \frac{1}{2}v - C$ whenever $\alpha \geq \alpha_0$. Since $\frac{1}{2}v - C = v = \frac{v}{2} - C \subseteq v - \text{int}(C)$, we have

$$\text{cl}((A_\alpha) \subseteq \varphi(x) + v - \text{int}(C), \quad \text{for all } \alpha \geq \alpha_0.$$

Conversely, assume that φ is not C-upper semicontinuous on X. Then there is $y_0 \in Y$ such that $\varphi^{-1}(y_0 - \text{int}(C))$ is not open in X. Hence there is $x_0 \in \varphi^{-1}(y_0 - \text{int}(C))$ such that every neighborhood of x_0 is not contained in $\varphi^{-1}(y_0 - \text{int}(C))$. Write $\varphi(x_0) = y_0 - v_0$ for some $v_0 \in \text{int}(C)$. Hence, there exists a net $\{x_\alpha\}_{\alpha\in\Lambda}$ in X such that $x_\alpha \to x$ and every $\varphi(x_\alpha)$ does not lie in $y_0 - \text{int}(C) = \varphi(x_0) + v_0 - \text{int}(C)$. Since the complement of $\varphi(x_0) + v_0 - \text{int}(C)$ is closed, for every $\alpha \in \Lambda$,

$$\text{cl}(A_\alpha) \cap (\varphi(x_0) + v_0 - \text{int}(C)) = \emptyset,$$

which is a contradiction. □

Proposition 2.28 *Let* X *and* Z *be Hausdorff topological vector spaces and* $f : X \to Z$ *be a vector-valued function. If* φ *is* C*-upper semicontinuous on* X, *then the real-valued function* $g : X \to \mathbb{R}$ *defined by* $g(x) = \psi(f(x))$ *is upper semicontinuous.*

Proof Since f is C-upper semicontinuous on X, from Proposition 2.27, we have that for each $w \in \text{int}(C)$ and for any net $\{x_\alpha\}_{\alpha\in\Lambda}$ in X converging to x, there exists $\alpha_0 \in \Lambda$ such that

$$\text{cl}(\{g(x_\beta) : \beta \geq \alpha\}) \subset g(x) + w - \text{int}(C), \quad \text{for all } \alpha \geq \alpha_0.$$

Let us consider a net $\{w_j\}_{j\in J} \subseteq \text{int}(C)$ such that $w_j \to 0$. Then from the C-upper semicontinuity of f, we deduce that for each $j \in J$, there exists $\alpha_0(j) \in J$ such that

$$f(x_\beta) - f(x) - w_j \in -\,\text{int}(C), \quad \text{for all } \beta \geq \alpha_0(j).$$

Hence, $\psi\left(f(x_\beta) - f(x) - w_j\right) < 0$ for all $\beta \geq \alpha_0(j)$. It follows that

$$\max_{\lambda\in B^*}\langle \lambda, f(x_\beta)\rangle \leq \max_{\lambda\in B^*}\langle \lambda, f(x)\rangle + \max_{\lambda\in B^*}\langle \lambda, w_j)\rangle, \text{ for all } \beta \geq \alpha_0(j).$$

Consequently,

$$g(x_\beta) \leq g(x) + \max_{\lambda\in B^*}\langle \lambda, w_j)\rangle, \quad \text{for all } \beta \geq \alpha_0(j),$$

from which it follows that $\limsup_{\beta} g(x_\beta) \leq g(x)$ since $w_j \to 0$. Therefore, g is upper semicontinuous. □

Proposition 2.29 *Let K be a nonempty subset of a topological vector space X and Y be a topological vector space with a partial ordering defined by a pointed closed convex cone C. Let $\varphi,\ \xi : K \to Y$ be C-lower semicontinuous functions. Then*

(a) *$\varphi + \xi$ is C-lower semicontinuous;*
(b) *$t\varphi$ is C-lower semicontinuous for $t > 0$.*

Proof We only prove part (a). Part (b) is left as an exercise. For a given $x_0 \in K$, for every neighborhood V of $\varphi(x_0) + \xi(x_0)$, we can find open neighborhoods V_1 and V_2 of $\varphi(x_0)$ and $\xi(x_0)$, respectively, such that $V_1 + V_2 \subseteq V$. Since φ and ξ are C-lower semicontinuous, we can find neighborhoods U_1 and U_2 of x_0 such that

$$\varphi(x) \in V_1 + C, \quad \text{for all } x \in U_1 \cap K$$

and

$$\xi(x) \in V_2 + C, \quad \text{for all } x \in U_2 \cap K.$$

Setting $U = U_1 + U_2$, we see that the function $\varphi + \xi$ is C-lower semicontinuous. □

Proposition 2.30 *Let K be a nonempty subset of a topological vector space X and Y be a topological vector space with a partial ordering defined by a pointed closed convex cone C. Let $\varphi : K \to Y$ be a C-lower semicontinuous function. Then the real-valued function $\xi \circ \varphi$ is lower semicontinuous for all $\xi \in C^*$, where C^* is the dual cone of C.*

Proof If ξ is a zero function in C^*, then this is obvious. So, suppose that $\xi \in C^*$ is not a zero function and let $x_0 \in K$. For any $\varepsilon > 0$, define

$$V = \{y \in \varphi(K) : |\xi(y)| < \varepsilon\}.$$

Since φ is C-lower semicontinuous, for any $x_0 \in K$, there exists an open neighborhood U of x_0 such that

$$\varphi(x) \in (\varphi(x_0) + V) + C, \quad \text{for all } x \in U \cap K.$$

It follows that, for any $x \in U \cap K$, there exists $y \in V$ such that $\varphi(x) \geq_C \varphi(x_0) + y$. Hence,

$$\xi \circ \varphi(x) \geq \xi \circ \varphi(x_0) - \varepsilon,$$

that is, $\xi \circ \varphi$ is lower semicontinuous. □

Proposition 2.31 *Let K be a nonempty subset of a topological vector space X and Y be a topological vector space with a partial ordering defined by a pointed closed convex cone C. Let $\varphi : K \to Y$ be a C-lower semicontinuous function at $x_0 \in K$, and for every $x \in K \setminus \{x_0\}$, $\mathbf{0} \not<_C \varphi(x)$. Then $\varphi(x_0) \not>_C \mathbf{0}$.*

Proof Suppose to the contrary that $\varphi(x_0) >_C \mathbf{0}$, that is, $\varphi(x_0) \in \text{int}(C)$. Then there is a neighborhood V of $\varphi(x_0)$ such that $V \subseteq \text{int}(C)$. We have $V + C \subseteq \text{int}(C)$. Since φ is C-lower semicontinuous at x_0, one can find a neighborhood U of x_0 such that $\varphi(U \cap K) \subseteq V + C$. Taking $x \in U \cap K$, $x \neq x_0$, we have $\varphi(x) \in \text{int}(C)$. This contradicts the assumption that $\varphi(x) \notin \text{int}(C)$ for all $x \in K \setminus \{x_0\}$. This completes the proof. □

Proposition 2.32 ([25, Proposition 1.8]) *Let Y be a Hausdorff topological vector space. Assume that the cone $C \subseteq Y$ has a closed convex bounded base. Then for any neighborhood V of the origin in Y, there exists a neighborhood W of the origin in Y such that*

$$(W + C) \cap (W - C) \subseteq V. \tag{2.26}$$

Theorem 2.6 *Let Y be a Hausdorff topological vector space, X a topological vector space and $\emptyset \neq K \subseteq X$. Assume that the cone $C \subseteq Y$ has a closed convex bounded base. Then the continuity in ordinary sense is equivalent to the C-continuity, that is, the function $\varphi : K \to Y$ is continuous (in ordinary sense) if and only if it is both C-lower semicontinuous as well as C-upper semicontinuous.*

Proof It is obvious that if $\varphi : K \to Y$ is continuous, that is, $\{\mathbf{0}\}$-continuous, then it is C-continuous for any cone C in Y.

Suppose that $\varphi : K \to Y$ is C-continuous at a point $x_0 \in X$. Then it is C-lower semicontinuous as well as C-upper semicontinuous at $x_0 \in X$. Let $V \subseteq Y$ be a neighborhood of $\varphi(x_0)$. Then we have to show that there exists a neighborhood $U \subseteq X$ of x_0 such that

$$\varphi(x) \in V, \quad \text{for all } x \in U \cap K. \tag{2.27}$$

By Proposition 2.32, for the neighborhood V, we can find a neighborhood W of $\varphi(x_0)$ in Y such that (2.26) holds. By the assumption of the theorem, for W, there are two neighborhoods $U_1 \subseteq X$ and $U_2 \subseteq X$ of x_0 such that

$$\varphi(x) \in W + C, \quad \text{for all } x \in U_1 \cap K$$

and

$$\varphi(x) \in W - C, \quad \text{for all } x \in U_2 \cap K.$$

This and (2.26) imply (2.27) for $U = U_1 \cap U_2$. □

Lemma 2.12 ([25, Theorem 7.2]) *Let K be a nonempty compact convex subset of a Hausdorff topological vector space. Let Y be a topological vector space with a solid pointed convex cone $C \subseteq Y$. If $\varphi : K \to Y$ is C-continuous, then $\bigcup_{x \in K}\{\varphi(x)\}$ is C-compact.*

Definition 2.27 Let X and Y be topological vector spaces, C be a pointed closed convex cone in Y and $e \in Y \setminus \{\mathbf{0}\}$. A function $\varphi : X \to Y$ is said to be

(a) *C-bounded below* if there exists $\alpha \in Y$ such that $\varphi(x) \subseteq \alpha + C$ for all $x \in X$;
(b) *(e, C)-lower semicontinuous* if for all $r \in \mathbb{R}$, the set $\{x \in X : \varphi(x) \in re - C\}$ is closed;
(c) *(e, C)-upper semicontinuous* if for all $r \in \mathbb{R}$, the set $\{x \in X : \varphi(x) \in re + C\}$ is closed;
(d) *(e, C)-continuous* if it is both (e, C)-lower semicontinuous as well as (e, C)-upper semicontinuous.

Remark 2.15 It is easy to see that the C-lower (respectively, C-upper) semicontinuity of φ implies the (e, C)-lower (respectively, (e, C)-upper) semicontinuity.

Remark 2.16

(a) For $Y = \mathbb{R}$, $C = \mathbb{R}_+$, $e = 1$, let $\{x \in X : \varphi(x) \in re - C\} = \{x \in X : \varphi(x) \leq r\}$ be a closed set for all $r \in \mathbb{R}$. This is the usual definition of lower semicontinuity for functionals $\varphi : X \to \mathbb{R}$.
(b) Luc [25] used the following notion of cone-continuity: φ is C-semicontinuous if the set $\{x \in X : \varphi(x) \in y - C\}$ is closed at every $y \in Y$. It is easy to see that C-semicontinuity implies (e, C)-lower semicontinuity for all $e \in Y$.

Example 2.29 Consider the mapping $\varphi : \mathbb{R} \to \mathbb{R}^2$ defined in Example 2.28 by

$$\varphi(x) := (\varphi_1(x), \varphi_2(x)) = \begin{cases} \left(\frac{1}{x}, x\right), & \text{for } x > 0, \\ (-1, -1), & \text{for } x \leq 0. \end{cases}$$

It is easy to see that φ is $\mathbb{R}^2_+$-bounded below. Let $e = (1, 1)$. Then φ is (e, C)-lower semicontinuous and (e, C)-upper semicontinuous with $C = \mathbb{R}^2_+$, and hence (e, C)-continuous.

Definition 2.28 A vector-valued function $\varphi : K \to Y$ is called *C-upper hemicontinuous* (respectively, *C-lower hemicontinuous*) if for all $x, y \in K$, the vector-valued function $t \mapsto \varphi(x + t(y - x))$ defined for $t \in [0, 1]$, is C-upper semicontinuous (respectively, C-lower semicontinuous) at $t = 0$.

φ is called *C-hemicontinuous* if it is C-upper semicontinuous as well as C-lower semicontinuous at $t = 0$.

Example 2.30

(a) The mapping $\varphi: \mathbb{R} \to \mathbb{R}^2$ defined in Example 2.28 is C-lower hemicontinuous. Because φ is not C-upper semicontinuous and φ maps from $\mathbb{R}$ to $\mathbb{R}^2$, the function cannot be C-upper hemicontinuous.
(b) The function $\varphi : K \to \mathbb{R}^2$, defined by $\varphi(x_1, x_2) := \left(x_1^2, x_2^2\right) - \left(\frac{1}{2}, \frac{1}{2}\right)$ on the set $K := \left\{(x_1, x_2) \in \mathbb{R}^2 : -1 < x_1 < 1, -1 < x_2 < 1\right\}$, is C-hemicontinuous.

Example 2.31 Let $f: \mathbb{R}^2 \to \mathbb{R}$ with

$$f(x_1, x_2) := \begin{cases} 0\,, & \text{if } (x_1, x_2) = 0\,, \\ \frac{x_1^2 x_2}{x_1^4 + x_2^2}\,, & \text{else.} \end{cases}$$

For $(x_1, x_2) \neq \mathbf{0}$, the function f is obviously continuous and therefore hemicontinuous as well. If one approaches $\mathbf{0}$ along the path $(x_1(t), x_2(t)) = \psi(t) = (t, \operatorname{sgn}(t) \cdot t^2)$, where

$$\operatorname{sgn}(t) := \begin{cases} 1\,, & \text{if } t > 0, \\ 0\,, & \text{if } t = 0, \\ -1\,, & \text{if } t < 0, \end{cases}$$

the corresponding function values are

$$f(\psi(t)) = \frac{t^2 \cdot \operatorname{sgn}(t) \cdot t^2}{t^4 + (\operatorname{sgn}(t))^2 t^4} = \frac{\operatorname{sgn}(t)}{2}\,,$$

which has a discontinuity at $t = 0$, see Fig. 2.10. If, on the other hand, one only considers straight lines approaching $\mathbf{0}$, i. e. $(x_1(t), x_2(t)) = \varphi(t; \alpha, \beta) = t(\alpha, \beta)$ with $\alpha^2 + \beta^2 > 0$, the corresponding function values are a continuous function in t:

$$f(\varphi(t; \alpha, \beta)) = \frac{\alpha^2 \beta t^3}{\alpha^4 t^4 + \beta^2 t^2} = \frac{\alpha^2 \beta t}{\alpha^4 t^2 + \beta^2}\,.$$

Therefore, the function f is hemicontinuous at $\mathbf{0}$.

Now we discuss cone continuity of set-valued maps.

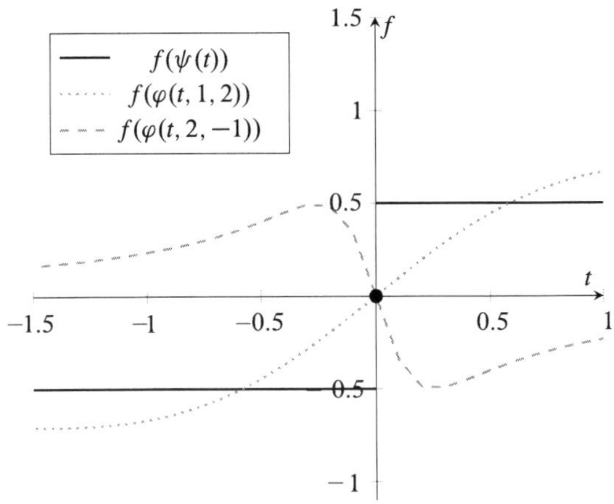

Fig. 2.10 Function values of the hemicontinuous function f from Example 2.31 along paths

Definition 2.29 (**C-Semicontinuity**) Let X be a topological space, Y be a topological vector space and C be a proper closed convex cone with $\operatorname{int}(C) \neq \emptyset$. A set-valued map $T : X \to 2^Y \setminus \{\emptyset\}$ is said to be:

(a) *C-lower semicontinuous* at $x \in X$ if for each open subset $\mathcal{V}$ of Y with $T(x) \cap \mathcal{V} \neq \emptyset$, there exists a neighborhood $\mathcal{U}$ of x such that

$$T(u) \cap \big(\mathcal{V} + \operatorname{int}(C)\big) \neq \emptyset, \quad \text{for all } u \in \mathcal{U}.$$

(b) *C-upper semicontinuous* at $x \in X$ if for each open subset $\mathcal{V}$ of Y with $\mathcal{V} \supseteq T(x)$, there exists a neighborhood $\mathcal{U}_x$ of x such that

$$T(u) \subseteq \mathcal{V} + \operatorname{int}(C), \quad \text{for all } u \in \mathcal{U}_x.$$

(c) *C-continuous* at $x \in X$ if it is C-lower semicontinuous as well as C-upper semicontinuous at x.

Georgiev and Tanaka [9] defined C-upper semicontinuity of a set-valued map in a different way.

Proposition 2.33 *Let X and Z be two topological spaces, and Y be a topological vector space with two proper solid convex cones C and D such that $C \supseteq D$. Suppose that $T : X \to 2^Y \setminus \{\emptyset\}$ is a set-valued map and $f : Z \to Y$ is a vector-valued function such that for fixed $x \in X$ and fixed $z \in Z$,*

(i) *T is D-upper semicontinuous at x;*
(ii) *f is D-continuous at z;*
(iii) *$T(x)$ is D-compact;*
(iv) *$T(x) + f(z) \subseteq \operatorname{int}(C)$.*

Then there exist two neighborhoods $\mathcal{U}_x$ of x and $\mathcal{V}_y$ of y such that

$$T(u) + f(v) \subseteq \operatorname{int}(C), \quad \text{for all } u \in \mathcal{U}_x \text{ and } v \in \mathcal{V}_y.$$

Also, if $C = D$, for each $\varepsilon \in \operatorname{int} D$ there exist $t > 0$ and neighborhoods $\mathcal{U}_x$ of x and $\mathcal{V}_y$ of y such that

$$T(u) + f(v) \subseteq t\varepsilon + \operatorname{int}(C), \quad \text{for all } u \in \mathcal{U}_x \text{ and } v \in \mathcal{V}_y.$$

Proof Let $A := T(x) + f(z)$. Then for each $y \in A$, there exists a positive number $t_z > 0$ such that $y - t_y\varepsilon \in \operatorname{int}(C)$. Note that $y - t_y\varepsilon + \operatorname{int} D \subseteq \operatorname{int}(C)$ and it is a neighborhood of y. Obviously, $\bigcup_{y \in A}\left(y - t_y\varepsilon + \operatorname{int} D\right) \supset A$. Hence, by condition (iii), there exist $y_1, y_2, \ldots, y_m \in A$ and corresponding positive numbers $t_{y_1}, t_{y_2}, \ldots, t_{y_m} > 0$ such that

$$\bigcup_{i=1}^{m}\left(y_i - t_{y_i}\varepsilon + \operatorname{int} D\right) \supset A$$

By condition (i), there exists a neighborhood $\mathcal{U}_x$ of x such that

$$T(u) + f(z) \subseteq \bigcup_{i=1}^{m}\left(y_i - t_{y_i}\varepsilon + \operatorname{int} D\right).$$

Since $y_1 - t_{y_1}\varepsilon, \ldots, y_n - t_{y_n}\varepsilon \in \operatorname{int}(C)$ are finitely many, there exists a positive number $\tau > 0$ such that

$$y_i - (t_{y_i} + 2\tau)\varepsilon \in \operatorname{int}(C), \quad i = 1, 2, \ldots, m. \tag{2.28}$$

By condition (ii), there exists a neighborhood $\mathcal{V}_y$ of y such that

$$f(v) \in f(z) - \tau\varepsilon + \operatorname{int} D, \quad \text{for all } v \in \mathcal{V}_y.$$

Note that $\operatorname{int}(C) + \operatorname{int} D \subseteq \operatorname{int}(C) + \operatorname{int}(C) = \operatorname{int}(C)$. Hence, we have

$$T(u) + f(v) \subseteq \operatorname{int}(C), \quad \text{for all } u \in \mathcal{U}_x \text{ and } z \in \mathcal{V}_z.$$

Also if $C = D$, by (2.28), we have

$$y_i - (t_{y_i} + \tau)\varepsilon \in \tau\varepsilon + \operatorname{int}(C), \quad i = 1, 2, \ldots, m.$$

Hence, we have

$$T(u) + f(v) \subseteq \tau\varepsilon + \operatorname{int}(C), \quad \text{for all } u \in \mathcal{U}_x \text{ and } y \in \mathcal{V}_y.$$

This completes the proof. □

Remark 2.17 If $T(x)$ is not C-compact, then Proposition 2.33 may not be true even if T has a constant value and f is continuous on Z.

Example 2.32 Let $X =]0, \infty[$, $Z = \mathbb{R}^2$, $Y = \mathbb{R}^2$, and $C = \mathbb{R}^2_+$. Suppose that $T : X \to 2^Y$ is defined by

$$T(x) := \left\{ (u, v) \in Y : v = \frac{1}{u} \right\},$$

and that $f : Z \to Y$ is defined by

$$f(z) = z.$$

Then T has constant values and f is continuous on Z. Also, we have

$$T(0) + f(\mathbf{0}_Z) \subset \text{int}(C),$$

where $\mathbf{0}_Z$ denotes the zero vector in Z. However for any neighborhood V of $\mathbf{0}_Z$, $V \cap -\text{int}(C) \neq \emptyset$. Let $\hat{z} \in V \cap -\text{int}(C)$. Then $T(0) + f(\hat{z}) \not\subset \text{int}(C)$. Hence, there is no neighborhood of $U \subset X$ of 0 and $V \subseteq Z$ of $\mathbf{0}_Z$ such that

$$T(u) + f(v) \subseteq \text{int}(C), \quad \text{for all } u \in U \text{ and } v \in V.$$

2.6 Nonlinear Scalarization Functions

The most useful and common practice to solve a vector optimization problem is to convert it into a scalar optimization problem. Gerstewitz [10] introduced a nonlinear scalarization function for such conversion. Pascoletti and Serafini [26] used this functional in multiobjective optimization. In the context of operator theory, this functional was mentioned by Krasnoselskiĭ [22] (see Rubinov [27]). Essential properties of such a functional were shown by Gerstewitz and Iwanow [11], Gerth and Weidner [12] and Göpfert et al. [14].

Several other mathematicians, namely, Chen et al. [3], Luc [25], etc., further used this scalarization function. Notice that Wierzbicki [32] was the first researcher to investigate nonconvex separation techniques.

Let Y be a topological vector space and e a fixed nonzero vector in Y. Furthermore, let D be a nonempty closed proper (i.e., $D \neq \{\mathbf{0}\}$ and $D \neq Y$) subset of Y. We assume that

$$D + [0, \infty) \cdot e \subset D, \tag{2.29}$$

which means that D contains the rays generated by e.

Now we introduce the following nonlinear scalarization function.

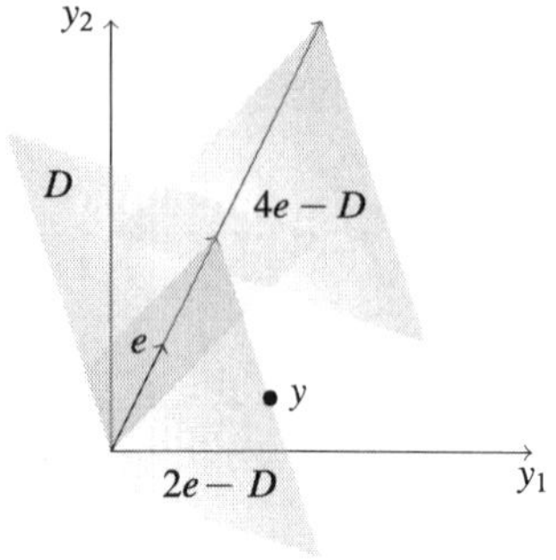

Fig. 2.11 Level sets of the functional $\xi_{e,D}$

Definition 2.30 For any closed set $D \subseteq Y$ and $e \in Y \setminus \{\mathbf{0}\}$ satisfying (2.29), a *nonlinear scalarization function* $\xi_{e,D} : Y \to \mathbb{R} \cup \{\pm\infty\}$ is defined by

$$\xi_{e,D}(y) = \inf\{t \in \mathbb{R} : y \in te - D\}, \quad \text{for all } y \in Y.$$

Applications of $\xi_{e,D}$ include coherent risk measures in financial mathematics (see, for instance, [17]). Furthermore, the functional $\xi_{e,D}$ can be used to separate nonconvex sets (see Theorem 2.8).

A visualization of the functional $\xi_{e,D}$ is given in Fig. 2.11.

Note that it is also possible to introduce a parameter α in the definition of the functional $\xi_{e,D}$. We have for a fixed $\alpha \in Y$

$$\begin{aligned}\xi_{e,D+\alpha}(y) &= \inf\{t \in \mathbb{R} : y \in te - (D + \alpha)\}\\ &= \inf\{t \in \mathbb{R} : y + a \in te - D\} = \xi_{e,D}(y + \alpha).\end{aligned}$$

The nonlinear scalarizing functional fulfills some important properties, some of which we need to define first.

Definition 2.31 Let Y be a topological vector space and C be a closed proper pointed convex cone in Y. A function $\xi : Y \to \mathbb{R}$ is said to be

(a) *monotone* with respect to C if for all $y_1, y_2 \in Y$,

$$y_1 \geq_C y_2 \quad \text{implies} \quad \xi(y_1) \geq \xi(y_2);$$

(b) *strictly monotone* with respect to C if for all $y_1, y_2 \in Y$,

$$y_1 >_C y_2 \quad \text{implies} \quad \xi(y_1) > \xi(y_2);$$

(c) *strongly monotone* with respect to C if for all $y_1, y_2 \in Y$,

$$y_1 \geq_{C_0} y_2 \quad \text{implies} \quad \xi(y_1) > \xi(y_2).$$

Of course, strong monotonicity implies strict monotonicity.

The nonlinear scalarization function satisfies the following important properties.

Theorem 2.7 ([12, 15]) *Let Y be a topological vector space, $D \subset Y$ a closed proper set and $B \subset Y$. Furthermore, let $e \in Y \setminus \{0\}$ be such that* (2.29) *is satisfied. Then the following properties hold for $\xi_{e,D}$:*

(a) *$\xi_{e,D}$ is lower semicontinuous.*
(b) *$\xi_{e,D}$ is convex $\Leftrightarrow$ D is convex.*
(c) *[for all $y \in Y$, for all $t > 0 : \xi_{e,D}(ty) = t\xi_{e,D}(y)$] $\Leftrightarrow$ D is a cone.*
(d) *$\xi_{e,D}$ is proper $\Leftrightarrow$ D does not contain lines parallel to e, i.e., for all $y \in Y$, there exists $r \in \mathbb{R}$ such that $y + re \notin D$.*
(e) *$\xi_{e,D}$ is monotone with respect to $B \Leftrightarrow D + B \subset D$.*
(f) *$\xi_{e,D}$ is subadditive $\Leftrightarrow D + D \subset D$.*
(g) *For all $y \in Y$ and all $r \in \mathbb{R} : \xi_{e,D}(y) \le r \Leftrightarrow y \in rk - D$. Equivalently, for all $y \in Y$ and all $r \in \mathbb{R} : \xi_{e,D}(y) > r \Leftrightarrow y \notin rk - D$.*
(h) *For all $y \in Y$ and all $r \in \mathbb{R} : \xi_{e,D}(y + rk) = \xi_{e,D}(y) + r$.*
(i) *$\xi_{e,D}$ is finite-valued $\Leftrightarrow$ D does not contain lines parallel to e and $\mathbb{R}e - D = Y$.*

Let furthermore $D + (0, \infty) \cdot e \subset \mathrm{int}(D)$. Then

(j) *For all $y \in Y$ and all $r \in \mathbb{R} : \xi_{e,D}(y) < r \Leftrightarrow y \in re - \mathrm{int}(D)$. Equivalently, for all $y \in Y$ and all $r \in \mathbb{R} : \xi_{e,D}(y) \ge r \Leftrightarrow y \notin re - \mathrm{int}(D)$.*
(k) *$\xi_{e,D}$ is continuous.*
(l) *For all $y \in Y$ and all $r \in \mathbb{R} : \xi_{e,D}(y) = r \Leftrightarrow y \in re - \mathrm{bd}(D)$.*
(m) *If $\xi_{e,D}$ is proper, then $\xi_{e,D}$ is monotone with respect to $B \Leftrightarrow D + B \subset D \Leftrightarrow \mathrm{bd}(D) + B \subset D$.*
(n) *If $\xi_{e,D}$ is finite-valued, then $\xi_{e,D}$ is strongly monotone with respect to $B \Leftrightarrow D + (B \setminus \{0\}) \subset \mathrm{int}(D) \Leftrightarrow \mathrm{bd}(D) + (B \setminus \{0\}) \subset \mathrm{int}(D)$.*
(o) *Suppose $\xi_{e,D}$ is proper. Then $\xi_{e,D}$ is subadditive $\Leftrightarrow D + D \subset D \Leftrightarrow \mathrm{bd}(D) + \mathrm{bd}(D) \subset D$.*

Proof

(a) To show that $\xi_{e,D}$ is lower semicontinuous, we prove that $\mathrm{epi}\,\xi_{e,D}$ is closed. In fact, for $D' := \{(y,t) \in Y \times \mathbb{R} : y \in te - D\}$, we observe that $D' \subseteq \mathrm{epi}\,\xi_{e,D} \subseteq \mathrm{cl}\,D'$. Notice that if D is closed, then D' is closed. Thus, if D is closed, we have $D' = \mathrm{epi}\,\xi_{e,D}$, and $\xi_{e,D}$ is lower semicontinuous.
(b) Let $\xi_{e,D}$ be convex, and choose $t_1 := \xi_{e,D}(y_1)$ and $t_2 := \xi_{e,D}(y_2)$, which means that $y_1 = t_1 e - d_1$, $y_2 = t_2 e - d_2$ for some $d_1, d_2 \in D$. We have

$$\lambda y_1 + (1-\lambda)y_2 = \lambda(t_1 e - d_1) + (1-\lambda)(t_2 e - d_2).$$

Because $\xi_{e,D}$ is convex, we have $\xi_{e,D}(\lambda y_1 + (1-\lambda)y_2) \le \lambda\xi_{e,D}(y_1) + (1-\lambda)\xi_{e,D}(y_2)$, which means that $\lambda y_1 + (1-\lambda)y_2 \in (\lambda t_1 + (1-\lambda)t_2)e - D$, taking into account (g). Thus,

$$\lambda y_1 + (1-\lambda)y_2 = \lambda(t_1 e - d_1) + (1-\lambda)(t_2 e - d_2) \in (\lambda t_1 + (1-\lambda)t_2)e - D,$$

and D is convex.

Conversely, let D be a closed convex set, $t_1 := \xi_{e,D}(y_1)$ and $t_2 := \xi_{e,D}(y_2)$. Then $y_1 = t_1 e - d_1$, $y_2 = t_2 e - d_2$ for some $d_1, d_2 \in D$. It follows for $\lambda \in [0, 1]$ that

$$\begin{aligned}\lambda y_1 + (1-\lambda)y_2 &= \lambda(t_1 e - d_1) + (1-\lambda)(t_2 e - d_2)\\ &= (\lambda t_1 + (1-\lambda)t_2 e) - \lambda d_1 - \lambda d_2\\ &\in (\lambda t_1 + (1-\lambda)t_2 e) - D.\end{aligned}$$

Thus, $\xi_{e,D}(\lambda y_1 + (1-\lambda)y_2) \leq \lambda t_1 + (1-\lambda)t_2 = \lambda \xi_{e,D}(y_1) + (1-\lambda)\xi_{e,D}(y_2)$, and thus $\xi_{e,D}$ is convex.

(c) Let $\xi_{e,D}(ty) = t\xi_{e,D}(y)$. It holds $y \in \xi_{e,D}(y)e - D$ and $ty \in \xi_{e,D}(ty) - D = t\xi_{e,D}(ty) - D$. Thus, D is a cone.

Conversely, let D be a cone. Then we have for $\bar{t} > 0$

$$\begin{aligned}\xi_{e,D}(\bar{t}y) &= \inf\{t \in \mathbb{R} : \bar{t}y \in te - D\}\\ &= \inf\{t \in \mathbb{R} : y \in \frac{1}{\bar{t}}te - D\}\\ &= \inf\{\bar{t} \cdot \tilde{t} \in \mathbb{R} : y \in \tilde{t}e - D\}\\ &= \bar{t} \inf\{\tilde{t} \in \mathbb{R} : y \in \tilde{t}e - D\}\\ &= \bar{t}\xi_{e,D}(y),\end{aligned}$$

with $\tilde{t} := \frac{1}{\bar{t}}t$.

(d) $\xi_{e,D}(y) = -\infty$ if and only if $y \in te - D$ for all $t \in \mathbb{R}$. This means $\{y - te : t \in \mathbb{R}\} \subseteq D$. So, $\xi_{e,D}$ is proper $\Leftrightarrow$ D does not contain lines parallel to e.

(e) Let $D + B \subseteq D$. Take $y_1, y_2 \in Y$ such that $y_1 \in y_2 - B$ holds. Let t be the smallest value such that $y_2 \in te - D$. Then $y_1 \in y_2 - B \subseteq te - D - B \subseteq te - D$. So, $\xi_{e,D}(y_1) \leq t = \xi_{e,D}(y_2)$. Now assume that $\xi_{e,D}$ is monotone with respect to B and take $y \in D$ and $b \in B$. $y \in D$ implies $-y \in -D$, and because of (g) with $r = 0$, we get $\xi_{e,D}(-y) \leq 0$. We have $(-y) - (-y - b) \in B$, which implies $\xi_{e,D}(-y - b) \leq \xi_{e,D}(-y) \leq 0$. Again by (g), we conclude with $-y - b \in -D$, thus $y + b \in D$.

(f) Let $\xi_{e,D}$ be subadditive, and choose $y_1, y_2 \in D$. Then by (g), we have $\xi_{e,D}(-y_1), \xi_{e,D}(-y_2) \leq 0$. Because $\xi_{e,D}$ is subadditive, we obtain $\xi_{e,D}(-y_1 - y_2) \leq \xi_{e,D}(-y_1) + \xi_{e,D}(-y_2) \leq 0$. Again by (g), we conclude $-y_1 - y_2 \in -D$, that is, $y_1 + y_2 \in D$.

(g) The implication $\Leftarrow$ is obvious from the definition of the functional $\xi_{e,D}$, while the inverse implication is immediate taking into account the closedness of D.

(h) We have $y \in \xi_{e,D}(y)e - D$ for some $y \in Y$. Adding $\bar{t}e$, we get $y + \bar{t}e \in (\xi_{e,D}(y) + \bar{t})e - D$, and thus $\xi_{e,D}(y + \bar{t}e) \leq \xi_{e,D}(y) + \bar{t}$. Conversely, consider

$y + \bar{t}e \in \xi_{e,D}(y + \bar{t}e)e - D$, so $y \in (\xi_{e,D}(y + \bar{t}e) - \bar{t})e - D$. Hence $\xi_{e,D}(y) \leq \xi_{e,D}(y + \bar{t}e) - \bar{t}$. In conclusion, we have $\xi_{e,D}(y + \bar{t}e) = \xi_{e,D}(y) + \bar{t}$.

(i) The assertion is deduced from (d) and $\operatorname{dom} \xi_{e,D} = \mathbb{R}e - D$.

(j) Let $t \in \mathbb{R}$ and choose $y \in te - \operatorname{int}(D)$. Then there exists some $\epsilon > 0$ such that $te - y - \epsilon e \in \operatorname{int}(D) \subset D$. This means $\xi_{e,D}(y) \leq t - \epsilon < t$. Conversely, choose $t \in \mathbb{R}$ and $y \in Y$ such that $\xi_{e,D}(y) < t$. Then there exists $\bar{t}$ such that $\bar{t} < t$, with $y \in \bar{t}e - D$. Then we obtain $y \in te - (D + (t - \bar{t})e) \subset te - \operatorname{int}(D)$.

(k) This assertion follows from (j).

(l) We obtain this result directly from (g) and (j).

(m) Let $\xi_{e,D}$ be proper and monotone with respect to B. Choose $y \in \operatorname{bd}(D)$ and $b \in B$. Then, by (l), we obtain $\xi_{e,D}(-y) = 0$. By the monotonicity assumption, we have $\xi_{e,D}(-y - b) \leq \xi_{e,D}(-y) = 0$. Then by (g), $y + b \in D$. This shows $\operatorname{bd}(D) + B \subset D$. Now let $\operatorname{bd}(D) + B \subset D$. We show that $\xi_{e,D}$ is monotone w. r. t. B. Choose $y_2 - y_1 \in B$. We know from (l) that $y_2 \in \xi_{e,D}(y_2)e - \operatorname{bd}(D)$, which means $\xi_{e,D}(y_2)e - y_2 \in \operatorname{bd}(D)$. Due to the assumption $\operatorname{bd}(D) + B \subset D$, we obtain that $\xi_{e,D}(y_2)e - y_2 + y_2 - y_1 \in D$, hence $\xi_{e,D}(y_2)e - y_1 \in D$. This means $\xi_{e,D}(y_1) \leq \xi_{e,D}(y_2)$. Thus, $\xi_{e,D}$ is monotone w. r. t. B. The remaining equivalence has been proved in (e).

(n) Let $\xi_{e,D}$ be finite-valued. Assume that $\xi_{e,D}$ is strictly monotone with respect to B. Take $y \in D$ and $b \in B \setminus \{\mathbf{0}\}$. It follows from (g) that $\xi_{e,D}(-y) \leq 0$. Because $\xi_{e,D}$ is strictly monotone with respect to B, we obtain from $(-y) - (-y - b) = b \in B \setminus \{\mathbf{0}\}$ that $\xi_{e,D}(-y - b) < \xi_{e,D}(-y) \leq 0$. Using (j), we obtain $y + b \in \operatorname{int}(D)$. Now let $\operatorname{bd}(D) + (B \setminus \{\mathbf{0}\}) \subset \operatorname{int}(D)$. Consider $y_2 - y_1 \in B \setminus \{\mathbf{0}\}$. Due to (l), we have $y_2 \in \xi_{e,D}(y_2)e - \operatorname{bd}(D)$, which means $\xi_{e,D}(y_2)e - y_2 \in \operatorname{bd}(D)$. Thus, by hypothesis, $\xi_{e,D}(y_2)e - y_2 + y_2 - y_1 \in \operatorname{int}(D)$, and thus $\xi_{e,D}(y_2)e - y_1 \in \operatorname{int}(D)$. That is $y_1 \in \xi_{e,D}(y_1)e - \operatorname{int}(D)$, implying $\xi_{e,D}(y_1) < \xi_{e,D}(y_2)$. The remaining implication follows since D is a closed set.

(o) Notice that the first part of this assertion was proven in (f). Let $\xi_{e,D}$ be proper, and let $\operatorname{bd}(D) + \operatorname{bd}(D) \subseteq D$. We consider $y_1, y_2 \in \operatorname{dom} \xi_{e,D}$. By (l), we obtain $y_i \in \xi_{e,D}(y_i)e - \operatorname{bd}(D)$, leading to

$$y_1 + y_2 \in (\xi_{e,D}(y_1) + \xi_{e,D}(y_2))\, e - \operatorname{bd}(D) - \operatorname{bd}(D) \subseteq (\xi_{e,D}(y_1) + \xi_{e,D}(y_2))\, e - D.$$

Thus, $\xi_{e,D}(y_1 + y_2) \leq \xi_{e,D}(y_1) + \xi_{e,D}(y_2)$. □

We summarize the properties of the functional $\xi_{e,B}$ for the case when $B = C$ is a given proper closed convex cone.

Corollary 2.1 ([15, Corollary 2.3.5.]) *Let $C \subset Y$ be a proper closed convex cone and $e \in \operatorname{int}(C)$. Then $\xi_{e,C}$, given in Definition 2.30, is finite-valued continuous, sublinear and strictly monotone with respect to C such that*

$$\xi_{e,C}(y) \leq r \Leftrightarrow y \in rk - C, \quad \textit{for all } y \in Y \textit{ and all } r \in \mathbb{R},$$

$$\xi_{e,C}(y) < r \Leftrightarrow y \in rk - \operatorname{int}(C), \quad \textit{for all } y \in Y \textit{ and all } r \in \mathbb{R}.$$

Lemma 2.13 ([25, Theorem 1.6]) *Let Y be a topological vector space which is partially ordered by a pointed convex cone C. Let X be a vector space with a nonempty subset K. A function $\varphi : K \to Y$ is C-quasiconvex (respectively, C-quasiconcave) if and only if the composite mapping $\xi_{e,D} \circ \varphi : K \to \mathbb{R}$ is $\mathbb{R}_+$-quasiconvex (respectively, $\mathbb{R}_+$-quasiconcave), where $\mathbb{R}_+ = \{x \in \mathbb{R} : x \geq 0\}$.*

Proof Let $\xi_{e,D} \circ \varphi$ be quasiconvex. Suppose that φ is not C-quasiconvex, that is, there exist some $\bar{\alpha} \in Y$, $x, y \in K$ and $t \in [0, 1]$ such that

$$\varphi(x), \varphi(y) \in \alpha - C \text{ and } \varphi(tx + (1-t)y) \notin \alpha - C.$$

By the definition of $\xi_{e,D}$ with $D := C - \alpha$, we have

$$\xi_{e,D}(\varphi(x)) = \inf\{t \in \mathbb{R} : \varphi(x) \in te - D\} = \inf\{t \in \mathbb{R} : \varphi(x) \in te - C + \alpha\} \leq 0$$

and

$$\xi_{e,D}(\varphi(y)) \leq 0,$$

while,

$$\xi_{e,D}(\varphi(tx + (1-t)y) > 0$$

which shows that $\xi_{e,D} \circ \varphi$ is not quasiconvex and so not $\mathbb{R}_+$-quasiconvex. The inverse implication of the assertion can be shown analogously. □

The following lemma relates (e, C)-lower (upper, respectively) semicontinuity of a vector-valued function φ (see Definition 2.27) to the lower (upper, respectively) semicontinuity of the composite mapping $\xi_{e,D} \circ \varphi$.

Lemma 2.14 ([30]) *Let Y be a topological vector space which is partially ordered by a pointed convex cone C. Let X be a vector space with a nonempty subset K. Consider a function $\varphi : K \to Y$ and the functional $\xi_{e,C}$ with $C \subset Y$ a closed convex pointed cone and $e \in \text{int}(C)$.*

(a) *If φ is (e, D)-lower semicontinuous and C-bounded below, then $\xi_{e,D} \circ \varphi$ is lower semicontinuous and bounded below.*
(b) *If φ is (e, D)-upper semicontinuous, then $\xi_{e,D} \circ \varphi$ is upper semicontinuous.*

We recall the following separation theorem for not necessarily convex sets.

Theorem 2.8 (Separation Theorem for Not Necessarily Convex Sets) [15, Theorem 2.3.6] *Let Y be a topological vector space, $D \subset Y$ be a closed proper set with nonempty interior, $A \subset Y$ be a nonempty set such that $A \cap (-\text{int}(D)) = \emptyset$ and $e \in Y$. Assume that one of the following two conditions holds:*

(i) *There is a cone $C \subset Y$ such that $e \in \text{int}(C)$ and $D + \text{int}(C) \subseteq D$;*
(ii) *D is convex, and (2.29) and $D + (0, \infty) \cdot e \subset \text{int}(D)$ are satisfied.*

Then $\xi_{e,D}$ *is a finite-valued continuous functional such that for all* $y \in A$, $d \in \operatorname{int}(D)$

$$\xi_{e,D}(y) \geq 0 > \xi_{e,D}(-d).$$

Moreover, $\xi_{e,D}(y) > 0$ *for every* $y \in \operatorname{int}(A)$.

Now we consider the nonlinear scalarizing functional with a variable ordering structure. Let X and Y be locally convex Hausdorff topological vector spaces. We consider a set-valued map $C : X \to 2^Y$ such that for each $x \in X$, $C(x)$ is a proper closed and convex cone with nonempty interior. Furthermore, let $e : X \to Y$ be a map. Suppose that for all $x \in X$, $e(x) \in \operatorname{int}(C(x))$. Let Y^* be the dual space of Y equipped with weak*-topology. Let $C^* : Y \to 2^{Y^*}$ be defined as

$$C^*(x) = \{\phi \in Y^* : \langle \phi, y \rangle \geq 0 \text{ for all } y \in C(x)\}, \quad \text{for all } x \in X.$$

Then the set

$$B^*(x) = \{\phi \in C^*(x) : \langle \phi, e(x) \rangle = 1\}$$

is a weak* compact base of the cone $C^*(x)$.

Definition 2.32 ([4]) The *nonlinear scalarization function* $\xi^v_{e,C} : X \times Y \to \mathbb{R}$ is defined as

$$\xi^v_{e,C}(x, y) = \inf\{\lambda \in \mathbb{R} : y \in \lambda e(x) - C(x)\}, \quad \text{for all } (x, y) \in Y \times Y.$$

Remark 2.18 Let P be a proper, closed and convex cone in Y with $\operatorname{int}(P) \neq \emptyset$, and $\hat{e} \in \operatorname{int}(P)$. If $C(x) = P$ is a constant set-valued map and $e(x) = \hat{e}$ a fixed vector in $\operatorname{int}(P)$ for all $x \in X$, then Definition 2.32 reduces to Definition 2.30.

In order to show that $\xi^v_{e,C}$ is well-defined, we first have to prove the following important proposition.

Proposition 2.34 ([12]) *For arbitrary* $x \in X$, *let* $C(x) \subset Y$ *be a solid cone. Then*

$$Y = \bigcup \{\operatorname{int}(C(x)) - \lambda e(x) : \lambda \in \mathbb{R}_+ \setminus \{0\}\},$$

for all $e(x) \in \operatorname{int}(C(x))$.

Proof For arbitrary $e(x) \in \operatorname{int}(C(x))$, we consider $V := \operatorname{int}(C(x)) - e(x)$. Of course, it holds $\mathbf{0} \in V$. Because $C(x)$ is a cone, we obtain

$$\lambda V = \lambda \operatorname{int}(C(x)) - \lambda e(x) \subseteq \operatorname{int}(C(x)) - \lambda e(x), \quad \text{for all } \lambda \in \mathbb{R}_+ \setminus \{0\}.$$

Consequently,

$$\bigcup \{\lambda V : \lambda \in \mathbb{R}_+ \setminus \{0\}\} \subseteq \bigcup \{\operatorname{int}(C(x)) - \lambda e(x) : \lambda \in \mathbb{R}_+ \setminus \{0\}\}$$

and because of $\mathbf{0} \in V$, we obtain

$$\bigcup\{\lambda V : \lambda \in \mathbb{R}_+\} \subseteq \bigcup\{\text{int}(C(x)) - \lambda e : \lambda \in \mathbb{R}_+ \setminus \{0\}\}.$$

By Proposition 1.1, we obtain

$$Y = \bigcup\{\lambda V : \lambda \in \mathbb{R}_+\},$$

and therefore,

$$Y = \bigcup\{\text{int}(C(x)) - \lambda e(x) : \lambda \in \mathbb{R}_+ \setminus \{0\}\}. \qquad \square$$

Example 2.33 For $Y = \mathbb{R}^2$, the constant cone $C = \mathbb{R}^2_+$ and fixed $e = (1,0) \notin \text{int}(C)$, we have

$$\bigcup\{\text{int}(\mathbb{R}^2_+) - \lambda(1,0) : \lambda \in \mathbb{R}_+ \setminus \{0\}\} = \{y \in \mathbb{R}^2 : y_2 > 0\} \neq \mathbb{R}^2.$$

Proposition 2.35 *The function* $\xi^v_{e,C} : X \times Y \to \mathbb{R}$ *is well defined.*

Proof We show that for any fixed $x \in X$ and $e(x) \in \text{int}(C(x))$, the set

$$A_{e(x)}(y) = \{t \in \mathbb{R} : y \in \alpha + te(x) - C(x)\}$$

is nonempty, closed and bounded from below for each $y \in Y$.

Since $\mathbf{0} \in e(x) - \text{int}(C(x))$, for each fixed $\alpha \in Y$, there is a positive number t such that

$$\frac{y}{t} \in e(x) - \text{int}(C(x)),$$

that is, $y \in te(x) - \text{int}(C(x)) \subseteq te(x) - C(x)$, and so $t \in A_{e(x)}(y)$. Thus, $A_{e(x)}(y)$ is nonempty for all $y \in Y$.

To show that $A_{e(x)}(y)$ is closed for all $y \in Y$, we let $\{t_n\} \subset \zeta_{e(x)}(y)$ be a sequence in $A_{e(x)}(y)$ such that $t_n \to t$ as $n \to \infty$. Then $y \in t_n e(x) - C(x)$ for all $n \in \mathbb{N}$, that is, $t_n e(x) - y \in C(x)$. Since $C(x)$ is closed, we have $te(x) - y \in C(x)$. Thus, $t \in A_{e(x)}(y)$ and so $A_{e(x)}(y)$ is closed.

Now we show that $A_{e(x)}$ is bounded from below. We have to show that for each $t \in \mathbb{R}$, there exists some $\lambda < t$ and $y \in \lambda e(x) - C(x)$. We proceed in two steps. At first we show that $\lambda \in A_{e(x)}(y)$ implies $\mu \in A_{e(x)}(y)$ for all $\mu > \lambda$. This is true because for $\lambda \in A_{e(x)}(y)$, we have

$$\begin{aligned} y \in \mu e(x) &= y - \lambda e(x) + (\lambda - \mu)e(x) \\ &\in -C(x) - \text{int}(C(x)) \subseteq -\text{int}(C(x)) \subseteq -C(x). \end{aligned} \tag{2.30}$$

Therefore, $\mu \in A_{e(x)}$.

Next we show that for all $y \in Y$, there is some $\lambda \in \mathbb{R}$ such that $\lambda \notin A_{e(x)}$. Now assume that there exists $y^0 \in Y$ with $\lambda \in A_{e(x)}(y^0)$ for all $\lambda \in \mathbb{R}$. Then due to (2.30), we have $y^0 \in -\operatorname{int}(C(x)) + \lambda e(x)$ for all $\lambda \in \mathbb{R}$. Hence, $\{y^0 - \lambda e(x) : \lambda \in \mathbb{R}\} \subseteq -\operatorname{int}(C(x))$. Because of Proposition 2.34, we have

$$Y = \bigcup \{\operatorname{int}(C(x)) - \lambda e(x) : \lambda \in \mathbb{R}_+ \setminus \{0\}\}.$$

Moreover, because Y is a vector space, we have $-y \in Y$. Therefore, for all $y \in Y$, there is some $c^0 \in \operatorname{int}(C(x))$ and some $\lambda^0 \in \mathbb{R}_+ \setminus \{0\}$ such that $-y = c^0 - \lambda^0 e(x)$. Hence,

$$\begin{aligned} y &= -c^0 + \lambda^0 e(x) + y^0 - y^0 \\ &= y^0 + \lambda^0 e(x) - c^0 - y^0 \\ &\in -\operatorname{int}(C(x)) - \operatorname{int}(C(x)) - y^0. \end{aligned}$$

Consequently, $y \in -C(x) - y^0$ for all $y \in Y$, and we obtain $Y \subseteq -C(x) - y^0$, which implies $Y = C(x)$, a contradiction. Thus, for all $y \in Y$ there is some $\lambda \in \mathbb{R}$ such that $\lambda \notin A_{e(x)}$.

In conclusion, this shows that for all $t \in \mathbb{R}$, there is some $\lambda_t < t$ such that $y \in \lambda_t e(x) - C(x)$. Hence, $\xi^v_{e,C}$ is well defined. □

The following result is important for the numerical treatment of the nonlinear scalarizing functional $\xi^v_{e,C}$.

Proposition 2.36 ([4]) *For any* $(x, y) \in X \times Y$,

$$\xi^v_{e,C}(x, y) = \max_{\phi \in B^*(x)} \frac{\langle \phi, y \rangle}{\langle \phi, e(x) \rangle},$$

where $B^*(x)$ *is a base for* $C^*(x)$.

Proof We first show that

$$\xi^v_{e,C}(x, y) = \sup_{\phi \in C^*(x) \setminus \{\mathbf{0}\}} \frac{\langle \phi, y \rangle}{\langle \phi, e(x) \rangle}.$$

Due to the definition of $\xi^v_{e,C}$, it holds $\xi^v_{e,C}(x, y)e(x) - y \in C(x)$. So, for any $\phi \in C^*(x) \setminus \{\mathbf{0}\} \subset C^*(x)$, we have $\langle \phi, \xi^v_{e,C}(x, y)e(x) - y \rangle \geq 0$, or equivalently, $\xi^v_{e,C}(x, y)\langle \phi, e(x) - y \rangle \geq 0$. Because $e(x) \in \operatorname{int}(C(x))$ and $\phi \in C^*(x) \setminus \{\mathbf{0}\}$, it holds $\langle \phi, e(x) \rangle > 0$. Thus, we obtain $\xi^v_{e,C}(x, y) \geq \frac{\langle \phi, y \rangle}{\langle \phi, e(x) \rangle}$, resulting in

$$\xi^v_{e,C}(x, y) \geq \sup_{\phi \in \operatorname{int} C^*(x) \setminus \{\mathbf{0}\}} \frac{\langle \phi, y \rangle}{\langle \phi, e(x) \rangle}.$$

Now let

$$\lambda_0 := \sup_{\phi \in C^*(x)\setminus\{\mathbf{0}\}} \frac{\langle \phi, y\rangle}{\langle \phi, e(x)\rangle}.$$

Then for any $\phi \in C^*(x) \setminus \{\mathbf{0}\}$, $\lambda_0 \geq \frac{\langle \phi, y\rangle}{\langle \phi, e(x)\rangle}$. Because $\langle \phi, e(x)\rangle > 0$, we have $\langle \phi, \lambda_0 e(x) - y\rangle \geq 0$. Then $\lambda_0 e(x) - y \in C(x)$. This means $y \in \lambda_0 e(x) - C(x)$, and by the definition of $\xi^v_{e,C}$, we immediately obtain $\xi^v_{e,C}(x, y) \leq \lambda_0$, i.e.,

$$\xi^v_{e,C}(x, y) \leq \sup_{\phi \in C^*(x)\setminus\{\mathbf{0}\}} \frac{\langle \phi, y\rangle}{\langle \phi, e(x)\rangle}.$$

Consequently, we get

$$\xi^v_{e,C}(x, y) = \sup_{\phi \in C^*(x)\setminus\{\mathbf{0}\}} \frac{\langle \phi, y\rangle}{\langle \phi, e(x)\rangle}.$$

Because $B^*(x)$ is a base for $C^*(x)$, for any $x \in Y$ and $\phi \in C^*(x) \setminus \{\mathbf{0}\}$, there exist some $\lambda > 0$ and $\varphi \in B^*(x)$ such that $\phi = \lambda\varphi$. So we obtain for all $x \in Y$,

$$\frac{\langle \phi, y\rangle}{\langle \phi, e(x)\rangle} = \frac{\langle \lambda\varphi, y\rangle}{\langle \lambda\varphi, e(x)\rangle} = \frac{\langle \varphi, y\rangle}{\langle \varphi, e(x)\rangle}.$$

Thus, we have

$$\xi^v_{e,C}(x, y) = \sup_{\phi \in C^*(x)\setminus\{\mathbf{0}\}} \frac{\langle \phi, y\rangle}{\langle \phi, e(x)\rangle} = \sup_{\phi \in B^*(x)} \frac{\langle \phi, y\rangle}{\langle \phi, e(x)\rangle}.$$

Finally, since $B^*(x)$ is weakly* compact, we conclude with

$$\xi^v_{e,C}(x, y) = \max_{\phi \in B^*(x)} \frac{\langle \phi, y\rangle}{\langle \phi, e(x)\rangle}.$$

□

The following results provide some important properties of the nonlinear scalarization function $\xi^v_{e,C}$.

Lemma 2.15 ([3]) *Let X and Y be locally convex Hausdorff topological vector spaces, $C : X \to 2^Y$ a set-valued map such that for all $x \in X$, $C(x)$ is a proper, pointed, closed and convex cone in Y with $\mathrm{int}(C(x)) \neq \emptyset$. Let $e : X \to Y$ be a vector-valued map such that for any $x \in X$, $e(x) \in \mathrm{int}(C(x))$. For each $\lambda \in \mathbb{R}$ and $(x, y) \in X \times Y$, we have*

(a) $\xi^v_{e,C}(x, y) < \lambda \Leftrightarrow y \in \lambda e(x) - \mathrm{int}(C(x))$;
(b) $\xi^v_{e,C}(x, y) \leq \lambda \Leftrightarrow y \in \lambda e(x) - C(x)$;

(c) $\xi^{v}_{e,C}(x,y) \geq \lambda \Leftrightarrow y \notin \lambda e(x) - \mathrm{int}(C(x))$;
(d) $\xi^{v}_{e,C}(x,y) > \lambda \Leftrightarrow y \notin \lambda e(x) - C(x)$;
(e) $\xi^{v}_{e,C}(x,y) = \lambda \Leftrightarrow y \in \lambda e(x) - \mathrm{bd}(C(x))$,

where $\mathrm{bd}(C(x))$ *denotes the topological boundary of* $C(x)$.

Proof We only prove part (a), as the remaining assertions can be proven in a similar way. Due to Proposition 2.36, it holds

$$\begin{aligned}
\xi^{v}_{e,C}(x,y) < \lambda \Leftrightarrow\ & \max_{\phi \in B^*(x)} \frac{\langle \phi, y\rangle}{\langle \phi, e(x)\rangle} < \lambda \\
\Leftrightarrow\ & \forall\, \phi \in B^*(x) : \ \langle \phi, y\rangle < \lambda \langle \phi, e(x)\rangle \\
\Leftrightarrow\ & \forall\, \phi \in B^*(x) : \ \langle \phi, \lambda e(x) - y\rangle > 0 \\
\Leftrightarrow\ & \forall\, \phi \in C^*(x) \setminus \{\mathbf{0}\} : \ \langle \phi, \lambda e(x) - y\rangle > 0 \\
\Leftrightarrow\ & \lambda e(x) - y \in \mathrm{int}(C(x)) \\
\Leftrightarrow\ & y \in \lambda e(x) - \mathrm{int}(C(x)).
\end{aligned}$$
□

Proposition 2.37 ([3]) *Let X and Y be a locally convex Hausdorff topological vector spaces. Then for any given $x \in X$,*

(a) $\xi^{v}_{e,C}(x,\cdot)$ *is positive homogeneous;*
(b) $\xi^{v}_{e,C}(x,\cdot)$ *is strictly monotone, that is, if $y_1 >_C y_2$, then $\xi^{v}_{e,C}(x,y_2) < \xi^{v}_{e,C}(x,y_1)$.*

Proof

(a) Let $\lambda > 0$. For $y \in Y$, we have

$$\begin{aligned}
\xi^{v}_{e,C}(x,\lambda y) &= \max_{\phi \in B^*(x)} \frac{\langle \phi, \lambda y\rangle}{\langle \phi, e(x)\rangle} \\
&= \lambda \max_{\phi \in B^*(x)} \frac{\langle \phi, y\rangle}{\langle \phi, e(x)\rangle} \\
&= \lambda \xi^{v}_{e,C}(x,y).
\end{aligned}$$

(b) Let $y_1 >_C y_2$ and set $r = \xi^{v}_{e,C}(x,y_1)$. Then by the definition of $\xi^{v}_{e,C}(x,y_1)$, we have

$$y_2 \in y_1 - \mathrm{int}(C) \subset re(x) - C(x) - \mathrm{int}(C(x)) \subset re(x) - \mathrm{int}(C(x)).$$

By Proposition 2.15 (a), we have

$$\xi^{v}_{e,C}(x,y_2) < r = \xi^{v}_{e,C}(x,y_1).$$
□

Proposition 2.38 ([3]) *For any fixed $x \in X$, and any $y_1, y_2 \in Y$, the following assertions hold:*

(a) $\xi^v_{e,C}(x, y_1 + y_2) \leq \xi^v_{e,C}(x, y_1) + \xi^v_{e,C}(x, y_2)$;
(b) $\xi^v_{e,C}(x, y_1 - y_2) \geq \xi^v_{e,C}(x, y_1) - \xi^v_{e,C}(x, y_2)$.

Proof

(a) It holds

$$\begin{aligned}\xi^v_{e,C}(x, y_1 + y_2) &= \max_{\phi \in B^*(x)} \frac{\langle \phi, y_1 + y_2 \rangle}{\langle \phi, e(x) \rangle} \\ &\leq \max_{\phi \in B^*(x)} \frac{\langle \phi, y_1 \rangle}{\langle \phi, e(x) \rangle} + \max_{\phi \in B^*(x)} \frac{\langle \phi, y_2 \rangle}{\langle \phi, e(x) \rangle} \\ &= \xi^v_{e,C}(x, y_1) + \xi^v_{e,C}(x, y_2).\end{aligned}$$

(b) It follows from (a) that

$$\xi^v_{e,C}(x, y_1) = \xi^v_{e,C}(x, y_1 - y_2 + y_2) \leq \xi^v_{e,C}(x, y_1 - y_2) + \xi^v_{e,C}(x, y_2).$$

Then $\xi^v_{e,C}(x, y_1) - \xi^v_{e,C}(x, y_2) \leq \xi^v_{e,C}(x, y_1 - y_2)$, and hence (b) holds. □

Remark 2.19 It is important to mention that there exist several other extensions of the nonlinear scalarizing functional $\xi_{e,D}$ introduced in Definition 2.30.

Hernández and Rodríguez-Marín [16] introduced the functional $Z_{e,C} : 2^Y \times 2^Y \to \mathbb{R} \cup \{+\infty\}$

$$Z_{e,C}(A, B) := \sup_{b \in B} \inf\{t \in \mathbb{R} : b \in te + A + C\},$$

where C is a closed pointed convex cone in Y and $\emptyset \neq A, B \in 2^Y$. Note that we have $Z_{e,C}(A, B) = \sup_{b \in B} \xi_{e,-(C+A)}(b)$ with the notations from Definition 2.30. Hernández and Rodríguez-Marín [16] used this functional in set-valued optimization for characterizing optimal solution sets of set optimization problems. They showed the equivalence

$$B \subseteq A + C \Leftrightarrow \text{ for some } e \in \operatorname{int}(C) : Z_{e,C}(A, B) \leq 0,$$

see [16, Thm. 3.10]. Thus, the functional $Z_{e,C}$ implicitly uses the lower set less order relation $A \preceq^l_C B :\Leftrightarrow B \subseteq A + C$ introduced by Kuroiwa [23, 24]. Moreover, Eichfelder [8, Sect. 5.2.1] (see also [6, 7]) studied the functional $\xi_{e,C-a} : Y \to \mathbb{R} \cup \{+\infty\}$

$$\xi_{e,C-a}(y) := \inf\{t \in \mathbb{R} : y \in te - (C(y) - a)\},$$

where $e \in \bigcap_{y \in Y} \operatorname{int}(C)(y) \setminus \{\mathbf{0}\}$ for a characterization of nondominated elements of a vector optimization problem equipped with a variable ordering structure.

2.7 Vector Conjugate

Let Y be a locally convex space with its dual Y^* and $C \subset Y$ be a proper closed solid convex cone with its dual C^*. Since $\operatorname{int}(C) \neq \emptyset$ and $C \neq Y$, we have $C^* \neq \{\mathbf{0}\}$, besides, C^* has a weakly* compact base, that is, there exists a convex, weak* compact set $B \subseteq C^*$ such that $\mathbf{0} \notin B$ and $C^* = \bigcup_{t \geq 0} tB$. We fix such a base and set

$$\sigma(u) := \max_{t \in B}(t, u), \quad \text{for all } u \in Y.$$

Then for all $u \in Y$,

$$\begin{aligned} u < 0 \Leftrightarrow \sigma(u) < 0; \quad & u \leq 0 \Leftrightarrow \sigma(u) \leq 0; \\ u \not< 0 \Leftrightarrow \sigma(u) \geq 0; \quad & u \not\leq 0 \Leftrightarrow \sigma(u) > 0. \end{aligned} \tag{2.31}$$

Let $g : X \to Y$ and $y \in X$. If a linear operator $l \in L(X, Y)$ satisfies

$$\langle l, z \rangle \not>_C g(z + y) - g(y), \quad \text{for all } z \in X,$$

then l is said to be a *weak subgradient of g at y* [34]. The set of all weak subgradients of g at y is denoted by $\partial^w g(y)$ and g is said to be *weakly subdifferentiable* at y [34] if $\partial^w g(y) \neq \emptyset$.

Let $A \subset Y$. Denote

$$\operatorname{Sup}_{\operatorname{int}(C)} A = \{u \in \operatorname{cl}(A) : (A - \{u\}) \cap \operatorname{int}(C) = \emptyset\}.$$

In the case that $Y = \mathbb{R}$ and $C = [0, \infty)$, we have

$$\operatorname{Sup}_{\operatorname{int}(C)} A = \begin{cases} \operatorname{Sup} A, & \text{if } \{u \in \operatorname{cl}(A) : (A - \{u\}) \cap \operatorname{int}(C) = \emptyset\} \neq \emptyset, \\ \infty, & \text{if } \{u \in \operatorname{cl}(A) : (A - \{u\}) \cap \operatorname{int}(C) = \emptyset\} = \emptyset. \end{cases}$$

Let $g : X \to Y$ and $l \in L(X, Y)$. The *vector conjugate function* [34], denoted by g^*_{sup}, of g at l is defined by

$$g^*_{sup}(l) = \operatorname{Sup}_{\operatorname{int}(C)} \{\langle l, y \rangle - g(y) : y \in X\}.$$

Let $y \in X$. The *vector biconjugate function*, denoted by g^{**}_{sup}, of g at y is defined by

$$g^{**}_{sup}(y) = \operatorname{Sup}_{\operatorname{int}(C)} \bigcup \{\langle l, y \rangle - g^*_{sup}(l) : l \in L(X, Y)\}.$$

Note that both g^*_{sup} and g^{**}_{sup} are set-valued maps and $g^*_{sup} : L(X,Y) \to 2^Y$, $g^{**}_{sup} : X \to 2^Y$. Throughout this section, we assume that $g^*_{sup}(l) \neq \emptyset$ and $g^{**}_{sup}(y) \neq \emptyset$.

Let $g : X \to Y$ and $y \in X$. g is said to be *externally stable at* y if $g(y) \in g^{**}_{sup}(y)$.

The external stability was introduced in [29] when the vector conjugate function is defined via the set of efficient points.

Lemma 2.16 *Let* $g : X \to Y$ *and* $y \in X$. *Then*

$$l \in \partial^w g(y) \quad \Leftrightarrow \quad \langle l, y\rangle - g(y) \in g^*_{sup}(l). \tag{2.32}$$

Proof It follows from the definitions of the vector conjugate function and the subgradient that $l \in \partial^w g(y)$ if and only if

$$\langle l, z\rangle \not>_C g(z+y) - g(y), \quad \text{for all } z \in X,$$

equivalently,

$$\langle l, z+y\rangle - g(z+y) \not>_C \langle l, y\rangle - g(y), \quad \text{for all } z \in X,$$

if and only if $\langle l, y\rangle - g(y) \in g^*_{sup}(l)$ as $y \in X$. □

Before we give an existence result for the weak subdifferential $\partial^w g(y)$, we need the following lemma.

Lemma 2.17 *Let* Y *be a locally convex space,* $B \subseteq Y$ *a set with nonempty interior and* $C \subset Y$ *a solid pointed cone. Let* $g : B \to Y$ *be* C*-convex and continuous at* $y \in \text{int}(B)$. *Then the epigraph of* g

$$\text{epi}(g) = \{(x,y) \in X \times Y : x \in B,\ y \in \{g(x)\} + C\}$$

has nonempty interior.

Proof Choose $z \in Y$ such that $z - g(y) \in \text{int}(C)$. Then $z - g(y) - 2M \subset C$ for some neighborhood M of $\mathbf{0}$ in Y. Since $y \in \text{int}(B)$ and g is continuous at y, there exists some neighborhood N of $\mathbf{0}$ in X such that $y+N \subset B$, and $g(y+N) \subset g(y)+M$. Then we have $z-g(y+N)-M \subset z-g(y)-M-M \subset C$. Therefore, $(y+N, z-M) \subset \text{epi}(g)$, and $\text{int}(\text{epi}(g))$ is nonempty. □

Theorem 2.9 *Let* Y *be a locally convex space,* $B \subseteq Y$ *a convex set with nonempty interior and* $C \subset Y$ *a solid pointed cone. Let* $g : B \to Y$ *be* C*-convex and continuous at* $y \in \text{int}(B)$. *Then there exists a weak subgradient* $l \in \partial^w g(y)$.

Proof Let $y \in \text{int}(B)$, and define $D := B - \{y\}$. Then $\mathbf{0} \in \text{int}(D)$. Let $\varphi(z) := g(y+z) - g(y)$. Then $\varphi(\mathbf{0}) = g(y) - g(y) = \mathbf{0}$, and φ is C-convex and continuous at $\mathbf{0}$. Therefore, the set $K := \{(z,x) \in (D \times Y) : x - \varphi(z) \in \text{int}(C)\}$ is nonempty (see Lemma 2.17) and convex (see Theorem 2.4). Since $(\mathbf{0}, \mathbf{0}) \notin K$, by a separation theorem for convex sets there exists nonzero $(-l_1, l_2) \in X^* \times Y^*$ with $-l_1(z) + l_2(x) \geq \mathbf{0}$ for all $(z,x) \in K$. If $l_1 = \mathbf{0}$, then $l_2(x) \geq \mathbf{0}$ for all $x \in Y$, contradicting

$(-l_1, l_2) \neq (\mathbf{0}, \mathbf{0})$. If $l_2 = \mathbf{0}$, then $-l_1(z) \geq \mathbf{0}$ for all $z \in D$. This, together with $\mathbf{0} \in \text{int}(D)$, shows that $l_1 = \mathbf{0}$ contradicting $(-l_1, l_2) \neq (\mathbf{0}, \mathbf{0})$; hence, $l_2 \neq \mathbf{0}$.

Since g is continuous at y, for all $z \in D$, we have $-l_1(z) + l_2(\varphi(z)) \geq \mathbf{0}$. A continuous linear mapping $l \in L(X, Y)$ is a weak subgradient of φ at $\mathbf{0}$ if for all $z \in D$, we have $\varphi(z) - l(z) \notin -\text{int}(C)$. If, for some $z \in D$, this does not hold and l is chosen to satisfy for all $z \in D$ the relation $l_2(l(z)) = l_1(z)$ (which is always possible since $l_2 \neq \mathbf{0}$), then we have $l_2(\varphi(z)) - l_1(z) < \mathbf{0}$ for this $z \in D$, a contradiction. Therefore, any $l \in L(X, Y)$ satisfying $l_2 l = l_1$ gives a weak subgradient to φ, and therefore to g. □

References

1. M. Bianchi, N. Hadjisavvas, S. Schaible, Vector equilibrium problems with generalized monotone bifunctions. J. Optim. Theory Appl. **92**(3), 527–542 (1997)
2. O. Chadli, Y. Chiang, S. Huang, Topological pseudomonotonicity and vector-valued equilibrium problems. J. Math. Anal. Appl. **270**, 435–450 (2002)
3. G.Y. Chen, X.X. Huang, X.Q. Yang, *Vector Optimization: Set-Valued and Variational Analysis* (Springer, Berlin/Heidelberg, 2005)
4. G.Y. Chen, X.Q. Yang, H. Yu, A nonlinear scalarization function and generalized vector quasi-equilibrium problems. J. Glob. Optim. **32**, 451–466 (2005)
5. M. Ehrgott, *Multicriteria Optimization*. Lecture Notes in Economics and Mathematical Sciences, vol. 491 (Springer, Berlin/Heidelberg, 2000)
6. G. Eichfelder, Variable Ordering Structures in Vector Optimization, Habilitation Thesis, University Erlangen-Nürnberg, 2011
7. G. Eichfelder, Variable ordering structures in vector optimization, in *Recent Developments in Vector Optimization*, ed. by Q.H. Ansari, J.-C. Yao (Springer, Heidelberg, 2012), pp. 95–126
8. G. Eichfelder, *Variable Ordering Structures in Vector Optimization* (Springer, Berlin/Heidelberg, 2014)
9. P.Gr. Georgiev, T. Tanaka, Vector-valued set valued variants of Ky Fan's inequality. J. Nonlinear Convex Anal. **1**(3), 245–254 (2000)
10. Chr. Gerstewitz (Tammer), Nichtkonvexe Dualität in der Vektoroptimierung. Wissensch. Zeitschr. TH Leuna-Mersebung. **25**, 357–364 (1983)
11. Chr. Gerstewitz (Tammer), E. Iwanow, Dualität für nichtkonvexe Vektoroptimierungsprobleme. Wiss. Z. Tech. Hochsch. Ilmenau **2**, 61–81 (1985)
12. Chr. Gerth (Tammer), P. Weidner, Nonconvex separation theorems and some applications in vector optimization. J. Optim. Theory Appl. **67**, 297–320 (1990)
13. X.H. Gong, K. Kimura, J.-C. Yao, Sensitivity analysis of strong vector equilibrium problems. J. Nonlinear Convex Anal. **9**(1), 83–94 (2008)
14. A. Göpfert, Chr. Tammer, C. Zălinescu, On the vectorial Ekeland's variational principle and minimal points in product spaces. Nonlinear Anal. **39**, 909–922 (2000)
15. A. Göpfert, Chr. Tammer, H. Riahi, C. Zălinescu, *Variational Methods in Partially Ordered Spaces* (Springer, Berlin/Heidelberg, 2003)
16. E. Hernández, L. Rodríguez-Marín, Nonconvex scalarization in set optimization with set-valued maps. J. Math. Anal. Appl. **325**(1), 1–18 (2007)
17. F. Heyde, Coherent risk measures and vector optimization, in *Multicriteria Decision Making and Fuzzy Systems: Theory, Methods and Applications*, ed. by K.-H. Küfer (Shaker-Verlag, Aachen, 2006), pp. 3–12
18. G. Jameson, *Ordered Linear Spaces* (Springer, Berlin, 1970)

19. V. Jeykumar, W. Oettli, M. Natividad, A solvability theorem for a class of quasiconvex mappings with applications to optimization. J. Math. Anal. Appl. **179**, 537–546 (1993)
20. A.A. Khan, Chr. Tammer, C. Zălinescu, *Set-valued Optimization: An Introduction with Applications* (Springer, Berlin/New York, 2015)
21. I.V. Konnov, J.-C. Yao, Existence of solutions for generalized vector equilibrium problems. J. Math. Anal. Appl. **233**, 328–335 (1999)
22. M.A. Krasnoselskiĭ, *Positive Solutions of Operator Equations*, ed. by L.F. Boron (P. Noordhoff Ltd., Groningen, 1964). Translated from the Russian by Richard E. Flaherty
23. D. Kuroiwa, The natural criteria in set-valued optimization, Sūrikaisekikenkyūsho Kōkyūroku. Res. Nonlinear Anal. Convex Anal., Kyoto **1031**, 85–90 (1997)
24. D. Kuroiwa, Some duality theorems of set-valued optimization with natural criteria, in *Proceedings of the International Conference on Nonlinear Analysis and Convex Analysis* (World Scientific, 1999), pp. 221–228
25. D.T. Luc, *Theory of Vector Optimization*. Lecture Notes in Economics and Mathematical Systems, vol. 319 (Springer, Berlin/ Heidelberg/New York, 1989)
26. A. Pascoletti, P. Serafini, Scalarizing vector optimization problems. J. Optim. Theory Appl. **42**, 499–524 (1984)
27. A.M. Rubinov, Sublinear operators and their applications. Russ. Math. Surv. **32**(4), 113–174 (1977). Akademiya Nauk SSSR i Moskovskoe Matematicheskoe Obshchestvo
28. P.H.M. Ruys, H.N. Weddepohl, Economic theory and duality, in *Convex Analysis and Mathematical Economics*, ed. by J. Kriens (Springer, Berlin/New York, 1979)
29. Y. Sawaragi, H. Nakayama, T. Tanino, *Theory of Multiobjective Optimization* (Academic Press Inc., Orlando, 1985)
30. Chr. Tammer, A generalization of ekellandz's variational priciple. Optimization **25**(2–3), 129–141 (1992)
31. T. Tanaka, Cone-convexity of vector-valued functions. Sci. Rep. Hirosaki Univ. **37**, 170–177 (1990)
32. A.P. Wierzbicki, Basis properties of scalarizing functionals for multiobjective optimization. Math. Opere Statist. Ser. Optim. **8**, 55–60 (1977)
33. K. Winkler, Aspekte mehrkriterieller Optimierung C(T)-wertiger Abbildungen, Dissertation, Martin-Luther-Universität Halle-Wittenberg, Germany, 2003
34. X.Q. Yang, Vector variational inequality and its duality. Nonlinear Anal. Theory Methods Appl. **21**, 869–877 (1993)
35. P.L. Yu, Cone convexity, cone extreme points, and nondominated solutions in decision problems with multiple objectives. J. Optim. Theory Appl. **14**, 319–377 (1974)

Chapter 3
Solution Concepts in Vector Optimization

Many applications require the optimization of multiple conflicting goals at the same time. Such a problem can be modeled as a vector optimization problem. The concept of vector optimization goes back to Edgeworth [4] and Pareto [16] who gave the definition of the standard optimality concept in multiobjective optimization. Vector optimization deals with the problem of finding efficient elements of a vector-valued function. In that sense, vector optimization generalizes the concept of scalar optimization. In scalar optimization, there is only one concept for efficiency which characterizes efficient elements, namely the solution which generates the smallest function value. But, due to the lack of a total order in $\mathbb{R}^\ell$ ($\ell \geq 2$), there are elements which cannot be compared, for example the vectors $(1, 0)$ and $(0, 1)$ are incomparable with respect to $\mathbb{R}^2_+$. Therefore, one has to define a different solution concept, and usually there does not exist one efficient solution, but a whole set of solutions. Moreover, different efficiency notions exist. For instance, we can define an element $\bar{x}$ for a vector optimization problem with an objective function $f : \mathbb{R}^n \to \mathbb{R}^\ell$ to be an efficient solution if there does not exist another element $x \in \mathbb{R}^n$ such that $f_i(x) \leq f_i(\bar{x})$ for all $i = 1, 2, \ldots, \ell$, and $f_j(x) < f_j(\bar{x})$ for some $j = 1, 2, \ldots, \ell$. A weaker notion is to define an element $\bar{x}$ to be a weakly efficient solution if there does not exist another element $x \in \mathbb{R}^n$ such that $f_i(x) < f_i(\bar{x})$ for all $i = 1, 2, \ldots, \ell$. Of course, it would be desirable to obtain a strongly (or an ideal) efficient solution $\bar{x}$ where $f_i(\bar{x}) \leq f_i(x)$ holds for all objectives $i = 1, 2, \ldots, \ell$ and for all $x \in \mathbb{R}^n$. The selection of the particular solution concept depends on the concrete application. In case the function f is linear, the reader may refer to Luc [15] for an introduction to multiobjective linear programming.

In this chapter, we discuss several solution concepts for a vector optimization problem. We present some existence results for solutions of vector optimization problems. Most of the definitions and results appearing in this chapter can be found in any standard book on vector optimization, see, for example, [3, 5, 11, 14, 17].

Q.H. Ansari et al., *Vector Variational Inequalities and Vector Optimization*, Vector Optimization, DOI 10.1007/978-3-319-63049-6_3

3.1 Optimality Notions

We have already defined the concept of minimal and maximal elements of a preordered set A which is not assumed to have a linear structure. If A is a subset of a preordered vector space, then Definition 2.7 is equivalent to the following definition.

Definition 3.1 ((Strong) Efficient Element) Let A be a nonempty subset of a preordered vector space Y with an ordering cone C.

(a) An element $\bar{y} \in A$ is called a *strong* (or *ideal*) *efficient element* or *strong* (or *ideal*) *minimal element* of the set A (with respect to C) if

$$A \subset \{\bar{y}\} + C, \quad \text{equivalently,} \quad \bar{y} \leq_C y \quad \text{for all } y \in A. \tag{3.1}$$

We denote by $\mathbb{SE}(A, C)$ the set of all strong efficient elements of A with respect to C.

(b) An element $\bar{y} \in A$ is called a *strong* (or *ideal*) *maximal element* of the set A if

$$A \subset \{\bar{y}\} - C, \quad \text{equivalently,} \quad y \leq_C \bar{y} \quad \text{for all } y \in A. \tag{3.2}$$

(c) An element $\bar{y} \in A$ is called an *efficient element* or *minimal element* of the set A (with respect to C) if

$$(\{\bar{y}\} - C) \cap A \subset \{\bar{y}\} + C. \tag{3.3}$$

The set of all efficient elements of A with respect to C is denoted by $\mathbb{E}(A, C)$.

(d) An element $\bar{y} \in A$ is called a *maximal element* of the set A if

$$(\{\bar{y}\} + C) \cap A \subset \{\bar{y}\} - C. \tag{3.4}$$

Since every (strong) maximal element of A is a (strong) efficient element with respect to the preordering induced by the convex cone $(-C)$, without loss of generality it is sufficient to study the (strong) efficiency notion.

If the ordering cone C is pointed, then the inclusions (3.3) and (3.4) can be replaced, respectively, by the following relations:

$$(\{\bar{y}\} - C) \cap A = \{\bar{y}\}, \quad \text{equivalently,} \quad y \leq_C \bar{y},\ y \in A \Rightarrow y = \bar{y}. \tag{3.5}$$

and

$$(\{\bar{y}\} + C) \cap A = \{\bar{y}\}, \quad \text{equivalently,} \quad \bar{y} \leq_C y,\ y \in A \Rightarrow y = \bar{y}. \tag{3.6}$$

In other words, an element $\bar{y} \in A$ is said to be an *efficient element* or *minimal element* (respectively, *maximal element*) of the set A if there is no $y \in A$ with $y \neq \bar{y}$ and $y \leq_C \bar{y}$ (respectively, $\bar{y} \leq_C y$).

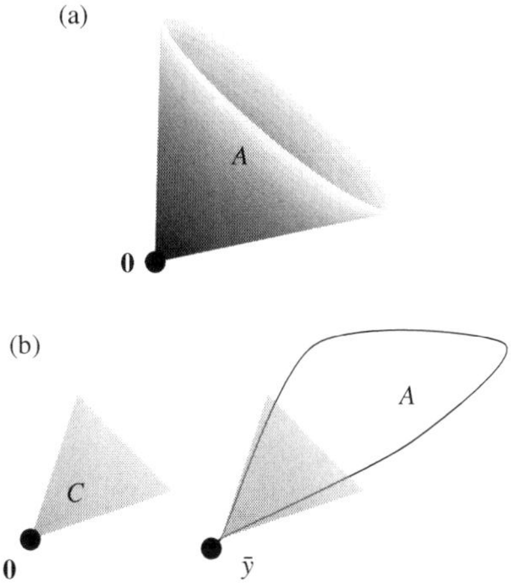

Fig. 3.1 (**a**) The point (0, 0, 0) is a strong efficient element of the set A, where the ordering cone is $C := \mathbb{R}^3_+$. (**b**) The element $\bar{y}$ is a strong efficient element of the set A with the given ordering cone C

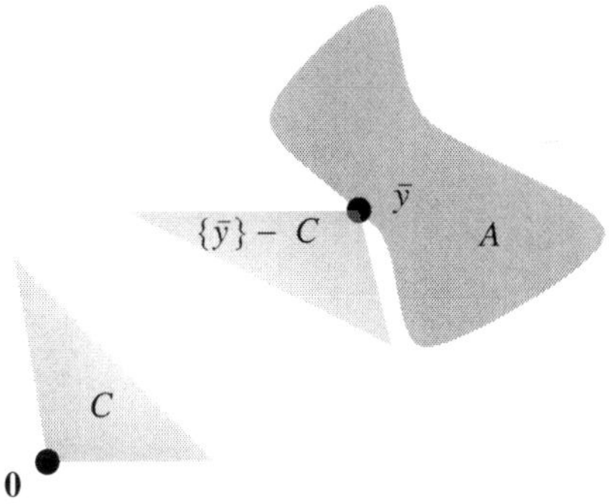

Fig. 3.2 Illustration of an efficient element $\bar{y}$ of the set A

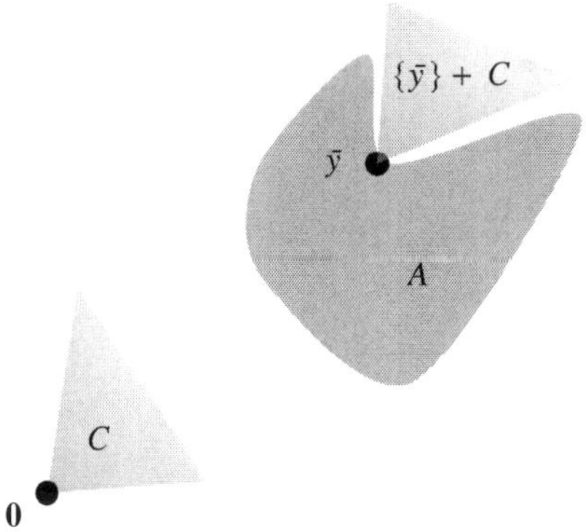

Fig. 3.3 Illustration of a maximal element $\bar{y}$ of the set A

Figures 3.1, 3.2 and 3.3 display a strong efficient element, an efficient element and a maximal element, respectively.

Example 3.1 Consider the set

$$A = \{(y_1, y_2) \in \mathbb{R}^2 : 2y_1^2 + y_2^2 \leq 3\}$$

Fig. 3.4 Efficient and maximal elements of the set $A := \{(y_1, y_2) \in \mathbb{R}^2 : 2y_1^2 + y_2^2 \leq 3\}$, where the ordering cone is given by $\mathbb{R}^2_+$

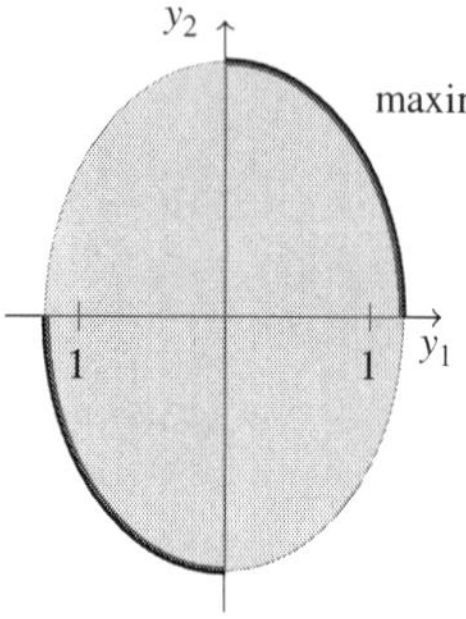

with the ordering cone $C = \mathbb{R}^2_+$ (see Fig. 3.4). The set of all maximal elements of A is

$$\left\{(y_1, y_2) \in \mathbb{R}^2 : y_1 \in \left[0, \sqrt{\frac{3}{2}}\right] \text{ and } y_2 = \sqrt{3 - 2y_1^2}\right\}$$

and the set of all efficient elements of A is

$$\left\{(y_1, y_2) \in \mathbb{R}^2 : y_1 \in \left[-\sqrt{\frac{3}{2}}, 0\right] \text{ and } y_2 = -\sqrt{3 - 2y_1^2}\right\}.$$

Remark 3.1

(a) For $C = \mathbb{R}_+$, which induces the usual ordering on $\mathbb{R}$, the concepts of efficient and strong efficient elements coincide, and agree with the usual definition of the minimum element of a set in $\mathbb{R}$.

(b) If A is a subset of a partially ordered vector space, then every strong efficient element of the set A is also an efficient element of A.

(c) Consider the cone $C = \mathbb{R}^2_+$, which induces the componentwise ordering in $\mathbb{R}^2$. The inequality $\bar{y} \leq_C y$ means y is above and to the right of $\bar{y}$. To say that $\bar{y} \in A \subseteq \mathbb{R}^2$ is a strong efficient element of a set A means that all other points of A lie above and to the right. To say that $\bar{y} \in A$ is an efficient element of a set A means that no other point of A lies to the left and below $\bar{y}$.

Later, we will see that to determine a solution of vector optimization problem is nothing else than to find the efficient element of the image set $f(A)$.

The following lemma shows that the efficient elements of a set A and the efficient elements of the set $A + C$ are closely related, where C is the ordering cone.

Lemma 3.1 *Let A be a nonempty subset of a preordered vector space Y with an ordering cone C.*

(a) *If the ordering cone C is pointed, then every efficient element of the set $A + C$ is also an efficient element of the set A.*

(b) *Every efficient element of the set A is also an efficient element of the set $A + C$.*

Proof

(a) Let $\bar{y} \in A + C$ be an arbitrary efficient element of the set $A + C$. Then we have $(\{\bar{y}\} - C) \cap (A + C) = \{\bar{y}\}$, because the cone C is assumed to be pointed. We suppose that $\bar{y} \notin A$. Then there exists an element $y \in A$ such that $y \neq \bar{y}$ and $y \in \{\bar{y}\} - C$. Consequently, we obtain

$$y \in (\{\bar{y}\} - C) \cap (A + C),$$

because $\{\bar{y}\} = (\{\bar{y}\} - C) \cap (A + C)$. But this contradicts the assumption that $\bar{y}$ is an efficient element of the set $A + C$. Hence, we obtain $\bar{y} \in A \subset A + C$, and therefore, $\bar{y}$ is also an efficient element of the set A.

(b) Let $\bar{y}$ be an efficient element of the set A. Then

$$(\{\bar{y}\} - C) \cap A \subset \{\bar{y}\} + C.$$

Now suppose that $\bar{y}$ is not an efficient element of the set $A + C$. Then we choose $y \in (\{\bar{y}\} - C) \cap (A + C) \neq \emptyset$. Assume that $y \notin \{\bar{y}\} + C$. Then there are elements $z \in A$ and $c \in C$ such that $y = z + c$. Consequently, we obtain $z = y - c \in \{\bar{y}\} - C$, and since $\bar{y}$ is an efficient element of the set A, we conclude $z \in \{\bar{y}\} + C$. But then we get $y \in \{\bar{y}\} + C$, a contradiction. □

Definition 3.2 (Properly Efficient Element) Let Y be an ordered normed space, A be a nonempty subset of Y and C be a pointed convex cone in Y which induces the partial ordering on Y.

(a) An element $\bar{y} \in A$ is called a *properly minimal element* or *properly efficient element* of the set A if $\bar{y}$ is an efficient element of the set A and the zero element $\mathbf{0}$ is an efficient element of the contingent cone $T(A + C, \bar{y})$. The set of all properly efficient elements of A with respect to C is denoted by $\mathbb{PE}(A, C)$.

(b) An element $\bar{y} \in A$ is called a *properly maximal element* of the set A if $\bar{y}$ is a maximal element of the set A and the zero element $\mathbf{0}$ is a maximal element of the contingent cone $T(A - C, \bar{y})$.

It is obvious that a properly efficient element of a set A is also an efficient element of A. For a visualization of an efficient element which is not properly efficient, consider the illustration in Fig. 3.5.

Definition 3.3 (Weakly Efficient Element) Let A be a nonempty subset of a preordered vector space Y with an ordering cone C which has a nonempty core.

(a) An element $\bar{y} \in A$ is called a *weakly efficient element* or *weakly minimal element* of the set A if

$$(\{\bar{y}\} - \operatorname{cor}(C)) \cap A = \emptyset. \tag{3.7}$$

The set of all weakly efficient elements of A with respect to C is denoted by $\mathbb{WE}(A, C)$.

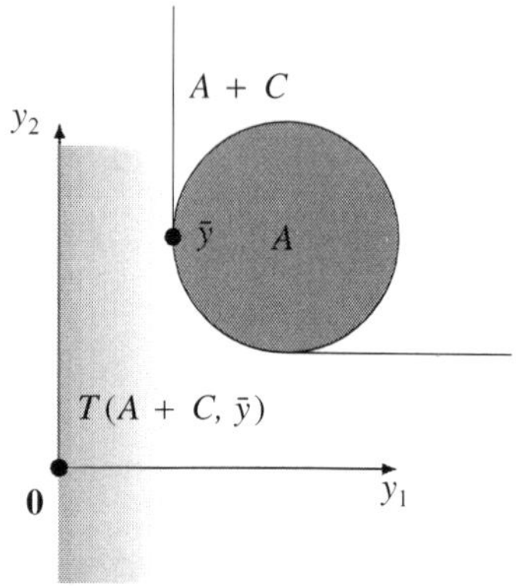

Fig. 3.5 Let $C = \mathbb{R}^2_+$. $\bar{y} \in A$ is an efficient element of the set A, but $\bar{y}$ is not properly efficient, as $\mathbf{0}$ is not an efficient element of the contingent cone $T(A + C, \bar{y})$

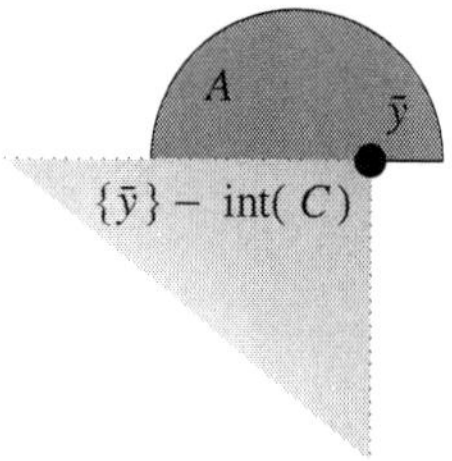

Fig. 3.6 A weakly efficient element $\bar{y}$ of a set A, where the ordering cone is given by $C = \mathbb{R}^2_+$

(b) An element $\bar{y} \in A$ is called a *weakly maximal element* of the set K if

$$(\{\bar{y}\} + \text{cor}(C)) \cap A = \emptyset. \tag{3.8}$$

In a more general setting, if A is a nonempty subset of a partially ordered topological vector space Y with a pointed closed convex solid cone C, then an element $\bar{y} \in A$ is called a *weakly minimal element* or *weakly efficient element* of the set A if $\bar{y}$ is an efficient element of A ordered by the cone $\tilde{C} = \{\mathbf{0}\} \cup \text{int}\, C$. An element $\bar{y} \in A$ is called a *weakly maximal element* of the set A if $\bar{y}$ is a maximal element of A ordered by the cone $\tilde{C} = \{\mathbf{0}\} \cup \text{int}\, C$.

A visualization of a weakly efficient element of a set A with the ordering cone $C = \mathbb{R}^2_+$ is given in Fig. 3.6.

The next lemma is similar to Lemma 3.1 and therefore, we omit the proof. It can be found in [11, pp. 110].

Lemma 3.2 *Let A be a nonempty subset of a preordered vector space Y with an ordering cone C which has a nonempty core. Then $\bar{y} \in A$ is a weakly efficient element of the set A if and only if it is a weakly efficient element of the set $A + C$.*

The following lemma gives the relationship between efficient and weakly efficient elements of a set.

Lemma 3.3 *Let A be a nonempty subset of a preordered vector space Y with an ordering cone C for which $C \neq Y$ and* $\text{cor}(C) \neq \emptyset$. *Then every efficient element of the set A is also a weakly efficient element of the set A.*

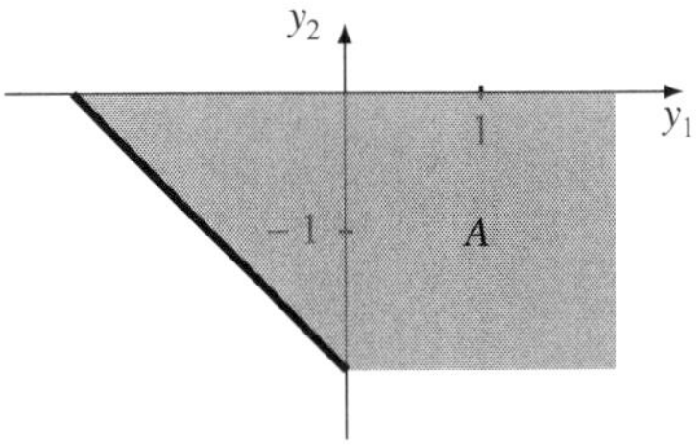

Fig. 3.7 The set of efficient elements is not equal to the set of weakly efficient element. Thus, we have $\mathbb{E}(A, C) \subsetneqq \mathbb{WE}(A, C)$ (see Example 3.2)

Proof Since $C \neq Y$, we have $\big(-\operatorname{cor}(C)\big) \cap C = \emptyset$. Therefore, for an arbitrary efficient element $\bar{y}$ of A, we have

$$\begin{aligned}\emptyset &= (\{\bar{y}\} - \operatorname{cor}(C)) \cap (\{\bar{y}\} + C)\\ &= (\{\bar{y}\} - \operatorname{cor}(C)) \cap (\{\bar{y}\} - C) \cap A\\ &= (\{\bar{y}\} - \operatorname{cor}(C)) \cap A\end{aligned}$$

that is, $\bar{y}$ is also a weakly efficient element of A. □

The following example shows that the converse of the above lemma is not true in general.

Example 3.2 Consider the set

$$A = \big\{(y_1, y_2) \in \mathbb{R}^2 : -y_1 - 2 \le y_2, \ -2 \le y_2 \le 0\big\}$$

in $Y = \mathbb{R}^2$ with the natural ordering cone $C = \mathbb{R}^2_+$ (see Fig. 3.7). There are no strong efficient elements of the set A. The set $\mathbb{E}(A, C)$ of all efficient elements of A is given by

$$\mathbb{E}(A, C) = \big\{(y_1, y_2) \in \mathbb{R}^2 : y_2 = -y_1 - 2, \ -2 \le y_2 \le 0\big\}.$$

The set $\mathbb{WE}(A, C)$ of all weakly efficient elements of A is

$$\mathbb{WE}(A, C) = \mathbb{E}(A, C) \cup \big\{(y_1, y_2) \in \mathbb{R}^2 : y_2 = -2, \ 0 \le y_1\big\}.$$

Consequently, we have $\mathbb{E}(A, C) \subsetneqq \mathbb{WE}(A, C)$.

Example 3.3 Consider the set

$$A = \big\{(y_1, y_2) \in \mathbb{R}^2 : y_1 - 2 \le y_2, \ -2 \le y_2 \le 0, \ y_1 \ge -2\big\}$$

in $Y = \mathbb{R}^2$ with the natural ordering cone $C = \mathbb{R}^2_+$. There is one strong efficient elements of the set A which is the only efficient element in A, namely, $(-2, -2)$. The set $\mathbb{WE}(A, C)$ of all weakly efficient elements of A is given by $\mathbb{WE}(A, C)$

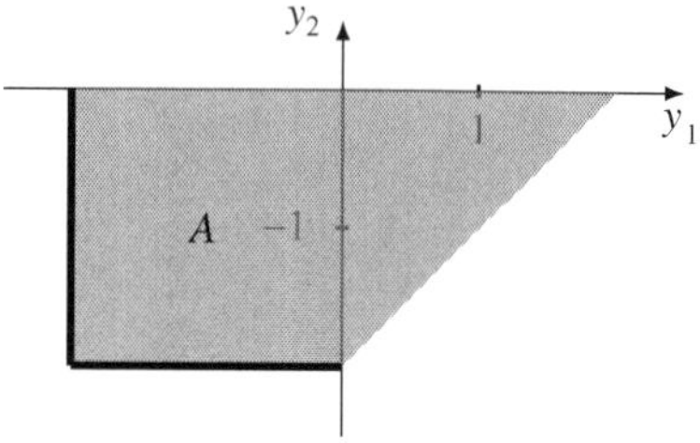

Fig. 3.8 There is only one efficient element $(-2, -2)$, and the set of weakly efficient element is given by $\{(y_1, y_2) \in \mathbb{R}^2 : -2 \le y_1 \le 0,\ y_2 = -2\} \cup \{(y_1, y_2) \in \mathbb{R}^2 : -2 \le y_2 \le 0,\ y_1 = -2\}$. Thus, we have $\mathbb{E}(A, C) \subsetneqq \mathbb{WE}(A, C)$

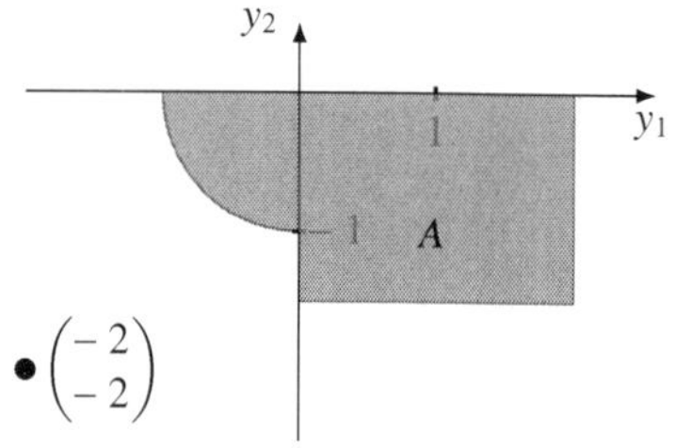

Fig. 3.9 The feasible objective region A given in Example 3.4

$= \big\{(y_1, y_2) \in \mathbb{R}^2 : -2 \le y_1 \le 0,\ y_2 = -2\big\} \cup \{(y_1, y_2) \in \mathbb{R}^2 : -2 \le y_2 \le 0,\ y_1 = -2\}$ (Fig. 3.8).

Example 3.4 Consider the set

$$A = \big\{(y_1, y_2) \in \mathbb{R}^2 : y_1^2 + y_2^2 \le 1,\ y_2 \le 0\big\} \bigcup \big\{(y_1, y_2) \in \mathbb{R}^2 : 0 \le y_1 \le 2,\ -1.5 \le y_2 \le 0\big\}$$

and

$$D = A \cup \{(-2, -2)\}$$

with the ordering cone $C = \mathbb{R}^2_+$ (see, Fig. 3.9). Then

$$\mathbb{SE}(D, C) = \mathbb{PE}(A, D) = \mathbb{E}(D, C) = \mathbb{WE}(D, C) = \{(-2, -2)\}.$$

We also have

$$\begin{aligned} \mathbb{SE}(A, C) &= \emptyset; \\ \mathbb{PE}(A, C) &= \{(y_1, y_2) \in \mathbb{R}^2 : y_1^2 + y_2^2 = 1,\ y_1, y_2 < 0\} \cup \{(0, -1.5)\}; \\ \mathbb{E}(A, C) &= \mathbb{PE}(A, C) \cup \{(-1, 0)\}; \\ \mathbb{WE}(A, C) &= \mathbb{E}(A, C) \cup \{(y_1, y_2) \in \mathbb{R}^2 : y_2 = -1.5,\ 0 \le y_1 \le 2\} \\ &\quad \cup \{(y_1, y_2) \in \mathbb{R}^2 : y_1 = 0,\ -1.5 \le y_2 \le -1\}. \end{aligned}$$

Consequently, we conclude

$$\mathbb{PE}(A, C) \subsetneq \mathbb{E}(A, C) \subsetneq \mathbb{WE}(A, C).$$

For the cone $C = (\mathbb{R}, 0) \subset \mathbb{R}^2$ which is not pointed, we have $\mathbb{SE}(D, C) = \emptyset$; $\mathbb{E}(D, C) = D$; $\mathbb{SE}(A, C) = \emptyset$; $\mathbb{E}(A, C) = A$.

Proposition 3.1 *Let A be a nonempty subset of a vector space Y with an ordering cone C for which $C \neq Y$. If $\mathbb{SE}(A, C) \neq \emptyset$, then*

$$\mathbb{SE}(A, C) = \mathbb{E}(A, C)$$

and $\mathbb{SE}(A, C)$ is a singleton whenever C is pointed.

Proof Let $\bar{y} \in \mathbb{SE}(A, C)$, that is, $\bar{y} \in \{y\} - C$ for all $y \in A$. Now suppose that $\bar{y} \notin \mathbb{E}(A, C)$. This means that $(\{\bar{y}\} - C) \cap A \not\subset \{\bar{y}\} + C$. Then there is an element $z \in A$ with $z \in \{\bar{y}\} - C$, but $z \notin \{\bar{y}\} + C$, a contradiction. This shows $\mathbb{SE}(A, C) \subseteq \mathbb{E}(A, C)$.

Conversely, let $\mathbb{SE}(A, C) \neq \emptyset$ and choose $y \in \mathbb{SE}(A, C)$, which means $y \in \{z\} - C$ for all $z \in A$. Then for each $\tilde{y} \in \mathbb{E}(A, C)$, $\tilde{y} \geq_C y$ implies $y \geq_C \tilde{y}$. From the transitivity of order, we have $z \geq_C \tilde{y}$ for all $z \in A$. This means that $\tilde{y} \in \mathbb{SE}(A, C)$ and hence $\mathbb{E}(A, C) \subseteq \mathbb{SE}(A, C)$.

Furthermore, we now assume that the cone C is pointed. Because the ordering $\geq_C$ is antisymmetric, for $y \in \mathbb{SE}(A, C)$ and $\tilde{y} \in \mathbb{E}(A, C)$, $y \geq_C \tilde{y}$ and $\tilde{y} \geq_C y$ yield $y = \tilde{y}$. Thus, $\mathbb{SE}(A, C)$ is a singleton. □

The following proposition follows from the definitions of a strong efficient element, an efficient element and a weakly efficient element. Therefore, we omit the proof which can be found in [14, pp. 43].

Proposition 3.2 *Let A and M be two subsets of a preordered vector space Y with an ordering cone C such that $M \subseteq A$. Then*

(a) $\mathbb{SE}(A, C) \cap M \subseteq \mathbb{SE}(M, C)$;
(b) $\mathbb{E}(A, C) \cap M \subseteq \mathbb{E}(M, C)$;
(c) $\mathbb{WE}(A, C) \cap M \subseteq \mathbb{WE}(M, C)$;

Figure 3.10 illustrates this result.

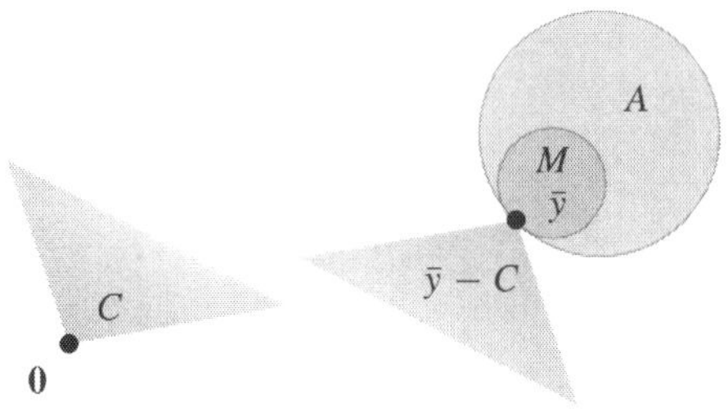

Fig. 3.10 The element $\bar{y}$ belongs to $\mathbb{E}(A, C) \cap M$, and by Proposition 3.2, we have $\bar{y} \in \mathbb{E}(M, C)$

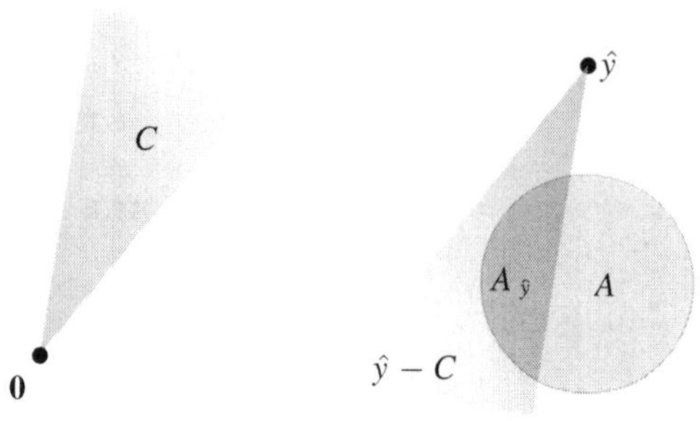

Fig. 3.11 A section $A_{\hat{y}}$ of A at $\hat{y}$

Definition 3.4 (Section of a Set) Let A be a subset of a preordered vector space Y with an ordering cone C. If for some $\hat{y} \in Y$, the set $A_{\hat{y}} = (\hat{y} - C) \cap A$ is nonempty, then $A_{\hat{y}}$ is called a *section* of y at $\hat{y}$ (see, Fig. 3.11).

The following proposition, which can be found in [14, Proposition 2.8] and [11, Lemma 6.2.], shows relations between efficient (strong efficient, weakly efficient) elements of a section $A_{\hat{y}}$ at some $\hat{y}$ and efficient (strong efficient, weakly efficient, respectively) elements of the set A.

Proposition 3.3 *Let A be a nonempty subset of a partially ordered vector space Y with an ordering cone C. For any $\hat{y} \in Y$, let the section $A_{\hat{y}}$ of A at $\hat{y}$ be nonempty.*

(a) *If $\mathbb{SE}(A, C) \neq \emptyset$, then $\mathbb{SE}(A_{\hat{y}}, C) \subseteq \mathbb{SE}(A, C)$.*
(b) $\mathbb{E}(A_{\hat{y}}, C) \subseteq \mathbb{E}(A, C)$.
(c) $\mathbb{WE}(A_{\hat{y}}, C) \subseteq \mathbb{WE}(A, C)$.

Proof

(a) Let $y \in \mathbb{SE}(A_{\hat{y}}, C)$. We prove that $A \subseteq \{y\} + C$ which implies that $y \in \mathbb{SE}(A, C)$. Let $z \in \mathbb{SE}(A, C)$. Then $A \subseteq \{z\} + C$ and, in particular,

$$y \in \{z\} + C. \tag{3.9}$$

This implies that $z \in A_{\hat{y}}$, and hence, $z \in \{y\} + C$. From the latter relation and (3.9), we have $z - y \in \ell(C) = C \cap (-C)$. Thus,

$$A \subseteq \{z\} + C = \{y\} + \{z\} - \{y\} + C \subseteq \{y\} + \ell(C) + C \subseteq \{y\} + C.$$

(b) Let $y \in \mathbb{E}(A_{\hat{y}}, C)$, i.e., $(\{y\} - C) \cap \left(A_{\hat{y}}\right) \subseteq \{y\} + C$. If there is some $\bar{y} \in A$ such that $\bar{y} \in \{y\} - C$, then $\bar{y} \in A_{\hat{y}}$. Hence, $y \in \{\bar{y}\} - C$ and so $y \in \mathbb{E}(A, C)$.
(c) Let $y \in \mathbb{WE}(A_{\hat{y}}, C)$, i.e., $(\{y\} - \mathrm{cor}(C)) \cap \left(A_{\hat{y}}\right) = \emptyset$. Suppose that $y \notin \mathbb{WE}(A, C)$. Then $(\{y\} - \mathrm{cor}(C)) \cap A \neq \emptyset$, and thus there exists $\bar{y} \in A$ such that $\bar{y} \in \{y\} - \mathrm{cor}(C)$. But then $\bar{y} \in A_{\hat{y}}$, a contradiction. □

It should be mentioned that a similar result for properly efficient elements is generally not true.

Example 3.5 Consider the set

$$A := \left\{(y_1, y_2) \in [0,1] \times [0,3] : y_2 \geq 1 - \sqrt{1-(y_1-1)^2},\ \forall y_1 \in [0,1]\right\}.$$

For $\hat{y} = (0,3)$, the element $(0,1)$ is properly efficient in the section $A_{\hat{y}} = \{\mathbf{0}\}\times[1,3]$, but $(0,1)$ is not properly efficient in A.

The following theorem provides the existence of efficient elements.

Theorem 3.1 ([14, Theorem 3.3]) *Let Y be a topological vector space, A be a nonempty subset of Y, and C be a convex correct cone in Y. Then $\mathbb{E}(A,C)$ is nonempty if and only if A has a nonempty C-complete section.*

Proof Let $\mathbb{E}(A,C) \neq \emptyset$. Then any point of $\mathbb{E}(A,C)$ will provide a C-complete section because no decreasing nets exist there.

Conversely, let A_y be a nonempty C-complete section of A. In view of Proposition 3.3, it is sufficient to show that $\mathbb{E}(A_y, C)$ is nonempty. First, we consider the set P consisting of decreasing nets from A. Since A is nonempty, so is P. Further, for two elements $a, b \in P$, we write $a \succ b$ if $\{b\} \subseteq \{a\}$ as two sets. It is clear that $\succ$ is a partial order in P. We claim that any chain in P has an upper bound. Indeed, let $\{a_\lambda\}_{\lambda\in\Gamma}$ be a chain in P and $\mathcal{B}$ denote the set of finite subsets B of Γ ordered by inclusion and let

$$a_B = \bigcup\{a_\lambda : \lambda \in B\}.$$

Now, set

$$a_o = \bigcup\{a_B : B \in \mathcal{B}\}.$$

Then, a_o is an element of P and $a_o \succ a_\lambda$ for each $\lambda \in \Gamma$, that is, a_o is an upper bound of the chain. Applying Zorn's Lemma, we obtain a maximal element, say, $a_* = \{x_\alpha : \alpha \in \Lambda\}$ forms a cover of A_y. With this in mind, remembering that a_* is a decreasing net in A_y, we arrive at a contradiction: A_y is not C-complete and the theorem is proven. Our last aim is to show that for each $\tilde{y} \subset A_y$, there is some $\alpha \in \Lambda$ such that $(y_\alpha - \operatorname{cl}\ C)^c$ contains $\tilde{y}$. If that is not the case, then $\tilde{y} \in y_\alpha - \operatorname{cl}\ C$ for each $\alpha \in \Lambda$. Since $\mathbb{E}(A_y, C) \neq \emptyset$, there is some $z \in A_y$ with $\tilde{y} >_C z$. Since C is correct, we obtain that $y_\alpha >_C z$ for all $\alpha \in \Lambda$. Adding z to the net a_*, we see that this net cannot be maximal, a contradiction. □

Theorem 3.2 ([14, Theorem 3.4]) *Let Y be a topological vector space, A be a nonempty subset of Y and C be a convex correct cone in Y. Then $\mathbb{E}(A,C)$ is nonempty if and only if A has a nonempty strongly C-complete section.*

Proof Clearly, if $\mathbb{E}(A,C)$ is nonempty, then any point of this set gives a required section. Now, let $A_{\hat{y}}$ denote the strongly C-complete section of A. If $\mathbb{E}(A_y, C) \neq \emptyset$, then by the same argument as in the proof of Theorem 3.1, we get a maximal net $\{y_\alpha\}_{\alpha\in\Lambda}$ in P and we prove that this net provides a cover of the second form in the definition. Indeed, if that is not true, then there is some $y \in A_{\hat{y}}$ such that $y \in y_\alpha - C$ for all $\alpha \in \Lambda$. Since $\mathbb{E}(A_{\hat{y}}, C) \neq \emptyset$, there is some $z \in A_{\hat{y}}$ with $y >_C z$. Since C is correct, we have

$$(\mathrm{cl}\ C) + C \setminus \ell(C) \subseteq C \setminus \ell(C),$$

and conclude that $y_\alpha >_C z$ for all $\alpha \in \Lambda$. So, the net a_* cannot be maximum. This contradiction completes the proof. □

Corollary 3.1 (Corley 1980, 1987) *Let C be an acute convex cone in a topological vector space Y and A be a C-semicompact set in Y. Then $\mathbb{E}(A, C)$ is nonempty.*

Proof By Lemma 2.10, A is $(\mathrm{cl}\, C)$-complete. In view of Theorem 3.1, the set $\mathbb{E}(A, \mathrm{cl}\, C)$ is nonempty. So, let $y \in \mathbb{E}(A, \mathrm{cl}\, C)$. Then $A \cap (\{y\} - \mathrm{cl}\, C) = \{y\}$. Since $C \subseteq \mathrm{cl}\, C$, $A \cap (\{y\} - C) \subseteq A \cap (\{y\} - \mathrm{cl}\, C)$. Consequently, $A \cap (\{y\} - C) = \{y\}$ and $y \in \mathbb{E}(A, C)$. Hence, $\mathbb{E}(A, C)$ is nonempty. □

Corollary 3.2 (Borwein 1983) *Let C be a closed convex cone in a topological vector space Y. Suppose that any one of the following conditions hold:*

(i) *A has a nonempty minorized closed section and C is Daniell;*
(ii) *A is closed and bounded, C is Daniell and Y is boundedly order complete;*
(iii) *A has a nonempty compact section.*

Then $\mathbb{E}(A, C)$ is nonempty.

Proof Due to Proposition 1.2, C is correct and in virtue of Lemma 2.10, the set A in the case (ii), or its section in the cases (i) and (iii) is C-complete. The result follows from Theorem 3.1 and Proposition 3.3. □

3.2 Solution Concepts

Let Y be a topological vector space with a pointed convex cone C, and let X be a vector space. We consider the following vector optimization problem (in short, VOP):

$$\begin{aligned} &\text{minimize}\, f(x), \\ &\text{subject to}\ x \in K, \end{aligned} \tag{3.10}$$

where

- $\emptyset \neq K \subseteq X$ is a feasible region
- $f : X \to Y$ is an objective function
- Y is the objective space
- x is a decision (variable) vector

- X is the decision variable space
- $\mathcal{Y} := f(K)$ is the feasible objective region

Let C be a convex cone generating the preorder in Y. A point $\bar{x} \in K$ is said to be

(a) a *strongly efficient solution* of VOP if $f(\bar{x}) \in \mathbb{SE}(\mathcal{Y}, C)$;
(b) an *efficient* or *Pareto efficient solution* of VOP if $f(\bar{x}) \in \mathbb{E}(\mathcal{Y}, C)$;
(c) a *weakly efficient* or *weakly Pareto efficient solution* of VOP if $f(\bar{x}) \in \mathbb{WE}(\mathcal{Y}, C)$.

If we choose $X = \mathbb{R}^n$ and $Y = \mathbb{R}^\ell$, then $f = (f_1, f_2, \ldots, f_\ell) : K \to \mathbb{R}^\ell$ is a multiobjective function. In that case, the VOP is the following:

$$\begin{aligned} &\text{minimize}\, f(x) = (f_1(x), f_2(x), \ldots, f_\ell(x)), \\ &\text{subject to}\, x \in K. \end{aligned} \tag{3.11}$$

Then corresponding efficient solutions can be defined, for instance, with respect to $C = \mathbb{R}^\ell_+$. In the next subsections, one by one, we shall discuss these solution concepts and some other solution concepts. We will also study the properties and existence of these solutions.

3.2.1 Efficient Solutions

Minimization of a vector-valued function f means that we look for the preimage of efficient elements of the set $\mathcal{Y} := f(K)$ with respect to the pointed convex cone $C \subset Y$. In practice, the minimal elements of the image set $f(K)$ do not play the central role but their preimages.

Definition 3.5 (Efficient Solution) An element $\bar{x} \in K$ is said to be an *efficient solution* or a *Pareto optimal solution* or *minimal solution* of VOP if $f(\bar{x})$ is an efficient element of the image set $f(K)$ with respect to C. In other words, $\bar{x} \in K$ is an efficient solution of VOP if

$$(\{f(\bar{x})\} - C) \cap \mathcal{Y} = \{f(\bar{x})\}.$$

The set of all efficient solutions $\bar{x} \in K$ is denoted by K_{eff} and it is called the *efficient solution set*. The set of all efficient elements $\bar{y} = f(\bar{x}) \in \mathcal{Y}$, where $\bar{x} \in K_{\text{eff}}$, is denoted by $\mathbb{E}(\mathcal{Y}, C)$ and it is called *nondominated set*. If there is no confusion which cone C is being used, we also denote $\mathcal{Y}_{\text{eff}} := \mathbb{E}(\mathcal{Y}, C)$. If $x, \tilde{x} \in K$ and $f(x) \leq_C f(\tilde{x})$, then we say that x dominates $\tilde{x}$ and $f(x)$ dominates $f(\tilde{x})$.

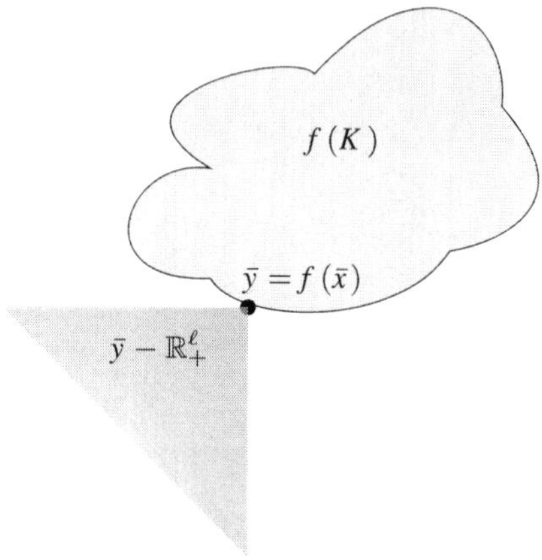

Fig. 3.12 Visualization of an efficient element $\bar{y}$

Remark 3.2 If $Y = \mathbb{R}^\ell$, $X = \mathbb{R}^n$ and $C = \mathbb{R}^\ell_+$, then an element $\bar{x} \in K$ is an efficient solution of VOP if there is no $x \in K$ such that

$$\begin{aligned} f_i(x) &\leq f_i(\bar{x}), \quad \text{for all } i \in \mathscr{F} = \{1, 2, \ldots, \ell\} \\ \text{and } f_i(x) &< f_i(\bar{x}), \quad \text{for some } i \in \mathscr{F} = \{1, 2, \ldots, \ell\}. \end{aligned}$$

When $\ell = 2, f(K) \subseteq \mathbb{R}^2$. In this case, the definition of an efficient solution says that when shifting the origin to a point $f(\bar{x})$, if $-\mathbb{R}^\ell_+ \setminus \{\mathbf{0}\}$ does not intersect the set $f(K)$, then $\bar{x}$ is an efficient solution of the VOP. See, Fig. 3.12 for an illustration.

The following are equivalent definitions of an efficient solution. Let $(\mathbb{R}^\ell, \leq_C)$ be an ordered space with ordering cone $C = \mathbb{R}^\ell_+$ which induces the componentwise ordering as $\bar{y} \leq_{C_0} y \Leftrightarrow \bar{y}_i \leq y_i$ for all $i = 1, 2, \ldots, \ell$ and $\bar{y} \neq y$. In particular, $\bar{x} \in K$ is an efficient solution of VOP if any one of the following equivalent conditions hold:

- there is no $x \in K$ such that $f(x) \leq_{C_0} f(\bar{x})$
- there is no $x \in K$ such that $f(x) - f(\bar{x}) \in -\mathbb{R}^\ell_+ \setminus \{\mathbf{0}\}$
- $(f(K) - f(\bar{x})) \cap \left(-\mathbb{R}^\ell_+ \setminus \{\mathbf{0}\}\right) = \emptyset$
- $f(x) - f(\bar{x}) \in \mathbb{R}^\ell \setminus \left\{-\mathbb{R}^\ell_+ \setminus \{\mathbf{0}\}\right\}$ for all $x \in K$
- $f(K) \cap \left(f(\bar{x}) - \mathbb{R}^\ell_+\right) = \{f(\bar{x})\}$
- there is no $f(x) \in f(K) \setminus \{f(\bar{x})\}$ such that $f(x) \in f(\bar{x}) - \mathbb{R}^\ell_+$
- $f(x) \leq_C f(\bar{x})$ for some $x \in K$ implies $f(x) = f(\bar{x})$

Example 3.6 Let $K := \left\{(x_1, x_2) \in \mathbb{R}^2 : -1 < x_1 < 1, -1 < x_2 < 1\right\}$ and let the objective function $f(x_1, x_2) := \left(x_1^2, x_2^2\right) - \left(\frac{1}{2}, \frac{1}{2}\right)$ be given. We consider the ordering cone $C = \mathbb{R}^2_+$. Then the only efficient solution of the corresponding VOP is $(x_1, x_2) = (0, 0)$ with objective function values $\left(-\frac{1}{2}, -\frac{1}{2}\right)$ (see Fig. 3.13).

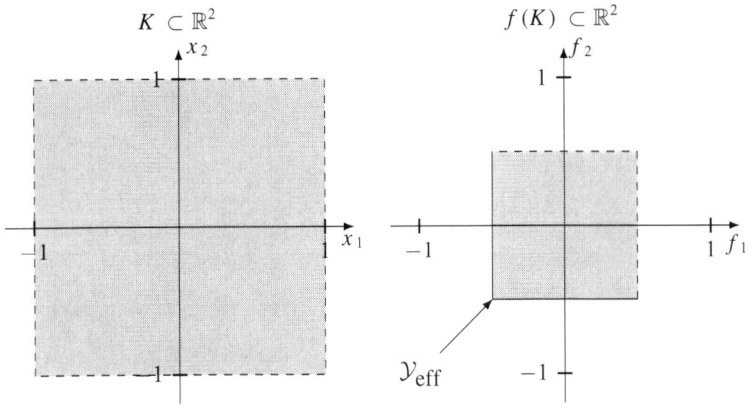

Fig. 3.13 *Left*: The feasible region K. *Right*: The feasible objective region. The only efficient solution is $(x_1, x_2) = (0, 0)$ with objective function values $\left(-\frac{1}{2}, -\frac{1}{2}\right)$ (see Example 3.6)

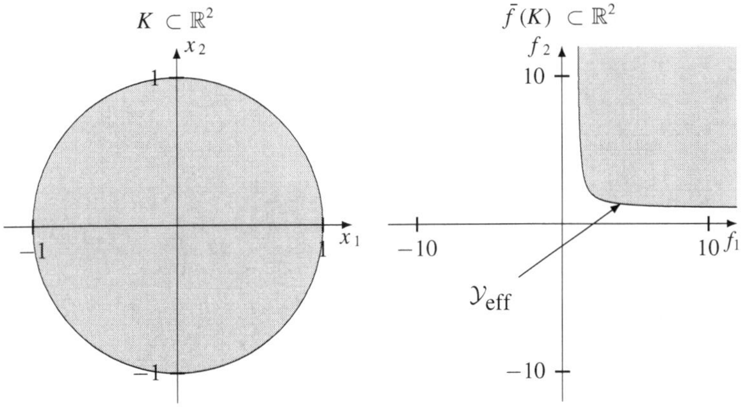

Fig. 3.14 *Left*: The feasible region K. *Right*: The feasible objective region with the nondominated set $\mathcal{Y}_{\text{eff}}$ (see Example 3.7)

Example 3.7 Let $K := \{(x_1, x_2) \in \mathbb{R}^2 : x_1^2 + x_2^2 \leq 1\}$ and consider the objective function $f(x_1, x_2) := \left(\frac{1}{x_1^2}, \frac{1}{x_2^2}\right)$. As in the previous example, we use the cone $C = \mathbb{R}^2_+$. Then the set of efficient solutions is given by $K_{\text{eff}} = \left\{(x_1, x_2) \in \mathbb{R}^2 : x_1^2 + x_2^2 = 1,\ x_1 \neq 0,\ x_2 \neq 0\right\}$ (see Fig. 3.14).

Let $\mathcal{Y} \subseteq \mathbb{R}^\ell$ and $\mathcal{Y}_{\text{eff}} = \{y \in \mathcal{Y} : \nexists\, \tilde{y} \in \mathcal{Y} \text{ such that } \tilde{y} \leq_C y\}$.

Next we collect important properties of the nondominated set $\mathcal{Y}_{\text{eff}}$, where $Y = \mathbb{R}^\ell$ and $C = \mathbb{R}^\ell_+$. The following proposition is a direct consequence of Lemma 3.1.

Proposition 3.4 *Let $\mathcal{Y} \subseteq \mathbb{R}^\ell$. Then $\mathcal{Y}_{eff} = \left(\mathcal{Y} + \mathbb{R}^\ell_+\right)_{eff}$*

Proposition 3.5 *Let $\mathcal{Y} \subseteq \mathbb{R}^\ell$. Every efficient point must lie on the boundary of $\mathcal{Y}$, that is, $\mathcal{Y}_{eff} \subseteq bd(\mathcal{Y})$.*

Proof Let $y \in \mathcal{Y}_{\text{eff}}$ and suppose $y \notin \text{bd}(\mathcal{Y})$. Then $y \in \text{int}(\mathcal{Y})$ and there exists a neighborhood $N_\varepsilon(y)$ of y with $N_\varepsilon(y) := y + N_\varepsilon(\mathbf{0}) \subseteq \mathcal{Y}$, where $N_\varepsilon(\mathbf{0})$ is an open ball with radius ε centered at the origin. Let $d \in \mathbb{R}^\ell_+$ such that $d \neq \mathbf{0}$. Then, we can choose some $\lambda > 0$ such that $\lambda d \in N_\varepsilon(\mathbf{0})$. Now, $y - \lambda d \in \mathcal{Y}$ with $\lambda d \in \mathbb{R}^\ell_+ \setminus \{\mathbf{0}\}$, that is, $y \notin \mathcal{Y}_{\text{eff}}$, a contradiction. Hence $y \in \text{bd}(\mathcal{Y})$. □

From Propositions 3.4 and 3.5, we can easily derive the following corollary.

Corollary 3.3 *Let $\mathcal{Y} \subseteq \mathbb{R}^\ell$. If $\mathcal{Y}$ is open or if $\mathcal{Y} + \mathbb{R}^\ell_+$ is open, then $\mathcal{Y}_{eff} = \emptyset$.*

Proposition 3.6 *Let $\mathcal{Y} \subseteq \mathbb{R}^\ell$. Then $(\mathcal{Y}_1 + \mathcal{Y}_2)_{eff} \subset (\mathcal{Y}_1)_{eff} + (\mathcal{Y}_2)_{eff}$.*

Proof Let $y \in (\mathcal{Y}_1 + \mathcal{Y}_2)_{\text{eff}}$. Then $y = y_1 + y_2$ for some $y_1 \in \mathcal{Y}_1$ and $y_2 \in \mathcal{Y}_2$. Assume that $y_1 \notin (\mathcal{Y}_1)_{\text{eff}}$. Then there exist $\tilde{y} \in \mathcal{Y}_1$ and $d \in \mathbb{R}^\ell_+ \setminus \{\mathbf{0}\}$ such that $y_1 = \tilde{y} + d$, and thus, $y = \tilde{y} + y_2 + d$ with $\tilde{y} + y_2 \in \mathcal{Y}_1 + \mathcal{Y}_2$. Hence $y \notin (\mathcal{Y}_1 + \mathcal{Y}_2)_{\text{eff}}$, a contradiction of our assumption. Thus, $y_1 \in (\mathcal{Y}_1)_{\text{eff}}$. Similarly, $y_2 \in (\mathcal{Y}_2)_{\text{eff}}$. Therefore, $y_1 + y_2 \in (\mathcal{Y}_1)_{\text{eff}} + (\mathcal{Y}_2)_{\text{eff}}$. □

Remark 3.3 We note that $(\mathcal{Y}_1)_{\text{eff}} + (\mathcal{Y}_2)_{\text{eff}} \subset (\mathcal{Y}_1 + \mathcal{Y}_2)_{\text{eff}}$ is not true in general, see the following example.

Example 3.8 Consider $\ell = 2$ and the sets

$$\mathcal{Y}_1 := \{(y_1, y_2) \in \mathbb{R}^2 : 2y_1 + y_2 \geq 0\}, \quad \mathcal{Y}_2 := \{(y_1, y_2) \in \mathbb{R}^2 : y_1 + 2y_2 \geq 0\}.$$

For both sets the nondominated set comprises the entire boundary. Nevertheless, the nondominated set of $\mathcal{Y}_1 + \mathcal{Y}_2 = \mathbb{R}^2$ is empty (cf. Fig. 3.15). So, the inclusion $(\mathcal{Y}_1)_{\text{eff}} + (\mathcal{Y}_2)_{\text{eff}} \subset (\mathcal{Y}_1 + \mathcal{Y}_2)_{\text{eff}}$ is not fulfilled in general.

The proof of the following proposition can be easily derived, and therefore, it is omitted.

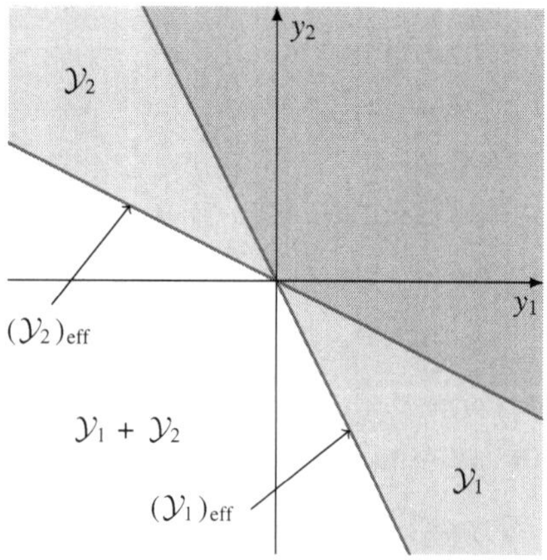

Fig. 3.15 Counterexample for Remark 3.3

Proposition 3.7 *Let $\mathcal{Y} \subseteq \mathbb{R}^\ell$. Then $(\alpha\mathcal{Y})_{eff} = \alpha\mathcal{Y}_{eff}$ for all $\alpha > 0$.*

3.2.2 Weakly and Strongly Efficient Solutions

The efficiency concept is the main optimality notion in vector optimization. But there are also other concepts being more weakly or more strongly formulated than the concept of efficiency. The Pareto optimality or efficiency is defined by using the componentwise ordering ($x \leq_{C_0} y \Leftrightarrow x_i \leq y_i$ for all $i = 1, 2, \ldots, \ell$ and $x \neq y$) on $\mathbb{R}^\ell$. When we replace it by the strict componentwise ordering ($x <_C y \Leftrightarrow x_i < y_i$ for all $i = 1, 2, \ldots, \ell$) on $\mathbb{R}^\ell$, we obtain the definition of weakly Pareto optimal solution of VOP.

Definition 3.6 (Weakly Efficient Solution) An element $\bar{x} \in K$ is said to be a *weakly efficient solution* or a *weakly Pareto optimal solution* or *weakly minimal solution* of VOP if $f(\bar{x})$ is a weakly efficient element of the image set $\mathcal{Y} = f(K)$ with respect to the pointed convex cone $C \subset Y$ with $\text{int}(C) \neq \emptyset$. In other words, $\bar{x}$ is a weakly efficient solution of VOP if

$$(\{f(\bar{x})\} - \text{int}(C)) \cap \mathcal{Y} = \emptyset.$$

The set of all weakly efficient solutions $\bar{x} \in K$ is denoted by $K_{\text{w-eff}}$ and it is called the *weakly efficient set*. The set of all weakly efficient elements $\bar{y} = f(\bar{x}) \in \mathcal{Y}$, where $\bar{x} \in K_{\text{w-eff}}$, is denoted by $\mathbb{WE}(f(K), C)$ and it is called *weakly nondominated set*. If there is no confusion about the selection of the cone C, we also use $\mathcal{Y}_{\text{w-eff}} := \mathbb{WE}(f(K), C)$.

Remark 3.4 If $\ell = 2$, $Y = \mathbb{R}^2$, $X = \mathbb{R}^n$, $C = \mathbb{R}^2_+$, then $f(K) \subseteq \mathbb{R}^2$. In this case, geometrically, a weakly efficient solution $\bar{x}$ of VOP is a point in $f(K)$ if we shift the origin to the point $\bar{x}$, then $f(K)$ does not intersect $-\text{int}(\mathbb{R}^2_+)$.

Remark 3.5 In case $Y = \mathbb{R}^\ell$, $X = \mathbb{R}^n$, $C = \mathbb{R}^\ell_+$, the definition of weakly efficient solutions reduces to the following: $\bar{x} \in K_{\text{w-eff}}$ if and only if there is no $y \subset K$ such that

$$f_i(y) < f_i(\bar{x}) \quad \text{for all } i \in \mathcal{F} = \{1, 2, \ldots, \ell\}.$$

Moreover, $\bar{x} \in K_{\text{w-eff}}$ if and only if there is no $x \in K$ such that $f(\bar{x}) - f(x) \in \text{int}(\mathbb{R}^\ell_+)$. Equivalently, $\bar{x} \in K_{\text{w-eff}}$ if and only if

$$(f(K) - f(\bar{x})) \cap \left(-\text{int}(\mathbb{R}^\ell_+)\right) = \emptyset.$$

Example 3.9 We reconsider Example 3.6, and we observe that the set of weakly efficient solutions is $K_{\text{w-eff}} = \{(x_1, x_2) \in \mathbb{R}^2 : x_1 = 0,\ x_2 \in [0, 1[\} \cup \{(x_1, x_2) \in \mathbb{R}^2 : x_2 = 0,\ x_1 \in [0, 1[\}$.

We know that every efficient solution is also weakly efficient. A converse statement is true under a certain convexity assumption on the set of feasible elements $\mathcal{Y}$. The following definition of a C-quasi-strictly convex set is given in [13].

Definition 3.7 (*C*-Quasi-Strict Convexity) Let $C \subset \mathbb{R}^\ell$ be a cone. A set $\mathcal{Y} \subset \mathbb{R}^\ell$ is *C-quasi-strictly convex* if $\mathcal{Y}$ is C-convex (that is, $\mathcal{Y} + C$ is a convex set) and if for each $y \in \mathbb{R}^\ell \setminus \mathcal{Y}_{\text{eff}}$, the set $y - C$ has a nonempty intersection with the relative interior relint$(\mathcal{Y} + C)$ of the set $\mathcal{Y} + C$.

It is easy to see that the following lemma holds.

Lemma 3.4 ([13]) *Let $C \subset \mathbb{R}^\ell_+$ be a pointed convex cone defining efficiency in $\mathbb{R}^\ell$. Let $\mathcal{Y} \subset \mathbb{R}^\ell$ be C-quasi-strictly convex and C-closed (that is, $\mathcal{Y}+C$ is closed). Then every weakly efficient solution is also efficient.*

The following stronger notion of strict convexity is also known.

Definition 3.8 (*C*-Strict Convexity) Let $C \subset \mathbb{R}^\ell$ be a cone. A set $\mathcal{Y} \subset \mathbb{R}^\ell$ is *C-strictly convex* if $\mathcal{Y}$ is C-convex (that is, $\mathcal{Y} + C$ is a convex set) with nonempty interior, and for each $y_1, y_2 \in \mathcal{Y}$ $(y_1 \neq y_2)$, the set $(y_1 + y_2)/2 - C$ has nonempty intersection with int$(\mathcal{Y})$.

Example 3.10 Let $Y = \mathbb{R}^2$ and $C = \mathbb{R}^2_+$.

(a) Let $\mathcal{Y}$ be a square. Because there exists weakly efficient solutions which are not efficient and $\mathcal{Y}+C$ is closed, $\mathcal{Y}$ is not C-quasi-strictly convex due to Lemma 3.4.
(b) Every C-strictly convex set is C-quasi-strictly convex. However, the converse implication is not generally true. For instance, the triangle with vertices $(1, 1)$, $(1, 0)$ and $(0, 1)$ is $\mathbb{R}^2_+$-quasi-strictly convex, but not $\mathbb{R}^2_+$-strictly convex.
(c) A circle is $\mathbb{R}^2_+$-strictly convex, and therefore $\mathbb{R}^2_+$-quasi-strictly convex.

Definition 3.9 (Strongly Efficient Solution) An element $\bar{x} \in K$ is said to be a *strongly efficient solution* or an *ideal efficient solution* if $f(\bar{x})$ is a strongly efficient element of the image set $\mathcal{Y} = f(K)$ with respect to the pointed convex cone $C \subset Y$. In other words, $\bar{x}$ is a strongly efficient solution of VOP if

$$\mathcal{Y} \subset \{f(\bar{x})\} + C.$$

The set of all strongly efficient solutions $\bar{x} \in K$ is denoted by $K_{\text{s-eff}}$ and it is called the *strongly efficient set*. The set of all strongly efficient elements $\bar{y} = f(\bar{x}) \in \mathcal{Y}$, where $\bar{x} \in K_{\text{s-eff}}$, is denoted by $\mathbb{SE}(\mathcal{Y}, C)$ and it is called *strongly nondominated set*. If there is no confusion about the choice of the cone C, we denote $\mathcal{Y}_{\text{s-eff}} := \mathbb{SE}(\mathcal{Y}, C)$.

Remark 3.6 We consider the case $Y = \mathbb{R}^\ell$, $X = \mathbb{R}^n$, $C = \mathbb{R}^\ell_+$. An element $\bar{x} \in K$ is a strongly efficient solution of VOP if $f(\bar{x})$ is a strongly efficient element of the image set $f(K)$ with respect to the componentwise ordering on $\mathbb{R}^\ell$, that is, if $f(\bar{x}) \leq_C f(y)$ for all $y \in K$. In other words, $f(\bar{x})$ is a strongly efficient solution of VOP if

$$f_i(\bar{x}) \leq f_i(y), \quad \text{for all } y \in K \text{ and } i \in \mathcal{F} = \{1, 2, \ldots, \ell\}.$$

Example 3.11 We revisit Example 3.6. We observe that $(x_1, x_2) = \left(-\frac{1}{2}, -\frac{1}{2}\right)$ is the only strongly efficient solution of the VOP.

It is clear that $\mathcal{Y}_{\text{s-eff}} \subset \mathcal{Y}_{\text{eff}} \subset \mathcal{Y}_{\text{w-eff}}$ and $K_{\text{s-eff}} \subset K_{\text{eff}} \subset K_{\text{w-eff}}$.

Example 3.12 Consider the VOP with feasible set

$$K = \left\{(x_1, x_2) \in \mathbb{R}^2 : 0 \leq x_1 \leq 2,\ 0 \leq x_2 \leq 3\right\}$$

and objective function $f : K \to \mathbb{R}^2$ defined as

$$f(x_1, x_2) = (x_1, x_2), \quad \text{for all } (x_1, x_2) \in K.$$

K describes a square in $\mathbb{R}^2$. Since f is the identity mapping, $f(K) = K$. The point $(0, 0)$ is the strongly efficient solution (in fact, $(0, 0)$ is the only efficient solution) whereas the set

$$\{(x_1, x_2) \in K : x_1 = 0 \text{ or } x_2 = 0\}$$

is the set of all weakly efficient solutions of VOP.

Example 3.13 Consider the VOP with feasible set

$$K = \left\{(x_1, x_2) \in \mathbb{R}^2 : x_1^2 + 2x_2^2 \leq 1\right\}$$

and objective function $f : K \to \mathbb{R}^2$ defined as

$$f(x_1, x_2) = (x_1, x_2), \quad \text{for all } (x_1, x_2) \in K,$$

see Fig. 3.16. Since f is the identity mapping, it holds $f(K) = K$. The set of efficient solutions is

$$\left\{(x_1, x_2) \in \mathbb{R}^2 : x_1 \in [-1, 0] \text{ and } x_2 = -\sqrt{\frac{1}{2} - \frac{x_1^2}{2}}\right\}.$$

But there is no strongly efficient solution.

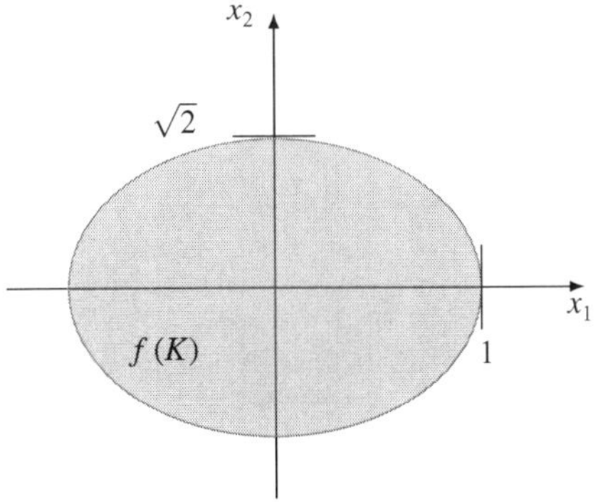

Fig. 3.16 The feasible solution set from Example 3.13

3.2.3 Properly Efficient Solutions

In this section, we present various types of proper efficiency for a solution x of a VOP. Here, we consider the following setting. Let $X = \mathbb{R}^n$, $Y = \mathbb{R}^\ell$, $\emptyset \neq K \subseteq \mathbb{R}^n$ be the feasible region, $f : X \to \mathbb{R}^\ell$ be the objective function, $\mathcal{Y} := f(K)$ be the feasible objective set, and $C \subset \mathbb{R}^\ell$ be a proper closed convex cone (unless otherwise noted). Most of the results in this section can be found in Sawaragi et al. [17].

Borwein [2] introduced the following sharper notion of a solution of a vector optimization problem, which we call proper efficiency in the sense of Borwein.

Definition 3.10 (Borwein 1977) An element $\bar{x} \in K$ is called *properly efficient in the sense of Borwein* if

$$T(f(K) + C, f(\bar{x})) \cap (-C) = \{\mathbf{0}\},$$

where $T(f(K) + C, f(\bar{x}))$ denotes the contingent cone of $f(K) + C$ at $f(\bar{x})$.

Proposition 3.8 ([17, Proposition 3.1.5]) *If an element $\bar{x} \in K$ is properly efficient in the sense of Borwein, then $\bar{x}$ is an efficient solution of* VOP.

Proof Let $\bar{x} \in K$ be properly efficient in the sense of Borwein, but suppose that $\bar{x}$ is not efficient for VOP. Then there exists a nonzero vector $c \in C$ such that $c = f(\bar{x}) - y$ for some $y \in \mathcal{Y}$. Let $c^k = \left(1 - \frac{1}{k}\right) c \in C$ and $t_k = k$ for $k = 1, 2, \ldots$. Then

$$y + c^k = f(\bar{x}) - c + \left(1 - \frac{1}{k}\right) c = f(\bar{x}) - \frac{1}{k} c \to f(\bar{x}), \quad \text{for } k \to +\infty,$$

and

$$t_k(y + c^k - f(\bar{x})) = k(-c + \left(1 - \frac{1}{k}\right) c) = -c \to -c, \quad \text{for } k \to +\infty.$$

Therefore, $T(f(K) + C, f(\bar{x})) \cap (-C) \neq \{\mathbf{0}\}$, and $\bar{x}$ is not a properly efficient solution in the sense of Borwein, a contradiction.

The converse assertion of Proposition 3.8 is not fulfilled, as the following example shows.

Example 3.14 In Example 3.13, the elements $(-1, 0)$ and $(0, -1)$ are efficient, but they are not properly efficient solutions in the sense of Borwein.

Definition 3.11 (Benson 1979) An element $\bar{x} \in K$ is called *properly efficient in the sense of Benson* if

$$\text{cl}\,(\text{cone}\,(f(K) + C - \{f(\bar{x})\})) \cap (-C) = \{\mathbf{0}\},$$

where cone $(f(K) + C - \{f(\bar{x})\})$ denotes the cone generated by $f(K) + C - \{f(\bar{x})\}$.

Remark 3.7 It holds $T(f(K) + C, f(\bar{x})) \subset \text{cl}\,(\text{cone}\,(f(K) + C - \{f(\bar{x})\}))$, which implies that proper efficiency in the sense of Benson strengthens the notion of proper efficiency in the sense of Borwein.

Before we show that a properly efficient solution in the sense of Borwein is, under appropriate assumptions, a properly efficient solution in the sense of Benson, we need the following lemma.

Lemma 3.5 *Let $S \subseteq \mathbb{R}^\ell$ be a convex set and let $y \in S$. Then*

$$T(S, y) = \text{cl}(\text{cone}(S - y)),$$

which is a closed convex cone.

Proof Due to its definition and because S is a convex set, $\text{cl}(\text{cone}(S - y))$ is a closed convex cone. As the relation $T(S, y) \subset \text{cl}(\text{cone}(S - y))$ is obvious, we only need to show that $\text{cone}(S - y) \subset T(S, y)$, because $T(S, y)$ is closed. Let $t \in \text{cone}(S - y)$. Then $t = \lambda(s - y)$ for some $\lambda \geq 0$ and $s \in S$. Let for $k = 1, 2, \ldots$

$$y^k = \left(1 - \frac{1}{k}\right) y + \frac{1}{k} s,$$

and $t_k = \lambda k \geq 0$. Then $t_k(y^k - y) = \lambda(s - y)$. Because S is a convex set, we obtain that $y^k \in S$, and

$$y^k \to y \quad \text{and} \quad t_k(y^k - y) \to t \quad \text{as} \quad k \to +\infty.$$

Hence, $t \in T(S, y)$, which completes the proof. □

Theorem 3.3 *If $\bar{x} \in K$ is a properly efficient solution in the sense of Benson, then it is also a properly efficient solution in the sense of Borwein. If K is a convex set and f is C-convex, then the converse statement is true as well.*

Proof If C is a convex cone and if f is C-convex on the convex set K, then $f(K) + C$ is a convex set. Then by Lemma 3.5, $T(S, y) = \text{cl}(\text{cone}(S - y))$. Therefore, proper efficiency in the sense of Benson is equivalent to the notion of proper efficiency in the sense of Borwein. □

The following example verifies that the converse statement of Theorem 3.3 is not true in general if K is not convex.

Example 3.15 Let $Y = \mathbb{R}^2$, $C = \mathbb{R}^2_+$, $K = \{(x_1, x_2) \in \mathbb{R}^2 : x_1 + x_2 \geq -1\} \cup \{(x_1, x_2) \in \mathbb{R}^2 : x_1 \geq 0\} \cup \{(x_1, x_2) \in \mathbb{R}^2 : x_2 \geq 0\}$, and $f : K \to \mathbb{R}^2$ with

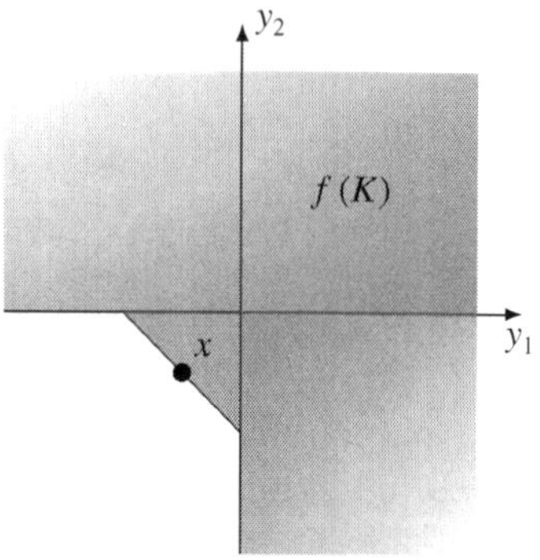

Fig. 3.17 The solution $x = (-0.5, -0.5)$ is properly efficient in the sense of Borwein, but not properly efficient in the sense of Benson (see Example 3.15)

$f(x_1, x_2) = (x_1, x_2)$. Then the solution $x = (-0.5, -0.5)$ is properly efficient in the sense of Borwein, but not properly efficient in the sense of Benson (see Fig. 3.17).

The following lemma will be used to show that if $f(K)$ is a polyhedral convex set, then any efficient solution is also properly efficient in the sense of Benson and Borwein.

Lemma 3.6 *Let $\beta_i \in \mathbb{R}$, $b^i \in \mathbb{R}^\ell$, $i = 1, 2, \ldots, n$ be given, and let $A \subset \mathbb{R}^\ell$ be a polyhedral convex set, i.e., $A := \{a \in \mathbb{R}^\ell : \langle b^i, a \rangle \leq \beta_i,\ i = 1, \ldots, \ell\}$, $\hat{a} \in A$, and $I(\hat{a}) := \{i \in \{1, 2, , \ldots, \ell\} : \langle b^i, \hat{a} \rangle = \beta_i\}$. Then*

$$T(A, \hat{a}) = \text{cone}(A - \hat{a}) = \{h \in \mathbb{R}^\ell : \langle b^i, h \rangle \leq 0 \, for \, i \in I(\hat{a})\}.$$

Theorem 3.4 *Let $f(K)$ be a polyhedral convex set, C a pointed closed convex cone in $\mathbb{R}^\ell$. Then any efficient solution of* VOP *is properly efficient in the sense of Borwein and in the sense of Benson.*

Proof Let $\bar{x} \in K$ be an efficient solution of (VOP). Suppose that $\bar{x} \in K$ is not properly efficient in the sense of Borwein or Benson. Then we can prove that there exists a nonzero vector $c \in C$ such that $-c \in T(f(K), f(\bar{x}))$, which means that

$$\langle b^i, -c \rangle \leq 0 \quad \text{for all } i \in I(f(\bar{x})).$$

Then for sufficiently small $\alpha > 0$,

$$\langle b^i, f(\bar{x} - \alpha c) \rangle \leq \beta_i, \quad \text{for all } i = 1, 2, \ldots, \ell.$$

Thus, $f(\bar{x}) - (f(\bar{x}) - \alpha c) \in C \setminus \{\mathbf{0}\}$ and $f(\bar{x}) - \alpha c \in f(K)$. But this means that $\bar{x}$ is not an efficient solution of VOP, a contradiction. □

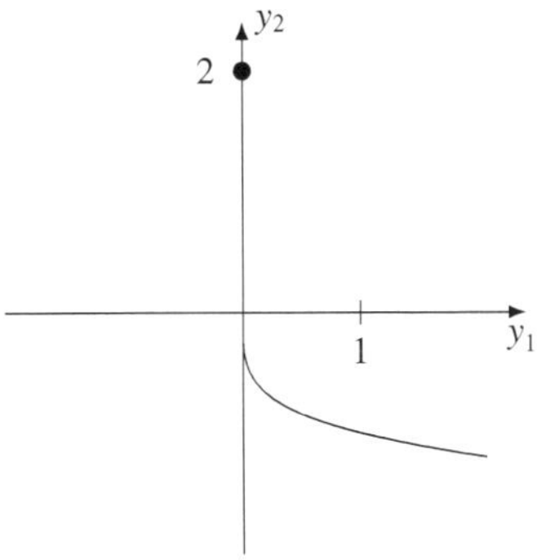

Fig. 3.18 The solution $x = (0, 2)$ is an efficient solution of VOP, but x does not belong to the closure of the set of properly efficient solutions in the sense of Benson (see Example 3.16)

Theorem 3.5 *Let C be a pointed closed convex cone in $\mathbb{R}^\ell$, $\mathcal{Y} = f(K)$ be C-closed (that is, $f(K) + C$ is closed) and $\mathcal{Y}^+ \cap (-C) = \{\mathbf{0}\}$, where*

$$\mathcal{Y}^+ := \big\{y \in \mathbb{R}^\ell : \text{there exist sequences } \{\alpha_k\} \subset \mathbb{R}, \{y^k\} \subset \mathcal{Y} \\ \text{such that } \alpha_k > 0, \alpha_k \to 0 \text{ and } \alpha_k y^k \to y\big\}. \qquad (3.12)$$

Then an efficient solution of VOP *belongs to the closure of the set of properly efficient solutions in the sense of Benson.*

Remark 3.8 Note that $\mathcal{Y}^+$ is an extension of the recession cone (see Definition 1.17).

The closedness assumption in Theorem 3.5 cannot be omitted, as the following example verifies.

Example 3.16 Let $Y = \mathbb{R}^2$, $C = \mathbb{R}^2_+$, $K = \{(x_1, x_2) \in \mathbb{R}^2 : x_1 = x_2^4,\ x_2 < 0\} \cup \{(0, 2)\}$, and $f : K \to \mathbb{R}^2$ with $f(x_1, x_2) = (x_1, x_2)$ (see Fig. 3.18). Note that $\mathcal{Y}^+ \cap (-C) = \{\mathbf{0}\}$ is satisfied, but $f(K)$ is not C-closed. Then the set of efficient solutions is the whole set K, but the set of properly efficient solutions in the sense of Benson is $\{(x_1, x_2) \in \mathbb{R}^2 : x_1 = x_2^4,\ x_2 < 0\}$. Therefore, the closure of the set of properly efficient solutions in the sense of Benson is $\{(x_1, x_2) \in \mathbb{R}^2 : x_1 = x_2^4,\ x_2 \leq 0\}$.

The converse assertion of the statement in Theorem 3.5 is generally not true, which can be seen in the following example.

Example 3.17 Consider $K = \{(x_1, x_2) \in \mathbb{R}^2 : x_2 = x_1, x_1 \leq 0\} \cup \{(x_1, x_2) \in \mathbb{R}^2 : x_1 = 0, -2 \leq x_2 \leq 0\}$, $C = \mathbb{R}^2_+$ and $f(x_1, x_2) = (x_1, x_2)$ (see Fig. 3.19). Note that $f(K)$ is C-closed and $\mathcal{Y}^+ \cap (-C) = \{\mathbf{0}\}$. Then the set of efficient solutions is equal to the set of properly efficient solutions in the sense of Benson, namely, $K_{\text{eff}} = \{(x_1, x_2) \in \mathbb{R}^2 : x_2 = x_1, x_1 < 0\} \cup \{(0, -2)\}$. Thus, $(0, 0)$ is not efficient. However, $(0, 0)$ belongs to the closure of the set of properly efficient solutions in the sense of Benson.

Another relation between efficient solutions of VOP and properly efficient solutions in the sense of Benson is given in the following corollary.

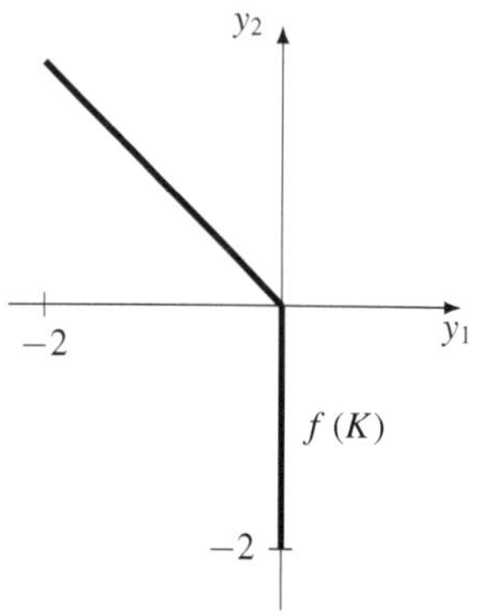

Fig. 3.19 The feasible solution set from Example 3.17

Corollary 3.4 *Let C be a pointed closed convex cone in $\mathbb{R}^\ell$ and $f(K)$ be a closed convex set or C-convex and C-closed. Then an efficient solution of* VOP *belongs to the closure of the set of properly efficient solutions in the sense of Benson.*

Definition 3.12 (Henig 1982)

(a) An element $\bar{x} \in K$ is called a *global properly efficient solution in the sense of Henig* if for some convex cone $C' \subset \mathbb{R}^\ell$ with $C \setminus \{\mathbf{0}\} \subset \operatorname{int}(C')$ it holds that

$$(\{f(\bar{x})\} - C') \cap \mathcal{Y} = \{\mathbf{0}\}.$$

(b) An element $\bar{x} \in K$ is called a *local properly efficient solution in the sense of Henig* if for every $\varepsilon > 0$, there exists a convex cone $C' \subset \mathbb{R}^\ell$ with $C \setminus \{\mathbf{0}\} \subset \operatorname{int}(C')$ it holds that

$$(\{f(\bar{x})\} - C') \cap ((\mathcal{Y} + C) \cap (\{f(\bar{x})\} + \varepsilon B)) = \{\mathbf{0}\},$$

where B denotes the closed unit ball in $\mathbb{R}^\ell$.

The following theorem states that global (local, respectively) proper efficiency in the sense of Henig reduces to proper efficiency in the sense of Borwein (Benson, respectively) if the cone C is closed and acute.

Theorem 3.6 *If C is closed and acute, then global proper efficiency in the sense of Henig is equivalent to proper efficiency in the sense of Benson. Moreover, under the assumption that C is closed and acute, then local proper efficiency in the sense of Henig is equivalent to proper efficiency in the sense of Borwein.*

Theorem 3.7 *A global properly efficient solution in the sense of Henig is also a local proper efficient solution in the sense of Henig. Conversely, if C is closed and acute, and if $\mathcal{Y}^+ \cap (-C) = \{\mathbf{0}\}$, then local proper efficiency in the sense of Henig implies global proper efficiency in the sense of Henig, where $\mathcal{Y}^+$ is defined by* (3.12).

Now we turn our attention to the special case where $Y = \mathbb{R}^\ell$ and the partial ordering is induced by the natural ordering cone $C = \mathbb{R}^\ell_+$. Geoffrion [10] introduced the following sharper notion of a solution of a vector optimization problem, known as properly Pareto optimal solution. The idea behind this kind of solution concept is to eliminate unbounded tradeoffs between various criteria.

Definition 3.13 (Geoffrion 1968) An element $\bar{x} \in K$ is said to be a *properly efficient solution in the sense of Geoffrion* or a *properly Pareto optimal solution in the sense of Geoffrion* or a *properly minimal solution in the sense of Geoffrion* of VOP if it is an efficient solution and there is a real number $\mu > 0$ such that for all $i \in \{1, 2, \ldots, \ell\}$ and every $y \in K$ with $f_i(y) < f_i(\bar{x})$, there exists at least one index $j \in \{1, 2, \ldots, \ell\}$ such that $f_j(y) > f_j(\bar{x})$ and

$$\frac{f_i(\bar{x}) - f_i(y)}{f_j(y) - f_j(\bar{x})} \leq \mu.$$

An efficient solution which is not a properly efficient solution in the sense of Geoffrion is called an *improperly efficient solution* in the sense of Geoffrion.

In most applications, improperly efficient solutions in the sense of Geoffrion are not desired because a possible improvement of one component leads to a drastic deterioration of another component, and a decision maker wishes to prevent the existence of unbounded trade-offs.

Definition 3.13 is a refinement of a notion of proper efficiency by Kuhn and Tucker [12]. Engau [9] extends Definition 3.13 to the case of an infinite number of objectives.

In Example 3.13, the set of all properly efficient solutions in the sense of Geoffrion is

$$\left\{ (x_1, x_2) \in \mathbb{R}^2 : x_1 \in (-1, 0) \text{ and } x_2 = -\sqrt{\frac{1}{2} - \frac{x_1^2}{2}} \right\},$$

thus, the points $(-1, 0)$ and $(0, -1)$ are efficient, but not properly efficient solutions in the sense of Geoffrion. This means that $(-1, 0)$ and $(0, -1)$ are improperly efficient solutions in the sense of Geoffrion.

Remark 3.9 In case $Y = \mathbb{R}^\ell$, $C = \mathbb{R}^\ell_+$, $K \subseteq \mathbb{R}^n$ is a convex set and $f_i : \mathbb{R}^n \to \mathbb{R}$, $i \in \{1, 2, \ldots, \ell\}$ are convex functions, it follows from Benson [1] and Ehrgott [5, Theorem 2.26] that Definition 3.10 and Definition 3.13 coincide.

Theorem 3.8 *Let $C = \mathbb{R}^\ell_+$. Then $\bar{x} \in K$ is a properly efficient solution in the sense of Geoffrion if and only if it is a properly efficient solution in the sense of Benson.*

Proof Let $\bar{x} \in K$ be a properly efficient solution in the sense of Geoffrion. Then $\bar{x}$ is efficient. Suppose that $\bar{x}$ is not a properly efficient solution in the sense of Benson. Then there exists a vector $d \neq \mathbf{0}$ such that

$$d \in \text{cl}(\text{cone}(\mathcal{Y} + \mathbb{R}^\ell_+ - \{f(\bar{x})\})) \cap (-\mathbb{R}^\ell_+).$$

Without loss of generality, we assume that $d_1 < -1$, $d_i \leq 0$ $(i = 2, 3, \ldots, \ell)$. Let $t_k(f(x^k) + r^k - \{f(\bar{x})\}) \to d$, where $r^k \in \mathbb{R}^\ell_+$, $t_k > 0$, and $x^k \in K$. By choosing a subsequence we can assume that $\tilde{I} := \{i : f_i(x^k) > f_i(\bar{x})\}$ is constant for all k, and nonempty since $\bar{x}$ is efficient. Choose a positive number M. Then there is some number k_0 such that for all k with $k \geq k_0$, we have

$$f_1(x^k) - f_1(\bar{x}) < -\frac{1}{2t_k}$$

and

$$f_i(x^k) - f_i(\bar{x}) \leq \frac{1}{2} M t_k \quad (i = 2, 3, \ldots, \ell).$$

Then for all $i \in \tilde{I}$, we have for $k \geq k_0$

$$0 < f_i(x^k) - f_i(\bar{x}) \leq \frac{1}{2Mt_k}$$

and

$$\frac{f_1(\bar{x}) - f_1(x^k)}{f_i(x^k) - f_i(\bar{x})} > \frac{\frac{1}{2t_k}}{\frac{1}{2Mt_k}} = M.$$

Therefore, $\bar{x}$ is not properly efficient in the sense of Geoffrion, a contradiction.

Conversely, let $\bar{x} \in K$ be properly efficient in the sense of Benson, and let $\bar{x}$ be an efficient solution. Assume that $\bar{x}$ is not properly efficient in the sense of Geoffrion. Let $\{M_k\}$ be an unbounded sequence of positive real numbers. Then, by reordering the objective functions (if necessary), we can assume that for each M_k there exists an $x^k \in K$ such that $f_1(x_k) < f_1(\bar{x})$ and

$$\frac{f_1(\bar{x}) - f_1(x_k)}{f_i(x^k) - f_i(\bar{x})} > M_k$$

for all $i = 2, 3, \ldots, \ell$ such that $f_i(x^k) > f_i(\bar{x})$. By choosing a subsequence of M_k (if necessary), we can assume that $\tilde{I} := \{i : f_i(x^k) > f_i(\bar{x})\}$ is constant and nonempty as $\bar{x}$ is efficient. For each k let $t_k := (f_1(\bar{x}) - f_1(\bar{x}))^{-1}$. Then t_k is positive for all k. Define

$$r_i^k := \begin{cases} 0, & \text{for } i = 1 \text{ or } i \in \tilde{I}, \\ f_i(\bar{x}) - f_i(x_k), & \text{for } i \neq 1,\ i \notin \tilde{I}. \end{cases}$$

Clearly, $r_k \in \mathbb{R}^\ell_+$, and

$$t_k(f_i(x^k)) + r_i^k - f_i(\bar{x}) = \begin{cases} -1, & \text{for } i = 1, \\ 0, & \text{for } i \neq 1,\ i \notin \tilde{I}, \end{cases}$$

$$0 < t_k(f_i(x^k) + r_i^k - f_i(\bar{x})) = t_k(f_i(x^k) - f_i(\bar{x})) < M_k^{-1} \quad \text{for } i \in \tilde{I}.$$

For $i = 1, 2, \ldots, \ell$, let $d_i := \lim_{k \to +\infty} t_k(f_i(x^k) + r_i^k - f_i(x))$. Then $d_1 = -1$ and $d_i = 0$ when $i \neq 1$ and $i \notin \tilde{I}$. Since $\{M_k\}$ is an unbounded sequence of positive real numbers, we have $d_i = 0$ for $i \in \tilde{I}$. Therefore,

$$d = (-1, 0, 0, \ldots, 0) \neq \mathbf{0} \in \text{cl}(\text{cone}(f(K) + \mathbb{R}^\ell_+ - \{f(\bar{x})\})) \cap -\mathbb{R}^\ell_+.$$

Hence $\bar{x}$ is not a properly efficient solution in the sense of Benson, in contradiction to the assumption. □

Furthermore, we introduce the following problem, which will be called VOP′. Let $f : \mathbb{R}^n \to \mathbb{R}^\ell$ and $g : \mathbb{R}^n \to \mathbb{R}^m$, and for all $i = 1, 2, \ldots, \ell$, $j = 1, 2, \ldots, m$, let f_i and g_j be continuously differentiable. Then VOP′ reads

$$\begin{aligned} &\text{minimize} f(x), \\ &\text{subject to } x \in K, \end{aligned}$$

where $K := \{x \in \mathbb{R}^n : g_1(x) \leq 0,\ g_2(x) \leq 0, \ldots, g_m(x) \leq 0\}$.

Definition 3.14 (Kuhn-Tucker 1951) An element $\bar{x} \in K \subseteq \mathbb{R}^n$ is a *properly efficient solution* of VOP′ *in the sense of Kuhn-Tucker* if it is efficient and if there is no $h \in \mathbb{R}^n$ such that

$$\begin{aligned} \langle \nabla f_i(\bar{x}), h \rangle &\leq 0 \quad \text{for all } i = 1, 2, \ldots, \ell, \\ \langle \nabla f_i(\bar{x}), h \rangle &< 0 \quad \text{for some } i = 1, 2, \ldots, \ell, \\ \text{and } \langle \nabla g_j(\bar{x}), h \rangle &\leq 0 \quad \text{for all } j \in J(\bar{x}) = \{j \in \{1, 2, \ldots, m\} : g_j(\bar{x}) = 0\}. \end{aligned}$$

Theorem 3.9 *Let the functions f_i and g_j be convex for all $i = 1, 2, \ldots, \ell$, $j = 1, 2, \ldots, m$. If $\bar{x} \in K$ is a properly efficient solution of* VOP′ *in the sense of Kuhn-Tucker, then it is also properly efficient in the sense of Geoffrion.*

The following example shows that the converse statement in Theorem 3.9 is not generally true.

Example 3.18 Let $K = \{(x_1, x_2) \in \mathbb{R}^2 : -x_1 \leq 0, -x_2 \leq 0, (x_1 - 1)^5 + x_2 \leq 0\}$, $f_1(x) = -3x_1 - 2x_2 + 3$, $f_2(x) = -x_1 - 3x_2$, and $C = \mathbb{R}^2_+$ (see Fig. 3.20). Then the element $x = (1, 0)$ is properly efficient in the sense of Geoffrion, but it is not a properly efficient solution in the sense of Kuhn-Tucker.

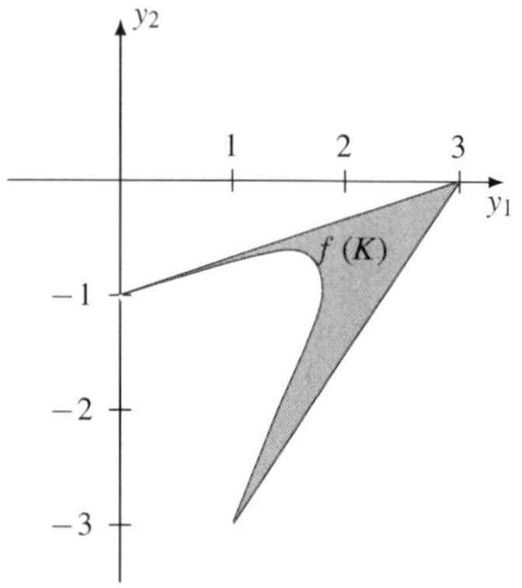

Fig. 3.20 The solution $x = (1, 0)$ with $f(x) = (0, -1)$ is properly efficient in the sense of Geoffrion, but not properly efficient in the sense of Kuhn-Tucker (see Example 3.18)

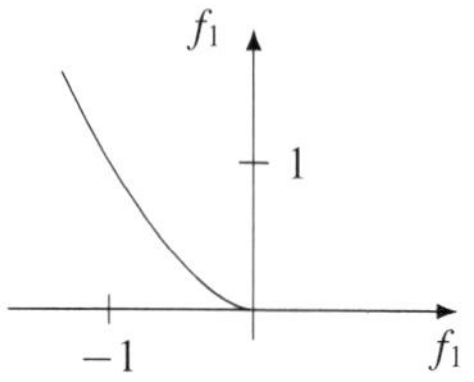

Fig. 3.21 $\bar{x} = 0$ with $f(\bar{x}) = (0, 0)$ is a properly efficient solution in the sense of Kuhn-Tucker, but not properly efficient in the sense of Geoffrion (see Example 3.19)

The following example shows that the assertion in Theorem 3.9 is not true if the objective functions are not convex.

Example 3.19 Let $K = \{x \in \mathbb{R} : -x \leq 0\}, f : K \to \mathbb{R}^2$ with $f_1 = -x^3$, which is not convex on K, and $f_2 = x^5$ (see Fig. 3.21). Then obviously $\bar{x} = 0$ is a properly efficient solution in the sense of Kuhn-Tucker. However, we show that $\bar{x} = 0$ is not properly efficient in the sense of Geoffrion. We have to show that for all $M > 0$, there exists an index i and elements $x \in K$ such that $f_i(x) < f_i(\bar{x})$, such that for all j with $f_j(\bar{x}) < f_j(x)$, we have

$$\frac{f_i(\bar{x}) - f_i(x)}{f_j(x) - f_j(\bar{x})} > M.$$

Let $i = 1$, and choose $x = \varepsilon$ for some $\varepsilon \in]0, 1[$. Then we have $f_1(x) = -\varepsilon^3 < 0 = f_1(\bar{x})$, as well as $f_2(x) = \varepsilon^5 > 0 = f_2(\bar{x})$. Moreover,

$$\frac{f_1(\bar{x}) - f_1(x)}{f_2(x) - f_2(\bar{x})} = \frac{\varepsilon^3}{\varepsilon^5} = \frac{1}{\varepsilon^2} \underset{\varepsilon \to 0}{\longrightarrow} +\infty.$$

Therefore, $\bar{x} = 0$ is not properly efficient in the sense of Geoffrion.

Figure 3.22 describes the discussed relations between the notions of proper efficiency.

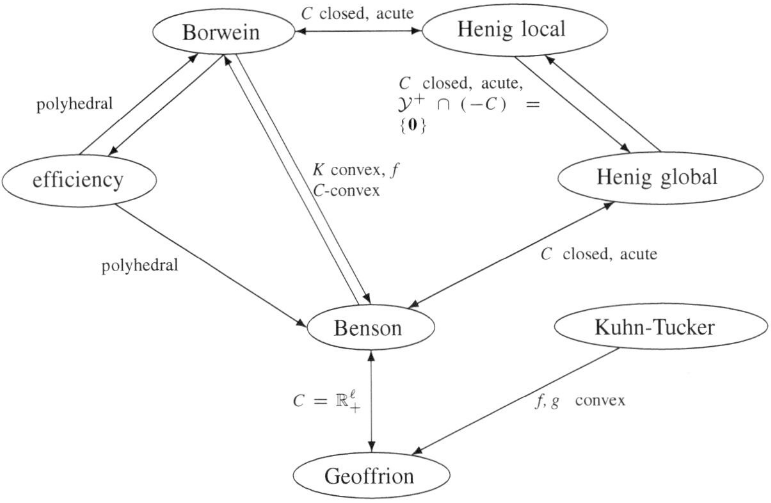

Fig. 3.22 Relations among different kinds of proper efficiency and efficiency

3.3 Existence of Solutions

For scalar optimization problems, Weierstrass's theorem guarantees the existence of extremal points for a continuous function $f : K \subset \mathbb{R}^n \to \mathbb{R}$ if K is compact. The well-known generalization assures the existence of a minimal point under the hypotheses that f is lower semicontinuous and the set K is compact. If we turn our attention to the image set $\mathcal{Y} = f(K)$, then we can state that $\mathcal{Y}$ has a minimal value when $\mathcal{Y} + \mathbb{R}_+$ is closed and bounded from below.

In this section, unless otherwise specified, we assume that $\mathcal{Y}$ is a subset of $\mathbb{R}^\ell$ and $C = \mathbb{R}^\ell_+$ or C is any proper convex pointed closed cone in $\mathbb{R}^\ell$.

Theorem 3.10 (Borwein 1983) *Suppose that there is some $\hat{y} \in \mathcal{Y}$ such that the section $\hat{Y} = \{y \in \mathcal{Y} : y \leq_C \hat{y}\} = (\hat{y} - C) \cap \mathcal{Y}$ is compact ("$\mathcal{Y}$ contains a compact section"). Then $\mathcal{Y}_{eff}$ is nonempty.*

Theorem 3.11 (Corley 1980) *If $\mathcal{Y}$ is a nonempty C-semicompact set, then $\mathcal{Y}_{eff} \neq \emptyset$.*

Corollary 3.5 (Hartley 1978) *If $\mathcal{Y} \subset \mathbb{R}^\ell$ is nonempty and C-compact, then $\mathcal{Y}_{eff} \neq \emptyset$.*

Proof The result follows from Theorem 3.1 and Proposition 2.12. □

Proposition 3.9 *Let $K \subset \mathbb{R}^n$ be a nonempty compact set and $f : \mathbb{R}^n \to \mathbb{R}^\ell$ be a C-upper semicontinuous vector-valued function. Then the nondominated set $\mathcal{Y}_{eff}$ is C-semicompact.*

Proof Let $\{(y_\alpha - C) : y_\alpha \in \mathcal{Y},\ \alpha \in \Lambda\}$ be an open cover of $\mathcal{Y}$. By C-semicontinuity of f, $\{f^{-1}((y_\alpha - C)^c) : y_\alpha \in \mathcal{Y},\ \alpha \in \Lambda\}$ is an open cover of K. Since K is compact, this open cover has a finite subcover of K. The image of such open subcover is a finite subcover of $\mathcal{Y}$ and whence $\mathcal{Y}$ is C-semicompact. □

Theorem 3.12 *Let $K \subset \mathbb{R}^n$ be a nonempty compact set and $f : \mathbb{R}^n \to \mathbb{R}^\ell$ be a C-upper semicontinuous vector-valued function. Then $K_{eff} \neq \emptyset$.*

Proof It follows directly from Proposition 3.9. □

Theorem 3.13 *If $\mathcal{Y} \subset \mathbb{R}^\ell$ is a nonempty and compact set, then $\mathcal{Y}_{w\text{-}eff} \neq \emptyset$.*

Proof Suppose that $\mathcal{Y}_{\text{w-eff}} = \emptyset$. Then for all $y \in \mathcal{Y}$, there is some $\tilde{y} \in \mathcal{Y}$ such that $y \in \tilde{y} + \text{int}(C)$. Taking the union over all $y \in \mathcal{Y}$, we obtain

$$\mathcal{Y} \subset \bigcup_{\tilde{y} \in \mathcal{Y}} (\tilde{y} + \text{int}(C))\,.$$

Since $\tilde{y} + \text{int}(C)$ is an open set, above inclusion implies that $\bigcup_{\tilde{y} \in \mathcal{Y}} (\tilde{y} + \text{int}(C))$ forms an open covering of $\mathcal{Y}$. Since $\mathcal{Y}$ is compact, there exists a finite subcover with

$$\mathcal{Y} \subset \bigcup_{i=1}^{n} (\tilde{y}_i + \text{int}(C))\,.$$

Choosing especially $\tilde{y}_i$ on the left hand side yields that for all $i = 1, 2, \ldots, k$, there is some $1 \le j \le k$ with $\tilde{y}_i \in \tilde{y}_j + \text{int}(C)$. In other words, for all i there is a j such that $\tilde{y}_j <_C \tilde{y}_i$. By transitivity of $<_C$ and because there are finitely many $\tilde{y}_i$ there is some i^* and a chain of inequalities such that

$$\tilde{y}_{i^*} <_C \tilde{y}_{i_1} <_C \cdots <_C \tilde{y}_{i_m} <_C \tilde{y}_{i^*}\,,$$

which is impossible. □

Corollary 3.6 *Let $K \subset \mathbb{R}^n$ be a nonempty and compact set, $C = \mathbb{R}^\ell_+$ and let $f : \mathbb{R}^n \to \mathbb{R}^\ell$ be continuous. Then $K_{w\text{-}eff} \neq \emptyset$.*

Proof The result follows from Theorem 3.13 and $K_{\text{eff}} \subset K_{\text{w-eff}}$ or from Theorem 3.12 and the fact that $f(K)$ is compact for compact K and continuous f.

□

Next we verify that the compactness assumption on K and the continuity of f is indeed necessary for the assertion in Corollary 3.6.

Example 3.20 In the following examples, we use the natural ordering cone $C = \mathbb{R}^2_+$.

(a) Consider the set $K := \{(x_1, x_2) \in \mathbb{R}^2 : 0 < x_1 \le 1, -\sqrt{x_1} < x_2 < \sqrt{x_1}\}$ and the objective function defined as $f(x_1, x_2) := (x_1, \frac{1}{2} - x_2^2)$. Then the set of weakly (in (a)): efficient solutions is empty, and therefore there do not exist any efficient solutions either. If, on the other hand, one considers the objective

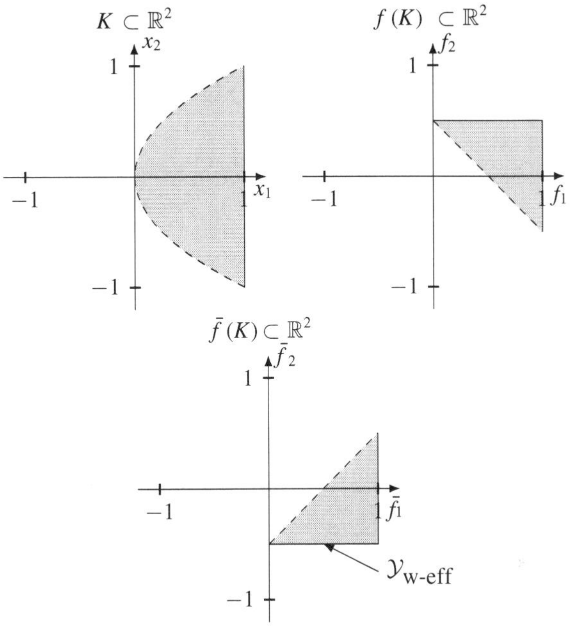

Fig. 3.23 The set of weakly efficient solutions of $f(K)$ is empty, whereas weakly efficient solutions exist for the set $\bar{f}(K)$ (see Example 3.20 (a))

$\bar{f}(x_1, x_2) := (x_1, -\frac{1}{2} + x_2^2)$ instead the set of weakly efficient solutions is not empty and given as $\{(x_1, x_2) \in \mathbb{R}^2 : 0 \leq x_1 \leq 1,\ x_2 = 0\}$. For an illustration, see Fig. 3.23.

(b) The same idea remains valid for the example $K := \{(x_1, x_2) \in \mathbb{R}^2 : -1 < x_1, x_2 < 1\}$ and the objective function $f(x_1, x_2) := (-x_1^2, -x_2^2) + (\frac{1}{2}, \frac{1}{2})$. However, if we choose $\bar{f}(x_1, x_2) := (x_1^2, x_2^2) - (\frac{1}{2}, \frac{1}{2})$, then there exist weakly efficient solutions (see Examples 3.6 and 3.9 and Fig. 3.24).

(c) To show that the continuity is necessary, we consider an example with the compact set $\mathbb{R}^1 \supset K := [0, 1]$ and the objective function

$$f(x) := \begin{cases} -x^{-1}, & \text{if } x \neq 0, \\ 1, & \text{otherwise.} \end{cases}$$

The image set of f is the open interval $(-\infty, 1]$ such that no weakly efficient solutions can be determined.

(d) This example can be extended to $\mathbb{R}^n$ in a straightforward manner. For the image space $\mathbb{R}^2$ and the choice $K := \{(x_1, x_2) : x_1^2 + x_2^2 \leq 1\}$ consider the objectives $f(x_1, x_2) := (-\frac{1}{x_1^2}, -\frac{1}{x_2^2})$ and $\bar{f}(x_1, x_2) := -f(x_1, x_2)$ (see Example 3.7). Again, we have an empty set of weakly efficient solutions for f, but in the latter case there are even efficient solutions. For an illustration, see Fig. 3.25.

(e) Let $K := \{(x_1, x_2) : x_1^2 + x_2^2 < 1\}$ the open unit ball in $\mathbb{R}^2$. We consider any linear mapping $f : \mathbb{R}^2 \to \mathbb{R}^2$ with $f(x_1, x_1) := A \cdot (x_1, x_2) + (b_1, b_2)$ for an

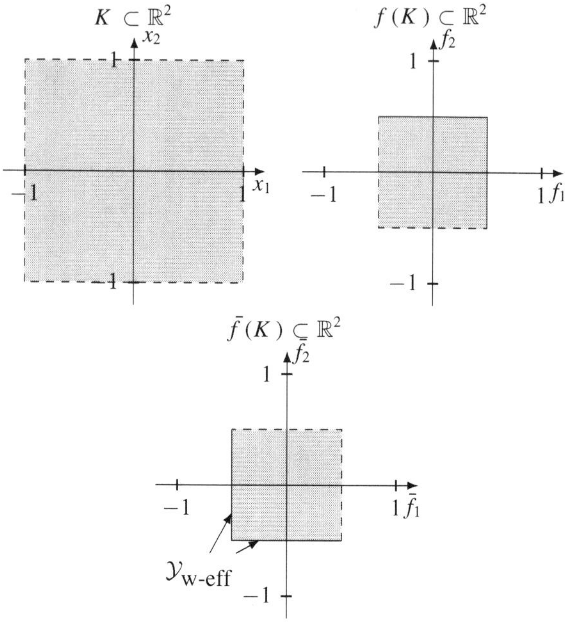

Fig. 3.24 Again, the set of weakly efficient solutions of $f(K)$ is empty, whereas weakly efficient solutions exist for the set $\bar{f}(K)$ (see Example 3.20 (b))

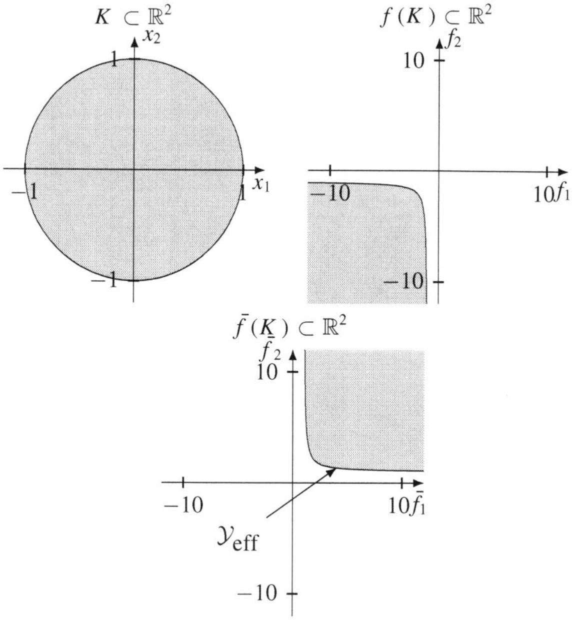

Fig. 3.25 The set of weakly efficient solutions for $f(K)$ is empty, but weakly efficient solutions of $\bar{f}(K)$ exist (see Example 3.20 (d))

invertible matrix $A \in \mathbb{R}^2 \times \mathbb{R}^2$ and $(b_1, b_2) \in \mathbb{R}^2$. The set of weakly efficient solutions is empty because K as well as the image set of K is an open set.

3.4 Optimality Notions for Variable Ordering Structures

It is well known from various applications that for modeling a problem as VOP, a fixed ordering cone is not sufficient for the description of efficient solutions. The motivation of introducing variable ordering cones is presented in [6–8] in the context of medical image registration.

Let X and Y be topological vector spaces and K be a nonempty subset of X. Let $C : K \to 2^Y$ be a set-valued map such that for all $x \in K$, $C(x)$ is a closed convex pointed cone. We further assume that $\text{int}(C(x)) \neq \emptyset$ wherever $\text{int}(C(x))$ is involved. Let $f : K \to Y$ be a vector-valued function. Recall the vector optimization problem (VOP):

$$\text{(VOP)} \qquad \text{minimize } f(x), \quad \text{subject to } x \in K.$$

Definition 3.15 A point $\bar{x} \in K$ is said to be a

(a) *dominated strong efficient solution* or *dominated strong Pareto solution* of VOP if

$$f(y) - f(\bar{x}) \in C(\bar{x}), \quad \text{for all } y \in K;$$

(b) *dominated efficient solution* or *dominated Pareto solution* of VOP if

$$f(y) - f(\bar{x}) \notin -C(\bar{x}) \setminus \{\mathbf{0}\}, \quad \text{for all } y \in K;$$

(c) *dominated weakly efficient solution* or *dominated weakly Pareto solution* of VOP if

$$f(y) - f(\bar{x}) \notin -\text{int}(C(\bar{x})), \quad \text{for all } y \in K;$$

(d) *dominated properly efficient solution (in the sense of Henig)* or *dominated properly Pareto solution (in the sense of Henig)* of VOP if there is a set-valued map $D : K \to 2^Y$ such that for all $x \in K$, $D(x)$ is a convex set and $C(x) \setminus \{\mathbf{0}\} \subseteq \text{int}(D(x))$, and

$$f(y) - f(\bar{x}) \notin -D(\bar{x}) \setminus \{\mathbf{0}\}, \quad \text{for all } y \in K.$$

Remark 3.10

(a) If C is a constant map, that is, for each $x \in K$, $C(x)$ is a fixed closed convex pointed cone with nonempty interior, then dominated strong efficient solution, dominated efficient solution, dominated weak efficient solution, and dominated properly efficient solution are called strong efficient solution (or strong Pareto solution), efficient solution (or Pareto solution), weak efficient solution (or weak

Pareto solution), and properly efficient solution (or properly Pareto solution), respectively.

(b) It is obvious that every (dominated) strong efficient solution is a (dominated) efficient solution, every (dominated) efficient solution is a (dominated) weak efficient solution and every (dominated) properly efficient solution is a (dominated) efficient solution.

Example 3.21 We revisit Example 3.6, that is, we consider the feasible set $K := \{(x_1, x_2) \in \mathbb{R}^2 : -1 < x_1 < 1, -1 < x_2 < 1\}$ and the objective function $f(x_1, x_2) := (x_1^2, x_2^2) - (\frac{1}{2}, \frac{1}{2})$. Let the variable ordering cone be given by

$$C(x) = \begin{cases} \mathbb{R}^2_+, & \text{if } x_2 = 0,\ x_1 \in [0, 1[, \\ \widetilde{C}, & \text{else,} \end{cases}$$

where $\widetilde{C} := \{(y_1, y_2) \in \mathbb{R}^2 : (y_1, y_2) = r_1 \cdot (1, 2) + r_2 \cdot (2, 1),\ r_1, r_2 \in \mathbb{R}_+\}$ (see, Fig. 3.26). Then $x = (0, 0)$ is not a dominated strong efficient solution, but it is a dominated efficient solution. The set of all dominated efficient solutions is $\{(x_1, x_2) \in K : x_1 = 0, x_2 \in [0, 1[\}$. The set of all dominated weakly efficient solutions is $\{(x_1, x_2) \in K : x_1 = 0, x_2 \in [0, 1[\} \bigcup \{(x_1, x_2) \in K : x_2 = 0, x_1 \in [0, 1[\}$, which equals the set of weakly efficient solutions if the fixed ordering cone $C = \mathbb{R}^2_+$ is chosen (see Example 3.6). The set of dominated properly efficient solutions is equal to the set of dominated efficient solutions.

Example 3.22 Consider the feasible set $K = \{(x_1, x_2) \in \mathbb{R}^2 : x_1^2 + x_2^2 \leq 1\}$, the objective function $f(x_1, x_2) = (x_1, x_2)$ and the ordering cone

$$C(x) = \begin{cases} \widetilde{C}, & \text{if } x_1 = 0,\ x_2 = -1, \\ \mathbb{R}^2_+, & \text{else,} \end{cases}$$

where $\widetilde{C} := \{(y_1, y_2) \in \mathbb{R}^2 : (y_1, y_2) = r_1 \cdot (0.25, 1) + r_2 \cdot (1, 0.5),\ r_1, r_2 \in \mathbb{R}_+\}$ (see, Fig. 3.27). It can be seen that $x = (0, -1)$ is a dominated properly efficient solution, but $x = (-1, 0)$ is not a dominated properly efficient solution. The set of dominated properly efficient solutions is $\{(x_1, x_2) \in K : x_1 < 0, x_2 = -\sqrt{1 - x_1^2}\} \bigcup \{(-1, 0)\}$.

Theorem 3.14 *Define $C_1 := \bigcap_{x \in K} C(x)$ and $C_2 := \bigcup_{x \in K} C(x)$, and assume that these cones are pointed and convex. Then we have the following assertions:*

(a) If $\bar{x} \in K$ is a dominated efficient solution of VOP, *then $\bar{x}$ is an efficient solution of* VOP, *where efficiency is defined by means of the cone C_1 in the sense of Definition 3.5.*

(b) If $\bar{x}$ is an efficient solution of VOP, *where efficiency is defined by means of the cone C_2 in the sense of Definition 3.5, then $\bar{x} \in K$ is a dominated efficient solution of* VOP.

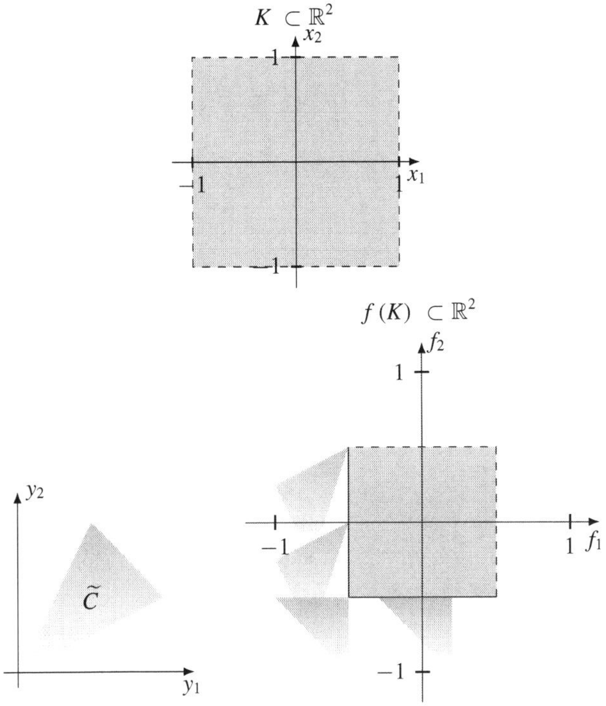

Fig. 3.26 The feasible region K, the cone $\widetilde{C}$ and the feasible region K with some attached variable ordering cones of Example 3.21

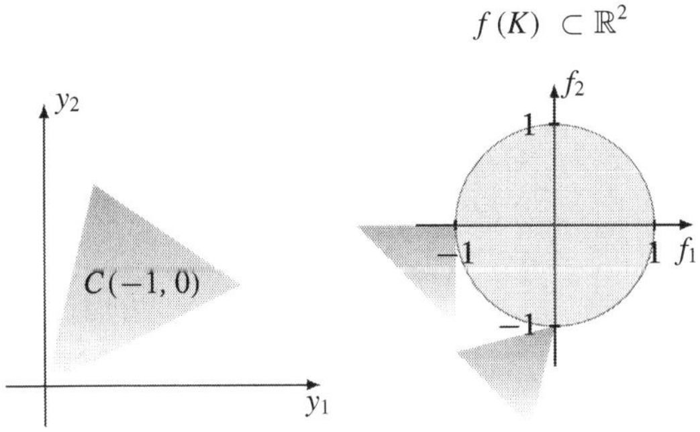

Fig. 3.27 Illustration of Example 3.22

Proof We only prove part (a), as (b) can be proven in a similar way. Let $\bar{x} \in K$ be a dominated efficient solution of VOP. Then, for all $y \in K$,

$$f(y) - f(\bar{x}) \notin -C(\bar{x}) \setminus \{\mathbf{0}\}.$$

Now suppose that $\bar{x}$ is not an efficient solution of VOP, where efficiency is defined by means of the cone C_2 in the sense of Definition 3.5. Then, $(f(\bar{x})-C_2)\cap\mathcal{Y} \neq \{f(\bar{x})\}$. This means that there is some $x \in K$ such that $f(x) \in \mathcal{Y}$ and

$$f(x) \in f(\bar{x}) - C_2 \setminus \{\mathbf{0}\} \subseteq f(\bar{x}) - C(\bar{x}) \setminus \{\mathbf{0}\},$$

contradicting the assumption. □

A similar result can be formulated for dominated weakly efficient solutions of VOP. We refer to [8] for corresponding results.

Example 3.23 Coming back to Example 3.21, Theorem 3.14 can be applied as follows. We have $C_1 = \mathbb{R}^2_+ \cap \widetilde{C} = \widetilde{C}$ and $C_2 = \mathbb{R}^2_+ \cup \widetilde{C} = \mathbb{R}^2$. The set of all dominated efficient solutions of VOP has been identified in Example 3.21 as $K_{\text{d-eff}} := \{(x_1, x_2) \in K : x_1 = 0, x_2 \in [0, 1[\}$. Therefore, we obtain with Theorem 3.14 (a) that every element in $K_{\text{d-eff}}$ is an efficient solution of VOP, where efficiency is defined by means of the cone C_1 in the sense of Definition 3.5. In order to apply Theorem 3.14 (b), we have to determine the efficient solution of VOP, by means of $C_2 = \mathbb{R}^2$ in the sense of Definition 3.5. The only efficient solution, which has already been found in Example 3.6, is $(x_1, x_2) = (0, 0)$. Therefore, we conclude that $(0, 0)$ is a dominated efficient solution of VOP.

Definition 3.16 Let K be a nonempty convex subset of X and $f : K \to Y$ be a Gâteaux differentiable function with Gâteaux derivative Df. Let $x \in K$ be an arbitrary element. Then f is said to be:

(a) *C_x-convex* if for all $y \in K$,

$$f(y) - f(x) - \langle Df(x), y - x\rangle \in C(x);$$

(b) *strictly C_x-convex* if for all $y \in K$, $y \neq x$,

$$f(y) - f(x) - \langle Df(x), y - x\rangle \in \operatorname{int}(C(x));$$

(c) *strongly C_x-pseudoconvex* if for all $y \in K$,

$$\langle Df(x), y - x\rangle \in C(x) \quad \text{implies} \quad f(y) - f(x) \in C(x);$$

(d) *C_x-pseudoconvex* if for all $y \in K$,

$$\langle Df(x), y - x\rangle \notin -C(x) \setminus \{\mathbf{0}\} \quad \text{implies} \quad f(y) - f(x) \notin -C(x) \setminus \{\mathbf{0}\};$$

(e) weakly *C_x-pseudoconvex* if for all $y \in K$,

$$\langle Df(x), y - x\rangle \notin -\operatorname{int}(C(x)) \quad \text{implies} \quad f(y) - f(x) \notin -\operatorname{int}(C(x)).$$

If $C(x)$ is a fixed closed convex pointed cone P with $\operatorname{int}(P) \neq \emptyset$, then C_x-convexity, strict C_x-convexity, strong C_x-pseudoconvexity, C_x-pseudoconvexity and weak C_x-pseudoconvexity are called P-convexity, strict P-convexity, strong P-pseudoconvexity, P-pseudoconvexity and weak P-pseudoconvexity, respectively.

Example 3.24 For the objective function $f(x_1, x_2) := \left(x_1^2, x_2^2\right) - \left(\frac{1}{2}, \frac{1}{2}\right)$ considered in Example 3.21, we observe that

$$\langle Df(x), y - x \rangle = (2x_1(y_1 - x_1), 2x_2(y_2 - x_2)),$$

and therefore, it holds

$$f(y) - f(x) - \langle Df(x), y - x \rangle = \left((y_1 - x_1)^2, (y_2 - x_2)^2\right).$$

One observes that f is C_x-convex for all $x = (x_1, x_2)$ with $x_2 = 0$, $x_1 \in [0, 1[$. If one chooses, for example, $x_1 = 0$, $x_2 = 0.5$. Then, for $y = (0, 0)$,

$$f(y) - f(x) - \langle Df(x), y - x \rangle = \left((y_1 - x_1)^2, (y_2 - x_2)^2\right) = (0, 0.25) \notin C(x).$$

Therefore, f is not C_x-convex for $x = (0, 0.5)$.

Example 3.25 Coming back to Example 3.22, we observe that the objective function $f(x_1, x_2) = (x_1, x_2)$ is C_x-convex, strongly C_x-convex, C_x-pseudoconvex, weakly C_x-pseudoconvex for all $x \in K$, but it is not strictly C_x-convex for any $x \in K$.

References

1. H.P. Benson, An improved definition of proper efficiency for vector maximization with respect to cones. J. Optim. Theory Appl. **71**, 232–241 (1979)
2. J. Borwein, Proper efficient points for maximization with respect to cones. SIAM J. Control Optim. **15**(1), 57–63 (1977)
3. V. Chankong, Y.Y. Haimes, *Multiobjective Decision Making: Theory and Methodology* (Elsevier Science Publishing Co., Inc., New York, 1983)
4. F.Y. Edgeworth, *Mathematical Psychics* (Kegan Paul, London, 1881)
5. M. Ehrgott, *Multicriteria Optimization*, Lecture Notes in Economics and Mathematical Sciences, vol. 491 (Springer, Berlin, 2000)
6. G. Eichfelder, Variable Ordering Structures in Vector Optimization, Habilitation Thesis, University Erlangen-Nürnberg, 2011
7. G. Eichfelder, Variable ordering structures in vector optimization, in *Recent Developments in Vector Optimization*, ed. by Q.H. Ansari, J.-C. Yao (Springer, Heidelberg, 2012), pp. 95–126
8. G. Eichfelder, *Variable Ordering Structures in Vector Optimization* (Springer, Berlin, 2014)
9. A. Engau, Definition and characterization of Geoffrion proper efficiency for real vector optimization with infinitely many criteria. J. Optim. Theory Appl. **165**, 439–457 (2015)
10. A.M. Geoffrion, Proper efficiency and the theory of vector maximization. J. Math. Anal. Appl. **22**, 618–630 (1968)

11. J. Jahn, *Vector Optimization: Theory, Applications, and Extensions* (Springer, Berlin, Heidelberg, 2004)
12. H.W. Kuhn, A.W. Tucker, Nonlinear programming, in *Proceedings of the second Berkeley Symposium on Mathematical Statistics and Probability, 1950* (University of California Press, Berkeley, 1951), pp. 481–492
13. D.T. Luc, Structure of the efficient point set. Proc. Amer. Math. Soc. **95**(3), 433–440 (1985)
14. D.T. Luc, *Theory of Vector Optimization*, Lecture Notes in Economics and Mathematical Systems, vol. 319 (Springer, Berlin, 1989)
15. D.T. Luc, *Multiobjective Linear Programming: An Introduction* (Springer, Berlin, 2016)
16. V. Pareto, *Manuale di economia politica* (Societa Editrice Libraria, Milano, 1906). English translation: V. Pareto, *Manual of Bibliography* 467 political economy, translated by A.S. Schwier, Augustus M. Kelley Publishers, New York (1971)
17. Y. Sawaragi, H. Nakayama, T. Tanino, *Theory of Multiobjective Optimization* (Academic Press Inc., Orlando, 1985)

Chapter 4
Classical Methods in Vector Optimization

In this chapter, we investigate solution procedures to obtain efficient solutions of a vector optimization problem.

The results that we are recalling in this chapter are standard and can be found in several books on vector optimization, for example in [10, 13, 32, 39, 44].

We briefly recall the vector optimization problem (in short, VOP): Let Y be a topological vector space with a nontrivial closed pointed convex cone C, and X be a vector space. Whenever we use $\text{int}(C)$, we assume that $\text{int}(C) \neq \emptyset$. We consider the problem

$$\begin{aligned} &\text{minimize}\, f(x), \\ &\text{subject to}\; x \in K, \end{aligned} \tag{4.1}$$

where

- $\emptyset \neq K \subseteq X$ is a *feasible region*
- $f : X \to Y$ is a *objective function*
- Y is the *objective space*
- x is a *decision (variable) vector*
- X is the *decision variable space*
- $\mathcal{Y} := f(K)$ is the *feasible objective region*

In the following sections, we show how VOP (4.1) can be converted into several scalar-valued minimization problems, and how solutions of the corresponding scalar-valued minimization problems relate to the efficient solutions of VOP. We present linear and nonlinear scalarization techniques. By means of these methods, we are able to characterize efficient solutions of VOP. The presented procedures are based on the scalarization of VOP, that is, on the principle of transforming VOP into a scalar optimization problem. The scalarization problem in these methods is formulated in a parameterizable way. By varying the parameter, different scalar optimization problems can be generated, and hence, several optimal solutions of

Q.H. Ansari et al., *Vector Variational Inequalities and Vector Optimization*,
Vector Optimization, DOI 10.1007/978-3-319-63049-6_4

such problems can be found. Because a scalarized problem is often easier to solve, there is a huge advantage of using scalarization techniques.

For the particular case where $X = \mathbb{R}^n$ and $Y = \mathbb{R}^\ell$, we present some well known and widely used methods, namely, the weighted sum method, ε-constraint method and the hybrid method.

For each given $\ell \in \mathbb{N}$, we denote by $\mathbb{R}^\ell_+$ the non-negative orthant of $\mathbb{R}^\ell$, that is,

$$\mathbb{R}^\ell_+ = \left\{x = (x_1, x_2, \dots, x_\ell) \in \mathbb{R}^\ell : x_i \geq 0, \text{ for } i = 1, 2, \dots, \ell\right\},$$

so that $\mathbb{R}^\ell_+$ has a nonempty interior with the topology induced in terms of convergence of vectors with respect to the Euclidean metric. That is,

$$\operatorname{int}(\mathbb{R}^\ell_+) = \left\{x = (x_1, x_2, \dots, x_\ell) \in \mathbb{R}^\ell : x_i > 0, \text{ for } i = 1, 2, \dots, \ell\right\}.$$

We denote by $\mathbb{T}^\ell_+$ and $\operatorname{int}(\mathbb{T}^\ell_+)$ the simplex of $\mathbb{R}^\ell_+$ and its relative interior, respectively, that is,

$$\mathbb{T}^\ell_+ = \left\{x = (x_1, x_2, \dots, x_\ell) \in \mathbb{R}^\ell_+ : \|x\| = \sum_{i=1}^{\ell} x_i = 1\right\},$$

and

$$\operatorname{int}(\mathbb{T}^\ell_+) = \left\{x = (x_1, x_2, \dots, x_\ell) \in \operatorname{int}(\mathbb{R}^\ell_+) : \|x\| = \sum_{i=1}^{\ell} x_i = 1\right\}.$$

e denotes the unit vector in $\mathbb{R}^\ell$, that is, $e = (1, 1, \dots, 1)$.

4.1 Linear Scalarization

This section deals with a linear scalarization of VOP. As we will see further on, such a method is beneficial for a convex VOP. The next theorem states how a linear scalarization of VOP relates to weakly efficient solutions of VOP.

Theorem 4.1 *Let Y be a topological vector space with a nontrivial pointed convex cone C, X be a vector space with $\emptyset \neq K \subseteq X$, and $f : X \to Y$ be a given function. If there exist a linear functional $W \in C^* \setminus \{\mathbf{0}\}$ and $\bar{x} \in K$ such that*

$$W\big(f(\bar{x})\big) \leq W\big(f(x)\big), \quad \textit{for all } x \in K,$$

then $\bar{x}$ is a weakly efficient solution of VOP.

Proof Suppose that $\bar{x}$ is not a weakly efficient solution of VOP. Then $(f(\bar{x}) - \text{int}(C)) \cap \mathcal{Y} \neq \emptyset$, and there is some element $\tilde{x}$ with $f(\tilde{x}) \in \mathcal{Y}$ such that $f(\tilde{x}) \in f(\bar{x}) - \text{int}(C)$. Hence, we conclude that for all $W \in C^* \setminus \{\mathbf{0}\}$,

$$W\big(f(\tilde{x})\big) < W\big(f(\bar{x})\big),$$

a contradiction to our supposition. □

As a particular case of the linear scalarization technique, we obtain the weighted sum method, which was presented by Gass and Saaty [18] and Zadeh [54]. The weighted sum method scalarizes a set of objectives into a single objective by pre-multiplying each objective with a user-supplied weight. In other words, the idea behind the weighted sum method is to associate each objective function with a weight coefficient and minimize the weighted sum of the objectives. In this way, the multiple objective functions are transformed into a single objective function. This method is the simplest approach and is probably the most widely used classical approach. Faced with multiple objectives, this method is the most convenient one that comes to mind.

Let $Y = \mathbb{R}^\ell$, $\emptyset \neq K \subset X = \mathbb{R}^n, f : X \to Y$, and let $W = (W_1, W_2, \ldots W_\ell) \in \mathbb{R}^\ell_+ \setminus \{\mathbf{0}\}$ be a vector of weight coefficients W_i. Then by multiplying each objective function f_i by the weight coefficient W_i and taking the sum of the resulting functions, we convert VOP into the following scalar optimization problem, called *weighted optimization problem* (in short, WOP):

$$\begin{aligned} &\text{minimize} \sum_{i=1}^{\ell} W_i f_i(x), \\ &\text{subject to } x \in K. \end{aligned} \tag{4.2}$$

The weight coefficient W_i $(i = 1, 2, \ldots, \ell)$ can be interpreted as some nonnegative weight or priority assigned to the ith objective criterion by the decision maker. Hypothetically, if $\ell = 3$ and $W_1 = 1$, $W_2 = 2$, $W_3 = 3$, then this means that the third objective criterion is 3 times important than the first objective criterion and 1.5 times more important in comparison with the second objective criterion, while second criterion is 2 times more important in comparison to the first objective criterion. Therefore, we can restrict the weights to belong to $\mathbb{T}^\ell_+ \setminus \{\mathbf{0}\}$, i.e., we replace W_1, W_2, W_3 by $\frac{1}{4}, \frac{2}{4} = \frac{1}{2}$ and $\frac{3}{4}$, respectively (see Remark 4.1).

A geometric interpretation of the weighted sum method can be easily seen by considering the scalar-valued function $g_W(x) := \sum_{i=1}^{\ell} W_i f_i(x)$. Since $\sum_{i=1}^{\ell} W_i f_i(x)$ defines a plane in the objective space characterized by its normal vector $W = (W_1, W_2, \ldots, W_\ell)$, each choice of the transformation parameter W induces a partition of the objective space into planes of identical g_W-values as shown in Fig. 4.1.

The following corollary follows from Theorem 4.1

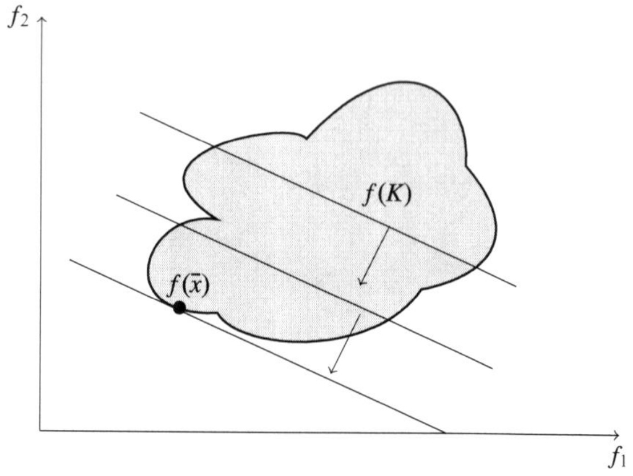

Fig. 4.1 Visualization of a feasible objective region $f(K)$ and some hyperplanes. If we minimize the weighted sum problem on the set $f(K)$, we obtain the efficient element $f(\bar{x})$

Corollary 4.1 *If* $W \in \mathbb{R}^{\ell}_{+} \setminus \{\mathbf{0}\}$, *then every solution of* WOP *is a weakly efficient solution of* VOP.

If the linear function belongs to the quasi-interior of the dual cone of C, then we can state the following connection between a linear scalarization and efficient solutions of VOP.

Theorem 4.2 *Let Y be a topological vector space with a nontrivial pointed convex cone C, X be a vector space with $\emptyset \neq K \subseteq X$, and $f : X \to Y$ be a given function. If there exist a linear functional $W \in C^{\#}$ and $\bar{x} \in K$ such that*

$$W\big(f(\bar{x})\big) \leq W\big(f(x)\big), \quad \textit{for all } x \in K,$$

then $\bar{x}$ is an efficient solution of VOP.

Proof Suppose that $\bar{x}$ is not an efficient solution of VOP. Then $(f(\bar{x}) - C) \cap \mathcal{Y} \neq \{f(\bar{x})\}$, and there is some element $\tilde{x}$ with $f(\tilde{x}) \in \mathcal{Y}$, $f(\tilde{x}) \neq f(\bar{x})$ such that $f(\tilde{x}) \in f(\bar{x}) - C$. Therefore, due to the definition of the quasi-interior of C^*, we obtain

$$W\big(f(\tilde{x})\big) < W\big(f(\bar{x})\big), \quad \text{for all } W \in C^{\#},$$

a contradiction to our supposition. □

For the weighted sum scalarization, we have the following corollary that is an immediate consequence of Theorem 4.2.

Corollary 4.2 *If* $W \in \operatorname{int}(\mathbb{R}^{\ell}_{+})$, *then every solution of* WOP *is an efficient solution of* VOP.

The next result relates unique solutions of a linear scalarization to efficient solutions of VOP.

Theorem 4.3 *Let Y be a topological vector space with a nontrivial pointed convex cone C, X be a vector space with $\emptyset \neq K \subseteq X$, and $f : X \to Y$ be a given function.*

If there exist a linear functional $W \in C^*$ *and* $\bar{x} \in K$ *such that*

$$W\big(f(\bar{x})\big) < W\big(f(x)\big), \quad \textit{for all } x \in K,$$

then $\bar{x}$ *is an efficient solution of* VOP.

Proof Suppose that $\bar{x}$ is not an efficient solution of VOP. Then $(f(\bar{x}) - C) \cap \mathcal{Y} \neq \{f(\bar{x})\}$. Thus, there exists some $\tilde{x}$ with $f(\tilde{x}) \in \mathcal{Y}$, $f(\tilde{x}) \neq f(\bar{x})$ such that $f(\tilde{x}) \in f(\bar{x}) - C$. Then we have

$$W\big(f(\tilde{x})\big) \leq W\big(f(\bar{x})\big), \quad \text{for all } W \in C^*,$$

in contradiction to the supposition. □

For the weighted sum scalarization, we obtain the following consequence of Theorem 4.3.

Corollary 4.3 *If* $W \in \mathbb{R}^{\ell}_{+}$ *and* $\bar{x}$ *is a unique solution of* WOP, *then it is an efficient solution of* VOP.

Example 4.1 Consider the following VOP:

$$\begin{aligned} &\text{minimize } f(x) = (x_1, x_2), \\ &\text{subject to } x \in K, \end{aligned} \tag{4.3}$$

where $K := \{(x_1, x_2) \in \mathbb{R}^2 : x_1^2 \leq x_2,\ x_2 \leq x_1 + 2\}$ (see Fig. 4.2). By choosing the weights $W_1 = W_2 = 1$, we obtain the following WOP:

$$\begin{aligned} &\text{minimize } x_1 + x_2, \\ &\text{subject to } x \in K. \end{aligned}$$

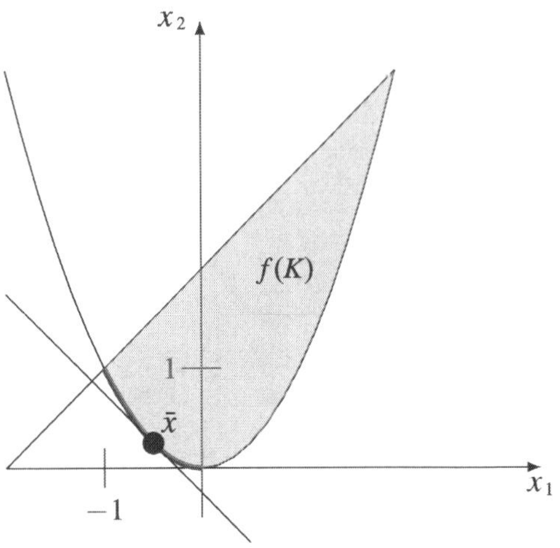

Fig. 4.2 $\bar{x}$ is a unique optimal solution of the weighted sum scalarization with weights $W_1 = W_2 = 1$, and thus $\bar{x}$ is an efficient solution of VOP (see Example 4.1)

The unique solution of this WOP is $\bar{x} = \left(-\frac{1}{2}, \frac{1}{4}\right)$. By Corollary 4.3, $\bar{x}$ is an efficient solution of VOP (4.3).

Remark 4.1 The assumption $W \in \mathbb{R}^{\ell}_{+} \setminus \{\mathbf{0}\}$ (respectively, $W \in \mathbb{R}^{\ell}_{+}$) in Corollary 4.1 (respectively, Corollary 4.3) can be replaced by the assumption $W \in \mathbb{T}^{\ell}_{+} \setminus \{\mathbf{0}\}$ (respectively, $W \in \mathbb{T}^{\ell}_{+}$), and $W \in \operatorname{int}(\mathbb{R}^{\ell}_{+})$ in Corollary 4.2 can be replaced by $W \in \operatorname{int}(\mathbb{T}^{\ell}_{+})$.

For the next results, we need the following separation theorem.

Theorem 4.4 ([32, Theorem 3.14.]) *Let A and B be nonempty convex subsets of a real vector space Y with* $\operatorname{cor}(A) \neq \emptyset$. *Then* $\operatorname{cor}(A) \cap B = \emptyset$ *if and only if there is a linear functional $W \in C^* \setminus \{\mathbf{0}\}$ and a real number α with $W(a) \leq W(b)$ for all $a \in A$ and for all $b \in B$ and $W(a) < \alpha$ for all $a \in \operatorname{cor}(A)$.*

The following theorem describes a necessary condition for weakly efficient solutions of VOP.

Theorem 4.5 *Let Y be a topological vector space with a nontrivial pointed convex cone C, X be a vector space with $\emptyset \neq K \subseteq X$, $f : X \to Y$ be a given function and the set $\mathcal{Y} + C$ be convex. Then for every weakly efficient solution $\bar{x} \in K$ of* VOP, *there exists a linear functional $W \in C^* \setminus \{\mathbf{0}\}$ such that*

$$W\big(f(\bar{x})\big) \leq W\big(f(x)\big), \quad \textit{for all } x \in K.$$

Proof Let $\bar{x} \in K$ be a weakly efficient solution of VOP. Then $f(\bar{x}) \in f(K) = \mathcal{Y}$ is a also a weakly efficient element of $\mathcal{Y}$. Thus, by Lemma 3.2 (b), $f(\bar{x})$ is also a weakly efficient element of the set $\mathcal{Y} + C$. Then

$$(f(\bar{x}) - \operatorname{int}(C)) \cap (\mathcal{Y} + C) = \emptyset. \tag{4.4}$$

Because $f(\bar{x}) - \operatorname{int}(C)$ and $\mathcal{Y} + C$ are convex sets, we can apply Theorem 4.4 and we observe that there exists a linear functional $W \in Y^* \setminus \{\mathbf{0}\}$ and a real number α with

$$W\big(f(\bar{x}) - c_1\big) \leq W\big(f(x) + c_2\big), \quad \text{for all } x \in K \text{ and all } c_1, c_2 \in C.$$

Because C is a cone, we obtain that $W \in C^* \setminus \{\mathbf{0}\}$, and by choosing $c_1 = c_2 = \mathbf{0}$, we conclude with

$$W\big(f(\bar{x})\big) \leq W\big(f(x)\big), \quad \text{for all } x \in K.$$

□

We apply Theorem 4.5 to the weighted sum scalarization.

Corollary 4.4 *Let the set $\mathcal{Y} + \mathbb{R}^{\ell}_{+}$ be convex. For every weakly efficient solution $\bar{x} \in K$ of* VOP, *there exists $W \in \mathbb{R}^{\ell}_{+} \setminus \{\mathbf{0}\}$ such that $\bar{x}$ is a solution of* WOP.

A necessary condition for efficient solutions is the following.

Theorem 4.6 *Let Y be a topological vector space with a nontrivial pointed convex cone C, X be a vector space with $\emptyset \neq K \subseteq X$, $f : X \to Y$ be a given function, and let the set $\mathcal{Y} + C$ be convex and have nonempty interior. Then for every efficient solution $\bar{x} \in K$ of* VOP, *there exists a linear functional $W \in C^* \setminus \{\mathbf{0}\}$ such that*

$$W\big(f(\bar{x})\big) \leq W\big(f(x)\big), \quad \text{for all } x \in K.$$

The proof of Theorem 4.6 is similar to the proof of Theorem 4.5, and therefore, it is omitted.

In the following corollary, we apply Theorem 4.6 to the weighted sum scalarization.

Corollary 4.5 *Let the set $\mathcal{Y} + \mathbb{R}^\ell_+$ be convex. For every efficient solution $\bar{x} \in K$ of* VOP, *there exists $W \in \mathbb{R}^\ell_+ \setminus \{\mathbf{0}\}$ such that $\bar{x}$ is a solution of* WOP.

The next result characterizes strongly efficient solutions of VOP.

Theorem 4.7 *Let Y be a topological vector space with a nontrivial pointed convex cone C, X be a vector space with $\emptyset \neq K \subseteq X$, and $f : X \to Y$ be a given function. Then $\bar{x} \in K$ is a strongly efficient solution for* VOP *if and only if for every linear functional $W \in C^*$, we have*

$$W\big(f(\bar{x})\big) \leq W\big(f(x)\big), \quad \text{for all } x \in K.$$

Proof Let $\bar{x} \in K$ be a strongly efficient solution of VOP. Then $f(K) = \mathcal{Y} \subset \{f(\bar{x})\} + C$. This means that for every $x \in K$, $f(x) \in \{f(\bar{x})\} + C$, and hence $W\big(f(\bar{x})\big) \leq W\big(f(x)\big)$ for all $x \in K$ and for all $W \in C^*$.

Conversely, suppose that $W\big(f(\bar{x})\big) \leq W\big(f(x)\big)$ holds true for all $x \in K$ and for all $W \in C^*$, but suppose that $\bar{x}$ is not strongly efficient for VOP. Then $\mathcal{Y} \not\subset f(\bar{x}) + C$, and thus there exists some $\tilde{x} \in K$ with $f(\tilde{x}) \notin \{f(\bar{x})\} + C$. But then there is some $W \in C^*$ such that $W(f(\tilde{x})) < W(f(\bar{x}))$, a contradiction. □

Corollary 4.6 *$\bar{x} \in K$ is a strongly efficient solution of* VOP *if and only if for all $x \in K$ and all $W \in \mathbb{R}^\ell_+$,*

$$\sum_{i=1}^{\ell} W_i f_i(\bar{x}) \leq \sum_{i=1}^{\ell} W_i f_i(x).$$

Now we turn our attention to VOP with $Y = \mathbb{R}^\ell$, $X = \mathbb{R}^n$ and $C = \mathbb{R}^\ell_+$. In the following theorem, we see that if all the weight coefficients in WOP are positive, then the optimal solution of WOP provides the properly efficient solution of VOP.

Theorem 4.8 ([20]) *Let $W \in \text{int}(\mathbb{T}^\ell_+)$. If $\bar{x}$ is a solution of* WOP, *then it is a properly efficient solution (in the sense of Geoffrion) of* VOP.

Proof In Corollary 4.2, we have already proved that $\bar{x}$ is an efficient solution of VOP. We shall show that $\bar{x}$ is a properly efficient solution with

$$\mu = (\ell - 1) \max_{i,j \in \mathscr{I}} \frac{W_j}{W_i}, \quad \text{for all } \ell \geq 2, \tag{4.5}$$

where $\mathscr{I} := \{1, 2, \ldots, \ell\}$.

Suppose that $\bar{x}$ is not a properly efficient solution of VOP. Then there exist $i \in \mathscr{I}$ and $y \in K$ such that $f_i(y) < f_i(\bar{x})$ and

$$f_i(\bar{x}) - f_i(y) > \mu \left(f_j(y) - f_j(\bar{x}) \right), \quad \text{for all } j \in \mathscr{I} \text{ for which } f_j(y) > f_j(\bar{x}).$$

Therefore, by the choice of μ, we have

$$f_i(\bar{x}) - f_i(y) > (\ell - 1) \frac{W_j}{W_i} \left(f_j(y) - f_j(\bar{x}) \right), \quad \text{for all } j \in \mathscr{I} \setminus \{i\}. \tag{4.6}$$

Multiplying (4.6) both sides by $\dfrac{W_i}{(\ell - 1)}$ for each $j \in \mathscr{I} \setminus \{i\}$, we get

$$\frac{W_i}{(\ell - 1)} \left(f_i(\bar{x}) - f_i(y) \right) > W_j \left(f_j(y) - f_j(\bar{x}) \right), \quad \text{for all } j \in \mathscr{I} \setminus \{i\}.$$

By taking the sum over all $j \neq i$, we obtain

$$W_i \left(f_i(\bar{x}) - f_i(y) \right) > \sum_{j=1,\, j \neq i}^{\ell} W_j \left(f_j(y) - f_j(\bar{x}) \right)$$

equivalently,

$$W_i \left(f_i(\bar{x}) - f_i(y) \right) > \sum_{j=1,\, j \neq i}^{\ell} W_j f_j(y) - \sum_{j=1,\, j \neq i}^{\ell} W_j f_j(\bar{x}).$$

This implies that

$$W_i f_i(\bar{x}) + \sum_{j=1,\, j \neq i}^{\ell} W_j f_j(\bar{x}) > W_i f_i(y) + \sum_{j=1,\, j \neq i}^{\ell} W_j f_j(y)$$

and thus,

$$\sum_{j=1}^{\ell} W_j f_j(\bar{x}) > \sum_{j=1}^{\ell} W_j f_j(y),$$

contradicting optimality of $\bar{x}$ for WOP. Thus, $\bar{x}$ is a properly efficient solution of VOP. □

Now we apply Theorem 4.8 to find properly efficient solutions (in the sense of Geoffrion) of vector optimization problems in the following example.

Example 4.2 Consider the following VOP:

$$\begin{aligned} &\text{minimize}\, f(x) = (x_1, x_2^2 - x_1), \\ &\text{subject to}\; x \in K, \end{aligned} \tag{4.7}$$

where $K := \{(x_1, x_2) \in \mathbb{R}^2 : 0 \leq x_1 \leq 2,\ 0 \leq x_2 \leq 2\}$ (see Fig. 4.2). By choosing the weights $W_1 = W_2 = \frac{1}{2}$, we obtain the following WOP:

$$\begin{aligned} &\text{minimize}\ \frac{1}{2}x_1 + \frac{1}{2}x_2^2, \\ &\text{subject to}\; x \in K. \end{aligned}$$

One solution of this WOP is $\bar{x} = (0, 0)$. Due to Theorem 4.8, $\bar{x}$ is a properly efficient solution of VOP (4.7). The set of all properly efficient solutions is given by $\{(x_1, x_2) : 0 \leq x_1 \leq 2, x_2 = 0\}$ (Fig. 4.3).

For further examples and exercises, we refer to [9].

Remark 4.2 One of the disadvantages of linear scalarization techniques, in particular the weighted sum method, is that all efficient solutions can only be found for convex problems, i.e., if the set $\mathcal{Y} + C$ is a convex set. The weighted sum method fails to find all efficient solutions of a nonconvex problem. This is the reason why we consider nonlinear scalarization methods in Sect. 4.2.

We state the following result to prove the necessary and sufficient conditions for a solution of VOP. In the following, we use the set $\mathcal{I} := \{1, 2, \ldots, \ell\}$.

Theorem 4.9 ([40]) *Let $K \subseteq \mathbb{R}^n$ be a nonempty convex set and for each $i \in \mathcal{I}$, $f_i : K \to \mathbb{R}$ be convex. If the system $f_i(x) < 0$ for all $i \in \mathcal{I}$ has no solution $x \in K$,*

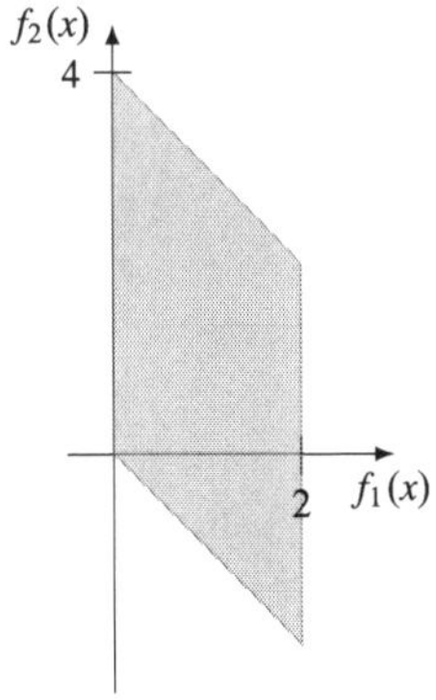

Fig. 4.3 $\bar{x} = (0, 0) = f(0, 0)$ is one efficient solution of the weighted sum scalarization with weights $W_1 = W_2 = 1$, and thus $\bar{x}$ is a properly efficient solution of VOP (see Example 4.2)

then there exist $\lambda_i \geq 0$, $i \in \mathscr{I}$ *with* $\sum_{i=1}^{\ell} \lambda_i = 1$ *such that*

$$\sum_{i=1}^{\ell} \lambda_i f_i(x) \geq 0, \quad \textit{for all } x \in K. \tag{4.8}$$

The following theorem shows that all properly efficient solutions (in the sense of Geoffrion) of a convex VOP can be found by means of WOP.

Theorem 4.10 ([20]) *Let* $K \subseteq \mathbb{R}^n$ *be a nonempty convex set and for each* $i \in \mathscr{I}$, $f_i : K \to \mathbb{R}$ *be convex. Then* $\bar{x}$ *is a properly efficient solution (in the sense of Geoffrion) of* VOP *if and only if there exists* $W \in \text{int}(\mathbb{R}^{\ell}_{+})$ *such that* $\bar{x}$ *is an optimal solution of* WOP.

Proof In view of Theorem 4.8, it is sufficient to prove the necessity condition. Let $\bar{x}$ be a properly efficient solution of VOP. Then, by definition, there exists a number $\mu > 0$ such that for all $i = 1, 2, \ldots, \ell$ and all $y \in K$ with $f_i(y) < f_i(\bar{x})$ at least one $j \in \mathscr{I}$ exists with $f_j(y) > f_j(\bar{x})$ and

$$\frac{f_i(\bar{x}) - f_i(y)}{f_j(y) - f_j(\bar{x})} \leq \mu. \tag{4.9}$$

We claim that for every $i \in \mathscr{I}$, the system

$$\begin{aligned} f_i(y) &< f_i(\bar{x}) \\ f_i(y) + \mu f_j(y) &< f_i(\bar{x}) + \mu f_j(\bar{x}), \quad \text{for all } j \in \mathscr{I} \text{ and } j \neq i \end{aligned} \tag{4.10}$$

has no solution.

Indeed, assume that for some $i \in \mathscr{I}$, the system (4.10) has a solution $y \in K$. If there is no $j \in \mathscr{I}$ with $f_j(y) > f_j(\bar{x})$, $\bar{x}$ cannot be a properly efficient solution. On the other hand, if there is some $j \in \mathscr{I}$ with $f_j(y) > f_j(\bar{x})$, we obtain from second inequality in (4.10) that

$$f_i(\bar{x}) - f_i(y) > \mu \left(f_j(y) - f_j(\bar{x}) \right)$$

contradicting with the inequality (4.9).

Theorem 4.9 implies that for the ith such system there exist $\lambda_j^{(i)} \geq 0$ for $j = 1, 2, \ldots, \ell$ such that $\sum_{i=1}^{\ell} \lambda_j^{(i)} = 1$ and for all $y \in K$, we have

$$\lambda_i^{(i)} f_i(y) + \sum_{j=1,\, j\neq i}^{\ell} \lambda_j^{(i)} \left(f_i(y) + \mu f_j(y) \right) \geq \lambda_i^{(i)} f_i(\bar{x}) + \sum_{j=1,\, j\neq i}^{\ell} \lambda_j^{(i)} \left(f_i(\bar{x}) + \mu f_j(\bar{x}) \right),$$

equivalently,

$$\lambda_i^{(i)} f_i(y) + \sum_{j=1,\, j\neq i}^{\ell} \lambda_j^{(i)} f_i(y) + \mu \sum_{j=1,\, j\neq i}^{\ell} \lambda_j^{(i)} f_j(y)$$

$$\geq \lambda_i^{(i)} f_i(\bar{x}) + \sum_{j=1,\, j\neq i}^{\ell} \lambda_j^{(i)} f_i(\bar{x}) + \mu \sum_{j=1,\, j\neq i}^{\ell} \lambda_j^{(i)} f_j(\bar{x}).$$

This implies that

$$\sum_{j=1}^{\ell} \lambda_j^{(i)} f_i(y) + \mu \sum_{j=1,\, j\neq i}^{\ell} \lambda_j^{(i)} f_j(y) \geq \sum_{j=1}^{\ell} \lambda_j^{(i)} f_i(\bar{x}) + \mu \sum_{j=1,\, j\neq i}^{\ell} \lambda_j^{(i)} f_j(\bar{x}). \tag{4.11}$$

Since $\sum_{j=1}^{\ell} \lambda_j^{(i)} = 1$, and $f_i(y)$ and $f_i(\bar{x})$ are independent of j, we have

$$\sum_{j=1}^{\ell} \lambda_j^{(i)} f_i(y) = f_i(y) \quad \text{and} \quad \sum_{j=1}^{\ell} \lambda_j^{(i)} f_i(\bar{x}) = f_i(\bar{x}).$$

Therefore, inequality (4.11) becomes

$$f_i(y) + \mu \sum_{j=1,\, j\neq i}^{\ell} \lambda_j^{(i)} f_j(y) \geq f_i(\bar{x}) + \mu \sum_{j=1,\, j\neq i}^{\ell} \lambda_j^{(i)} f_j(\bar{x}), \quad \text{for all } y \in K. \tag{4.12}$$

The inequality (4.12) holds for all $i = 1, 2, \ldots, \ell$. Taking the sum over i up to ℓ, we obtain

$$\sum_{i=1}^{\ell} f_i(y) + \mu \sum_{i=1}^{\ell} \sum_{j=1,\, j\neq i}^{\ell} \lambda_j^{(i)} f_j(y) \geq \sum_{i=1}^{\ell} f_i(\bar{x}) + \mu \sum_{i=1}^{\ell} \sum_{j=1,\, j\neq i}^{\ell} \lambda_j^{(i)} f_i(\bar{x}),$$

for all $y \in K$, which implies that

$$\sum_{i=1}^{\ell} \left(1 + \mu \sum_{j=1,\, j\neq i}^{\ell} \lambda_j^{(i)} \right) f_i(y) \geq \sum_{i=1}^{\ell} \left(1 + \mu \sum_{j=1,\, j\neq i}^{\ell} \lambda_j^{(i)} \right) f_i(\bar{x}),$$

for all $y \in K$. Since $\sum_{k=1}^{\ell} \lambda_k^{(i)} = 1$, we have

$$\sum_{i=1}^{\ell}\left(1+\mu\sum_{j=1,\,j\neq i}^{\ell}\frac{\lambda_j^{(i)}}{\sum_{k=1}^{\ell}\lambda_k^{(i)}}\right)f_i(y) \geq \sum_{i=1}^{\ell}\left(1+\mu\sum_{j=1,\,j\neq i}^{\ell}\frac{\lambda_j^{(i)}}{\sum_{k=1}^{\ell}\lambda_k^{(i)}}\right)f_i(\bar{x}), \tag{4.13}$$

for all $y \in K$. Let

$$t_i = 1+\mu\sum_{j=1,\,j\neq i}^{\ell}\frac{\lambda_j^{(i)}}{\sum_{k=1}^{\ell}\lambda_k^{(i)}}.$$

Then $t_i \geq 0$ for all i, and the inequality (4.13) can be written as

$$\sum_{i=1}^{\ell} t_i f_i(y) \geq \sum_{i=1}^{\ell} t_i f_i(\bar{x}), \quad \text{for all } y \in K.$$

Hence, $\bar{x}$ is an optimal solution of WOP. □

4.2 Nonlinear Scalarization Method

Here we present the concept of nonlinear scalarization introduced by Gerth and Weidner [22]. Several scalarization concepts known from the literature may be obtained by a variation of parameters involved in this prominent scalarization method (see Sect. 2.6). For instance, Klamroth et al. [33] showed how robust and stochastic scalar optimization problems can be characterized by using a nonlinear scalarizing functional. Let Y be a topological vector space, $e \in Y \setminus \{\mathbf{0}\}$ and D be a nonempty closed proper (i.e., $D \neq \{\mathbf{0}\}$ and $D \neq Y$) subset of Y satisfying

$$D + [0, +\infty[\cdot e \subset D. \tag{4.14}$$

We use the functional $\xi_{e,D}$ given in Definition 2.30, namely the functional $\xi_{e,D} : Y \to \mathbb{R} \cup \{\pm\infty\}$ defined by

$$\xi_{e,D}(y) = \inf\{t \in \mathbb{R} : y \in te - D\}, \quad \text{for all } y \in Y. \tag{4.15}$$

Now we formulate the problem of minimizing the functional $\xi_{e,D}$ as

$$\begin{aligned} &\text{minimize } \xi_{e,D}(f(x)),\\ &\text{subject to } f(x) \in f(K), x \in K. \end{aligned} \tag{4.16}$$

Notice that the functional $\xi_{e,D}$ operates in the objective space Y of the multiobjective function $f : X \to Y$, where X is a vector space. When searching for efficient solutions $x \in K \subseteq X$ of f, the functional $\xi_{e,D}$ can be used to scalarize f. Since the functional's well-studied monotonicity properties allow for connections to vector-valued optimization problems, $\xi_{e,D}$ may be used to gain efficient solutions of the set $f(K) = \mathcal{Y}$.

Figure 4.4 visualizes the functional $\xi_{e,D}$ for a cone $D \subset \mathbb{R}^2_+$ and a given vector $e \in \text{int}(D)$. We can see that the set $-D$ is moved along the line $\mathbb{R}\cdot e$ up until y belongs to $te - D$. The functional $\xi_{e,D}$ assigns the smallest value t such that the property $y \in te - D$ is fulfilled. By a variation of the vector e that satisfy (4.14) all efficient solutions of a vector optimization problem without any convexity assumptions can be found.

The scalarizing functional $\xi_{e,D}$ was used in [22] to prove nonconvex separation theorems. Applications of $\xi_{e,D}$ include coherent risk measures in financial mathematics (see, for instance, [29]). Many properties of $\xi_{e,D}$ were studied in [22, 25, 52].

In Theorem 2.7, we already mentioned that the functional $\xi_{e,D}$ fulfills certain monotonicity properties under quite general assumptions. The theorem below, which appeared in [22, 31], relates functionals that are monotone with respect to a cone C to the nondominated and weakly nondominated set of a VOP.

Theorem 4.11 *Let Y be a real preordered vector space whose ordering is defined by a convex cone C.*

(a) *If there exists a functional $\xi : Y \to \mathbb{R}$ which is monotone with respect to C, where for all $y \in f(K) \setminus \{\bar{y}\} : \xi(\bar{y}) < \xi(y)$, then $\bar{y} \in \mathbb{E}(f(K), C)$.*

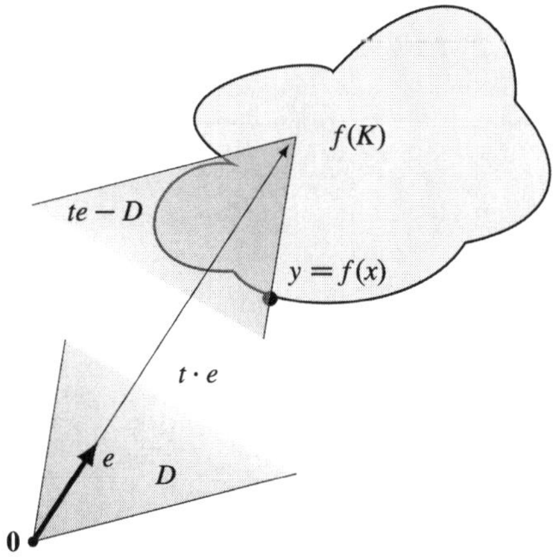

Fig. 4.4 Visualization of the functional $\xi_e(y) = \min\{t \in \mathbb{R} : y \in te - D\}$

(b) *If there exists a functional $\xi : Y \to \mathbb{R}$ which is strictly monotone with respect to C, where for all $y \in f(K) : \xi(\bar{y}) \leq \xi(y)$, then $\bar{y} \in \mathbb{WE}(f(K), C)$.*

Proof

(a) Let ξ be monotone with respect to C and $\xi(\bar{y}) < \xi(y)$ for all $y \in f(K) \setminus \{\bar{y}\}$. Suppose that $\bar{y} \notin \mathbb{E}(f(K), C)$. Then $(\{\bar{y}\} - C) \cap f(K) \not\subseteq \{\bar{y}\} + C$. Thus, there exists some $y \in f(K)$, such that $y \in \{\bar{y}\} - C$ and $y \notin \{\bar{y}\} + C$. Then by the monotonicity of z with respect to C, we obtain $\xi(y) \leq \xi(\bar{y})$. This contradicts the unique minimality of the element $\bar{y}$.
(b) Let ξ be strictly monotone with respect to C and $\xi(\bar{y}) \leq \xi(y)$ for all $y \in f(K)$, but suppose that $\bar{y} \notin \mathbb{WE}(f(K), C)$. Then there exists some $y \in f(K)$ with $y \in \{\bar{y}\} - \text{int}(C)$. But then we immediately obtain $\xi(y) < \xi(\bar{y})$, a contradiction.

□

Thus, we have the following corollary, which follows from the monotonicity properties of the functional $\xi_{e,D}$ for a cone $C = D$ (see Theorem 2.7).

Corollary 4.7 *Let $C \subset Y$ be a closed convex pointed cone in a real partially ordered topological vector space Y, $K \subseteq X$ a nonempty set in a vector space X and $f : X \to Y$ be a given function. Furthermore, let $e \in Y \setminus \{\mathbf{0}\}$ and $D = C$ such that* (4.14) *is fulfilled. If $f(\bar{x})$ is a solution of the scalarized problem* (4.16)*, then $f(\bar{x}) \in \mathbb{WE}(f(K), C)$. Furthermore, if $f(\bar{x})$ is a unique solution of the scalarized problem* (4.16)*, then $f(\bar{x}) \in \mathbb{E}(f(K), C)$.*

The following theorem shows that all weakly efficient solutions of VOP can be found by means of the nonlinear scalarizing functional.

Theorem 4.12 *Let Y be a topological vector space, $B \subset Y$ be a closed set with nonempty interior and $\mathbf{0} \in bd(B)$. Assume that there exists a cone $C \subset Y$ with nonempty interior such that $B + \text{int}(C) \subset B$. Then $\bar{y} \in \mathbb{WE}(f(K), B)$ if and only if $\bar{y} \in f(K)$ and there exists a continuous function $\xi : Y \to \mathbb{R}$ such that for all $y \in \text{int}(B)$ and for all $\hat{y} \in f(K)$,*

$$\xi(\bar{y} - y) < 0 = \xi(\bar{y}) \leq \xi(\hat{y}). \tag{4.17}$$

Proof Let $e \in \text{int}(C)$, $A := f(K)$ and $D := C - \bar{y}$. Then the inequality (4.17) follows from Theorem 2.8 for $\xi := \xi_{e,D}$. Conversely, let (4.17) be satisfied. Then $f(K) \cap (\bar{y} - \text{int}(B)) = \emptyset$, which means that $\bar{y} \in \mathbb{WE}(f(K), B)$. □

In case $Y = \mathbb{R}^\ell$, and by a specific selection of the set D and the vector $e \in \mathbb{R}^\ell \setminus \{\mathbf{0}\}$, we are able to express the objective function in the weighted sum scalarization method by means of the functional $\xi_{e,D}$.

Theorem 4.13 *Let $Y = \mathbb{R}^\ell$ and $X = \mathbb{R}^n$. Choose weights $W \in \mathbb{T}^\ell_+ \setminus \{\mathbf{0}\}$ and let $D := \left\{y \in \mathbb{R}^\ell : \sum_{i=1}^\ell W_i y_i \geq 0\right\}$, $e := \mathbf{1}_\ell = (1, 1, \ldots, 1)$. Then $\xi_{e,D}(f(x)) = \sum_{i=1}^\ell W_i f_i(x)$ for every $x \in K$.*

Proof It holds for all $x \in K$,

$$\begin{aligned}
\xi_{e,D}(f(x)) &= \min\{t \in \mathbb{R} : f(x) \in te - D\} \\
&= \min\left\{t \in \mathbb{R} : \sum_{i=1}^{\ell} W_i \cdot (f(x) - te)_i \leq 0\right\} \\
&= \min\left\{t \in \mathbb{R} : \sum_{i=1}^{\ell} W_i \cdot f_i(x) \leq t \cdot \underbrace{\sum_{i=1}^{\ell} W_i}_{=1}\right\} \\
&= \sum_{i=1}^{\ell} W_i f_i(x).
\end{aligned}$$

□

Note that the selection of the cone D in and the vector e in Theorem 4.13 is crucial and if these parameters are changed, Theorem 4.13 may not hold anymore. If $\sum_{i=1}^{\ell} W_i \neq 1$, then the solutions of WOP and (4.16) for the parameters e and D given in Theorem 4.13 are the same, but the objective function values may differ.

Due to the monotonicity properties of the functional $\xi_{e,D}$, we are able to prove relations between solutions of the weighted sum method and efficient solutions (see Corollaries 4.1 and 4.2).

Corollary 4.8 *If* $W \in \mathbb{R}^{\ell}_{+} \setminus \{\mathbf{0}\}$, *then every solution of* WOP *is a weakly efficient solution of* VOP. *Furthermore, every unique solution of* WOP *is an efficient solution of* VOP.

Proof With the choice of D and e as in Theorem 4.13, the assertions follow directly from the fact that D is a closed proper set, and together with $D + \mathbb{R}^{\ell}_{+} \subseteq D$, Theorem 2.7 implies that the functional $\xi_{e,D}$ is monotone with respect to $\mathbb{R}^{\ell}_{+}$ and strictly monotone with respect to $\mathbb{R}^{\ell}_{+}$. Therefore, the assertions follow with Theorems 4.11 and 4.13. □

4.2.1 ε-Constraint Method

Now we again turn our attention to the finite dimensional case. We consider the selection $Y = \mathbb{R}^{\ell}$ for the objective space and $X = \mathbb{R}^n$ as the decision variable space. Furthermore, we define efficiency of VOP by means of the natural ordering cone $C = \mathbb{R}^{\ell}_{+}$. Then we consider the ε-constraint method, introduced by Haimes et al. [27] in 1971, which is probably best known technique to solve VOP besides the weighted sum method. In the ε-constraint method, only one of the original objective functions is selected to be optimized and all other objective functions are converted into constraints by setting an upper bound to each of them. The problem to be solved is now the following *ε-constraint optimization problem* (in short, ε-COP):

$$\begin{aligned} &\text{minimize } f_k(x), \\ &\text{subject to } f_j(x) \leq \varepsilon_j, \quad \text{for all } j = 1, 2, \ldots, \ell,\ j \neq k, \\ &x \in K, \end{aligned} \tag{4.18}$$

where $\varepsilon = (\varepsilon_1, \varepsilon_2, \ldots, \varepsilon_\ell) \in \mathbb{R}^\ell$ and $k \in \{1, 2, \ldots, \ell\}$. The component ε_k is irrelevant for (4.18), but the convention is to include it as it will be convenient later.

The following theorem provides the relationship between an optimal solution of ε-COP and a weakly efficient solution of VOP.

Theorem 4.14 *Every solution of ε-COP is a weakly efficient solution of* VOP. *In other words, if $\bar{x} \in K$ is a solution of* ε-COP *for some k, then it is a weakly efficient solution of* VOP.

Proof Let $\bar{x} \in K$ be a solution of ε-COP but not a weakly efficient solution of VOP. Then there exists some $y \in K$ such that $f_i(y) < f_i(\bar{x})$ for all $i = 1, 2, \ldots, \ell$. In particular, $f_j(y) < f_j(\bar{x})$ for all $j = 1, 2, \ldots, \ell,\ j \neq k$. Since $f_j(y) < f_j(\bar{x}) \leq \varepsilon_j$ for all $j \neq k$, y is feasible for ε-COP, but then $\bar{x}$ is not an optimal solution of ε-COP, contradicting our assumption that $\bar{x}$ is an optimal solution of ε-COP. Thus, $\bar{x}$ is a weakly efficient solution of VOP. □

The following result relates optimal solutions of ε-COP to efficient solutions of VOP.

Theorem 4.15 *If $\bar{x} \in K$ is a unique optimal solution of* ε-COP, *then it is an efficient solution of* VOP.

Proof Let $\bar{x} \in K$ be a unique solution of ε-COP for some k but not an efficient solution of VOP. Then there exists some point $y \in K$ such that $f_j(y) \leq f_j(\bar{x})$ for all $j = 1, 2, \ldots, \ell$ and $f_k(y) < f_k(\bar{x})$ for at least one k. The uniqueness of $\bar{x}$ implies that for all $y \in K$ with $f_j(y) \leq f_j(\bar{x}) = \varepsilon_j, j \neq k$, we have $f_k(\bar{x}) < f_k(y)$. So, we have a contradiction with the preceding inequalities, and x must be an efficient solution of VOP. □

Theorem 4.16 *The vector $\bar{x} \in K$ is an efficient solution of* VOP *if and only if there exists an $\bar{\varepsilon} \in \mathbb{R}^\ell$ such that $\bar{x}$ is a solution of $\bar{\varepsilon}$-*COP *for every $k = 1, 2, \ldots, \ell$.*

Proof Let $\bar{\varepsilon} = f(\bar{x})$. Assume that $\bar{x}$ is an efficient solution of VOP but not a solution of $\bar{\varepsilon}$-COP for some k. Then there exists $y \in K$ such that $f_k(y) < f_k(\bar{x})$ and $f_j(y) \leq \bar{\varepsilon}_j = f_j(\bar{x})$ for all $j \neq k$, that is, $\bar{x}$ is not an efficient solution of VOP.

Conversely, let $\bar{x} \in K$ be a solution of $\bar{\varepsilon}$-COP. Then for all $k = 1, 2, \ldots, \ell$, there is no $y \in K$ such that $f_k(y) < f_k(\bar{x})$ and $f_j(y) \leq f_j(\bar{x}) = \bar{\varepsilon}_j$ when $j \neq k$. So, $\bar{x}$ is an efficient solution of VOP. □

The following theorem shows that it is possible to characterize the ε-constraint method by means of the functional $\xi_{e,D}$. Note that here, the set D is not a cone.

Theorem 4.17 *Let*

$$e = (e_1, , e_2, \ldots, e_\ell), \text{ where } e_j = \begin{cases} 1, & \text{for } j = k, \\ 0, & \text{for } j \neq k, \end{cases} \tag{4.19}$$

and

$$D = \mathbb{R}^\ell_+ - \bar{c}, \text{ with } \bar{c} = (\bar{c}_1, \bar{c}_2, \ldots, \bar{c}_\ell),\ \bar{c}_j = \begin{cases} 0, & \text{for } j = k, \\ \varepsilon_j, & \text{for } j \neq k, \end{cases} \tag{4.20}$$

Then $\bar{x} \in K$ *is a solution of* ε-COP *if and only if* $f(\bar{x})$ *is a solution of problem* (4.16).

Proof It holds for all $x \in K$,

$$\begin{aligned} \xi_{e,D}(f(x)) &= \min\{t \in \mathbb{R} : f(x) \in te - D\} \\ &= \min\{t \in \mathbb{R} : f(x) \in te - \mathbb{R}^\ell_+ - \bar{c}\} \\ &= \min\{f_k(x) : f_j(x) \leq \varepsilon_j,\ j = 1, 2, \ldots, \ell,\ j \neq k\}. \end{aligned}$$

□

The assertions in Theorems 4.14 and 4.15 can be verified by using the nonlinear scalarizing functional $\xi_{e,D}$.

Theorem 4.18 *Every solution of* ε-COP *is a weakly efficient solution of* VOP. *Furthermore, every unique solution of* ε-COP *is an efficient solution of* VOP.

Proof Let e and D be given by (4.19) and (4.20), respectively. Then we have $D + \mathbb{R}^\ell_+ \subseteq D$, thus, by Theorem 2.7 (e), $\xi_{e,D}$ is monotone with respect to $\mathbb{R}^\ell_+$. Then Theorem 4.11 yields the first assertion. Now we show directly that $\xi_{e,D}(\cdot)$ is strictly monotone with respect to $\mathbb{R}^\ell_+$. Consider $t \in \mathbb{R}$, $y \in te - \text{int}(D)$. Then $te - y \in \text{int}(D)$. Consequently, there exists an $s > 0$ such that $te - y - se \in \text{int}(D) \subset D$. We deduce $\xi_{e,D}(y) \leq t - s < t$, and thus

$$te - \text{int}(D) \subset \{y \in \mathbb{R}^\ell : \xi_{e,D}(y) < t\}. \tag{4.21}$$

Furthermore, for $y_1 \in y_2 - \text{int}(\mathbb{R}^\ell_+)$, it holds

$$\begin{aligned} y_1 \in y_2 - \text{int}(\mathbb{R}^\ell_+) &\subset \xi_{e,D}(y_2)e - D - \text{int}(\mathbb{R}^\ell_+) \\ &\subset \xi_{e,D}(y_2)e - \text{int}(D) \\ &\subset \{y \in \mathbb{R}^\ell : \xi_{e,D}(y) < \xi_{e,D}(y_2)\} \quad \text{because of (4.21).} \end{aligned}$$

We conclude that $\xi_{e,D}(y_1) < \xi_{e,D}(y_2)$ and thus $\xi_{e,D}$ is strictly monotone with respect to $\mathbb{R}^\ell_+$. Then Theorem 4.11 gives the second assertion. □

The following results provide the relationship between the weighted sum method and the ε-constraint method.

Theorem 4.19 *Let $\bar{x} \in K$ be an optimal solution of* WOP *and* $W = (W_1, W_2, \ldots, W_\ell) \in \mathbb{R}^\ell_+$ *be the corresponding weight vector.*

(a) *If $W_k > 0$, then $\bar{x}$ is a solution of ε-COP for f_k as the objective function and $\varepsilon_j = f_j(\bar{x})$ for $j = 1, 2, \ldots, \ell$, $j \neq k$.*
(b) *If $\bar{x}$ is a unique solution of* WOP, *then $\bar{x}$ is a solution of ε-COP when $\varepsilon_j = f_j(\bar{x})$ for $j = 1, 2, \ldots, \ell$, $j \neq k$ and for every f_k, $k = 1, 2, \ldots, \ell$, as the objective function.*

Proof Let $\bar{x} \in K$ be a solution of WOP for some weight vector $W = (W_1, W_2, \ldots, W_\ell) \in \mathbb{R}^\ell_+$.

(a) Let $W_k > 0$. Since $\bar{x}$ is an optimal solution of WOP, we have

$$\sum_{i=1}^{\ell} W_i f_i(y) \geq \sum_{i=1}^{\ell} W_i f_i(\bar{x}), \quad \text{for all } y \in K. \tag{4.22}$$

Assume that $\bar{x} \in K$ is not a solution of ε-COP. Then there exists a point $\hat{y} \in K$ such that $f_k(\hat{y}) < f_k(\bar{x})$ and $f_j(\hat{y}) \leq f_j(\bar{x})$ when $j = 1, 2, \ldots, \ell$, $j \neq k$. Since $W_k > 0$ and $W_j \geq 0$ when $k \neq j$, we have

$$0 > W_k \left(f_k(\hat{y}) - f_k(\bar{x})\right) + \sum_{j=1,\, j\neq k}^{\ell} W_j \left(f_j(\hat{y}) - f_j(\bar{x})\right),$$

equivalently,

$$0 > \sum_{j=1}^{\ell} W_j \left(f_j(\hat{y}) - f_j(\bar{x})\right),$$

a contradiction of inequality (4.22). Thus, $\bar{x}$ is an optimal solution of ε-COP.
(b) If $\bar{x} \in K$ is a unique solution of WOP, then

$$\sum_{i=1}^{\ell} W_i f_i(\bar{x}) < \sum_{i=1}^{\ell} W_i f_i(y), \quad \text{for all } y \in K. \tag{4.23}$$

If there is some objective function f_k such that $\bar{x}$ does not solve ε-COP when f_k is to be minimized, then we can find a $\hat{y} \in K$ such that $f_k(\hat{y}) < f_k(\bar{x})$ and $f_j(\hat{y}) \leq f_j(\bar{x})$ when $j \neq k$. Therefore, for any $W \in \mathbb{R}^\ell_+$, we obtain

$$\sum_{i=1}^{\ell} W_i f_i(\hat{y}) \leq \sum_{i=1}^{\ell} W_i f_i(\bar{x}),$$

a contradiction to inequality (4.23). Thus, $\bar{x}$ is a solution of ε-COP for all f_k to be minimized. □

Theorem 4.20 *Let $K \subseteq \mathbb{R}^n$ be a nonempty convex set and for each $i \in \mathscr{I}$, let $f_i : K \to \mathbb{R}$ be convex. If $\bar{x} \in K$ is a solution of* ε-COP *for some k, then there exists $W \in \mathbb{R}^{\ell}_{+} \setminus \{\mathbf{0}\}$ such that $\bar{x} \in K$ is a solution of* WOP.

Proof Suppose that $\bar{x}$ is a solution of ε-COP for some k. Then there is no $\hat{y} \in K$ satisfying $f_k(\hat{y}) < f_k(\bar{x})$ and $f_j(\hat{y}) \leq f_j(\bar{x}) \leq \varepsilon_j$ for $j \neq k$. Using the convexity of f_i and Theorem 4.9, we conclude that there is a $W \in \mathbb{R}^{\ell}_{+} \setminus \{\mathbf{0}\}$ such that

$$\sum_{i=1}^{\ell} W_i \left(f_i(\hat{y}) - f_i(\bar{x})\right) \geq 0, \quad \text{for all } y \in K.$$

Since $W \in \mathbb{R}^{\ell}_{+} \setminus \{\mathbf{0}\}$, we have

$$\sum_{i=1}^{\ell} W_i f_i(\hat{y}) \geq \sum_{i=1}^{\ell} W_i f_i(\bar{x}), \quad \text{for all } y \in K$$

and W is the desired weight vector. □

The following corollary relates efficient solutions of VOP to solutions of WOP by means of the ε-constraint scalarization (see Corollary 4.4).

Corollary 4.9 *Let $K \subseteq \mathbb{R}^n$ be a nonempty convex set and for each $i \in \mathscr{I}$, let $f_i : K \to \mathbb{R}$ be convex. If $\bar{x} \in K$ is an efficient solution of* VOP, *then there exists a weight vector $W \in \mathbb{T}^{\ell}_{+} \setminus \{\mathbf{0}\}$ such that $\bar{x}$ is a solution of* WOP.

Proof Since $\bar{x}$ is an efficient solution of VOP, by Theorem 4.16, it is an optimal solution of ε-COP for every objective function f_k to be minimized. Since each f_i is convex, by Theorem 4.20 we obtain the desired result. □

4.2.2 *Hybrid Method*

By combining the weighted sum method and the ε-constraint method, Corley [11] and Wendell and Lee [53] described another scalarization method which characterizes the optimal solutions of VOP. It is called hybrid method by Chankong and Haimes [10]. We describe the *hybrid method* as to solve the following problem:

$$\begin{aligned}
&\text{minimize} \sum_{i=1}^{\ell} W_i f_i(x), \\
&\text{subject to } f_j(x) \leq \varepsilon_j, \quad \text{for all } j = 1, 2, \ldots, \ell, \\
&x \in K,
\end{aligned} \tag{4.24}$$

where $W_i > 0$ for all $i = 1, 2, \ldots, \ell$ and $\varepsilon = (\varepsilon_1, \varepsilon_2, \ldots, \varepsilon_\ell) \in \mathbb{R}^\ell$. The problem (4.24) is called *hybrid problem.*

The following theorem shows that all efficient solutions of VOP can be found by means of the hybrid problem.

Theorem 4.21 *If $\bar{x} \in K$ is a solution of the hybrid problem for any upper bound vector $\varepsilon \in \mathbb{R}^\ell$, then it is an efficient solution of* VOP. *Conversely, if $\bar{x} \in K$ is an efficient solution of* VOP, *then it is a solution of the hybrid problem for $\varepsilon = f(\bar{x})$.*

Proof Let $\bar{x} \in K$ be an optimal solution of the hybrid problem for any upper bound vector $\varepsilon \in \mathbb{R}^\ell$ and for some $W_i > 0$, $i = 1, 2, \ldots, \ell$. Then $f_j(\bar{x}) \leq \varepsilon_j$ for all $j = 1, 2, \ldots, \ell$. Suppose that $\bar{x}$ is not an efficient solution of VOP. Then there exists some $y \in K$ such that $f_i(y) \leq f_i(\bar{x})$ for all $i = 1, 2, \ldots, \ell$ and $f_j(y) < f_j(\bar{x})$ for at least one j. Since $W_i > 0$ for all $i = 1, 2, \ldots, \ell$, we have

$$\sum_{i=1}^{\ell} W_i f_i(y) < \sum_{i=1}^{\ell} W_i f_i(\bar{x})$$

and

$$f_i(y) \leq f_i(\bar{x}) \leq \varepsilon_i, \quad \text{for all } i = 1, 2, \ldots, \ell.$$

The last inequality shows that y is a feasible point for the hybrid problem. Then $\bar{x}$ is not an optimal solution of the hybrid problem, a contradiction to our assumption. Hence, $\bar{x}$ is an efficient solution of VOP.

We prove the converse part by showing that if $\bar{x}$ is not an optimal solution of the hybrid problem, then it is not an efficient solution of VOP. Assume that $\bar{x} \in K$ is not an optimal solution of the hybrid problem, where $\varepsilon = f(\bar{x})$. Let $\hat{x}$, $\hat{x} \neq \bar{x}$, be an optimal solution of the hybrid problem. Then $\hat{x} \in K$ and $f_j(\hat{x}) \leq \varepsilon_j$ for all $j = 1, 2, \ldots, \ell$. Since $\varepsilon = f(\bar{x})$, we have $f_j(\hat{x}) \leq \varepsilon_j = f_j(\bar{x})$ for all $j = 1, 2, \ldots, \ell$. Also,

$$\sum_{i=1}^{\ell} W_i f_i(\hat{x}) < \sum_{i=1}^{\ell} W_i f_i(\bar{x}) \tag{4.25}$$

and because $W_i > 0$ for all $i = 1, 2, \ldots \ell$,

$$f_j(\hat{x}) < f_j(\bar{x}), \quad \text{for some } j = 1, 2, \ldots, \ell. \tag{4.26}$$

The inequalities (4.25) and (4.26) imply that $f_i(\hat{x}) \leq f_i(\bar{x})$ for all $i = 1, 2, \ldots, \ell$ and $f_j(\hat{x}) < f_j(\bar{x})$ for at least one j. Thus, $\bar{x}$ is a not an efficient solution of VOP. □

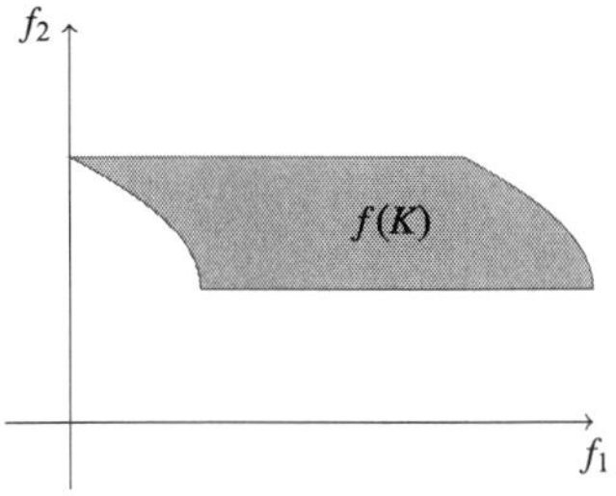

Fig. 4.5 The set of feasible elements in Example 4.3

Example 4.3 Consider the following VOP:

$$\text{minimize}\, f(x) = (x_1^2 - (1-x_2)^2, x_2),$$
$$\text{subject to}\, x \in K,$$

where $K := \{(x_1, x_2) \in \mathbb{R}^2 : 1 \le x_1 \le 2,\ 1 \le x_2 \le 2\}$. The set of feasible elements is illustrated in Fig. 4.5. As can be seen, the problem is nonconvex, and therefore, the weighted-sum method would not be able to yield all efficient solutions of VOP. Only the solutions $(x_1, x_2) = (\sqrt{2}, 0)$ and $(x_1, x_2) = (1, 2)$ are found by the weighted sum method. The hybrid method, however, is able to find all efficient solutions of VOP. The set of efficient solutions is $\{(x_1, x_2) \in K : 1 \le x_1 \le 2, x_2 = 1\}$.

We present the procedure to obtain all optimal solutions of VOP which is based on Theorem 4.21. First one has to solve the hybrid problem for the parametric solution $x_0(\varepsilon)$. Then equate

$$(f_1(x_0(\varepsilon)), f_2(x_0(\varepsilon)), \ldots, f_\ell(x_0(\varepsilon))) = \varepsilon. \tag{4.27}$$

For any solution ε of (4.27), x_0 is a minimal point by the first part of Theorem 4.21. The second part guarantees that all minimal points can be achieved in this way. The hybrid problem can be solved by using any appropriate method.

If the hybrid problem is slightly modified, it can be expressed via the nonlinear scalarizing functional $\xi_{e,D}$. To this end, we introduce the following problem.

$$\begin{aligned} &\text{minimize} \sum_{i=1}^{\ell} W_i f_i(x), \\ &\text{subject to}\, f_j(x) \le \varepsilon_j, \quad \text{for all } j \in P, \\ &x \in K, \end{aligned} \tag{4.28}$$

where $W_i > 0$ for all $i = 1, 2, \ldots, \ell$, $\varepsilon = (\varepsilon_1, \varepsilon_2, \ldots, \varepsilon_\ell) \in \mathbb{R}^\ell$ and $P \subsetneq \{1, 2, \ldots, \ell, \}$. Then we have the following theorem.

Theorem 4.22 ([52]) *Let $Y = \mathbb{R}^\ell$, $X = \mathbb{R}^n$. For the choice $e := (e_1, e_2, \dots, e_\ell)$ with*

$$e_i := \begin{cases} 0, & \text{for all } i \in P, \\ 1, & \text{if } i \notin P, \end{cases}$$

$\bar{\varepsilon} := (\bar{\varepsilon}_1, \bar{\varepsilon}_2, \dots, \bar{\varepsilon}_\ell)$ with

$$\bar{\varepsilon}_i := \begin{cases} \varepsilon_i, & \text{for all } i \in P, \\ 0, & \text{if } i \notin P, \end{cases}$$

and $D := \left\{y \in \mathbb{R}^\ell : y_i \geq 0 \; \forall i \in P, \; \sum_{i=1}^{\ell} W_i y_i \geq 0\right\} - \bar{\varepsilon}$, problem (4.28) *and problem* (4.16) *have the same optimal solutions.*

Remark 4.3 It is clear that the hybrid problem (4.24) cannot be characterized by means of the functional $\xi_{e,D}$, because e would need to be the zero vector. But then the functional $\xi_{e,D}$ would not be defined. This issue is resolved in problem (4.28) by the particular choice of the set P.

4.2.3 Application: A Unified Approach to Uncertain Optimization

In this section, we show that the nonlinear scalarizing functional $\xi_{D,e}$ is a useful tool to represent a wide range of uncertain optimization problems. Specifically, we show that many different concepts of robustness and of stochastic programming can be described as special cases of the general nonlinear scalarization method described in Sect. 4.2 (see also Sect. 2.6) by choosing the involved parameters and sets appropriately. This leads to a unifying concept which can be used to handle robust and stochastic optimization problems as well as to derive new concepts of robustness. We introduce multiple objective (deterministic) counterparts for uncertain optimization problems and discuss their relations to well-known scalar robust optimization problems by using the nonlinear scalarization concept. Finally, we mention some relations between robustness and coherent risk measures. In this section, we follow the explanations conducted in [33].

Since most real world optimization problems (OPs) are contaminated with uncertain data, it is very important to include uncertainty into the optimization model. One way of dealing with such optimization problems is described in the concept of robustness: Instead of assuming that all data are known, one allows different scenarios for the input parameters and looks for a solution that works well in every uncertain scenario. Robust optimization is an active field of research, we refer to Ben-Tal et al. [5] and Kouvelis and Yu [37] for an extensive collection of results and applications for the most prominent concepts. Several other concepts of robustness were introduced more recently, e.g. the concept of light robustness by Fischetti and Monaci [16] or the concept of recovery-robustness in Liebchen et al. [38]. A scenario-based approach is suggested in Goerigk and Schöbel [23]. Moreover, there are many works devoted to uncertain discrete optimization. Various applications of robust approaches to discrete uncertain optimization can be found, for example, in [8, 26, 28].

In all these approaches, the uncertain optimization problem is replaced by a deterministic version, called the *robust counterpart* of the uncertain problem.

Another prominent way of dealing with uncertain optimization is the field of stochastic programming; for an introduction we refer to Birge and Louveaux [7]. Different from robust optimization, stochastic programming assumes some knowledge about the probability distribution of the uncertain data. The objective usually is to find a feasible solution (or a solution that is feasible with a certain probability) that optimizes the expected value of some objective or cost function.

In this section, we will link the different concepts of robustness and of stochastic programming, that usually have been considered fundamentally different, in a very general and unifying framework. Assuming that the set of scenarios is finite, we will show that all of the considered uncertain optimization problems have a (deterministic) counterpart in this framework. Our analysis is based on two closely related concepts: First, we will use *nonlinear scalarizing functionals* to describe the counterpart of uncertain optimization problems, and second, we will relate the solutions of this functional (and likewise of the respective uncertain optimization problems) to the efficient set of a *multiple objective counterpart*. This will lay the ground for a thorough analysis of the interrelations and also the differences between established concepts in robust optimization and stochastic programming. By providing additional trade-off information between alternative efficient solutions, the multiple objective counterpart can facilitate the decision making process when deciding for a most preferred robust solution.

For specific robustness concepts, the connection between uncertain scalar optimization problems and an associated (deterministic) multiple objective counterpart were observed by several authors. Kouvelis and Sayin [36, 45] use this relation to develop efficient solution methods for bi- and multiple objective discrete optimization problems based on algorithms that were originally developed to solve uncertain scalar optimization problems. They focus on two classical robustness concepts that will be referred to as strict robustness and deviation robustness in below, see also [35]. Perny et al. [42] use a multiple objective counterpart to introduce a robustness measure based on the Lorenz dominance rule in the context of minimum spanning tree and shortest path problems. From the stochastic programming perspective, a multiple objective counterpart for a two-stage stochastic programming problem was introduced in Gast [19] and used to interrelate stochastic programming models with the concept of recoverable robustness, see Stiller [49]. A critical analysis is given in Hites et al. [30] who give a qualitative description of the similarities and differences between the two modeling paradigms. They conclude that from a modeling perspective, a multiple objective counterpart can in general not be used to represent an uncertain scalar optimization problem. However, as will be seen below, there certainly is a strong relation from a theoretical point of view. This reveals interesting properties of alternative solutions of scalar uncertain optimization problems.

In a first step, we will treat uncertain scalar optimization problems as special cases of a nonlinear scalarizing functional. Many methods for scalarization were suggested in the literature that are special cases of a nonlinear scalarization concept introduced by Gerstewitz (Tammer) [21], see also Gerth and Weidner [22], Pascoletti and Serafini [41], Göpfert et al. [24, 25]. This scalarization method includes, for instance, weighted-sums, Chebyshev- and ε-constraint-scalarization (see Sect. 4.2). We show that this scalarization method includes a variety of different models from robust optimization and stochastic programming as specifications. So it is possible to get a unified approach for different types of optimization models including uncertainties. Moreover, the well-studied properties of this scalarization method allow the establishment of relations to multiple objective optimization problems (MOPs).

Let Y be a topological vector space, $e \in Y \setminus \{\mathbf{0}\}$ and let $\mathcal{F}, D$ be proper subsets of Y. We assume that D is closed and

$$D + [0, +\infty[\cdot e \subset D. \tag{4.29}$$

Now we formulate the problem

$$\begin{aligned} &\text{minimize } \xi_{e,D}(y), \\ &\text{subject to } y \in \mathcal{F}, \end{aligned} \qquad (P_{e,D,\mathcal{F}})$$

where $\xi_{e,D}$ is defined by (4.15).

Based on the interpretation of uncertain scalar optimization problems by means of the nonlinear scalarizing functional $\xi_{e,D}$, we will in a second step formulate multiple objective counterparts, whose efficient set comprises optimal solutions of the considered uncertain scalar optimization problems.

We now formulate an optimization problem with uncertainties.

Throughout this section, let $\mathcal{U} := \{\zeta_1, \zeta_2, \dots, \zeta_q\}$ be a finite uncertainty set, i.e., $\zeta \in \mathcal{U}$ can take on q different values. One could think of ζ being real numbers or real vectors. Furthermore, let $f : \mathbb{R}^n \times \mathcal{U} \to \mathbb{R}$, $F_i : \mathbb{R}^n \times \mathcal{U} \to \mathbb{R}$, $i = 1, 2, \dots, m$. Then an *uncertain scalar optimization problem* (uncertain OP) is defined as a parametrized optimization problem

$$Q(\zeta), \quad \zeta \in \mathcal{U}, \tag{4.30}$$

where for a given $\zeta \in \mathcal{U}$ the optimization problem $(Q(\zeta))$ is given by

$$\begin{aligned} &\text{minimize } f(x, \zeta), \\ &\text{subject to } F_i(x, \zeta) \leq 0, \ i = 1, 2, \dots, m, \\ &x \in \mathbb{R}^n. \end{aligned} \qquad (Q(\zeta))$$

When the uncertain OP (4.30) has to be solved, it is not known which value $\zeta \in \mathcal{U}$ is going to be realized. We call $\hat{\zeta} \in \mathcal{U}$ the *nominal value*, i.e., the value of ζ that we believe is true today. $(Q(\hat{\zeta}))$ is called the *nominal problem*. Throughout this section, we assume that the minima of the problems to be proposed in the following sections exist. The results in this section are based on Klamroth et al. [33].

4.2.3.1 Strict Robustness

Our first approach to uncertain optimization is a concept called *strict robustness*, which has been first mentioned by Soyster [48] and then formalized and extensively analyzed by Ben-Tal et al. [5] in numerous publications, see e.g. [3, 15] for early contributions and [5] for a detailed collection of results. The idea of this concept is twofold: On the one hand, the worst possible objective function value is minimized in order to get a solution that is "good enough" even in the worst case scenario. On the other hand, all constraints have to be satisfied for *every* scenario $\zeta \in \mathcal{U}$. Thus, this concept is extremely convertive and would suit a risk-averse decision maker. The *strictly robust counterpart* of the uncertain optimization problem $(Q(\zeta), \zeta \in \mathcal{U})$ is defined by

$$\begin{aligned} &\text{minimize } \rho_{RC}(x) = \max_{\zeta\in\mathcal{U}} f(x,\zeta), \\ &\text{subject to } \forall \zeta \in \mathcal{U}: \ F_i(x,\zeta) \leq 0, \ i = 1,2,\ldots,m, \\ &x \in \mathbb{R}^n. \end{aligned} \tag{RC}$$

A feasible solution of the problem (RC) is called *strictly robust*. The set of strictly robust solutions is denoted as

$$\mathfrak{A} := \{x \in \mathbb{R}^n : \forall \zeta \in \mathcal{U} \text{ such that } F_i(x,\zeta) \leq 0, \ i = 1,2,\ldots,m\}. \tag{4.31}$$

In the following theorem, we present how (RC) can be expressed using the nonlinear scalarizing functional $\xi_{e,D}$ (cf. [33, 34]).

Theorem 4.23 *Consider*

$$\mathfrak{A}_1 := \mathfrak{A}, \tag{4.32}$$

$$D_1 := \mathbb{R}^q_+, \tag{4.33}$$

$$e_1 = 1_q := (1,1,\ldots,1), \tag{4.34}$$

$$\mathcal{F}_1 = \left\{(f(x,\zeta_1),(f(x,\zeta_2),\ldots,f(x,\zeta_q)) : x \in \mathfrak{A}_1\right\}. \tag{4.35}$$

For $e = e_1$, $D = D_1$, condition (4.29) is satisfied and with $\mathcal{F} = \mathcal{F}_1$, problem $(P_{e,D,\mathcal{F}})$ is equivalent to problem (RC) in the following sense:

$$\begin{aligned}\min\{\xi_{e_1,D_1}(y) : y \in \mathcal{F}_1\} &= \xi_{e_1,D_1}(y^*)\\ &= \min\{\rho_{RC}(x) : x \in \mathfrak{A}_1\}\\ &= \rho_{RC}(x^*),\end{aligned}$$

where $y^* = (f(x^*, \zeta_1), f(x^*, \zeta_2), \ldots, f(x^*, \zeta_q))$.

Proof Since $D_1 + [0, \infty) \cdot e_1 = \mathbb{R}^q_+ + [0, \infty) \cdot 1_q \subset \mathbb{R}^q_+ = D_1$, condition (4.29) is satisfied. Since $e_1 \in \text{int}(\mathbb{R}^q_+)$ and $D_1 = \mathbb{R}^q_+$ is closed, the infimum in the definition of $\xi_{e,D}$ is finite and attained such that we can replace the infimum by a minimum, and we have for $C = \mathbb{R}^q_+$:

$$\begin{aligned}\min_{y\in\mathcal{F}_1}\xi_{e_1,D_1}(y) &= \min_{y\in\mathcal{F}_1}\min\{t \in \mathbb{R} : y \in te_1 - D_1\}\\ &= \min_{y\in\mathcal{F}_1}\min\{t \in \mathbb{R} : y - te_1 \in -D_1\}\\ &= \min_{x\in\mathfrak{A}_1}\min\left\{t \in \mathbb{R} : (f(x,\zeta_1), f(x,\zeta_2), \ldots, f(x,\zeta_q)) - t\cdot(1,1,\ldots,1) \leq_C 0_q\right\}\\ &= \min_{x\in\mathfrak{A}_1}\min\left\{t \in \mathbb{R} : (f(x,\zeta_1), f(x,\zeta_2), \ldots, f(x,\zeta_q)) \leq_C t\cdot(1,1,\ldots,1)\right\}\\ &= \min\left\{\max_{\zeta\in\mathcal{U}} f(x,\zeta) : x \in \mathfrak{A}_1\right\}\\ &= \min\{\rho_{RC}(x) : x \in \mathfrak{A}_1\}.\end{aligned}$$

□

Note that the selection of $e_1 = 1_q$ means that every objective function $f(x, \zeta)$, $\zeta \in \mathcal{U}$, is treated in the same way, i.e., no objective function is preferred to another one.

Remark 4.4 Since D_1 is a proper closed convex cone and $e_1 \in \text{int}(D_1)$, the functional ξ_{e_1,D_1} is continuous, finite-valued, monotone with respect to $\mathbb{R}^q_+$, strictly monotone with respect to $\mathbb{R}^q_+$ and sublinear, taking into account Corollary 2.1.

Remark 4.5 The concept of strict robustness is described by the Chebyshev scalarization with the origin as reference point as a special case of functional $\xi_{e,D}$. Theorem 4.23 shows that $\xi_{e,D}$ can be interpreted as a max-ordering problem as defined in multiple objective optimization, see [13]. This relationship was also observed by Kouvelis and Sayin [36, 45] where it was used to determine the nondominated set of discrete bicriteria optimization problems.

4.2.3.2 Deviation Robustness

We now introduce another prominent robustness concept, called *deviation-robustness* or *min max regret robustness*. In contrast to the concept of strict robustness, the function to be minimized is $\max_{\zeta\in\mathcal{U}}(f(x,\zeta) - f^*(\zeta))$, where $f^*(\zeta) \in \mathbb{R}$ is the optimal value of problem $(Q(\zeta))$ for the parameter $\zeta \in \mathcal{U}$. This robustness concept has a long tradition in many applications such as scheduling or location theory, mostly if no uncertainty in the constraints is present. We refer to [35] for numerous applications of this approach. We formulate the *deviation-robust counterpart* of (4.30) as

$$\begin{aligned}
&\text{minimize } \rho_{dRC}(x) = \max_{\zeta\in\mathcal{U}}(f(x,\zeta) - f^*(\zeta)) \\
&\text{subject to } \forall\zeta \in \mathcal{U} : F_i(x,\zeta) \le 0,\ i = 1,2,\dots,m, \qquad (dRC)\\
&x \in \mathbb{R}^n.
\end{aligned}$$

Now let

$$f^* := (f^*(\zeta_1), f^*(\zeta_2), \dots, f^*(\zeta_q)) \tag{4.36}$$

be the vector consisting of the individual minimizers for the respective scenarios which can be interpreted as an *ideal solution vector*. The relation to the ideal point of the multiple objective counterpart of problem (4.30) will be discussed later but can already be noted here. Then we have the following theorem:

Theorem 4.24 *Consider*

$$\mathfrak{A}_2 := \mathfrak{A}, \tag{4.37}$$

$$D_2 := \mathbb{R}^q_+ - f^*, \tag{4.38}$$

$$e_2 := 1_q, \tag{4.39}$$

$$\mathcal{F}_2 = \{(f(x,\zeta_1), f(x,\zeta_2), \dots, f(x,\zeta_q)) : x \in \mathfrak{A}_2\}. \tag{4.40}$$

For $e = e_2$, $D = D_2$, condition (4.29) is satisfied and with $\mathcal{F} = \mathcal{F}_2$, problem $(P_{e,D,\mathcal{F}})$ is equivalent to problem (dRC) in the following sense:

$$\begin{aligned}
\min\{\xi_{e_2,D_2}(y) : y \in \mathcal{F}_2\} &= \xi_{e_2,D_2}(y^*) \\
&= \min\{\rho_{dRC}(x) : x \in \mathfrak{A}_2\} \\
&= \rho_{dRC}(x^*),
\end{aligned}$$

where $y^* = (f(x^*,\zeta_1), f(x^*,\zeta_2), \dots, f(x^*,\zeta_q))$.

Proof Since $D_2 + [0,\infty)\cdot e_2 = (\mathbb{R}^q_+ - f^*) + [0,\infty)\cdot 1_q \subset \mathbb{R}^q_+ - f^* = D_2$, condition (4.29) is satisfied. Moreover, for $C = \mathbb{R}^q_+$,

$$\begin{aligned}\min_{y\in\mathcal{F}_2} \xi_{e_2,D_2}(y) &= \min_{y\in\mathcal{F}_2} \min\{t\in\mathbb{R} : y \in te_2 - D_2\}\\ &= \min_{x\in\mathfrak{A}_2} \min\{t\in\mathbb{R} : (f(x,\zeta_1), f(x,\zeta_2),\dots,f(x,\zeta_q))\\ &\qquad\qquad - (f^*(\zeta_1), f^*(\zeta_2),\dots,f^*(\zeta_q)) \leq_C t\cdot 1_q\}\\ &= \min\{\max_{\zeta\in\mathcal{U}}(f(x,\zeta) - f^*(\zeta)) : x\in\mathfrak{A}_2\}\\ &= \min\{\rho_{dRC}(x) : x \in \mathfrak{A}_2\}.\end{aligned}$$

□

Alternatively, we could have taken $\widetilde{D_2} := D_1 = \mathbb{R}^q_+$ and we could have minimized ξ_{e_1,D_1} over the set $\widetilde{\mathcal{F}_2} := \{(f(x,\zeta_1),\dots,f(x,\zeta_q)) - (f^*(\zeta_1),\dots,f^*(\zeta_q)) : x \in \mathfrak{A}_2\}$. Therefore, one can observe that (*dRC*) is a shifted version of (*RC*). This means, whenever we have a finite uncertainty set and f^* is known beforehand, the concepts of strict robustness and of deviation robustness can be solved within the same complexity. However, for general uncertainty sets, (*dRC*) is usually harder to solve than (*RC*).

Remark 4.6 Using Theorem 2.7 and the fact that $D_2 + (0,+\infty)\cdot e_2 \subset \mathrm{int}(D_2)$, we can conclude that the functional ξ_{e_2,D_2} is continuous, finite-valued, convex, monotone with respect to $\mathbb{R}^q_+$, and strictly monotone with respect to $\mathbb{R}^q_+$. Note that since D_2 is not a cone, Corollary 2.1 cannot be applied.

Remark 4.7 Similar to the case of strict robustness, the concept of deviation robustness can be described by the Chebyshev scalarization, however, not with the origin as reference point but with the ideal point f^* defined in (4.36) as reference point. This shows once again the close relationship between these two robustness concepts, see also Kouvelis and Sayin [36, 45].

4.2.3.3 Reliable Robustness

Sometimes it is difficult to find a point x that satisfies all constraints $F_i(x,\zeta) \leq 0$ for all $\zeta \in \mathcal{U}$, or it is simply not useful for x to satisfy the constraints at the cost of minimality of f. We therefore introduce the concept of reliable robustness, where the constraints are allowed to differ from the original problem. Instead of having hard constraints $F_i(x,\zeta) \leq 0$ for all $\zeta \in \mathcal{U}$, we now allow the constraints to satisfy an infeasibility tolerance $\delta_i \in \mathbb{R}_+$ in order to achieve soft constraints $F_i(x,\zeta) \leq \delta_i$. Nevertheless, the original constraints for the nominal value $\hat{\zeta}$ should be fulfilled, i.e., $F_i(x,\hat{\zeta}) \leq 0,\ i = 1,2,\dots,m$. Then the *reliably robust counterpart* of (4.30) proposed by Ben-Tal and Nemirovski [4], is defined by

$$\begin{aligned} &\text{minimize } \rho_{rRC}(x) = \max_{\zeta \in \mathcal{U}} f(x, \zeta), \\ &\text{subject to } F_i(x, \hat{\zeta}) \leq 0, \; i = 1, 2, \ldots, m, \\ &\forall \zeta \in \mathcal{U} : \; F_i(x, \zeta) \leq \delta_i, \; i = 1, 2, \ldots, m, \\ &x \in \mathbb{R}^n. \end{aligned} \tag{rRC}$$

A feasible solution of (rRC) is called *reliably robust*. If $\delta_i = 0$ for all $i = 1, 2, \ldots, m$, the reliably robust OP (rRC) is equivalent to the strictly robust OP (RC). Strict robustness is therefore a special case of reliable robustness.

Theorem 4.25 *Consider*

$$\begin{aligned} \mathfrak{A}_3 := \{x \in \mathbb{R}^n : F_i(x, \hat{\zeta}) \leq 0, \\ \forall \zeta \in \mathcal{U} : F_i(x, \zeta) \leq \delta_i, i = 1, 2, \ldots, m\}, \end{aligned} \tag{4.41}$$

$$D_3 := \mathbb{R}^q_+, \tag{4.42}$$

$$e_3 := 1_q, \tag{4.43}$$

$$\mathcal{F}_3 = \{(f(x, \zeta_1), f(x, \zeta_2), \ldots, f(x, \zeta_q)) : x \in \mathfrak{A}_3\}. \tag{4.44}$$

For $e = e_3$, $D = D_3$, condition (4.29) is satisfied and with $\mathcal{F} = \mathcal{F}_3$, problem ($P_{e,D,\mathcal{F}}$) is equivalent to problem (rRC) in the following sense:

$$\begin{aligned} \min\{\xi_{e_3,D_3}(y) : y \in \mathcal{F}_3\} &= \xi_{e_3,D_3}(y^*) \\ &= \min\{\rho_{rRC}(x) : x \in \mathfrak{A}_3\} \\ &= \rho_{rRC}(x^*), \end{aligned}$$

where $y^* = (f(x^*, \zeta_1), f(x^*, \zeta_2), \ldots, f(x^*, \zeta_q))$.

Because $\rho_{rRC} = \rho_{RC}$ and only the set $\mathcal{F}_3$ of feasible points differs from the concept of strict robustness, the proof is left out.

Remark 4.8 Since $\xi_{e_3,D_3} = \xi_{e_1,D_1}$, the functional ξ_{e_3,D_3} is again continuous, finite-valued, monotone with respect to $\mathbb{R}^q_+$, strictly monotone with respect to $\mathbb{R}^q_+$ and sublinear, taking into account Corollary 2.1.

Remark 4.9 The concept of reliable robustness is - similar to strict robustness - described by the Chebyshev scalarization with the origin as reference point and on the basis of a relaxed feasible set, as a special case of functional $\xi_{e,D}$.

4.2.3.4 Light Robustness

When we examine a variation of the constraints $F_i(x, \zeta) \leq \delta_i$, where F_i, δ_i, $i = 1, 2, \ldots, m$, are defined as in the concept of reliable robustness, it could be useful to minimize these tolerances, which the concept of light robustness describes. It was introduced in 2008 by Fischetti and Monaci in [16] for linear programs with the Γ-uncertainty set introduced by Bertsimas and Sim [6] and generalized to general uncertain robust optimization problems by Schöbel [46]. Let z^* be the optimal value of the nominal problem $(Q(\hat{\zeta}))$, which we assume to be positive, i.e., $z^* > 0$. We want the nominal value $f(x, \hat{\zeta})$ to be bounded by $(1 + \gamma)z^*$, where $\gamma \geq 0$. Then a solution of the *lightly robust counterpart* of (4.30)

$$
\begin{aligned}
&\text{minimize } \rho_{lRC}(\delta, x) = \sum_{i=1}^{m} w_i \delta_i, \\
&\text{subject to } F_i(x, \hat{\zeta}) \leq 0, \ i = 1, 2, \ldots, m, \\
&f(x, \hat{\zeta}) \leq (1 + \gamma) z^*, \\
&\forall \zeta \in \mathcal{U} : \ F_i(x, \zeta) \leq \delta_i, \ i = 1, 2, \ldots, m, \\
&x \in \mathbb{R}^n, \\
&\delta_i \in \mathbb{R}_+, \ i = 1, 2, \ldots, m,
\end{aligned}
\tag{lRC}
$$

where $w_i \geq 0$, $i = 1, 2, \ldots, m$, $\sum_{i=1}^m w_i = 1$, is called *lightly robust*.

Theorem 4.26 *Consider*

$$
D_4 := \left\{ (\delta_1, \delta_2, \ldots, \delta_m) : \sum_{i=1}^{m} w_i \delta_i \geq 0, \ \delta_i \in \mathbb{R}, \ i = 1, 2, \ldots, m \right\}, \tag{4.45}
$$

$$
e_4 := 1_m, \tag{4.46}
$$

$$
\begin{aligned}
\mathcal{F}_4 = &\{(\delta_1, \delta_2, \ldots, \delta_m) : \exists x \in \mathbb{R}^n : F_i(x, \hat{\zeta}) \leq 0, \ f(x, \hat{\zeta}) \leq (1 + \gamma) z^*, \\
&\forall \zeta \in \mathcal{U} : F_i(x, \zeta) \leq \delta_i, \ \delta_i \in \mathbb{R}_+, \ i = 1, 2, \ldots, m\}.
\end{aligned}
\tag{4.47}
$$

For $e = e_4$, $D = D_4$, condition (4.29) is satisfied and with $\mathcal{F} = \mathcal{F}_4$, problem $(P_{e,D,\mathcal{F}})$ is equivalent to problem (lRC) in the following sense:

$$
\begin{aligned}
\min\{\xi_{e_4,D_4}(y) : y \in \mathcal{F}_4\} &= \xi_{e_4,D_4}(y^*) \\
&= \min\{\rho_{lRC}(\delta) : \delta \in \mathcal{F}_4\} \\
&= \rho_{lRC}(\delta^*),
\end{aligned}
$$

where $y^* = \delta^* = (\delta_1^*, \delta_2^*, \ldots, \delta_m^*)$.

Proof In this case, $D_4 + [0, +\infty) \cdot e_4 = \{(\delta_1, \delta_2, \dots, \delta_m) \in \mathbb{R}^m : \sum_{i=1}^m w_i\delta_i \geq 0\} + [0, +\infty) \cdot 1_m \subset D_4$, and (4.29) is satisfied in $\mathbb{R}^m$. Moreover,

$$\begin{aligned}
\min_{y\in\mathcal{F}_4} \xi_{e_4,D_4}(y) &= \min_{y\in\mathcal{F}_4} \min\,\{t \in \mathbb{R} : y \in te_4 - D_4\} \\
&= \min_{y\in\mathcal{F}_4} \min\,\{t \in \mathbb{R} : y - te_4 \in -D_4\} \\
&= \min_{\delta\in\mathcal{F}_4} \min \left\{t \in \mathbb{R} : \sum_{i=1}^m w_i(\delta_i - t) \leq 0\right\} \\
&= \min_{\delta\in\mathcal{F}_4} \min \left\{t \in \mathbb{R} : \sum_{i=1}^m w_i\delta_i \leq t \cdot \sum_{i=1}^m w_i\right\} \\
&= \min \left\{\sum_{i=1}^m w_i\delta_i : \delta \in \mathcal{F}_4\right\} \\
&= \min\,\{\rho_{lRC}(\delta) : \delta \in \mathcal{F}_4\}\,.
\end{aligned}$$

□

Remark 4.10 Note that D_4 is a proper closed convex cone with $e_4 \in \operatorname{int}(D_4)$ and Corollary 2.1 implies that the functional ξ_{e_4,D_4} is continuous, finite-valued, monotone with respect to $\mathbb{R}^m_+$, strictly monotone with respect to $\mathbb{R}^m_+$ and sublinear.

Remark 4.11 The concept of light robustness can be interpreted as a weighted sum approach with upper bound constraints in the dimension defined by the number of constraints of the uncertain OP, including the original constraints in the weighted objective function.

4.2.3.5 Stochastic Programming

Stochastic programming models are conceptually different from robust optimization models in the sense that they take information on the probability distribution of the uncertain data into account. For an introduction to stochastic programming we refer to [7, 47]. We focus on two-stage stochastic programming models in the following, see Beale [2], Dantzig [12] and Tintner [51] for early references. Two-stage stochastic programming models allow for a later correction of a solution x selected in stage 1 of the decision process by a recourse action u when the realization of the random data is known. Note that since we assumed that the scenario set $\mathcal{U}$ is finite, each scenario $\zeta_k \in \mathcal{U}$ now has an associated probability $p_k \geq 0$, $k = 1, \dots, q$, $\sum_{q=1}^k p_k = 1$. In this situation, a *two-stage stochastic counterpart* can be formulated as

$$\begin{aligned}
&\text{minimize } \rho_{SP}(x) = \mathbb{E}[Q(x, \zeta)] = \sum_{k=1}^q p_k Q(x, \zeta_k), \\
&\text{subject to } x \in X.
\end{aligned} \tag{4.48}$$

Here, X denotes the feasible set of the first-stage problem which could, for example, be defined based on the nominal scenario as $X = \{x \in \mathbb{R}^n : F_i(x, \hat{\zeta}) \leq 0, i = 1, \ldots, m\}$, or as the set of strictly robust solutions $X = \mathfrak{A}$, see (4.31). The objective is to minimize the expectation of the overall cost $Q(x, \zeta)$ that involves, for given $x \in X$ and known $\zeta \in \mathcal{U}$, an optimal recourse action u, i.e., an optimal solution of the second-stage problem

$$\begin{aligned} &\text{minimize } f(x, u, \zeta) = Q(x, \zeta), \\ &\text{subject to } u \in \mathcal{G}(x, \zeta). \end{aligned} \tag{4.49}$$

The second-stage objective function $f(x, u, \zeta)$ and the feasible set $\mathcal{G}(x, \zeta)$ of the second-stage problem are both parametrized with respect to the stage 1 solution $x \in X$ and the scenario $\zeta \in \mathcal{U}$.

In the light of the uncertain optimization problem (4.30), we can assume that the objective function f in (4.30) depends both on the first-stage and the second-stage variables, i.e., on the nominal cost and the cost of the recourse action. We hence consider the following specification of problem (4.48):

$$\begin{aligned} &\text{minimize } \rho_{SP}(x, u) = \sum_{k=1}^{q} p_k f(x, u_k, \zeta_k), \\ &\text{subject to } \forall \zeta_k \in \mathcal{U} : \; F_i(x, \zeta_k) - \delta_k(u_k) \leq 0, \; i = 1, 2, \ldots, m, \\ &x \in \mathbb{R}^n, \\ &u_k \in \mathcal{G}(x, \zeta_k), \; k = 1, 2, \ldots, q, \end{aligned} \tag{SP}$$

with compensations $\delta_k : \mathbb{R}^{\bar{n}} \to \mathbb{R}$ that depend on the second-stage decisions $u_k \in \mathbb{R}^{\bar{n}}$, $k = 1, \ldots, q$.

Remark 4.12 If we set $\mathcal{G}(x, \zeta) = \emptyset$ in the two-stage stochastic programming formulation (*SP*), we obtain a static model as a special case in which the second-stage variables $u \in \mathbb{R}^{\bar{n}q}$ can be omitted. Since this model plays a special role in the comparison of the different robustness and stochastic programming concepts below, we include it here for the sake of completeness. The static stochastic programming problem is defined by:

$$\begin{aligned} &\text{minimize } \rho_{sSP}(x) = \sum_{k=1}^{q} p_k f(x, \zeta_k), \\ &\text{subject to } \forall \zeta_k \in \mathcal{U} : \; F_i(x, \zeta_k) \leq 0, \; i = 1, 2, \ldots, m, \\ &x \in \mathbb{R}^n. \end{aligned} \tag{sSP}$$

Theorem 4.27 *Let*

$$\begin{aligned}\mathfrak{A}_5 :=&\{(x,u) := (x,u_1,\ldots,u_q) \in \mathbb{R}^{n\times \bar{n}^q} \;:\; \forall \zeta_k \in \mathcal{U} :\\ &F_i(x,\zeta_k) - \delta_k(u_k) \leq 0,\; i = 1,2,\ldots,m\},\end{aligned} \tag{4.50}$$

$$D_5 := \{(y_1,\ldots,y_q) : \sum_{k=1}^{q} p_k y_k \geq 0,\; y_k \in \mathbb{R},\; k = 1,2,\ldots,q\}, \tag{4.51}$$

$$e_5 := 1_q, \tag{4.52}$$

$$\mathcal{F}_5 =\{(f(x,u_1,\zeta_1),(f(x,u_2,\zeta_2),\ldots,f(x,u_q,\zeta_q)) : (x,u) \in \mathfrak{A}_5\}. \tag{4.53}$$

For $e = e_5$, $D = D_5$, condition (4.29) is satisfied and with $\mathcal{F} = \mathcal{F}_5$, problem $(P_{e,D,\mathcal{F}})$ is equivalent to problem (SP) in the following sense:

$$\begin{aligned}\min\{\xi_{e_5,D_5}(y) : y \in \mathcal{F}_5\} &= \xi_{e_5,D_5}(y^*)\\ &= \min\{\rho_{SP}(x,u) : (x,u) \in \mathfrak{A}_5\}\\ &= \rho_{SP}(x^*,u^*),\end{aligned}$$

where $y^* = (f(x^*,u_1^*,\zeta_1),f(x^*,u_2^*,\zeta_2),\ldots,f(x^*,u_q^*,\zeta_q))$.

Proof We have $D_5 + [0,+\infty)\cdot e_5 = \left\{(y_1,\ldots,y_q) \in \mathbb{R}^q : \sum_{k=1}^q p_k y_k \geq 0\right\} + [0,+\infty)\cdot 1_m \subset D_5$, thus (4.29) is satisfied. Moreover,

$$\begin{aligned}\min_{y\in\mathcal{F}_5} \xi_{e_5,D_5}(y) &= \min_{y\in\mathcal{F}_5} \min\{t \in \mathbb{R} : y \in te_5 - D_5\}\\ &= \min_{y\in\mathcal{F}_5} \min\{t \in \mathbb{R} : y - te_5 \in -D_5\}\\ &= \min_{y\in\mathcal{F}_5} \min\left\{t \in \mathbb{R} : \sum_{k=1}^{q} p_k(y_k - t) \leq 0\right\}\\ &= \min_{y\in\mathcal{F}_5} \min\left\{t \in \mathbb{R} : \sum_{k=1}^{q} p_k y_k \leq t\cdot \underbrace{\sum_{k=1}^{q} p_k}_{=1}\right\}\\ &= \min\left\{\sum_{k=1}^{q} p_k y_k : y \in \mathcal{F}_5\right\}\\ &= \min\{\rho_{SP}(x,u) : (x,u) \in \mathfrak{A}_5\}.\end{aligned}$$

□

Remark 4.13 D_5 is a proper closed convex cone with $e_5 \in \text{int}(D_5)$ and Corollary 2.1 implies that the functional ξ_{e_5,D_5} is continuous, finite-valued, monotone with respect to $\mathbb{R}^q_+$, strictly monotone with respect to $\mathbb{R}^q_+$ and sublinear.

Remark 4.14 Similar to the case of light robustness, the above formulated two-stage stochastic programming problem can be interpreted as a weighted sums approach, however, in this case with a relaxed feasible set. This relation was also observed by Gast [19] in the multiple objective context. Note that in the special case of the static model (*sSP*), the feasible set is in fact identical to the set of strictly robust solutions $\mathfrak{A}$ (and not relaxed), see (4.31).

4.2.3.6 Properties of the Nonlinear Functional for the Description of Robustness and Stochastic Programming

The properties of the nonlinear scalarizing functional $\xi_{e,D}$ that were used for the description of the respective robust and stochastic programming counterparts are summarized in the following corollary (see Theorem 2.7 and Corollary 2.1).

Corollary 4.10 *The following properties hold for* $i = 1, 2, 3, 5$ *(*$i = 1$*: strict robustness,* $i = 2$*: deviation robustness,* $i = 3$*: reliable robustness,* $i = 5$*: stochastic programming): The corresponding functional* ξ_{e_i,D_i} *is continuous, finite-valued, convex, monotone with respect to* $\mathbb{R}^q_+$ *and strictly monotone with respect to* $\mathbb{R}^q_+$*, and the following properties hold:*

$$\forall y \in \mathcal{F}_i,\ \forall r \in \mathbb{R} : \xi_{e_i,D_i}(y) \leq r \Leftrightarrow y \in re_i - D_i, \tag{P1}$$

$$\forall y \in \mathcal{F}_i,\ \forall r \in \mathbb{R} : \xi_{e_i,D_i}(y + re_i) = \xi_{e_i,D_i}(y) + r. \tag{P2}$$

$$\forall y \in \mathcal{F}_i,\ \forall r \in \mathbb{R} : \xi_{e_i,D_i}(y) = r \Leftrightarrow y \in re_i - \partial D_i, \tag{P3}$$

$$\forall y \in \mathcal{F}_i,\ \forall r \in \mathbb{R} :\ \xi_{e_i,D_i}(y) < r \Leftrightarrow y \in re_i - \text{int}(D_i). \tag{P4}$$

For $i = 1, 3, 5$, z^{B_i,k_i} *is even sublinear. For* $i = 4$ *(light robustness), the properties* (P1)–(P4) *are fulfilled, and* ξ_{e_4,D_4} *is continuous, sublinear and finite-valued. Additionally,* ξ_{e_i,D_i} *is monotone with respect to* $\mathbb{R}^m_+$ *and strictly monotone with respect to* $\mathbb{R}^m_+$.

4.2.3.7 New Concepts for Robustness

As we have shown in Sect. 4.2, the functional $\xi_{e,D}$ contains many scalarizations of VOP as special cases which are well known in the literature (see [50]), for instance the weighted Chebyshev scalarization or weighted sum scalarization, see [52] for details. These scalarizations can be regarded in the context of robustness. Specifically, one can develop new concepts for robustness that fit the specific needs of a decision-maker. In order to illustrate this point we exemplary use the well-known *ε-constraint scalarization method* (see [13, 14, 27]). This scalarization leads to the ε-constraint method in multiple objective optimization. In the following we analyze which type of robust counterpart is defined by this scalarization.

Note that a similar analysis can be done for every variation of the parameters D, e and for every feasible set $\mathcal{F}$ in $(P_{e,D,\mathcal{F}})$, i.e., many other possibilities for defining a robust counterpart of an uncertain OP may be derived.

Let us first define the ε-constraint scalarization. To this end, let some $k \in \{1, 2, \ldots, q\}$ and some real values $\varepsilon_l \in \mathbb{R}$, $l = 1, 2, \ldots, q, l \neq k$ be given. Then the ε-constraint scalarization is given by

$$e_6 = (e_6^1, e_6^2, \ldots, e_6^q) \text{ where } e_6^j = \begin{cases} 1, & \text{for } j = k, \\ 0, & \text{for } j \neq k, \end{cases} \tag{4.54}$$

$$D_6 := \mathbb{R}^q_+ - \bar{b}, \text{ with } \bar{b} = (\bar{b}^1, \bar{b}^2, \ldots, \bar{b}^q), \bar{b}^j = \begin{cases} 0, & \text{for } j = k, \\ \varepsilon_j, & \text{for } j \neq k, \end{cases} \tag{4.55}$$

$$\mathcal{F}_6 = \left\{(f(x, \zeta_1), f(x, \zeta_2), \ldots, f(x, \zeta_q)) : x \in \mathfrak{A}\right\}. \tag{4.56}$$

Then the following reformulation holds.

Theorem 4.28 *Let* $\varepsilon := (\varepsilon_1, \varepsilon_2, \ldots, \varepsilon_q) \in \mathbb{R}^q$ *and* $k \in \{1, 2, \ldots, q\}$. *Then for* $e = e_6$, $D = D_6$, (4.29) *holds and with* $\mathcal{F} = \mathcal{F}_6$, *problem* $(P_{e,D,\mathcal{F}})$ *is equivalent to*

$$\begin{aligned} &\textit{minimize } \rho_{\varepsilon RC}(x) = f(x, \zeta_k), \\ &\textit{subject to } \forall \zeta \in \mathcal{U}: \ F_i(x, \zeta) \leq 0, \ i = 1, 2, \ldots, m, \\ &x \in \mathbb{R}^n, \\ &f(x, \zeta_j) \leq \varepsilon_j, \ j \in \{1, 2, \ldots, q\}, \ j \neq k. \end{aligned} \tag{εRC}$$

Theorem 4.28 shows that the nonlinear scalarizing functional $z^{B,k}$ can be formulated as (εRC). Let us call (εRC) the *ε-constraint robust counterpart* of (4.30). We now discuss its meaning for robust optimization. Contrary to the other robustness concepts, the parameter k_6 symbolizes that only a single objective function is minimized. This means, the decision maker picks one particular objective function that is to be minimized subject to the constraints that were also used in strict and deviation-robustness. Additionally, the former objective functions $f(x, \zeta_j)$, $j \in \{1, 2, \ldots, q\}$, $j \neq k$, are moved to the constraints. This concept makes sense if a solution is required with a given nominal quality for every scenario $\zeta_j, j = 1, 2, \ldots, q, j \neq k$ while finding the best possible for the remaining scenarios k. Applying this concept, one question immediately arises: How can a decision-maker be sure how to pick the upper bounds ε_j for these constraints? If the bounds ε_j are chosen too small, the set of feasible solutions of (εRC) may be empty, or the objective function value of $f(x, \zeta_k)$ may decrease in an undesired manner. On the other hand, if the bounds ε_j are chosen too large, the quality for the other scenarios decreases.

Note that we could have included the constraint $f(x, \zeta_j) \leq \varepsilon_j, j \in \{1, 2, \ldots, q\}$, $j \neq k$ in the set of feasible points $\widetilde{\mathcal{F}_6}$, and we would have obtained $\widetilde{\mathcal{F}_6} = \{(f(x, \zeta_1), f(x, \zeta_2), \ldots, f(x, \zeta_q)) : x \in \mathbb{R}^n : f(x, \zeta_j) \leq \varepsilon_j, j \in \{1, 2, \ldots, q\}, j \neq k, \forall \zeta \in \mathcal{U} : F_i(x, \zeta) \leq 0, i = 1, 2, \ldots, m\}$. Then we could have used $\widetilde{D_6} = \mathbb{R}^q_+$ instead of D_6, but the set of feasible points $\widetilde{\mathcal{F}_6}$ would have been smaller and possibly harder to deal with.

Corollary 4.11 *The functional ξ_{e_6, D_6} is lower semi-continuous, convex, proper, monotone with respect to $\mathbb{R}^q_+$ and strictly monotone with respect to $\mathbb{R}^q_{+}$, and the properties* (P1) *and* (P2) *from Corollary 4.10 hold for $i = 6$.*

4.2.3.8 Multiple Objective Counterpart Problems and Relations to Scalar Robust Optimization and Stochastic Programming

In this section we propose a new concept, namely to replace an uncertain (scalar) OP $(Q(\zeta), \zeta \in \mathcal{U})$, as introduced in (4.30), by its (deterministic) multiple objective counterpart. The idea is that every scenario $\zeta_l \in \mathcal{U}, l = 1, 2, \ldots, q$ yields its own objective function $h_l(x) := f(x, \zeta_l)$, the only exception being the case of light robustness where the roles of objective and constraints are reversed. Following the example of the different robustness concepts discussed above, the multiple objective counterparts formulated below can be distinguished with respect to the solution set $\mathfrak{A}$, i.e., the way in which the (uncertain) constraints are handled. To simplify the following analysis, in the case of stochastic programming we focus on the static model (*sSP*).

Let $h : \mathbb{R} \to \mathbb{R}^q$ be defined by

$$h(x) := \left(h_1(x), h_2(x), \ldots, h_q(x)\right) = \left(f(x, \zeta_1), f(x, \zeta_2), \ldots, f(x, \zeta_q)\right). \tag{4.57}$$

Recall from (4.31) that

$$\mathfrak{A}_1 = \mathfrak{A}_2 = \mathfrak{A}_5 = \mathfrak{A} = \{x \in \mathbb{R}^n : \forall \zeta \in \mathcal{U} : F_i(x, \zeta) \leq 0, i = 1, 2, \ldots, m\}.$$

Then the *multiple objective strictly robust counterpart* to $(Q(\zeta), \zeta \in \mathcal{U})$ is defined by

$$\begin{aligned} &\text{minimize } h(x), \\ &\text{subject to } x \in \mathfrak{A}_1, \end{aligned} \tag{RC'}$$

where (weakly) efficient elements are defined by means of the natural ordering cone in $\mathbb{R}^q$ and $h[\mathfrak{A}_1] = \mathcal{F}_1$ (see (4.35)).

Similarly, recall from (4.41) that

$$\mathfrak{A}_3 := \{x \in \mathbb{R}^n : F_i(x, \hat{\zeta}) \leq 0, \ \forall \zeta \in \mathcal{U} : \ F_i(x, \zeta) \leq \delta_i, \ i = 1, 2, \ldots, m\}.$$

We propose the *multiple objective reliably robust counterpart* to $(Q(\zeta), \zeta \in \mathcal{U})$ as

$$\begin{aligned} &\text{minimize } h(x), \\ &\text{subject to } x \in \mathfrak{A}_3, \end{aligned} \tag{rRC'}$$

and (weakly) efficient elements are defined again by means of the natural ordering cone in $\mathbb{R}^q$. We have $h[\mathfrak{A}_3] = \mathcal{F}_3$ (see formula (4.44)).

Now let us introduce a multiple objective counterpart that corresponds to the lightly robust counterpart (lRC). Let $\mathcal{F}_4$ be defined by (4.47), i.e.,

$$\begin{aligned} \mathcal{F}_4 =& \{(\delta_1, \delta_2, \ldots, \delta_m) : \exists x \in \mathbb{R}^n : F_i(x, \hat{\zeta}) \leq 0, \, f(x, \hat{\zeta}) \leq (1 + \gamma) z^*, \\ & \forall \zeta \in \mathcal{U} : F_i(x, \zeta) \leq \delta_i, \ \delta_i \in \mathbb{R}, \ i = 1, 2, \ldots, m\}. \end{aligned}$$

We propose the *multiple objective lightly robust counterpart* to $(Q(\zeta), \zeta \in \mathcal{U})$ by

$$\begin{aligned} &\text{minimize } \delta, \\ &\text{subject to } \delta \in \mathcal{F}_4. \end{aligned} \tag{lRC'}$$

Here, (weakly) efficient elements are defined by means of the natural ordering cone in $\mathbb{R}^m$.

Using Theorem 4.11 together with Corollaries 4.10 and 4.11, we can conclude that problem $(P_{e_i, D_i, \mathcal{F}_i})$, $i = 1, 2, 5, 6$ ($(P_{e_3, D_3, \mathcal{F}_3})$, $(P_{e_4, D_4, \mathcal{F}_4})$, respectively) is a scalarization of the multiple objective counterpart (RC') ((rRC'), (lRC'), respectively), and the following corollary holds due to the monotonicity properties of ξ_{e_i, D_i}, $i = 1, \ldots, 6$.

Corollary 4.12 *For $i = 1, 2, 5, 6$ ($i = 1$: strict robustness, $i = 2$: deviation robustness, $i = 5$: static stochastic programming, $i = 6$: ε-constraint robustness), we have:*

$$\begin{aligned} [\forall y \in \mathcal{F}_i \setminus \{y^*\} : \xi_{e_i, D_i}(y^*) < \xi_{e_i, D_i}(y)] &\Rightarrow y^* \in \mathbb{E}(h[\mathfrak{A}_1], \mathbb{R}^q_+), \\ [\forall y \in \mathcal{F}_i : \xi_{e_i, D_i}(y^*) \leq \xi_{e_i, D_i}(y)] &\Rightarrow y^* \in \mathbb{WE}(h[\mathfrak{A}_1], \mathbb{R}^q_+). \end{aligned}$$

Concerning reliably robustness ($i = 3$), it holds

$$\begin{aligned} [\forall y \in \mathcal{F}_3 \setminus \{y^*\} : \xi_{e_3, D_3}(y^*) < \xi_{e_3, D_3}(y)] &\Rightarrow y^* \in \mathbb{E}(h[\mathfrak{A}_3], \mathbb{R}^q_+), \\ [\forall y \in \mathcal{F}_3 : \xi_{e_3, D_3}(y^*) \leq \xi_{e_3, D_3}(y)] &\Rightarrow y^* \in \mathbb{WE}(h[\mathfrak{A}_3], \mathbb{R}^q_+). \end{aligned}$$

For light robustness ($i = 4$)*, we conclude*

$$[\forall y \in \mathcal{F}_4 \setminus \{y^*\} : \xi_{e_4,D_4}(y^*) < \xi_{e_4,D_4}(y)] \Rightarrow y^* \in \mathbb{E}(\mathcal{F}_4, \mathbb{R}^m_+),$$
$$[\forall y \in \mathcal{F}_4 : \xi_{e_4,D_4}(y^*) \leq \xi_{e_4,D_4}(y)] \Rightarrow y^* \in \mathbb{WE}(\mathcal{F}_4, \mathbb{R}^m_+).$$

Because problem $(P_{e_i,D_i,\mathcal{F}_i})$, $i = 1, 2, 5, 6$, with the nonlinear scalarizing objective functional ξ_{e_i,D_i} is a scalarization of the multiple objective counterpart (RC') and $(P_{e_1,D_1,\mathcal{F}_1})$ $((P_{e_2,D_2,\mathcal{F}_2})$, $(P_{e_5,D_5,\mathcal{F}_5})$, $(P_{e_6,D_6,\mathcal{F}_6})$, respectively) is equivalent to the optimization problem (RC) $((dRC)$, (sSP), (εRC), respectively), we can conclude: If x^* solves (RC) $((dRC)$, (sSP), (εRC), respectively), then $y^* = (f(x^*, \zeta_1), f(x^*, \zeta_2), \ldots, f(x^*, \zeta_q))$ is weakly efficient for (RC'). If x^* is the unique solution of problem (RC) $((dRC)$, (sSP), (εRC), respectively), then $y^* = (f(x^*, \zeta_1), f(x^*, \zeta_2), \ldots, f(x^*, \zeta_q))$ is efficient for (RC').

Remark 4.15 The (weakly) efficient set of the multiple objective strictly robust counterpart (RC') thus comprises optimal solutions of the (scalar) strictly robust counterpart (RC) (which are obtained by Chebyshev scalarization with the origin as reference point), the deviation-robust counterpart (dRC) (Chebyshev scalarization with the ideal point as reference point), the static stochastic programming equivalent (sSP) (weighted sums scalarization), and the ε-constraint robust counterpart (εRC) (ε-constraint scalarization).

Analogously, similar results also hold for the other concepts of robustness: $(P_{e_3,D_3,\mathcal{F}_3})$ is equivalent to the reliably robust counterpart (rRC). Therefore, we conclude: If x^* solves (rRC), then $y^* = (f(x^*, \zeta_1), f(x^*, \zeta_2), \ldots, f(x^*, \zeta_q))$ is weakly efficient for (rRC'). If x^* is the unique solution of problem (rRC), then $y^* = (f(x^*, \zeta_1), f(x^*, \zeta_2), \ldots, f(x^*, \zeta_q))$ is efficient for (rRC'). Concerning light robustness, we can conclude: If $\delta^* = (\delta_1^*, \delta_2^*, \ldots, \delta_m^*)$ is a (unique) solution of the lightly robust counterpart (lRC), then δ^* is (efficient) weakly efficient for the corresponding multiple objective counterpart (lRC').

4.2.3.9 Application: Risk Measures and Robustness in Financial Theory

The functional $\xi_{e,D}$ is an important tool in the field of financial mathematics (see Heyde [29]). As already mentioned before, it can be used as a coherent risk measure of an investment. For a better understanding of the topic, we now introduce coherent risk measures and their relation to robustness.

Let Y be a vector space of random variables, and let Ω be a set of elementary events (a set of all possible states of the future). Then a future payment of an investment is a random variable $y : \Omega \to \mathbb{R}$. Positive payments in the future are wins, negative ones are losses. If no investment is being done, then y takes on the value zero. In order to valuate such an investment, we need to valuate random variables by comparing them. To do that, we introduce an ordering relation that is induced by a set $D \subseteq Y$. Artzner et al. [1] axioms for a cone $D \subseteq Y$ of random variables that represent acceptable investments:

(i) $\{y \in Y : y(\omega) \geq 0 \ (\omega \in \Omega)\} \subset D, D \cap \{y \in Y : y(\omega) < 0 \ (\omega \in \Omega)\} = \emptyset$,
(ii) $D + D \subset D$.

In financial terms, the cone property says that every nonnegative multiple of an acceptable investment is again acceptable. Furthermore, axiom (i) means that every investment with almost sure nonnegative results will be accepted and every investment with almost sure negative results is not acceptable. The convexity property in axiom (ii) means that merging two acceptable investments together results again in an acceptable investment. However, in some applications the cone property of D and axiom (ii) are not useful, especially, if the investor does not want to lose more than a certain amount of money. In this case Föllmer and Schied [17] replace the cone property and axiom (ii) by a convexity axiom.

Cones $D \subset Y$ satisfying the axioms (i) and (ii) of acceptable investments can be used in order to introduce a preference relation on Y. The decision maker prefers y_1 to y_2 (changing from y_2 to y_1 is an acceptable risk) if and only if $y_1 - y_2$ is an element of D, i.e.,

$$y_1 \geq_D y_2 \Leftrightarrow y_1 - y_2 \in D.$$

The smallest cone D satisfying the axioms (i) and (ii) is $D = \{y \in Y : y(\omega) \geq 0 \ (\omega \in \Omega)\}$. A decision maker using this particular cone D of acceptable investments is risk-averse, i.e., he only accepts investments with nonnegative payments.

The functional $\xi_{e,D}$ can be used to describe risks associated with investments. Artzner et al. [1] introduced coherent risk measures, i.e., functionals $\mu : Y \to \mathbb{R} \cup \{+\infty\}$, where Y is the vector space of random variables, that satisfy the following properties:

(a) $\mu(y + tk) = \mu(y) - t$,
(b) $\mu(0) = 0$ and $\mu(\lambda y) = \lambda \mu(y)$ for all $y \in Y$ and $\lambda > 0$,
(c) $\mu(y_1 + y_2) \leq \mu(y_1) + \mu(y_2)$ for all $y_1, y_2 \in Y$,
(d) $\mu(y_1) \leq \mu(y_2)$ if $y_1 \geq y_2$.

The following interpretation of the properties (a)–(d) is to mention: The translation property (a) means that the risk would be mitigated by an additional safe investment with a corresponding amount, especially, it holds

$$\mu(y + \mu(y)e) = 0.$$

The positive homogeneity of the risk measure in (b) means that double risk must be secured by double risk capital; the subadditivity in (c) means that a diversification of risk should be recompensed and finally, the monotonicity of the risk measure in (d) means that higher risk needs more risk capital.

A risk measure may be negative. In this case it can be interpreted as a maximal amount of cash that could be given away such that the reduced result remains acceptable.

It can be shown that

$$\mu(y) = \inf\{t \in \mathbb{R} : y + te \in D\}$$

is a coherent risk measure. Obviously, we have (cf. Heyde [29])

$$\mu(y) = \xi_{e,D}(-y).$$

A risk measure induces a set D_μ of acceptable risks (dependent on μ)

$$D_\mu = \{y \in Y : \mu(y) \leq 0\}.$$

Based on our investigation of robustness by means of the functional $\xi_{e,D}$, the following interpretation of coherent risk measures is now possible: If $Y = \mathbb{R}^q$ (there are q states of the future), $D_1 = \mathbb{R}^q_+$ (that is, indeed, the smallest set that satisfies axioms (i) and (ii)) and $e_1 = 1_q$, then

$$\mu(y) = \xi_{e_1,D_1}(-y) = \max_{\xi \in \mathcal{U}}(-f(x,\zeta)) = -\min_{\zeta \in \mathcal{U}} f(x,\zeta)$$

is a coherent risk measure. Specifically, the risk measure $\max_{\zeta \in \mathcal{U}}(-f(x,\zeta))$ is the objective function of the strictly robust counterpart (*RC*) with negative values of f. Because $\mu(y) = -\min_{\zeta \in \mathcal{U}} f(x,\zeta)$, negative payments f of an investment in the future result in a positive risk measure, and positive payments result in a negative risk measure. This seems very reasonable since negative payments (losses) are riskier than investments with only positive payments (bonds). The above approach can analogously be used for other concepts of robustness (assuming that the cone D satisfies (i) and (ii)).

Interrelations between robustness and coherent risk measures have also been studied by Quaranta and Zaffaroni [43]: They minimized the conditional value at risk (which is a coherent risk measure) of a portfolio of shares using concepts of robust optimization.

References

1. P. Artzner, F. Delbean, J.-M. Eber, D. Heath, Coherent measures of risk. Math. Financ. **9**, 203–228 (1999)
2. E.M.L. Beale, On minimizing a convex function subject to linear inequalities. J. R. Stat. Soc. Ser. B **17**, 173–184 (1955)
3. A. Ben-Tal, A. Nemirovski, Robust convex optimization. Math. Oper. Res. **23**(4), 769–805 (1998)
4. A. Ben-Tal, A. Nemirovski, Robust solutions of linear programming problems contaminated with uncertain data. Math. Program. **88**, 411–424 (2000)
5. A. Ben-Tal, L. El Ghaoui, A. Nemirovski, *Robust Optimization* (Princeton University Press, Princeton, Oxford, 2009)

6. D. Bertsimas, M. Sim, The price of robustness. Oper. Res. **52**(1), 35–53 (2004)
7. J.R. Birge, F.V. Louveaux, *Introduction to Stochastic Programming* (Springer, New York, 1997)
8. C. Buchheim, F. Liers, L. Sanità, An exact algorithm for robust network design, in *Network Optimization*, ed. by J. Pahl, T. Reiners, S. Voß. Lecture Notes in Computer Science, vol. 6701 (Springer, Berlin, Heidelberg, 2011)
9. S. Chandra, Jayadeva, A. Mehra, *Numerical Optimization with Applications* (Narosa Publishing House, New Delhi, 2009)
10. V. Chankong, Y.Y. Haimes, *Multiobjective Decision Making: Theory and Methodology* (Elsevier Science Publishing Co., Inc., New York, NY, 1983)
11. H.W. Corley, An existence result for maximization with respect to cone. J. Optim. Theory Appl. **31**(2), 277–281 (1980)
12. G.B. Dantzig, Linear programming under uncertainty. Manage. Sci. **1**, 197–206 (1955)
13. M. Ehrgott, *Multicriteria Optimization*. Lecture Notes in Economics and Mathematical Sciences, vol. 491 (Springer, Berlin, Heidelberg, 2000)
14. G. Eichfelder, *Adaptive Scalarization Methods in Multiobjective Optimization* (Springer, Berlin, Heidelberg, 2008)
15. L. El Ghaoui, H. Lebret, Robust solutions to least-squares problems with uncertain data. SIAM J. Matrix Anal. Appl. **18**, 1034–1064 (1997)
16. M. Fischetti, M. Monaci, Light robustness, in *Robust and Online Large-Scale Optimization*, ed. by R.K. Ahuja, R. Moehring, C. Zaroliagis. Lecture Notes in Computer Science, vol. 5868 (Springer, Berlin, Heidelberg, 2009), pp. 61–84
17. H. Föllmer, A. Schied, *Stochastic Finance* (Walter de Gruyter, Berlin, 2004)
18. S. Gass, T. Saaty, The computational algorithm for the parametric objective function. Naval Res. Logist. Quart. **2**, 39–45 (1995)
19. S. Gast, *Network Flow Problems with Uncertain Input Data in the Context of Supply Chain Management Applications* (University of Wuppertal, Wuppertal, 2010)
20. A.M. Geoffrion, Proper efficiency and the theory of vector maximization. J. Math. Anal. Appl. **22**, 618–630 (1968)
21. C. Gerstewitz (Tammer), Nichtkonvexe Dualität in der Vektoroptimierung. Wissenschaftliche Zeitschrift TH Leuna-Mersebung. **25**, 357–364 (1983)
22. C. Gerth (Tammer), P. Weidner, Nonconvex separation theorems and some applications in vector optimization. J. Optim. Theory Appl. **67**, 297–320 (1990)
23. M. Goerigk, A. Schöbel, A scenario-based approach for robust linear optimization, in *Proceedings of the 1st International ICST Conference on Practice and Theory of Algorithms in (Computer) Systems (TAPAS)*. Lecture Notes in Computer Science (Springer, New York, 2011), pp. 139–150
24. A. Göpfert, C. Tammer, C. Zălinescu, A new minimal point theorem in product spaces. Z. Anal. Anwendungen **18**, 767 770 (1999)
25. A. Göpfert, C. Tammer, H. Riahi, C. Zălinescu, *Variational Methods in Partially Ordered Spaces* (Springer, Berlin, Heidelberg, 2003)
26. C. Gottschalk, A.M.C.A. Koster, F. Liers, B. Peis, D. Schmand, A. Wierz, Robust Flows Over Time: Models and Complexity Results (2017). arXiv:1608.06520v2 [cs.DM]
27. Y. Haimes, L.S. Lasdon, D.A. Wismer, On a bicriterion formulation of the problems of integrated system identification and system optimization. IEEE Trans. Syst. Man Cyberbet. **1**, 296–297 (1971)
28. A. Heidt, H. Helmke, M. Kapolke, F. Liers, A. Martin, Robust runway scheduling under uncertain conditions. J. Air Transp. Manag. **56**(A), 28–37 (2016)
29. F. Heyde, Coherent risk measures and vector optimization, in *Multicriteria Decision Making and Fuzzy Systems: Theory, Methods and Applications*, ed. by K.-H. Küfer (Shaker-Verlag, Herzogenrath, 2006), pp. 3–12
30. R. Hites, Y. De Smet, N. Risse, M. Salazar-Neumann, P. Vincke, About the applicability of MCDA to some robustness problems. Eur. J. Oper. Res. **174**, 322–332 (2006)
31. J. Jahn, Scalarization in vector optimization. Math. Program. **29**, 203–218 (1984)

32. J. Jahn, *Vector Optimization: Theory, Applications, and Extensions* (Springer, Berlin, Heidelberg, 2004)
33. K. Klamroth, E. Köbis, A. Schöbel, C. Tammer, A unified approach for different concepts of robustness and stochastic programming via nonlinear scalarizing functionals. Optimization **62**(5), 649–671 (2013)
34. E. Köbis, C. Tammer, Relations between strictly robust optimization problems and a nonlinear scalarization method. AIP Conf. Proc. **1479**, 2371–2374 (2012)
35. I.V. Konnov, J.-C. Yao, On the generalized vector variational inequality problem. J. Math. Anal. Appl. **206**, 42–58 (1997)
36. P. Kouvelis, S. Sayin, Algorithm robust for the bicriteria discrete optimization problem. Ann. Oper. Res. **147**, 71–85 (2006)
37. P. Kouvelis, G. Yu, *Robust Discrete Optimization and Its Applications* (Kluwer Academic Publishers, New York, 1997)
38. C. Liebchen, M. Lübbecke, R.H. Möhring, S. Stiller, The concept of recoverable Robustness, linear programming recovery, and railway applications, in *Robust and Online Large-Scale Optimization*, ed. by R.K. Ahuja, R.H. Möhring, C.D. Zaroliagis. Lecture Note on Computer Science, vol. 5868 (Springer, New York, 2009)
39. D.T. Luc, *Theory of Vector Optimization*. Lecture Notes in Economics and Mathematical Systems, vol. 319 (Springer, Berlin, Heidelberg, New York, 1989)
40. O. Mangasarian, *Nonlinear Programming* (McGraw Hill, New York, 1969)
41. A. Pascoletti, P. Serafini, Scalarizing vector optimization problems. J. Optim. Theory Appl. **42**, 499–524 (1984)
42. P. Perny, O. Spanjaard, L.-X. Storme, A decision-theoretic approach to robust optimization. Ann. Oper. Res. **147**, 317–341 (2006)
43. A.G. Quaranta, A. Zaffaroni, Robust optimization of conditional value at risk and portfolio selection. J. Bank. Financ. **32**, 2046–2056 (2008)
44. Y. Sawaragi, H. Nakayama, T. Tanino, *Theory of Multiobjective Optimization* (Academic, Orlando, 1985)
45. S. Sayin, P. Kouvelis, The multiobjective discrete optimization problem: a weighted min-max two-stage optimization approach and a bicriteria algorithm. Manage. Sci. **51**, 1572–1581 (2005)
46. A. Schöbel, Generalized light robustness and the trade-off between robustness and nominal quality. Math. Meth. Oper. Res. **80**(2), 161–191 (2014)
47. A. Shapiro, D. Dentcheva, A. Ruszczyński, *Lectures on Stochastic Programming: Modeling and Theory*. MOS-SIAM Series on Optimization, vol. 9 (SIAM, Philadelphia, 2009)
48. A.L. Soyster, Convex programming with set-inclusive constraints and applications to inexact linear programming. Oper. Res. **21**, 1154–1157 (1973)
49. S. Stiller, *Extending Concepts of Reliability: Network Creation Games, Real-time Scheduling, and Robust Optimization* (TU, Berlin, 2009)
50. C. Tammer, K. Winkler, A new scalarization approach and applications in multicriteria D.C. optimization. J. Nonlinear Convex Anal. **4**(3), 365–380 (2003)
51. G. Tintner, Stochastic linear programming with applications to agricultural economics, in *Proceedings of the 2nd Symposium on Linear Programming*, ed. by H.A. Antosiewicz (National Bureau of Standards, Washington, DC, 1955), pp. 197–228
52. P. Weidner, Ein Trennungskonzept und seine Anwendung auf Vektoroptimierungsverfahren. Dissertation B, Martin-Luther-Universität Halle-Wittenberg (1990)
53. R.E. Wendell, D.N. Lee, Efficiency in multiple objective optimization problems. Math. Program. **12**, 406–414 (1977)
54. L. Zadeh, Optimality and non-scalar-valued performance ceriteria. IEEE Trans. Autom. Control **8**, 59–60 (1963)

Chapter 5
Vector Variational Inequalities

The theory of vector variational inequalities began with the pioneer work of F. Giannessi [10] in 1980 where he extended the classical variational inequality for vector-valued functions in the setting of finite dimensional spaces. He also provided some applications to alternative theorems, quadratic programs and complementarity problems. Since then, a large number of papers have appeared in the literature on different aspects of vector variational inequalities. These references are gathered in the bibliography. Later, it is proved that the theory of vector variational inequalities is a powerful tool to study vector optimization problems. In this chapter, we give an introduction to vector variational inequalities, existence theory of their solutions and some applications to vector optimization problems.

5.1 Formulations and Preliminary Results

Let X and Y be topological vector spaces and K be a nonempty subset of X. Let $T : K \to \mathcal{L}(X, Y)$ be an operator and $C : K \to 2^Y$ be a set-valued map such that for all $x \in K$, $C(x)$ is a closed convex pointed cone. We denote by $\text{int}(C(x))$ the interior of $C(x)$. We further assume that $\text{int}(C(x)) \neq \emptyset$ wherever $\text{int}(C(x))$ is involved. For every $l \in \mathcal{L}(X, Y)$, the value of l at x is denoted by $\langle l, x \rangle$. The *vector variational inequality problems* (in short, VVIPs) are defined as follows:

- *Strong Vector Variational Inequality Problem* (in short, SVVIP): Find $\bar{x} \in K$ such that

$$\langle T(\bar{x}), y - \bar{x} \rangle \in C(\bar{x}), \quad \text{for all } y \in K. \tag{5.1}$$

- *Vector Variational Inequality Problem* (in short, VVIP): Find $\bar{x} \in K$ such that

$$\langle T(\bar{x}), y - \bar{x} \rangle \notin -C(\bar{x}) \setminus \{\mathbf{0}\}, \quad \text{for all } y \in K. \tag{5.2}$$

Q.H. Ansari et al., *Vector Variational Inequalities and Vector Optimization*, Vector Optimization, DOI 10.1007/978-3-319-63049-6_5

- *Weak Vector Variational Inequality Problem* (in short, WVVIP): Find $\bar{x} \in K$ such that

$$\langle T(\bar{x}), y - \bar{x} \rangle \notin -\operatorname{int}(C(\bar{x})), \quad \text{for all } y \in K. \tag{5.3}$$

Whenever we are more specific, we write Stampacchia strong vector variational inequality problem, Stampacchia vector variational inequality problem, and Stampacchia weak vector variational inequality problem instead of strong vector variational inequality problem, vector variational inequality problem and weak vector variational inequality problem, respectively.

When $Y = \mathbb{R}$ and $C(x) = \mathbb{R}_+$ for all $x \in K$, then above mentioned vector variational inequality problems reduce to the classical variational inequality problem discussed in Chap. 1.

We denote by Sol(SVVIP)^d, Sol(VVIP)^d and Sol(WVVIP)^d the set of solutions of SVVIP, VVIP and WVVIP, respectively. If for all $x \in K$, $C(x) = P$ is a fixed closed convex pointed cone with $\operatorname{int}(P) \neq \emptyset$, then the solution sets of SVVIP, VVIP and WVVIP are denoted by Sol(SVVIP), Sol(VVIP) and Sol(WVVIP), respectively. It is clear that Sol(SVVIP)^d $\subseteq$ Sol(VVIP)^d $\subseteq$ Sol(WVVIP)^d, but the converse need not be true as illustrated by the following example.

Example 5.1 ([28]) Let $X = Y = \mathbb{R}^2$, $K = \{x = (x_1, x_2) \in \mathbb{R}^2 : x_1 \geq 0\}$, $C(x) = \mathbb{R}^2_+$ for all $x \in K$ and $f_1, f_2 : \mathbb{R}^2 \to \mathbb{R}^2$ be defined by

$$f_1(x) = (x_1 - 1, x_2) \quad \text{and} \quad f_2(x) = \left(\frac{1}{2}x_1, x_2 - 1\right), \quad \text{for all } x = (x_1, x_2) \in \mathbb{R}^2.$$

Let $T : K \to \mathcal{L}(X, Y)$ be defined by

$$T(x) = (f_1(x), f_2(x)), \quad \text{for all } x = (x_1, x_2) \in K.$$

Then

$$\begin{aligned}
\text{Sol(WVVIP)} &= \left\{x = (x_1, x_2) \in K : \langle T(x), y - x \rangle \notin -\operatorname{int}\left(\mathbb{R}^2_+\right) \text{ for all } y \in K\right\} \\
&= \left\{ x = (x_1, x_2) \in K : \left((x_1 - 1)(y_1 - x_1) + \frac{1}{2}x_1(y_2 - x_2), \right.\right. \\
&\qquad \left.\left. x_2(y_1 - x_1) + (x_2 - 1)(y_2 - x_2) \right) \notin -\operatorname{int}\left(\mathbb{R}^2_+\right) \text{ for all } y \in K \right\} \\
&= \left\{ x = (x_1, x_2) \in K : 0 \leq x_1 \leq 1,\ x_2 = 2 + \frac{2}{x_1 - 2} \right\}.
\end{aligned}$$

For $\bar{x} = (0, 1) \in \mathrm{Sol(WVVIP)}$, we have

$$\begin{aligned}\langle T(\bar{x}), y - \bar{x}\rangle &= \{(\langle f_1(\bar{x}), y - \bar{x}\rangle, \langle f_2(\bar{x}), y - \bar{x}\rangle) : y_1 \geq 0,\ y_2 \in \mathbb{R}\}\\ &= \{(-y_1 + y_2 - 1, 0) : y_1 \geq 0,\ y_2 \in \mathbb{R}\}\\ &= \mathbb{R} \times \{0\}.\end{aligned}$$

This means that there exists $y \in K$ such that

$$\langle T(\bar{x}), y - \bar{x}\rangle \in -\mathbb{R}^2_+ \setminus \{\mathbf{0}\},$$

that is, $\bar{x} = (0, 1) \notin \mathrm{Sol(VVIP)}$. Similarly, it can be easily verified that $\hat{x} = (1, 0) \in \mathrm{Sol(WVVIP)}$ but $\hat{x} \notin \mathrm{Sol(VVIP)}$.

Definition 5.1 A subset K of X is said to be a *strictly convex body* if $\mathrm{int}(K) \neq \emptyset$ and for arbitrary given points $x, y \in K$, $x \neq y$,

$$\{x_\lambda := \lambda x + (1 - \lambda)y : \lambda \in]0, 1[\} \subset \mathrm{int}(K).$$

Yen and Lee [28] provided the following criterion under which Sol(VVIP) and Sol(WVVIP) are equal.

Theorem 5.1 *Let $K \subseteq X$ be a strictly convex body, $Y = \mathbb{R}^\ell$ and C be a fixed closed convex cone in Y with* $\mathrm{int}(C) \neq \emptyset$. *For each $x \in K$, let the linear operator $v \mapsto \langle T(x), v\rangle$ from X to $\mathbb{R}^\ell$ be surjective. Then* Sol(VVIP) *and* Sol(WVVIP) *are equal.*

Proof Suppose contrary that there exists $\bar{x} \in K$ such that $\bar{x} \in \mathrm{Sol(WVVIP)}$ but $\bar{x} \notin \mathrm{Sol(VVIP)}$. Then there exists $y \in K$ such that

$$\langle T(\bar{x}), y - \bar{x}\rangle \in -C \setminus \{\mathbf{0}\}. \tag{5.4}$$

In particular, $y \neq \bar{x}$. Since K is a strictly convex body, $y_\lambda := \lambda y + (1 - \lambda)\bar{x} \in \mathrm{int}(K)$ for all $\lambda \subset]0, 1[$. Then from (5.4), we have

$$\langle T(\bar{x}), y_\lambda - \bar{x}\rangle = \lambda\langle T(\bar{x}), y - \bar{x}\rangle + (1 - \lambda)\langle T(\bar{x}), \bar{x} - \bar{x}\rangle \in -C \setminus \{\mathbf{0}\}. \tag{5.5}$$

Let $\varepsilon > 0$ be such that $B_\varepsilon(y_\lambda) \subseteq K$, where $B_\varepsilon(y_\lambda)$ denotes the open ball with center at y_λ and radius ε. By assumption, the linear operator $v \mapsto \langle T(\bar{x}), v\rangle$ is surjective, so it is an open mapping (see Theorem C.10). Since $B_\varepsilon(y_\lambda) - \bar{x}$ is a neighborhood of $y_\lambda - \bar{x}$, $\langle T(\bar{x}), (B_\varepsilon(y_\lambda) - y)\rangle := \{\langle T(\bar{x}), z - \bar{x}\rangle : z \in B_\varepsilon(y_\lambda)\}$ must be a neighborhood of $z_\lambda := \langle T(\bar{x}), y_\lambda - \bar{x}\rangle$. Let $\rho > 0$ be such that

$$B_\rho(z_\lambda) \subset \langle T(\bar{x}), (B_\varepsilon(y_\lambda) - y)\rangle. \tag{5.6}$$

Since $\text{int}(C) \neq \emptyset$, from (5.5) and Remark 1.7, it follows that $B_\rho(z_\lambda) \cap (-\text{int}(C) \setminus \{\mathbf{0}\}) \neq \emptyset$. Hence, (5.6) implies that there exists a point $z \in B_\varepsilon(y_\lambda) \subset K$ such that

$$\langle T(\bar{x}), z - \bar{x} \rangle \in -\text{int}(C) \setminus \{\mathbf{0}\},$$

contradicting our assumption that $\bar{x} \in \text{Sol(WVVIP)}$. □

The following problems are closely related to the above mentioned vector variational inequality problems, known as *Minty vector variational inequality problems*.

- *Minty Strong Vector Variational Inequality Problem* (in short, MSVVIP): Find $\bar{x} \in K$ such that

$$\langle T(y), y - \bar{x} \rangle \in C(\bar{x}), \quad \text{for all } y \in K. \tag{5.7}$$

- *Minty Vector Variational Inequality Problem* (in short, MVVIP): Find $\bar{x} \in K$ such that

$$\langle T(y), y - \bar{x} \rangle \notin -C(\bar{x}) \setminus \{\mathbf{0}\}, \quad \text{for all } y \in K. \tag{5.8}$$

- *Minty Weak Vector Variational Inequality Problem* (in short, MWVVIP): Find $\bar{x} \in K$ such that

$$\langle T(y), y - \bar{x} \rangle \notin -\text{int}(C(\bar{x})), \quad \text{for all } y \in K. \tag{5.9}$$

When $Y = \mathbb{R}$ and $C(x) = \mathbb{R}_+$ for all $x \in K$, then the above mentioned Minty vector variational inequality problems reduce to the Minty variational inequality problem discussed in Chap. 1.

We denote by Sol(MSVVIP)^d, Sol(MVVIP)^d and Sol(MWVVIP)^d the set of solutions of MSVVIP, MVVIP and MWVVIP, respectively. It is clear that $\text{Sol(MSVVIP)}^d \subseteq \text{Sol(MVVIP)}^d \subseteq \text{Sol(MWVVIP)}^d$.

The following example shows that the converse is not true in general.

Example 5.2 Consider $X = \mathbb{R}$, $Y = \mathbb{R}^2$, $K = [0, 1]$ $C(x) = \mathbb{R}^2_+$ for all $x \in K$ and the operator $T(y) = \left(y^2, -y\right)$. Then the whole interval $[0, 1]$ is a solution of MWVVIP, but there do not exist any solutions of MSVVIP.

If for all $x \in K$, $C(x) = P$ a fixed closed convex pointed cone with $\text{int}(P) \neq \emptyset$, then the solution sets of MSVVIP, MVVIP and MWVVIP are denoted by Sol(MSVVIP), Sol(MVVIP) and Sol(MWVVIP), respectively.

The following examples due to Charitha et al. [6] show that the solution set of (Stampacchia) vector variational inequality problems is not equal to the solution set of Minty vector variational inequality problems.

Example 5.3 Let $K = [-1, 1]$, $C(x) = \mathbb{R}^2_+$ for all $x \in K$ and $T : \mathbb{R} \to \mathbb{R}^2$ be defined by

$$T(x) = (-x, x^2), \quad \text{for all } x \in \mathbb{R}.$$

Then

$$\begin{aligned}\text{Sol(WVVIP)} &= \bigcap_{y\in[-1,1]} \left\{x \in [-1,1] : \left(-x(y-x), x^2(y-x)\right) \notin -\operatorname{int}\left(\mathbb{R}^2_+\right)\right\} \\ &= \bigcap_{y\in[-1,1]} \left\{x \in [-1,1] : \max\left\{-x(y-x), x^2(y-x)\right\} \geq 0\right\} \\ &= \{-1\} \cup [0,1].\end{aligned}$$

We further have

$$\begin{aligned}\text{Sol(MWVVIP)} &= \bigcap_{y\in[-1,1]} \left\{x \in [-1,1] : \left(-y(y-x), y^2(y-x)\right) \notin -\operatorname{int}\left(\mathbb{R}^2_+\right)\right\} \\ &= \bigcap_{y\in[-1,1]} \left\{x \in [-1,1] : \max\left\{-y(y-x), y^2(y-x)\right\} \geq 0\right\} \\ &= \{-1\}.\end{aligned}$$

In the above example the solution set Sol(WVVIP) is disconnected and Sol(MWVVIP) $\subseteq$ Sol(WVVIP). However, this may not be true in general as the following example shows.

Example 5.4 Let $K = [-1,1]$, $C(x) = \mathbb{R}^2_+$ for all $x \in K$ and $T : \mathbb{R} \to \mathbb{R}^2$ be defined by

$$T(x) = \begin{cases} (-1, x - \frac{1}{2}), & \text{if } x \leq 0, \\ (1, x - \frac{1}{2}), & \text{if } x > 0. \end{cases}$$

Then Sol(WVVIP) $=]0, \frac{1}{2}]$ and Sol(WMVVIP) $= [0, \frac{1}{2}]$.

The following example, due to Giannessi [11], shows that Sol(MVVIP) $\not\subseteq$ Sol(VVIP). In general, $\text{Sol(MVVIP)}^d \subseteq \text{Sol(VVIP)}^d$ does not hold.

Example 5.5 Let $X = \mathbb{R}$, $Y = \mathbb{R}^2$, $K = [-1,0]$, $C(x) = \mathbb{R}^2_+$ for all $x \in K$ and $T : K \to \mathbb{R}^2$ be defined by $T(x) = (1, 2x) \in \mathbb{R}^2$. Then it can be easily checked that $\bar{x} = 0$ is a solution of MVVIP but not a solution of VVIP.

We next discuss the conditions under which the solution set of Stampacchia vector variational inequality problems is equal to the solution set of Minty vector variational inequality problems.

Definition 5.2 Let K be a nonempty convex subset of X and $x \in K$ be an arbitrary element. The operator $T : K \to \mathcal{L}(X, Y)$ is said to be:

(a) *strongly C_x-upper sign continuous* if for all $y \in K$ and all $\lambda \in]0, 1[$,

$$\langle T(x + \lambda(y - x)), y - x\rangle \in C(x) \text{ implies } \langle T(x), y - x\rangle \in C(x);$$

(b) *C_x-upper sign continuous* if for all $y \in K$ and all $\lambda \in]0,1[$,

$$\langle T(x+\lambda(y-x)), y-x\rangle \notin -C(x)\setminus\{\mathbf{0}\}$$
$$\text{implies } \langle T(x), y-x\rangle \notin -C(x)\setminus\{\mathbf{0}\};$$

(c) *weakly C_x-upper sign continuous* if for all $y \in K$ and all $\lambda \in]0,1[$,

$$\langle T(x+\lambda(y-x)), y-x\rangle \notin -\operatorname{int}(C(x))$$
$$\text{implies } \langle T(x), y-x\rangle \notin -\operatorname{int}(C(x)).$$

Definition 5.3 Let K be a nonempty convex subset of X. An operator $T : K \to \mathcal{L}(X,Y)$ is said to be *υ-hemicontinuous* if for every $x, y \in K$ and $\lambda \in [0,1]$, the mapping $\lambda \mapsto \langle T(x+\lambda(y-x)), y-x\rangle$ is continuous at 0^+.

Remark 5.1 It is easy to see that the υ-hemicontinuity of T implies strong C_x-upper sign continuity provided that the set-valued map C is closed; as well as weak C_x-upper sign continuity of T provided that the set-valued map $W : K \to 2^Y$ defined by $W(x) = Y \setminus \{-\operatorname{int}(C(x))\}$, is closed.

If $X = Y = \mathbb{R}$ and $K = C(x) = [0,\infty[$ for all $x \in K$, then any positive mapping $T : K \to \mathcal{L}(X,Y)$ is weakly C_x-upper sign continuous while it is not υ-hemicontinuous. In this case, the concept of weak C_x-upper sign continuity reduces to the upper sign continuity introduced by Hadjisavvas [13].

Example 5.6 Let $X = \mathbb{R}$, $K = [0,1]$, $Y = \mathbb{R}^2$, $C(x) = \mathbb{R}^2_+$ for all $x \in K$ the constant natural ordering cone, and $T(x) = (1, 2|x|)$. Then the operator T is strongly C_x-upper sign continuous, C_x-upper sign continuous as well as weakly C_x-upper sign continuous.

Lemma 5.1 *Let K be a nonempty convex subset of X and $T : K \to \mathcal{L}(X,Y)$ be an operator. Then*

(a) $\text{Sol(MSVVIP)}^d \subseteq \text{Sol(SVVIP)}^d$ *if T is strongly C_x-upper sign continuous;*
(b) $\text{Sol(MVVIP)}^d \subseteq \text{Sol(VVIP)}^d$ *if T is C_x-upper sign continuous;*
(c) $\text{Sol(MWVVIP)}^d \subseteq \text{Sol(WVVIP)}^d$ *if T is weakly C_x-upper sign continuous.*

Proof (a) Let $\bar{x} \in \text{Sol(MSVVIP)}^d$. Then

$$\langle T(y), y-\bar{x}\rangle \in C(\bar{x}), \quad \text{for all } y \in K.$$

Since K is convex, $\bar{x}+\lambda(y-\bar{x}) \in K$ for all $\lambda \in]0,1[$. Therefore, in particular, we have

$$\langle T(\bar{x}+\lambda(y-\bar{x})), \bar{x}+\lambda(y-\bar{x})-\bar{x}\rangle \in C(\bar{x}),$$

equivalently,

$$\lambda \langle T(\bar{x} + \lambda(y - \bar{x})), y - \bar{x}\rangle \in C(\bar{x}).$$

Since $C(\bar{x})$ is a convex cone, we have

$$\langle T(\bar{x} + \lambda(y - \bar{x})), y - \bar{x}\rangle \in C(\bar{x}).$$

By strong C_x-upper sign continuity of T, we obtain

$$\langle T(\bar{x}), y - \bar{x}\rangle \in C(\bar{x}).$$

Hence, $\bar{x} \in \text{Sol(SVVIP)}^d$.

Since $W(x) = Y \setminus \{-C(x) \setminus \{\mathbf{0}\}\}$ and $W(x) = Y \setminus \{-\operatorname{int}(C(x))\}$ are cones, the proof of the part (b) and (c) lies on the lines of the proof of part (a). □

In view of Remark 5.1 and Lemma 5.1, we have the following result.

Lemma 5.2 *Let K be a nonempty convex subset of X and $T : K \to \mathcal{L}(X, Y)$ be v-hemicontinuous. Then*

(a) $\text{Sol(MSVVIP)}^d \subseteq \text{Sol(SVVIP)}^d$ *provided that the set-valued map C is closed;*
(b) $\text{Sol(MWVVIP)}^d \subseteq \text{Sol(WVVIP)}^d$ *provided that the set-valued map $W : K \to 2^Y$ defined by $W(x) = Y \setminus \{-\operatorname{int}(C(x))\}$, is closed.*

As we have seen in Examples 5.4 and 5.5 that Sol(MVVIP) $\nsubseteq$ Sol(VVIP) and Sol(MWVVIP) $\nsubseteq$ Sol(WVVIP). Therefore, Giannessi [11] showed that this insufficient character of VVIP can be weakened by performing a perturbation on VVIP as follows: find $\bar{x} \in K$ such that there exists $\varepsilon \in]0, 1[$ and

$$\langle T(\bar{x} + \lambda(y - \bar{x}), y - \bar{x}\rangle \notin -C(\bar{x}) \setminus \{\mathbf{0}\}, \tag{5.10}$$

for all $y \in K$ and all $\lambda \in]0, \varepsilon[$. It is called *perturbed vector variational inequality problem* (in short, PVVIP).

In a similar way, we define perturbed strong vector variational inequality problem and perturbed weak vector variational inequality problem as follows:

- *Perturbed Strong Vector Variational Inequality Problem* (in short, PSVVIP): Find $\bar{x} \in K$ such that

$$\langle T(\bar{x} + \lambda(y - \bar{x}), y - \bar{x}\rangle \in C(\bar{x}), \tag{5.11}$$

  for all $y \in K$ and all $\lambda \in]0, \varepsilon[$.
- *Perturbed Weak Vector Variational Inequality Problem* (in short, PWVVIP): Find $\bar{x} \in K$ such that

$$\langle T(\bar{x} + \lambda(y - \bar{x}), y - \bar{x}\rangle \notin -\operatorname{int}(C(\bar{x})), \tag{5.12}$$

  for all $y \in K$ and all $\lambda \in]0, \varepsilon[$.

The set of solutions of PSVVIP, PVVIP and PWVVIP are denoted by $\mathrm{Sol(PSVVIP)}^d$, $\mathrm{Sol(PVVIP)}^d$ and $\mathrm{Sol(PWVVIP)}^d$, respectively.

Proposition 5.1 *Let K be a nonempty convex subset of X. Then* $\mathrm{Sol(MVVIP)}^d \subseteq \mathrm{Sol(PVVIP)}^d$, $\mathrm{Sol(MSVVIP)}^d \subseteq \mathrm{Sol(PSVVIP)}^d$ *and* $\mathrm{Sol(MWVVIP)}^d \subseteq \mathrm{Sol(PWVVIP)}^d$.

Proof Let $\bar{x} \in K$ be a solution of MVVIP. Then

$$\langle T(y), y - \bar{x} \rangle \notin -C(\bar{x}) \setminus \{\mathbf{0}\}, \quad \text{for all } y \in K. \tag{5.13}$$

Since K is convex, $x_\lambda = \bar{x} + \lambda(z - \bar{x}) \in K$ for all $z \in K$ and all $\lambda \in [0, 1]$. Taking $y = x_\lambda$ with $\varepsilon \in]0, 1[$ and $\lambda \in]0, \varepsilon[$, we have

$$\langle T(x_\lambda), x_\lambda - \bar{x} \rangle \notin -C(\bar{x}) \setminus \{\mathbf{0}\}.$$

Since $x_\lambda - \bar{x} = \lambda(z - \bar{x})$, we have

$$\langle T(x_\lambda), z - \bar{x} \rangle \notin -C(\bar{x}) \setminus \{\mathbf{0}\}, \quad \text{for all } z \in K \text{ and } \lambda \in]0, \varepsilon[.$$

Thus, $\bar{x} \in \mathrm{Sol(PVVIP)}^d$.

Similarly, we can easily show that $\mathrm{Sol(MSVVIP)}^d \subseteq \mathrm{Sol(PSVVIP)}^d$ and $\mathrm{Sol(MWVVIP)}^d \subseteq \mathrm{Sol(PWVVIP)}^d$. □

We introduce the following set-valued maps:

- $S^S(y) = \{x \in K : \langle T(x), y - x \rangle \in C(x)\}$;
- $M^S(y) = \{x \in K : \langle T(y), y - x \rangle \in C(x)\}$;
- $S(y) = \{x \in K : \langle T(x), y - x \rangle \notin -C(x) \setminus \{\mathbf{0}\}\}$;
- $M(y) = \{x \in K : \langle T(y), y - x \rangle \notin -C(x) \setminus \{\mathbf{0}\}\}$;
- $S^W(y) = \{x \in K : \langle T(x), y - x \rangle \notin -\operatorname{int}(C(x))\}$;
- $M^W(y) = \{x \in K : \langle T(y), y - x \rangle \notin -\operatorname{int}(C(x))\}$.

From the above definition of set-valued maps, the following result can be easily derived.

Proposition 5.2

(a) $\mathrm{Sol(SVVIP)}^d = \bigcap_{y \in K} S^S(y)$ *and*

$\mathrm{Sol(MSVVIP)}^d = \bigcap_{y \in K} M^S(y)$;

(b) $\text{Sol(VVIP)}^d = \bigcap_{y \in K} S(y)$ *and* $\text{Sol(MVVIP)}^d = \bigcap_{y \in K} M(y)$;

(c) $\text{Sol(WVVIP)}^d = \bigcap_{y \in K} S^W(y)$ *and* $\text{Sol(MWVVIP)}^d = \bigcap_{y \in K} M^W(y)$.

Proposition 5.3

(a) *If the set-valued map* $C : K \to 2^Y$ *is closed, then for each* $y \in K$, $M^S(y)$ *is a closed set.*

(b) *If the set-valued map* $W : K \to 2^Y$, *defined by* $W(x) = Y \setminus \{-\operatorname{int}(C(x))\}$, *is closed, then for each* $y \in K$, $M^W(y)$ *is a closed set.*

(c) *If the set-valued map* $W : K \to 2^Y$, *defined by* $W(x) = Y \setminus \{-\operatorname{int}(C(x))\}$, *is concave, then for each* $y \in K$, $M^W(y)$ *is a convex set.*

Proof

(a) Let $y \in K$ be arbitrary and $\{x_n\}$ be a sequence in $M^S(y)$ such that $x_n \to x \in K$. Then $\langle T(y), y - x_n \rangle \in C(x_n)$. Since $T(y) \in \mathcal{L}(X, Y)$,

$$\langle T(y), y - x_n \rangle \to \langle T(y), y - x \rangle$$

and $\langle T(y), y - x \rangle \in C(x)$ because C is a closed set-valued map. Hence, $x \in M^S(y)$, and thus, $M^S(y)$ is closed.

(b) Let $y \in K$ be arbitrary and $\{x_n\}$ be a sequence in $M^W(y)$ such that $x_n \to x \in K$. Then $\langle T(y), y - x_n \rangle \in W(x_n)$. Since $T(y) \in \mathcal{L}(X, Y)$,

$$\langle T(y), y - x_n \rangle \to \langle T(y), y - x \rangle$$

and $\langle T(y), y - x \rangle \in W(x)$ because W is a closed set-valued map. Hence, $x \in M^W(y)$, and thus, $M^W(y)$ is closed.

(c) Let $x_1, x_2 \in M^W(y)$. Then

$$\langle T(y), y - x_1 \rangle \in W(x_1) \quad \text{and} \quad \langle T(y), y - x_2 \rangle \in W(x_2).$$

By concavity of W, for all $\lambda \in [0, 1]$, we have

$$\begin{aligned} \langle T(y), y - (\lambda x_1 + (1-\lambda)x_2) \rangle &= \lambda \langle T(y), y - x_1 \rangle + (1-\lambda) \langle T(y), y - x_2 \rangle \\ &\in \lambda W(x_1) + (1-\lambda) W(x_2) \\ &\subseteq W(\lambda x_1 + (1-\lambda)x_2). \end{aligned}$$

Therefore, $\lambda x_1 + (1-\lambda)x_2 \in M^W(y)$, and hence, $M^W(y)$ is convex. □

Remark 5.2 In addition to the hypothesis of Proposition 5.3 (a) (respectively, (b)), if we further assume that the operator $T : K \to \mathcal{L}(X, Y)$ and the pairing $\langle ., . \rangle$

are continuous, then $S^S(y)$ (respectively, $S^W(y)$) is closed for all $y \in K$. We note that the pairing $\langle .,. \rangle : \mathcal{L}(X,Y) \times X \to Y$ is continuous if X and Y are Hausdorff topological vector spaces and $\mathcal{L}(X,Y)$ is equipped with σ-topology (see, Lemma C.1).

Proposition 5.4 *Let K be a nonempty convex subset of X. The set-valued maps S and S^W are KKM-maps.*

Proof Let $\hat{x}$ be a point in the convex hull of any finite subset $\{y_1, y_2, \ldots, y_m\}$ of K. Then $\hat{x} = \sum_{i=1}^m \lambda_i y_i$ for some nonnegative real number λ_i, $1 \leq i \leq m$, with $\sum_{i=1}^m \lambda_i = 1$. If $\hat{x} \notin \bigcup_{i=1}^m S(y_i)$, then

$$\langle T(\hat{x}), y_i - \hat{x} \rangle \in -C(\hat{x}) \setminus \{\mathbf{0}\}, \quad \text{for each } i = 1, 2, \ldots, m.$$

Since $-C(\hat{x})$ is a convex cone and $\lambda_i \geq 0$ with $\sum_{i=1}^m \lambda_i = 1$, we have

$$\sum_{i=1}^m \lambda_i \langle T(\hat{x}), y_i - \hat{x} \rangle \in -C(\hat{x}) \setminus \{\mathbf{0}\}.$$

It follows that

$$\begin{aligned} \mathbf{0} = \langle T(\hat{x}), \hat{x} - \hat{x} \rangle &= \left\langle T(\hat{x}), \sum_{i=1}^m \lambda_i y_i - \sum_{i=1}^m \lambda_i \hat{x} \right\rangle \\ &= \left\langle T(\hat{x}), \sum_{i=1}^m \lambda_i (y_i - \hat{x}) \right\rangle = \sum_{i=1}^m \lambda_i \langle T(\hat{x}), y_i - \hat{x} \rangle \in -C(\hat{x}) \setminus \{\mathbf{0}\}. \end{aligned}$$

Thus, $\mathbf{0} \in -C(\hat{x}) \setminus \{\mathbf{0}\}$, a contradiction. Therefore, we must have

$$\operatorname{co}(\{y_1, y_2, \ldots, y_m\}) \subseteq \bigcup_{i=1}^m S(y_i),$$

and hence, S is a KKM map on K.

By using the similar argument, we can easily prove that S^W is a KKM map on K. □

Remark 5.3 The above argument cannot be applied for S^S. In general, S^S is not a KKM map. Consider, for instance, $X = \mathbb{R}$, $K = [0, 1]$, $Y = \mathbb{R}^2$, $C(x) = \mathbb{R}^2_+$ for all $x \in K$, and $T(y) = (-1, y)$. Then for $y_1 = 0$, $y_2 = 1$, we have $S^S(y_1) = \{y \in K : (y, -y^2) \in \mathbb{R}^2_+\}$ and $S^S(y_2) = \{y \in K : (y - 1, y \cdot (1 - y)) \in \mathbb{R}^2_+\}$. It follows that $y = 0.5 \in \operatorname{co}(\{y_1, y_2\})$, but $y \notin S^S(y_1) \cup S^S(y_2)$.

5.2 Existence Results for Solutions of Vector Variational Inequalities Under Monotonicity

Let X and Y be topological vector spaces such that the pairing $\langle .,. \rangle$ is continuous, and K be a nonempty subset of X. Let $T : K \to \mathcal{L}(X, Y)$ be an operator and $C : K \to 2^Y$ be a set-valued map such that for all $x \in K$, $C(x)$ is a closed convex pointed cone. We further assume that $\text{int}(C(x)) \neq \emptyset$ wherever $\text{int}(C(x))$ is involved.

Definition 5.4 Let $x \in K$ be an arbitrary element. An operator $T : K \to \mathcal{L}(X, Y)$ is said to be:

(a) *C_x-monotone* on K if for every $y \in K$, we have

$$\langle T(x) - T(y), x - y \rangle \in C(x);$$

(b) *strongly C_x-pseudomonotone* on K if for every $y \in K$, we have

$$\langle T(x), y - x \rangle \in C(x) \quad \text{implies} \quad \langle T(y), y - x \rangle \in C(x);$$

(c) *C_x-pseudomonotone* on K if for every $y \in K$, we have

$$\langle T(x), y - x \rangle \notin -C(x) \setminus \{\mathbf{0}\} \quad \text{implies} \quad \langle T(y), y - x \rangle \notin -C(x) \setminus \{\mathbf{0}\};$$

(d) *weakly C_x-pseudomonotone* on K if for every $y \in K$, we have

$$\langle T(x), y - x \rangle \notin -\text{int}(C(x)) \quad \text{implies} \quad \langle T(y), y - x \rangle \notin -\text{int}(C(x)).$$

If $C(x) = P$ is a fixed closed convex pointed cone with $\text{int}(P) \neq \emptyset$, then C_x-monotone operator, strongly C_x-pseudomonotone operator, C_x-pseudomonotone operator and weakly C_x-pseudomonotone operator are called P-monotone operator, strongly P-pseudomonotone operator, P-pseudomonotone and weakly P-pseudomonotone operator, respectively.

It can be argued directly from Definition 5.4 that

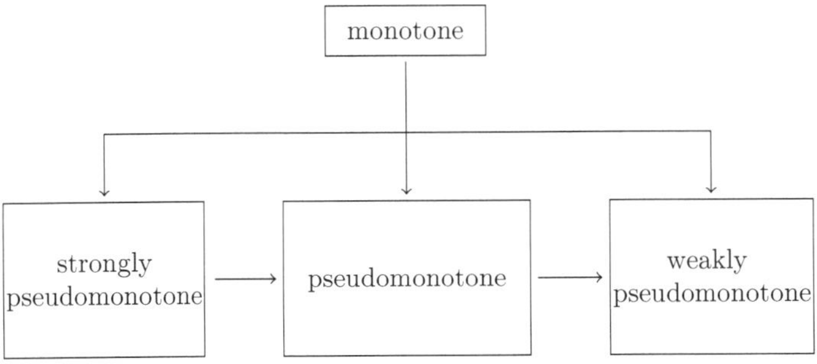

Example 5.7 Let $X = K = \mathbb{R}$, $Y = \mathbb{R}^2$ and $C(x) = \mathbb{R}^2_+$ for all $x \in K$. Let $T : K \to \mathcal{L}(X, Y)$ be defined by

$$T(x)(t) = \langle T(x), t\rangle = (x, x^2)t, \quad \text{for all } x \in X \text{ and all } t \in X.$$

If $\langle T(y), y-x\rangle \in -\operatorname{int}\left(\mathbb{R}^2_+\right)$, then $(y, y^2)(y-x) = (y(y-x), y^2(y-x)) \in -\operatorname{int}\left(\mathbb{R}^2_+\right)$, and so, $y(y-x) < 0$ and $y^2(y-x) < 0$. This implies that $y < x$ and $y > 0$, and thus $x > 0$. Hence, $\langle T(x), y-x\rangle = (x, x^2)(y-x) = (x(y-x), x^2(y-x)) \in -\operatorname{int}\left(\mathbb{R}^2_+\right)$. This shows that T is weakly C_x-pseudomonotone.

The following example shows that weakly C_x-pseudomonotonicity does not imply C_x-monotonicity.

Example 5.8 Let $X = Y = \mathbb{R}^2$, $K = [-1, 1] \times [-1, 1] \subset \mathbb{R}^2$. The vectors of $\mathbb{R}^2$ are denoted by $x = (x_1, x_2)$. Let $C : K \to 2^Y$ be defined by

$$C(x) = \begin{cases} C_1 = \{(y_1, y_2) \in Y : y_1 \geq 0, y_2 \geq 0\}, & \text{if } x_1 \geq 0, x_2 \geq 0, x \neq (0,0), \\ C_2 = \{(y_1, y_2) \in Y : y_1 \leq 0, y_2 \geq 0\}, & \text{if } x_1 < 0, x_2 > 0, x = (0,0), \\ C_3 = \{(y_1, y_2) \in Y : y_1 \leq 0, y_2 \leq 0\}, & \text{if } x_1 \leq 0, x_2 \leq 0, x \neq (0,0), \\ C_4 = \{(y_1, y_2) \in Y : y_1 \geq 0, y_2 \leq 0\}, & \text{if } x_1 > 0, x_2 < 0, \end{cases}$$

and let $T : K \to \mathcal{L}(X, Y)$ be defined by

$$\langle T(x), y\rangle = \left(\int_0^{x_1} y_1 t dt, \int_0^{x_2} y_2 t dt\right) = \frac{1}{2}\left(x_1^2 y_1, x_2^2 y_2\right).$$

Then T is weakly C_x-pseudomonotone on K, but not C_x-monotone.

From the definition of different kinds of pseudomonotonicity of T, we have following lemma.

Lemma 5.3 *Let K be a nonempty convex subset of X and $x \in K$ be an arbitrary element. Then the following statements hold.*

(a) Sol(SVVIP)^d $\subseteq$ Sol(MSVVIP)^d *if $T : K \to \mathcal{L}(X, Y)$ is strongly C_x-pseudomonotone.*
(b) Sol(VVIP)^d $\subseteq$ Sol(MVVIP)^d *if $T : K \to \mathcal{L}(X, Y)$ is C_x-pseudomonotone.*
(c) Sol(WVVIP)^d $\subseteq$ Sol(MWVVIP)^d *if $T : K \to \mathcal{L}(X, Y)$ is weakly C_x-pseudomonotone.*

The following existence results for a solution of vector variational inequalities are the extension of the classical existence result for a solution of scalar variational inequality problem due to Browder [5] and Hartman and Stampacchia [14].

Theorem 5.2 *Let K be a nonempty compact convex subset of X and $W : K \to 2^Y$ be a set-valued map defined by $W(x) = Y \setminus \{-\operatorname{int}(C(x))\}$ such that the graph $\mathcal{G}(W)$ of W is closed in $K \times Y$. For each $x \in K$, suppose that $T : K \to \mathcal{L}(X, Y)$ is weakly C_x-pseudomonotone and weakly C_x-upper sign continuous. Then* Sol(WVVIP)^d *is nonempty.*

Proof By Lemmas 5.1 (c) and 5.3 (c), Sol(MWVVIP)^d = Sol(WVVIP)^d. Hence,

$$\bigcap_{y \in K} S^W(y) = \bigcap_{y \in K} M^W(y).$$

By Proposition 5.4, S^W is a KKM map, and so is M^W because $S^W(y) \subseteq M^W(y)$ for all $y \in K$ by weakly C_x-pseudomonotonicity of T. By Proposition 5.3 (b), $M^W(y)$ is a closed subset of a compact set K, and hence $M^W(y)$ is compact for all $y \in K$. By Fan-KKM Lemma 1.14, we have

$$\bigcap_{y \in K} S^W(y) = \bigcap_{y \in K} M^W(y) \neq \emptyset.$$

Hence, Sol(WVVIP)^d is nonempty. □

The following example due to Ansari et al. [4] illustrates Theorem 5.2.

Example 5.9 Let $X = \mathbb{R}$, $Y = \mathbb{R}^2$, $K = [0, 1]$ and $C(x) = \mathbb{R}^2_+$ for all $x \in K$. Let $T : K \to \mathcal{L}(X, Y)$ be defined by

$$T(x)(t) = \langle T(x), t \rangle = (x, x^2)t, \quad \text{for all } x \in X \text{ and all } t \in X.$$

Then T is weakly C_x-upper sign continuous and weakly C_x-pseudomonotone. Note that

$$\langle T(x), y - x \rangle = (x, x^2)(y - x) = (x(y - x), x^2(y - x)).$$

It is easy to see that the set $\{x \in K : \langle T(y), y - x \rangle \notin -\operatorname{int}(C(x))\} = [0, y]$ is closed. Since K is compact, all the conditions of Theorem 5.2 hold. Therefore, Sol(WVVIP) is nonempty. It can be easily checked that $x = 0$ is the only solution of WVVIP as well as of MWVVIP, that is, Sol(WVVIP) = Sol(MWVVIP) = $\{0\}$.

Analogously, we have the following existence result for a solution of VVIP in the setting of a compact convex subset of a topological vector space.

Theorem 5.3 *Let K be a nonempty compact convex subset of X. For each $x \in K$, suppose that $T : K \to \mathcal{L}(X, Y)$ is C_x-pseudomonotone and C_x-upper sign continuous such that the set*

$$M(y) = \{x \in K : \langle T(y), y - x \rangle \notin -C(x) \setminus \{\mathbf{0}\}\}$$

is closed. Then Sol(VVIP)^d *is nonempty.*

Remark 5.4 Since the mapping S^S is not a KKM mapping, we cannot use a similar argument for SVVIP.

Fang and Huang [9] considered the following definition of pseudomonotonicity to establish the existence of solutions of SVVIP.

Definition 5.5 Let $C : K \to 2^Y$ be a set-valued map such that for each $x \in K$, $C(x)$ is a closed convex pointed cone in Y, and let $x \in K$ be an arbitrary element. An operator $T : K \to \mathcal{L}(X, Y)$ is said to be C_x*-pseudomonotone*$_+$ if for all $y \in K$,

$$\langle T(x), y - x\rangle \notin -C(x) \setminus \{\mathbf{0}\} \quad \text{implies} \quad \langle T(y), y - x\rangle \in C(x).$$

If $C(x) = P$ is a fixed closed pointed convex cone in Y for all $x \in X$, then C_x-monotonicity$_+$ is called P*-monotonicity*.

It is easy to see that the C_x-pseudomonotonicity$_+$ is stronger than the C_x-pseudomonotonicity.

Example 5.10 ([9]) Let $X = Y = \mathbb{R}^2$, $K = C(x) = \mathbb{R}^2_+$ for all $x \in K$, and $T : K \to \mathcal{L}(X, Y)$ be defined by

$$T(x) = ((x_1 + 1, x_2 + 2), (x_1 + 1, x_2 + 2)), \quad \text{for all } x = (x_1, x_2) \in K.$$

Let $x = (x_1, x_2)$, $y = (y_1, y_2) \in K$ and

$$\begin{aligned}\langle T(x), y - x\rangle &= ((x_1 + 1)[y_1 + y_2 - (x_1 + x_2)], (x_2 + 2)[y_1 + y_2 - (x_1 + x_2)]) \\ &\notin -\mathbb{R}^2_+ \setminus \{\mathbf{0}\}.\end{aligned}$$

The above relation implies that $y_1 + y_2 \geq x_1 + x_2$. It follows that

$$\begin{aligned}\langle T(y), y - x\rangle &= ((y_1 + 1)[y_1 + y_2 - (x_1 + x_2)], (y_2 + 2)[y_1 + y_2 - (x_1 + x_2)]) \\ &\in \mathbb{R}^2_+.\end{aligned}$$

Hence, T is C_x-pseudomonotone$_+$.

The next result provides the existence of a solution of SVVIP under C_x-pseudomonotonicity$_+$ of T.

Theorem 5.4 *Let K be a nonempty compact convex subset of X and $C : K \to 2^Y$ be a closed set-valued map such that for each $x \in K$, $C(x)$ is a closed convex pointed cone in Y. For each $x \in K$, suppose that $T : K \to \mathcal{L}(X, Y)$ is C_x-pseudomonotone$_+$ and strongly C_x-upper sign continuous. Then* Sol(SVVIP)^d *is nonempty.*

Proof By C_x-pseudomonotonicity$_+$ of T, we have

$$\text{Sol(VVIP)}^d \subseteq \text{Sol(MSVVIP)}^d.$$

By Lemma 5.1, $\text{Sol(MSVVIP)}^d \subseteq \text{Sol(SVVIP)}^d \subseteq \text{Sol(VVIP)}^d$. Therefore,

$$\text{Sol(VVIP)}^d \subseteq \text{Sol(MSVVIP)}^d \subseteq \text{Sol(VVIP)}^d,$$

and thus, $\text{Sol(VVIP)}^d = \text{Sol(MSVVIP)}^d$, that is,

$$\bigcap_{y \in K} S(y) = \bigcap_{y \in K} M^S(y).$$

By Proposition 5.4, S is a KKM map. By C_x-pseudomonotonicity$_+$ of T, $S(y) \subseteq M^S(y)$ for all $y \in K$. Thus, M^S is a KKM map. By Proposition 5.3 (a), $M^S(y)$ is a closed subset of a compact set K, and hence $M^S(y)$ is compact for all $y \in K$. By Fan-KKM Lemma 1.14, we have

$$\bigcap_{y \in K} S(y) = \bigcap_{y \in K} M^S(y) \neq \emptyset.$$

Hence, Sol(SVVIP)^d is nonempty. □

Remark 5.5 Fang and Huang [9] proved Theorem 5.4 in the setting of a closed bounded convex subset of a reflexive Banach space X but for a fixed closed convex pointed cone.

From now onward, we present existence results for a solution of SVVIP, VVIP and WVVIP in the setting of Banach spaces.

Theorem 5.5 *Let X and Y be Banach spaces and K be a nonempty weakly compact convex subset of X. Let the set-valued map $W : K \to 2^Y$ be defined by $W(x) = Y \setminus \{-\operatorname{int}(C(x))\}$ such that the graph $\mathcal{G}(W)$ of W is weakly closed in $K \times Y$. For each $x \in K$, suppose that $T : K \to \mathcal{L}(X, Y)$ is weakly C_x-pseudomonotone and v-hemicontinuous on K. Then* Sol(WVVIP)^d *is nonempty.*

Proof By Lemmas 5.2 (b) and 5.3 (c),

$$\bigcap_{y \in K} S^W(y) = \bigcap_{y \in K} M^W(y).$$

Following the similar argument as in the proof of Theorem 5.2 and Proposition 5.4, we see that M^W is a KKM map.

We claim that for each $y \in K$, $M^W(y)$ is a weakly closed subset of K. For any $y \in K$, let $\{x_\alpha\}$ be a net in $M^W(y)$ such that x_α converges weakly to $\hat{x} \in K$. Since $x_\alpha \in M^W(y)$, we have

$$\langle T(y), y - x_\alpha \rangle \in Y \setminus \{-(C(x_\alpha))\}, \quad \text{for all } \alpha.$$

Since $T(y) \in \mathcal{L}(X, Y)$, by Theorem C.11, $T(y)$ is continuous from the weak topology of X to the weak topology of Y. We achieve that the net $\{\langle T(y), y - x_\alpha \rangle\}$ converges weakly to $\{\langle T(y), y - \hat{x} \rangle\} \in Y$. So, we obtain that $(x_\alpha, \langle T(y), y - x_\alpha \rangle)$ converges weakly to $(\hat{x}, \langle T(y), y - \hat{x} \rangle) \in \mathcal{G}(W)$, since $\mathcal{G}(W)$ is weakly closed. Therefore,

$$\langle T(y), y - \hat{x} \rangle \in W(\hat{x}) = Y \setminus \{-\operatorname{int}(C(\hat{x}))\},$$

so that $\hat{x} \in M^W(y)$. Consequently, $M^W(y)$ is a weakly closed subset of K.

Since K is weakly compact subset of a Banach space X and $M^W(y)$ is a weakly closed subset of K, we have that $M^W(y)$ is a weakly compact subset of K for all $y \in K$. Rest of the proof follows on the lines of the proof of Theorem 5.2. □

Example 5.11 In continuation of Example 5.8, for any $x = (x_1, x_2) \in K$, $y = (y_1, y_2) \in K$, it is clear that

$$\begin{aligned} t \mapsto & \langle T(ty + (1-t)x), y - x \rangle \\ &= \left(\int_0^{ty_1+(1-t)x_1} (y_1 - x_1) t dt, \int_0^{ty_2+(1-t)x_2} (y_2 - x_2) t dt \right) \\ &= \tfrac{1}{2} \left((y_1 - x_1)[ty_1 + (1-t)x_1]^2, (y_2 - x_2)[ty_2 + (1-t)x_2]^2 \right) \end{aligned}$$

is continuous at 0^+; so, T is a v-hemicontinuous operator. By Theorem 5.5, the WVVIP for the corresponding T and K has a solution. Indeed, it can be verified that $\bar{x} = (0, 1) \in K$ is a solution of WVVIP, since $C(\bar{x}) = C_1$ (as in Example 5.8) and

$$\begin{aligned} \langle T(\bar{x}), y - \bar{x} \rangle &= \left(\int_0^0 (y_1 - 0) t dt, \int_0^1 (y_2 - 1) t dt \right) \\ &= \frac{1}{2} (0, (y_2 - 1)) \notin -\operatorname{int}(C(\bar{x})), \quad \text{for all } y = (y_1, y_2) \in K. \end{aligned}$$

By Proposition 5.4, $S(y) := \{x \in K : \langle T(x), y - x \rangle \notin -C(x) \setminus \{\mathbf{0}\}\}$ is a KKM map. Therefore, analogously to Theorem 5.5, we have the following existence result for a solution of VVIP.

Theorem 5.6 *Let X and Y be Banach spaces and K be a nonempty weakly compact convex subset of X. For each $x \in K$, suppose that $T : K \to \mathcal{L}(X, Y)$ is C_x-pseudomonotone and v-hemicontinuous such that the set*

$$M(y) = \{x \in K : \langle T(y), y - x \rangle \notin -C(x) \setminus \{\mathbf{0}\}\}$$

is weakly closed. Then Sol(VVIP)^d *is nonempty.*

By using the similar argument as in the proof of Theorems 5.4 and 5.5, we can easily prove the following result on the existence of a solution of SVVIP.

Theorem 5.7 *Let X and Y be Banach spaces, K be a nonempty weakly compact convex subset of X and $C : K \to 2^Y$ be a closed set-valued map such that for each $x \in K$, $C(x)$ is a closed convex pointed cone in Y. For each $x \in K$, suppose that $T : K \to \mathcal{L}(X, Y)$ is C_x-pseudomonotone$_+$ and v-hemicontinuous on K. Then* Sol(SVVIP)^d *is nonempty.*

Remark 5.6

(a) Since every nonempty, closed, bounded and convex subset of a reflexive Banach space is weakly compact, we can consider K as a nonempty closed, bounded and convex subset of a reflexive Banach space X in Theorems 5.5, 5.3 and 5.7.
(b) When $C(x)$ is a fixed closed convex pointed cone with nonempty interior and T is a C_x-monotone operator, then Theorem 5.5 is considered in [21].
(c) When $Y = \mathbb{R}$, $\mathcal{L}(X, Y) = X^*$ and $C(x) = \mathbb{R}_+$ for all $x \in K$, then Theorems 5.5, 5.3 and 5.7 are same and established in [24].

Let X and Y be Banach spaces. A mapping $g : X \to Y$ is said to be *completely continuous* if the weak convergence of any sequence $\{x_n\}$ to x in X implies the strong convergence of $\{g(x_n)\}$ to $g(x)$ in Y.

We denote by $\mathcal{L}_c(X, Y)$ the set of all completely continuous mappings from X to Y. Obviously, $\mathcal{L}_c(X, Y) \subseteq \mathcal{L}(X, Y)$.

Huang and Fang [15] proved Theorem 5.5 for $T : K \to \mathcal{L}_c(X, Y)$.

Definition 5.6 The operator $T : K \to \mathcal{L}(X, Y)$ is said to satisfy the *L-condition* on K if and only if the following condition holds:

$$\sum_{i=1}^{m} \lambda_i \langle T(y_i), y_i \rangle - \sum_{i=1}^{m} \lambda_i \langle T(y_i), \hat{x} \rangle \in C(\hat{x}),$$

for any finite subset $\{y_1, y_2, \ldots, y_m\}$ of K, $\hat{x} = \sum_{i=1}^{m} \lambda_i y_i$, with $\lambda_i \geq 0$, $1 \leq i \leq m$, and $\sum_{i=1}^{m} \lambda_i = 1$.

We note that the particular form of the above condition is used in [25].

We derive the following existence results for solutions of WVVIP and MWVVIP under L-condition.

Proposition 5.5 *Let X, Y, K, C, W, and $\mathcal{G}(W)$ be the same as in Theorem 5.5. Suppose that $T : K \to \mathcal{L}(X, Y)$ satisfies L-condition on K. Then* MWVVIP *has a solution.*

Proof We first claim that M^W is a KKM map on K. To this end, let $\hat{x} \in \text{co}(\{y_1, y_2, \ldots, y_m\}) \subseteq K$. Then $\hat{x} = \sum_{i=1}^{m} \lambda_i y_i$ for some nonnegative real number λ_i, $1 \leq i \leq m$, with $\sum_{i=1}^{m} \lambda_i = 1$. If $\hat{x} \notin \bigcup_{i=1}^{m} M^W(y_i)$, then

$$\langle T(y_i), y_i - \hat{x} \rangle \in -\text{int}(C(\hat{x})), \quad \text{for each } i = 1, 2, \ldots, m.$$

Since $-C(\hat{x})$ is a convex cone and $\lambda_i \geq 0$ with $\sum_{i=1}^{m} \lambda_i = 1$, we have

$$\sum_{i=1}^{m} \lambda_i \langle T(y_i), y_i - \hat{x} \rangle \in -\text{int}(C(\hat{x})). \tag{5.14}$$

Moreover, since T satisfies L-condition on K,

$$\sum_{i=1}^{m} \lambda_i \langle T(y_i), y_i\rangle - \sum_{i=1}^{m} \lambda_i \langle T(y_i), \hat{x}\rangle \in C(\hat{x}). \tag{5.15}$$

It follows from (5.14) and (5.15) that $\mathbf{0} \in \text{int}(C(\hat{x}))$, a contradiction. Hence,

$$\hat{x} \in \bigcup_{i=1}^{m} M^W(y_i),$$

and so M^W is a KKM map on K. From the proof of Theorem 5.5, we derive that $M^W(y)$ is a weakly compact subset of K for all $y \in K$. By Fan-KKM Lemma 1.14, we have

$$\bigcap_{y\in K} M^W(y) \neq \emptyset,$$

that is, Sol(MWVVIP)^d is nonempty. □

Since $-C(x)$ is a convex cone, analogously we can prove the following result.

Proposition 5.6 *Let X, Y, K and C be the same as in Theorem 5.3. Suppose that $T : K \to \mathcal{L}(X, Y)$ satisfies L-condition on K such that the set*

$$M(y) = \{x \in K : \langle T(y), y - x\rangle \notin -C(x) \setminus \{\mathbf{0}\}\}$$

is weakly closed. Then MVVIP *has a solution.*

In Theorem 5.5, we can replace the weak C_x-pseudomonotonicity by L-condition.

Theorem 5.8 *Let X, Y, K, C, W, and $\mathcal{G}(W)$ be the same as in Theorem 5.5. Suppose that $T : K \to \mathcal{L}(X, Y)$ is v-hemicontinuous and satisfies L-condition on K. Then* WVVIP *has a solution.*

Proof The result follows from Proposition 5.5 and Lemma 5.2 (b). □

Theorem 5.9 *Let X, Y, K and C be the same as in Theorem 5.3. Suppose that $T : K \to \mathcal{L}(X, Y)$ is v-hemicontinuous and satisfies L-condition on K. Then* VVIP *has a solution.*

Proof The result follows from Proposition 5.6 and Lemma 5.2 (a). □

To consider the unbounded case, we need the following coercivity conditions.

Definition 5.7 The operator $T : K \to \mathcal{L}(X, Y)$ is said to satisfy

(a) *strongly v-coercive condition* C_1 if there exist a weakly compact subset B of X and $\tilde{y} \in B \cap K$ such that

$$\langle T(x), \tilde{y} - x\rangle \notin C(x), \quad \text{for all } x \in K \setminus B;$$

(b) *strongly υ-coercive condition* C_2 if there exist a weakly compact subset B of X and $\tilde{y} \in B \cap K$ such that

$$\langle T(\tilde{y}), \tilde{y} - x \rangle \notin C(x), \quad \text{for all } x \in K \setminus B;$$

(c) *υ-coercive condition* C_1 if there exist a weakly compact subset B of X and $\tilde{y} \in B \cap K$ such that

$$\langle T(x), \tilde{y} - x \rangle \in -C(x) \setminus \{\mathbf{0}\}, \quad \text{for all } x \in K \setminus B;$$

(d) *υ-coercive condition* C_2 if there exist a weakly compact subset B of X and $\tilde{y} \in B \cap K$ such that

$$\langle T(\tilde{y}), \tilde{y} - x \rangle \in -C(x) \setminus \{\mathbf{0}\}, \quad \text{for all } x \in K \setminus B;$$

(e) *weakly υ-coercive condition* C_1 if there exist a weakly compact subset B of X and $\tilde{y} \in B \cap K$ such that

$$\langle T(x), \tilde{y} - x \rangle \in -\operatorname{int}(C(x)), \quad \text{for all } x \in K \setminus B;$$

(f) *weakly υ-coercive condition* C_2 if there exist a weakly compact subset B of X and $\tilde{y} \in B \cap K$ such that

$$\langle T(\tilde{y}), \tilde{y} - x \rangle \in -\operatorname{int}(C(x)), \quad \text{for all } x \in K \setminus B.$$

We now present some existence results for solutions of WVVIP and VVIP defined on a closed (not necessarily bounded) convex subset K of a Banach space X.

Theorem 5.10 *Let X and Y be Banach spaces and K be a nonempty closed convex subset of X. Let $W : K \to 2^Y$ be defined by $W(x) = Y \setminus \{-\operatorname{int}(C(x))\}$ such that the graph $\mathcal{G}(W)$ of W is weakly closed in $K \times Y$. Suppose that $T : K \to \mathcal{L}(X, Y)$ is weakly C_x-pseudomonotone and υ-hemicontinuous on K. In addition, assume that T satisfies weakly υ-coercive condition C_1. Then* WVVIP *has a solution.*

Proof Let $\tilde{y} \in K$, and let the weakly compact subset B of X satisfy weakly υ-coercive condition C_1.

We claim that the weak closure $\overline{S^W(\tilde{y})}^w$ of $S^W(\tilde{y})$ is a weakly compact subset of K. If $S^W(\tilde{y}) \nsubseteq B$, then there exists $x \in S^W(\tilde{y})$ such that $x \in K \setminus B$. It follows that

$$\langle T(x), \tilde{y} - x \rangle \notin -\operatorname{int}(C(x)),$$

which contradicts weakly υ-coercive condition C_1. Therefore, we have $S^W(\tilde{y}) \subseteq B$; hence, $\overline{S^W(\tilde{y})}^w$ is a weakly compact subset of K. By Fan-KKM Lemma 1.14, we have

$$\bigcap_{y\in K} \overline{S^W(\tilde{y})}^{w} \neq \emptyset.$$

Again, by weak C_x-pseudomonotonicity of T, $S^W(y) \subseteq M^W(y)$ for all $y \in K$, and so that $\overline{S^W(y)}^{w} \subseteq \overline{M^W(y)}^{w} = M^W(y)$ for all $y \in K$, since $M^W(y)$ is weakly closed as proved in Theorem 5.5. Consequently, $\bigcap_{y\in K} M^W(y) \neq \emptyset$. As in Theorem 5.5,

$$\bigcap_{y\in K} S^W(y) = \bigcap_{y\in K} M^W(y) \neq \emptyset.$$

Hence, Sol(WVVIP)^d is nonempty. □

Example 5.12 ([30]) In continuation of Example 5.11, we consider the following unbounded set

$$K = \{(x_1, x_2) \in X : x_1 - 1 \leq x_2 \leq x_1 + 1 \text{ and } x_2 \geq 0\} \subset \mathbb{R}^2.$$

Let the operator T and the mapping C be the same as in Example 5.11. Consider the compact set $B = [-1, 2] \times [0, 1] \subset \mathbb{R}^2$ and $\tilde{y} = (-1, 0) \in B \cap K$. We have

$$\begin{aligned}\langle T(x), \tilde{y} - x\rangle &= \left(\int_0^{x_1} (-1 - x_1)tdt, \int_0^{x_2} (-x_2)tdt\right)\\ &= \frac{1}{2}\left((-1 - x_1)x_1^2, -x_2^3\right) \in -\operatorname{int}(C(x)),\end{aligned}$$

for all $x \in K \setminus B$, where $x = (x_1, x_2)$. Hence, B and $\tilde{y}$ satisfy weakly v-coercive condition C_1. By Theorem 5.10, we know that there exists a solution $\bar{x} \in K$ of WVVIP. Indeed, we can verify that $\bar{x} = (-1, 0) \in K$ is a solution of WVVIP.

Remark 5.7 When $Y = \mathbb{R}$, $\mathcal{L}(X, Y) = X^*$ and $C(x) = \mathbb{R}_+$ for all $x \in K$, then Theorem 5.10 is considered in [27].

Analogously, we have the following existence result for a solution of VVIP without boundedness assumption on the underlying set.

Theorem 5.11 *Let X and Y be Banach spaces and K be a nonempty closed convex subset of X. For each $x \in K$, suppose that $T : K \to \mathcal{L}(X, Y)$ is C_x-pseudomonotone and v-hemicontinuous on K such that the set*

$$G(y) = \{x \in K : \langle T(y), y - x\rangle \notin -C(x) \setminus \{\mathbf{0}\}\}$$

is weakly closed. In addition, assume that T satisfies v-coercive condition C_1. Then VVIP *has a solution.*

The following result provides the existence of a solution of MWVVIP defined on a noncompact set, but under weakly υ-coercivity condition C_2.

Proposition 5.7 *Let X, Y, K, C, W, and $\mathcal{G}(W)$ be the same as in Theorem 5.10. Suppose that $T : K \to \mathcal{L}(X, Y)$ satisfies L-condition and weakly υ-coercive condition C_2. Then* MWVVIP *has a solution.*

Proof Let $\tilde{y} \in K$ and weakly compact subset $B \subset X$ satisfy υ-coercive condition C_2. By employing the same argument as in Proposition 5.5 and Theorem 5.5, respectively, we see that M^W is a KKM map on K and $M^W(y)$ is a weakly compact subset of K for all $y \in K$. Now, we claim that $M^W(\tilde{y}) \subseteq B$. If there exists $x \in M^W(\tilde{y})$ such that $x \in K \setminus B$, then

$$\langle T(\tilde{y}), \tilde{y} - x \rangle \notin -\operatorname{int}(C(x)),$$

which contradicts υ-coercive condition C_2. Therefore, $M^W(\tilde{y}) \subseteq B$; hence, $M^W(\tilde{y})$ is a weakly compact subset of K. By Fan-KKM Lemma 1.14, we have

$$\bigcap_{y \in K} M^W(y) \neq \emptyset;$$

so, $\mathrm{Sol(MWVVIP)}^d$ is nonempty. □

By using a similar argument as in the proof of Proposition 5.7 and Proposition 5.6, we obtain the following existence result for a solution of MVVIP defined over a noncompact set, but under the υ-coercive condition C_2.

Proposition 5.8 *Let X, Y, K, C and M^W be the same as in Theorem 5.11. Suppose that $T : K \to \mathcal{L}(X, Y)$ satisfies L-condition and weakly υ-coercive condition C_2. Then* MVVIP *has a solution.*

If we add the υ-hemicontinuity in Proposition 5.7 and 5.8, we derive the existence results for a solution of WVVIP and VVIP defined over a noncompact set by solving the corresponding WMVVIP and MVVIP, respectively.

Theorem 5.12 *Let X, Y, K, C, W, and $\mathcal{G}(W)$ be the same as in Proposition 5.7. Suppose that $T : K \to \mathcal{L}(X, Y)$ is υ-hemicontinuous such that both L-condition and weakly υ-coercive condition C_2 hold on K. Then* WVVIP *has a solution.*

Proof The result follows from Proposition 5.7 and Lemma 5.2 (b). □

Analogously, by using Proposition 5.8 and Lemma 5.2 (a), we derive the following existence result for a solution of VVIP.

Theorem 5.13 *Let X, Y, K and C be the same as in Proposition 5.8. Suppose that $T : K \to \mathcal{L}(X, Y)$ is υ-hemicontinuous such that both L-condition and υ-coercive condition C_2 hold on K and the set*

$$M(y) = \{x \in K : \langle T(y), y - x \rangle \notin -C(x) \setminus \{\mathbf{0}\}\}$$

is weakly closed. Then VVIP *has a solution.*

In Theorem 5.10, if we replace weakly v-coercive condition C_1 by weakly v-coercive condition C_2, then the existence theorem for a solution of WVVIP can be obtained as follows.

Theorem 5.14 *Let X, Y, K, C, W, and $\mathcal{G}(W)$ be the same as in Proposition 5.7. For each $x \in K$, suppose that $T : K \to \mathcal{L}(X, Y)$ is weakly C_x-pseudomonotone and v-hemicontinuous on K, and assume in addition that T satisfies weakly v-coercive condition C_2 on K. Then* WVVIP *has a solution.*

Proof Let $\tilde{y} \in K$, and let weakly compact subset B of X be such that T satisfies weakly v-coercive condition C_2. By the same argument as in the proof of Theorem 5.5, we can derive that M^W is a KKM map on K and, for each $y \in K$, $M^W(y)$ is a weakly compact subset of K. It can be checked that $M^W(\tilde{y}) \subseteq (K \cap D)$ is a weakly compact subset of K. By Fan-KKM Lemma 1.14, we get

$$\bigcap_{y \in K} S^W(y) = \bigcap_{y \in K} M^W(y) \neq \emptyset.$$

Therefore, Sol(WVVIP)^d is nonempty. □

By using Theorem 5.3, we have following existence result for a solution of VVIP.

Theorem 5.15 *Let X, Y, K and C be the same as in Theorem 5.3. For each $x \in K$, suppose that $T : K \to \mathcal{L}(X, Y)$ is C_x-pseudomonotone and v-hemicontinuous on K such that the set*

$$M(y) = \{x \in K : \langle T(y), y - x \rangle \notin -C(x) \setminus \{\mathbf{0}\}\}$$

is weakly closed. Further, assume that T satisfies v-coercive condition C_2 on K. Then VVIP *has a solution.*

Now we use the scalarized coercivity condition to derive the existence results for a solution of WVVIP. We set

$$C_+ = \text{co}(\{C(x) : x \in K\}),$$

and note that

$$C_+^* = \{l \in Y^* : \langle l, y \rangle \geq 0 \text{ for all } y \in C_+\},$$

$$\text{int}(C_+^*) = \{l \in Y^* : \langle l, y \rangle > 0 \text{ for all } y \in \text{int}(C_+)\},$$

where Y^* denotes the topological dual of Y. Let $s \in \text{int}(C_+^*)$ and $T : K \to \mathcal{L}(X, Y)$ be an operator. We define the operator $T_s : K \to X^*$ by

$$\langle T_s(x), y \rangle = \langle s, \langle T(x), y \rangle \rangle, \quad \text{for all } x \in K \text{ and all } y \in X,$$

where X^* denotes the topological dual of X.

In order to solve WVVIP over an unbounded domain, we need the following coercivity conditions.

Definition 5.8 Let X and Y be Banach spaces and K be a nonempty closed convex subset of X. Let $C : K \to 2^Y$ be a set-valued map such that C_+ is a closed convex pointed cone with $\text{int}(C_+) \neq \emptyset$. An operator $T : K \to \mathcal{L}(X, Y)$ is said to be:

(a) *v-coercive* if there exist $x_0 \in K$ and $s \in C_+^* \setminus \{\mathbf{0}\}$ such that

$$\frac{\langle T_s(x), x - x_0 \rangle}{\|x - x_0\|} \to \infty, \quad \text{as } x \in K, \ \|x\| \to \infty.$$

(b) *weakly v-coercive* if there exist $x_0 \in K$ and $s \in C_+^* \setminus \{\mathbf{0}\}$ such that

$$\langle T_s(x), x - x_0 \rangle \to \infty, \quad \text{as } x \in K, \ \|x\| \to \infty.$$

Remark 5.8

(a) It is clear that if T is v-coercive, then it is weakly v-coercive.
(b) If $Y = \mathbb{R}$, $C(x) = \mathbb{R}_+$ for all $x \in K$ and $T : K \to X^*$, Definition 5.8 collapses to the following definitions:

(i) T is *coercive* if

$$\frac{\langle T(x), x - x_0 \rangle}{\|x\|} \to \infty, \quad \text{whenever } x \in K \text{ and } \|x\| \to \infty.$$

(ii) T is *weakly coercive* if

$$\langle T(x), x - x_0 \rangle \to \infty, \quad \text{whenever } x \in K \text{ and } \|x\| \to \infty.$$

(c) Suppose that T is v-coercive, then

$$\langle s, \langle T(x) - T(x_0), x - x_0 \rangle \rangle \to \infty.$$

Thus, the definition of v-coerciveness of T is equivalent to one used in [7]. If $C : K \to 2^Y$ is a constant mapping and $s \in \text{int}(C_+^*)$, then the definition of v-coerciveness of T is considered in [8].

Recall that a mapping $H : X \to X^*$ is said to be *pseudomonotone* if for every pair of points $x \in K$, $y \in K$, we have

$$\langle H(x), y - x \rangle \geq 0 \quad \text{implies} \quad \langle H(y), y - x \rangle \geq 0.$$

Proposition 5.9 *Let X and Y be Banach spaces and K be a nonempty closed convex subset of X. Let $C : K \to 2^Y$ be a set-valued map such that C_+ is a closed convex* *pointed cone with* $\text{int}(C_+) \neq \emptyset$. *Suppose that* $T : K \to \mathcal{L}(X,Y)$ *is weakly* C_+-*pseudomonotone and there exists* $s \in \text{int}(C_+^*)$ *such that*

$$\langle s, y \rangle \geq 0, \quad \textit{for all } y \notin -\text{int}(C_+). \tag{5.16}$$

Then the operator T_s *is pseudomonotone on* K.

Proof Let $x, y \in K$ be such that $\langle T_s(x), y - x \rangle \geq 0$. If $\langle T(x), y - x \rangle \in -\text{int}(C_+)$, then since $s \in \text{int}(C_+^*)$,

$$-\langle T_s(x), y - x \rangle > 0,$$

which contradicts to the assumption. Thus, we have

$$\langle T(x), y - x \rangle \notin -\text{int}(C_+).$$

By weak C_+-pseudomonotonicity of T, $\langle T(y), y - x \rangle \notin -\text{int}(C_+)$. From (5.16), it follows that

$$\langle s, \langle T(y), y - x \rangle \rangle \geq 0, \quad \text{that is,} \quad \langle T_s(y), y - x \rangle \geq 0.$$

Therefore, T_s is pseudomonotone on K. □

Remark 5.9 Proposition 5.16 also holds in the setting of Hausdorff topological vector spaces (see [30]).

Under the assumption of weak v-coercivity of T, we have the following existence theorem for a solution of WVVIP.

Theorem 5.16 *Let* X, Y, K, C, *and* W *be the same as in Theorem 5.10; in addition, let* X *be reflexive and* C_+ *be closed convex. Suppose that* $T : K \to \mathcal{L}(X,Y)$ *is weakly* C_+-*pseudomonotone,* v-*hemicontinuous, and is weakly* v-*coercive with respect to an* $s \in \text{int}(C_+^*)$ *on* K *such that* $\langle s, y \rangle \geq 0$ *for all* $y \notin -\text{int}(C_+)$. *Then* WVVIP *has a solution.*

Proof For $s \in \text{int}(C_+^*)$, suppose that there exists an $\bar{x} \in K$ which is a solution of the following variational inequality problem, denoted by $(\text{VIP})_s$: find $\bar{x} \in K$ such that

$$\langle T_s(\bar{x}), y - \bar{x} \rangle \geq 0, \quad \text{for all } y \in K.$$

If

$$\langle T(\bar{x}), y - \bar{x} \rangle \in -\text{int}(C(\bar{x})), \quad \text{for some } y \in K,$$

then

$$\langle T(\bar{x}), y - \bar{x} \rangle \in -\text{int}(C_+);$$

thus, $\langle T_s(\bar{x}), y-\bar{x}\rangle < 0$, which contradicts the fact that $\bar{x}$ solves $(\mathrm{VIP})_s$. Consequently, we have

$$\langle T(\bar{x}), y - \bar{x}\rangle \notin -\operatorname{int}(C(\bar{x})), \quad \text{for all } y \in K,$$

that is, $\bar{x}$ is a solution of WVVIP. So. it is sufficient to prove that there exists a solution of $(\mathrm{VIP})_s$ for some $s \in \operatorname{int}(C_+^*)$ where s satisfies the weak υ-coercivity of the operator T.

Let B_r denote the closed ball (under the norm) of X with center at zero and radius r. In the special case where $Y = \mathbb{R}$, $C(x) = \mathbb{R}_+$ for all $x \in K \cap B_r$, Proposition 5.9 and Theorem 1.41, guarantee the existence of a solution x_r for $(\mathrm{VIP})_s$, that is,

$$x_r \in K \cap B_r, \qquad \langle T_s(x_r), y - x_r\rangle \geq 0, \quad \text{for all } y \in K \cap B_r.$$

Choose $r \geq \|x_0\|$, where x_0 satisfies the weak υ-coercivity of T. Then

$$\langle T_s(x_r), x_0 - x_r\rangle \geq 0. \tag{5.17}$$

We observe that $\{x_r : r > 0\}$ must be bounded. Otherwise, we can choose r large enough so that the weak υ-coercivity of T implies that

$$\langle T_s(x_r), x_0 - x_r\rangle < 0,$$

which contradicts (5.17). Therefore, there exists r such that $\|x_r\| < r$. Now, for each $x \in K$, we can choose $\varepsilon > 0$ small enough such that $x_r + \varepsilon(x - x_r) \in K \cap B_r$. Then

$$\langle T_s(x_r), x_r + \varepsilon(x - x_r) - x_r\rangle \geq 0.$$

Dividing by ε on both sides of the above inequality, we obtain

$$\langle T_s(x_r), x - x_r\rangle \geq 0, \quad \text{for all } x \in K,$$

which shows that x_r is a solution of $(\mathrm{VIP})_s$. By the above observation, x_r is a solution of WVVIP. □

We note that it is possible to impose other topologies on X so that the conclusion of theorems in this section still hold. For example, we may consider the weakest topology τ on X such that for all $l \in \mathcal{L}(X, Y)$, l is continuous. With suitable modification of the statements of the results of this section, we can obtain some existence results for a solution of WVVIP.

As an application of above results, we provide some existence results for solutions of generalized vector complementarity problems.

Let X and Y be Banach spaces and K be a convex pointed cone in X. Let $C : K \to 2^Y$ be a set-valued map such that for all $x \in K$, $C(x)$ is a closed convex pointed cone with $\operatorname{int}(C(x)) \neq \emptyset$. Let $T : K \to \mathcal{L}(X, Y)$ be an operator. The *generalized vector complementarity problem* (in short, GVCP) is to find $\bar{x} \in K$ such that

$$\langle T(\bar{x}), \bar{x} \rangle \notin \text{int}(C(\bar{x})) \quad \text{and} \quad \langle T(\bar{x}), y \rangle \notin -\text{int}(C(\bar{x})), \text{ for all } y \in K. \tag{5.18}$$

Such problem is an extension of the vector complementarity problem (in short, VCP) studied in [8], where Y is an ordered Banach space induced by a constant positive cone P. When $Y = \mathbb{R}$, GVCP coincides with the *generalized complementarity problem* (in short, GCP), that is, to find $\bar{x} \in K$ such that

$$\langle T(\bar{x}), \bar{x} \rangle = 0 \quad \text{and} \quad T(\bar{x}) \in K^*, \tag{5.19}$$

where K^* is the dual cone of K. This problem was introduced by Karamardian [17]. He proved that both variational inequality problem and complementarity problem have the same solution if the underlying set K is a closed convex cone; see [17, Lemma 3.1]. Chen and Yang [8] showed the equivalence between WVVIP and GVCP with underlying space Y being an ordered Banach space; See [8, Proposition 4.2], and its remark.

Ansari et al. [4], Huang and Fang [16] and Yin et al. [29] considered more general forms of vector complementarity problem and established some existence results for solutions of such problems by proving their equivalence with corresponding vector variational inequality problems.

The following lemma gives the equivalence relation between WVVIP and GVCP.

Lemma 5.4 *Let X and Y be Banach spaces and K be a closed convex pointed cone in X. Let $C : K \to 2^K$ be a set-valued map such that for all $x \in K$, $C(x)$ is a closed convex pointed cone in X. Let $T : K \to \mathcal{L}(X, y)$ be a nonlinear operator. Then the following statements hold.*

(a) *If $\bar{x}$ is a solution of* WVVIP, *then it is a solution of* GVCP.
(b) *If $\bar{x}$ is a solution of* GVCP, *and we assume in addition that*

$$\langle T(\bar{x}), \bar{x} \rangle \in -C(\bar{x}), \tag{5.20}$$

then $\bar{x}$ is a solution of WVVIP.

Proof

(a) Since $\bar{x}$ is a solution of WVVIP, we have

$$\langle T(\bar{x}), y - \bar{x} \rangle \notin -\text{int}(C(\bar{x})), \quad \text{for all } y \in K. \tag{5.21}$$

First, we take $y = \mathbf{0}$. Then (5.21) becomes

$$\langle T(\bar{x}), \bar{x} \rangle \notin \text{int}(C(\bar{x})). \tag{5.22}$$

Next, for each $z \in K$, replace y by $z + \bar{x} \in K$ in (5.21), we have

$$\langle T(\bar{x}), z \rangle = \langle T(\bar{x}), (z + \bar{x}) - \bar{x} \rangle \notin -\operatorname{int}(C(\bar{x})). \tag{5.23}$$

Relations (5.22) and (5.23) show that $\bar{x}$ is a solution of GVCP.

(b) If $\bar{x}$ solves GVCP, then

$$\langle T(\bar{x}), y \rangle \notin -\operatorname{int}(C(\bar{x})), \quad \text{for all } y \in K. \tag{5.24}$$

By (5.20) and (5.24), we have

$$\langle T(\bar{x}), y - \bar{x} \rangle \notin -\operatorname{int}(C(\bar{x})), \quad \text{for all } y \in K.$$

Thus, $\bar{x}$ is a solution of WVVIP. □

For all $x \in K$, when $C(x)$ is a fixed closed convex pointed cone with nonempty interior, Lemma 5.4 is considered in [8].

By Lemma 5.4 and by existence results for a solution of WVVIP, we have the following existence results for a solution of GVCP.

Theorem 5.17 *Let X and Y be Banach spaces and K be a closed convex pointed cone in X. Let C and W be the same as in Theorem 5.10. Suppose that $T : K \to \mathcal{L}(X, Y)$ is v-hemicontinuous on K. Then* GVCP *has a solution under each of the following conditions:*

(i) *T is weakly C-pseudomonotone and satisfy v-coercive condition C_1;*
(ii) *T satisfies both L-condition and v-coercive condition C_2 on K;*
(iii) *T is weakly C-pseudomonotone and satisfies v-coercive condition C_2.*

Proof

(i) The result follows from Lemma 5.4 and Theorem 5.10.
(ii) The result follows from Lemma 5.4 and Theorem 5.12.
(iii) The result follows from Lemma 5.4 and Theorem 5.14. □

Under the assumption of the weak v-coercivity of T, we have the following existence result for a solution of GVCP.

Theorem 5.18 *Let X, Y, K, C, and W be the same as in Theorem 5.16. Suppose that $T : K \to \mathcal{L}(X, Y)$ is weakly C_+-pseudomonotone and v-hemicontinuous. Then* GVCP *has a solution under each of the following conditions:*

(i) *T is weakly v-coercive with respect to an $s \in \operatorname{int}(C_+^*)$ on K such that $\langle s, y \rangle \geq 0$ for all $y \notin -\operatorname{int}(C_+)$;*
(ii) *T is v-coercive with respect to an $s \in \operatorname{int}(C_+^*)$ on K such that $\langle s, y \rangle \geq 0$ for all $y \notin -\operatorname{int}(C_+)$.*

Proof

(i) The result follows from Lemma 5.4 (i) and Theorem 5.16.
(ii) It clearly follows from (i). □

5.3 Existence Results for Solutions of Vector Variational Inequalities Without Monotonicity

The following example due to Huang and Fang [15] shows that the operator T in WVVIP is not C_x-monotone, even then the corresponding WVVIP has a solution.

Example 5.13 Let $X = \mathbb{R}$, $K = [-\pi/2, \pi/2]$, $Y = \mathbb{R}^2$ and $C(x) = \mathbb{R}^2_+$ for all $x \in K$. Let $T : K \to \mathcal{L}(X, Y)$ be defined by

$$T(x) = \left(\sin x \cos x, \sin^2 x - x\right), \quad \text{for all } x \in K.$$

Then T is continuous, but not C_x-monotone on K. However, 0 is a solution of the corresponding WVVIP.

The above example motivates us to present some existence results for solutions of vector variational inequality problems without any kind of monotonicity assumption.

Theorem 5.19 *Let X and Y be Hausdorff topological vector spaces and K be a nonempty convex subset of X. Let $C : K \to 2^Y$ be a set-valued map such that for all $x \in K$, $C(x)$ is a closed convex pointed cone with* $\text{int}(C(x)) \neq \emptyset$, *and $W : K \to 2^Y$ be defined by $W(x) = Y \setminus \{-\text{int}(C(x))\}$ such that the graph of W is closed in $K \times Y$. Let $T : K \to \mathcal{L}(X, Y)$ be an operator such that for all $y \in K$, the mapping $x \mapsto \langle T(x), y - x\rangle$ is continuous. Assume that for a nonempty compact convex subset $D \subseteq K$ with each $x \in K$, there exists $y \in D$ such that $\langle T(x), y - x\rangle \in -\text{int}(C(x))$. Then there exists a solution $\bar{x} \in K$ of* WVVIP.

Proof Let $A := \{(x, y) \in K \times K : \langle T(x), y - x\rangle \notin -\text{int}(C(x))\}$. Then, it is clear that $(x, x) \in A$ for all $x \in K$. We show that for all $y \in K$, the set $A_y := \{x \in K : (x, y) \in A\}$ is closed. To this end, let $\{x_\alpha\}$ be a net in A_y converging to some $x \in K$. Since $(x_\alpha, y) \in A$ for each α, we have

$$\langle T(x_\alpha), y - x_\alpha\rangle \notin -\text{int}(C(x_\alpha)),$$

equivalently,

$$\langle T(x_\alpha), y - x_\alpha\rangle \in Y \setminus \{-\text{int}(C(x_\alpha))\}.$$

By assumption $\langle T(x_\alpha), y - x_\alpha \rangle$ converges to $\langle T(x), y - x \rangle$. Since the graph of W is closed in $K \times Y$, we have

$$\langle T(x), y - x \rangle \in Y \setminus \{-\operatorname{int}(C(x))\}, \quad \text{equivalently,} \quad \langle T(x), y - x \rangle \notin -\operatorname{int}(C(x)).$$

Thus, $x \in A_y$, and consequently, A_y is closed.

Finally, we show that for each $x \in K$, the set $A_x := \{y \in K : (x, y) \notin A\}$ is convex. To this end, let $y_1, y_2 \in A_x$ and $\lambda_1 \geq 0, \lambda_2 \geq 0$ with $\lambda_1 + \lambda_2 = 1$. As $C(x)$ is a convex cone, we have

$$\langle T(x), \lambda_1(y_1 - x) \rangle \in -\operatorname{int}(C(x)) \quad \text{and} \quad \langle T(x), \lambda_2(y_2 - x) \rangle \in -\operatorname{int}(C(x))$$

which imply by the convexity of $\operatorname{int}(C(x))$ that

$$\langle T(x), \lambda_1 y_1 + \lambda_2 y_2 - (\lambda_1 + \lambda_2)x \rangle = \langle T(x), \lambda_1 y_1 + \lambda_2 y_2 - x \rangle \in -\operatorname{int}(C(x));$$

hence, $\lambda_1 y_1 + \lambda_2 y_2 \in A_x$, and therefore, A_x is convex.

By invoking Lemma 1.17, there exists $\bar{x} \in K$ such that $\{\bar{x}\} \times K \subset A$. This implies that $\bar{x} \in K$ and

$$\langle T(\bar{x}), y - \bar{x} \rangle \notin -\operatorname{int}(C(\bar{x})), \quad \text{for all } y \in K$$

from which the result follows. □

Lai and Yao [18] proved above theorem in the setting of Banach spaces X and Y and compactness assumption on K.

In a similar manner as the proof of Theorem 5.19, we derive the following existence result for a solution of VVIP.

Theorem 5.20 *Let X and Y be Hausdorff topological vector spaces, K be a nonempty convex subset of X and $C : K \to 2^Y$ be a set-valued map such that for all $x \in K$, $C(x)$ is a closed convex pointed cone. Let $T : K \to \mathcal{L}(X, Y)$ be an operator such that for all $y \in K$, the set $\{x \in K : \langle T(x), y - x \rangle \notin -C(x) \setminus \{\mathbf{0}\}\}$ is closed. Assume that for a nonempty compact convex subset $D \subseteq K$ with each $x \in K$, there exists $y \in D$ such that $\langle T(x), y - x \rangle \in -C(x) \setminus \{\mathbf{0}\}$. Then there exists a solution $\bar{x} \in K$ of* VVIP.

Proof Let $A = \{(x, y) \in K \times K : \langle T(x), y - x \rangle \notin -C(x) \setminus \{\mathbf{0}\}\}$. Then it is clear that $(x, x) \in A$ for all $x \in K$. Since $C(x)$ is a convex cone, by using the similar argument as in the proof of Theorem 5.19, we can show that for each $x \in K$, the set $A_x := \{y \in K : (x, y) \notin A\}$ is convex. Rest of the proof lies on the lines of the proof of Theorem 5.19. □

Fang and Huang [9] proved Theorem 5.20 for a fixed cone C and in the setting of a compact convex subset K of a Banach space X, but by using the method of partition of unity. They presented the following example to show that the condition "the set $\{x \in K : \langle T(x), y - x \rangle \notin -C \setminus \{\mathbf{0}\}$ is closed for all $y \in K$" is nontrivial.

Example 5.14 Let $X = \mathbb{R}$, $K = [0, 1]$, $Y = \mathbb{R}^2$, $C(x) = \mathbb{R}^2_+$ for all $x \in K$ and $f_1, f_2 : K \to \mathbb{R}$ be continuous increasing functions such that $f_1(x) > 0$ and $f_2(x) > 0$ for all $x \in K$, and

$$T(x) = (f_1(x), f_2(x)), \quad \text{for all } x \in K.$$

Then, obviously, $T : K \to \mathcal{L}(X, Y)$ is continuous and C_x-monotone. It can be easily seen that for each $y \in K$, $\{x \in K : \langle T(x), y - x\rangle \in -\mathbb{R}^2_+ \setminus \{\mathbf{0}\}\} =]y, 1]$ and so is open in K, and hence, its complement is closed.

The following result provides the existence of a solution of SVVIP without any kind of monotonicity assumption.

Theorem 5.21 *Let X and Y be Hausdorff topological vector spaces, K be a nonempty convex subset of X and $C : K \to 2^Y$ be a closed set-valued map such that for all $x \in K$, $C(x)$ is a closed convex pointed cone. Let $T : K \to \mathcal{L}(X, Y)$ be an operator such that the mapping $x \mapsto \langle T(x), y - x\rangle$ is continuous, and for all $x \in K$, the set $\{y \in K : \langle T(x), y - x\rangle \notin C(x)\}$ is convex. Assume that for a nonempty compact convex subset $D \subseteq K$ with each $x \in K$, there exists $y \in D$ such that $\langle T(x), y - x\rangle \notin C(x)$. Then there exists a solution $\bar{x} \in K$ of* SVVIP.

Proof Let $A := \{(x, y) \in K \times K : \langle T(x), y - x\rangle \in C(x)\}$. Then it is clear that $(x, x) \in A$ for all $x \in K$. Since C is a closed set-valued map, its graph must be closed. Therefore, by using similar argument as in the proof of Theorem 5.19, we can show that for each $y \in K$, the set $A_y := \{x \in K : (x, y) \in A\}$ is closed. Rest of the proof lies on the lines of the proof of Theorem 5.19. □

Remark 5.10

(a) It can be easily seen that if the set K is compact in Theorems 5.19 – 5.21, then the last assumption in these theorems is satisfied.
(b) If we consider $\mathcal{L}(X, Y)$ as a topological vector space under the σ-topology and $T : K \to \mathcal{L}(X, Y)$ is a continuous operator, then the mapping $x \mapsto \langle T(x), y - x\rangle$ is continuous in Theorems 5.19 and 5.21.

    Indeed, by Lemma C.1, $\langle ., .\rangle$ is continuous. Since T is continuous, for each $y \in K$, the mapping $x \mapsto \langle T(x), y - x\rangle$ is continuous as it is a composition of two continuous functions.

Corollary 5.1 *Let X and Y be Banach spaces and K be a nonempty convex subset of X. Let $T : K \to \mathcal{L}(X, Y)$ be a continuous operator, $C : K \to 2^Y$ be a set-valued map such that for all $x \in K$, $C(x)$ is a closed convex pointed cone with* $\text{int}(C(x)) \neq \emptyset$, *and $W : K \to 2^Y$ be defined by $W(x) = Y \setminus \{-\text{int}(C(x))\}$ such that the graph $\mathcal{G}(W)$ of W is closed in $K \times Y$. Assume that for a nonempty compact convex subset $D \subseteq K$ with each $x \in K$, there exists $y \in D$ such that $\langle T(x), y - x\rangle \in -\text{int}(C(x))$. Then there exists a solution $\bar{x} \in K$ of* WVVIP.

Proof In view of Theorem 5.19, it suffices to check that for each $y \in K$, the mapping $x \mapsto \langle T(x), y - x\rangle$ is continuous. To this end, let $y \in K$ be arbitrary but fixed and let

$T_y : K \to Y$ be defined by

$$T_y(x) = \langle T(x), y - x \rangle, \quad \text{for all } x \in K.$$

Let $\{x_\alpha\}$ be any net in K converging to some $x \in K$. By the assumption, we have

$$\|T(x_\alpha) - T(x)\|_{\mathcal{L}(X,Y)} \to 0.$$

Since the net $\{x_\alpha\}$ is convergent, it is bounded. Therefore,

$$|\langle T(x_\alpha) - T(x), y - x_\alpha \rangle| \leq \|T(x_\alpha) - T(x)\|_{\mathcal{L}(X,Y)} \|y - x_\alpha\|_X \to 0,$$

and hence, $\langle T(x_\alpha) - T(x), y - x_\alpha \rangle$ converges to $\mathbf{0}$ in Y.

On the other hand, as $T(x) \in \mathcal{L}(X,Y)$, we have $\langle T(x), y - x_\alpha \rangle$ converges to $\langle T(x), y - x \rangle$ in Y. Consequently, the net

$$T_y(x_\alpha) = \langle T(x_\alpha), y - x_\alpha \rangle = \langle T(x_\alpha) - T(y), y - x_\alpha \rangle + \langle T(y), y - x_\alpha \rangle$$

converges to $\langle T(x), y - x \rangle = T_y(x)$ in Y. Hence, the operator T_y is continuous. Thus, the result follows from Theorem 5.19. □

By using Theorem 5.21 and argument similar to Corollary 5.1, we have the following result.

Corollary 5.2 *Let X and Y be Banach spaces and K be a nonempty convex subset of X. Let $T : K \to \mathcal{L}(X,Y)$ be a continuous operator such that for all $x \in K$, the set $\{y \in K : \langle T(x), y - x \rangle \notin C(x)\}$ is convex. Let $C : K \to 2^Y$ be a closed set-valued map such that for all $x \in K$, $C(x)$ is a closed convex pointed cone. Assume that for a nonempty compact convex subset $D \subseteq K$ with each $x \in K$, there exists $y \in D$ such that $\langle T(x), y - x \rangle \notin C(x)$. Then there exists a solution $\bar{x} \in K$ of* SVVIP.

5.4 Applications to Vector Optimization

Since the invention of vector variational inequalities in 1980 by F. Giannessi, the theory of vector variational inequalities became a powerful tool to study vector optimization problems (VOP); See, for example, [1–3, 11, 12, 19, 20, 22, 23, 26] and the references therein. In most of the papers appeared in the literature, the VOP is studied by using Stampacchia vector variational inequality problems. In 1998, Giannessi [11] established the necessary and sufficient conditions for a point to be an efficient solution of VOP for differentiable and convex functions are that the point to be a solution of Minty vector variational inequality problem. It was the first paper in which a direct application of Minty vector variational inequality problem to vector optimization problems is given. Later, Yang et al. [26] extended the results of Giannessi [11] for differentiable but pseudoconvex functions.

In this section, we present some necessary and sufficient conditions in terms of Stampacchia vector variational inequality problems or Minty vector variational inequality problems, for an efficient, weak efficient, properly efficient (in different sense) solutions of vector optimization problems

5.4.1 *Relations Between Vector Variational Inequalities and Vector Optimization*

Let X and Y be topological vector spaces and K be a nonempty subset of X. Let $C : K \to 2^Y$ be a set-valued map such that for all $x \in K$, $C(x)$ is a closed convex pointed cone. We further assume that $\operatorname{int}(C(x)) \neq \emptyset$ wherever $\operatorname{int}(C(x))$ is involved. Let $f : K \to Y$ be a vector-valued function. Recall the vector optimization problem (VOP):

$$\text{minimize} f(x), \quad \text{subject to } x \in K. \tag{VOP}$$

Definition 5.9 Let K be a nonempty convex subset of X and $f : K \to Y$ be a Gâteaux differentiable function with Gâteaux derivative Df. Let $x \in K$ be an arbitrary element. Then f is said to be:

(a) C_x-*convex* if for all $y \in K$,

$$f(y) - f(x) - \langle Df(x), y - x \rangle \in C(x);$$

(b) *strictly* C_x-*convex* if for all $y \in K$,

$$f(y) - f(x) - \langle Df(x), y - x \rangle \in \operatorname{int}(C(x));$$

(c) *strongly* C_x-*pseudoconvex* if for all $y \in K$,

$$\langle Df(x), y - x \rangle \in C(x) \quad \text{implies} \quad f(y) - f(x) \in C(x);$$

(d) C_x-*pseudoconvex* if for all $y \in K$,

$$\langle Df(x), y - x \rangle \notin -C(x) \setminus \{\mathbf{0}\} \quad \text{implies} \quad f(y) - f(x) \notin -C(x) \setminus \{\mathbf{0}\};$$

(e) weakly C_x-*pseudoconvex* if for all $y \in K$,

$$\langle Df(x), y - x \rangle \notin -\operatorname{int}(C(x)) \quad \text{implies} \quad f(y) - f(x) \notin -\operatorname{int}(C(x)).$$

If $C(x)$ is a fixed closed convex pointed cone P with $\operatorname{int}(P) \neq \emptyset$, then C_x-convexity, strictly C_x-convexity, strongly C_x-pseudoconvexity, C_x-pseudoconvexity and weakly C_x-pseudoconvexity are called P-convexity, strictly P-convexity, strongly P-pseudoconvexity, P-pseudoconvexity and weakly P-pseudoconvexity, respectively.

The following results provide the relation between the solutions of VVIP and VOP.

Theorem 5.22 *Let K be a nonempty convex subset of X and $f : K \to Y$ be a Gâteaux differentiable vector-valued function with Gâteaux derivative Df.*

(a) *If $\bar{x} \in K$ is a dominated strongly efficient solution of* VOP, *then it is a solution of* SVVIP (5.1) *with $T(\bar{x}) = Df(\bar{x})$.*
(b) *If for all $x \in K$, f is strongly C_x-pseudoconvex and $\bar{x}$ is a solution of* SVVIP (5.1) *with $T(\bar{x}) = Df(\bar{x})$, then $\bar{x}$ is a dominated strongly efficient solution of* VOP.

Proof

(a) Let $\bar{x} \in K$ be a dominated strongly efficient solution of VOP. Then

$$f(y) - f(\bar{x}) \in C(\bar{x}), \quad \text{for all } y \in K.$$

Since K is convex, $\bar{x} + \lambda(y - \bar{x}) \in K$ for all $\lambda \in [0, 1]$, and therefore, we have

$$f(\bar{x} + \lambda(y - \bar{x})) - f(\bar{x}) \in C(\bar{x}), \quad \text{for all } \lambda \in [0, 1].$$

Since $C(x)$ is a closed cone for all $x \in K$, we have

$$\frac{f(\bar{x} + \lambda(y - \bar{x})) - f(\bar{x})}{\lambda} \in C(\bar{x}), \quad \text{for all } \lambda \in]0, 1].$$

By the definition of Gâteaux derivative, we have

$$\langle Df(\bar{x}), y - \bar{x} \rangle = \lim_{\lambda \to 0+} \frac{f(\bar{x} + \lambda(y - \bar{x})) - f(\bar{x})}{\lambda} \in C(\bar{x}).$$

Thus, $\bar{x} \in K$ is a solution of SVVIP (5.1) with $T(\bar{x}) = Df(\bar{x})$.

(b) Let $\bar{x}$ be a solution of SVVIP (5.1) with $T(\bar{x}) = Df(\bar{x})$. Then

$$\langle Df(\bar{x}), y - \bar{x} \rangle \in C(\bar{x}), \quad \text{for all } y \in K.$$

By using strong C_x-pseudoconvexity of f, we obtain the desired conclusion. □

If $\bar{x} \in K$ is a dominated efficient solution of VOP, then $\bar{x}$ may not be a solution of VVIP (5.2) with $T(\bar{x}) = Df(\bar{x})$ as shown in the following example.

Example 5.15 Let $X = \mathbb{R}$, $K = [-1, 0]$, $Y = \mathbb{R}^2$ and $C(x) = \mathbb{R}^2_+$ for all $x \in K$. Let $f : K \to \mathcal{L}(X, Y)$ be defined by

$$f(x) = (x, x^2), \quad \text{for all } x \in K.$$

Then every $\bar{x} \in K$ is an efficient solution of VOP. But $\bar{x} = 0$ is not a solution of VVIP (5.2) with $T(\bar{x}) = Df(\bar{x})$. Indeed, let $\bar{x} = 0$. Then, for $y = -1$, we have

$$\langle Df(\bar{x}), y - x \rangle = (-1, 0) \in -\mathbb{R}^2_+ \setminus \{\mathbf{0}\}.$$

Hence, $\bar{x} = 0$ is not a solution of VVIP (5.2) with $T(\bar{x}) = Df(\bar{x})$.

However, by using the C_x-pseudoconvexity, we obtain the following result.

Theorem 5.23 *Let K be a nonempty convex subset of X and $f : K \to Y$ be a Gâteaux differentiable vector-valued function with Gâteaux derivative Df. If for all $x \in K$, f is C_x-pseudoconvex and $\bar{x} \in K$ is a solution of* VVIP (5.2) *with $T(\bar{x}) = Df(\bar{x})$, then it is a dominated efficient solution of* VOP.

Theorem 5.24 *Let K be a nonempty convex subset of X and $f : K \to Y$ be a Gâteaux differentiable vector-valued function with Gâteaux derivative Df.*

(a) *If $\bar{x} \in K$ is a dominated weakly efficient solution of* VOP, *then it is a solution of* WVVIP (5.3) *with $T(\bar{x}) = Df(\bar{x})$.*
(b) *If for all $x \in K$, T is weakly C_x-pseudoconvex and $\bar{x} \in K$ is a solution of* WVVIP (5.3) *with $T(\bar{x}) = Df(\bar{x})$, then it is a dominated weakly efficient solution of* VOP.

Proof

(a) Let $\bar{x} \in K$ be a dominated weakly efficient solution of VOP. Then,

$$f(y) - f(\bar{x}) \notin -\operatorname{int}(C(\bar{x})), \quad \text{for all } y \in K.$$

Since K is convex, $\bar{x} + \lambda(y - \bar{x}) \in K$ for all $\lambda \in [0, 1]$, and therefore, we have

$$f(\bar{x} + \lambda(y - \bar{x})) - f(\bar{x}) \notin -\operatorname{int}(C(\bar{x})), \quad \text{for all } \lambda \in [0, 1],$$

that is,

$$f(\bar{x} + \lambda(y - \bar{x})) - f(\bar{x}) \in W(\bar{x}) = Y \setminus \{-\operatorname{int}(C(\bar{x}))\}, \quad \text{for all } \lambda \in [0, 1].$$

Since $W(x)$ is a closed cone for all $x \in K$, we have

$$\frac{f(\bar{x} + \lambda(y - \bar{x})) - f(\bar{x})}{\lambda} \in W(\bar{x}), \quad \text{for all } \lambda \in]0, 1].$$

Then

$$\langle Df(\bar{x}), y - \bar{x} \rangle = \lim_{\lambda \to 0+} \frac{f(\bar{x} + \lambda(y - \bar{x})) - f(\bar{x})}{\lambda} \in W(\bar{x}),$$

and hence,

$$\langle Df(\bar{x}), y - \bar{x}\rangle \notin -\operatorname{int}(C(\bar{x})), \quad \text{for all } y \in K.$$

Therefore, $\bar{x} \in K$ is a solution of WVVIP (5.3) with $T(\bar{x}) = Df(\bar{x})$.

(b) It directly follows from the definition of weakly C_x-pseudoconvexity of f. □

Theorem 5.25 *Let K be a nonempty convex subset of X and $f : K \to Y$ be Gâteaux differentiable with Gâteaux derivative Df such that $(-f)$ is C_x-strictly convex for all $x \in K$, that is,*

$$f(y) - f(x) - \langle Df(x), y - x\rangle \in -\operatorname{int}(C(x)), \quad \textit{for all } y \in K. \tag{5.25}$$

Then every dominated weak efficient solution of VOP is a solution of VVIP (5.2) *with $T(\bar{x}) = Df(\bar{x})$.*

Proof Assume that $\bar{x}$ is a dominated weakly efficient solution of VOP but not a solution of VVIP (5.2) with $T(\bar{x}) = Df(\bar{x})$. Then there exists $y \in K$ such that

$$\langle Df(\bar{x}), y - \bar{x}\rangle \in -C(\bar{x}) \setminus \{\mathbf{0}\}. \tag{5.26}$$

Combining (5.25) and (5.26), we obtain

$$f(y) - f(\bar{x}) \in -\operatorname{int}(C(\bar{x})),$$

a contradiction to our assumption that $\bar{x}$ is a dominated weak efficient solution of VOP. Hence, $\bar{x}$ is a solution of VVIP with $T(\bar{x}) = Df(\bar{x})$. □

Corollary 5.3 *Let K be a nonempty convex subset of X and $f : K \to Y$ be Gâteaux differentiable with Gâteaux derivative Df such that $(-f)$ is C_x-strictly convex for all $x \in K$. Then every dominated efficient solution of* VOP *is a solution of* VVIP (5.2) *with $T(\bar{x}) = Df(\bar{x})$.*

Theorem 5.26 *Let K be a nonempty convex subset of X. For all $x \in K$, let $f : K \to \mathbb{R}^\ell$ be strictly C_x-convex function. Then every dominated weakly efficient solution of* VOP *is a dominated efficient solution of* VOP.

Proof Assume that $\bar{x}$ is a dominated weakly efficient solution of VOP, but not dominated efficient solution of VOP. Then there exists $y \in K$ such that

$$f(y) - f(\bar{x}) \in -C(\bar{x}) \setminus \{\mathbf{0}\}. \tag{5.27}$$

Since f is strictly C_x-convex, we have

$$\langle Df(\bar{x}), y - \bar{x}\rangle - f(y) + f(\bar{x}) \in -\operatorname{int}(C(\bar{x})). \tag{5.28}$$

By combining (5.27) and (5.28), we obtain

$$\langle Df(\bar{x}), y - \bar{x} \rangle \in - \operatorname{int}(C(\bar{x})).$$

Thus, $\bar{x}$ is not a solution of WVVIP with $T(\bar{x}) = Df(\bar{x})$. By using Theorem 5.24 (a), we see that $\bar{x}$ is not a dominated weak efficient solution of VOP, a contradiction to our assumption. □

Finally, we present the relationship between a solution of VVIP (5.2) and a dominated properly efficient solution (in the sense of Henig) of VOP.

Theorem 5.27 *Let K be a nonempty convex subset of X and $f : K \to Y$ be Gâteaux differentiable with Gâteaux derivative Df. If $\bar{x} \in K$ is a dominated properly efficient solution (in the sense of Henig) of* VOP, *then it is a solution of* VVIP (5.2) *with $T(\bar{x}) = Df(\bar{x})$.*

Proof Since $\bar{x} \in K$ is a dominated properly efficient solution of VOP, there is a set-valued map $D : K \to 2^Y$ such that for all $x \in K$, $D(x)$ is a convex cone and $C(x) \setminus \{\mathbf{0}\} \subseteq \operatorname{int}(D(x))$, and

$$f(y) - f(\bar{x}) \notin -D(\bar{x}) \setminus \{\mathbf{0}\}, \quad \text{for all } y \in K.$$

Since $-\operatorname{int}(D(x)) \subseteq -D(x) \setminus \{\mathbf{0}\}$, we have

$$f(y) - f(\bar{x}) \notin - \operatorname{int}(D(\bar{x})), \quad \text{for all } y \in K.$$

As in the proof of Theorem 5.24 (a), it follows that

$$\langle Df(\bar{x}), y - \bar{x} \rangle \notin - \operatorname{int}(D(\bar{x})), \quad \text{for all } y \in K.$$

Since $C(x) \setminus \{\mathbf{0}\} \subseteq \operatorname{int}(D(x))$ for all $x \in K$, we obtain the desired result. □

5.4.2 *Relations Between Vector Variational Inequalities and Vector Optimization in Finite Dimensional Spaces*

Throughout this section, unless otherwise specified, we assume that K is a nonempty convex subset of $\mathbb{R}^n$.

Let us write down the formulation of (Stampacchia) vector variational inequality problems and Minty vector variational inequality problem in finite dimensional setting.

For each $i = 1, 2, \ldots, \ell$, let $T_i : K \to \mathbb{R}^n$ be a vector-valued function such that $T = (T_1, T_2, \ldots, T_\ell) : K \to \mathbb{R}^{\ell \times n}$ is a matrix-valued function. For abbreviation, we put

$$\langle T(x), v\rangle_\ell := (\langle T_1(x), v\rangle, \ldots, \langle T_\ell(x), v\rangle), \quad \text{for all } x \in K \text{ and all } v \in \mathbb{R}^n.$$

The (Stampacchia) vector variational inequality problems and Minty vector variational inequality problems can be written in the following forms:

- (VVIP): Find $\bar{x} \in K$ such that

$$\langle T(\bar{x}), y - \bar{x}\rangle_\ell \notin -\mathbb{R}^\ell_+ \setminus \{\mathbf{0}\}, \quad \text{for all } y \in K. \tag{5.29}$$

- (WVVIP): Find $\bar{x} \in K$ such that

$$\langle T(\bar{x}), y - \bar{x}\rangle_\ell \notin -\operatorname{int}\left(\mathbb{R}^\ell_+\right), \quad \text{for all } y \in K, \tag{5.30}$$

- (MVVIP): Find $\bar{x} \in K$ such that

$$\langle T(y), y - \bar{x}\rangle_\ell \notin -\mathbb{R}^\ell_+ \setminus \{\mathbf{0}\}, \quad \text{for all } y \in K. \tag{5.31}$$

- (MWVVIP): Find $\bar{x} \in K$ such that

$$\langle T(y), y - \bar{x}\rangle_\ell \notin -\operatorname{int}\left(\mathbb{R}^\ell_+\right), \quad \text{for all } y \in K. \tag{5.32}$$

If $T(x) = \nabla f(x) = (\nabla f_1(x), \ldots, \nabla f_\ell(x))$, then the inclusion relations (5.29) and (5.30), (5.31) and (5.32) reduce to the following inclusion relations, respectively:

$$\left(\langle \nabla f_1(\bar{x}), y - \bar{x}\rangle, \ldots, \langle \nabla f_\ell(y), y - \bar{x}\rangle\right) \notin -\mathbb{R}^\ell_+ \setminus \{\mathbf{0}\}, \quad \text{for all } y \in K,$$

$$\left(\langle \nabla f_1(\bar{x}), y - \bar{x}\rangle, \ldots, \langle \nabla f_\ell(y), y - \bar{x}\rangle\right) \notin -\operatorname{int}\left(\mathbb{R}^\ell_+\right), \quad \text{for all } y \in K,$$

$$\left(\langle \nabla f_1(y), y - \bar{x}\rangle, \ldots, \langle \nabla f_\ell(y), y - \bar{x}\rangle\right) \notin -\mathbb{R}^\ell_+ \setminus \{\mathbf{0}\}, \quad \text{for all } y \in K,$$

and

$$\left(\langle \nabla f_1(y), y - \bar{x}\rangle, \ldots, \langle \nabla f_\ell(y), y - \bar{x}\rangle\right) \notin -\operatorname{int}\left(\mathbb{R}^\ell_+\right), \quad \text{for all } y \in K.$$

If each component $f_i : K \to \mathbb{R}$ of a vector-valued function $f = (f_1, f_2, \ldots, f_\ell) : K \to \mathbb{R}^\ell$ is convex, then f is $\mathbb{R}^\ell_+$-convex. Therefore, all the results in the previous subsection hold if we consider each f_i is convex.

As we have seen in Example 5.15, an efficient solution of VOP may not be a solution of VVIP (5.2) with $T(\bar{x}) = Df(\bar{x})$. However, Giannessi [11] showed that every efficient solution of VOP is a solution of MVVIP (5.8) with $T(\bar{x}) = Df(\bar{x})$ and vice-versa if the underlying objective function is convex. It is further generalized by Yang et al. [26] for pseudoconvex functions and obtained the following result.

Theorem 5.28 (Giannessi) *For each $i \in \mathscr{I} = \{1, 2, \ldots, \ell\}$, let $f_i : K \to \mathbb{R}$ be pseudoconvex and differentiable on an open set containing K such that for all $x \in K$, $\nabla f(x) = (\nabla f_1(x), \ldots, \nabla f_\ell(x)) = T(x)$. Then $\bar{x} \in K$ is an efficient solution of* VOP *if and only if it is a solution of* MVVIP (5.31).

Proof Let $\bar{x} \in K$ be a solution of MVVIP but not an efficient solution of VOP. Then there exists $z \in K$ such that

$$f(z) - f(\bar{x}) = (f_1(z) - f_1(\bar{x}), \ldots, f_\ell(z) - f_\ell(\bar{x})) \in -\mathbb{R}^\ell_+ \setminus \{\mathbf{0}\}. \tag{5.33}$$

Since K is convex, $z(\lambda) := \lambda z + (1 - \lambda)\bar{x} \in K$ for all $\lambda \in [0, 1]$. Also since each f_i is pseudoconvex, it follows from Theorem 1.21 that each f_i is both semistrictly quasiconvex and quasiconvex. Thus, by (5.33), we have

$$f(z(\lambda)) - f(\bar{x}) = (f_1(z(\lambda)) - f_1(\bar{x}), \ldots, f_\ell(z(\lambda)) - f_\ell(\bar{x})) \\ \in -\mathbb{R}^\ell_+ \setminus \{\mathbf{0}\}, \quad \text{for all } \lambda \in]0, 1[, \tag{5.34}$$

that is,

$$f(z(\lambda)) - f(z(0)) = (f_1(z(\lambda)) - f_1(z(0)), \ldots, f_\ell(z(\lambda)) - f_\ell(z(0))) \in -\mathbb{R}^\ell_+ \setminus \{\mathbf{0}\}, \tag{5.35}$$

for all $\lambda \in]0, 1[$, and so,

$$f_i(z(\lambda)) - f_i(z(0)) \leq 0, \quad \text{for all } \lambda \in]0, 1[, \tag{5.36}$$

with strict inequality holds for some k such that $1 \leq k \leq \ell$.

By Mean Value Theorem, there exist $\lambda_i \in]0, 1[$ such that

$$\frac{d}{d\lambda} f_i(z(\lambda_i)) = \frac{f_i(z(\lambda)) - f_i(z(0))}{\lambda}.$$

Thus,

$$\langle \nabla f_i(z(\lambda_i)), z - \bar{x} \rangle = \frac{f_i(z(\lambda)) - f_i(z(0))}{\lambda} \leq 0, \tag{5.37}$$

for all $\lambda \in]0, 1[$ and for each $i = 1, 2, \ldots, \ell$ with strict inequality holds for some $1 \leq k \leq \ell$. Thus,

$$\big(\langle \nabla f_1(z(\lambda_1)), z - \bar{x} \rangle, \ldots, \langle \nabla f_\ell(z(\lambda_\ell)), z - \bar{x} \rangle\big) \in -\mathbb{R}^\ell_+ \setminus \{\mathbf{0}\},$$

that is,

$$\langle \nabla f_i(z(\lambda_i)), z - \bar{x} \rangle \leq 0, \quad \text{for each } i = 1, 2, \ldots, \ell, \tag{5.38}$$

and one of which becomes a strict inequality.

Suppose that $\lambda_1, \lambda_2, \ldots, \lambda_\ell$ are all equal. Then by (5.38) and the fact that $z(\lambda_i) - \bar{x} = \lambda_i(z - \bar{x})$, we have

$$\langle \nabla f_i(z(\lambda_i)), z(\lambda_i) - \bar{x} \rangle = \lambda_i \langle \nabla f_i(z(\lambda_i)), z - \bar{x} \rangle \leq 0,$$

for all $\lambda_i \in]0, 1[$ and for each $i \in \mathscr{I}$ with one of which becomes a strict inequality. Therefore,

$$\big(\langle \nabla f_1(z(\lambda_1)), z(\lambda_1) - \bar{x} \rangle, \ldots, \langle \nabla f_\ell(z(\lambda_\ell)), z(\lambda_\ell) - \bar{x} \rangle\big) \in -\mathbb{R}^\ell_+ \setminus \{\mathbf{0}\}$$

which contradicts to our supposition that $\bar{x}$ is a solution of MVVIP (5.31).

Consider the case when $\lambda_1, \lambda_2, \ldots, \lambda_\ell$ are not equal. Let $\lambda_1 \neq \lambda_2$.

If $\lambda_1 < \lambda_2$, then by inequality (5.38) and using the fact that $z(\lambda_2) - z(\lambda_1) = (\lambda_2 - \lambda_1)(z - \bar{x})$, we have

$$\langle \nabla f_2(z(\lambda_2)), z(\lambda_2) - z(\lambda_1) \rangle = (\lambda_2 - \lambda_1)\langle \nabla f_2(z(\lambda_2)), z - \bar{x} \rangle \leq 0, \tag{5.39}$$

with strict inequality for $k = 2$. Since each f_i is pseudoconvex, by Theorem 1.27, ∇f_i is pseudomonotone, and thus, we have

$$\langle \nabla f_2(z(\lambda_1)), z(\lambda_2) - z(\lambda_1) \rangle \leq 0$$

with strict inequality for $k = 2$. Thus,

$$\langle \nabla f_2(z(\lambda_1)), z - \bar{x} \rangle \leq 0,$$

with strict inequality for $k = 2$.

If $\lambda_1 > \lambda_2$, then again by using the same argument as above, we have

$$\langle \nabla f_1(z(\lambda_1)), z(\lambda_1) - z(\lambda_2) \rangle = (\lambda_1 - \lambda_2)\langle \nabla f_1(z(\lambda_1)), z - \bar{x} \rangle \leq 0,$$

with strict inequality for $k = 1$. Since each ∇f_i is pseudomonotone, we have

$$\langle \nabla f_1(z(\lambda_2)), z(\lambda_1) - z(\lambda_2) \rangle \leq 0,$$

and therefore,

$$\langle \nabla f_1(z(\lambda_2)), z - \bar{x}) \rangle \leq 0,$$

with strict inequality for $k = 1$.

For the case $\lambda_1 \neq \lambda_2$, let $\bar{\lambda} = \min\{\lambda_1, \lambda_2\}$. Then, we have

$$\langle \nabla f_i(z(\bar{\lambda})), z - \bar{x} \rangle \leq 0, \quad \text{for } i = 1, 2.$$

By continuing this process, we can find $\lambda^* \in]0,1[$ such that

$$\langle \nabla f_i(z(\lambda^*)), z - \bar{x} \rangle \leq 0, \quad \text{for } i = 1, 2, \ldots, \ell,$$

with strict inequality for some k such that $1 \leq k \leq \ell$. Multiplying above inequality by $-\lambda$ and using the fact that $\bar{x} - z(\lambda^*) = -\lambda(z - \bar{x})$, we obtain

$$\langle \nabla f_i(z(\lambda^*)), \bar{x} - z(\lambda^*) \rangle = -\lambda \langle \nabla f_i(z(\lambda^*)), z - \bar{x} \rangle \geq 0, \quad \text{for } i = 1, 2, \ldots, \ell,$$

with strict inequality for some k such that $1 \leq k \leq \ell$, that is,

$$\big(\langle \nabla f_1(z(\lambda^*)), z(\lambda^*) - \bar{x} \rangle, \ldots, \langle \nabla f_\ell(z(\lambda^*)), z(\lambda^*) - \bar{x} \rangle \big) \in -\mathbb{R}^\ell_+ \setminus \{\mathbf{0}\}$$

which contradicts to our supposition that $\bar{x}$ is a solution of MVVIP (5.31).

Conversely, let $\bar{x} \in K$ be a solution of VOP. Assume contrary that there exists $\hat{y} \in K$ such that

$$\langle F(\hat{y}), \hat{y} - \bar{x} \rangle_\ell = \big(\langle \nabla f_1(\hat{y}), \hat{y} - \bar{x} \rangle, \ldots, \langle \nabla f_\ell(\hat{y}), \hat{y} - \bar{x} \rangle \big) \in -\mathbb{R}^\ell_+ \setminus \{\mathbf{0}\}, \tag{5.40}$$

that is,

$$\langle \nabla f_i(\hat{y}), \bar{x} - \hat{y} \rangle \geq 0, \quad \text{for each } i = 1, 2, \ldots, \ell,$$

with one of which inequality becomes strict. Since each f_i is pseudoconvex, we deduce that for each $i \in \mathscr{I}$,

$$f_i(\bar{x}) \geq f_i(\hat{y}). \tag{5.41}$$

Let $j \in \mathscr{I}$ be such that $\langle \nabla f_i(\hat{y}), \bar{x} - \hat{y} \rangle > 0$. Since each f_i is pseudoconvex, therefore by Theorem 1.21, it is quasiconvex as well as semistrictly quasiconvex. Thus, by Theorem 1.18, we have

$$f_j(\bar{x}) > f_j(\hat{y}). \tag{5.42}$$

Combining (5.41) and (5.42), we have

$$(f_1(\hat{y}) - f_1(\bar{x}), \ldots, f_\ell(\hat{y}) - f_\ell(\bar{x})) \in -\mathbb{R}^\ell_+ \setminus \{\mathbf{0}\}$$

which is a contradiction to our assumption. □

References

1. Q.H. Ansari, A.H. Siddiqi, A generalized vector variational-like inequality and optimization over an efficient set, in *Functional Analysis with Current Applications in Science, Technology and Industry*, ed. by M. Brokate, A.H. Siddiqi. Pitman Research Notes in Mathematics, vol. **377** (Addison Wesley Longman Limited, Essex, 1998), pp. 177–191
2. Q.H. Ansari, J.C. Yao, Systems of generalized variational inequalities and their applications. Appl. Anal. **76**, 203–217 (2000)
3. Q.H. Ansari, J.C. Yao, On nondifferentiable and nonconvex vector optimization problems. J. Optim. Theory Appl. **106**, 487–500 (2000)
4. Q.H. Ansari, A.P. Farajzadeh, S. Schaible, Existence of solutions of vector variational inequalities and vector complementarity problems. J. Glob. Optim. **45**, 297–307 (2009)
5. F.E. Browder, Nonlinear monotone operators and convex sets in Banach spaces. Bull. Am. Math. Soc. **71**, 780–785 (1965)
6. C. Charitha, J. Dutta, C.S. Lalitha, Gap functions for vector variational inequalities. Optimization **64**(7), 1499–1520 (2015)
7. G.Y. Chen, Existence of solution for a vector variational inequality: an extension of Hartmann-Stampacchia theorem. J. Optim. Theory Appl. **74**, 445–456 (1992)
8. G.Y. Chen, X.Q. Yang, Vector complementarity problem and its equivalence with weak minimal element in ordered spaces. J. Math. Anal. Appl. **153**, 136–158 (1990)
9. Y.P. Fang, N-J. Huang, Strong vector variational inequalities in Banach spaces. Appl. Math. Lett. **19**, 362–368 (2006)
10. F. Giannessi, Theorems of alternative, quadratic programs and complementarity problems, in *Variational Inequalities and Complementarity Problems*, ed. by R.W. Cottle, F. Giannessi, J.L. Lions (John Wiley and Sons, New York, 1980), pp. 151–186
11. F. Giannessi, On Minty variational principle, in *New Trends in Mathematical Programming*, ed. by F. Giannessi, S. Komlòski, T. Rapcsáck (Kluwer Academic Publisher, Dordrech, Holland, 1998), pp. 93–99
12. F. Giannessi (ed.), *Vector Variational Inequality and Vector Equilibria: Mathematical Theories* (Kluwer Academic Publishers, Dordrechet, Boston, 2000)
13. N. Hadjisavvas, Continuity amd maximality properties of pseudomonotone operators. J. Convex Anal. **10**(2), 459–469 (2003)
14. P. Hartmann, G. Stampacchia, On some nonlinear elliptic differential functional equations. Acta Math. **115**, 271–310 (1966)
15. N.-J. Huang, Y.-P. Fang, On vector variational inequalities in reflexive Banach spaces. J. Global Optim. **32**, 495–505 (2005)
16. N.-J. Huang, Y.-P. Fang, Strong vector F-complementarity problem and least element problem of feasible set. Nonlinear Anal. **61**, 901–918 (2005)
17. S. Karamardian, Generalized complementarity problem. J. Optim. Theory Appl. **8**, 161–168 (1971)
18. T.C. Lai, J.C. Yao, Existence results for VVIP. Appl. Math. Lett. **9**(3), 17–19 (1996)
19. G.M. Lee, On relations between vector variational inequality and vector optimization problem, in *Progress in Optimization, II: Contributions from Australasia*, ed. by X.Q. Yang, A.I. Mees, M.E. Fisher, L.S. Jennings (Kluwer Academic Publisher, Dordrecht, Holland, 2000), pp. 167–179
20. G.M. Lee, D.S. Kim, Existence of solutions for vector optimization problems. J. Math. Anal. Appl. **220**, 90–98 (1998)
21. G.M. Lee, D.S. Kim, B.S. Lee, S.J. Cho, Generalized vector variational inequality and fuzzy extension. Appl. Math. Lett. **6**, 47–51 (1993)
22. G.M. Lee, D.S. Kim, B.S. Lee, N.D. Yen, Vector variational inequality as a tool for studing vector optimization. Nonlinear Anal. **34**, 745–765 (1998)
23. G. Ruiz-Garzón, R. Osuna-Gómez, A. Rufián-Lizana, Relationships between vector variational-like inequality and optimization problems. Eur. J. Oper. Res. **157**, 113–119 (2004)

24. S. Schaible, J.C. Yao, On the equivalence of nonlinear complementarity problems. Math. Program. **70**, 191–200 (1995)
25. X.Q. Yang, Vector variational inequality and its duality. Nonlinear Anal. Theory Methods Appl. **21**, 869–877 (1993)
26. X.M. Yang, X.Q. Yang, K.L. Teo, Some remarks on the Minty vector variational inequalities. J. Optim. Theory Appl. **121**(1), 193–201 (2004)
27. J.C. Yao, Variational inequalities with generalized monotone operators. Math. Oper. Res. **19**, 691–705 (1994)
28. N.D. Yen, G.M. Lee, On monotone and strongly monotone vector variational inequalities, in *Vector Variational Inequality and Vector Equilibria: Mathematical Theories*, ed. by F. Giannessi (Kluwer Academic Publishers, Dordrechet, Boston, 2000), pp. 467–478
29. H. Yin, C.X. Xu, Z.X. Zhang, The F-complemenatrity problems and its equivalence with least element problem. Acta Math. Sin. **44**, 679–686 (2001)
30. S.J. Yu, J.C. Yao, On vector variational inequalities. J. Optim. Theory Appl. **89**(3), 749–769 (1996)

Chapter 6
Linear Scalarization of Vector Variational Inequalities

This chapter deals with linear scalarization techniques for vector variational inequality problems and Minty vector variational inequality problems. Such concepts are important for deriving numerical algorithms for solving vector variational inequalities.

For each given $\ell \in \mathbb{N}$, we denote by $\mathbb{R}^{\ell}_{+}$ the non-negative orthant of $\mathbb{R}^{\ell}$, that is,

$$\mathbb{R}^{\ell}_{+} = \left\{x = (x_1, x_2, \ldots, x_\ell) \in \mathbb{R}^{\ell} : x_i \geq 0, \quad \text{for } i = 1, 2, \ldots, \ell\right\},$$

so that $\mathbb{R}^{\ell}_{+}$ has a nonempty interior with the topology induced in terms of convergence of vectors with respect to the Euclidean metric. That is,

$$\operatorname{int}(\mathbb{R}^{\ell}_{+}) = \left\{x = (x_1, x_2, \ldots, x_\ell) \in \mathbb{R}^{\ell} : x_i > 0, \quad \text{for } i = 1, 2, \ldots, \ell\right\}.$$

We denote by $\mathbb{T}^{\ell}_{+}$ and $\operatorname{int}(\mathbb{T}^{\ell}_{+})$ the simplex of $\mathbb{R}^{\ell}_{+}$ and its relative interior, respectively, that is,

$$\mathbb{T}^{\ell}_{+} = \left\{x = (x_1, x_2, \ldots, x_\ell) \in \mathbb{R}^{\ell}_{+} : \|x\| = \sum_{i=1}^{\ell} x_i = 1\right\},$$

and

$$\operatorname{int}(\mathbb{T}^{\ell}_{+}) = \left\{x = (x_1, x_2, \ldots, x_\ell) \in \operatorname{int}(\mathbb{R}^{\ell}_{+}) : \|x\| = \sum_{i=1}^{\ell} x_i = 1\right\}.$$

e denotes the unit vector in $\mathbb{R}^{\ell}$, that is, $e = (1, 1, \ldots, 1)$.

Q.H. Ansari et al., *Vector Variational Inequalities and Vector Optimization*, Vector Optimization, DOI 10.1007/978-3-319-63049-6_6

Let K be a nonempty convex subset of $\mathbb{R}^n$. For each $i = 1, 2, \ldots, \ell$, let $T_i : K \to \mathbb{R}^n$ be a vector-valued function such that $T = (T_1, T_2, \ldots, T_\ell) : K \to \mathbb{R}^{\ell \times n}$ is a matrix-valued function. For abbreviation, we put

$$\langle T(x), v \rangle_\ell := (\langle T_1(x), v \rangle, \ldots, \langle T_\ell(x), v \rangle), \quad \text{for all } x \in K \text{ and all } v \in \mathbb{R}^n.$$

Let us define the vector variational inequality problem and weak vector variational inequality problem in these settings.

- *Vector Variational Inequality Problem* (in short, FVVIP): Find $\bar{x} \in K$ such that for all $y \in K$

$$\langle T(\bar{x}), y - \bar{x} \rangle_\ell := (\langle T_1(\bar{x}), y - \bar{x} \rangle, \ldots, \langle T_\ell(\bar{x}), y - \bar{x} \rangle) \notin -\mathbb{R}^\ell_+ \setminus \{\mathbf{0}\}. \tag{6.1}$$

- *Weak Vector Variational Inequality Problem* (in short, FWVVIP): Find $\bar{x} \in K$ such that for all $y \in K$

$$\langle T(\bar{x}), y - \bar{x} \rangle_\ell := (\langle T_1(\bar{x}), y - \bar{x} \rangle, \ldots, \langle T_\ell(\bar{x}), y - \bar{x} \rangle) \notin -\text{int}(\mathbb{R}^\ell_+). \tag{6.2}$$

We denote the solution set of FVVIP and FWVVIP by Sol(FVVIP) and Sol(FWVVIP), respectively.

Let $W = (W_1, W_2, \ldots, W_\ell) \in \mathbb{R}^\ell_+ \setminus \{\mathbf{0}\}$ be arbitrary. The *weighted variational inequality problem* (in short, WVIP) consists of finding $\bar{x} \in K$ w. r. t. the weight vector $W = (W_1, W_2, \ldots, W_\ell) \in \mathbb{R}^\ell_+ \setminus \{\mathbf{0}\}$ such that

$$W \cdot \langle T(\bar{x}), y - \bar{x} \rangle_\ell := \sum_{i=1}^{\ell} W_i \langle T_i(\bar{x}), y - \bar{x} \rangle \geq 0, \quad \text{for all } y \in K. \tag{6.3}$$

The solution set of WVIP is denoted by Sol(WVIP).

If $W \in \mathbb{T}^\ell_+$, then the solution of WVIP is called *normalized*. The set of normalized solutions of WVIP is denoted by Sol(WVIP)$_n$.

The following lemmas show the relationship among Sol(FVVIP), Sol(FWVVIP) and Sol(WVIP).

Lemma 6.1 *For any given weight vector $W = (W_1, W_2, \ldots, W_\ell) \in \text{int}(\mathbb{R}^\ell_+)$ (respectively, $W = (W_1, W_2, \ldots, W_\ell) \in \mathbb{R}^\ell_+ \setminus \{\mathbf{0}\}$), Sol(WVIP) $\subseteq$ Sol(FVVIP) (respectively, Sol(WVIP) $\subseteq$ Sol(FWVVIP)).*

Proof Let $\bar{x} \in$ Sol(WVIP) w. r. t. the weight vector $W \in \text{int}(\mathbb{R}^\ell_+)$ (respectively, $W \in \mathbb{R}^\ell_+ \setminus \{\mathbf{0}\}$) but $\bar{x} \notin$ Sol(FVVIP) (respectively, $\bar{x} \notin$ Sol(FWVVIP)). Then, there would exist some $y \in K$ such that

$$\langle T(\bar{x}), y - \bar{x} \rangle_\ell = \angle T_1(\bar{x}, y - \bar{x} \rangle, \ldots, \langle T_\ell(\bar{x}, y - \bar{x} \rangle \in -\mathbb{R}^\ell_+ \setminus \{\mathbf{0}\}, \quad \text{for all } y \in K$$

$$\Big(\langle T(\bar{x}), y - \bar{x} \rangle_\ell = \angle T_1(\bar{x}, y - \bar{x} \rangle, \ldots, \langle T_\ell(\bar{x}, y - \bar{x} \rangle \in -\text{int}(\mathbb{R}^\ell_+), \quad \text{for all } y \in K \Big).$$

Since $W \in \text{int}(\mathbb{R}^{\ell}_{+})$ (respectively, $W \in \mathbb{R}^{\ell}_{+} \setminus \{\mathbf{0}\}$), we have

$$W \cdot \langle T(\bar{x}), \bar{x} - y \rangle_{\ell} = \sum_{i=1}^{\ell} W_i \langle T_i(\bar{x}), y - \bar{x} \rangle > 0,$$

that is,

$$W \cdot \langle T(\bar{x}), y - \bar{x} \rangle_{\ell} < 0,$$

which contradicts our assumption that $\bar{x} \in K$ is a solution of WVIP. Hence, $\bar{x} \in K$ is a solution of FVVIP (respectively, FWVVIP). □

Lemma 6.2 *If $\bar{x}$ is a solution of FWVVIP, then there exists a weight vector $W = (W_1, W_2, \ldots, W_{\ell}) \in \mathbb{R}^{\ell}_{+} \setminus \{\mathbf{0}\}$ such that $\bar{x} \in$ Sol(WVIP) w. r. t. W.*

Proof Let $\bar{x} \in$ Sol(FWVVIP). Then,

$$\{\langle T(\bar{x}), y - \bar{x} \rangle_{\ell} : y \in K\} \cap \{-\text{int}(\mathbb{R}^{\ell}_{+})\} = \emptyset.$$

So, by a separation theorem, there exists $W \in \mathbb{R}^{\ell}_{+} \setminus \{\mathbf{0}\}$ such that

$$\inf_{y \in K} W \cdot \langle T(\bar{x}), y - \bar{x} \rangle_{\ell} \geq \sup_{v \in -\text{int}(\mathbb{R}^{\ell}_{+})} W \cdot v.$$

This implies that $W \in \mathbb{R}^{\ell}_{+} \setminus \{\mathbf{0}\}$. Then, the right-hand side of the above inequality is 0, and therefore, $W \cdot \langle T(\bar{x}), y - \bar{x} \rangle_{\ell} \geq 0$ for all $y \in K$. Hence, $\bar{x} \in K$ is a solution of WVIP. □

By combining Lemma 6.1 and L:5.6.2, we have the following relations in terms of Cheng [2] and Lee et al. [4].

Remark 6.1

(a) $$\bigcup_{W \in \text{int}(\mathbb{R}^{\ell}_{+})} \text{Sol(WVIP)} \subseteq \text{Sol(VVIP)} \subseteq \text{Sol(WVVIP)} = \bigcup_{W \in \mathbb{R}^{\ell}_{+} \setminus \{\mathbf{0}\}} \text{Sol(WVIP)}.$$

(b) Since the solution set Sol(WVIP) of the WVIP w. r. t. the weight vector $W \in \mathbb{R}^{\ell}_{+} \setminus \{\mathbf{0}\}$ is equal to the solution set of the WVIP w. r. t. the weight vector αW, for any $\alpha > 0$, the above inclusion can be rewritten as

$$\bigcup_{W \in \text{int}(\mathbb{T}^{\ell_i}_{+})} \text{Sol(WVIP)} \subseteq \text{Sol(VVIP)} \subseteq \text{Sol(WVVIP)} = \bigcup_{W \in \mathbb{T}^{\ell_i}_{+}} \text{Sol(WVIP)}.$$

(c) Cheng [2] and Lee et al. [4] studied the nonemptyness, compactness, convexity and connectedness of the solution set Sol(WVIP).

As we have seen in Chap. 5, the Minty vector variational inequalities are useful to establish the existence of a solution for (Stampacchia) vector variational inequalities and also have their own importance while dealing with vector optimization problems. Therefore, we consider the following Minty weighted variational inequality problem.

Let $W = (W_1, W_2, \ldots, W_\ell) \in \mathbb{R}^\ell_+ \setminus \{\mathbf{0}\}$ be arbitrary. The *Minty weighted variational inequality problem* (in short, MWVIP) consists in finding $\bar{x} \in K$ w. r. t. the weight vector $W = (W_1, W_2, \ldots, W_\ell) \in \mathbb{R}^\ell_+ \setminus \{\mathbf{0}\}$ such that

$$W \cdot \langle T(y), y - \bar{x} \rangle_\ell := \sum_{i=1}^{\ell} W_i \langle T_i(y), y - \bar{x} \rangle \geq 0, \quad \text{for all } y \in K. \tag{6.4}$$

The solution set of MWVIP is denoted by Sol(MWVIP).

It can be easily seen that the solution set Sol(MWVIP) is convex for every $W \in \mathbb{R}^\ell_+ \setminus \{\mathbf{0}\}$.

The following lemmas show that the relationship among Sol(FVVIP), Sol(FWVVIP) and Sol(WVIP).

Lemma 6.3 *For any given weight vector* $W = (W_1, W_2, \ldots, W_\ell) \in \operatorname{int}(\mathbb{R}^\ell_+)$ *(respectively,* $W = (W_1, W_2, \ldots, W_\ell) \in \mathbb{R}^\ell_+ \setminus \{\mathbf{0}\}$*), Sol(MWVIP)* $\subseteq$ *Sol(FMVVIP) (respectively, Sol(MWVIP)* $\subseteq$ *Sol(FMWVVIP)).*

Proof Let $\bar{x} \in$ Sol(MWVIP) w. r. t. the weight vector $W \in \operatorname{int}(\mathbb{R}^\ell_+)$ (respectively, $W \in \mathbb{R}^\ell_+ \setminus \{\mathbf{0}\}$) but $\bar{x} \notin$ Sol(FVVIP) (respectively, $\bar{x} \notin$ Sol(FWVVIP)). Then, there would exist some $y \in K$ such that

$$\langle T(\bar{x}), y - \bar{x} \rangle_\ell \in -\mathbb{R}^\ell_+ \setminus \{\mathbf{0}\}, \quad \text{for all } y \in K$$

$$\Big(\langle T(\bar{x}), y - \bar{x} \rangle_\ell \in -\operatorname{int}(\mathbb{R}^\ell_+), \quad \text{for all } y \in K \Big).$$

Since $W \in \operatorname{int}(\mathbb{R}^\ell_+)$ (respectively, $W \in \mathbb{R}^\ell_+ \setminus \{\mathbf{0}\}$), we have

$$W \cdot \langle T(\bar{x}), \bar{x} - y \rangle_\ell = \sum_{i=1}^{\ell} W_i \langle T_i(\bar{x}), y - \bar{x} \rangle > 0,$$

that is,

$$W \cdot \langle T(\bar{x}), y - \bar{x} \rangle_\ell < 0,$$

which contradicts our assumption that $\bar{x} \in K$ is a solution of WVIP. Hence, $\bar{x} \in K$ is a solution of FVVIP (respectively, FWVVIP). □

In general, we have

$$\bigcup_{W \in \operatorname{int}(\mathbb{R}^\ell_+)} \operatorname{Sol(WMVIP)} \subseteq \operatorname{Sol(MVVIP)} \subseteq \operatorname{Sol(MWVVIP)}$$

$$= \bigcup_{W \in \mathbb{R}^\ell_+ \setminus \{\mathbf{0}\}} \operatorname{Sol(MWVIP)}.$$

Definition 6.1 Let $W = (W_1, W_2, \dots, W_n) \in \mathbb{R}^\ell_+ \setminus \{\mathbf{0}\}$ be a weight vector. A matrix-valued function $T = (T_1, T_2, \dots T_\ell) : K \to \mathbb{R}^{\ell \times n}$ is said to be

(a) *weighted monotone w. r. t. the weight vector W* if for all $x, y \in K$, we have

$$W \cdot \langle T(x) - T(y), x - y \rangle_\ell \geq 0,$$

and *weighted strictly monotone w. r. t. the weight vector W* if the inequality is strict for all $x \neq y$;

(b) *weighted pseudomonotone w. r. t. the weight vector W* if for all $x, y \in K$, we have

$$W \cdot \langle T(x), y - x \rangle_\ell \geq 0 \;\Rightarrow\; W \cdot \langle T(y), y - x \rangle_\ell \geq 0,$$

and *weighted strictly pseudomonotone w. r. t. the weight vector W* if the second inequality is strict for all $x \neq y$;

(c) *weighted maximal pseudomonotone w. r. t. the weight vector W* if it is weighted pesudomonotone and for all $x, y \in K$, we have

$$W \cdot \langle T(z), z - x \rangle_\ell \leq 0 \;\; \forall z \in \,]x, y] \;\Rightarrow\; W \cdot \langle T(x), y - x \rangle \geq 0, \tag{6.5}$$

and *weighted maximal strictly pseudomonotone w. r. t. the weight vector W* if it is weighted strictly pseudomonotone and (6.5) holds.

It can easily seen that if each T_i is monotone, then T is weighted monotone w. r. t. the any weight vector $W \in \mathbb{R}^\ell_+ \setminus \{\mathbf{0}\}$.

Definition 6.2 Let $W = (W_1, W_2, \dots, W_n) \in \mathbb{R}^\ell_+ \setminus \{\mathbf{0}\}$ be a weight vector. A matrix-valued function $T = (T_1, T_2, \dots, T_\ell) : K \to \mathbb{R}^{\ell \times n}$ is said to be *weighted hemicontinuous w. r. t. the weight vector W* if for all $x, y \in K$ and $\lambda \in [0, 1]$, the mapping $\lambda \mapsto \sum_{i=1}^{\ell} W_i \cdot \langle T_i(x + \lambda(y - x)), y - x \rangle$ is continuous.

If each T_i is continuous, then T is continuous, and hence, T is hemicontinuous.

Proposition 6.1 *Let $W = (W_1, W_2, \dots, W_n) \in \mathbb{R}^\ell_+ \setminus \{\mathbf{0}\}$ be a weight vector and $T = (T_1, T_2, \dots, T_\ell) : K \to \mathbb{R}^{\ell \times n}$ be weighted hemicontinuous and weighted pseudomonotone w. r. t. the weight vector $W = (W_1, W_2, \dots, W_n) \in \mathbb{R}^\ell_+ \setminus \{\mathbf{0}\}$. Then, T is weighted maximal pseudomonotone w. r. t. the same weight vector $W = (W_1, W_2, \dots, W_n) \in \mathbb{R}^\ell_+ \setminus \{\mathbf{0}\}$.*

Proof Assume that for all $x, y \in K$,

$$W \cdot \langle T(z), z - x \rangle_\ell \geq 0, \quad \text{for all } z \in \,]x, y].$$

Then,

$$W \cdot \langle T(x + \lambda(y - x)), y - (x + \lambda(y - x)) \rangle_\ell \geq 0, \quad \text{for all } \lambda \in \,]0, 1]$$

which implies that

$$W \cdot \langle T(x + \lambda(y - x)), y - x \rangle_\ell \geq 0, \quad \text{for all } \lambda \in \,]0, 1].$$

By the weighted hemicontinuity of T, we have

$$W \cdot \langle T(y), y - x \rangle_\ell \geq 0.$$

Hence, T is weighted maximal pseudomonotone w. r. t. the weight vector W. □

The following lemma can be viewed as a generalization of the Minty lemma (see, [3]).

Lemma 6.4 *Let* $W = (W_1, W_2, \ldots, W_n) \in \mathbb{R}^\ell_+ \setminus \{\mathbf{0}\}$ *be a weight vector and* $T = (T_1, T_2, \ldots, T_\ell) : K \to \mathbb{R}^{\ell \times n}$ *be weighted maximal pseudomonotone w. r. t. the weight vector* $W = (W_1, W_2, \ldots, W_n) \in \mathbb{R}^\ell_+ \setminus \{\mathbf{0}\}$. *Then, Sol(WVIP) = Sol(MWVIP).*

Proof It is obvious that Sol(WVIP) $\subseteq$ Sol(MWVIP) by the weighted pseudomonotonicity of T.

Let $x \in$ Sol(MWVIP), then

$$W \cdot \langle T(y), y - \bar{x} \rangle_\ell \geq 0, \quad \text{for all } y \in K.$$

Since K is convex, we have $]\bar{x}, y] \subseteq K$, and therefore,

$$W \cdot \langle T(z), z - \bar{x} \rangle_\ell \geq 0, \quad \text{for all } z \in \,]\bar{x}, y].$$

By the weighted maximal pseudomonotonicity of T, we have

$$W \cdot \langle T(\bar{x}), y - \bar{x} \rangle_\ell \geq 0, \quad \text{for all } y \in K.$$

This shows that $\bar{x} \in$ Sol(WVIP), and hence, Sol(WVIP) = Sol(MWVIP). □

Remark 6.2 In view of Proposition 6.1 and Lemma 6.4, we have that if T is weighted hemicontinuous and weighted pseudomonotone w. r. t. the same weight vector $W = (W_1, W_2, \ldots, W_n) \in \mathbb{R}^\ell_+ \setminus \{\mathbf{0}\}$, then Sol(WVIP) = Sol(MWVIP).

However, Charitha et al. [1] proved that Sol(WVIP) = Sol(MWVIP) if each T_i, $i = 1, 2, \ldots, \ell$ is continuous and monotone.

We now have some existence results for solutions of WVIP.

Theorem 6.1 *Let $W = (W_1, W_2, \ldots, W_n) \in \mathbb{R}^\ell_+ \setminus \{\mathbf{0}\}$ be a weight vector and K be a nonempty convex subset of $\mathbb{R}^n$. Let $T = (T_1, T_2, \ldots, T_\ell) : K \to \mathbb{R}^{\ell \times n}$ be weighted maximal pseudomonotone w. r. t. $W = (W_1, W_2, \ldots, W_n) \in \mathbb{R}^\ell_+ \setminus \{\mathbf{0}\}$. Assume that there exist a nonempty, closed and compact subset D of K and $\tilde{y} \in D$ such that for each $x \in K \setminus D$, $W \cdot \langle T(x), y - x\rangle_\ell < 0$. Then, there exists a solution $\bar{x} \in K$ of WVIP.*

Proof For each $x \in K$, define set-valued maps $F, G : K \to 2^K$ by

$$F(x) = \{y \in K : W \cdot \langle T(y), y - x\rangle_\ell < 0\}$$

and

$$G(x) = \{y \in K : W \cdot \langle T(x), y - x\rangle_\ell < 0\}.$$

Then, it is clear that for each $x \in K$, $G(x)$ is convex. By weighted pseudomonotonicity of T, we have $F(x) \subseteq G(x)$ for all $x \in K$.

For each $y \in K$, the complement of $F^{-1}(y)$ in K is

$$[F^{-1}(y)]^c = \{x \in K : W \cdot \langle T(y), y - x\rangle_\ell \geq 0\}$$

is closed in K, and hence, $F^{-1}(y)$ is open in K. Therefore, $F^{-1}(y)$ is compactly open.

Assume that for all $x \in K$, $F(x)$ is nonempty. Then all the conditions of Theorem 1.36 are satisfied, and therefore, there exists $\hat{x} \in K$ such that $\hat{x} \in G(\hat{x})$. It follows that

$$0 = W \cdot \langle T(\tilde{x}), \tilde{x} - \tilde{x}\rangle < 0,$$

a contradiction. Hence, there exists $\bar{x} \in K$ such that $F(\bar{x}) = \emptyset$. This implies that for all $y \in K$,

$$W \cdot \langle T(y), y - \bar{x}\rangle_\ell \geq 0,$$

that is, there exists $\bar{x} \in K$ w. r. t. the weight vector $W = (W_1, W_2, \ldots, W_\ell) \in \mathbb{R}^\ell_+ \setminus \{\mathbf{0}\}$ such that

$$W \cdot \langle T(y), y - \bar{x}\rangle_\ell \leq 0, \quad \text{for all } y \in K.$$

By Lemma 6.4, $\bar{x} \in K$ is a solution of WVIP. □

Remark 6.3 In view of Remark 6.2, the assumption that T is weighted maximal monotone in Theorem 6.1 can be replaced by weighted hemicontinuous and weighted pseudomonotone w. r. t. W.

Remark 6.4 In Theorem 6.1, if T is weighted maximal strictly pseudomonotone w. r. t. W, then solution of WVIP is unique.

Indeed, assume that there exist two solutions x' and x'' of WVIP. Then, we have

$$W \cdot \langle T(x''), x' - x'' \rangle_\ell \geq 0.$$

By the weighted strictly pseudomonotonicity of T, we have

$$W \cdot \langle T(x'), x' - x'' - \rangle_\ell > 0, \quad \text{i.e.} \quad W \cdot \langle T(x'), x'' - x' \rangle_\ell < 0,$$

that is, x' is not a solution of WVIP, a contradiction.

Now we present the definition of weighted B-pseudomonotonicity.

Definition 6.3 Let $W = (W_1, W_2, \dots, W_n) \in \mathbb{R}^\ell_+ \setminus \{\mathbf{0}\}$ be a weight vector. A matrix-valued function $T = (T_1, T_2, \dots, T_\ell) : K \to \mathbb{R}^{\ell \times n}$ is said to be *weighted B-pseudomonotone w. r. t. the weight vector* W if for each $x \in K$ and every sequence $\{x_m\}_{m \in \mathbb{N}}$ in K converging to x with

$$\limsup_{m \to \infty} W \cdot \langle T(x_m), x - x_m \rangle_\ell \geq 0,$$

we have

$$\limsup_{m \to \infty} W \cdot \langle T(x_m), y - x_m \rangle \leq W \cdot \langle T(x), y - x \rangle, \quad \text{for all } y \in K.$$

Theorem 6.2 *Let $W = (W_1, W_2, \dots, W_n) \in \mathbb{R}^\ell_+ \setminus \{\mathbf{0}\}$ be a weight vector and K be a nonempty convex subset of $\mathbb{R}^n$. Let $T = (T_1, T_2, \dots, T_\ell) : K \to \mathbb{R}^{\ell \times n}$ be weighted B-pseudomonotone* w. r. t. *W such that for each $A \in \mathscr{F}(K)$, $x \mapsto W \cdot \langle T(x), y - x \rangle_\ell$ is lower semicontinuous on* co*A. Assume that there exist a nonempty, closed and compact subset D of K and $\tilde{y} \in D$ such that for all $x \in K \setminus D$, $W \cdot \langle T(x), \tilde{y} - x \rangle_\ell < 0$. Then, there exists a solution $\bar{x} \in K$ of WVIP.*

Proof For each $x \in K$, let $G : K \to 2^K$ be defined by

$$G(x) = \{y \in K : W \cdot \langle T(x), y - x \rangle_\ell < 0\}.$$

Then, for all $x \in K$, $G(x)$ is convex. Let $A \in \mathscr{F}(K)$, then for all $y \in \text{co}A$,

$$[G^{-1}(y)]^c \cap \text{co}A = \{x \in \text{co}A : W \cdot \langle T(x), y - x \rangle_\ell \geq 0\}$$

is closed in co$A$ by lower semicontinuity of the map $x \mapsto W \cdot \langle T(x), y - x \rangle_\ell$ on co$A$. Hence $G^{-1}(y) \bigcap \text{co}A$ is open in co$A$.

Suppose that $x, y \in \text{co}A$ and $\{x_m\}_{m \in \mathbb{N}}$ is a sequence in K converging to x such that

$$W \cdot \langle T(x_m), (\alpha x + (1 - \alpha) y) - x_m \rangle_\ell \geq 0, \quad \text{for all} \quad m \in \mathbb{N} \text{ and all } \alpha \in [0, 1].$$

For $\alpha = 0$, we have

$$W \cdot \langle T(x_m), x - x_m \rangle_\ell \geq 0, \quad \text{for all } m \in \mathbb{N},$$

and therefore,

$$\limsup_{m\to\infty} W \cdot \langle T(x_m), x - x_m \rangle_\ell \geq 0.$$

By the weighted B-pseudomonotonicity of T, we have

$$\limsup_{m\to\infty} W \cdot \langle T(x_m), y - x_m \rangle_\ell \leq W \cdot \langle T(x), y - x \rangle_\ell. \tag{6.6}$$

For $\alpha = 1$, we have

$$W \cdot \langle T(x_m), y - x_m \rangle \geq 0, \quad \text{for all } m \in \mathbb{N},$$

and therefore,

$$\limsup_{m\to\infty} W \cdot \langle T(x_m), y - x_m \rangle_\ell \geq 0. \tag{6.7}$$

From (6.6) and (6.7), we get

$$W \cdot \langle T(x), y - x \rangle_\ell \geq 0,$$

and thus $y \notin G(x)$.

Assume that for all $x \in K$, $G(x)$ is nonempty. Then all the conditions of Theorem 1.36 are satisfied. The rest of the proof follows the lines of the proof of Theorem 6.1. □

Some existence results for solutions of WVIP have been studied in [4] under strong monotonicity and Lipschitz continuity of each T_i and in [1] under continuity and monotonicity each T_i.

References

1. C. Charitha, J. Dutta, C.S. Lalitha, Gap functions for vector variational inequalities. Optimization **64**(7), 1499–1520 (2015)
2. Y. Cheng, On the connectedness of the solution set for the weak vector variational inequality. J. Math. Anal. Appl. **260**, 1–5 (2001)
3. D. Kinderlehrer, G. Stampacchia, *An Introduction to Variational Inequalities and Their Applications* (Academic, New York, 1980)
4. G.M. Lee, D.S. Kim, B.S. Lee, N.D. Yen, Vector variational inequality as a tool for studying vector optimization. Nonlinear Anal. **34**, 745–765 (1998)

Chapter 7
Nonsmooth Vector Variational Inequalities

We have seen in Chap. 5 that if the objective function of a vector optimization problem is smooth (that is, differentiable), then its solution, namely, weak efficient solution, strong efficient solution, efficient solution, properly efficient solution, can be characterized by the corresponding vector variational inequality problems. If the objective function is not smooth but it has some kind of directional derivative, namely, (upper or lower) Dini directional derivative, Clarke directional derivative, Dini-Hadamard directional derivative, etc., then the vector variational inequality problems studied in Chap. 5 would not be useful, and therefore, we need to define different kinds of vector variational inequality problems by means of bifunctions, called nonsmooth vector variational inequality problems. In the formulation of nonsmooth vector variational inequality problems, we consider different kinds of directional derivatives as a bifunction. For a comprehensive study of different kinds of directional derivatives and nonsmooth (scalar) variational inequalities, we refer the recent book [2]. Some recent papers on this topic are [1, 5, 7–9].

In this chapter, we define different kinds of nonsmooth vector variational inequality problems by means of a bifunction. Several existence results for solutions of these nonsmooth vector variational inequality problems are studied. We give some relations among different kinds of solutions of nonsmooth vector optimization problems and nonsmooth vector variational inequality problems.

7.1 Formulations and Preliminary Results

Throughout the section, unless otherwise specified, we assume that K is a nonempty convex subset of $\mathbb{R}^n$ and $C = \mathbb{R}^\ell_+$. Let $h = (h_1, h_2, \ldots, h_\ell) : K \times \mathbb{R}^n \to \mathbb{R}^\ell$ be a vector-valued function such that for each fixed $x \in K$, $h(x; d)$ is positively homogeneous in d, that is, $h(x; \alpha d) = \alpha h(x; d)$ for all $\alpha > 0$.

Q.H. Ansari et al., *Vector Variational Inequalities and Vector Optimization*, Vector Optimization, DOI 10.1007/978-3-319-63049-6_7

We consider the following nonsmooth vector variational inequality problems:

- *Strong h-Vector Variational Inequality Problem* (h-SVVIP): Find $\bar{x} \in K$ such that

$$h(\bar{x}; y - \bar{x}) = \big(h_1(\bar{x}; y - \bar{x}), \ldots, h_\ell(\bar{x}; y - \bar{x})\big) \in C, \quad \text{for all } y \in K. \tag{7.1}$$

- *h-Vector Variational Inequality Problem* (h-VVIP): Find $\bar{x} \in K$ such that

$$h(\bar{x}; y-\bar{x}) = \big(h_1(\bar{x}; y-\bar{x}), \ldots, h_\ell(\bar{x}; y-\bar{x})\big) \notin -C \setminus \{\mathbf{0}\}, \quad \text{for all } y \in K. \tag{7.2}$$

- *Weak h-Vector Variational Inequality Problem* (h-WVVIP): Find $\bar{x} \in K$ such that

$$h(\bar{x}; y-\bar{x}) = \big(h_1(\bar{x}; y-\bar{x}), \ldots, h_\ell(\bar{x}; y-\bar{x})\big) \notin -\text{int}(C), \quad \text{for all } y \in K. \tag{7.3}$$

As we have seen in Chap. 5, the Minty vector variational inequality problems are closely related to the (Stampacchia) vector variational inequality problems, therefore, we also consider the following Minty nonsmooth vector variational inequality problems.

- *Minty Strong h-Vector Variational Inequality Problem* (h-MSVVIP): Find $\bar{x} \in K$ such that

$$h(y; \bar{x} - y) = \big(h_1(y; \bar{x} - y), \ldots, h_\ell(y; \bar{x} - y)\big) \in -C, \quad \text{for all } y \in K. \tag{7.4}$$

- *Minty h-Vector Variational Inequality Problem* (h-MVVIP): Find $\bar{x} \in K$ such that

$$h(y; \bar{x}-y) = \big(h_1(y; \bar{x}-y), \ldots, h_\ell(y; \bar{x}-y)\big) \notin C \setminus \{\mathbf{0}\}, \quad \text{for all } y \in K. \tag{7.5}$$

- *Minty Weak h-Vector Variational Inequality Problem* (h-MWVVIP): Find $\bar{x} \in K$ such that

$$h(y; \bar{x} - y) = \big(h_1(y; \bar{x} - y), \ldots, h_\ell(y; \bar{x} - y)\big) \notin \text{int}(C), \quad \text{for all } y \in K. \tag{7.6}$$

The set of solutions of h-SVVIP, h-VVIP, h-WVVIP, h-MSVVIP, h-MVVIP and h-MWVVIP are denoted by Sol(h-SVVIP), Sol(h-VVIP), Sol(h-WVVIP), Sol(h-MSVVIP), Sol(h-MVVIP) and Sol(h-MWVVIP), respectively.

Let $f = (f_1, f_2, \ldots, f_\ell) : \mathbb{R}^n \to \mathbb{R}^\ell$ be avector-valued function and

$$D^+ f(x; d) = \big(D^+ f_1(x; d), \ldots, D^+ f_\ell(x; d)\big),$$

where $D^+ f_i(x; d)$ denotes the upper Dini directional derivative of f_i at x in the direction d.

When $h(x; \cdot) = D^+ f(x; \cdot)$, then h-SVVIP, h-VVIP, h-WVVIP, h-MSVVIP, h-MVVIP and h-MWVVIP become the following nonsmooth vector variational inequality problems.

- (D^+-SVVIP): Find $\bar{x} \in K$ such that

$$D^+f(\bar{x}; y - \bar{x}) \in C, \quad \text{for all } y \in K. \tag{7.7}$$

- (D^+-VVIP): Find $\bar{x} \in K$ such that

$$D^+f(\bar{x}; y - \bar{x}) \notin -C \setminus \{\mathbf{0}\}, \quad \text{for all } y \in K. \tag{7.8}$$

- (D^+-WVVIP): Find $\bar{x} \in K$ such that

$$D^+f(\bar{x}; y - \bar{x}) \notin -\text{int}(C), \quad \text{for all } y \in K. \tag{7.9}$$

- (D^+-MSVVIP): Find $\bar{x} \in K$ such that

$$D^+f(y; \bar{x} - y) \in -C, \quad \text{for all } y \in K. \tag{7.10}$$

- (D^+-MVVIP): Find $\bar{x} \in K$ such that

$$D^+f(y; \bar{x} - y) \notin C \setminus \{\mathbf{0}\}, \quad \text{for all } y \in K. \tag{7.11}$$

- (D^+-MWVVIP): Find $\bar{x} \in K$ such that

$$D^+f(y; \bar{x} - y) \notin \text{int}(C), \quad \text{for all } y \in K. \tag{7.12}$$

Similarly, we can define D_+-SVVIP, D_+-VVIP, D_+-WVVIP, D_+-MSVVIP, D_+-MVVIP and D_+-MWVVIP by considering $D_+f(x; \cdot)$ in place of $h(x; \cdot)$ in h-SVVIP, h-VVIP, h-WVVIP, h-MSVVIP, h-MVVIP and h-MWVVIP, respectively.

If we consider (upper or lower) Dini directional derivative as a bifunction $h(x; d)$, with x referring to a point in $\mathbb{R}^n$ and d referring to a direction from $\mathbb{R}^n$, then (7.1), (7.2), (7.3), (7.4), (7.5) and (7.6) are equivalent to (7.7), (7.8), (7.9), (7.10), (7.11) and (7.12), respectively. In general, if we treat any generalized directional derivative as a bifunction $h(x; d)$ with x referring to a point in $\mathbb{R}^n$ and d referring to a direction from $\mathbb{R}^n$, then the corresponding nonsmooth vector variational inequality problems can be defined in the same way.

Definition 7.1 A vector-valued bifunction $h = (h_1, h_2, \ldots, h_\ell) : K \times \mathbb{R}^n \to \mathbb{R}^\ell$ is said to be:

(a) *strongly C-pseudomonotone* if for all $x, y \in K$,

$$h(x; y - x) \in C \quad \text{implies} \quad h(y; x - y) \in -C;$$

(b) *C-pseudomonotone* if for all $x, y \in K$,

$$h(x; y - x) \notin -C \setminus \{\mathbf{0}\} \quad \text{implies} \quad h(y; x - y) \notin C \setminus \{\mathbf{0}\};$$

(c) *weakly C-pseudomonotone* if for all $x, y \in K$,

$$h(x; y - x) \notin -\mathrm{int}(C) \quad \text{implies} \quad h(y; x - y) \notin \mathrm{int}(C);$$

(d) *C-properly subodd* if

$$h(x; d_1) + h(x; d_2) + \cdots + h(x; d_m) \in C,$$

for every $d_i \in \mathbb{R}^n$ with $\sum_{i=1}^m d_i = \mathbf{0}$ and for all $x \in K$.

If $m = 2$, the definition of proper suboddness reduces to the definition of suboddness.

Example 7.1 The function $h : \mathbb{R} \times \mathbb{R} \to \mathbb{R}^2$, with $h(x, d) = (x, -x - d)$ is strongly $\mathbb{R}^2_+$-pseudomonotone, $\mathbb{R}^2_+$-pseudomonotone and weakly $\mathbb{R}^2_+$-pseudomonotone, but h is not $\mathbb{R}^2_+$-properly subodd.

Definition 7.2 A vector-valued bifunction $h = (h_1, h_2, \ldots, h_\ell) : K \times \mathbb{R}^n \to \mathbb{R}^\ell$ is said to be *C-upper sign continuous* (respectively, *strongly C-upper sign continuous* and *weakly C-upper sign continuous*) if for all $x, y \in K$ and $\lambda \in]0, 1[$,

$$h(x + \lambda(y - x); x - y) \notin C \setminus \{\mathbf{0}\} \quad \text{implies} \quad h(x; x - y) \notin C \setminus \{\mathbf{0}\}$$

$$\Big(\text{respectively,} \quad h(x + \lambda(y - x); x - y) \in -C \quad \text{implies} \quad h(x; x - y) \in -C$$

$$\text{and} \quad h(x + \lambda(y - x); x - y) \notin \mathrm{int}(C) \quad \text{implies} \quad h(x; x - y) \notin \mathrm{int}(C)\Big).$$

Definition 7.3 A vector-valued bifunction $h = (h_1, h_2, \ldots, h_\ell) : K \times \mathbb{R}^n \to \mathbb{R}^\ell$ is said to be *v-hemicontinuous* if for each fixed $d \in \mathbb{R}^n$ and for all $x, y \in K$,

$$\lim_{\lambda \to 0+} h(x + \lambda(y - x)); d) = h(x; d).$$

It can be easily seen that if each component h_i, $i = 1, 2, \ldots, \ell$, of h is hemicontinuous, that is,

$$\lim_{\lambda \to 0+} h_i(x + \lambda(y - x)); d) = h_i(x; d),$$

then h is v-hemicontinuous.

Remark 7.1 If h is v-hemicontinuous, then it is strongly C-upper sign continuous and weakly C-upper sign continuous as C and $\mathbb{R}^\ell \setminus \{\mathrm{int}(C)\}$ are closed sets.

Example 7.2 The function $h : \mathbb{R} \times \mathbb{R} \to \mathbb{R}^2$, which is defined by $h(x; d) = \big(|x| \cdot x^2 \cdot d, \exp(x) \cdot d\big)$, is strongly $\mathbb{R}^2_+$-upper sign continuous.

The following result provides the relationship between nonsmooth vector variational inequality problems and Minty nonsmooth vector variational inequality problems.

Lemma 7.1 *Let $h : K \times \mathbb{R}^n \to \mathbb{R}^\ell$ be C-pseudomonotone (respectively, strongly C-pseudomonotone and weakly C-pseudomonotone) and C-upper sign continuous (respectively, strong C-upper sign continuous and weakly C-upper sign continuous) such that for each fixed $x \in K$, $h(x;\cdot)$ is C-properly subodd and positively homogeneous. Then $\bar{x} \in K$ is a solution of h-*VVIP *(respectively, h-*SVVIP *and h-*WVVIP*) if and only if it is a solution of h-*MVVIP *(respectively, h-*MSVVIP *and h-* MWVVIP*).*

Proof The C-pseudomonotonicity of h implies that every solution of h-VVIP is a solution of h-MVVIP.

Conversely, let $\bar{x} \in K$ be a solution of h-MVVIP. Then

$$h(y; \bar{x} - y) \notin C \setminus \{\mathbf{0}\}, \quad \text{for all } y \in K. \tag{7.13}$$

Since K is convex, we have $y_\lambda := \bar{x} + \lambda(y - \bar{x}) \in K$ for all $\lambda \in]0, 1[$, therefore, (7.13) becomes

$$h(y_\lambda; \bar{x} - y_\lambda) \notin C \setminus \{\mathbf{0}\}.$$

Since $\bar{x} - y_\lambda = \lambda(\bar{x} - y)$ and $h(x;\cdot)$ is positively homogeneous, we have

$$h(y_\lambda; \bar{x} - y) \notin C \setminus \{\mathbf{0}\}.$$

Thus, the C-upper sign continuity and the C-proper suboddness of h imply that $\bar{x} \in K$ is a solution of h-VVIP.

Similarly, we can prove Sol(h-SVVIP) = Sol(h-MSVVIP) and Sol(h-WVVIP) = Sol(h-MWVVIP). □

Example 7.3 Let $X = \mathbb{R}$, $K = [0, 1]$, $Y = \mathbb{R}^2$, and $C = \mathbb{R}^2_+$. Consider the function $h(x; d) = \left(x^2 d, |x| d\right)$. Note that h is strongly $\mathbb{R}^2_+$-pseudomonotone, strongly $\mathbb{R}^2_+$-upper sign continuous, $\mathbb{R}^2_+$-properly subodd and positive homogeneous in the second variable. The element $\bar{x} = 0$ is the only solution of the strong h-vector variational inequality problem h-SVVIP as well as the only solution of the Minty strong h-vector variational inequality problem h-MSVVIP.

In general, Sol(h-SVVIP) $\neq$ Sol(h-MSVVIP), Sol(h-VVIP) $\neq$ Sol(h-MVVIP) and Sol(h-WVVIP)) $\neq$ Sol(h-MWVVIP).

To overcome this deficiency, we define the following perturbed h-vector variational inequality problems.

- *ε-Perturbed Strong h-Vector Variational Inequality Problem* (ε-h-PSVVIP): Find $\bar{x} \in K$ for which there exists $\bar{\varepsilon} \in]0, 1[$ such that

$$h(\bar{x} + \varepsilon(y - \bar{x}); y - \bar{x}) \in -C, \quad \text{for all } y \in K \text{ and all } \varepsilon \in]0, \bar{\varepsilon}[. \tag{7.14}$$

- *ε-Perturbed h-Vector Variational Inequality Problem* (ε-h-PVVIP): Find $\bar{x} \in K$ for which there exists $\bar{\varepsilon} \in]0, 1[$ such that

$$h(\bar{x} + \varepsilon(y - \bar{x}); y - \bar{x}) \notin C \setminus \{\mathbf{0}\}, \quad \text{for all } y \in K \text{ and all } \varepsilon \in]0, \bar{\varepsilon}[. \tag{7.15}$$

- *ε-Perturbed Weak h-Vector Variational Inequality Problem* (ε-h-PWVVIP): Find $\bar{x} \in K$ for which there exists $\bar{\varepsilon} \in]0, 1[$ such that

$$h(\bar{x} + \varepsilon(y - \bar{x}); y - \bar{x}) \notin \text{int}(C), \quad \text{for all } y \in K \text{ and all } \varepsilon \in]0, \bar{\varepsilon}[. \tag{7.16}$$

Proposition 7.1 *Let $h = (h_1, h_2, \ldots, h_\ell) : K \times \mathbb{R}^n \to \mathbb{R}^\ell$ be C-pseudomonotone (respectively, strongly C-pseudomonotone and weakly C-pseudomonotone) and C-properly subodd such that it is positively homogeneous in the second argument. Then $\bar{x} \in K$ is a solution of* ε-h-PVVIP *(respectively,* ε-h-PSVVIP *and* ε-h-PWVVIP*) if and only if it is a solution of* h-MVVIP *(respectively,* h-MSVVIP *and* h- MWVVIP*).*

Proof Let $\bar{x}$ be a solution of h-MVVIP. Then

$$h(y; \bar{x} - y) \notin C \setminus \{\mathbf{0}\}, \quad \text{for all } y \in K. \tag{7.17}$$

Since K is convex, we have

$$x_\varepsilon := \bar{x} + \varepsilon(z - \bar{x}) \in K, \quad \text{for all } z \in K \text{ and all } \varepsilon \in [0, 1].$$

Taking $y = x_\varepsilon$ with $\bar{\varepsilon} = 1$ and $\varepsilon \in]0, \bar{\varepsilon}[$ in (7.17), we have

$$h(x_\varepsilon; \bar{x} - x_\varepsilon) \notin C \setminus \{\mathbf{0}\}.$$

Since $\bar{x} - x_\varepsilon = \varepsilon(\bar{x} - z)$ and $h(x; .\cdot)$ is positively homogeneous, we have

$$h(x_\varepsilon; \bar{x} - z) \notin C \setminus \{\mathbf{0}\}, \quad \text{for all } z \in K \text{ and all } \varepsilon \in]0, \bar{\varepsilon}[\,. \tag{7.18}$$

Since $(z - \bar{x}) + (\bar{x} - z) = \mathbf{0}$ and h is C-properly subodd, we have

$$h(x_\varepsilon; z - \bar{x}) + h(x_\varepsilon; \bar{x} - z) \in C. \tag{7.19}$$

Combining (7.18) and (7.19), we obtain

$$h(x_\varepsilon; z - \bar{x}) \notin C \setminus \{\mathbf{0}\}, \quad \text{for all } z \in K \text{ and all } \varepsilon \in]0, \bar{\varepsilon}[.$$

Therefore, $\bar{x} \in K$ is a solution of ε-h-PVVIP.

Conversely, suppose that $\bar{x} \in K$ is a solution of ε-h-PVVIP, but not a solution of h-MVVIP. Then there exists z $\in K$ such that

$$h(z; \bar{x} - z) \in C \setminus \{\mathbf{0}\}.$$

Since K is convex, we have

$$x_\varepsilon := \bar{x} + \varepsilon(z - \bar{x}) \in K, \quad \text{for all } \varepsilon \in [0, 1].$$

Since $x_\varepsilon - z = (1 - \varepsilon)(\bar{x} - z)$ and $h(x; \cdot)$ is positively homogeneous, we have

$$h(z; \bar{x} - z) = \frac{1}{1 - \varepsilon} h(z; x_\varepsilon - z) \in C \setminus \{\mathbf{0}\}, \quad \text{for all } \varepsilon \in]0, 1[;$$

thus,

$$h(z; x_\varepsilon - z) \in C \setminus \{\mathbf{0}\}, \quad \text{for all } \varepsilon \in]0, 1[.$$

By C-pseudomonotonicity of h, we obtain

$$h(x_\varepsilon; z - x_\varepsilon) \in -C \setminus \{\mathbf{0}\}, \quad \text{for all } \varepsilon \in]0, 1[.$$

Since $z - x_\varepsilon = (1 - \varepsilon)(z - \bar{x})$ and $h(x; \cdot)$ is positively homogeneous, we have

$$h(x_\varepsilon; z - \bar{x}) \in C \setminus \{\mathbf{0}\}, \quad \text{for all } \varepsilon \in]0, 1[,$$

which contradicts our supposition that $\bar{x}$ is a solution of ε-h-PVVIP.

The rest of the part can be proved in a similar way. □

7.2 Existence Results for Solutions of Nonsmooth Vector Variational Inequalities

We first present an existence result for a solution of h-VVIP without using any kind of monotonicity.

Theorem 7.1 *Let K be a nonempty compact convex subset of $\mathbb{R}^n$. Let $h = (h_1, h_2, \ldots, h_\ell) : K \to \mathbb{R}^\ell$ be a vector-valued function such $h(x; \mathbf{0}) = \mathbf{0}$ and $h(x; \cdot)$ is positively homogeneous for each fixed $x \in K$, and the set $\{x \in K : h(x; y - x) \in -C \setminus \{\mathbf{0}\}\}$ is open in K for every fixed $y \in K$. Then h-VVIP has a solution $\bar{x} \in K$.*

Proof Suppose that h-VVIP has no solution. Then for every $\bar{x} \in K$, there exists $y \in K$ such that

$$h(\bar{x}; y - \bar{x}) \in -C \setminus \{\mathbf{0}\}. \tag{7.20}$$

For every $y \in K$, define the set N_y by

$$N_y = \{x \in K : h(x; y - x) \in -C \setminus \{\mathbf{0}\}\}. \tag{7.21}$$

By assumption, the set N_y is open in K for each $y \in K$. Therefore, from (7.20), $\{N_y : y \in K\}$ is an open cover of K. Since K is compact, there exists a finite subset $\{y_1, y_2, \ldots, y_k\}$ of K such that

$$K = \bigcup_{i=1}^{k} N_{y_i}.$$

Thus, there exists a continuous partition of unity $\{\beta_1, \beta_2, \ldots, \beta_k\}$ subordinated to $\{N_{y_1}, N_{y_2}, \ldots, N_{y_k}\}$ such that for all $x \in K$,

- $\beta_j(x) \geq 0, j = 1, 2, \ldots, k$
- $\sum_{j=1}^{k} \beta_j(x) = 1$
- $\beta_j(x) = 0$ whenever $x \notin N_{y_j}$, and $\beta_j(x) > 0$ whenever $x \in N_{y_j}$

Let $p : K \to \mathbb{R}^n$ be defined by

$$p(x) = \sum_{j=1}^{k} \beta_j(x) y_j, \quad \text{for all } x \in K.$$

Since each β_i is continuous, we have p is continuous. Let $\Delta = \mathrm{co}(\{y_1, y_2, \ldots, y_k\}) \subset K$. Then Δ is a simplex of the finite dimensional space and p maps Δ into itself. By Brouwer's Fixed Point Theorem 1.39, there exists $\hat{x} \in \Delta$ such that $p(\hat{x}) = \hat{x}$.

Define $q : K \to \mathbb{R}^\ell$ by

$$q(x) = h(x; x - p(x)) = \sum_{j=1}^{k} \beta_j h(x; x - y_j), \quad \text{for all } x \in K. \tag{7.22}$$

For any given $x \in K$, let $J = \{j : x \in N_{y_j}\} = \{j : \beta_j(x) > 0\}$. Obviously, J is nonempty. It follows from (7.21) and (7.22) that

$$q(x) = \sum_{j \in J} \beta_j(x) h(x; y_j - x) \in -C \setminus \{\mathbf{0}\}, \quad \text{for all } x \in K.$$

Since $\hat{x} \in \Delta \subset K$ is a fixed of p, from (7.21), we have

$$q(\hat{x}) = h(\hat{x}; \hat{x} - \hat{x}) = \mathbf{0} \in -C \setminus \{\mathbf{0}\},$$

a contradiction. Hence, h-VVIP has a solution $\bar{x} \in K$. □

The following result provides the existence of a solution of h-MWVVIP and h-WVVIP in the setting of compact convex set but under weakly C-pseudomonotonicity.

Theorem 7.2 *Let $K \subseteq \mathbb{R}^n$ be a nonempty, convex and compact set and $h = (h_1, h_2, \ldots, h_\ell) : K \to \mathbb{R}^\ell$ be a positively homogeneous in the second argument, C-properly subodd and weakly C-pseudomonotone vector-valued function such that for all $i \in \mathscr{I} = \{1, 2, \ldots, \ell\}$ and for each fixed $x \in K$, $h_i(x; \cdot)$ is continuous. Then* h-MWVVIP *has a solution $\bar{x} \in K$.*

Furthermore, if h is weakly C-upper sign continuous, then $\bar{x} \in K$ is a solution of h-WVVIP.

Proof For all $y \in K$, we define two set-valued maps $S, M : K \to 2^K$ by

$$S(y) = \{x \in K : h(x; y - x) \notin -\text{int}(C)\}$$

and

$$M(y) = \{x \in K : h(y; x - y) \notin \text{int}(C)\}.$$

We show that S is a KKM map. Let $\hat{x} \in \text{co}\left(\{y_1, y_2 \ldots, y_p\}\right)$, then $\hat{x} = \sum_{k=1}^p \lambda_k y_k$ with $\lambda_k \geq 0$ and $\sum_{k=1}^p \lambda_k = 1$. If $\hat{x} \notin \bigcup_{k=1}^p S(y_k)$, then

$$h(\hat{x}; y_k - \hat{x}) \in -\text{int}(C), \quad \text{for all } k = 1, 2, \ldots, p.$$

Since $-C$ is a convex cone and $\lambda_k \geq 0$ with $\sum_{k=1}^p \lambda_k = 1$, we have

$$\sum_{k=1}^p \lambda_k h(\hat{x}; y_k - \hat{x}) \in -\text{int}(C). \tag{7.23}$$

Since

$$\sum_{k=1}^p \lambda_k (y_k - \hat{x}) = \sum_{k=1}^p \lambda_k y_k - \sum_{k=1}^p \lambda_k \hat{x} = \hat{x} - \hat{x} = \mathbf{0},$$

by C-proper suboddness of h, we have

$$\sum_{k=1}^p h(\hat{x}; \lambda_k (y_k - \hat{x})) \in C.$$

By positive homogenuity of h, we have

$$\sum_{k=1}^p \lambda_k h(\hat{x}; y_k - \hat{x}) \in C,$$

which contradicts (7.23). Therefore, $\text{co}\left(\{y_1, y_2, \ldots, y_p\}\right) \subseteq \bigcup_{k=1}^{p} S(y_k)$. Hence, S is a KKM map.

The weak C-pseudomonotonicity of h implies that $S(y) \subseteq M(y)$ for all $y \in K$; hence, M is a KKM map.

We claim that $M(y)$ is a closed set in K for all $y \in K$. Indeed, let $\{x_m\}$ be a sequence in $M(y)$ which converges to $x \in K$. Then

$$h(y; x_m - y) \notin \text{int}(C), \quad \text{that is,} \quad h(y; x_m - y) \in \mathbb{R}^\ell \setminus \{\text{int}(C)\}.$$

Since $\mathbb{R}^\ell \setminus \{\text{int}(C)\}$ is a closed set and each $h_i(x; \cdot)$ is continuous, we have $h(y; x-y) \notin \text{int}(C)$, and hence, $x \in M(y)$. Thus, $M(y)$ is closed in K.

Further, since K is compact, it follows that $M(y)$ is compact for all $y \in K$. Then, by Fan-KKM Lemma 1.14,

$$\bigcap_{y \in K} M(y) \neq \emptyset,$$

that is, there exists $\bar{x} \in K$ such that

$$h(y; \bar{x} - y) \notin \text{int}(C), \quad \text{for all } y \in K.$$

Thus, $\bar{x} \in K$ is a solution of h-MVVIP.

By Lemma 7.1, $\bar{x} \in K$ is a solution of h-WVVIP. □

Definition 7.4 A vector-valued function $h = (h_1, h_2, \ldots, h_\ell) : K \to \mathbb{R}^\ell$ is said to be *C-pseudomonotone$_+$* if for all $x, y \in K$,

$$h(x; y - x) \notin -C \setminus \{\mathbf{0}\} \quad \text{implies} \quad h(y; x - y) \in -C.$$

Clearly, C-pseudomonotonicity$_+$ is stronger than C-pseudomonotonicity.

Example 7.4 Let $X = \mathbb{R}$, $K = [-2, 2]$, $Y = \mathbb{R}^2$, and $C = \mathbb{R}^2_+$. The function $h : K \times \mathbb{R} \to \mathbb{R}^2$, defined by $h(x; d) = \left(x \cdot d, x^2 \cdot d\right)$, is $\mathbb{R}^2_+$-pseudomonotone, but not $\mathbb{R}^2_+$-pseudomonotone$_+$.

We have the following existence result for a solution of h-SVVIP under C-pseudomonotonicity$_+$ assumption.

Theorem 7.3 *Let $K \subseteq \mathbb{R}^n$ be a nonempty, convex and compact set and $h = (h_1, h_2, \ldots, h_\ell) : K \to \mathbb{R}^\ell$ be a positively homogeneous in the second argument, C-properly subodd and C-pseudomonotone$_+$ vector-valued function such that for all $i \in \mathscr{I} = \{1, 2, \ldots, \ell\}$ and for each fixed $x \in K$, $h_i(x; \cdot)$ is continuous. Furthermore, if h is strongly C-upper sign continuous, then h-*VVIP *has a solution.*

Proof For all $y \in K$, we define set-valued maps $S, M : K \to 2^K$ by

$$S(y) = \{x \in K : h(x; y - x) \notin -C \setminus \{\mathbf{0}\}\}$$

and

$$M(y) = \{x \in K : h(y; x - y) \in -C\}.$$

By C-pseudomonotonicity$_+$, Sol(h-VVIP) $\subseteq$ Sol(h-MSVVIP). From Lemma 7.1, Sol(h-MSVVIP) $\subseteq$ Sol(h-SVVIP) $\subseteq$ Sol(h-VVIP). Thus, Sol(h-VVIP) $\subseteq$ Sol(h-MSVVIP) $\subseteq$ Sol(h-VVIP), and hence, Sol(h-VVIP) $=$ Sol(h-MSVVIP), that is,

$$\bigcap_{y \in K} S(y) = \bigcap_{y \in K} M(y).$$

We prove that S is a KKM map. Let $\hat{x} \in \text{co}\left(\{y_1, y_2, \ldots, y_p\}\right)$, then $\hat{x} = \sum_{k=1}^{p} \lambda_k y_k$ with $\lambda_k \geq 0$ and $\sum_{k=1}^{p} \lambda_k = 1$. If $\hat{x} \notin \bigcup_{k=1}^{p} S(y_k)$, then

$$h(\hat{x}; y_k - \hat{x}) \in -C \setminus \{\mathbf{0}\}, \quad \text{for all } k = 1, 2, \ldots, p.$$

Since $-C$ is a convex cone and $\lambda_k \geq 0$ with $\sum_{k=1}^{p} \lambda_k = 1$, we have

$$\sum_{k=1}^{p} \lambda_k h(\hat{x}; y_k - \hat{x}) \in -C \setminus \{\mathbf{0}\}. \tag{7.24}$$

Since

$$\sum_{k=1}^{p} \lambda_k (y_k - \hat{x}) = \sum_{k=1}^{p} \lambda_k y_k - \sum_{k=1}^{p} \lambda_k \hat{x} = \hat{x} - \hat{x} = \mathbf{0},$$

by C-proper suboddness of h, we have

$$\sum_{k=1}^{p} h(\hat{x}; \lambda_k (y_k - \hat{x})) \in C.$$

The positive homogenuity of h in the second argument implies that

$$\sum_{k=1}^{p} \lambda_k h(\hat{x}; y_k - \hat{x}) \in C,$$

which contradicts (7.24). Therefore, $\text{co}\left(\{y_1, y_2, \ldots, y_p\}\right) \subseteq \bigcup_{k=1}^{p} S(y_k)$. Hence, S is a KKM map.

By C-pseudomonotonicity$_+$ of h implies that $S(y) \subseteq M(y)$ for all $y \in K$; hence, M is a KKM map. Since $-C$ is closed and each $h_i(x; \cdot)$ is continuous, it can be easily seen that $M(y)$ is closed subset of a compact set K, and hence, compact. Therefore, by Fan-KKM Lemma 1.14,

$$\bigcap_{y \in K} S(y) = \bigcap_{y \in K} M(y) \neq \emptyset,$$

that is, there exists a solution of h-VVIP. □

Definition 7.5 Let K be a nonempty convex subset of $\mathbb{R}^n$. A vector-valued function $h = (h_1, h_2, \ldots, h_\ell) : K \to \mathbb{R}^\ell$ is said to be

(a) *strongly proper C-quasimonotone*$_*$ if for every finite set $\{y_1, y_2, \ldots, y_p\}$ in K and $x \in \mathrm{co}\left(\{y_1, y_2, \ldots, y_p\}\right)$, there exists $i \in \{1, 2, \ldots, p\}$ such that $h(x; y_i - x) \in C$;
(b) *proper C-quasimonotone*$_*$ if for every finite set $\{y_1, y_2, \ldots, y_p\}$ in K and $x \in \mathrm{co}\left(\{y_1, y_2, \ldots, y_p\}\right)$, there exists $i \in \{1, 2, \ldots, p\}$ such that $h(x; y_i - x) \notin -C \setminus \{\mathbf{0}\}$;
(c) *weakly proper C-quasimonotone*$_*$ if for every finite set $\{y_1, y_2, \ldots, y_p\}$ in K and $x \in \mathrm{co}\left(\{y_1, y_2, \ldots, y_p\}\right)$, there exists $i \in \{1, 2, \ldots, p\}$ such that $h(x; y_i - x) \notin -\mathrm{int}(C)$.

Example 7.5 Let $X = \mathbb{R}$, $K = [0, 1]$, $Y = \mathbb{R}^2$, and $C = \mathbb{R}^2_+$. The function $h : K \times \mathbb{R} \to \mathbb{R}^2$, defined by $h(x; d) = \left(x, \sqrt{x} \cdot |d|\right)$, is strongly proper C-quasimonotone$_*$.

Theorem 7.4 *Let $K \subseteq \mathbb{R}^n$ be a nonempty, convex and compact set and $h = (h_1, h_2, \ldots, h_\ell) : K \to \mathbb{R}^\ell$ be a C-properly subodd and strongly proper C-quasimonotone$_*$ vector-valued function such that for all $i \in \mathscr{I} = \{1, 2, \ldots, \ell\}$ and for each fixed $x \in K$, $h_i(x; \cdot)$ is continuous and $h(x; \mathbf{0}) = \mathbf{0}$ for all $x \in K$. Then h-SVVIP has a solution $\bar{x} \in K$.*

Proof For all $y \in K$, we define a set-valued map $S : K \to 2^K$ by

$$S(y) = \{x \in K : h(x; y - x) \in C\}.$$

Since $h(x; y-y) = h(x; \mathbf{0}) = \mathbf{0} \in C$ for each $y \in K$, $y \in S(y)$, and thus, $S(y) \neq \emptyset$. We show that S is a KKM map. Let $\hat{x} \in \mathrm{co}\left(\{y_1, y_2, \ldots, y_p\}\right)$ such that $\hat{x} \notin \bigcup_{k=1}^p S(y_k)$. This implies that

$$h(\hat{x}; y_k - \hat{x}) \notin C, \quad \text{for all } k = 1, 2, \ldots, p.$$

This contradicts the strong proper C-quasimonotonicity$_*$ of h. Hence, S is a KKM map.

Since $h(x; \cdot)$ is continuous and C is closed, it can be easily seen that $S(y)$ is a closed subset of a compact set K, and hence, $S(y)$ is compact for all $y \in K$. Then, by Fan-KKM Lemma 1.14,

$$\bigcap_{y \in K} S(y) \neq \emptyset,$$

that is, there exists $\bar{x} \in K$ such that

$$h(\bar{x}; y - \bar{x}) \in C, \quad \text{for all } y \in K.$$

Thus, $\bar{x} \in K$ is a solution of h-SVVIP. □

Similarly, we can prove the following results.

Theorem 7.5 *Let $K \subseteq \mathbb{R}^n$ be a nonempty, convex and compact set and $h = (h_1, h_2, \ldots, h_\ell) : K \to \mathbb{R}^\ell$ be a C-properly subodd and weakly proper C-quasimonotone$_*$ vector-valued function such that for all $i \in \mathscr{I} = \{1, 2, \ldots, \ell\}$ and for each fixed $x \in K$, $h_i(x; \cdot)$ is continuous and $h(x; \mathbf{0}) = \mathbf{0}$ for all $x \in K$. Then* h-WVVIP *has a solution $\bar{x} \in K$.*

Theorem 7.6 *Let $K \subseteq \mathbb{R}^n$ be a nonempty, convex and compact set and $h = (h_1, h_2, \ldots, h_\ell) : K \to \mathbb{R}^\ell$ be a C-properly subodd and proper C-quasimonotone$_*$ vector-valued function such that for all $i \in \mathscr{I} = \{1, 2, \ldots, \ell\}$ and for each fixed $x \in K$, the set $\{x \in K : h(x; y - x) \notin -C \setminus \{\mathbf{0}\}\}$ is closed and $h(x; \mathbf{0}) = \mathbf{0}$ for all $x \in K$. Then* h-VVIP *has a solution $\bar{x} \in K$.*

7.3 Nonsmooth Vector Variational Inequalities and Nonsmooth Vector Optimization

The optimization problem may have a nonsmooth objective function. Therefore, Crespi et al. [3, 4] introduced Minty variational inequality for scalar-valued functions defined by means of lower Dini directional derivative. More recently, the same authors extended their formulation to the vector case in [5]. They have also established the relations between a Minty Vector Variational Inequality problem (in short, MVVIP) and solutions of VOP (both ideal and weak efficient but not efficient) solution. Crespi et al. [5] used the scalarization method to obtain their results. The similar VVIP is also considered by Lalitha and Mehta [8] and proved some existence results. They also provided some relationships between the solutions of VOP and this kind of VVIP.

In this section, we propose some relations between vector optimization and vector variational inequalities when the objective functions are not necessarily smooth.

Throughout this section, unless otherwise specified, we assume that K is nonempty convex subset of $\mathbb{R}^n$, $C = \mathbb{R}^\ell_+$ and $h = (h_1, h_2, \ldots, h_\ell) : K \times \mathbb{R}^n \to \mathbb{R}^\ell$ is a vector-valued function.

Definition 7.6 A vector-valued function $f = (f_1, f_2, \ldots, f_\ell) : K \to \mathbb{R}^\ell$ is said to be

(a) *C-h-convex* if for all $x, y \in K$,

$$f(y) - f(x) - h(x; y - x) \in C;$$

(b) *strictly C-h-convex* if for all $x, y \in K$, $x \neq y$,

$$f(y) - f(x) - h(x; y - x) \in \operatorname{int}(C);$$

(c) *strongly C-h-pseudoconvex* if for all $x, y \in K$,

$$f(y) - f(x) \notin C \quad \text{implies} \quad h(x; y - x) \notin C,$$

equivalently,

$$h(x; y - x) \in C \quad \text{implies} \quad f(y) - f(x) \in C;$$

(d) *C-h-pseudoconvex* if for all $x, y \in K$,

$$f(y) - f(x) \in -C \setminus \{\mathbf{0}\} \quad \text{implies} \quad h(x; y - x) \in -C \setminus \{\mathbf{0}\};$$

equivalently,

$$h(x; y - x) \notin C \setminus \{\mathbf{0}\} \quad \text{implies} \quad f(y) - f(x) \notin -C \setminus \{\mathbf{0}\};$$

(e) *weakly C-h-pseudoconvex* if for all $x, y \in K$,

$$f(y) - f(x) \in -\operatorname{int}(C) \quad \text{implies} \quad h(x; y - x) \in -\operatorname{int}(C),$$

equivalently,

$$h(x; y - x) \notin -\operatorname{int}(C) \quad \text{implies} \quad f(y) - f(x) \notin -\operatorname{int}(C).$$

Obviously, strictly C-h-convexity implies C-h-convexity, and C-h-convexity implies C-h-pseudoconvexity.

If $h(x; d) = D^+f(x; d)$ (respectively, $D_+f(x; d)$) upper (respectively, lower) Dini directional derivative of a function f at x in the direction d, then C-h-convexity is called C-D^+-convexity (respectively, C-D_+-convexity), and so on.

Example 7.6 Let $X = \mathbb{R}$, $K = [0, 1]$, $Y = \mathbb{R}^2$, and $C = \mathbb{R}^2_+$. Let the function $h : K \times \mathbb{R} \to \mathbb{R}^2$, be given as $h(x; d) = \left(-x^2, -|x| - |d|\right)$. Furthermore, let $f : K \to \mathbb{R}^2$ be defined by $f = \left(x^2, |x|\right)$. Then f is C-h-convex, but not strictly C-h-convex. Moreover, f is strongly C-h-pseudoconvex and C-h-pseudoconvex, but f is not weakly C-h-pseudoconvex.

The following result provides the relation among the weakly efficient solution of VOP and the solutions of h-WVVIP and (D^+-WVVIP).

Theorem 7.7 *Let $f : K \to \mathbb{R}^\ell$ be a vector-valued function. Then the following statements hold.*

(a) *Every strongly efficient solution of* VOP *is a solution of* D^+-SVVIP (7.7).

(b) *If f is strongly C-h-pseudoconvex, then every solution of h-*SVVIP *is a strongly efficient solution of* VOP.

Proof

(a) Let $\bar{x}$ be a strongly efficient solution of VOP. Then

$$f(y) - f(\bar{x}) \in C, \quad \text{for all } y \in K.$$

Since K is a convex set, we have $\bar{x} + \lambda(y - \bar{x}) \in K$ for all $\lambda \in [0, 1]$; thus,

$$\frac{f(\bar{x} + \lambda(y - \bar{x})) - f(\bar{x})}{\lambda} \in C, \quad \text{for all } \lambda \in]0, 1[.$$

Taking the limit sup as $\lambda \downarrow 0$, we obtain

$$D^+ f(\bar{x}; y - \bar{x}) = \limsup_{\lambda \downarrow 0} \frac{f(\bar{x} + \lambda(y - \bar{x})) - f(\bar{x})}{\lambda} \in C, \quad \text{for all } y \in K.$$

Hence, $\bar{x}$ is a solution of D^+-SVVIP (7.7).

(b) Assume that $\bar{x} \in K$ is a solution of D^+-SVVIP (7.7) but not a strongly efficient solution of VOP. Then there exists $y \in K$ such that

$$f(y) - f(\bar{x}) \notin C.$$

Since f is strongly C-h-pseudoconvex, we have

$$h(\bar{x}; y - \bar{x}) \notin C,$$

contradicting our assumption that $\bar{x}$ is a solution of h-SVVIP. □

Since $\mathbb{R}^\ell \setminus \{-\operatorname{int}(C)\}$ is a closed convex cone, in a similar way, we can prove the following result.

Theorem 7.8 *Let $f : K \to \mathbb{R}^\ell$ be a vector-valued function. Then the following statements hold.*

(a) *Every weakly efficient solution of* VOP *is a solution of D^+-*WVVIP (7.9).

(b) *If f is weakly C-h-pseudoconvex, then every solution of h-*WVVIP *is a weakly efficient solution of* VOP.

Theorem 7.9 *Let $f : K \to \mathbb{R}^\ell$ be a C-h-pseudoconvex vector-valued function. Then every solution of h-VVIP is an efficient of VOP.*

Proof It lies on the lines of the proof of Theorem 7.7 (b), therefore, we omit it. □

Theorem 7.10 *Let* $f : K \to \mathbb{R}^\ell$ *be a vector-valued function such that* $(-f)$ *is strictly* C*-*h*-convex, that is,*

$$-(f(y) - f(x)) - h(\bar{x}; y - \bar{x}) \in \operatorname{int}(C), \quad \text{for all } x, y \in K. \tag{7.25}$$

Then every weakly efficient solution of VOP *is a solution of* h-VVIP.

Proof Assume that $\bar{x}$ is a weakly efficient solution of VOP but not a solution of h-VVIP. Then there exists $y \in K$ such that

$$h(\bar{x}; y - \bar{x}) \in C \setminus \{\mathbf{0}\}. \tag{7.26}$$

Combining (7.26) and (7.25), we obtain

$$f(y) - f(\bar{x}) \in -\operatorname{int}(C),$$

a contradiction to our assumption that $\bar{x}$ is a weakly efficient solution of VOP. Hence, $\bar{x}$ is a solution of h-VVIP. □

Theorem 7.11 *If* $f : K \to \mathbb{R}^\ell$ *is strictly* C*-*D^+*-convex function, then every weakly efficient solution of* VOP *is an efficient solution of* VOP.

Proof Assume that $\bar{x}$ is a weak efficient solution of VOP, but not an efficient solution of VOP. Then there exists $y \in K$ such that

$$f(y) - f(\bar{x}) \in -C \setminus \{\mathbf{0}\}. \tag{7.27}$$

Since f is strictly C-D^+-convex, we have

$$D^+f(\bar{x}; y - \bar{x}) - f(y) + f(\bar{x}) \in -\operatorname{int}(C). \tag{7.28}$$

By combining (7.27) and (7.28), we obtain

$$D^+f(\bar{x}; y - \bar{x}) \in -\operatorname{int}(C).$$

Thus, $\bar{x}$ is not a solution of D^+-WVVIP (7.9). By using Theorem 7.8 (a), we see that $\bar{x}$ is not a weak efficient solution of VOP, a contradiction to our assumption. □

Theorem 7.12 *For each* $i = 1, 2, \ldots, \ell$*, let* $f_i : K \to \mathbb{R}$ *be* D^+*-pseduoconvex and lower semicontinuous. If* $\bar{x} \in K$ *is a solution of* D^+-VVIP*, then it is an efficient solution of* VOP.

Proof Suppose that $\bar{x} \in K$ is not an efficient solution of VOP. Then there exists $\hat{y} \in K$ such that $f_i(\hat{y}) < f_i(\bar{x})$ for some i and $f_j(\hat{y}) \leq f_j(\bar{x})$ for all $j \neq i$. By D^+-pseudoconvexity of f_i, $D^+f_i(\bar{x}; \hat{y} - \bar{x}) < 0$. By Remark 1.18 (b), f_j is quasiconvex for all $j \neq i$, and hence, $D^+f_j(\bar{x}; \hat{y} - \bar{x}) \leq 0$ for all $j \neq i$. Thus, $\bar{x} \in K$ is not a solution of D^+-VVIP. □

The following result, due to Ansari and Lee [1], provides the relationship between a solution of an h-MVVIP and an efficient solution of VOP. It can be treated as a nonsmooth version of Theorem 5.28.

Theorem 7.13 *For each $i \in \mathscr{I}$, let $f_i : K \to \mathbb{R}$ be upper semicontinuous and D^+-pseudoconvex. For each $i \in \mathscr{I} = \{1, 2, \ldots, \ell\}$ and for all $x \in K$, let $h_i(x; \cdot)$ be positively homogeneous and subodd such that $h_i(x; \cdot) \leq D^+ f_i(x; \cdot)$. Then $\bar{x} \in K$ is a solution of* h-MVVIP *if and only if it is an efficient solution of* VOP.

Proof Let $\bar{x} \in K$ be a solution of h-MVVIP but not an efficient solution of VOP. Then there exists $z \in K$ such that

$$f(\bar{x}) - f(z) \in C \setminus \{\mathbf{0}\}. \tag{7.29}$$

Set $z(\lambda) := \lambda \bar{x} + (1 - \lambda) z$ for all $\lambda \in [0, 1]$. Since K is convex, $z(\lambda) \in K$ for all $\lambda \in [0, 1]$. Also since each f_i is D^+-pseudoconvex, it follows from Lemma 1.4 that f_i is quasiconvex and semistrictly quasiconvex. By using quasiconvexity, semistrictly quasiconvexity and (7.29), we get

$$f_i(\bar{x}) - f_i(z(\lambda)) \in C \setminus \{\mathbf{0}\}, \quad \text{for all } \lambda \in]0, 1[.$$

That is,

$$f_i(\bar{x}) \geq f_i(z(\lambda)), \quad \text{for all } \lambda \in]0, 1[\text{ and all } i = 1, 2, \ldots, \ell, \tag{7.30}$$

with strict inequality holds in (7.30) for some k such that $1 \leq k \leq \ell$.

By Diewert Mean-Value Theorem 1.31, there exists $\alpha_i \in]0, 1[$ such that

$$f_i(z(\lambda)) - f_i(\bar{x}) \geq D^+ f_i\left(z(\alpha_i); z(\lambda) - \bar{x}\right), \quad \text{for all } \lambda \in]0, 1[\text{ and all } i \in \mathscr{I}. \tag{7.31}$$

Combining inequalities (7.30) and (7.31), we obtain

$$D^+ f_i(z(\alpha_i); z(\lambda) - \bar{x}) \leq 0, \quad \text{for all } \lambda \in]0, 1[\text{ and all } i = 1, 2, \ldots, \ell,$$

with strict inequality holds for some k such that $1 \leq k \leq \ell$. Since, for each fixed $x \in K$, $h_i(x; \cdot) \leq D^+ f_i(x; \cdot)$, we have

$$h_i(z(\alpha_i); z(\lambda) - \bar{x}) \leq 0, \quad \text{for all } \lambda \in]0, 1[\text{ and all } i = 1, 2, \ldots, \ell,$$

where strict inequality holds for some k such that $1 \leq k \leq \ell$. By using the positive homogeneity of h_i in the second argument, we get

$$h_i(z(\alpha_i); z(\lambda) - \bar{x}) = h_i(z(\alpha_i); \lambda \bar{x} + (1 - \lambda) z - \bar{x}) = (1 - \lambda) h_i(z(\alpha_i); z - \bar{x})$$

and so,

$$h_i(z(\alpha_i); z - \bar{x}) \leq 0, \quad \text{for all } i = 1, 2, \ldots, \ell, \tag{7.32}$$

where strict inequality holds for some k such that $1 \leq k \leq \ell$. By the suboddness of h_i in the second argument, we have

$$h_i(z(\alpha_i); \bar{x} - z) \geq 0, \quad \text{for all } i = 1, 2, \ldots, \ell, \tag{7.33}$$

where strict inequality holds for some k such that $1 \leq k \leq \ell$.

Suppose that $\alpha_1, \ldots, \alpha_\ell$ are all equal. Then by (7.33), the positive homogeneity of h_i in the second argument, and the fact that

$$\bar{x} - z(\alpha_i) = (1 - \alpha_i)(\bar{x} - z),$$

we have

$$h_i(z(\alpha_i); \bar{x} - z(\alpha_i)) \geq 0, \quad \text{for all } i = 1, 2, \ldots, \ell,$$

where strict inequality holds for some i, that is,

$$\big(h_1(z(\alpha_1); \bar{x} - z(\alpha_1)), \ldots, h_\ell(z(\alpha_\ell); \bar{x} - z(\alpha_\ell))\big) \in C \setminus \{\mathbf{0}\},$$

which contradicts to our assumption that $\bar{x}$ is a solution of (h-MVVIP).

Consider the case when $\alpha_1, \alpha_2, \ldots, \alpha_\ell$ are not equal. Let $\alpha_1 \neq \alpha_2$.

If $\alpha_1 < \alpha_2$, then by the positive homogeneity and the suboddness of $h_i(x; \cdot)$, we get

$$h_1(z(\alpha_1); z(\alpha_2) - z(\alpha_1)) = (\alpha_2 - \alpha_1) h_1(z(\alpha_1); \bar{x} - z),$$

and by using (7.33), we obtain

$$h_1(z(\alpha_1); z(\alpha_2) - z(\alpha_1)) = (\alpha_2 - \alpha_1) h_1(z(\alpha_1); \bar{x} - z) \geq 0, \tag{7.34}$$

where strict inequality holds for $k = 1$.

Since each f_i is D^+-pseudoconvex and $h_i(x; \cdot) \leq D^+ f_i(x; \cdot)$, by Lemma 1.5, f_i is h_i-pseudoconvex; further, by Lemma 1.7 (b), h_i is pseudomonotone. Therefore, we have

$$h_1(z(\alpha_2); z(\alpha_2) - z(\alpha_1)) \geq 0, \tag{7.35}$$

where strict inequality holds for $k = 1$ by virtue of Lemma 1.6. The positive homogeneity of $h_i(x; \cdot)$ implies that

$$(\alpha_2 - \alpha_1) h_1(z(\alpha_2); \bar{x} - z) \geq 0,$$

where strict inequality holds for $k = 1$. Since $\alpha_2 - \alpha_1 > 0$, we have

$$h_1(z(\alpha_2); \bar{x} - z) \geq 0,$$

with strict inequality for $k = 1$.

If $\alpha_1 > \alpha_2$, then by the positive homogeneity and the suboddness of $h_i(x; \cdot)$, we get

$$h_2(z(\alpha_2); z(\alpha_1) - z(\alpha_2)) = (\alpha_1 - \alpha_2) h_2(z(\alpha_2); \bar{x} - z),$$

and by using (7.33), we obtain

$$h_2(z(\alpha_2); z(\alpha_1) - z(\alpha_2)) = (\alpha_1 - \alpha_2) h_2(z(\alpha_2); \bar{x} - z) \geq 0, \tag{7.36}$$

with strict inequality for $k = 2$.

As above, each h_i is pseudomonotone; therefore,

$$h_2(z(\alpha_1); z(\alpha_1) - z(\alpha_2)) \geq 0, \tag{7.37}$$

with strict inequality for $k = 2$ by virtue of Lemma 1.6 Again as above, by using the positive homogeneity of $h_i(x; \cdot)$, we get

$$h_2(z(\alpha_1); \bar{x} - z) \geq 0,$$

with strict inequality for $k = 2$.

For the case $\alpha_1 \neq \alpha_2$, let $\bar{\alpha} = \max\{\alpha_1, \alpha_2\}$. Then we have

$$h_i(z(\bar{\alpha}); \bar{x} - z) \geq 0, \quad \text{for all } i = 1, 2.$$

By continuing this process, we can find $\alpha^* \in]0, 1[$ such that

$$h_i(z(\alpha^*); \bar{x} - z) \geq 0, \quad \text{for all } i = 1, 2, \ldots, \ell,$$

with strict inequality for some k such that $1 \leq k \leq \ell$. By multiplying the above inequality by $1 - \alpha^*$, we obtain

$$h_i(z(\alpha^*); \bar{x} - z(\alpha^*)) \geq 0, \quad \text{for all } i = 1, 2, \ldots, \ell,$$

with strict inequality for some k such that $1 \leq k \leq \ell$. Thus,

$$\left(h_1(z(\alpha^*); \bar{x} - z(\alpha^*)), \ldots, h_\ell(z(\alpha^*); \bar{x} - z(\alpha^*))\right) \in C \setminus \{\mathbf{0}\},$$

which contradicts our supposition that $\bar{x}$ is a solution of h-MVVIP.

Conversely, suppose that $\bar{x} \in K$ is an efficient solution of VOP, but not a solution of h-MVVIP. Then there exists $z \in K$ such that

$$h(z; \bar{x} - z) \in C \setminus \{\mathbf{0}\},$$

that is,

$$h_i(z; \bar{x} - z) \geq 0, \quad \text{for all } i \in \mathscr{I},$$

with strict inequality holds for some i. Since $h_i(z; \ ) \leq D^+ f_i(z; \cdot)$ for all $i \in \mathscr{I}$,

$$D^+ f_i(z; \bar{x} - z) \geq 0, \quad \text{for all } i \in \mathscr{I},$$

with strict inequality holds for some i. Since each f_i is D^+-pseudoconvex, we have

$$f_i(\bar{x}) \geq f_i(z), \quad \text{for all } i \in \mathscr{I}.$$

Let $j \in \mathscr{I}$ be such that $D^+ f_j(z; \bar{x} - z) > 0$. Since f_j is upper semicontinuous and D^+-pseudoconvex, it follows from Lemma 1.4 that f_j is quasiconvex; hence, it follows from Theorem 4 in [6] that $f_j(\bar{x}) > f_j(z)$. Thus, $f(z) - f(\bar{x}) \in -C \setminus \{\mathbf{0}\}$; hence, $\bar{x}$ is not an efficient solution of VOP. This contradiction proves our result. □

The following result gives the relation between a solution of h-VVIP and a properly efficient solution (in the sense of Henig) of VOP.

Theorem 7.14 *If $\bar{x} \in K$ is a properly efficient solution (in the sense of Henig) of* VOP, *then it is a solution of* D^+-VVIP.

Proof Since $\bar{x} \in K$ is a properly efficient solution (in the sense of Henig) of VOP, there is convex cone D in $\mathbb{R}^\ell$ such that $C \setminus \{\mathbf{0}\} \subseteq \text{int}(D)$, and

$$f(y) - f(\bar{x}) \notin -D \setminus \{\mathbf{0}\}, \quad \text{for all } y \in K.$$

Since $-\text{int}(D) \subseteq -D \setminus \{\mathbf{0}\}$, we have

$$f(y) - f(\bar{x}) \notin -\text{int}(D), \quad \text{for all } y \in K.$$

Since K is a convex set, we have $\bar{x} + \lambda(y - \bar{x}) \in K$ for all $\lambda \in [0, 1]$; thus,

$$\frac{f(\bar{x} + \lambda(y - \bar{x})) - f(\bar{x})}{\lambda} \in \mathbb{R}^\ell \setminus \{-\text{int}(D)\}, \quad \text{for all } \lambda \in]0, 1[.$$

Since $\mathbb{R}^\ell \setminus \{-\text{int}(D)\}$ is a closed convex cone, by taking the limit sup as $\lambda \downarrow 0$, we obtain

$$D^+ f(\bar{x}; y - \bar{x}) = \limsup_{\lambda \downarrow 0} \frac{f(\bar{x} + \lambda(y - \bar{x})) - f(\bar{x})}{\lambda} \in \mathbb{R}^\ell \setminus \{-\text{int}(D)\}, \quad \text{for all } y \in K.$$

Therefore,

$$\langle D^+ f(\bar{x}), y - \bar{x} \rangle \notin -\operatorname{int}(D), \quad \text{for all } y \in K.$$

Since $C \setminus \{\mathbf{0}\} \subseteq \operatorname{int}(D)$ for all $x \in K$, we obtain the result. □

Theorem 7.15 ([10]) *If $\bar{x} \in K$ is a properly efficient solution (in the sense of Geoffrion) of* VOP *and for each $i = 1, 2, \ldots, \ell$, $D^+ f_i(\bar{x}; \cdot)$ is finite on $K - \bar{x}$, then $\bar{x} \in K$ is a solution of D^+*-VVIP.

Proof Let $\bar{x} \in K$ be a properly efficient solution (in the sense of Geoffrion) of VOP. Suppose on contrary that there exists $d \in \mathbb{R}^n$ such that $d \in K - \bar{x}$, $D^+ f_i(\bar{x}; d) < 0$ and $D^+ f_j(\bar{x}; d) \leq 0$, $j \neq i$. We choose $v \in K$ such that $d = v - \bar{x}$. Since K is convex, we can choose a sequence $\{t_n\}$ of positive real numbers such that $t_n \downarrow 0$, $\bar{x} + t_n(v - \bar{x}) \in K$ for all n and

$$D^+ f_i(\bar{x}; d) = \lim_{n \to \infty} \frac{f_i(\bar{x} + t_n(v - \bar{x})) - f_i(\bar{x})}{t_n}.$$

Since $D^+ f_i(\bar{x}; d) < 0$,

$$\lim_{n \to \infty} \frac{f_i(\bar{x} + t_n(v - \bar{x})) - f_i(\bar{x})}{t_n} < 0,$$

and hence, there exists a natural number N such that for all $n \geq N$,

$$\frac{1}{t_n} [f_i(\bar{x} + t_n(v - \bar{x})) - f_i(\bar{x})] < 0,$$

that is, $f_i(\bar{x} + t_n(v - \bar{x})) < f_i(\bar{x})$. Since $\bar{x}$ is an efficient solution VOP, choosing a subsequence of the sequence $\{\bar{x} + t_n(v - \bar{x})\}$, if necessary, we may assume that

$$I := \{j : f_j(\bar{x} + t_n(v - \bar{x})) > f_j(\bar{x})\}$$

is constant for all $n \geq N$. So, for all $j \in I$, we have

$$\limsup_{n \to \infty} \frac{f_j(\bar{x} + t_n(v - \bar{x})) - f_j(\bar{x})}{t_n} \geq 0.$$

Since $D^+ f_j(\bar{x}; d) \leq 0$ for all $j \in I$, we have

$$\limsup_{n \to \infty} \frac{f_j(\bar{x} + t_n(v - \bar{x})) - f_j(\bar{x})}{t_n} = 0,$$

for all $j \in I$. So, choosing a subsequence of $\{t_n\}$, if necessary, we may assume that

$$\lim_{n\to\infty} \frac{f_j(\bar{x}+t_n(v-\bar{x}))-f_j(\bar{x})}{t_n} = 0,$$

for all $j \in I$. So, for all $j \in I$, we have

$$\frac{f_i(\bar{x})-f_i(\bar{x}+t_n(v-\bar{x}))}{f_j(\bar{x}+t_n(v-\bar{x}))-f_j(\bar{x})} = \frac{\frac{1}{t_n}[f_i(\bar{x})-f_i(\bar{x}+t_n(v-\bar{x}))]}{\frac{1}{t_n}[f_j(\bar{x}+t_n(v-\bar{x}))-f_j(\bar{x})]} \to +\infty \text{ as } n \to \infty,$$

which contradicts to the proper efficiency (in the sense of Geoffrion) of $\bar{x}$. □

Remark 7.2 Since proper efficiency in the sense of Benson and in the sense of Geoffrion are equivalent when $C = \mathbb{R}^\ell_+$, Theorem 7.15 also holds for proper efficiency in the sense of Benson.

Corollary 7.1 ([11, Theorem 4]) *For each $i = 1, 2, \ldots, \ell$, let $f_i : K \to \mathbb{R}$ be convex and differentiable. If $\bar{x} \in K$ is a properly efficient solution (in the sense of Benson) of* VOP, *then it is a solution of* ∇-VVIP.

The following example shows that the Corollary 7.1 cannot be extended to efficient solutions of VOP even though each f_i is convex.

Example 7.7 Let $K = [-1, 0]$ and $f(x) = (x, x^2)$. Then $\bar{x} = 0$ is an efficient solution of VOP, but it is not a solution of the following ∇-VVIP: Find $\bar{x} \in K$ such that for all $y \in K$,

$$(\langle \nabla f_1(\bar{x}), y-\bar{x}\rangle, \langle \nabla f_2(\bar{x}), y-\bar{x}\rangle) = (y-\bar{x}, 2\bar{x}(y-\bar{x})) \notin -\mathbb{R}^2_+ \setminus \{\mathbf{0}\}.$$

We notice that $\bar{x} = 0$ is not a properly efficient solution (in the sense of Benson) of VOP.

References

1. Q.H. Ansari, G.M. Lee, Nonsmooth vector optimization problems and Minty vector variational inequalities. J. Optim. Theory Appl. **145**(1), 1–16 (2010)
2. Q.H. Ansari, C.S. Lalitha, M. Mehta, *Generalized Convexity, Nonsmooth Variational Inequalities and Nonsmooth Optimization* (CRC Press/Taylor & Francis Group, Boca Raton/London/New York, 2014)
3. G.P. Crespi, I. Ginchev, M. Rocca, Minty variational inequalities, increase along rays property and optimization. J. Optim. Theory Appl. **123**(3), 479–496 (2004)
4. G.P. Crespi, I. Ginchev, M. Rocca, Existence of solutions and star-shapedness in Minty variational inequalities. J. Glob. Optim. **32**, 485–494 (2005)
5. G.P. Crespi, I. Ginchev, M. Rocca, Some remarks on the Minty vector variational inequalities. J. Math. Anal. Appl. **345**(3), 165–175 (2008)

6. W.E. Diewert, Alternative characterizations of six kinds of quasiconcavity in the nondifferentiable case with applications to nonsmooth programming, in *Generalized Concavity in Optimization and Economics*, ed. by S. Schaible, W.T. Ziemba (Academic Press, New York, 1981), pp. 51–93
7. Y.-P. Fang, R. Hu, A non-smooth version of Minty variational principle. Optimization **58**, 401–412 (2009)
8. C.S. Lalitha, M. Mehta, Vector variational inequalities with cone-pseudomonotone bifunctions. Optimization **54**(3), 327–338 (2005)
9. C.S. Lalitha, M. Mehta, On vector variational inequality problem in terms of bifunctions. Aust. J. Math. Anal. Appl. **3**(2), 1–11 (2005). Article ID 11
10. G.M. Lee, M.H. Kim, Remarks on relations between vector variational inequality and vector optimization. Nonlinear Anal. **47**, 627–635 (2001)
11. X.Q. Yang, On some equivalent conditions of vector variational inequalities, in *Vector Variational Inequality and Vector Equilibria: Mathematical Theories*, ed. by F. Giannessi (Kluwer Academic Publishers, Dordrechet/Boston, 2000), pp. 423–432

Chapter 8
Generalized Vector Variational Inequalities

When the objective function involved in the vector optimization problem is not necessarily differentiable, then the method to solve VOP via corresponding vector variational inequality problems is no longer valid. We need to generalize the vector variational inequality problems for set-valued maps. There are several ways to generalize vector variational inequality problems discussed in Chap. 5. The main objective of this chapter is to generalize the vector variational inequality problems for set-valued maps and to present the existence results for such generalized vector variational inequality problems with or without monotonicity assumption. We also present some relations between a generalized vector variational inequality problem and a vector optimization problem with a nondifferentiable objective function. Several results of this chapter also hold in the setting of Hausdorff topological vector spaces, but for the sake of convenience, our setting is Banach spaces.

8.1 Formulations and Preliminaries

When the map T involved in the formulation of vector variational inequality problems and Minty vector variational inequality problems is a set valued map, then the vector variational inequality problems and Minty vector variational inequality problems, discussed in Chap. 5, are called (more precisely, Stampacchia) generalized vector variational inequality problems and Minty generalized vector variational inequality problems, respectively.

Let X and Y be Banach spaces and K be a nonempty convex subset of X. Let $T : K \to 2^{\mathcal{L}(X,Y)}$ be a set-valued map with nonempty values, and $C : K \to 2^Y$ be a set-valued map such that for all $x \in K$, $C(x)$ is a closed convex pointed cone. We also assume that $\text{int}(C(x)) \neq \emptyset$ wherever $\text{int}(C(x))$ the interior of the set $C(x)$ is involved in the formulation of a problem. For every $l \in \mathcal{L}(X, Y)$, the value of l at x is denoted by $\langle l, x \rangle$.

Q.H. Ansari et al., *Vector Variational Inequalities and Vector Optimization*, Vector Optimization, DOI 10.1007/978-3-319-63049-6_8

We consider the following *generalized vector variational inequality problems* (SGVVIP) and *Minty generalized vector variational inequality problems* (MGVVIP).

$(\text{GSVVIP})_s$: Find $\bar{x} \in K$ such that there exists $\bar{\zeta} \in T(\bar{x})$ satisfying

$$\langle \bar{\zeta}, y - \bar{x} \rangle \in C(\bar{x}), \quad \text{for all } y \in K. \tag{8.1}$$

$(\text{GSVVIP})_w$: Find $\bar{x} \in K$ such that for all $y \in K$, there exists $\bar{\zeta} \in T(\bar{x})$ satisfying

$$\langle \bar{\zeta}, y - \bar{x} \rangle \in C(\bar{x}). \tag{8.2}$$

$(\text{MGSVVIP})_g$: Find $\bar{x} \in K$ such that for all $y \in K$ and all $\xi \in T(y)$, we have

$$\langle \xi, y - \bar{x} \rangle \in C(\bar{x}). \tag{8.3}$$

$(\text{MGSVVIP})_w$: Find $\bar{x} \in K$ such that for all $y \in K$, there exists $\xi \in T(y)$ satisfying

$$\langle \xi, y - \bar{x} \rangle \in C(\bar{x}). \tag{8.4}$$

$(\text{GVVIP})_g$: Find $\bar{x} \in K$ such that for all $\bar{\zeta} \in T(\bar{x})$, we have

$$\langle \bar{\zeta}, y - \bar{x} \rangle \notin -C(\bar{x}) \setminus \{\mathbf{0}\}, \quad \text{for all } y \subset K. \tag{8.5}$$

$(\text{GVVIP})_s$: Find $\bar{x} \in K$ such that there exists $\bar{\zeta} \in T(\bar{x})$ satisfying

$$\langle \bar{\zeta}, y - \bar{x} \rangle \notin -C(\bar{x}) \setminus \{\mathbf{0}\}, \quad \text{for all } y \in K. \tag{8.6}$$

$(\text{GVVIP})_w$: Find $\bar{x} \in K$ such that for all $y \in K$, there exists $\bar{\zeta} \in T(\bar{x})$ satisfying

$$\langle \bar{\zeta}, y - \bar{x} \rangle \notin -C(\bar{x}) \setminus \{\mathbf{0}\}. \tag{8.7}$$

$(\text{MGVVIP})_g$: Find $\bar{x} \in K$ such that for all $y \in K$ and all $\xi \in T(y)$, we have

$$\langle \xi, y - \bar{x} \rangle \notin -C(\bar{x}) \setminus \{\mathbf{0}\}. \tag{8.8}$$

(MGVVIP)$_w$: Find $\bar{x} \in K$ such that for all $y \in K$, there exists $\xi \in T(y)$ satisfying

$$\langle \xi, y - \bar{x} \rangle \notin -C(\bar{x}) \setminus \{\mathbf{0}\}. \tag{8.9}$$

(GWVVIP)$_g$: Find $\bar{x} \in K$ such that for all $\bar{\zeta} \in T(\bar{x})$, we have

$$\langle \bar{\zeta}, y - \bar{x} \rangle \notin -\operatorname{int}(C(\bar{x})), \quad \text{for all } y \in K. \tag{8.10}$$

(GWVVIP)$_s$: Find $\bar{x} \in K$ such that there exists $\bar{\zeta} \in T(\bar{x})$ satisfying

$$\langle \bar{\zeta}, y - \bar{x} \rangle \notin -\operatorname{int}(C(\bar{x})), \quad \text{for all } y \in K. \tag{8.11}$$

(GWVVIP)$_w$: Find $\bar{x} \in K$ such that for all $y \in K$, there exists $\bar{\zeta} \in T(\bar{x})$ satisfying

$$\langle \bar{\zeta}, y - \bar{x} \rangle \notin -\operatorname{int}(C(\bar{x})). \tag{8.12}$$

(MGWVVIP)$_g$: Find $\bar{x} \in K$ such that for all $y \in K$ for all $\xi \in T(y)$, we have

$$\langle \xi, y - \bar{x} \rangle \notin -\operatorname{int}(C(\bar{x})). \tag{8.13}$$

(MGWVVIP)$_w$: Find $\bar{x} \in K$ such that for all $y \in K$, there exists $\xi \in T(y)$ satisfying

$$\langle \xi, y - \bar{x} \rangle \notin -\operatorname{int}(C(\bar{x})). \tag{8.14}$$

In (GSVVIP)$_w$, (GVVIP)$_w$, and (GWVVIP)$_w$, $\bar{\zeta} \in T(\bar{x})$ depends on $y \in K$; Also, in (MGSVVIP)$_w$, (MGVVIP)$_w$, and (MGWVVIP)$_w$, $\xi \in T(y)$ depends on $y \in K$.

We denote by Sol(GSVVIP)^d_g, Sol(GSVVIP)^d_s, Sol(GSVVIP)^d_w, Sol(MGSVVIP)^d_g, Sol(MGSVVIP)^d_w, Sol(GVVIP)^d_g, Sol(GVVIP)^d_s, Sol(GVVIP)^d_w, Sol(MGVVIP)^d_g, Sol(MGVVIP)^d_w, Sol(GWVVIP)^d_g, Sol(GWVVIP)^d_s, Sol(GWVVIP)^d_w, Sol(MGWVVIP)^d_g, and Sol(MGWVVIP)^d_w, the set of solutions of (GSVVIP)$_g$, (GSVVIP)$_s$, (GSVVIP)$_w$, (MGSVVIP)$_g$, (MGSVVIP)$_w$, (GVVIP)$_g$, (GVVIP)$_s$, (GVVIP)$_w$, (MGVVIP)$_g$, (MGVVIP)$_w$, (GWVVIP)$_g$, (GWVVIP)$_s$, (GWVVIP)$_w$, (MGWVVIP)$_g$, and (MGWVVIP)$_w$, respectively.

If for all $x \in K$, $C(x) = D$ is a fixed closed convex pointed cone with $\operatorname{int}(D) \neq \emptyset$, then the solution set of (GSVVIP)$_g$, (GSVVIP)$_s$, (GSVVIP)$_w$, (MGSVVIP)$_g$, (MGSVVIP)$_w$, (GVVIP)$_g$, (GVVIP)$_s$, (GVVIP)$_w$, (MGVVIP)$_g$, (MGVVIP)$_w$, (GWVVIP)$_g$, (GWVVIP)$_s$, (GWVVIP)$_w$, (MGWVVIP)$_g$, and (MGWVVIP)$_w$, are denoted by Sol(GSVVIP)$_g$, Sol(GSVVIP)$_s$, Sol(GSVVIP)$_w$, Sol(MGSVVIP)$_g$, Sol(MGSVVIP)$_w$, Sol(GVVIP)$_g$, Sol(GVVIP)$_s$, Sol(GVVIP)$_w$, Sol(MGVVIP)$_g$,

Sol(MGVVIP)$_w$, Sol(GWVVIP)$_g$, Sol(GWVVIP)$_s$, Sol(GWVVIP)$_w$, Sol(MGWVVIP)$_g$, and Sol(MGWVVIP)$_w$, respectively

Remark 8.1 It is clear that

(a) $\text{Sol(GSVVIP)}_g^d \subseteq \text{Sol(GSVVIP)}_s^d \subseteq \text{Sol(GSVVIP)}_w^d$;
(b) $\text{Sol(MGSVVIP)}_g^d \subseteq \text{Sol(MGSVVIP)}_w^d$;
(c) $\text{Sol(GVVIP)}_g^d \subseteq \text{Sol(GVVIP)}_s^d \subseteq \text{Sol(GVVIP)}_w^d$;
(d) $\text{Sol(MGVVIP)}_g^d \subseteq \text{Sol(MGVVIP)}_w^d$;
(e) $\text{Sol(GWVVIP)}_g^d \subseteq \text{Sol(GWVVIP)}_s^d \subseteq \text{Sol(GWVVIP)}_w^d$;
(f) $\text{Sol(MGWVVIP)}_g^d \subseteq \text{Sol(MGWVVIP)}_w^d$;
(g) $\text{Sol(GSVVIP)}_g^d \subseteq \text{Sol(GVVIP)}_g^d \subseteq \text{Sol(GWVVIP)}_g^d$;
(h) $\text{Sol(GSVVIP)}_s^d \subseteq \text{Sol(GVVIP)}_s^d \subseteq \text{Sol(GWVVIP)}_s^d$;
(i) $\text{Sol(SGVVIP)}_w^d \subseteq \text{Sol(GVVIP)}_w^d \subseteq \text{Sol(GWVVIP)}_w^d$;
(j) $\text{Sol(MGSVVIP)}_g^d \subseteq \text{Sol(MGVVIP)}_g^d \subseteq \text{Sol(MGWVVIP)}_g^d$;
(k) $\text{Sol(MGSVVIP)}_w^d \subseteq \text{Sol(MGVVIP)}_w^d \subseteq \text{Sol(MGWVVIP)}_w^d$.

Definition 8.1 Let K be a nonempty convex subset of X and $x \in K$ be an arbitrary element. The set-valued map $T : K \to 2^{\mathcal{L}(X,Y)}$ is said to be

(a) *strongly generalized C_x-upper sign continuous* if for all $y \in K$,

$$\begin{aligned}&\text{there exists } \xi_\lambda \in T(x+\lambda(y-x)) \text{ for } \lambda \in]0,1[\text{ such that}\\ &\langle \xi_\lambda, y-x\rangle \in C(x) \text{ implies that there exists } \zeta \in T(x)\\ &\text{such that } \langle \zeta, y-x\rangle \in C(x);\end{aligned}$$

(b) *strongly generalized C_x-upper sign continuous$_+$* if for all $y \in K$,

$$\begin{aligned}&\text{there exists } \xi_\lambda \in T(x+\lambda(y-x)) \text{ for } \lambda \in]0,1[\text{ such that}\\ &\langle \xi_\lambda, y-x\rangle \in C(x) \text{ implies that } \langle \zeta, y-x\rangle \in C(x) \text{ for all } \zeta \in T(x);\end{aligned}$$

(c) *strongly generalized C_x-upper sign continuous$^+$* if for all $y \in K$,

$$\begin{aligned}&\text{for all } \xi_\lambda \in T(x+\lambda(y-x)) \text{ for } \lambda \in]0,1[\text{ such that } \langle \xi_\lambda, y-x\rangle \in C(x)\\ &\text{implies that there exists } \zeta \in T(x) \text{ such that } \langle \zeta, y-x\rangle \in C(x);\end{aligned}$$

(d) *strongly generalized C_x-upper sign continuous$_+^+$* if for all $y \in K$,

$$\begin{aligned}&\text{for all } \xi_\lambda \in T(x+\lambda(y-x)) \text{ for } \lambda \in]0,1[\text{ such that } \langle \xi_\lambda, y-x\rangle \in C(x)\\ &\text{implies that } \langle \zeta, y-x\rangle \in C(x) \text{ for all } \zeta \in T(x);\end{aligned}$$

(e) *generalized C_x-upper sign continuous* if for all $y \in K$,

$$\text{there exists } \xi_\lambda \in T(x+\lambda(y-x)) \text{ for } \lambda \in]0,1[\text{ such that}$$
$$\langle \xi_\lambda, y-x \rangle \notin -C(x)\setminus\{\mathbf{0}\} \text{ implies that there exists } \zeta \in T(x) \text{ such that}$$
$$\langle \zeta, y-x \rangle \notin -C(x)\setminus\{\mathbf{0}\};$$

(f) *generalized C_x-upper sign continuous$_+$* if for all $y \in K$,

$$\text{there exists } \xi_\lambda \in T(x+\lambda(y-x)) \text{ for } \lambda \in]0,1[\text{ such that}$$
$$\langle \xi_\lambda, y-x \rangle \notin -C(x)\setminus\{\mathbf{0}\} \text{ implies that } \langle \zeta, y-x \rangle \notin -C(x)\setminus\{\mathbf{0}\}$$
$$\text{for all } \zeta \in T(x);$$

(g) *generalized C_x-upper sign continuous$^+$* if for all $y \in K$,

$$\text{for all } \xi_\lambda \in T(x+\lambda(y-x)) \text{ for } \lambda \in]0,1[\text{ such that}$$
$$\langle \xi_\lambda, y-x \rangle \notin -C(x)\setminus\{\mathbf{0}\} \text{ implies that there exists } \zeta \in T(x) \text{ such that}$$
$$\langle \zeta, y-x \rangle \notin -C(x)\setminus\{\mathbf{0}\};$$

(h) *generalized C_x-upper sign continuous$_+^+$* if for all $y \in K$,

$$\text{for all } \xi_\lambda \in T(x+\lambda(y-x)) \text{ for } \lambda \in]0,1[\text{ such that}$$
$$\langle \xi_\lambda, y-x \rangle \notin -C(x)\setminus\{\mathbf{0}\} \text{ implies that } \langle \zeta, y-x \rangle \notin -C(x)\setminus\{\mathbf{0}\}$$
$$\text{for all } \zeta \in T(x);$$

(i) *weakly generalized C_x-upper sign continuous* if for all $y \in K$,

$$\text{there exists } \xi_\lambda \in T(x+\lambda(y-x)) \text{ for } \lambda \in]0,1[\text{ such that}$$
$$\langle \xi_\lambda, y-x \rangle \notin -\operatorname{int}(C(x)) \text{ implies that there exists } \zeta \in T(x) \text{ such that}$$
$$\langle \zeta, y-x \rangle \notin -\operatorname{int}(C(x));$$

(j) *weakly generalized C_x-upper sign continuous$_+$* if for all $y \in K$,

$$\text{there exists } \xi_\lambda \in T(x+\lambda(y-x)) \text{ for } \lambda \in]0,1[\text{ such that}$$
$$\langle \xi_\lambda, y-x \rangle \notin -\operatorname{int}(C(x)) \text{ implies that } \langle \zeta, y-x \rangle \notin -\operatorname{int}(C(x))$$
$$\text{for all } \zeta \in T(x);$$

(k) *weakly generalized C_x-upper sign continuous*$^+$ if for all $y \in K$,

$$\text{for all } \xi_\lambda \in T(x + \lambda(y - x)) \text{ for } \lambda \in]0, 1[\text{ such that}$$
$$\langle \xi_\lambda, y - x\rangle \notin -\operatorname{int}(C(x)) \text{ implies that there exists } \zeta \in T(x)$$
$$\text{such that } \langle \zeta, y - x\rangle \notin -\operatorname{int}(C(x));$$

(l) *weakly generalized C_x-upper sign continuous*$_+^+$ if for all $y \in K$,

$$\text{for all } \xi_\lambda \in T(x + \lambda(y - x)) \text{ for } \lambda \in]0, 1[\text{ such that}$$
$$\langle \xi_\lambda, y - x\rangle \notin -\operatorname{int}(C(x)) \text{ implies that } \langle \zeta, y - x\rangle \notin -\operatorname{int}(C(x))$$
$$\text{for all } \zeta \in T(x).$$

Example 8.1 Let $X = \mathbb{R}$, $Y = \mathbb{R}^2$, $K =]0, 1]$ and $C(x) = \mathbb{R}^2_+$ for all $x \in K$. Consider the map $T(x) := \{(y_1, y_2) \in \mathbb{R}^2 : |y_1| \leq x,\ |y_2| \leq x\}$. Then T is strongly generalized C_x-upper sign continuous, strongly generalized C_x-upper sign continuous$^+$, generalized C_x-upper sign continuous, generalized C_x-upper sign continuous$^+$, weakly generalized C_x-upper sign continuous, and weakly generalized C_x-upper sign continuous$^+$. However, T is not strongly generalized C_x-upper sign continuous$_+$, strongly generalized C_x-upper sign continuous$_+^+$, generalized C_x-upper sign continuous$_+$, generalized C_x-upper sign continuous$_+^+$, weakly generalized C_x-upper sign continuous$_+$, or weakly generalized C_x-upper sign continuous$_+^+$ (Fig. 8.1).

Definition 8.2 Let K be a nonempty convex subset of X. A set-valued map $T : K \to 2^{\mathcal{L}(X,Y)}$ is said to be *generalized v-hemicontinuous* if for all $x, y \in K$, the set-valued map $F : [0, 1] \to 2^Y$, defined by $F(\lambda) = \langle T(x + \lambda(y - x)), y - x\rangle$, is upper semicontinuous at 0^+, where $\langle T(x + \lambda(y - x)), y - x\rangle = \{\langle \zeta, y - x\rangle : \zeta \in T(x + \lambda(y - x))\}$.

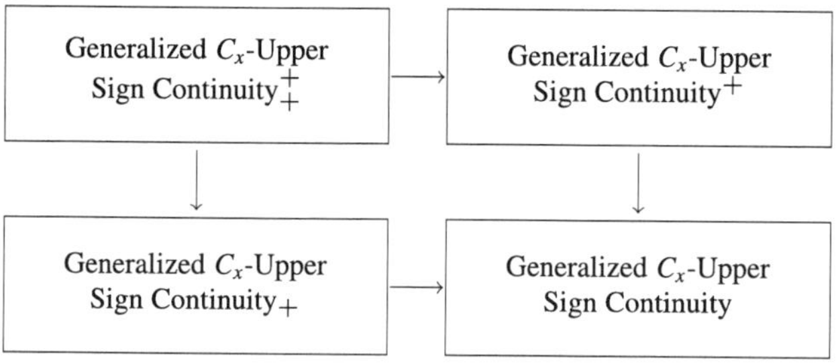

Fig. 8.1 Relations among different kinds of generalized C_x-upper sign continuities. The similar diagram also holds for weak as well as for strong cases

Lemma 8.1 *Let K be a nonempty convex subset of X and $x \in K$ be an arbitrary element. If the set-valued map $T : K \to 2^{\mathcal{L}(X,Y)}$ is generalized υ-hemicontinuous, then it is strongly generalized C_x-upper sign continuous as well as weakly generalized C_x-upper sign continuous.*

Proof Let x be an arbitrary but fixed element. Suppose to the contrary that T is not weakly generalized C_x-upper sign continuous. Then for some $y \in K$ and all $\xi_\lambda \in T(x + \lambda(y - x))$, $\lambda \in]0, 1[$, we have

$$\langle \xi_\lambda, y - x \rangle \notin -\operatorname{int}(C(x)) \tag{8.15}$$

implies

$$\langle \zeta, y - x \rangle \in -\operatorname{int}(C(x)), \quad \text{for all } \zeta \in T(x).$$

Since T is generalized υ-hemicontinuous, the set-valued map $F : [0, 1] \to 2^Y$, defined in Definition 8.2, is upper semicontinuous at 0^+, and $F(0) = \langle T(x), y-x \rangle \subseteq -\operatorname{int}(C(x))$, we have that there exists an open neighborhood $V =]0, \delta[\subseteq [0, 1]$ such that $F(\lambda) = \langle T(x + \lambda(y - x)), y - x \rangle \subseteq -\operatorname{int}(C(x))$ for all $\lambda \in]0, \delta[$, that is, for all $\xi_\lambda \in T(x + \lambda(y - x))$ and all $\lambda \in]0, \delta[$, we have $\langle \xi_\lambda, y - x \rangle \in -\operatorname{int}(C(x))$, a contradiction of (8.15). Hence, T is weakly generalized C_x-upper sign continuous.

Since $W(x) = Y \setminus \{C(x)\}$ is an open set for all $x \in K$, the proof for strong case is similar, and therefore, we omit it. □

Remark 8.2 The generalized υ-hemicontinuity does not imply the generalized C_x-upper sign continuity.

Definition 8.3 Let K be a nonempty convex subset of X and $T : K \to 2^{\mathcal{L}(X,Y)}$ be a set-valued map with nonempty compact values. Then T is said to be *$\mathscr{H}$-hemicontinuous* if for all $x, y \in K$, the set-valued map $F : [0, 1] \to 2^Y$, defined by $F(\lambda) = \mathscr{H}(T(x + \lambda(y - x)), T(x))$, is $\mathscr{H}$-continuous at 0^+, where $\mathscr{H}$ denotes the Hausdorff metric on the family of all nonempty closed bounded subsets of $\mathcal{L}(X, Y)$.

Lemma 8.2 *Let K be a nonempty convex subset of X and $x \in K$ be an arbitrary element. If the set-valued map $T : K \to 2^{\mathcal{L}(X,Y)}$ is nonempty compact valued and $\mathscr{H}$-hemicontinuous, then it is strongly generalized C_x-upper sign continuous$^+$ as well as weakly generalized C_x-upper sign continuous$^+$.*

Proof Let x be an arbitrary but fixed element and suppose that T is strongly generalized C_x-upper sign continuous$^+$. Let $x_\lambda := x + \lambda(y - x)$ for all $y \in K$ and $\lambda \in]0, 1[$. Assume that for all $y \in K$ and all $\xi_\lambda \in T(x_\lambda)$, $\lambda \in]0, 1[$, we have

$$\langle \xi_\lambda, y - x \rangle \in C(x).$$

Since $T(x_\lambda)$ and $T(x)$ are compact, from Lemma 1.13, it follows that for each fixed $\xi_\lambda \in T(x_\lambda)$, there exists $\zeta_\lambda \in T(x)$ such that

$$\|\xi_\lambda - \zeta_\lambda\| \le \mathscr{H}(T(x_\lambda), T(x)).$$

Since $T(x)$ is compact, without loss of generality, we may assume that $\zeta_\lambda \to \zeta \in T(x)$ as $\lambda \to 0^+$. Since T is $\mathscr{H}$-hemicontinuous, $\mathscr{H}(T(x_\lambda), T(x)) \to 0$ as $\lambda \to 0^+$. Thus,

$$\begin{aligned} \|\xi_\lambda - \zeta\| &\le \|\xi_\lambda - \zeta_\lambda\| + \|\zeta_\lambda - \zeta\| \\ &\le \mathscr{H}(T(x_\lambda), T(x)) + \|\zeta_\lambda - \zeta\| \to 0 \text{ as } \lambda \to 0^+. \end{aligned}$$

This implies that $\xi_\lambda \to \zeta \in T(x)$. Since $C(x)$ is closed, we have that there exists $\zeta \in T(x)$ such that $\langle \zeta, y - x \rangle \in C(x)$ for all $y \in K$. Hence, T is strongly generalized C_x-upper sign continuous$^+$.

Since $W(x) = Y \setminus \{-\operatorname{int}(C(x))\}$ is closed for all $x \in K$, by using the similar argument, it is easy to show that T is weakly generalized C_x-upper sign continuous$^+$. □

Remark 8.3 The $\mathscr{H}$-hemicontinuity does not imply the generalized C_x-upper sign continuity$^+$.

Lemma 8.3 *Let K be a nonempty convex subset of X and $T : K \to 2^{\mathcal{L}(X,Y)}$ be a set-valued map with nonempty values. Then*

(a) $\text{Sol(MGSVVIP)}_w^d \subseteq \text{Sol(GSVVIP)}_s^d$ *if T is strongly generalized C_x-upper sign continuous;*
(b) $\text{Sol(MGVVIP)}_w^d \subseteq \text{Sol(GVVIP)}_s^d \subseteq \text{Sol(GVVIP)}_w^d$ *if T is generalized C_x-upper sign continuous;*
(c) $\text{Sol(MGWVVIP)}_w^d \subseteq \text{Sol(GVVIP)}_s^d \subseteq \text{Sol(GWVVIP)}_w^d$ *if T is weakly generalized C_x-upper sign continuous.*

Proof (a) Let $\bar{x} \in \text{Sol(MGSVVIP)}_w^d$. Then for all $y \in K$, there exists $\xi \in T(y)$ such that

$$\langle \xi, y - \bar{x} \rangle \in C(\bar{x}).$$

Since K is convex, for all $\lambda \in]0,1[$, $y_\lambda := x + \lambda(y - \bar{x}) \in K$. Therefore, for $y_\lambda \in K$, there exists $\xi_\lambda \in T(y_\lambda)$ such that

$$\langle \xi_\lambda, \bar{x} + \lambda(y - \bar{x}) - \bar{x} \rangle \in C(\bar{x}),$$

equivalently,

$$\lambda \langle \xi_\lambda, y - \bar{x} \rangle \in C(\bar{x}).$$

Since $C(x)$ is a convex cone, we have

$$\langle \xi_\lambda, y - \bar{x} \rangle \in C(\bar{x}).$$

By strong generalized C_x-upper sign continuity of T, there exists $\bar{\zeta} \in T(\bar{x})$ such that

$$\langle \bar{\zeta}, y - \bar{x} \rangle \in C(\bar{x}), \quad \text{for all } y \in K.$$

Hence, $\bar{x} \in \text{Sol(GSVVIP)}_s^d$.

Since $W(x) = Y \setminus \{-C(x) \setminus \{\mathbf{0}\}\}$ and $W(x) = Y \setminus \{-\text{int}(C(x))\}$ are cones, the proof of the part (b) and (c) lies on the lines of the proof of part (a). □

Similarly, we can prove the following lemma.

Lemma 8.4 *Let K be a nonempty convex subset of X and $T : K \to 2^{\mathcal{L}(X,Y)}$ be a set-valued map with nonempty values. Then*

(a) $\text{Sol(MGSVVIP)}_g^d \subseteq \text{Sol(GSVVIP)}_g^d$ *if T is strongly generalized C_x-upper sign continuous$_+^+$;*
(b) $\text{Sol(MGSVVIP)}_g^d \subseteq \text{Sol(GSVVIP)}_s^d$ *if T is strongly generalized C_x-upper sign continuous$^+$;*
(c) $\text{Sol(MGSVVIP)}_w^d \subseteq \text{Sol(GSVVIP)}_g^d$ *if T is strongly generalized C_x-upper sign continuous$_+$;*
(d) $\text{Sol(MGVVIP)}_g^d \subseteq \text{Sol(GVVIP)}_g^d$ *if T is generalized C_x-upper sign continuous$_+^+$;*
(e) $\text{Sol(MGVVIP)}_g^d \subseteq \text{Sol(GVVIP)}_s^d$ *if T is generalized C_x-upper sign continuous$^+$;*
(f) $\text{Sol(MGVVIP)}_w^d \subseteq \text{Sol(GVVIP)}_g^d$ *if T is generalized C_x-upper sign continuous$_+$;*
(g) $\text{Sol(MGWVVIP)}_g^d \subseteq \text{Sol(GWVVIP)}_g^d$ *if T is weakly generalized C_x-upper sign continuous$_+^+$;*
(h) $\text{Sol(MGWVVIP)}_g^d \subseteq \text{Sol(GWVVIP)}_s^d$ *if T is weakly generalized C_x-upper sign continuous$^+$;*
(i) $\text{Sol(MGWVVIP)}_w^d \subseteq \text{Sol(GWVVIP)}_g^d$ *if T is weakly generalized C_x-upper sign continuous$_+$.*

We introduce the following set-valued maps:

- $S_g^S(y) = \{x \in K : \forall \zeta \in T(x) \text{ satisfying } \langle \zeta, y - x \rangle \in C(x)\}$;
- $S_w^S(y) = \{x \in K : \exists \zeta \in T(x) \text{ satisfying } \langle \zeta, y - x \rangle \in C(x)\}$;
- $M_g^S(y) = \{x \in K : \forall \xi \in T(y) \text{ satisfying } \langle \xi, y - x \rangle \in C(x)\}$;
- $M_w^S(y) = \{x \in K : \exists \xi \in T(y) \text{ satisfying } \langle \xi, y - x \rangle \in C(x)\}$;
- $S_g(y) = \{x \in K : \forall \zeta \in T(x) \text{ satisfying } \langle \zeta, y - x \rangle \notin -C(x) \setminus \{\mathbf{0}\}\}$;
- $S_w(y) = \{x \in K : \exists \zeta \in T(x) \text{ satisfying } \langle \zeta, y - x \rangle \notin -C(x) \setminus \{\mathbf{0}\}\}$;

- $M_g(y) = \{x \in K : \forall \xi \in T(y) \text{ satisfying } \langle \xi, y - x \rangle \notin -C(x) \setminus \{\mathbf{0}\}\}$;
- $M_w(y) = \{x \in K : \exists \xi \in T(y) \text{ satisfying } \langle \xi, y - x \rangle \notin -C(x) \setminus \{\mathbf{0}\}\}$;
- $S_g^W(y) = \{x \in K : \forall \zeta \in T(x) \text{ satisfying } \langle \zeta, y - x \rangle \notin -\operatorname{int}(C(x))\}$;
- $S_w^W(y) = \{x \in K : \exists \zeta \in T(x) \text{ satisfying } \langle \zeta, y - x \rangle \notin -\operatorname{int}(C(x))\}$;
- $M_g^W(y) = \{x \in K : \forall \xi \in T(y) \text{ satisfying } \langle \xi, y - x \rangle \notin -\operatorname{int}(C(x))\}$;
- $M_w^W(y) = \{x \in K : \exists \xi \in T(y) \text{ satisfying } \langle \xi, y - x \rangle \notin -\operatorname{int}(C(x))\}$.

From the above definition of set-valued maps, the following result can be easily derived.

Proposition 8.1

(a) $\text{Sol(GSVVIP)}_g^d = \bigcap_{y \in K} S_g^S(y)$ *and* $\text{Sol(GSVVIP)}_w^d = \bigcap_{y \in K} S_w^S(y)$;

(b) $\text{Sol(MGSVVIP)}_g^d = \bigcap_{y \in K} M_g^S(y)$ *and* $\text{Sol(MGSVVIP)}_w^d = \bigcap_{y \in K} M_w^S(y)$;

(c) $\text{Sol(GVVIP)}_g^d = \bigcap_{y \in K} S_g(y)$ *and* $\text{Sol(GVVIP)}_w^d = \bigcap_{y \in K} S_w(y)$;

(d) $\text{Sol(MGVVIP)}_g^d = \bigcap_{y \in K} M_g(y)$ *and* $\text{Sol(MGVVIP)}_w^d = \bigcap_{y \in K} M_w(y)$;

(e) $\text{Sol(GWVVIP)}_g^d = \bigcap_{y \in K} S_g^W(y)$ *and* $\text{Sol(GWVVIP)}_w^d = \bigcap_{y \in K} S_w^W(y)$;

(f) $\text{Sol(MGWVVIP)}_g^d = \bigcap_{y \in K} M_g^W(y)$ *and* $\text{Sol(MGWVVIP)}_w^d = \bigcap_{y \in K} M_w^W(y)$.

Proposition 8.2

(a) *If the set-valued map $C : K \to 2^Y$ is closed, then for each $y \in K$, $M_g^S(y)$ is a closed set.*
(b) *If the set-valued map $W : K \to 2^Y$, defined by $W(x) = Y \setminus \{-\operatorname{int}(C(x))\}$, is closed, then for each $y \in K$, $M_g^W(y)$ is a closed set.*
(c) *If K is compact and the set-valued map $T : K \to 2^{\mathcal{L}(X,Y)}$ is nonempty compact valued and the set-valued map $C : K \to 2^Y$ is closed, then for each $y \in K$, $M_w^S(y)$ is a closed set.*
(d) *If K is compact and the set-valued map $T : K \to 2^{\mathcal{L}(X,Y)}$ is nonempty compact valued and the set-valued map $W : K \to 2^Y$, defined by $W(x) = Y \setminus \{-\operatorname{int}(C(x))\}$, is closed, then for each $y \in K$, $M_w^W(y)$ is a closed set.*

(e) *If the set-valued map* $T : K \to 2^{\mathcal{L}(X,Y)}$ *is lower semicontinuous and the set-valued map* $C : K \to 2^Y$ *is closed, then for each* $y \in K$, $S_g^S(y)$ *is a closed set.*
(f) *If the set-valued map* $T : K \to 2^{\mathcal{L}(X,Y)}$ *is lower semicontinuous and the set-valued map* $W : K \to 2^Y$*, defined by* $W(x) = Y \setminus \{-\operatorname{int}(C(x))\}$*, is closed, then for each* $y \in K$, $S_g^W(y)$ *is a closed set.*
(g) *If the set-valued map* $W : K \to 2^Y$*, defined by* $W(x) = Y \setminus \{-\operatorname{int}(C(x))\}$*, is concave, then for each* $y \in K$, $M_g^W(y)$ *is a convex set.*
(h) *If the set-valued map* $C : K \to 2^Y$ *is concave, then for each* $y \in K$, $M_g^S(y)$ *is a convex set.*

Proof The proof of part (a) is similar to that of (b), therefore, we prove only part (b).

(b) For any fixed $y \in K$, let $\{x_m\}$ be a sequence in $M_g^W(y)$ such that $\{x_m\}$ converges to $x \in K$. Since $x_m \in M_g^W(y)$, for all $\xi \in T(y)$, we have

$$\langle \xi, y - x_m \rangle \in W(x_m) = Y \setminus \{-\operatorname{int}(C(x_m))\}, \quad \text{for all } m.$$

Since $\xi \in \mathcal{L}(X, Y)$, ξ is continuous, and so, the sequence $\{\langle \xi, y - x_m \rangle\}$ converges to $\langle \xi, y - x \rangle \in Y$. Since W is closed, so its graph $\mathcal{G}(W)$ is closed, and therefore, we have $(x_m, \langle \xi, y - x_m \rangle)$ converges to $(x, \langle \xi, y - x \rangle) \in \mathcal{G}(W)$. Thus,

$$\langle \xi, y - x \rangle \in W(x) = Y \setminus \{-\operatorname{int}(C(x))\},$$

so that $x \in M_g^W(y)$. Consequently, $M_g^W(y)$ is a closed subset of K.

The proof of part (c) is similar to that of (d), therefore, we prove only part (d).

(d) For any fixed $y \in K$, let $\{x_m\}$ be a sequence in $M_w^W(y)$ such that $\{x_m\}$ converges to $x \in K$. Since $x_m \in M_w^W(y)$, there exists $\xi_m \in T(y)$ such that

$$\langle \xi_m, y - x_m \rangle \in W(x_m) = Y \setminus \{-\operatorname{int}(C(x_m))\}, \quad \text{for all } m.$$

Since $T(y)$ is compact, we may assume that $\{\xi_m\}$ converges to some $\xi \in T(y)$. Besides, since K is compact, $\{x_m\}$ is bounded. Therefore, $\langle \xi_m - \xi, y - x_m \rangle$ converges to $\mathbf{0}$, but $\langle \xi, y - x_m \rangle$ converges to $\langle \xi, y - x \rangle \in Y$ due to $\xi \in \mathcal{L}(X, Y)$. Hence, $\langle \xi_m, y - x_m \rangle$ converges to $\langle \xi, y - x \rangle \in Y$. Therefore, $(x_m, \langle \xi_m, y - x_m \rangle)$ converges to $(x, \langle \xi, y - x \rangle) \in \mathcal{G}(W)$ since $\mathcal{G}(W)$ is closed. Thus, for $\xi \in T(y)$,

$$\langle \xi, y - x \rangle \in W(x) = Y \setminus \{-\operatorname{int}(C(x))\},$$

so that $x \in M_w^W(y)$. Consequently, $M_w^W(y)$ is a closed subset of K.

The proof of part (f) is similar to that of (e), therefore, we prove only part (e).

(e) For any fixed $y \in K$, let $\{x_m\}$ be a sequence in $S_g^S(y)$ converging to $x \in K$. By lower semicontinuity (see Lemma 1.9) of T, for any $\zeta \in T(x)$, there exists $\zeta_m \in T(x_m)$ for all m such that the sequence $\{\zeta_m\}$ converges to $\zeta \in \mathcal{L}(X, Y)$. Since

$x_m \in S_g^S(y)$ for all m, we have

$$\langle \zeta_m, y - x_m \rangle \in C(x_m).$$

Moreover,

$$\begin{aligned}
\| \langle \zeta_m, y - x_m \rangle - \langle \zeta, y - x \rangle \| &= \| \langle \zeta_m, y - x_m \rangle - \langle \zeta_m, x \rangle + \langle \zeta_m, x \rangle - \langle \zeta, y - x \rangle \| \\
&= \| \langle \zeta_m, x - x_m \rangle + \langle \zeta_m, y - x \rangle - \langle \zeta, y - x \rangle \| \\
&= \| \langle \zeta_m, x - x_m \rangle + \langle \zeta_m - \zeta, y - x \rangle \| \\
&\leq \| \zeta_m \| \, \| x - x_m \| + \| \zeta_m - \zeta \| \, \| y - x \|.
\end{aligned}$$

Since $\{\zeta_m\}$ is bounded in $\mathcal{L}(X, Y)$, $\{\langle \zeta_m, y - x_m \rangle\}$ converges to $\langle \zeta, y - x \rangle$. By the closedness of C, we have $\langle \zeta, y - x \rangle \in C(x)$. Hence, $x \in S_g^S(y)$, and therefore, $S_g^S(y)$ is closed.

(g) Let $y \in K$ be any fixed element and let $x_1, x_2 \in M_g^W(y)$. Then for all $\xi \in T(y)$, we have

$$\langle \xi, y - x_1 \rangle \in W(x_1) \quad \text{and} \quad \langle \xi, y - x_2 \rangle \in W(x_2).$$

By concavity of W, for all $\lambda \in [0, 1]$, we have

$$\begin{aligned}
\langle \xi, y - (\lambda x_1 + (1 - \lambda) x_2) \rangle &= \lambda \, \langle \xi, y - x_1 \rangle + (1 - \lambda) \, \langle \xi, y - x_2 \rangle \\
&\in \lambda W(x_1) + (1 - \lambda) W(x_2) \\
&\subseteq W(\lambda x_1 + (1 - \lambda) x_2).
\end{aligned}$$

Therefore, $\lambda x_1 + (1 - \lambda) x_2 \in M_g^W(y)$, and hence, $M_g^W(y)$ is convex.

Similarly, we can prove part (h). □

Remark 8.4 The set-valued maps S_g, S_w, M_g, and M_w fail to have the property that $S_g(y)$, $S_w(y)$, $M_g(y)$, and $M_w(y)$ are closed for all $y \in K$.

Example 8.2 Consider $X = Y = \mathbb{R}$, $K =]0, 1]$, $C(x) = R_+$ for all $x \in K$ and $T(x) = [0, 1]$. Then the set

$$\begin{aligned}
S_g(y) &= \{x \in K : \forall \zeta \in T(x) \text{ satisfying } \langle \zeta, y - x \rangle \notin -C(x) \setminus \{\mathbf{0}\}\} \\
&= \{x \in]0, 1] : x \leq y\}
\end{aligned}$$

is not closed.

Proposition 8.3 *Let K be a nonempty convex subset of X. The set-valued maps S_w and S_w^W are KKM-maps.*

Proof Let $\hat{x}$ be in the convex hull of any finite subset $\{y_1, y_2, \dots, y_p\}$ of K. Then $\hat{x} = \sum_{i=1}^{p} \lambda_i y_i$ for some nonnegative real number λ_i, $1 \leq i \leq p$, with $\sum_{i=1}^{p} \lambda_i = 1$. If $\hat{x} \notin \bigcup_{i=1}^{p} S_w(y_i)$, then for all $\zeta \in T(\hat{x})$, we have

$$\langle \zeta, y_i - \hat{x} \rangle \in -C(\hat{x}) \setminus \{\mathbf{0}\}, \quad \text{for each } i = 1, 2, \dots, p.$$

Since $-C(\hat{x})$ is a convex cone and $\lambda_i \geq 0$ with $\sum_{i=1}^{p} \lambda_i = 1$, we have

$$\sum_{i=1}^{p} \lambda_i \langle \zeta, y_i - \hat{x} \rangle \in -C(\hat{x}) \setminus \{\mathbf{0}\}.$$

It follows that

$$\begin{aligned} \mathbf{0} = \langle \zeta, \hat{x} - \hat{x} \rangle &= \left\langle \zeta, \sum_{i=1}^{p} \lambda_i y_i - \sum_{i=1}^{p} \lambda_i \hat{x} \right\rangle \\ &= \left\langle \zeta, \sum_{i=1}^{p} \lambda_i (y_i - \hat{x}) \right\rangle = \sum_{i=1}^{p} \lambda_i \langle \zeta, y_i - \hat{x} \rangle \in -C(\hat{x}) \setminus \{\mathbf{0}\}. \end{aligned}$$

Thus, we have $\mathbf{0} \in -C(\hat{x}) \setminus \{\mathbf{0}\}$, a contradiction. Therefore, we must have

$$\operatorname{co}(\{y_1, y_2, \dots, y_p\}) \subseteq \bigcup_{i=1}^{p} S_w(y_i),$$

and hence, S_w is a KKM map on K.

Since $-C(x)$ is a convex cone, by using the similar argument, we can easily prove that S_w^W is a KKM map on K. □

Remark 8.5 The above argument cannot be applied for S_g^S and S_w^S. In general, S_g^S and S_w^S are not KKM maps.

Example 8.3 Let $X = K = \mathbb{R}$, $Y = \mathbb{R}^2$ and let the operator $T : K \to 2^{\mathcal{L}(X,Y)}$ be the single valued map $T(x) := (x, \; x)$. Then the sets S_g^S and S_w^S coincide, and it can be easily seen that they are not KKM maps: Consider, for instance, the points $y_1 = 0$ and $y_2 = 1$. Then $S_g^S(y_1) = S_w^S(y_1) = \{0\}$ and $S_g^S(y_2) = S_w^S(y_2) = \{0, 1\}$. However, $\frac{1}{2} \in \operatorname{co}(y_1, y_2)$ and $S_g^S\left(\frac{1}{2}\right) = S_w^S\left(\frac{1}{2}\right) = \left\{0, \frac{1}{2}\right\}$, but $\frac{1}{2} \notin \{0, 1\}$.

8.2 Existence Results under Monotonicity

Let X and Y be Banach spaces and K be a nonempty convex subset of X. Let $T : K \to 2^{\mathcal{L}(X,Y)}$ be a set-valued map with nonempty values, and $C : K \to 2^Y$ be a set-valued map such that for all $x \in K$, $C(x)$ is a closed convex pointed cone with $\operatorname{int}(C(x)) \neq \emptyset$.

Definition 8.4 Let $x \in K$ be an arbitrary element. A set-valued map $T : K \to 2^{\mathcal{L}(X,Y)}$ is said to be

(a) *strongly generalized C_x-monotone* on K if for every $y \in K$ and for all $\zeta \in T(x)$, $\xi \in T(y)$, we have

$$\langle \zeta - \xi, x - y \rangle \in C(x);$$

(b) *strongly generalized C_x-monotone$^+$* on K if for every $y \in K$ and for all $\zeta \in T(x)$, there exists $\xi \in T(y)$ such that

$$\langle \zeta - \xi, x - y \rangle \in C(x);$$

(c) *strongly generalized C_x-monotone$_+$* on K if for every $y \in K$ and for all $\xi \in T(y)$, there exists $\zeta \in T(x)$ such that

$$\langle \zeta - \xi, x - y \rangle \in C(x);$$

(d) *strongly generalized C_x-pseudomonotone* on K if for every $y \in K$ and for all $\zeta \in T(x)$ and $\xi \in T(y)$, we have

$$\langle \zeta, y - x \rangle \in C(x) \quad \text{implies} \quad \langle \xi, y - x \rangle \in C(x);$$

(e) *strongly generalized C_x-pseudomonotone$^+$* on K if for every $y \in K$ and for all $\zeta \in T(x)$, we have

$$\langle \zeta, y - x \rangle \in C(x) \quad \text{implies} \quad \langle \xi, y - x \rangle \in C(x), \ \text{for some } \xi \in T(y);$$

(f) *strongly generalized C_x-pseudomonotone$_+$* on K if for every $y \in K$, we have for some $\zeta \in T(x)$,

$$\langle \zeta, y - x \rangle \in C(x) \ \text{implies} \ \langle \xi, y - x \rangle \in C(x), \ \text{for all } \xi \in T(y).$$

Definition 8.5 Let $x \in K$ be an arbitrary element. A set-valued map $T : K \to 2^{\mathcal{L}(X,Y)}$ is said to be

(a) *generalized C_x-monotone* on K if for every $y \in K$ and for all $\zeta \in T(x)$, $\xi \in T(y)$, we have

$$\langle \zeta - \xi, x - y \rangle \notin -C(x) \setminus \{\mathbf{0}\};$$

(b) *generalized C_x-monotone$^+$* on K if for every $y \in K$ and for all $\zeta \in T(x)$, there exists $\xi \in T(y)$ such that

$$\langle \zeta - \xi, x - y \rangle \notin -C(x) \setminus \{\mathbf{0}\};$$

(c) *generalized C_x-monotone$_+$* on K if for every $y \in K$ and for all $\xi \in T(y)$, there exists $\zeta \in T(x)$ such that

$$\langle \zeta - \xi, x - y \rangle \notin -C(x) \setminus \{\mathbf{0}\};$$

(d) *generalized C_x-pseudomonotone* on K if for every $y \in K$ and for all $\zeta \in T(x)$ and $\xi \in T(y)$, we have

$$\langle \zeta, y - x \rangle \notin -C(x) \setminus \{\mathbf{0}\} \quad \text{implies} \quad \langle \xi, y - x \rangle \notin -C(x) \setminus \{\mathbf{0}\};$$

(e) *generalized C_x-pseudomonotone$^+$* on K if for every $y \in K$ and for all $\zeta \in T(x)$, we have

$$\langle \zeta, y - x \rangle \notin -C(x) \setminus \{\mathbf{0}\} \text{ implies } \langle \xi, y - x \rangle \notin -C(x) \setminus \{\mathbf{0}\},$$

for some $\xi \in T(y)$;

(f) *generalized C_x-pseudomonotone$_+$* on K if for every $y \in K$, we have

$$\text{for some } \zeta \in T(x), \ \ \langle \zeta, y - x \rangle \notin -C(x) \setminus \{\mathbf{0}\}$$
$$\text{implies } \langle \xi, y - x \rangle \notin -C(x) \setminus \{\mathbf{0}\}, \quad \text{for all } \xi \in T(y).$$

When we replace $C(x) \setminus \{\mathbf{0}\}$ by $\text{int}(C(x))$ in the above definitions, then T is called *weakly generalized C_x-monotone*, *weakly generalized C_x-monotone$^+$*, *weakly generalized C_x-monotone$_+$*, *weakly generalized C_x-pseudomonotone*, *weakly generalized C_x-pseudomonotone$^+$*, and *weakly generalized C_x-pseudomonotone$_+$*, respectively.

The following example shows that the weakly generalized C_x-pseudomonotonicity does not imply weakly generalized C_x-monotonicity.

Example 8.4 Let $X = Y = \mathbb{R}$, $C(x) = [0, \infty)$ for all $x \in X$, and let $T : \mathbb{R} \to 2^{\mathbb{R}}$ be defined as $T(x) =]-\infty, x]$ for all $x \in \mathbb{R}$. Then it is easy to see that T is weakly generalized C_x-pseudomonotone but not weakly generalized C_x-monotone.

From the above definition, we have the following diagram (Fig. 8.2).

The implications in the following lemma follow from the definition of different kinds of monotonicities, and therefore, we omit the proof.

Lemma 8.5 *Let K be a nonempty subset of X and $T : K \to 2^{\mathcal{L}(X,Y)}$ be a set-valued map with nonempty values. Then*

(a) $\text{Sol(GSVVIP)}_w^d \subseteq \text{Sol(MGSVVIP)}_w^d$ *if T is strongly generalized C_x-pseudomonotone$^+$;*
(b) $\text{Sol(GSVVIP)}_w^d \subseteq \text{Sol(MGSVVIP)}_g^d$ *if T is strongly generalized C_x-pseudomonotone$_+$;*
(c) $\text{Sol(GVVIP)}_w^d \subseteq \text{Sol(MGVVIP)}_w^d$ *if T is generalized C_x-pseudomonotone$^+$;*
(d) $\text{Sol(GVVIP)}_w^d \subseteq \text{Sol(MGVVIP)}_g^d$ *if T is generalized C_x-pseudomonotone$_+$;*

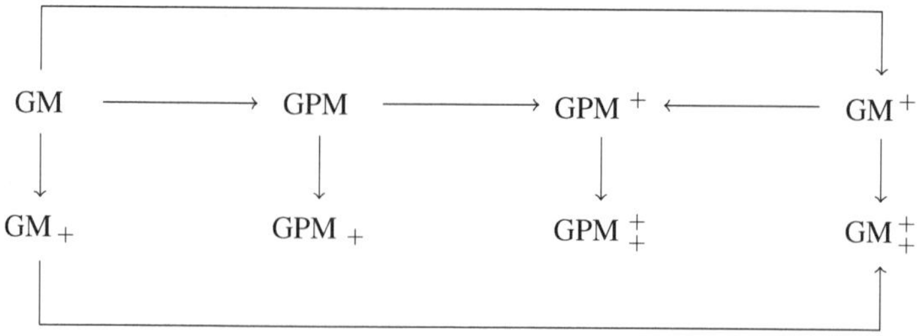

Fig. 8.2 Relations among different kinds of generalized C_x-monotonicity. GM and GPM stand for generalized C_x-monotonicity and generalized C_x-pseudomonotonicity, respectively

(e) $\text{Sol(GWVVIP)}_w^d \subseteq \text{Sol(MGWVVIP)}_w^d$ *if* T *is weakly generalized* C_x-*pseudomonotone*$^+$;

(f) $\text{Sol(GWVVIP)}_w^d \subseteq \text{Sol(MGWVVIP)}_g^d$ *if* T *is weakly generalized* C_x-*pseudomonotone*$_+$.

Next we give the first result on the existence of a solution of $(\text{GWVVIP})_w$.

Theorem 8.1 *Let* X *and* Y *be Banach spaces and* K *be a nonempty compact convex subset of* X. *Let* $C : K \to 2^Y$ *be a set-valued map such that for each* $x \in K$, $C(x)$ *is a proper, closed and convex (not necessarily pointed) cone with* $\text{int}(C(x)) \neq \emptyset$; *and let* $W : K \to 2^Y$ *be defined by* $W(x) = Y \setminus \{-\text{int}(C(x))\}$, *such that the graph* $\mathcal{G}(W)$ *of* W *is closed in* $X \times Y$. *Let* $x \in K$ *be arbitrary and suppose that* $T : K \to 2^{\mathcal{L}(X,Y)}$ *is weakly generalized* C_x-*pseudomonotone*$_+$ *and weakly generalized* C_x-*upper sign continuous*$^+$ *on* K. *Then there exists a solution of* $(\text{GWVVIP})_w$.

Proof Define set-valued maps $S_w^W, M_g^W : K \to 2^K$ by

$$S_w^W(y) = \{x \in K : \exists \zeta \in T(x) \text{ satisfying } \langle \zeta, y - x\rangle \notin -\text{int}(C(x))\},$$

and

$$M_g^W(y) = \{x \in K : \forall \xi \in T(y) \text{ satisfying } \langle \xi, y - x\rangle \notin -\text{int}(C(x))\},$$

for all $y \in K$. Then by Proposition 8.3, S_w^W is a KKM map on K. By generalized C_x-pseudomonotonicity$_+$ of T, $S_w^W(y) \subseteq M_g^W(y)$ for all $y \in K$. Since S_w^W is a KKM map, so is M_g^W. Also,

$$\bigcap_{y \in K} S_w^W(y) \subseteq \bigcap_{y \in K} M_g^W(y).$$

by Lemma 8.4 (i),

$$\bigcap_{y \in K} M_g^W(y) \subseteq \bigcap_{y \in K} S_w^W(y),$$

and thus,

$$\bigcap_{y\in K} S_w^W(y) = \bigcap_{y\in K} M_g^W(y).$$

By Proposition 8.2 (b) and the assumption that the graph $\mathcal{G}(W)$ of W is closed, $M_g^W(y)$ is closed for all $y \in K$. Since K is compact, so is $M_g^W(y)$ for all $y \in K$. By Fan-KKM Lemma 1.14, we have

$$\bigcap_{y\in K} S_w^W(y) = \bigcap_{y\in K} M_g^W(y) \neq \emptyset.$$

Hence, there exists $\bar{x} \in K$ such that for all $y \in K$, there exists $\bar{\zeta} \in T(\bar{x})$ satisfying

$$\langle \bar{\zeta}, y - \bar{x} \rangle \notin -\operatorname{int}(C(\bar{x})).$$

The proof of theorem is complete. □

Remark 8.6 We note that the assumptions of Theorem 8.1 imply that, in case of an infinite-dimensional space Y, the cone $C(x)$ cannot be pointed for each $x \in K$. Indeed, the assumptions imply that $Y \setminus \{-\operatorname{int}(C(x))\}$ is closed for each $x \in K$; hence $\operatorname{int}(C(x))$ is open. Since Y is infinite-dimensional, $\operatorname{int}(C(x))$ contains a whole straight line. That is, there exist $y, z \in Y$ such that $y + tz$, $y - tz \in \operatorname{int}(C(x))$ for all $t \in \mathbb{R}$. By convexity, $\mathbf{0} \in C(x)$ which gives $(1/t)y + z$, $(1/t)y + z \in C(x)$ for all $t > 1$. Since $C(x)$ is closed, $z \in C(x)$ and $-z \in C(x)$. Consequently, $C(x)$ cannot be pointed.

Analogously to Theorem 8.1, we have the following existence result for a solution of $(\text{GVVIP})_w$.

Theorem 8.2 *Let X, Y, K, C and W be the same as in Theorem 8.1. Let $x \in K$ be arbitrary and suppose that $T : K \to 2^{\mathcal{L}(X,Y)}$ is generalized C_x-pseudomonotone$_+$ and generalized C_x-upper sign continuous$^+$ on K such that the set $M_g^W(y) = \{x \in K : \forall \xi \in T(y)$ satisfying $\langle \xi, y - x \rangle \notin -\operatorname{int}(C(x))\}$ is closed for all $y \in K$. Then there exists a solution of* $(\text{GVVIP})_w$.

Remark 8.7 Theorem 8.1 and 8.2 also hold when K is nonempty weakly compact convex subset of a Banach space X.

Since S_w^S is not a KKM map, the argument similar to Theorem 8.2 cannot be used for proving the existence of a solution of $(\text{GSVVIP})_w$. Therefore, we define the following concept of pseudomonotonicity.

Definition 8.6 Let $x \in K$ be an arbitrary element. A set-valued map $T : K \to 2^{\mathcal{L}(X,Y)}$ is said to be *generalized C_x-pseudomonotone** on K if for every $y \in K$ and for all $\zeta \in T(x)$ and $\xi \in T(y)$, we have

$$\langle \zeta, y - x \rangle \notin -C(x) \setminus \{\mathbf{0}\} \quad \text{implies} \quad \langle \xi, y - x \rangle \in C(x).$$

We use the above definition of pseudomonotonicity and establish the following existence result for a solution of $(\text{GSVVIP})_w$.

Theorem 8.3 *Let X, Y, K and C be the same as in Theorem 8.1. In addition, we assume that the graph of C is closed. Let $x \in K$ be arbitrary and suppose that $T : K \to 2^{\mathcal{L}(X,Y)}$ is generalized C_x-pseudomonotone* and strongly generalized upper sign continuous$^+$ on K. Then there exists a solution of* $(\text{GSVVIP})_w$.

Proof Define set-valued maps $S_w, M_g^S : K \to 2^K$ by

$$S_w(y) = \{x \in K : \exists \zeta \in T(x) \text{ satisfying } \langle \zeta, y - x \rangle \notin -C(x) \setminus \{\mathbf{0}\}\},$$

and

$$M_g^S(y) = \{x \in K : \forall \xi \in T(y) \text{ satisfying } \langle \xi, y - x \rangle \in C(x)\},$$

for all $y \in K$. Then by Proposition 8.3, S_w is a KKM map on K. By generalized C_x-pseudomonotonicity* of T, $S_w(y) \subseteq M_g^S(y)$ for all $y \in K$. Since S_w is a KKM map, so is M_g^M. Also,

$$\bigcap_{y \in K} S_w(y) \subseteq \bigcap_{y \in K} M_g^S(y).$$

By using strongly generalized C_x-upper sign continuity$^+$ of T and Lemma 8.4 (b), we have

$$\begin{aligned}\bigcap_{y \in K} M_g^S(y) &= \text{Sol}(\text{MGSVVIP})_g^d \subseteq \text{Sol}(\text{GSVVIP})_s^d \\ &\subseteq \text{Sol}(\text{GVVIP})_s^d \subseteq \text{Sol}(\text{GVVIP})_w^d \\ &= \bigcap_{y \in K} S_w(y),\end{aligned}$$

and thus,

$$\bigcap_{y \in K} S_w(y) = \bigcap_{y \in K} M_g^S(y).$$

Since the graph $\mathcal{G}(C)$ of C is closed and K is compact, we have that $M_g^S(y)$ is compact for all $y \in K$. By Fan-KKM Lemma 1.14, we have

$$\bigcap_{y \in K} S_w(y) = \bigcap_{y \in K} M_g^S(y) \neq \emptyset.$$

Hence, there exists $\bar{x} \in K$ such that for all $y \in K$, there exists $\zeta \in T(\bar{x})$ satisfying

$$\langle \zeta, y - \bar{x} \rangle \notin -C(x) \setminus \{\mathbf{0}\}.$$

This completes the proof. □

To give the existence results for solutions of $(\mathrm{GWVVIP})_w$ defined on a closed (not necessarily bounded) convex subset K of a Banach space X, we need the following coercivity conditions.

Definition 8.7 The set-valued map $T : K \to 2^{\mathcal{L}(X,Y)}$ is said to be

(a) *weakly generalized* v*-coercive* on K if there exist a compact subset B of X and $\tilde{y} \in B \cap K$ such that for every $\zeta \in T(x)$,

$$\langle \zeta, \tilde{y} - x \rangle \in -\operatorname{int}(C(x)), \quad \text{for all } x \in K \setminus B. \tag{8.16}$$

(b) *generalized* v*-coercive* on K if there exist a compact subset B of X and $\tilde{y} \in B \cap K$ such that for every $\zeta \in T(x)$,

$$\langle \zeta, \tilde{y} - x \rangle \in -C(x) \setminus \{\mathbf{0}\}, \quad \text{for all } x \in K \setminus B. \tag{8.17}$$

Theorem 8.4 *Let X, Y, C, W and $\mathcal{G}(W)$ be the same as in Theorem 8.1, and K be a nonempty closed convex subset of X. Let $x \in K$ be an arbitrary element and suppose that $T : K \to 2^{\mathcal{L}(X,Y)}$ is weakly generalized C_x-pseudomonotone$_+$, weakly generalized C_x-upper sign continuous$^+$ and weakly generalized v-coercive on K and it has nonempty values. Then* $(\mathrm{GWVVIP})_w$ *has a solution.*

Proof Let S_w^W and M_g^W be the set-valued maps defined as in the proof of Theorem 8.1. Choose a compact subset B of X and $\tilde{y} \in B \cap K$ such that for every $\zeta \in T(x)$, (8.16) holds.

We claim that the closure $\operatorname{cl}(S_w^W(\tilde{y}))$ of $S_w^W(\tilde{y})$ is a compact subset of K. If $S_w^W(\tilde{y}) \nsubseteq B$, then there exists $x \in S_w^W(\tilde{y})$ such that $x \in K \setminus B$. It follows that, for some $\zeta \in T(x)$,

$$\langle \zeta, \tilde{y} - x \rangle \notin -\operatorname{int}(C(x)),$$

which contradicts (8.16). Therefore, we have $S_w^W(\tilde{y}) \subseteq B$; hence, $\operatorname{cl}(S_w^W(\tilde{y}))$ is a compact subset of K.

As in the proof of Theorem 8.1, by Fan-KKM Lemma 1.14, we have

$$\bigcap_{y \in K} \operatorname{cl}(S_w^W(\tilde{y})) \neq \emptyset.$$

Again, as in the proof of Theorem 8.1, $M_g^W(y)$ is closed for all $y \in K$. By weakly generalized C_x-pseudomonotonicity$_+$ of T, $S_w^W(y) \subseteq M_g^W(y)$ for all $y \in K$. Therefore,

$$\operatorname{cl}(S_w^W(\tilde{y})) \subseteq M_g^W(y), \quad \text{for all } y \in K.$$

Consequently,

$$\bigcap_{y\in K} M_g^W(y) \neq \emptyset.$$

Furthermore, as in the proof of Theorem 8.1, we have

$$\bigcap_{y\in K} S_w^W(y) = \bigcap_{y\in K} M_g^W(y) \neq \emptyset.$$

Hence, $(\text{GWVVIP})_w$ has a solution. □

Analogous to Theorem 8.4, we can prove the following existence result for a solution of $(\text{GVVIP})_w$.

Theorem 8.5 *Let X, Y, C, W and $\mathcal{G}(W)$ be the same as in Theorem 8.2, and K be a nonempty closed convex subset of X. Let $x \in K$ be an arbitrary element and suppose that $T : K \to 2^{\mathcal{L}(X,Y)}$ is nonempty valued, generalized C_x-pseudomonotone$_+$, generalized C_x-upper sign continuous$^+$ and generalized v-coercive on K such that the set*

$$M_g(y) = \{x \in K : \forall \xi \in T(y) \text{ satisfying } \langle \xi, y - x\rangle \notin -C(x) \setminus \{\mathbf{0}\}\}$$

is closed for all $y \in K$. Then $(\text{GVVIP})_w$ has a solution.

Definition 8.8 The set-valued map $T : K \to 2^{\mathcal{L}(X,Y)}$ is said to be

(a) *weakly generalized d-coercive* on K if there exist a point $\tilde{y}$ and a number $d > 0$ such that for every $\zeta \in T(x)$,

$$\langle \zeta, \tilde{y} - x\rangle \in -\operatorname{int}(C(x)), \text{ if } x \in K \text{ and } \|\tilde{y} - x\| > d;$$

(b) *generalized d-coercive* on K if there exist a point $\tilde{y}$ and a number $d > 0$ such that for every $\zeta \in T(x)$,

$$\langle \zeta, \tilde{y} - x\rangle \in -C(x) \setminus \{\mathbf{0}\}, \text{ if } x \in K \text{ and } \|\tilde{y} - x\| > d.$$

Now we present an existence theorem for a solution of problem $(\text{GWVVIP})_w$ under weakly generalized C_x-pseudomonotonicity$_+$ assumption.

Theorem 8.6 *Let X, Y, C, W and $\mathcal{G}(W)$ be the same as in Theorem 8.1. Let K be a nonempty convex subset of X and $T : K \to 2^{\mathcal{L}(X,Y)}$ be a weakly generalized C_x-pseudomonotone$_+$, weakly generalized C_x-upper sign continuous$^+$ on K with nonempty compact values. Suppose that at least one of the following assumptions holds:*

(i) *K is weakly compact.*
(ii) *X is reflexive, K is closed, and T is generalized d-coercive on K.*

Then $(\text{GWVVIP})_w$ has a solution.

Proof Let S_w^W be the set-valued map defined as in the proof of Theorem 8.1. Define a set-valued map M_w^W by

$$M_w^W(y) = \{x \in K : \exists \xi \in T(y) \text{ satisfying } \langle \xi y - x\rangle \notin -\operatorname{int}(C(x))\},$$

for all $y \in K$. In order to prove the theorem under assumptions (i) it suffices to follow the proof of Theorem 8.1.

As in the proof of Theorem 8.1, S_w^W is a KKM map. By weakly generalized C_x-pseudomonotone$_+$, $S_w^W(y) \subseteq M_w^W(y)$ for all $y \in K$, and so M_w^W is a KKM-map. As in the proof of Proposition 8.2 (d), we can easily show that $M_w^W(y)$ is weakly closed for all $y \in K$.

Let us now consider the case (ii). Let B_r denote the closed ball (under the norm) of X with center at origin and radius r. If $K \cap B_r \neq \emptyset$, part (i) guarantees the existence of a solution x_r for the following problem, denoted by $(\text{GWVVIP})_w^r$:

$$\begin{aligned}
&\text{find } x_r \in K \cap B_r \text{ such that for all } y \in K \cap B_r,\\
&\text{there exists } \zeta_r \in T(x_r) \text{ satisfying } \langle \zeta_r, y - x_r\rangle \notin -\operatorname{int}(C(x_r)).
\end{aligned}$$

We observe that $\{x_r : r > 0\}$ must be bounded. Otherwise, we can choose r large enough so that $r \geq \|\tilde{y}\|$ and $d < \|\tilde{y} - x_r\|$, where $\tilde{y}$ satisfies the weakly generalized d-coercivity of T. It follows that, for every $\zeta_r \in T(x_r)$,

$$\langle \zeta_r, y_0 - x_r\rangle \in -\operatorname{int}(C(x_r)),$$

that is, x_r is not a solution of problem $(\text{GWVVIP})_w^r$, a contradiction. Therefore, there exist r such that $\|x_r\| < r$. Choose for any $x \in K$. Then we can choose $\varepsilon > 0$ small enough such that $x_r + \varepsilon(x - x_r) \in K \cap B_r$. If we suppose that for every $\zeta_r \in T(x_r)$,

$$\langle \zeta_r, x - x_r\rangle \in -\operatorname{int}(C(x_r)),$$

then

$$\langle \zeta_r, x_r + \varepsilon(x - x_r) - x_r\rangle = \varepsilon\langle \zeta_r, x - x_r\rangle \in -\operatorname{int}(C(x_r)),$$

that is, x_r is not a solution of $(\text{GWVVIP})_w^r$. Thus, x_r is a solution of $(\text{GWVVIP})_w$. □

Analogous to Theorem 8.6, we have the following existence result for a solution of $(\text{GVVIP})_w$.

Theorem 8.7 *Let X, Y, C, W and $\mathcal{G}(W)$ be the same as in Theorem 8.2. Let K be a nonempty convex subset of X and $T : K \to 2^{\mathcal{L}(X,Y)}$ be nonempty valued, generalized C_x-pseudomonotone$_+$ and generalized C_x-upper sign continuous$^+$ on K such that the set*

$$S_g^M(y) = \{x \in K : \forall \xi \in T(y) \text{ satisfying } \langle \xi, y - x\rangle \notin -C(x) \setminus \{\mathbf{0}\}\}$$

is weakly closed for all $y \in K$. Suppose that at least one of the following assumptions holds:

(i) *K is weakly compact.*
(ii) *X is reflexive, K is closed, and T is generalized d-coercive on K.*

Then (GVVIP)$_w$ *has a solution.*

In order to derive the existence results for solution of (GWVVIP)$_w$ and (GWVVIP)$_s$ by the way of solving an appropriate Stampacchia generalized (scalar) variational inequality problem (in short, GVIP), we use the following scalarization technique.

Let $s \in Y^*$ and $T : K \to 2^{\mathcal{L}(X,Y)}$ be a set-valued map with nonempty values. We define a set-valued map $T_s : K \to 2^{X^*}$ by

$$\langle T_s(x), y\rangle = \langle s, T(x), y\rangle, \quad \text{for all } x \in K \text{ and } y \in X.$$

Also, set

$$H(s) = \{y \in Y : \langle s, y\rangle \geq 0\}.$$

Then for all $s \in Y^*$, $H(s)$ is a closed convex cone in Y.

Recall that a set-valued map $Q : X \to 2^{X^*}$ is said to be *generalized pseudomonotone* on X if for every pair of points $x, y \in X$ and for all $u \in Q(x)$, $v \in Q(y)$, we have

$$\langle u, y - x\rangle \geq 0 \quad \text{implies} \quad \langle v, y - x\rangle \geq 0.$$

Also, a set-valued map $Q : X \to 2^{X^*}$ is said to be *generalized pseudomonotone*$^+$ on X if for every pair of points $x, y \in X$ and for all $u \in Q(x)$, we have

$$\langle u, y - x\rangle \geq 0 \quad \text{implies} \quad \langle v, y - x\rangle \geq 0, \quad \text{for some } v \in Q(y).$$

Obviously, every generalized pseudomonotone set-valued map is generalized pseudomonotone$^+$.

Proposition 8.4 *Let X and Y be Banach spaces and K be a nonempty closed convex subset of X. Suppose that $T : K \to 2^{\mathcal{L}(X,Y)}$ is strongly generalized $H(s)$-pseudomonotone (respectively, strongly generalized $H(s)$-pseudomonotone$^+$) for some $s \in Y^* \setminus \{\mathbf{0}\}$. Then the mapping T_s is generalized pseudomonotone (respectively, generalized pseudomonotone$^+$) on K.*

Proof For any $x, y \in K$, let

$$\langle \zeta_s, y - x\rangle \geq 0, \quad \text{for all } \zeta_s \in T_s(x). \tag{8.18}$$

Then $\langle s, \langle \zeta, y - x\rangle\rangle \geq 0$ for all $\zeta \in T(x)$. Therefore, $\langle \zeta, y - x\rangle \in H(s)$ for all $\zeta \in T(x)$. If T is strongly generalized $H(s)$-pseudomonotone, then we must have

$\langle \xi, y - x \rangle \in H(s)$ for all $\xi \in T(y)$, and thus $\langle s, \langle \xi, y - x \rangle \rangle \geq 0$ for all $\xi \in T(y)$. Hence, for all $\xi_s \in T_s(y)$,

$$\langle \xi_s, y - x \rangle \geq 0, \tag{8.19}$$

that is, T_s is generalized pseudomonotone on K. Analogously, if T is strongly generalized $H(s)$-pseudomonotone$^+$, (8.18) implies (8.19) for some $\xi_s \in T_s(y)$ and T_s is generalized pseudomonotone$^+$ on K. □

Theorem 8.8 *Let X and Y be Banach spaces and K be a nonempty compact convex subset of X. Let $C : K \to 2^Y$ be defined as in Theorem 8.1 such that $C_+^* \setminus \{\mathbf{0}\} \neq \emptyset$. Let $x \in K$ be arbitrary and suppose that $T : K \to 2^{\mathcal{L}(X,Y)}$ is weakly generalized C_x-upper sign continuous and weakly generalized $H(s)$-pseudomonotone on K for some $s \in C_+^*$ where $H(s) \neq Y$, and has nonempty values. Then the following statements hold.*

(a) *There exists a solution of* (GWVVIP)$_w$.
(b) *If for each $x \in K$, the set $T(x)$ is convex and weakly compact in $\mathcal{L}(X, Y)$, then there exists a solution of* (GWVVIP)$_s$.

Proof

(a) Since $H(s) \neq Y$, we note that int $H(s) = s^{-1}((0, \infty))$. To see this, consider the following argument. It is clear that $s^{-1}((0, \infty)) \subset \text{int}(H(s))$.

Conversely, let $y \in \text{int}(H(s))$. Then there exists $r > 0$ such that $B_r(y) \subset H(s)$, where $B_r(y)$ denotes the ball with center at y and radius r. Hence, $\langle s, y + rz \rangle \geq 0$ for all $\|z\| < 1$. If $\langle s, y \rangle = 0$, then from the above inequality we conclude that $\langle s, w \rangle \geq 0$ for all $w \in Y$ or $Y \subset H(s)$ which is a contradiction. Therefore, $\langle s.w \rangle > 0$ and $y \in s^{-1}((0, \infty))$. Consequently, $\text{int}(H(s)) = s^{-1}((0, \infty))$.

As $s \in C_+^* \setminus \{\mathbf{0}\}$, the mapping T_s is generalized pseudomonotone on K due to Proposition 8.4. Beside, since T is weakly generalized C_x-upper sign continuous, so is T_s. Now, in the special case where $Y = \mathbb{R}$, $C(x) = \mathbb{R}_+$ for all $x \in K$. Theorem 8.1 guarantees the existence of a solution $\bar{x} \in K$ of (GVIP)$_w^s$, that is, for all $y \in K$, there exists $\zeta_s \in T_s(\bar{x})$ satisfying

$$\langle \zeta_s, y - \bar{x} \rangle \geq 0. \tag{8.20}$$

Consequently, for every $y \in K$, there exists $\bar{\zeta} \in T(\bar{x})$ such that

$$\langle s, \langle \bar{\zeta}, y - \bar{x} \rangle \rangle \geq 0,$$

hence, $\langle \bar{\zeta}, y - \bar{x} \rangle \notin -\text{int}(H(s))$. Since $s \in C_+^*$, $-\text{int}(H(s)) \supseteq -\text{int}(C_+) \supseteq -\text{int}(C(\bar{x}))$, so that

$$\langle \bar{\zeta}, y - \bar{x} \rangle \notin -\text{int}(C(\bar{x})).$$

Therefore, $\bar{x}$ is a solution of (GWVVIP)$_w$.

(b) Let, in addition, the set $T(\bar{x})$ be convex and compact. Then $T_s(\bar{x})$ is obviously convex in X^*. We show that $T_s(\bar{x})$ is also compact.

Let $\{z_\alpha\}$ be a net in $T_s(\bar{x})$. Then there exists a net $\{\zeta_\alpha\}$ in $T(\bar{x})$ such that

$$\langle z_\alpha, x\rangle = \langle s, \langle \zeta_\alpha, x\rangle\rangle, \quad \text{for all } x \in X.$$

Since $T(\bar{x})$ is compact, there exists a subnet of $\{\zeta_\alpha\}$ which is converging to some $\zeta \in T(\bar{x})$. Without loss of generality, we suppose that ζ_α converges to ζ. Fix any $x \in X$. Then we can define

$$\langle l, u\rangle = \langle s, \langle u, x\rangle\rangle, \quad \text{for all } u \in \mathcal{L}(X, Y),$$

hence, $l \in \mathcal{L}(X, Y)^*$. Therefore, there exists $\bar{z} \in X^*$ such that

$$\lim_\alpha \langle z_\alpha, x\rangle = \lim_\alpha \langle l, \zeta_\alpha\rangle = \langle l, \zeta\rangle = \langle s, \langle \zeta, x\rangle\rangle = \langle \bar{z}, x\rangle,$$

that is, $\bar{z} \in T_s(\bar{x})$. Thus, $T_s(\bar{x})$ is compact set in X^*.

By (8.20) and the well known minimax theorem [4], we have

$$\max_{\zeta_s \in T_s(\bar{x})} \min_{y \in K} \langle \zeta_s, y - \bar{x}\rangle = \min_{y \in K} \max_{\zeta_s \in T_s(\bar{x})} \langle \zeta_s, y - \bar{x}\rangle \geq 0.$$

Hence, there exists $\zeta_s \in T_s(\bar{x})$ such that

$$\langle \zeta_s, y - \bar{x}\rangle \geq 0, \quad \text{for all } y \in K,$$

that is, there exists $\zeta \in T(\bar{x})$ such that

$$\langle s, \langle \zeta, y - \bar{x}\rangle\rangle \geq 0, \quad \text{for all } y \in K.$$

Analogously, it follows that

$$\langle \zeta, y - \bar{x}\rangle \notin -\operatorname{int}(C(\bar{x})), \quad \text{for all } y \in K.$$

Therefore, $\bar{x}$ is a strong solution of (GWVVIP)$_w$. □

In order to solve (GWVVIP)$_w$ with an unbounded domain, we need the following coercivity conditions. We first note that

$$C_+^* = \{l \in Y^* : \langle l, y\rangle \geq 0 \text{ for all } y \in C_+\},$$

and

$$\operatorname{int}(C_+^*) = \{l \in Y^* : \langle l, y\rangle > 0 \text{ for all } y \in C_+\},$$

where $C_+ = \operatorname{co}(\{C(x) : x \in K\})$.

Definition 8.9 Let X and Y be Banach spaces and K be a nonempty closed convex subset of X. Let $C : K \to 2^Y$ be a set-valued map such that $C_+^* \setminus \{\mathbf{0}\} \neq \emptyset$. A set-valued map $T : K \to 2^{\mathcal{L}(X,Y)}$ is said to be

(a) *generalized* v*-coercive* if there exist $x_0 \in K$ and $s \in C_+^* \setminus \{\mathbf{0}\}$ such that

$$\inf_{\zeta \in T_s(x)} \frac{\langle \zeta, x - x_o \rangle}{\|x - x_0\|} \to \infty, \quad \text{as } x \in K,\ \|x\| \to \infty.$$

(b) *weakly generalized* v*-coercive* if there exist $y \in K$ and $s \in C_+^* \setminus \{\mathbf{0}\}$ such that

$$\inf_{\zeta \in T_s(x)} \langle \zeta, x - y \rangle \to \infty, \quad \text{as } x \in K,\ \|x\| \to \infty.$$

It is clear that if T is generalized v-coercive, then it is weakly generalized v-coercive.

Under the assumption of the weak generalized v-coercivity of T, we have the following existence theorem for solutions of (GWVVIP)$_w$ and (GWVVIP)$_s$.

Theorem 8.9 *Let X, Y and C be the same as in Theorem 8.8 and, in addition, X be reflexive. Let K be a nonempty convex closed subset of X. Suppose that $T : K \to 2^{\mathcal{L}(X,Y)}$ is weakly generalized $H(s)$-upper sign continuous, weakly generalized $H(s)$-pseudomonotone, and weakly generalized v-coercive with respect to an $s \in C_+^* \setminus \{\mathbf{0}\}$ on K, where $H(s) \neq Y$, and has nonempty values. Then the following statements hold.*

(a) *There exists a solution of* (GWVVIP)$_w$.
(b) *If, for each $x \in K$, the set $T(x)$ is convex and weakly compact in $\mathcal{L}(X, Y)$, there exists a solution of* (GWVVIP)$_s$.

Proof If, for the given $s \in C_+^* \setminus \{\mathbf{0}\}$, there exists $\bar{x} \in K$ which is a solution of (GVIP)$_w$, that is, for all $y \in K$, there exists $\zeta_s \in T_s(\bar{x})$ satisfying

$$\langle \zeta_s, y - \bar{x} \rangle \geq 0.$$

Then as in the proof of Theorem 8.8, assertions (a) and (b) are true. So, for the proof of this theorem, it is sufficient to prove that there exists a solution of (GVIP)$_w$.

Let B_r denote the closed ball (under the norm) of X with center at origin and radius r. In the special case where $Y = \mathbb{R}$, $C(x) = \mathbb{R}_+$ for all $x \in K \cap B_r$, Proposition 8.4 and Theorem 8.1 with Remark 8.7 guarantee the existence of a solution x_r for the following problem, denoted by (GVIP)$_w^r$:

$$\text{Find } x_r \in K \cap B_r \text{ such that for all } y \in K \cap B_r,$$

$$\text{there exists } \zeta_s \in T_s(\bar{x}) \text{ satisfying } \langle \zeta_s, y - \bar{x} \rangle \geq 0,$$

if $K \cap B_r \neq \emptyset$. Choose $r \geq \|x_0\|$, where x_0 satisfies the weak generalized υ-coercivity of T. Then for some $\zeta_s' \in T_s(\bar{x})$, we have

$$\langle \zeta_s', y - \bar{x} \rangle \geq 0. \tag{8.21}$$

We observe that $\{x_r : r > 0\}$ must be bounded. Otherwise, we can choose r large enough so that the weak generalized υ-coercivity of T yields

$$\langle \zeta_s, x_0 - \bar{x} \rangle < 0, \quad \text{for all } \zeta_s \in T_s(\bar{x}),$$

which contradicts (8.21). Therefore, there exists r such that $\|x_0\| < r$. Now, for each $x \in K$, we can choose $\varepsilon > 0$ small enough such that $x_r + \varepsilon(x - x_r) \in K \cap B_r$. Then

$$\langle \zeta_s, x_r + \varepsilon(x - x_r) - \bar{x} \rangle \geq 0, \quad \text{for some } \zeta_s \in T_s(\bar{x}).$$

Dividing by ε on both sides of the above inequality, we obtain

$$\langle \zeta_s, x - x_r \rangle \geq 0, \quad \text{for all } x \in K,$$

which shows that x_r is s solution of $(\text{GVIP})_w^s$ and the result follows. □

We now obtain similar results in the case of weak generalized $H(s)$-pseudomonotonicity.

Theorem 8.10 *Let X, Y and C be the same as in Theorem 8.8. Let K be a nonempty convex subset of X and $T : K \to 2^{\mathcal{L}(X,Y)}$ be a weakly generalized $H(s)$-upper sign continuous, weakly generalized $H(s)$-pseudomonotone mapping with nonempty compact values on K with respect to $s \in C_+^* \setminus \{\mathbf{0}\}$ where $H(s) \neq Y$. Suppose that at least one of the following conditions hold:*

(i) *K is weakly compact.*
(ii) *K is closed, T is weakly υ-coercive on K with respect to the same $s \in C_+^* \setminus \{\mathbf{0}\}$, and X is reflexive.*

Then the following statements hold.

(a) *There exists a solution of* $(\text{GWVVIP})_w$.
(b) *If, for each $x \in K$, the set $T(x)$ is convex, there exists a solution of* $(\text{GWVVIP})_s$.

Proof We first note that, in case (i), the existence of a solution to the $(\text{GVIP})_w$ defined in (8.20) is guaranteed by Theorem 8.6 (a). In addition, under assumptions of (ii), the set $T_s(x)$ is also convex and sequential compact. Therefore, in order to prove this theorem it suffices to follow the proofs of Theorems 8.8 and 8.9 with the corresponding modifications, respectively. □

Remark 8.8 Let X and Y be Banach spaces and K be a closed convex pointed cone in X. Let $C : K \to 2^Y$ be such that for all $x \in K$, $C(x)$ is a closed convex pointed cone with $\text{int}(C(x)) \neq \emptyset$. Let $T : K \to 2^{\mathcal{L}(X,Y)}$ be a set-valued map with nonempty values. The *generalized vector complementarity problem* (in short, GVCP) is to find

$(\bar{x}, \bar{\zeta}) \in K \times T(\bar{x})$ such that

$$\langle \bar{\zeta}, \bar{x} \rangle \notin \text{int}(C(\bar{x})) \text{ and } \langle \bar{\zeta}, y \rangle \notin -\text{int}(C(\bar{x})), \quad \text{for all } y \in K.$$

It can be shown that if (GWVVIP)$_s$ has a solution, then (GVCP) has a solution. Then by using Theorems 8.9 and 8.10, we can derive existence results for solutions of (GVCP). For further details, we refer [5].

Definition 8.10 Let $x \in K$ be an arbitrary element. A set-valued map $T : K \to 2^{\mathcal{L}(X,Y)}$ is said to be

(a) *generalized C_x-quasimonotone* on K if for every $y \in K$ and for all $\zeta \in T(x)$ and all $\xi \in T(y)$, we have

$$\langle \zeta, y - x \rangle \notin -C(x) \quad \text{implies} \quad \langle \xi, y - x \rangle \notin -\text{int}(C(x));$$

(b) *generalized C_x-quasimonotone*$^+$ on K if for every $y \in K$ and for all $\zeta \in T(x)$, we have

$$\langle \zeta, y - x \rangle \notin -C(x) \text{ implies } \langle \xi, y - x \rangle \notin -\text{int}(C(x)), \text{ for some } \xi \in T(y).$$

Daniilidis and Hadjisavvas [2] established some existence results for a solution of (GWVVIP)$_w$ under generalized C_x-quasimonotonicity or generalized C_x-quasimonotonicity$^+$.

Now we establish some existence results for solutions of (GSVVIP)$_s$, (GVVIP)$_s$ and (GWVVIP)$_s$.

Definition 8.11 Let $T : K \to 2^{\mathcal{L}(X,Y)}$ be a set-valued map. A single-valued map $f : K \to \mathcal{L}(X, Y)$ is said to be a *selection* of T if for all $x \in K, f(x) \in T(x)$. It is called *continuous selection* if, in addition, f is continuous

Lemma 8.6 *If u is a selection of T, then every solution of* SVVIP (5.1), VVIP (5.2) *and* WVVIP (5.3) *(all these defined by means of f) is a solution of* (GSVVIP)$_s$, (GVVIP)$_s$ *and* (GWVVIP)$_s$, *respectively.*

Proof Assume that $\bar{x} \in K$ is a solution of SVVIP (5.1), that is,

$$\langle f(\bar{x}), y - \bar{x} \rangle \in C(x), \quad \text{for all } y \in K.$$

Let $\bar{\zeta} = f(\bar{x})$. Then, $\bar{\zeta} \in T(\bar{x})$ such that

$$\langle \bar{\zeta}, y - \bar{x} \rangle \in C(x), \quad \text{for all } y \in K.$$

Thus, $\bar{x} \in K$ is a solution of (GSVVIP)$_s$.

Similarly, we can prove the other cases. □

Lemma 8.7 *Let $f : K \to \mathcal{L}(X,Y)$ be a selection of $T : K \to 2^{\mathcal{L}(X,Y)}$ and $x \in K$ be an arbitrary element. If T is (respectively, strongly and weakly) generalized C_x-pseudomonotone, then f is (respectively, strongly and weakly) C_x-pseudomonotone.*

Theorem 8.11 *Let X and Y be Banach spaces and K be a nonempty compact convex subset of X. Let $C : K \to 2^Y$ be a set-valued map such that for each $x \in K$, $C(x)$ is a proper closed convex (not necessarily pointed) cone with $\operatorname{int}(C(x)) \neq \emptyset$; and let $W : K \to 2^Y$ be defined by $W(x) = Y \setminus \{-\operatorname{int}(C(x))\}$, such that the graph $\mathcal{G}(W)$ of W is closed in $X \times Y$. For arbitrary $x \in K$, suppose that $T : K \to 2^{\mathcal{L}(X,Y)}$ is nonempty valued, weakly generalized C_x-pseudomonotone$_+$ and has continuous selection f on K. Then there exists a solution of* (GWVVIP)$_s$.

Proof By the hypothesis, there is a continuous function $f : K \to \mathcal{L}(X,Y)$ such that $f(x) \in T(x)$ for all $x \in K$. From Lemma 8.7, f is weakly C_x-pseudomonotone. Then all the conditions of Theorem 5.2 are satisfied. Hence, there exists a solution of the following WVVIP: Find $\bar{x} \in K$ such that

$$\langle f(\bar{x}), y - \bar{x}\rangle \notin -\operatorname{int}(C(\bar{x})), \quad \text{for all } y \in K.$$

By Lemma 8.6, $\bar{x}$ is a solution of (GWVVIP)$_s$.

Similarly, by using Lemmas 8.6 and 8.7, and Theorem 5.3, we can establish the following result.

Theorem 8.12 *Let X and Y be Banach spaces and K be a nonempty compact convex subset of X. Let $C : K \to 2^Y$ be a set-valued map such that for each $x \in K$, $C(x)$ is a proper closed convex (not necessarily pointed) cone with $\operatorname{int}(C(x)) \neq \emptyset$; and let $W : K \to 2^Y$ be defined by $W(x) = Y \setminus \{-\operatorname{int}(C(x))\}$, such that the graph $\mathcal{G}(W)$ of W is closed in $X \times Y$. Let $x \in K$ be arbitrary and suppose that $T : K \to 2^{\mathcal{L}(X,Y)}$ is nonempty valued, generalized C_x-pseudomonotone$_+$ and has continuous selection f on K such that the set*

$$M_g^W(y) = \{x \in K : \forall \xi \in T(y) \textit{ satisfying } \langle \xi, y - x\rangle \notin -\operatorname{int}(C(x))\}$$

is closed for all $y \in K$. Then there exists a solution of (GVVIP)$_s$.

Remark 8.9 If K is compact and $T : K \to 2^{\mathcal{L}(X,Y)}$ is continuous, then T has a continuous selection, see, for example [3].

8.3 Existence Results Without Monotonicity

Let X and Y be two Banach spaces, $K \subset X$ be a nonempty, closed and convex set, and $C \subset Y$ be a closed, convex and pointed cone with $\operatorname{int}(C) \neq \emptyset$.

Recall that a mapping $g : X \to Y$ is said to be *completely continuous* if the weak convergence of x_n to x in X implies the strong convergence of $g(x_n)$ to $g(x)$ in Y.

Definition 8.12 Let K be a nonempty, closed and convex subset of a Banach space X and Y be a Banach space ordered by a closed, convex and pointed cone C with $\text{int}(C) \neq \emptyset$. A set-valued map $T : K \to 2^{\mathcal{L}(X,Y)}$ is said to be

(a) *completely semicontinuous* if for each $y \in K$,

$$\{x \in K : \langle \zeta, y - x \rangle \in -\text{int}(C) \text{ for all } \zeta \in T(x)\}$$

is open in K with respect to the weak topology of X;

(b) *strongly semicontinuous* if for each $y \in K$,

$$\{x \in K : \langle \zeta, y - x \rangle \in -\text{int}(C) \text{ for all } \zeta \in T(x)\}$$

is open in K with respect to the norm topology of X.

Remark 8.10

(a) Let K be a nonempty, bounded, closed and convex subset of a reflexive Banach space X and Y be a Banach space ordered by a closed, convex and pointed cone C with $\text{int}(C) \neq \emptyset$. Let $T : K \to \mathcal{L}(X, Y)$ be completely continuous. Then T is completely semicontinuous.
(b) Let K be a nonempty, compact and convex subset of a Banach space X and Y be a Banach space ordered by a closed, convex and pointed cone C with $\text{int}(C) \neq \emptyset$. Let $T : K \to \mathcal{L}(X, Y)$ be continuous. Then T is strongly semicontinuous.
(c) When $X = \mathbb{R}^n$, complete continuity is equivalent to continuity, and complete semicontinuity is equivalent to strong semicontinuity.

Next we state and prove the existence result for a solution of $(\text{GWVVIP})_s$ with $C(x)$ is a fixed pointed solid closed convex cone in Y.

Theorem 8.13 *Let K be a nonempty, bounded closed and convex subset of a reflexive Banach space X and Y be a Banach space ordered by a proper closed convex and pointed cone C with* $\text{int}(C) \neq \emptyset$. *Let $T : K \to 2^{\mathcal{L}(X,Y)}$ be a completely semicontinuous set-valued map with nonempty values. Then there exists a solution of* $(\text{GWVVIP})_s$ *for a fixed pointed solid closed convex cone C in Y, that is, there exist $\bar{x} \in K$ and $\zeta \in T(\bar{x})$ such that*

$$\langle \zeta, y - \bar{x} \rangle \notin -\text{int}(C), \quad \textit{for all } y \in K.$$

Proof Suppose that the conclusion is not true. Then for each $\hat{x} \in K$, there exists $y \in K$ such that

$$\langle \hat{\zeta}, y - \hat{x} \rangle \in -\text{int}(C), \quad \text{for all } \hat{\zeta} \in T(\hat{x}). \tag{8.22}$$

For every $y \in K$, define the set N_y as

$$N_y = \{x \in K : \langle \zeta, y - x \rangle \in -\text{int}(C) \text{ for all } \zeta \in T(x)\}.$$

Since T is completely semicontinuous, the set N_y is open in K with respect to the weak topology of X for every $y \in K$.

We assert that $\{N_y : y \in K\}$ is an open cover of K with respect to the weak topology of X. Indeed, first it is easy to see that

$$\bigcup_{y \in K} N_y \subseteq K.$$

Second, for each $\hat{x} \in K$, by (8.22) there exists $y \in K$ such that $\hat{x} \in N_y$. Hence $\hat{x} \in \bigcup_{y \in K} N_y$. This shows that $K \subseteq \bigcup_{y \in K} N_y$. Consequently,

$$K = \bigcup_{y \in K} N_y.$$

So, the assertion is valid.

The weak compactness of K implies that there exists a finite set of elements $\{y_1, y_2, \ldots, y_m\} \subseteq K$ such that $K = \bigcup_{i=1}^{m} N_{y_i}$. Hence there exists a continuous (with respect to the weak topology of X) partition of unity $\{\beta_1, \beta_2, \ldots, \beta_m\}$ subordinated to $\{N_{y_1}, N_{y_2}, \ldots, N_{y_m}\}$ such that $\beta_j(x) \geq 0$ for all $x \in K, j = 1, 2, \ldots, m, \sum_{j=1}^{m} \beta_j(x) = 1$ for all $x \in K$, and

$$\beta_j(x) \begin{cases} = 0, & \text{whenever } x \notin N_{y_j}, \\ > 0, & \text{whenever} x \in N_{y_j}. \end{cases}$$

Let $p : K \to X$ be defined by

$$p(x) = \sum_{j=1}^{m} \beta_j(x) y_j, \quad \text{for all } x \in K. \tag{8.23}$$

Since β_i is continuous with respect to the weak topology of X for each i, p is continuous with respect to the weak topology of X. Let $\Delta := \text{co}(\{y_1, y_2, \ldots, y_m\}) \subseteq K$. Then Δ is a simplex of a finite dimensional space and p maps Δ into itself. By Brouwer's Fixed Point Theorem 1.39, there exists $\tilde{x} \in \Delta$ such that $p(\tilde{x}) = \tilde{x}$. For any given $x \in K$, let

$$k(x) = \left\{j : x \in N_{y_j}\right\} = \left\{j : \beta_j(x) > 0\right\}.$$

Obviously, $k(x) \neq \emptyset$.

Since $\tilde{x} \in \Delta \subseteq K$ is a fixed point of p, we have $p(\tilde{x}) = \sum_{j=1}^{m} \beta_j(\tilde{x})y_j$ and hence by the definition of N_y, we derive for each $\tilde{\zeta} \in T(\tilde{x})$

$$\begin{aligned} \mathbf{0} &= \langle \tilde{\zeta}, \tilde{x} - \tilde{x} \rangle \\ &= \langle \tilde{\zeta}, \tilde{x} - p(\tilde{x}) \rangle \\ &= \left\langle \tilde{\zeta}, \tilde{x} - \sum_{j=1}^{m} \beta_j(\tilde{x})y_j \right\rangle \\ &= \sum_{j \in k(x_0)} \beta_j(x_0) \langle \tilde{\zeta}, \tilde{x} - y_j \rangle \in \operatorname{int}(C) \end{aligned}$$

which leads to a contradiction. Therefore, there exist $\bar{x} \in K$ and $\zeta \in T(\bar{x})$ such that

$$\langle \zeta, y - \bar{x} \rangle \notin -\operatorname{int}(C), \quad \text{for all } y \in K.$$

This completes the proof. □

The proof of the following result can be easily derived on the lines of the proof of Theorem 8.13.

Theorem 8.14 *Let K be a nonempty, compact and convex subset of a Banach space X and Y be a Banach space ordered by a proper closed convex and pointed cone C with $\operatorname{int}(C) \neq \emptyset$. Let $T : K \to 2^{\mathcal{L}(X,Y)}$ be strongly semicontinuous with nonempty values. Then there exist $\bar{x} \in K$ and $\zeta \in T(\bar{x})$ such that*

$$\langle \zeta, y - \bar{x} \rangle \notin -\operatorname{int}(C), \quad \textit{for all } y \in K.$$

Now we establish an existence theorem for a solution of $(\text{GWVVIP})_g$ under lower semicontinuity assumption on the underlying set-valued map T.

Theorem 8.15 *Let X and Y be Hausdorff topological vector spaces, K be a nonempty convex subset of X and the set-valued map $T : K \to 2^{\mathcal{L}(X,Y)}$ be lower semicontinuous such that the set*

$$A_x := \{y \in K : \langle \zeta, y - x \rangle \in -\operatorname{int}(C(x)) \textit{ for all } \zeta \in T(x)\}$$

is convex for all $x \in K$. Let the set-valued map $W : K \to 2^Y$, defined by $W(x) = Y \setminus \{-\operatorname{int}(C(x))\}$ for all $x \in K$, be closed. Assume that for a nonempty compact convex set $D \subset K$ with each $x \in D \setminus K$, there exists $y \in D$ such that for any $\zeta \in T(x)$, $\langle \zeta, y - x \rangle \in -\operatorname{int}(C(x))$. Then $(\text{GWVVIP})_g$ has a solution.

Proof Let

$$A = \{(x, y) \in K \times K : \langle \zeta, y - x \rangle \notin -\operatorname{int}(C(x)) \text{ for all } \zeta \in T(x)\}.$$

Clearly, $(x, x) \in A$ for all $x \in K$. For each fixed $y \in K$, let

$$\begin{aligned} A_y &:= \{x \in K : (x, y) \in A\} \\ &= \{x \in K : \langle \zeta, y - x \rangle \notin -\operatorname{int}(C(x)) \text{ for all } \zeta \in T(x)\}. \end{aligned}$$

Then by Proposition 8.2 (f), A_y is closed. By hypothesis, for each fixed $y \in K$, the set $A_x := \{y \in K : (x, y) \notin A\}$ is convex.

By Lemma 1.17, there exists $\bar{x} \in K$ such that $\{\bar{x}\} \times K \subset A$, that is, $\bar{x} \in K$ such that $\langle \xi, \bar{x} - y \rangle \notin -\operatorname{int}(C(\bar{x}))$, for all $\xi \in T(\bar{x})$ and $y \in K$. □

8.4 Generalized Vector Variational Inequalities and Optimality Conditions for Vector Optimization Problems

Throughout this section, unless otherwise specified, we assume that K is a nonempty convex subset of $\mathbb{R}^n$ and $f = (f_1, f_2, \ldots, f_\ell) : \mathbb{R}^n \to \mathbb{R}^\ell$ be a vector-valued function. The subdifferential of a convex function f_i is denoted by ∂f_i.

Corresponding to K and ∂f_i, the (Stampacchia) generalized vector variational inequality problems and Minty generalized vector variational inequality problems are defined as follows:

$(\mathrm{GVVIP})^\ell_g$: Find $\bar{x} \in K$ such that for all $y \in K$ and all $\bar{\zeta}_i \in \partial f_i(\bar{x})$, $i \in \mathscr{I} = \{1, 2, \ldots, \ell\}$,

$$\langle \bar{\zeta}, y - \bar{x} \rangle_\ell := \big(\langle \bar{\zeta}_1, y - \bar{x} \rangle, \ldots, \langle \bar{\zeta}_\ell, y - \bar{x} \rangle \big) \notin -\mathbb{R}^\ell_+ \setminus \{\mathbf{0}\}. \tag{8.24}$$

$(\mathrm{GVVIP})^\ell_s$: Find $\bar{x} \in K$ such that there exist $\bar{\zeta}_i \in \partial f_i(\bar{x})$, $i \in \mathscr{I} = \{1, 2, \ldots, \ell\}$, such that for all $y \in K$

$$\langle \bar{\zeta}, y - \bar{x} \rangle_\ell := \big(\langle \bar{\zeta}_1, y - \bar{x} \rangle, \ldots, \langle \bar{\zeta}_\ell, y - \bar{x} \rangle \big) \notin -\mathbb{R}^\ell_+ \setminus \{\mathbf{0}\}. \tag{8.25}$$

$(\mathrm{GVVIP})^\ell_w$: Find $\bar{x} \in K$ such that for all $y \in K$, there exist $\bar{\zeta}_i \in \partial f_i(\bar{x})$, $i \in \mathscr{I} = \{1, 2, \ldots, \ell\}$, satisfying

$$\langle \bar{\zeta}, y - \bar{x} \rangle_\ell := \big(\langle \bar{\zeta}_1, y - \bar{x} \rangle, \ldots, \langle \bar{\zeta}_\ell, y - \bar{x} \rangle \big) \notin -\mathbb{R}^\ell_+ \setminus \{\mathbf{0}\}. \tag{8.26}$$

$(\mathrm{MGVVIP})^\ell_g$: Find $\bar{x} \in K$ such that for all $y \in K$ and all $\xi_i \in \partial f_i(y)$, $i \in \mathscr{I} = \{1, 2, \ldots, \ell\}$,

$$\langle \xi, y - \bar{x} \rangle_\ell := \big(\langle \xi_1, y - \bar{x} \rangle, \ldots, \langle \xi_\ell, y - \bar{x} \rangle \big) \notin -\mathbb{R}^\ell_+ \setminus \{\mathbf{0}\}. \tag{8.27}$$

(MGVVIP)$^\ell_w$: Find $\bar{x} \in K$ such that for all $y \in K$, there exist $\xi_i \in \partial f_i(y)$, $i \in \mathscr{I} = \{1, 2, \ldots, \ell\}$,

$$\langle \xi, y - \bar{x} \rangle_\ell := \left(\langle \xi_1, y - \bar{x} \rangle, \ldots, \langle \xi_\ell, y - \bar{x} \rangle \right) \notin -\mathbb{R}^\ell_+ \setminus \{\mathbf{0}\}. \tag{8.28}$$

(GWVVIP)$^\ell_g$: Find $\bar{x} \in K$ such that for all $y \in K$ and all $\bar{\zeta}_i \in \partial f_i(\bar{x})$, $i \in \mathscr{I} = \{1, 2, \ldots, \ell\}$,

$$\langle \bar{\zeta}, y - \bar{x} \rangle_\ell := \left(\langle \bar{\zeta}_1, y - \bar{x} \rangle, \ldots, \langle \bar{\zeta}_\ell, y - \bar{x} \rangle \right) \notin -\text{int}\left(\mathbb{R}^\ell_+ \right). \tag{8.29}$$

(GWVVIP)$^\ell_s$: Find $\bar{x} \in K$ such that there exist $\bar{\zeta}_i \in \partial f_i(\bar{x})$, $i \in \mathscr{I} = \{1, 2, \ldots, \ell\}$, such that for all $y \in K$

$$\langle \bar{\zeta}, y - \bar{x} \rangle_\ell := \left(\langle \bar{\zeta}_1, y - \bar{x} \rangle, \ldots, \langle \bar{\zeta}_\ell, y - \bar{x} \rangle \right) \notin -\text{int}\left(\mathbb{R}^\ell_+ \right). \tag{8.30}$$

(GWVVIP)$^\ell_w$: Find $\bar{x} \in K$ such that for all $y \in K$, there exist $\bar{\zeta}_i \in \partial f_i(\bar{x})$, $i \in \mathscr{I} = \{1, 2, \ldots, \ell\}$, satisfying

$$\langle \bar{\zeta}, y - \bar{x} \rangle_\ell := \left(\langle \bar{\zeta}_1, y - \bar{x} \rangle, \ldots, \langle \bar{\zeta}_\ell, y - \bar{x} \rangle \right) \notin -\text{int}\left(\mathbb{R}^\ell_+ \right). \tag{8.31}$$

(MGWVVIP)$^\ell_g$: Find $\bar{x} \in K$ such that for all $y \in K$ and all $\xi_i \in \partial f_i(y)$, $i \in \mathscr{I} = \{1, 2, \ldots, \ell\}$,

$$\langle \xi, y - \bar{x} \rangle_\ell := \left(\langle \xi_1, y - \bar{x} \rangle, \ldots, \langle \xi_\ell, y - \bar{x} \rangle \right) \notin -\text{int}\left(\mathbb{R}^\ell_+ \right). \tag{8.32}$$

(MGWVVIP)$^\ell_w$: Find $\bar{x} \in K$ such that for all $y \in K$, there exist $\xi_i \in \partial f_i(y)$, $i \in \mathscr{I} = \{1, 2, \ldots, \ell\}$, such that

$$\langle \xi, y - \bar{x} \rangle_\ell := \left(\langle \xi_1, y - \bar{x} \rangle, \ldots, \langle \xi_\ell, y - \bar{x} \rangle \right) \notin -\text{int}\left(\mathbb{R}^\ell_+ \right). \tag{8.33}$$

We denote the solution sets of the above mentioned problems (GVVIP)$^\ell_g$, (GVVIP)$^\ell_s$, (GVVIP)$^\ell_w$, (MGVVIP)$^\ell_g$, (MGVVIP)$^\ell_w$, (GWVVIP)$^\ell_g$, (GWVVIP)$^\ell_s$, (GWVVIP)$^\ell_w$, (MGWVVIP)$^\ell_g$ and (MGWVVIP)$^\ell_w$ by Sol(GVVIP)$^\ell_g$, Sol(GVVIP)$^\ell_s$, Sol(GVVIP)$^\ell_w$, Sol(MGVVIP)$^\ell_g$, Sol(MGVVIP)$^\ell_w$, Sol(GWVVIP)$^\ell_g$, Sol(GWVVIP)$^\ell_s$, Sol(GWVVIP)$^\ell_w$, Sol(MGWVVIP)$^\ell_g$ and Sol(MGWVVIP)$^\ell_w$, respectively.

As in Remark 8.1, we have

(a) Sol(GVVIP)$^\ell_g$ $\subseteq$ Sol(GVVIP)$^\ell_s$ $\subseteq$ Sol(GVVIP)$^\ell_w$;
(b) Sol(GWVVIP)$^\ell_g$ $\subseteq$ Sol(GWVVIP)$^\ell_s$ $\subseteq$ Sol(GWVVIP)$^\ell_w$;
(c) Sol(GVVIP)$^\ell_g$ $\subseteq$ Sol(GWVVIP)$^\ell_g$;
(d) Sol(GVVIP)$^\ell_s$ $\subseteq$ Sol(GWVVIP)$^\ell_s$;

(e) $\text{Sol(GVVIP)}_w^\ell \subseteq \text{Sol(GWVVIP)}_w^\ell$;
(f) $\text{Sol(MGVVIP)}_g^\ell \subseteq \text{Sol(MGWVVIP)}_g^\ell$;
(g) $\text{Sol(MGVVIP)}_w^\ell \subseteq \text{Sol(MGWVVIP)}_w^\ell$.

The following example shows that $\text{Sol(GVVIP)}_w^\ell \subseteq \text{Sol(GVVIP)}_s^\ell$ may not be true.

Example 8.5 [7] Let $K = \left\{(x_1, x_2) \in \mathbb{R}^2 : x_1 \leq 0, \; -\sqrt{-x_1} \leq x_2 \leq 0\right\}$ and

$$f_1(x_1, x_2) = \sqrt{x_1^2 + x_2^2} + x_2, \quad \text{for all } (x_1, x_2) \in K,$$
$$f_2(x_1, x_2) = x_2, \quad \text{for all } (x_1, x_2) \in K.$$

If $(x_1, x_2) = (0, 0)$, then

$$\begin{aligned}\partial f_1(x_1, x_2) &= \{(\zeta_1, \zeta_2) \in \mathbb{R}^2 : \zeta_1^2 + \zeta_2^2 \leq 1\} + \{(0, 1)\} \\ &= \{(\zeta_1, \zeta_2) \in \mathbb{R}^2 : \zeta_1^2 + (\zeta_2 - 1)^2 \leq 1\}.\end{aligned}$$

If $(x_1, x_2) \neq (0, 0)$, then

$$\partial f_1(x_1, x_2) = \left\{\left(\frac{x_1}{\sqrt{x_1^2 + x_2^2}}, \frac{x_2}{\sqrt{x_1^2 + x_2^2}} + 1\right)\right\}.$$

It can be easily checked that for all $(\zeta_1, \zeta_2) \in \partial f_1(0, 0)$, there exists $(x_1, x_2) \in K$ such that

$$(\zeta_1 x_1 + \zeta_2 x_2, x_2) \in -\mathbb{R}_+^2 \setminus \{\mathbf{0}\},$$

and that for all $(x_1, x_2) \in K$, there exists $(\xi_1, \xi_2) \in \partial f_1(0, 0)$ such that

$$(\xi_1 x_1 + \xi_2 x_2, x_2) \notin -\mathbb{R}_+^2 \setminus \{\mathbf{0}\}.$$

Hence, $(0, 0) \in \text{Sol(GVVIP)}_w^\ell$, but $(0, 0) \notin \text{Sol(GVVIP)}_s^\ell$.

Moreover, $\text{Sol(GVVIP)}_s^\ell = \left\{(x, -\sqrt{-x}) : x < 0\right\}$ and $\text{Sol(GVVIP)}_w^\ell = \left\{(x, -\sqrt{-x}) : x \leq 0\right\}$.

Proposition 8.5 *For each $i \in \mathscr{I} = \{1, 2, \ldots, \ell\}$, let $f_i : K \to \mathbb{R}$ be convex. Then* $\text{Sol(GVVIP)}_w^\ell \subseteq \text{Sol(MGVVIP)}_g^\ell \subseteq \text{Sol(MGVVIP)}_w^\ell$.

Proof Let $\bar{x} \in K$ be a solution of (GVVIP)_w^ℓ. Then for all $y \in K$, there exist $\bar{\zeta}_i \in \partial f_i(\bar{x})$, $i = 1, 2, \ldots, \ell$, such that

$$\left(\langle \bar{\zeta}_1, y - \bar{x} \rangle, \ldots, \langle \bar{\zeta}_\ell, y - \bar{x} \rangle\right) \notin -\mathbb{R}_+^\ell \setminus \{\mathbf{0}\}. \tag{8.34}$$

Since each f_i is convex, ∂f_i, $i \in \mathscr{I}$, is monotone, and therefore, we have

$$\langle \xi_i - \bar{\zeta}_i, y - \bar{x} \rangle \geq 0, \quad \text{for all } \xi_i \in \partial f_i(y) \text{ and for each } i \in \mathscr{I}. \tag{8.35}$$

From (8.34) and (8.35), it follows that for all $y \in K$ and all $\xi_i \in \partial f_i(y)$, $i \in \mathscr{I}$,

$$\big(\langle \xi_1, y - \bar{x} \rangle, \ldots, \langle \xi_\ell, y - \bar{x} \rangle \big) \notin -\mathbb{R}^\ell_+ \setminus \{\mathbf{0}\}.$$

Thus, $\bar{x} \in K$ is a solution of (MGVVIP)$^\ell_g$. □

The converse of the above proposition may not be true, that is, Sol(MGVVIP)$^\ell_g \not\subseteq$ Sol(GVVIP)$^\ell_w$.

Example 8.6 Let $K =]-\infty, 0]$ and $f_1(x) = x, f_2(x) = x^2$. Since $(x, 0) \in -\mathbb{R}^2_+ \setminus \{\mathbf{0}\}$ for all $x \in]-\infty, 0[$, we have $0 \notin$ Sol(GVVIP)$^\ell_w$.

But, since $(x, 2x^2) \notin -\mathbb{R}^2_+ \setminus \{\mathbf{0}\}$, we have $0 \in$ Sol(MGVVIP)$^\ell_g$. Moreover, we can easily verify that Sol(GVVIP)$^\ell_w =]-\infty, 0[$ and Sol(MGVVIP)$^\ell_g =]-\infty, 0]$.

The following result provides the relationship between the solutions of (MGWVVIP)$^\ell_g$ and (GWVVIP)$^\ell_g$.

Theorem 8.16 *For each $i \in \mathscr{I} = \{1, 2, \ldots, \ell\}$, let $f_i : K \to \mathbb{R}$ be convex. Then $\bar{x} \in K$ is a solution* (GWVVIP)$^\ell_w$ *if and only if it is a solution of* (MGWVVIP)$^\ell_w$.

Proof Let $\bar{x} \in K$ be a solution of (GWVVIP)$^\ell_w$. Then for any $y \in K$, there exist $\bar{\zeta}_i \in \partial f_i(\bar{x})$, $i = 1, 2, \ldots, \ell$, such that

$$\Big(\langle \bar{\zeta}_1, y - \bar{x} \rangle, \ldots, \langle \bar{\zeta}_\ell, y - \bar{x} \rangle \Big) \notin -\text{int}\left(\mathbb{R}^\ell_+\right). \tag{8.36}$$

Since each f_i is convex, ∂f_i $(i \in \mathscr{I})$ is monotone, and therefore, we have

$$\langle \xi_i - \bar{\zeta}_i, y - \bar{x} \rangle \geq 0, \quad \text{for all } y \in K,\ \xi_i \in \partial f_i(y) \text{ and for each } i \in \mathscr{I}. \tag{8.37}$$

From (8.36) and (8.36), it follows that for any $y \in K$ and any $\xi_i \in \partial f_i(y)$, $i \in \mathscr{I}$,

$$\Big(\langle \xi_1, y - \bar{x} \rangle, \ldots, \langle \xi_\ell, y - \bar{x} \rangle \Big) \notin -\text{int}\left(\mathbb{R}^\ell_+\right).$$

Thus, $\bar{x} \in K$ is a solution of (MGWVVIP)$^\ell_g$. Since Sol(MGWVVIP)$^\ell_g \subseteq$ Sol(MGWVVIP)$^\ell_w$, $\bar{x} \in K$ is a solution of Sol(MGWVVIP)$^\ell_w$.

Conversely, let $\bar{x} \in K$ be a solution of (MGWVVIP)$^\ell_w$. Consider any $y \in K$ and any sequence $\{\alpha_m\} \searrow 0$ with $\alpha_m \in]0, 1]$. Since K is convex, $y_m := \bar{x} + \alpha_m(y - \bar{x}) \in K$. Since $\bar{x} \in K$ is a solution of (MGWVVIP)$^\ell_w$, there exist $\xi^m_i \in \partial f_i(y_m)$, $i \in \mathscr{I}$, such that

$$(\langle \xi^m_1, y_m - \bar{x} \rangle), \ldots, \langle \xi^m_\ell, \eta(y_m, \bar{x}) \rangle) \notin -\text{int}\left(\mathbb{R}^\ell_+\right).$$

Since each f_i is convex and so it is locally Lipschitz (see Theorem 1.16), and hence, there exists $k > 0$ such that for sufficiently large m and for all $i \in \mathscr{I}$, $\|\xi_i^m\| \leq k$. So, we can assume that the sequence $\{\xi_i^m\}$ converges to $\bar{\zeta}_i$ for each $i \in \mathscr{I}$. Since the set-valued map $y \mapsto \partial f_i(y)$ is closed (see Lemma 1.8), $\xi_i^m \in \partial f_i(y_m)$ and $y_m \to \bar{x}$ as $m \to \infty$, we have $\bar{\zeta}_i \in \partial f_i(\bar{x})$ for each $i \in \mathscr{I}$. Therefore, for any $y \in K$, there exist $\bar{\zeta}_i \in \partial f_i(\bar{x})$, $i \in \mathscr{I}$, such that

$$\left(\langle \bar{\zeta}_1, y - \bar{x}\rangle, \ldots, \langle \bar{\zeta}_\ell, y - \bar{x}\rangle\right) \notin -\text{int}\left(\mathbb{R}^\ell_+\right).$$

Hence, $\bar{x} \in K$ is a solution of (GWVVIP)$^\ell_w$. □

Next theorem provides the necessary and sufficient conditions for an efficient solution of VOP.

Theorem 8.17 ([6]) *For each $i \in \mathscr{I} = \{1, 2, \ldots, \ell\}$, let $f_i : K \to \mathbb{R}$ be convex. Then $\bar{x} \in K$ is an efficient solution of VOP if and only if it is a solution of* (MGVVIP)$^\ell_w$.

Proof Let $\bar{x} \in K$ be a solution of (MGVVIP)$^\ell_w$ but not an efficient solution of VOP. Then there exists $z \in K$ such that

$$\left(f_1(z) - f_1(\bar{x}), \ldots, f_\ell(z) - f_\ell(\bar{x})\right) \in -\mathbb{R}^\ell_+ \setminus \{\mathbf{0}\}. \tag{8.38}$$

Set $z(\lambda) := \lambda z + (1 - \lambda)\bar{x}$ for all $\lambda \in [0, 1]$. Since K is convex, $z(\lambda) \in K$ for all $\lambda \in [0, 1]$. Since each f_i is convex, we have

$$f_i(z(\lambda)) = f_i(\lambda z + (1 - \lambda)\bar{x}) \leq \lambda f_i(z) + (1 - \lambda) f_i(\bar{x}), \quad \text{for each } i = 1, 2, \ldots, \ell,$$

that is,

$$f_i(\bar{x} + \lambda(z - \bar{x})) - f_i(\bar{x}) \leq \lambda[f_i(z) - f_i(\bar{x})],$$

for all $\lambda \in [0, 1]$ and for each $i = 1, 2, \ldots, \ell$. In particular, for $\lambda \in]0, 1[$, we have

$$\frac{f_i(z(\lambda)) - f_i(\bar{x})}{\lambda} \leq f_i(z) - f_i(\bar{x}), \quad \text{for each } i = 1, 2, \ldots, \ell. \tag{8.39}$$

By Lebourg's Mean Value Theorem 1.32, there exist $\lambda_i \in]0, 1[$ and $\xi_i \in \partial f_i(z(\lambda_i))$ such that

$$\langle \xi_i, z - \bar{x}\rangle = f_i(z(\lambda)) - f_i(\bar{x}), \quad \text{for each } i = 1, 2, \ldots, \ell. \tag{8.40}$$

By combining (8.39)–(8.40), we obtain

$$\langle \xi_i, z - \bar{x}\rangle \leq f_i(z) - f_i(\bar{x}), \quad \text{for each } i = 1, 2, \ldots, \ell. \tag{8.41}$$

Suppose that $\lambda_1, \lambda_2, \ldots, \lambda_\ell$ are all equal. Then it follows from (8.38) and (8.41) that $\bar{x}$ is not a solution of $(\text{MGVVIP})_w^\ell$. This contradicts to the fact the $\bar{x}$ is a solution of $(\text{MGVVIP})_w^\ell$.

Consider the case when $\lambda_1, \lambda_2, \ldots, \lambda_\ell$ are not equal. Let $\lambda_1 \neq \lambda_2$. Then from (8.41), we have

$$\langle \xi_1, z - \bar{x} \rangle \leq f_1(z) - f_1(\bar{x}) \tag{8.42}$$

and

$$\langle \xi_2, z - \bar{x} \rangle \leq f_2(z) - f_2(\bar{x}). \tag{8.43}$$

Since f_i and f_2 are convex, ∂f_1 and ∂f_2 are monotone, that is,

$$\langle \xi_1 - \xi_2^*, z(\lambda_1) - z(\lambda_2) \rangle \geq 0, \quad \text{for all } \xi_2^* \in \partial f_1(z(\lambda_2)), \tag{8.44}$$

and

$$\langle \xi_1^* - \xi_2, z(\lambda_1) - z(\lambda_2) \rangle \geq 0, \quad \text{for all } \xi_1^* \in \partial f_2(z(\lambda_1)). \tag{8.45}$$

If $\lambda_1 > \lambda_2$, then by (8.44), we obtain

$$0 \leq \langle \xi_1 - \xi_2^*, z(\lambda_1) - z(\lambda_2) \rangle = (\lambda_1 - \lambda_2)\langle \xi_1 - \xi_2^*, z - \bar{x} \rangle,$$

and so,

$$\langle \xi_1 - \xi_2^*, z - \bar{x}) \rangle \geq 0 \quad \Leftrightarrow \quad \langle \xi_1, z - \bar{x} \rangle \geq \langle \xi_2^*, z - \bar{x} \rangle.$$

From (8.42), we have

$$\langle \xi_2^*, z - \bar{x} \rangle \leq f_1(z) - f_1(\bar{x}), \quad \text{for all } \xi_2^* \in \partial f_1(z(\lambda_2)).$$

If $\lambda_1 < \lambda_2$, then by (8.45), we have

$$0 \leq \langle \xi_1^* - \xi_2, z(\lambda_1) - z(\lambda_2) \rangle = (\lambda_1 - \lambda_2)\langle \xi_1^* - \xi_2, z - \bar{x} \rangle,$$

and so,

$$\langle \xi_1^* - \xi_2, z - \bar{x} \rangle \leq 0 \quad \Leftrightarrow \quad \langle \xi_1^*, z - \bar{x} \rangle \leq \langle \xi_2, z - \bar{x} \rangle.$$

From (8.43), we obtain

$$\langle \xi_1^*, z - \bar{x} \rangle \leq f_2(z) - f_2(\bar{x}), \quad \text{for all } \xi_1^* \in \partial f_2(z(\lambda_1)).$$

Therefore, for the case $\lambda_1 \neq \lambda_2$, let $\bar{\lambda} = \min\{\lambda_1, \lambda_2\}$. Then, we can find $\bar{\xi}_i \in \partial f_i(z(\bar{\lambda}))$ such that

$$\langle \bar{\xi}_i, z - \bar{x} \rangle \leq f_i(z) - f_i(\bar{x}), \quad \text{for all } i = 1, 2.$$

By continuing this process, we can find $\lambda^* \in]0, 1[$ and $\xi_i^* \in \partial f_i(z(\lambda^*))$ such that $\lambda^* = \min\{\lambda_1, \lambda_2, \ldots, \lambda_\ell\}$ and

$$\langle \xi_i^*, z - \bar{x} \rangle \leq f_i(z) - f_i(\bar{x}), \quad \text{for each } i = 1, 2, \ldots, \ell. \tag{8.46}$$

From (8.38) and (8.46), we have $\xi_i^* \in \partial f_i(z(\lambda^*))$, $i = 1, 2, \ldots, \ell$, and

$$\big(\langle \xi_1^*, z - \bar{x} \rangle, \ldots, \langle \xi_\ell^*, z - \bar{x} \rangle\big) \in -\mathbb{R}_+^\ell \setminus \{\mathbf{0}\}.$$

By multiplying above inclusion by $-\lambda^*$, we obtain

$$\big(\langle \xi_1^*, z(\lambda^*) - \bar{x} \rangle, \ldots, \langle \xi_\ell^*, z(\lambda^*) - \bar{x} \rangle\big) \in -\mathbb{R}_+^\ell \setminus \{\mathbf{0}\}.$$

which contradicts to our supposition that $\bar{x}$ is a solution of $(\text{MGVVIP})_w^\ell$.

Conversely, suppose that $\bar{x} \in K$ is an efficient solution of VOP. Then we have

$$\big(f_1(y) - f_1(\bar{x}), \ldots, f_\ell(y) - f_\ell(\bar{x})\big) \notin -\mathbb{R}_+^\ell \setminus \{\mathbf{0}\}, \quad \text{for all } y \in K. \tag{8.47}$$

Since each f_i is convex, we deduce that

$$\langle \xi_i, \bar{x} - y \rangle \leq f_i(\bar{x}) - f_i(y), \quad \text{for all } y \in K,\ \xi_i \in \partial f_i(y) \text{ and } i \in \mathcal{I}.$$

Also, we obtain

$$\langle \xi_i, y - \bar{x} \rangle \geq f_i(y) - f_i(\bar{x}), \quad \text{for all } y \in K,\ \xi_i \in \partial f_i(y) \text{ and } i \in \mathcal{I}. \tag{8.48}$$

From (8.47) and (8.48), it follows that $\bar{x}$ is a solution of $(\text{MGVVIP})_w^\ell$. □

Theorem 8.17 is extended for Dini subdifferential by Al-Homidan and Ansari [1].

Theorem 8.18 *[6] For each* $i \in \mathcal{I} = \{1, 2, \ldots, \ell\}$, *let* $f_i : K \to \mathbb{R}$ *be convex. If* $\bar{x} \in K$ *is a solution* $(\text{GVVIP})_w^\ell$, *then it is an efficient solution of* VOP *and hence a solution of* $(\text{MGVVIP})_w^\ell$.

Proof Since $\bar{x} \in X$ is a solution of $(\text{GVVIP})_w^\ell$, for any $y \in K$, there exist $\bar{\zeta}_i \in \partial f_i(\bar{x})$, $i = 1, 2, \ldots, \ell$, such that

$$\big(\langle \bar{\zeta}_1, y - \bar{x} \rangle, \ldots, \langle \bar{\zeta}_\ell, y - \bar{x} \rangle\big) \notin -\mathbb{R}_+^\ell \setminus \{\mathbf{0}\}. \tag{8.49}$$

Since each f_i is convex, we have

$$\langle \bar{\zeta}_i, y - \bar{x} \rangle \leq f_i(y) - f_i(\bar{x}) \quad \text{for any } y \in K \text{ and all } i \in \mathcal{I}. \tag{8.50}$$

By combining (8.49) and (8.50), we obtain

$$\left(f_1(y) - f_1(\bar{x}), \ldots, f_\ell(y) - f_\ell(\bar{x})\right) \notin -\mathbb{R}^\ell_+ \setminus \{\mathbf{0}\}, \quad \text{for all } y \in K.$$

Thus, $\bar{x} \in K$ is an efficient solution of VOP. □

From Theorem 8.18, we see that $(\text{GVVIP})^\ell_w$ is a sufficient optimality condition for an efficient solution of VOP. However, it is not, in general, a necessary optimality condition for an efficient solution of VOP.

Example 8.7 Let $K = [-1, 0]$ and $f(x) = (x, x^2)$. Consider the following differentiable convex vector optimization problem:

$$\text{minimize } f(x), \quad \text{subject to } x \in K, \tag{VOP}$$

Then $\bar{x} = 0$ is an efficient solution of VOP and $\bar{x} = 0$ is a solution of the following (MVVIP): Find $\bar{x} \in K$ such that for all $y \in K$,

$$\left(\langle \nabla f_1(y), y - \bar{x}\rangle, \langle \nabla f_2(y), y - \bar{x}\rangle\right) = \left(y - \bar{x}, 2y(y - \bar{x})\right) \notin -\mathbb{R}^2_+ \setminus \{\mathbf{0}\}.$$

However, $\bar{x} = 0$ is not a solution of the following (VVIP): Find $\bar{x} \in K$ such that for all $y \in K$,

$$\left(\langle \nabla f_1(\bar{x}), y - \bar{x}\rangle, \langle \nabla f_2(\bar{x}), y - \bar{x}\rangle\right) = \left(y - \bar{x}, 2\bar{x}(y - \bar{x})\right) \notin -\mathbb{R}^2_+ \setminus \{\mathbf{0}\}.$$

The following result presents the equivalence between the solution of $(\text{GWVVIP})^\ell_w$ and a weakly efficient solution of VOP.

Theorem 8.19 *For each $i \in \mathcal{I} = \{1, 2, \ldots, \ell\}$, let $f_i : K \to \mathbb{R}$ be convex. If $\bar{x} \in K$ is a weakly efficient solution of* VOP *if and only if it is a solution of* $(\text{GWVVIP})^\ell_w$.

Proof Suppose that $\bar{x}$ is a solution of $(\text{GWVVIP})^\ell_w$ but not a weakly efficient solution of VOP. Then there exists $y \in K$ such that

$$\left(f_1(y) - f_1(\bar{x}), \ldots, f_\ell(y) - f_\ell(\bar{x})\right) \in -\text{int}\left(\mathbb{R}^\ell_+\right). \tag{8.51}$$

Since each f_i, $i \in \mathcal{I}$, is convex, we have

$$\langle \zeta_i, y - \bar{x}\rangle \leq f_i(y) - f_i(\bar{x}), \quad \text{for all } \zeta_i \in \partial f_i(\bar{x}). \tag{8.52}$$

Combining (8.51) and (8.52), we obtain

$$\left(\langle \zeta_1, y - \bar{x}\rangle, \ldots, \langle \zeta_\ell, y - \bar{x}\rangle\right) \in -\text{int}\left(\mathbb{R}^\ell_+\right), \quad \text{for all } \zeta_i \in \partial f_i(\bar{x})$$

which contradicts to our supposition that $\bar{x}$ is a solution of $(\text{GWVVIP})^\ell_w$.

Conversely, assume that $\bar{x} \in K$ is a weakly efficient solution of VOP but not a solution of $(\text{GWVVIP})^\ell_w$. Then by Theorem 8.16, $\bar{x}$ is not a solution of

$(\text{MGWVVIP})_w^\ell$. Thus, there exist $y \in K$ and $\xi_i \in \partial f_i(y)$, $i \in \mathscr{I}$, such that

$$\left(\langle \xi_1, y-\bar{x}\rangle, \ldots, \langle \xi_\ell, y, \bar{x}\rangle\right) \in -\text{int}\left(\mathbb{R}_+^\ell\right). \tag{8.53}$$

By convexity of f_i, $i \in \mathscr{I}$, we have

$$0 > \langle \xi_i, y-\bar{x}\rangle \geq f_i(y) - f_i(\bar{x}). \tag{8.54}$$

From (8.53) and (8.54), we then have

$$\left(f_1(y)-f_1(\bar{x}), \ldots, f_\ell(y)-f_\ell(\bar{x})\right) \in -\text{int}\left(\mathbb{R}_+^\ell\right).$$

which contradicts to our assumption that $\bar{x}$ is a weakly efficient solution of VOP. □

The following example shows that the weakly efficient solution of VOP may not be a solution of $(\text{GWVVIP})_g^\ell$.

Example 8.8 ([7]) Let $K =]-\infty, 0]$ and

$$f_1(x) = x, \quad f_2(x) = \begin{cases} x^2, & x < 0 \\ x, & x \geq 0. \end{cases}$$

Then sol$(\text{GWVVIP})_g^\ell =]-\infty, 0[$, but the set of weakly efficient solution of VOP is $]-\infty, 0]$.

The relations between a properly efficient solution in the sense of Geoffrion and a solution of $(\text{GVVIP})_w^\ell$ is studied in [6].

References

1. S. Al-Homidan, Q.H. Ansari, Relations between generalized vector variational-like inequalities and vector optimization problems. Taiwan. J. Math. **16**(3), 987–998 (2012)
2. A. Daniilidis, N. Hadjisavvas, Existence theorems for vector variational inequalities. Bull. Aust. Math. Soc. **54**, 473–481 (1996)
3. X.P. Ding, W.K. Kim, K.K. Tan, A selection theorem and its applications. Bull. Aust. Math. Soc. **46**, 205–212 (1992)
4. H. Kneser, Sur le théorème fondamentale de la thérie des jeux. C. R. Acad. Sci. Paris **234**, 2418–2420 (1952)
5. I.V. Konnov, J.-C. Yao, On the generalized vector variational inequality problem. J. Math. Anal. Appl. **206**, 42–58 (1997)
6. G.M. Lee, On relations between vector variational inequality and vector optimization problem, in *Progress in Optimization, II: Contributions from Australasia*, ed. by X.Q. Yang, A.I. Mees, M.E. Fisher, L.S. Jennings (Kluwer Academic Publisher, Dordrecht, 2000), pp. 167–179
7. G.M. Lee, K.B. Lee, Vector variational inequalities for nondifferentiable convex vector optimization problems. J. Glob. Optim. **32**, 597–612 (2005)

Chapter 9
Vector Equilibrium Problems

Motivated by various applications of multi-criteria decision making, extensions of scalar equilibrium problems, discussed in Chap. 1, for the vector case were proposed. Among them, the most investigated problems are vector optimization and vector saddle point. In 1956, Blackwell [27] considered matrix games with vector payoffs and proved the existence theorem for such problems. Since then, many researchers studied vector non-cooperative games in finite- and infinite-dimensional spaces. In 1980, Giannessi [51] extended the variational inequality problem for vector-valued functions known as "vector variational inequality problem" (in short, VVIP) with further applications. Vector equilibrium problems (in short, VEPs) can be viewed as further and natural extension of the previous concepts. It is a unified model of several known problems, namely, vector variational inequality problems, vector optimization problems, vector saddle point problems and Nash equilibrium problems for vector-valued functions. The theory of VVIPs and VEPs has been developing extensively since the early ninety's. In particular, a number of various kinds of these problems were proposed and the corresponding existence results both on bounded and on unbounded sets were established. The mathematical theory of VEPs is presented in this chapter.

9.1 Introduction

Let K be a nonempty convex subset of a topological vector space X and Y be an ordered topological vector space with a proper closed convex cone C such that $\text{int}(C) \neq \emptyset$. Throughout the chapter, we denote by $\mathbf{0}$ the zero element of a vector space. Let $f : K \times K \to Y$ be a bifunction such that $f(x, x) \geq_C \mathbf{0}$ for all $x \in K$. There are several possible ways to extend the equilibrium problem (in short, EP) for vector-valued bifunctions. Here, we present main three formulations of such extensions, are called *vector equilibrium problems* (in short, VEPs).

Q.H. Ansari et al., *Vector Variational Inequalities and Vector Optimization*,
Vector Optimization, DOI 10.1007/978-3-319-63049-6_9

The *weak vector equilibrium problem* (in short, WVEP) is to find $\bar{x} \in K$ such that

$$f(\bar{x}, y) \not<_C \mathbf{0}, \quad \text{for all } y \in K. \tag{9.1}$$

The *vector equilibrium problem* (in short, VEP) is to find $\bar{x} \in K$ such that

$$f(\bar{x}, y) \not\leq_{C_0} \mathbf{0}, \quad \text{for all } y \in K. \tag{9.2}$$

The *strong vector equilibrium problem* (in short, SVEP) is to find $\bar{x} \in K$ such that

$$f(\bar{x}, y) \geq_C \mathbf{0}, \quad \text{for all } y \in K. \tag{9.3}$$

The sets of solutions of problems WVEP, VEP, and SVEP are denoted by $\mathbb{S}(\text{WVEP}(f, C))$, $\mathbb{S}(\text{VEP}(f, C))$, and $\mathbb{S}(\text{SVEP}(f, C))$, respectively.

It is clear that

$$\mathbb{S}(\text{WVEP}(f, C)) \subset \mathbb{S}(\text{VEP}(f, C)) \subset \mathbb{S}(\text{SVEP}(f, C)),$$

but the converse need not be true.

Other problems closely related to VEPs are the following problems, called *Minty vector equilibrium problems* (in short, MVEPs) or *dual vector equilibrium problems* (in short, DVEPs).

The *Minty weak vector equilibrium problem* (in short, MWVEP) is to find $\bar{x} \in K$ such that

$$f(y, \bar{x}) \not>_C \mathbf{0}, \quad \text{for all } y \in K. \tag{9.4}$$

The *Minty vector equilibrium problem* (in short, MVEP) is to find $\bar{x} \in K$ such that

$$f(y, \bar{x}) \not\geq_{C_0} \mathbf{0}, \quad \text{for all } y \in K. \tag{9.5}$$

The *Minty strong vector equilibrium problem* (in short, MSVEP) is to find $\bar{x} \in K$ such that

$$f(y, \bar{x}) \leq_C \mathbf{0}, \quad \text{for all } y \in K. \tag{9.6}$$

The sets of solutions of MWVEP, MVEP, and MSVEP are denoted by $\mathbb{S}(\text{MWVEP}(f, C))$, $\mathbb{S}(\text{MVEP}(f, C))$, and $\mathbb{S}(\text{MSVEP}(f, C))$, respectively.

Obviously,

$$\mathbb{S}(\text{MWVEP}(f, C)) \subset \mathbb{S}(\text{MVEP}(f, C)) \subset \mathbb{S}(\text{MSVEP}(f, C)),$$

but the converse assertions are not true.

The examples of VEPs are vector optimization problems, vector variational inequality problems, vector saddle point problems, noncooperative equilibrium problem with vector payoff, etc.

Example 9.1

(a) **(Vector optimization Problem)** Let $\varphi : K \to Y$ be a vector-valued function. Set

$$f(x,y) = \varphi(x) - \varphi(y), \quad \text{for all } x, y \in K.$$

Then the above mentioned VEPs reduce to the corresponding VOPs, and provide, respectively, weak efficient, efficient, and strong efficient solutions of VOP.

(b) **(Vector Variational Inequalities)** Let $T : K \to \mathcal{L}(X,Y)$ be a nonlinear operator. Set

$$f(x,y) = \langle T(x), y - x \rangle, \quad \text{for all } x, y \in K.$$

Then VEPs are equivalent to the corresponding VVIPs.

(c) **(Vector Saddle Point Problems)** Let K_1 and K_2 be nonempty subsets of X and $L : K_1 \times K_2 \to Y$ be a vector-valued function. The *weak regular saddle point problem* [70] is to find $\bar{x} = (\bar{x}_1, \bar{x}_2) \in K_1 \times K_2$ such that

$$L(\bar{x}_1, y_2) \not>_C L(y_1, \bar{x}_2), \quad \text{for all } (y_1, y_2) \in K_1 \times K_2. \tag{9.7}$$

If we replace $\not>_C$ by $\not\geq_{C_0}$ (respectively, by $\leq_C$) in (9.7), then above problem is called *regular saddle point problem* (respectively, *strong regular saddle point problem*).

The *weak vector saddle point problem* is to find $\bar{x} = (\bar{x}_1, \bar{x}_2) \in K_1 \times K_2$ such that

$$L(\bar{x}_1, y_2) \not>_C L(\bar{x}_1, \bar{x}_2) \not>_C L(y_1, \bar{x}_2), \quad \text{for all } (y_1, y_2) \in K_1 \times K_2. \tag{9.8}$$

If we replace $\not>_C$ by $\not\geq_{C_0}$ (respectively, by $\leq_C$) in (9.8), then the weak vector saddle point problem is called *vector saddle point problem* (respectively, *strong vector saddle point problem*).

Clearly, every solution of the regular saddle point problem (respectively, weak regular saddle point problem) is a solution of saddle point problem (respectively, weak saddle point problem), but converse is not true since $\not>_C$ is not transitive. Set $K = K_1 \times K_2$ and define $f : K \times K \to Y$ by

$$f((x_1, x_2), (y_1, y_2)) = L(y_1, x_2) - L(x_1, y_2),$$

for all $(x_1, x_2), (y_1, y_2) \in K_1 \times K_2$. Then VEPs are equivalent to the corresponding regular saddle point problems.

(d) **(Noncooperative Vector Equilibrium Problem)** For each $i = 1, 2, \ldots, m$, let K_i be a nonempty subset of a topological vector space X_i, $K := \prod_{i=1}^m K_i$ and $X = \prod_{i=1}^m X_i$. For each $i = 1, 2, \ldots, m$, let $\varphi_i : K \to Y$ be a vector-valued function. For each $x = (x_1, x_2, \ldots, x_m) \in K$, we denote by $x^i = (x_1, x_2, \ldots, x_{i-1}, x_{i+1}, \ldots, x_m) \in \prod_{j \neq i} K_j$ and write $x = (x_i, x^i) \in K_i \times \prod_{j \neq i} K_j = \prod_{i=1}^m K_i$. The *noncooperative vector equilibrium problem* [71] or *Nash equilibrium problem with vector payoff* is to find $\bar{x} = K$ such that for each $i = 1, 2, \ldots, m$,

$$\varphi_i(\bar{x}_i, \bar{x}^i) \nleq_{C_0} \varphi_i(y_i, \bar{x}^i), \quad \text{for all } y_i \in K_i. \tag{9.9}$$

If we replace $\nleq_{C_0}$ in (9.9) by $\not>_C$ (respectively, by $\leq_C$), then noncooperative vector equilibrium problem is called *noncooperative weak vector equilibrium problem* (respectively, *noncooperative strong vector equilibrium problem*).

If we consider

$$f(x, y) = \sum_{i=1}^m \left[\varphi_i(y_i, x^i) - \varphi_i(x_i, x^i) \right], \quad \text{for all } x, y \in K,$$

then noncooperative vector equilibrium problem (respectively, noncooperative weak vector equilibrium problem and noncooperative strong vector equilibrium problem) is equivalent to VEP (respectively, WVEP and SVEP).

When $f(x, y) = g(x, y) + h(x, y)$ for all $x, y \in K$ with $g, h : K \times K \to Y$ such that $g(x, x) \geq_C \mathbf{0}$ and $h(x, x) = \mathbf{0}$ for all $x \in K$, then WVEP can be written as to find $\bar{x} \in K$ such that

$$g(\bar{x}, y) + h(\bar{x}, y) \not<_C \mathbf{0}, \quad \text{for all } y \in K. \tag{9.10}$$

The set of solutions of WVEP (9.10) is denoted by $\mathbb{S}(\text{WVEP}(g, h, C))$. Let $l : K \times K \to Y$ be a vector-valued bifunction. We also consider the following more general problem known as *implicit weak vector variational problem* (for short, IWVVP) which contains WVEP (9.1) and (9.10) as special cases: Find $\bar{x} \in K$ such that

$$l(\bar{x}, \bar{x}) + g(\bar{x}, \bar{x}) \not>_C l(\bar{x}, y) + g(\bar{x}, y), \quad \text{for all } y \in K. \tag{9.11}$$

Similarly, we can define implicit vector variational problem and implicit strong vector variational problem by replacing $\not>_C$ by $\nleq_{C_0}$ and by $\leq_C$, respectively, in (9.11).

Let $T, G : K \to \mathcal{L}(X, Y)$ be nonlinear operators. Set

$$l(x, y) = \langle T(x), y - x \rangle \quad \text{and} \quad g(x, y) = \langle G(x), y - x \rangle, \quad \text{for all } x, y \in K.$$

Then IWVVP reduces to the problem of finding $\bar{x} \in K$ such that

$$\langle T(\bar{x}) + G(\bar{x}), y - \bar{x} \rangle \not<_C \mathbf{0}, \quad \text{for all } y \in K. \tag{9.12}$$

It is known as *strongly nonlinear weak vector variational inequality problem* (in short, SNWVVIP) considered and studied in [3]. For $Y = \mathbb{R}$ and $C = \mathbb{R}_+$, SNWVVIP is studied by Mosco [81].

When $K = X$, $Y = \mathbb{R}$ and $C = \mathbb{R}_+$, IWVVP reduces to the problem of finding $\bar{x} \in X$ such that

$$l(\bar{x}, \bar{x}) + g(\bar{x}, \bar{x}) \leq l(\bar{x}, y) + g(\bar{x}, y), \quad \text{for all } y \in X, \tag{9.13}$$

which is known as the *implicit variational problem* (for short, IVP). It includes variational and quasi-variational inequalities [18], fixed point problem and saddle point problem, Nash equilibrium problem of non-cooperative games as special cases. The existence of solutions of IVP was studied by Mosco [81], while Dolcetta and Matzeu [37] discussed its duality and applications.

9.2 Existence Results

Throughout this section, unless otherwise specified, we assume that K is a nonempty convex subset of a topological vector space X and (Y, C) is an ordered topological vector space with a proper, closed and convex cone C. Whenever $\text{int}(C)$ is involved in the formulation of a problem, we further assume that $\text{int}(C) \neq \emptyset$.

Definition 9.1 Let K be a nonempty subset of X. A vector-valued bifunction $f : K \times K \to Y$ is said to be:

(a) *C-monotone* if for all $x, y \in K$, we have

$$f(x, y) + f(y, x) \leq_C \mathbf{0};$$

(b) *C-pseudomonotone* if for all $x, y \in K$, we have

$$f(x, y) >_C \mathbf{0} \quad \text{implies} \quad f(y, x) <_C \mathbf{0},$$

equivalently,

$$f(y, x) \not<_C \mathbf{0} \quad \text{implies} \quad f(x, y) \not>_C \mathbf{0};$$

(c) *strictly C-pseudomonotone* if for all $x, y \in K$, $x \neq y$, we have

$$f(x, y) \not<_C \mathbf{0} \quad \text{implies} \quad f(y, x) <_C \mathbf{0},$$

equivalently,

$$f(y, x) \not<_C \mathbf{0} \quad \text{implies} \quad f(x, y) <_C \mathbf{0};$$

(d) *C-quasimonotone* if for all $x, y \in K$, we have

$$f(x,y) >_C \mathbf{0} \quad \text{implies} \quad f(y,x) \leq_C \mathbf{0},$$

equivalently,

$$f(y,x) \nleq_C \mathbf{0} \quad \text{implies} \quad f(x,y) \ngtr_C \mathbf{0};$$

(e) *maximal C-pseudomonotone* if it is C-pseudomonotone and, for all $x, y \in K$, we have

$$f(u,x) \ngtr_C \mathbf{0} \text{ for all } u \in]x,y] \quad \text{implies} \quad f(x,y) \nless_C \mathbf{0},$$

where $]x,y]$ denotes the line segment joining x and y with the endpoint y.

Obviously, C-monotonicity implies C-pseudomonotonicity and C-pseudomonotonicity implies strictly C-pseudomonotonicity as well as C-quasimonotonicity. The reverse assertions may not be true.

Lemma 9.1 *If $f : K \times K \to Y$ is C-pseudomonotone, C-hemicontinuous in the first argument and explicitly C-quasiconvex in the second argument such that $f(x,x) \geq_C \mathbf{0}$ for all $x \in K$, then it is maximal C-pseudomonotone.*

Proof We first prove that for all $x, y \in K$ and all $u \in]x,y]$,

$$f(u,x) \ngtr_C \mathbf{0} \quad \text{implies} \quad f(u,y) \nless_C \mathbf{0}.$$

Let $f(u,x) \ngtr_C \mathbf{0}$ for all $u \in]x,y]$. Assume contrary that $f(\hat{z},y) <_C \mathbf{0}$ for some $\hat{z} \in]x,y]$.

We consider two cases:

CASE 1. If $f(\hat{z},x) \geq_C \mathbf{0}$, then $f(\hat{z},x) >_C f(\hat{z},y)$. By explicitly C-quasiconvexity, we have $f(\hat{z},\hat{z}) <_C f(\hat{z},x)$. Then $f(\hat{z},\hat{z}) \geq_C \mathbf{0}$ implies that $f(\hat{z},x) >_C \mathbf{0}$ which contradicts to our supposition that $f(\hat{z},x) \ngtr_C \mathbf{0}$ for all $\hat{z} \in]x,y]$.

CASE 2. If $f(\hat{z},x) \ngeq_C \mathbf{0}$, then by Lemma 2.5, there exists $c \ngeq_C \mathbf{0}$ such that $f(\hat{z},y) \leq_C c$ and $f(\hat{z},x) \leq_C c$. From C-quasiconvexity of $f(x,.)$, we deduce that $f(\hat{z},\hat{z}) \leq_C c$ implying that $c \geq_C \mathbf{0}$. This contradicts the fact that $c \ngeq_C \mathbf{0}$.

Hence, $f(u,y) \nless_C \mathbf{0}$ for all $z \in]x,y]$. By C-hemicontinuity of $f(.,y)$, we have $f(x,y) \nless_C \mathbf{0}$. Hence, f is maximal C-pseudomonotone. □

Lemma 9.2 *Let $f : K \times K \to Y$ be maximal C-pseudomonotone. Then* $\mathbb{S}(\text{WVEP}(f,C)) = \mathbb{S}(\text{MWVEP}(f,C))$.

Proof The inclusion $\mathbb{S}(\text{WVEP}(f,C)) \subseteq \mathbb{S}(\text{MWVEP}(f,C))$ directly follows from the C-pseudomonotonicity of f.

Let $\bar{x} \in \mathbb{S}(\text{MWVEP}(f,C))$, then

$$f(y,\bar{x}) \ngtr_C \mathbf{0}, \quad \text{for all } y \in K.$$

Now fix $y \in K$ arbitrarily. Then $]\bar{x}, y] \subseteq K$, and therefore,

$$f(u, \bar{x}) \not\geq_C \mathbf{0}, \quad \text{for all } u \in]\bar{x}, y].$$

By the definition of maximal C-pseudomonotone of f, we have $f(\bar{x}, y) \not<_C \mathbf{0}$. Since $y \in K$ was arbitrary, $\bar{x} \in \mathbb{S}(\text{MWVEP}(f, C))$. □

From Lemma 9.1 and 9.2, we have the following result.

Lemma 9.3 *Let $f : K \times K \to Y$ be C-pseudomonotone, C-hemicontinuous in the first argument and explicitly C-quasiconvex in the second argument such that $f(x, x) \geq_C \mathbf{0}$ for all $x \in K$. Then $\mathbb{S}(\text{WVEP}(f, C)) = \mathbb{S}(\text{MWVEP}(f, C))$.*

9.2.1 Existence Results for Solution of Weak Vector Equilibrium Problems

Theorem 9.1 *Let $f : K \times K \to Y$ be satisfy the following conditions:*

(i) *For all $x \in K$, $f(x, x) \geq_C \mathbf{0}$;*
(ii) *For all $y \in K$, $G(y) := \{x \in K : f(y, x) \not\geq_C \mathbf{0}\}$ is closed in K;*
(iii) *For all $x \in K$, $H(x) := \{y \in K : f(x, y) <_C \mathbf{0}\}$ is convex;*
(iv) *f is maximal C-pseudomonotone;*
(v) *There exist a nonempty closed and compact subset D of K and an element $\tilde{y} \in D$ such that $f(z, \tilde{y}) <_C \mathbf{0}$ for all $z \in K \setminus D$.*

Then there exists a solution $\bar{x} \in D$ of WVEP (9.1).

Proof Define set-valued maps $S, T : K \to 2^K$ by

$$S(y) = \{x \in K : f(x, y) \not<_C \mathbf{0}\}, \quad \text{for all } y \in K,$$

and

$$T(y) - \text{cl}_K S(y), \quad \text{for all } y \in K.$$

By (i), $S(y)$ is nonempty for all $y \in K$ and so is $T(y)$. S is a KKM-map, that is, for every finite subset $\{y_1, y_2, \ldots, y_m\}$ of K there holds $\text{co}(\{y_1, y_2, \ldots, y_m\}) \subseteq \bigcup_{i=1}^m S(y_i)$. Indeed, assume contrary that $x \in \text{co}(\{y_1, y_2, \ldots, y_m\})$, but $x \notin S(y_i)$ for all $i = 1, 2, \ldots, m$. Then $y_i \in H(x)$ for all $i = 1, 2, \ldots, m$. By condition (iii), $f(x, x) <_C \mathbf{0}$ which contradicts condition (i). Hence S is a KKM-map.

Since $S(y) \subseteq T(y)$ for all $y \in K$, T is a KKM-map with closed values and $S(\tilde{y})$ is contained in the compact set D by condition (v). Thus $\text{cl}_K(S(\tilde{y})) \subseteq D$ because D is closed in K, also $\text{cl}_K(S(\tilde{y})) = T(\tilde{y})$. Therefore, $T(\tilde{y}) \subseteq D$. By Fan-KKM Lemma 1.14, there exists $\bar{x} \in D$ such that $\bar{x} \in T(y)$ for all $y \in K$. By condition (ii), $G(y)$ is closed in K for all $y \in K$, and from the C-pseudomonotonicity of f, we have $S(y) \subseteq G(y)$, and thus, $T(y) \subseteq \mathrm{cl}_K(S(y)) \subseteq G(y)$. Therefore, we obtain that $\bar{x} \in G(y)$ for all $y \in K$, that is,

$$f(y, \bar{x}) \not>_C \mathbf{0}, \quad \text{for all } y \in K.$$

By Lemma 9.2, $\bar{x} \in D$ is a solution of WVEP (9.1). □

Remark 9.1

(a) The condition (ii) of Theorem 9.1 holds if for all $x \in K$, the function $y \mapsto g(x, y)$ is C-lower semicontinuous on K; See, Proposition 2.25.
(b) The condition (iii) of Theorem 9.1 holds if for all $x \in K$, the function $y \mapsto f(y, x)$ is C-quasiconvex; See, Proposition 2.15.

From Theorem 9.1 and Lemma 9.3, we obtain the following result.

Theorem 9.2 *Let the bifunction $f : K \times K \to Y$ satisfy the following conditions:*

(i) *For all $x \in K$, $f(x, x) \not>_C \mathbf{0}$;*
(ii) *For all $x \in K$, the function $y \mapsto f(x, y)$ is C-hemicontinuous and C-lower semicontinuous on K;*
(iii) *For all $x \in K$, the function $y \mapsto f(x, y)$ is explicitly C-quasiconvex on K;*
(iv) *f is C-pseudomonotone;*
(v) *There exist a nonempty closed and compact subset D of K and an element $\tilde{y} \in D$ such that $f(z, \tilde{y}) <_C \mathbf{0}$ for all $z \in K \setminus D$.*

Then there exists a solution $\bar{x} \in D$ of WVEP (9.1).

Proof Define set-valued maps $S, M : K \to 2^K$ by

$$S(y) = \{x \in K : f(x, y) \not<_C \mathbf{0}\}, \quad \text{for all } y \in K,$$

and

$$M(y) = \{x \in K : f(y, x) \not>_C \mathbf{0}\}, \quad \text{for all } y \in K.$$

By (i) and C-hemicontinuity of $y \mapsto f(x, y)$, $M(y)$ is nonempty and closed for all $y \in K$. In view of Remark 9.1, by the same argument as in the proof of Theorem 9.1, S is a KKM-map. By C-pseudomonotonicity of f, $S(y) \subseteq M(y)$ for all $y \in K$. Hence M is a KKM-map with closed values. By condition (v), $M(\tilde{y})$ is contained in the compact set D, and thus, $M(\tilde{y})$ is compact. Therefore, by Fan-KKM Lemma 1.14, there exists $\bar{x} \in D$ such that $\bar{x} \in S(y)$ for all $y \in K$, that is,

$$f(y, \bar{x}) \not>_C \mathbf{0}, \quad \text{for all } y \in K.$$

By Lemma 9.2, $\bar{x} \in D$ is a solution of WVEP (9.1). □

Bianchi, Hadjisavvas and Schaible [23, 60, 61] established some existence results for a solution of WVEP (9.1) under C-quasimonotonicity.

Now we present some existence results for a solution of WVEP (9.1) without any kind of C-monotonicity.

Theorem 9.3 *Let K be a nonempty convex subset of a topological vector space X. Let $f : K \times K \to Y$ be a vector-valued bifunction such that the following conditions hold.*

(i) *For all $y \in K$, $x \mapsto f(x,y)$ is C-upper semicontinuous on each nonempty compact subset of K;*
(ii) *For all $A \in \mathscr{F}(X)$ and each $x \in \mathrm{co}(A)$, $y \mapsto f(y,x)$ is C-quasiconvex;*
(iii) *For all $x \in K$, $f(x,x) \not<_C \mathbf{0}$;*
(iv) *There exist a nonempty closed compact subset $D \subseteq K$ and $\tilde{y} \in D$ such that $f(x,\tilde{y}) <_C \mathbf{0}$ for all $x \in K \setminus D$.*

Then there exists a solution $\bar{x} \in D \subseteq K$ of WVEP (9.1).

Proof For all $y \in K$, define

$$G(y) = \{x \in D : f(x,y) \not<_C \mathbf{0}\}.$$

Since for each $y \in K$, $x \mapsto f(x,y)$ is C-upper semicontinuous on each nonempty compact subset of K, we have that each $G(y)$ is closed. Since every element $\bar{x} \in \bigcap_{y\in K} G(y)$ is a solution of WVEP (9.1), we have to prove that $\bigcap_{y\in K} G(y) \neq \emptyset$. Since D is compact, it is sufficient to show that the family $\{G(y)\}_{y\in K}$ has the finite intersection property.

Let $\{y_1, y_2, \ldots, y_m\}$ be a finite subset of K. Let us note that $A := \mathrm{co}(\{y_1, y_2, \ldots, y_m\})$ is a compact convex subset of K. We define a set-valued map $S : A \to 2^A$ by

$$S(y) = \{x \in A : f(x,y) \not<_C \mathbf{0}\}, \quad \text{for all } y \in A.$$

By condition (iii), $S(y)$ is nonempty.

In view of condition (ii), it can be easily seen that S is a KKM-map. Clearly, $S(y) \subseteq G(y)$ for all $y \in A$.

We note that for each $y \in A$, $\mathrm{cl}_A(S(y))$ is closed in A and is therefore also compact. By Fan-KKM Lemma 1.14, $\bigcap_{y\in A} \mathrm{cl}_A(S(y)) \neq \emptyset$. We can choose $\bar{x} \in \bigcap_{y\in A} \mathrm{cl}_A(S(y))$, and note that $\tilde{y} \in A$ and $S(\tilde{y}) \subseteq D$ by (iv). Thus, $\bar{x} \in \mathrm{cl}_A(S(\tilde{y})) \subseteq \mathrm{cl}_K(S(\tilde{y})) = \mathrm{cl}_D(S(\tilde{y})) \subseteq D$. Since $\bar{x} \in \bigcap_{j=1}^m \mathrm{cl}_A(S(y_j))$ and for each $j = 1, 2, \ldots, m$,

$$\begin{aligned} \mathrm{cl}_A(S(y_j)) &= \mathrm{cl}_A\left(\{x \in A : f(x,y_j) \not<_C \mathbf{0}\}\right) \\ &= \{x \in A : f(x,y_j) \not<_C \mathbf{0}\}, \end{aligned}$$

we have $f(\bar{x}, y_j) \not<_C \mathbf{0}$ for all $j = 1, 2, \ldots, m$, and hence, $\bar{x} \in \bigcap_{j=1}^m G(y_j)$. Therefore, $\{G(y)\}_{y\in K}$ has the finite intersection property and the proof is finished. □

Definition 9.2 Let $g, h : K \times K \to Y$ be vector-valued bifunctions. Then g is called *C-pseudomonotone with respect to h* if for all $x, y \in K$, we have

$$g(x, y) + h(x, y) \not<_C \mathbf{0} \quad \text{implies} \quad g(y, x) - h(x, y) \not>_C \mathbf{0}.$$

The following existence results for a solution of WVEP (9.10) under C-pseudomonotonicity is established by Bianchi et al. [23] (see also [3] for moving cone).

Theorem 9.4 *Let $g, h : K \times K \to Y$ be satisfy the following conditions:*

(i) *For all $x \in K$, $g(x, x) \geq_C \mathbf{0}$;*
(ii) *For all $y \in K$, the function $x \mapsto g(x, y)$ is C-hemicontinuous;*
(iii) *For all $x \in K$, the function $y \mapsto g(x, y)$ is C-convex and C-lower semicontinuous;*
(iv) *g is C-pseudomonotone w. r. t. h;*
(v) *For all $x \in K$, $h(x, x) = \mathbf{0}$;*
(vi) *For all $y \in K$, the function $x \mapsto h(x, y)$ is C-upper semicontinuous;*
(vii) *For all $x \in K$, the function $y \mapsto h(x, y)$ is C-convex;*
(viii) *There exist a nonempty closed compact subset $D \subseteq K$ and $\tilde{y} \in D$ such that $g(z, \tilde{y}) + h(z, \tilde{y}) <_C \mathbf{0}$ for all $x \in K \setminus D$.*

Then WVEP (9.10) *has a solution.*

Proof For all $y \in K$, define two set-valued maps $S, M : K \to 2^K$ by

$$S(y) = \{x \in K : g(x, y) + h(x, y) \not<_C \mathbf{0}\}$$

and

$$M(y) = \{x \in K : g(y, x) - h(x, y) \not>_C \mathbf{0}\}.$$

Then by conditions (i) and (v), $S(y)$ is nonempty for all $y \in K$. Since $g(x, .)$ and $h(x, .)$ are C-convex, $S(y)$ is convex. It can be easily proved that S is a KKM-map (see the proof of Theorem 9.1). Let $T : K \to 2^K$ be defined as $T(y) = \text{cl}_K S(y)$ for all $y \in K$. Then T is also a KKM-map with closed values and by condition (viii), $T(\tilde{y})$ is contained in the compact set D. Then by Fan-KKM Lemma 1.14, there exists $\bar{x} \in D$ such that $\bar{x} \in T(y)$ for all $y \in K$. Since the sum of two C-lower semicontinuous functions is C-lower semicontinuous, we have $M(y)$ is closed in K for all $y \in K$. By C-pseudomonotonicity of g w. r. t. h, $T(y) \subseteq M(y)$ for all $y \in K$. Therefore, we obtain that $\bar{x} \in M(y)$ for all $y \in K$, that is,

$$g(y, \bar{x}) - h(\bar{x}, y) \not>_C \mathbf{0}, \quad \text{for all } y \in K.$$

For a fixed $y \in K$, we set $y_t = ty + (1 - t)\bar{x}$ for $t \in]0, 1[$. Then

$$g(y_t, \bar{x}) - h(\bar{x}, y_t) \not>_C \mathbf{0}, \quad \text{for all } t \in]0, 1[,$$

and consequently,

$$(1-t)g(y_t,\bar{x}) + tg(y_t,y) \not>_C (1-t)h(\bar{x},y_t) + tg(y_t,y).$$

Since $g(x,x) \geq_C \mathbf{0}$ for all $x \in K$ and $g(x,.)$ is C-convex, we have

$$\mathbf{0} \leq_C g(y_t,y_t) \leq_C tg(y_t,y) + (1-t)g(y_t,\bar{x}).$$

Note that for $a, b \in Y$ with $a \not>_C b$ and $a \geq_C \mathbf{0}$, we have $b \not<_C \mathbf{0}$. Therefore, we conclude that

$$tg(y_t,y) + (1-t)h(\bar{x},y_t) \not<_C \mathbf{0}.$$

From conditions (v) and (vii), we have $h(\bar{x},y_t) \leq_C t\,h(\bar{x},y)$, and thus,

$$g(y_t,y) + (1-t)h(\bar{x},y) \not<_C \mathbf{0}, \quad \text{for all } t \in]0,1[.$$

By condition (ii), we deduce that $g(\bar{x},y) + h(\bar{x},y) \not<_C \mathbf{0}$. Since y was arbitrary, the result is completed. □

Next, we present an existence result for a solution of WVEP (9.10) without any kind of C-monotonicity assumption.

Theorem 9.5 *Let K be a nonempty convex subset of a Hausdorff topological vector space X. Assume that the bifunctions $g, h : K \times K \to Y$ satisfy the following conditions:*

(i) *For all $x \in K$, $h(x,x) = \mathbf{0}$;*
(ii) *For all $x, y \in K$, $g(x,y) + h(x,y) <_C \mathbf{0}$ implies $h(y,x) <_C \mathbf{0}$;*
(iii) *For each fixed $y \in K$, $x \mapsto h(x,y)$ is C-quasiconcave and C-upper semicontinuous on K.*
(iv) *For each fixed $y \in K$, $x \mapsto g(x,y)$ is C-upper semicontinuous on K.*
(v) *There exist a nonempty compact convex subset B of K and a nonempty compact subset D of K such that for each $x \in K \setminus D$, there exists $\tilde{y} \in B$ satisfying $g(x,\tilde{y}) + h(x,\tilde{y}) <_C \mathbf{0}$.*

Then WVEP (9.10) *has a solution.*

Proof Assume that the conclusion of this theorem is not true. Then for each $x \in K$, the set

$$\{y \in K : g(x,y) + h(x,y) <_C \mathbf{0}\} \neq \emptyset.$$

For all $x \in K$, define two set-valued maps $S, T : K \to 2^K$ by

$$S(x) = \{y \in K : g(x,y) + h(x,y) <_C \mathbf{0}\}$$

and

$$T(x) = \{y \in K : h(y, x) >_C \mathbf{0}\}.$$

Clearly, for all $x \in K$, $S(x) \neq \emptyset$. Let $\{y_1, y_2, \ldots, y_m\}$ be a finite subset of $S(x)$ and $\lambda_i \geq 0$ for all $i = 1, 2, \ldots, m$ with $\sum_{i=1}^{m} \lambda_i = 1$. Then

$$g(x, y_i) + h(x, y_i) <_C \mathbf{0}, \quad \text{for all } i = 1, 2, \ldots, m.$$

By condition (ii), we have $h(y_i, x) >_C \mathbf{0}$ for all $i = 1, 2, \ldots, m$. Since $h(., x)$ is C-quasiconcave, we have $h\left(\sum_{i=1}^{m} \lambda_i y_i, x\right) >_C \mathbf{0}$, and hence, $\sum_{i=1}^{m} \lambda_i y_i \in T(x)$. Therefore, $\text{co}(S(x)) \subseteq T(x)$, for all $x \in K$.

Since $g(., y)$ and $h(., y)$ are C-upper semicontinuous and so is $g(., y) + h(., y)$. Therefore, the complement of $S^{-1}(y)$ in K,

$$[S^{-1}(y)]^c = \{x \in K : g(x, y) + h(x, y) \not<_C \mathbf{0}\}$$

is closed in K. Hence $S^{-1}(y)$ is open in K. Since $S(x) \neq \emptyset$ for all $x \in K$, we have

$$K = \bigcup_{y \in K} S^{-1}(y) = \bigcup_{y \in K} \text{int}_K S^{-1}(y).$$

By condition (v), for each $x \in K \setminus D$, there exists $\tilde{y} \in B$ such that

$$x \in S^{-1}(\tilde{y}) = \text{int}_K S^{-1}(\tilde{y}).$$

Hence, all the conditions of Theorem 1.35 are satisfied, and therefore, there exists $\bar{x} \in K$ such that $\bar{x} \in T(\bar{x})$, that is, $h(\bar{x}, \bar{x}) >_C \mathbf{0}$, which is a contradiction. This completes the proof. □

For further existence results for solutions of WVEP (9.10), we refer [3, 10, 23, 64, 94] and the references therein.

9.2.2 *Existence Results for Strong Vector Equilibrium Problems*

In this section, we present some existence results for a solution of the strong vector equilibrium problem (in short, SVEP) (9.3).

Flores-Bazán and Flores-Bazán [48] obtained the following abstract result in the setting of finite dimensional Euclidean space $\mathbb{R}^n$ but for moving cone.

Theorem 9.6 *Let K be a nonempty compact convex subset of a Hausdorff topological vector space X, Y be a topological vector space and W be a nonempty subset of Y. Let $f : K \times K \to Y$ be a vector-valued bifunction satisfying the following conditions:*

(i) *For all $x, \in K$, $f(x,x) \in W$;*
(ii) *For all $x, \in K$, $f(x,y) \in W$ implies $f(y,x) \in -W$;*
(iii) *For all $x, \in K$, the set $\{y \in K : f(x,y) \in -W\}$ is closed;*
(iv) *For all $x, \in K$, the set $\{y \in K : f(x,y) \notin W\}$ is convex;*
(v) *For all $x, y, \in K$, $f(x,y) \in -W$ implies $f(y,x) \in W$.*

Then the solution set of the following abstract vector equilibrium problem (in short, AVEP*):*

$$\text{find } \bar{x} \in K \text{ such that } f(\bar{x}, y) \in W, \quad \text{for all } y \in K, \tag{9.14}$$

and that of the following dual abstract vector equilibrium problem (in short, DAVEP*):*

$$\text{find } \bar{x} \in K \text{ such that } f(y, \bar{x}) \in -W, \quad \text{for all } y \in K \tag{9.15}$$

are nonempty and both coincide, and they are closed.

Proof We first find $\bar{x} \in K$ such that

$$\bar{x} \in \bigcap_{y \in K} \{x \in K : f(y,x) \in -W\}.$$

To this end, we set

$$S(y) = \{x \in K : f(y,x) \in -W\}.$$

By condition (ii), for each $y \in K$, $S(y)$ is a closed subset of a compact set K, and hence it is compact. By using similar argument as in the proof of Theorem 9.1 and using conditions (i) and (iv), we conclude that S is a KKM mapping. Hence, by Fan-KKM Lemma 1.14, there exists $\bar{x} \in K$ such that $\bar{x} \in \bigcap_{y \in K} S(y)$, that is, $f(y, \bar{x}) \in -W$ for all $y \in K$, in other words, the second problem has a solution. By condition (v), we conclude that such a solution is also a solution of the first problem. Since every solution of the first problem is a solution of the second problem by condition (ii), we conclude that both solution sets coincides. The closedness is a consequence of condition (iii). □

By using Theorem 9.6, Flores-Bazán and Flores-Bazán [48] characterized the solution set of SVEP (9.3) in terms of its asymptotic cone.

Farajzadeh et al. [46] further considered abstract vector equilibrium problems and proved the following result without compactness assumption on the set K.

Theorem 9.7 *Let X and Y be Hausdorff topological vector spaces, K be a nonempty convex subset of X, and W be a nonempty set in Y. Let $f : K \times K \to Y$ be a vector-valued bifunction such that the following conditions hold:*

(i) *For all finite subsets A of K and all $y \in \operatorname{co}(A)$, there exists $y \in A$ such that $f(x, y) \in W$;*
(ii) *For all $y \in K$, the set $\{x \in K : f(x, y) \in W\}$ is closed;*
(iii) *There exists a nonempty compact subset B of K and a nonempty compact convex subset D of K such that for each $x \in K \setminus B$, there exists $\tilde{y} \in D$ such that $f(x, \tilde{y}) \notin W$.*

Then the solution set of AVEP (9.14) *is nonempty and compact.*

Proof Define a set-valued mapping $S : K \to 2^K$ by

$$S(y) = \{x \in K : f(x, y) \in W\}.$$

It can be easily seen that S is a KKM mapping by using condition (i). By applying conditions (ii) and (iii), we deduce that $\bigcap_{y \in D} S(y)$ is a closed subset of B. Then, by Fan-KKM Lemma 1.14, $\bigcap_{x \in K} S(x) \neq \emptyset$. This means that the solution set of AVEP (9.14) is nonempty. By (ii), the solution set of AVEP (9.14) is closed subset of the compact set B and hence compact. □

Farajzadeh et al. [46] also used the scalarization method to prove the existence of a solution of AVEP (9.14).

Definition 9.3 Let $y \in K$ be any fixed element and $f : K \times K \to Y$ be a vector-valued bifunction. A function $x \mapsto f(x, y)$ is said to be *C-upper sign continuous* if for every $x \in K$,

$$f(u, y) \geq_C \mathbf{0}, \quad \text{for all } u \in]x, y[\quad \Rightarrow f(x, y) \geq_C \mathbf{0},$$

where $]x, y[$ denotes the open line segment joining x and y.

If $x \mapsto f(x, y)$ is C-hemicontinuous, that is, the restriction of f to line segments in K is C-continuous, then $x \mapsto f(x, y)$ is C-upper sign continuous. Even this fact is true when $x \mapsto f(x, y)$ is C-upper hemicontinuous.

Definition 9.4 A bifunction $f : K \times K \to Y$ is said to be

(a) *strong C-pseudomonotone* if for all $x, y \in K$,

$$f(x, y) \not\leq_C \mathbf{0} \quad \Rightarrow \quad f(y, x) \leq_{C_0} \mathbf{0};$$

(b) *strong C-quasimonotone* if for all $x, y \in K$,

$$f(x, y) \not\leq_C \mathbf{0} \quad \Rightarrow \quad f(y, x) \leq_C \mathbf{0};$$

(c) *strong properly C-quasimonotone* if for every finite set $\{x_1, x_2, \ldots, x_m\} \subseteq K$ and for all $x \in \text{co}(\{x_1, x_2, \ldots, x_m\})$, there exists $i \in \{1, 2, \ldots, m\}$ such that $f(x_i, x) \leq_C \mathbf{0}$.

It is clear from the definition that strong C-pseudomonotonicity of f implies strong C-quasimonotonicity. But in general there is no relationship between strong properly C-quasimonotonicity and strong C-quasimonotonicity or strong C-pseudomonotonicity.

The following proposition is a vector version of Proposition 1 in [20]. It provides a criteria for the strong properly C-quasimonotonicity of a bifunction.

Proposition 9.1 *Let $f : K \times K \to Y$ be a vector-valued bifunction such that $f(x,x) = \mathbf{0}$ for all $x \in K$. If one of the following conditions holds:*

(i) *the set $\{x \in K : f(x,y) \not\leq_C \mathbf{0}\}$ is convex, or*
(ii) *the set $\{y \in K : f(x,y) \leq_{C_0} \mathbf{0}\}$ is convex and f is strong C-pseudomonotone,*

then, f is strong properly C-quasimonotone.

Proof Assume contrary that f is not strong properly C-quasimonotone. Then there exist a finite set $\{x_1, x_2, \ldots, x_m\} \subseteq K$ and $\tilde{x} \in \text{co}(\{x_1, x_2, \ldots, x_m\})$ such that $f(x_i, \tilde{x}) \not\leq_C \mathbf{0}$ for all $i = 1, 2, \ldots, m$.

Suppose that (i) holds. Then we have $f(\tilde{x}, \tilde{x}) \not\leq_C \mathbf{0}$, and so $\mathbf{0} = f(\tilde{x}, \tilde{x}) \not\leq_C \mathbf{0}$ a contradiction. Hence f is strong properly C-quasimonotone.

Suppose that (ii) holds. Then by C-pseudomonotonicity of f, we have $f(\tilde{x}, x_i) \leq_{C_0} \mathbf{0}$. By the first condition in (ii), we deduce $\mathbf{0} = f(\tilde{x}, \tilde{x}) \not\leq_C \mathbf{0}$ a contradiction. Then f is strong properly C-quasimonotone. □

Remark 9.2

(a) For each fixed $x \in K$, if the mapping $y \mapsto f(x,y)$ is C-convex, then the set $\{y \in K : f(x,y) \leq_{C_0} \mathbf{0}\}$ is convex for all $x \in K$.

Indeed, let $\lambda \in]0,1[$ and $f(x, y_i) \leq_{C_0} \mathbf{0}$ for all $i = 1, 2$. Since C is a pointed convex cone, we have

$$\lambda f(x, y_1) + (1 - \lambda) f(x, y_2) \leq_{C_0} \mathbf{0}. \tag{9.16}$$

By (9.16) and C-convexity of $y \mapsto f(x,y)$, we obtain

$$f(x, ty + (1-t)z) \in C \backslash \{\mathbf{0}\} + C \subseteq C \backslash \{\mathbf{0}\},$$

that is, $f(x, ty + (1-t)z) \geq_{C_0} \mathbf{0}$.

(b) If $Y \backslash (-C)$ is convex and f is C-concave in the first variable, then (i) of Proposition 9.1 holds. The proof is straight forward by using $Y \backslash (-C) + C \subseteq Y \backslash (-C)$.

The sets of local solutions for SVEP (9.3) and MSVEP (9.6) are denoted by $\mathbb{S}_{K,loc}$ and $\mathbb{S}^M_{K,loc}$, respectively, and defined as follows:

$$\mathbb{S}_{K,loc} = \{x \in K : \text{ there exists an open neighborhood } V \text{ of } x \text{ such that } f(x,y) \geq_C \mathbf{0} \text{ for all } y \in V \cap K\},$$

$$\mathbb{S}^M_{K,loc} = \{x \in K : \text{ there exists an open neighborhood } V \text{ of } x \text{ such that } f(y,x) \leq_C \mathbf{0} \text{ for all } y \in V \cap K\}.$$

Obviously, $\mathbb{S}(\text{SVEP}(f,C)) \subseteq \mathbb{S}_{K,loc}$ and $\mathbb{S}(\text{MSVEP}(f,C)) \subseteq \mathbb{S}^M_{K,loc}$.

The following lemma provides a relationship between the solution sets $\mathbb{S}^M_{K,loc}$ and $\mathbb{S}(\text{SVEP}(f,C))$. Furthermore, it is a vector version of Lemma 2.1 in [21].

Lemma 9.4 *Let K be a nonempty convex subset of X and $f : K \times K \to Y$ be a vector-valued bifunction such that the following conditions hold:*

(i) *$f(x,x) \geq_C \mathbf{0}$ for all $x \in K$;*
(ii) *For each fixed $y \in K$, the mapping $x \mapsto f(x,y)$ is C-upper sign continuous;*
(iii) *If $f(x,y) \not\geq_C \mathbf{0}$ and $f(x,z) \leq_C \mathbf{0}$, then $f(x,u) \not\geq_C \mathbf{0}$ for all $u \in]y,z[$.*

Then $\mathbb{S}^M_{K,loc} \subseteq \mathbb{S}(\text{SVEP}(f,C))$.

Proof Let $z \in \mathbb{S}^M_{K,loc}$. In order to show that $z \in \mathbb{S}(\text{SVEP}(f,C))$, we assume contrary that there exists $y \in K$ such that

$$f(z,y) \not\geq_C \mathbf{0}. \tag{9.17}$$

From the definition of $S^M_{K,loc}$, there exists an open neighborhood V of z such that $f(v,z) \leq_C \mathbf{0}$ for all $v \in K \cap V$. Since $V - z$ is a neighborhood of $\mathbf{0}$, there exists $t_0 \in]0,1[$ such that $t(y-z) \in V - z$ for all $0 < t \leq t_0$. Let $\bar{y} := z + t_0(y-z)$ and $y_t := (1-t)z + t\bar{y} \in [z,\bar{y}]$ for $t \in [0,1]$. Then $y_t \in K \cap V$, since $y_t = (1-t)z + tz + t\,t_0(y-z) = z + t\,t_0(y-z)$ and $t\,t_0(y-z) \in V - z$. Hence (9.17) implies that $f(y_t,z) \leq_C \mathbf{0}$ and by condition (i), $f(z,z) = \mathbf{0}$. Now we will show that $f(u,\bar{y}) \geq_C \mathbf{0}$ for all $u \in]z,\bar{y}[$. Indeed, if $f(u,\bar{y}) \not\geq_C \mathbf{0}$ for some $u \in]z,\bar{y}[$, then as $f(u,z) \leq_C \mathbf{0}$, we deduce from (iii) that $f(u,v) \not\geq_C \mathbf{0}$ for all $v \in]z,\bar{y}[$ and in particular $f(u,u) = \mathbf{0} \not\geq_C \mathbf{0}$. Hence $\mathbf{0} \notin C$ which contradicts the fact that $\mathbf{0} \in C$ since C is a pointed cone. Therefore, $f(u,\bar{y}) \geq_C \mathbf{0}$ for all $u \in]z,\bar{y}[$. Thus by (ii), we have

$$f(z,\bar{y}) \in C. \tag{9.18}$$

Since $f(z,z) = \mathbf{0}$ and $f(z,y) \not\geq_C \mathbf{0}$, it follows from (iii) that $f(z,\bar{y}) \not\geq_C \mathbf{0}$ which contradicts (9.18). □

The following example shows that the condition (iii) in Lemma 9.4 is essential.

Example 9.2 Let $X = Y = \mathbb{R}$, $K = [-1, 1]$, $C = [0, \infty)$ and $f : [-1, 1] \times [-1, 1] \to \mathbb{R}$ be defined as

$$f(x, y) = \begin{cases} 0, & \text{if } (x, y) \in \{0\} \times \left[-\frac{1}{2}, \frac{1}{2}\right] \text{ or } x = y, \\ -1, & \text{otherwise.} \end{cases}$$

It is clear that $f(x, x) = 0$ for all $x \in K$, and if $f(u, y) \geq 0$ for all $u \in]x, y[$, then $u = 0$ for all $u \in]x, y[$, which is impossible. This shows that the mapping $x \mapsto f(x, y)$ is C-upper sign continuous for each fixed $y \in K$. Since $f\left(\frac{1}{4}, \frac{1}{3}\right) < 0$ and $f\left(\frac{3}{4}, \frac{1}{3}\right) < 0$, we can easily see that $f\left(u, \frac{1}{3}\right) < 0$ does not hold for all $u \in]\frac{1}{4}, \frac{3}{4}[$, for example, take $u = \frac{1}{3} \in]\frac{1}{4}, \frac{3}{4}[$, and so the example does not fulfill the condition (iii) of Lemma 9.4. Moreover, the conclusion of Lemma 9.4 is not true for this example, since $x_0 = 0$ is a solution of SVEP(f, C) defined over $\left[-\frac{1}{2}, \frac{1}{2}\right]$ and $\mathbb{S}(\text{SVEP}(f, C)) = \emptyset$.

The following example illustrates that $\mathbb{S}^M_{K,loc}$ is a singleton, while the set $\mathbb{S}(\text{SVEP}(f, C))$ is uncountable.

Example 9.3 Let $X = \mathbb{R}$, $K = [0, 1]$, $C = \{(x, y) \in \mathbb{R}^2 : x \geq 0,\ y \geq 0\}$, $Y = \mathbb{R}^2$, and $f : K \times K \to \mathbb{R}^2$ be defined by $f(x, y) = (x, y)$.

It is easy to verify that f satisfies all the assumptions of Lemma 9.4 and $S^M_{K,loc} = \{0\}$, $\mathbb{S}(\text{SVEP}(f, C)) = K$.

Remark 9.3

(a) If f is C-convex in the second variable, then condition (ii) in Lemma 9.4 holds. Indeed, let $f(x, y) \not\leq_C \mathbf{0}$ and $f(x, z) \leq_C \mathbf{0}$. Since $Y \backslash C$ and $-C$ are cone, we have $tf(x, u) \not\leq_C \mathbf{0}$ and $(1 - t)f(x, z) \leq_C \mathbf{0}$ and also from $(Y \backslash C) - C \subseteq Y \backslash C$, we get

$$tf(x, u) + (1 - t)f(x, z) \not\leq_C \mathbf{0}. \tag{9.19}$$

Since $(Y \backslash C) - C \subseteq Y \backslash C$, we obtain

$$f(x, ty + (1 - t)z) \not\leq_C \mathbf{0}, \quad \text{for all } t \in [0, 1].$$

This shows that condition (ii) of Lemma 9.4 holds.

(b) Lemma 9.4 improves and extends Lemma 2.4 in [63] and Lemma 2.1 in [21] to vector-valued bifunctions.

(c) If for all $x, y \in K$, $f(x, y) \not\leq_{C_0} \mathbf{0}$ implies $f(y, x) \leq_C \mathbf{0}$, then we obtain the inclusion $\mathbb{S}(\text{SVEP}(f, C)) \subseteq \mathbb{S}(\text{MSVEP}(f, C))$. Therefore, under this assumption, we have $\mathbb{S}(\text{SVEP}(f, C)) = S^M_{K,loc} = \mathbb{S}(\text{SVEP}(f, C))$. Thus, if $Y = \mathbb{R}$ and $C = [0, \infty)$, we deduce Proposition 2.5 in [63]. Moreover, if f is C-quasimonotone and $f(x, y) = \mathbf{0}$ implies $f(y, x) = \mathbf{0}$, then $\mathbb{S}(\text{SVEP}(f, C)) \subseteq \mathbb{S}(\text{MSVEP}(f, C))$. Hence we obtain the C-quasimonotone version of Proposition 2.5 in [63] for the vector case.

We have the following existence result for a solution of $\mathbb{S}(\text{MSVEP}(f, C))$.

Theorem 9.8 *Let K be a nonempty convex subset of a Hausdorff topological vector space X and $f : K \times K \to Y$ be a vector-valued bifunction such that $f(x, x) = \mathbf{0}$ for all $x \in K$. Assume that the following conditions hold:*

(i) *For each $A \in \mathscr{F}(K)$ and for all $x \in \operatorname{co}(A)$, there exists $y \in A$ such that $f(y, x) \leq_C \mathbf{0}$;*
(ii) *For all $x \in K$, the set $\{y \in K : f(x, y) \leq_C \mathbf{0}\}$ is closed in K;*
(iii) *There exist a nonempty compact subset D and a nonempty compact convex subset B of K such that for all $y \in K \setminus D$, there exists $\tilde{x} \in B$ such that $f(\tilde{x}, y) \nleq_C \mathbf{0}$.*

Then the solution set $\mathbb{S}(\text{MSVEP}(f, C))$ of MSVEP (9.6) is nonempty and compact.

Proof Define a set-valued mapping $T : K \to 2^K$ by

$$T(x) = \{y \in K : f(x, y) \leq_C \mathbf{0}\}, \quad \text{for all } x \in K.$$

Then T is a KKM mapping. Indeed, let $A \in \mathscr{F}(K)$ and $x \in \operatorname{co}(A)$. Then by (i), there is a $y \in A$ such that $f(y, x) \leq_C \mathbf{0}$ and so $x \in T(y)$. This shows that $x \in \operatorname{co}(A)$. Since $f(x, x) = \mathbf{0}$, we have $x \in T(x)$. Consequently, $\operatorname{co}(A) \subseteq \bigcup_{x \in A} T(x)$. So T is a KKM mapping. It is obvious that the solution set $\mathbb{S}(\text{MSVEP}(f, C))$ equals to the set $\bigcap_{x \in K} T(x)$. Since T is a KKM mapping and B is compact and convex, then by Lemma 1.14, we deduce that $\bigcap_{x \in B} T(x)$ is nonempty. By (ii) and (iii), it is a closed subset of D and so is compact. □

The following lemma provides a relationship among strong C-quasimonotonicity, strong properly C-quasimonotonicity and $\mathbb{S}^M_{K,loc}$. Also, it is a vector version of Lemma 4.2 in [21] without assuming that f is quasiconvex in the second variable.

Lemma 9.5 *Let K be a nonempty convex subset of X and $f : K \times K \to Y$ be strong C-quasimonotone bifunction. If for each $x \in K$, the set $\{y \in K : f(x, y) \leq_C \mathbf{0}\}$ is closed in K and convex, then either f is strong properly C-quasimonotone or $\mathbb{S}^D_{K,loc} \neq \emptyset$.*

Proof If f is not strong properly C-quasimonotone, then there exist a finite set $\{x_1, x_2, \ldots, x_m\} \subseteq K$ and $\bar{x} \in \operatorname{co}(\{x_1, x_2, \ldots, x_m\})$ such that $f(x_i, \bar{x}) \leq_C \mathbf{0}$ for all $i = 1, 2, \ldots, m$. Since the set $A_i = \{y \in K : f(x_i, y) \leq_C \mathbf{0}\}$ is closed in K for all $i = 1, 2, \ldots, m$ and $\bar{x} \notin \bigcup_{i=1}^m A_i$, there exists an open neighborhood V of $\bar{x}$ such that

$$f(x_i, y) \nleq_C \mathbf{0}, \quad \text{for all } y \in K \cap V,\ i = 1, 2, \ldots, m,$$

and so by strong C-quasimonotonicity of f, we have

$$f(y, x_i) \leq_C \mathbf{0}, \quad \text{for all } i = 1, 2, \ldots, m \text{ and } y \in K \cap V.$$

Since the set $\{y \in K : f(x,y) \leq_C \mathbf{0}\}$ is convex for all $x \in K$, we get

$$f(y,\bar{x}) \leq_C \mathbf{0}, \quad \text{for all } y \in K \cap V.$$

Therefore, $\bar{x} \in \mathbb{S}^M_{K,loc}$. □

We now present the following existence result for a solution of SVEP under strong C-quasimonotonicity assumption.

Proposition 9.2 *Let K be a nonempty convex subset of X and $f : K \times K \to Y$ be a strong C-quasimonotone bifunction satisfying the conditions of Lemma 9.4 and 9.5 and condition* (iii) *of Theorem 9.8. Then* $\mathbb{S}(\text{SVEP}(f, C)) \neq \emptyset$.

Proof If f is strong C-properly quasimonotone, then the result deduces from Theorem 9.8 through Lemma 9.4. Otherwise, we obtain the result from Lemma 9.4 and 9.5. □

Next we present existence result for a solution of SVEP without strong C-quasimonotonicity assumption.

Proposition 9.3 *Let K be a nonempty convex subset of X. Assume that $f : K \times K \to Y$ satisfies condition* (iii) *of Lemma 9.4. Let $x_0 \in K$ be a local solution of* SVEP *in a neighborhood V of x_0. If there exists $\bar{y} \in K \cap \text{int}(V)$ such that $f(x_0,\bar{y}) \leq_C \mathbf{0}$, then $x_0 \in \mathbb{S}(\text{SVEP}(f, C))$.*

Proof Assume contrary that there exists $z \in K$ such that $f(x_0, z) \not\leq_C \mathbf{0}$. Then by condition (iii) of Lemma 9.4, note that $f(x_0, \bar{y}) \leq_C \mathbf{0}$, we have

$$f(x_0, u) \not\leq_C \mathbf{0}, \quad \text{for all } u \in]\bar{y}, z[,$$

and so this is a contradiction by using $\bar{y} \in K \cap \text{int}(V)$ and x_0 is a local solution of SVEP in the neighborhood V of x_0. □

Rest of this subsection, we assume that X is a Hausdorff locally bounded topological vector space, K is a nonempty unbounded convex subset of X and $f : K \times K \to Y$ is a vector-valued bifunction.

Assumption 9.5 *There exists an open bounded neighborhood V of $\mathbf{0}$ such that*

$$\forall x \in K\backslash\overline{V}, \ \exists y \in K \cap \overline{V} \text{ satisfying } f(x,y) \not\leq_C \mathbf{0}.$$

This coercivity condition extends the coercivity condition considered in [21] in the setting of reflexive Banach spaces.

The next result extends Proposition 2.2 in [21] and it provides a necessary condition for boundedness of the solution set of SVEP.

Proposition 9.4 *Let $f : K \times K \to Y$ be a vector-valued bifunction such the following conditions hold:*

(i) $f(x,x) = \mathbf{0}$ *for all $x \in K$;*
(ii) *For each fixed $y \in K$, the set $\{x \in K : f(x,y) \geq_C \mathbf{0}\}$ is convex;*

(iii) *If* $f(x, y) \not\geq_C \mathbf{0}$ *and* $f(x, z) \leq_C \mathbf{0}$, *then* $f(x, u) \not\geq_C \mathbf{0}$ *for all* $u \in]y, z[$.

If the solution set $\mathbb{S}(\text{SVEP}(f, C))$ *of* SVEP (9.3) *is nonempty and bounded, then Assumption 9.5 holds.*

Proof Suppose that the Assumption 9.5 does not hold. Let V be an arbitrary open bounded, balanced neighborhood of $\mathbf{0}$ and $x_0 \in \mathbb{S}(\text{SVEP}(f, C))$. Consider positive integer m_0 such that $x_0 \in m_0 V$. Let $m > m_0$ and $W_m = \underbrace{V + V + \cdots + V}_{m}$. Since Assumption 9.5 is not true and W_m is bounded, there exists $x_m \in K \setminus W_m$ such that

$$f(x_m, y) \geq_C \mathbf{0}, \quad \text{for all } y \in K \cap W_m. \tag{9.20}$$

Since $(m-1)V \subseteq W_m$ and $x_m \notin W_m$, we have

$$t_0 = \sup\{t \in [0, 1] : x_0 + t(x_m - x_0) \in (m-1)V\} < 1. \tag{9.21}$$

Therefore, for all positive number t' with $t_0 + t' < 1$, we deduce that

$$z_m = x_0 + (t_0 + t')(x_m - x_0) \notin (m-1)V. \tag{9.22}$$

We claim that $z_m \in W_m \cap K$. Indeed, we can choose small positive number t such that $t(x_m - x_0) \in V$ and $t < 2(1 - t_0)$. By (9.21), there exists t_1 such that $t_0 - \frac{t}{2} < t_1$ and

$$z_m = x_0 + (t_1 + t)(x_m - x_0) \in (m-1)V + V \subseteq W_m.$$

Since $x_0 \in \mathbb{S}(\text{SVEP}(f, C))$, the convexity of the set $\{x \in K : f(x, y) \geq_C \mathbf{0}\}$ and (9.20) imply that

$$f(z_m, y) \in C, \quad \text{for all } y \in K \cap W_m. \tag{9.23}$$

By (9.23) and Proposition 9.3, we obtain $z_m \in \mathbb{S}(\text{SVEP}(f, C))$. Hence, the sequence $\{z_m\}$ is unbounded, which contradicts the boundedness of $\mathbb{S}(\text{SVEP}(f, C))$. □

Now we give necessary and sufficient conditions for nonemptiness of the solution set of SVEP (9.3).

Theorem 9.9 *Let* $f : K \times K \to Y$ *be a strong C-pseudomonotone bifunction such that the following conditions hold:*

(i) *For all* $x \in K$, $f(x, x) = \mathbf{0}$;
(ii) *For all* $y \in K$, *the mapping* $x \mapsto f(x, y)$ *is C-upper sign continuous;*
(iii) *For each* $x \in K$, *the set* $\{y \in K : f(x, y) \leq_C \mathbf{0}\}$ *is convex and closed in* K;
(iv) *If* $f(x, y) \not\geq_C \mathbf{0}$ *and* $f(x, z) \leq_C \mathbf{0}$, *then* $f(x, u) \not\geq_C \mathbf{0}$ *for all* $u \in]y, z[$.

If $\mathbb{S}(\text{SVEP}(f, C))$ *is nonempty and bounded, then Assumption 9.5 holds. Moreover, if Assumption 9.5 holds with bounded open neighborhood* V *and* $f|\text{co}(K \cap W)$, *the restriction of f to* co$(K \cap W)$, *satisfies conditions* (i) *and* (iii) *of Theorem 9.8 for every bounded neighborhood W with* $\text{int}(W) \supseteq \overline{V}$. *Then* $\mathbb{S}(\text{SVEP}(f, C))$ *is nonempty, compact and convex.*

Proof Suppose that the Assumption 9.5 does not hold. Let V be a bounded open balanced neighborhood of $\mathbf{0}$, $x_0 \in \mathbb{S}(\text{SVEP}(f, C))$ and m_0 positive integer such that $x_0 \in m_0 V$. Let $m > m_0$ and $W_m = \underbrace{V + V + \cdots + V}_{m}$. Since Assumption 9.5 does not hold and W_m is a bounded neighborhood of $\mathbf{0}$, there exists $x_m \in K \setminus W_m$ such that

$$f(x_m, y) \geq_C \mathbf{0}, \quad \text{for all } y \in K \cap W_m. \tag{9.24}$$

Since C is pointed, we have $C \cap (-C \setminus \{\mathbf{0}\}) = \emptyset$, and therefore,

$$f(x_m, y) \not\leq_{C_0} \mathbf{0}, \quad \text{for all } y \in K \cap W_m. \tag{9.25}$$

The strong C-pseudomonotonicity of f implies that

$$f(y, x_m) \leq_C \mathbf{0}, \quad \text{for all } y \in K \cap W_m. \tag{9.26}$$

Since f is strong C-pseudomonotone and $x_0 \in \mathbb{S}(\text{SVEP}(f, C))$, we have

$$f(y, x_0) \leq_C \mathbf{0}, \quad \text{for all } y \in K. \tag{9.27}$$

Since $(m-1)\, V \subseteq W_m$ and $x_m \notin W_m$, we have

$$t_0 = \sup\{t \in [0, 1] : x_0 + t(x_m - x_0) \in (m-1)\, V\} < 1.$$

Therefore, for all positive number t' such that $t_0 + t' < 1$, we deduce

$$z_m = x_0 + (t_0 + t')(x_m - x_0) \notin (m-1)\, V.$$

By using (9.26) and (9.26), we obtain

$$f(y, z_m) \leq_C \mathbf{0}, \quad \text{for all } y \in W_m \cap K,$$

as the set $\{y \in K : f(x, y) \leq_C\}$ is convex. Consequently, $z_m \in \mathbb{S}^M_{K,loc}$. Therefore, by Lemma 9.4, we have $z_m \in \mathbb{S}(\text{SVEP}(f, C))$. Hence, the sequence $\{z_m\}$ is unbounded, which contradicts the boundedness of $\mathbb{S}(\text{SVEP}(f, C))$.

Conversely, let Assumption 9.5 hold with an open neighborhood V and W be an open bounded balanced neighborhood of $\mathbf{0}$ containing V. Since the mapping $f|\text{co}(K \cap W)$ satisfies all the conditions of Theorem 9.8, there exists a solution $\bar{x}$ of SVEP defined over co$(K \cap W)$. If $\bar{x}$ is an element of W, then by Proposition 9.3 $\bar{x} \in \mathbb{S}(\text{SVEP}(f, C))$. Otherwise, by Assumption 9.5, there exists $y \in \overline{V}$ such that

$f(\bar{x}, y) \not\geq_C \mathbf{0}$. Since $f(\bar{x}, \bar{x}) = \mathbf{0} \in -C$, by condition (iv), we have, $f(\bar{x}, u) \not\geq_C \mathbf{0}$ for all $u \in]\bar{x}, y[$ which contradicts that $\bar{x}$ is a solution of SVEP defined over $\text{co}(K \cap W)$. Therefore, $\bar{x} \in \mathbb{S}(\text{SVEP}(f, C))$.

Now we show that $\mathbb{S}(\text{SVEP}(f, C))$ is a compact subset of K. To see this, let $x_\alpha \in \mathbb{S}(\text{SVEP}(f, C))$ and $x_\alpha \to x$. Then $f(x_\alpha, y) \geq_C \mathbf{0}$ for all $y \in K$ and all α. Since C is pointed, we have by $f(x_\alpha, y) \not\leq_{C_0} \mathbf{0}$ for all $y \in K$ and all α. The strong C-pseudomonotonicity of f implies that $f(y, x_\alpha) \leq_C \mathbf{0}$ for all $y \in K$ and all α. Since $x_\alpha \to x$ and the set $\{y \in K : f(x, y) \leq_C \mathbf{0}\}$ is closed in K for all $x \in K$, we get $f(y, x) \geq_C \mathbf{0}$ for all $y \in K$. This means that $x \in \mathbb{S}(\text{MSVEP}(f, C))$ and by Lemma 9.4, we have $x \in \mathbb{S}(\text{SVEP}(f, C))$ as $\mathbb{S}(\text{MSVEP}(f, C)) \subseteq \mathbb{S}^M_{K,loc}$. Consequently, $\mathbb{S}(\text{SVEP}(f, C))$ is closed in K. It follows from condition (iii) of Theorem 9.8 that $\mathbb{S}(\text{SVEP}(f, C))$ is a subset of the compact subset D of K. Moreover, strong C-pseudomonotonicity of f and convexity of the set $\{y \in K : f(x, y) \leq_C \mathbf{0}\}$ for all $x \in K$ imply that $\mathbb{S}(\text{SVEP}(f, C))$ is convex. □

Assumption 9.5 entails the boundedness of the solution set but the following coercivity condition allows that the solution set to be unbounded.

Assumption 9.6 *There exists an open bounded neighborhood V of $\mathbf{0}$ such that for all $x \in K \setminus \overline{V}$ and for all $W \in \mathcal{B}$ with $W \supseteq \overline{V}$ containing x, there exists $y \in \text{int}(W \cap K)$ satisfying $f(x, y) \leq_C \mathbf{0}$, where $\mathcal{B}$ is a base at $\mathbf{0}$ consists of neighborhoods of $\mathbf{0}$ for topological vector space X.*

Theorem 9.10 *Let $f : K \times K \to Y$ be a strong C-pseudomonotone bifunction such that $f(x, x) = \mathbf{0}$ for all $x \in K$. If $\mathbb{S}(\text{SVEP}(f, C))$ is nonempty, then the Assumption 9.6 holds. Moreover, if f satisfies conditions* (ii) *and* (iii) *of Lemma 9.4, Assumption 9.6 with bounded open neighborhood V of $\mathbf{0}$, and conditions* (i)–(iii) *of Theorem 9.8 hold for $f|\text{co}(K \cap W)$ and for every $W \in \mathcal{B}$ with $\text{int}(W \supseteq V)$, then $\mathbb{S}(\text{SVEP}(f, C))$ is nonempty.*

Proof Suppose that $x_0 \in \mathbb{S}(\text{SVEP}(f, C))$. Then $f(x_0, y) \geq_C \mathbf{0}$ for all $y \in K$. Since C is pointed, we have $f(x_0, y) \not\leq_{C_0} \mathbf{0}$ for all $y \in K$. The strong C-pseudomonotonicity of f implies that $f(y, x_0) \leq_C \mathbf{0}$ for all $y \in K$. Then the Assumption 9.6 trivially holds for $y = x_0$ and for every $x \in K \setminus V$, where V is an arbitrary bounded open neighborhood of $\mathbf{0}$ such that $x_0 \in V$.

To see the converse, let $W \in \mathcal{B}$. By Theorem 9.8, there exists a solution $\bar{x}$ of $\text{SVEP}(f, C)$ defined over $\text{co}(K \cap W)$. In the case that $\bar{x} \in \overline{V}$, by our assumption $\overline{V} \subset \text{int}(W)$ and Proposition 9.3, we obtain $\bar{x} \in \mathbb{S}(\text{SVEP}(f, C))$. If $\bar{x} \notin \overline{V}$, by Assumption 9.6, there exist $y \in \text{int}(W \cap K)$ such that $f(\bar{x}, y) \leq_C \mathbf{0}$. Proposition 9.3 implies that $\bar{x} \in \mathbb{S}(\text{SVEP}(f, C))$. □

We deal with the strong C-quasimonotone bifunctions and establish the following existence results for a solution of SVEP in the presence of Assumption 9.5 and 9.6, respectively. The first theorem is a vector version of Theorem 4.1 in [21].

Theorem 9.11 *Let $f : K \times K \to Y$ be a strong C-quasimonotone bifunction satisfying the conditions of Lemma 9.4 and for each $x \in K$, the set $\{y \in K : f(x, y) \leq_C \mathbf{0}\}$ is closed and convex. If the set $\mathbb{S}(\text{SVEP}(f, C))$ is bounded and $\mathbb{S}(\text{MSVEP}(f, C))$ is nonempty, then the Assumption 9.5 holds. Moreover, if f*

satisfies Assumption 9.5 with an open bounded balanced neighborhood V of $\mathbf{0}$, and the condition (iii) *of Theorem 9.8 for $f|\text{co}(K \cap W)$ and for every $W \in \mathcal{B}$ with* $\text{int}(W) \supseteq \overline{V}$. *Then $\mathbb{S}(\text{SVEP}(f, C))$ is nonempty.*

Proof There exists an integer $m_0 > 1$ such that the set $\mathbb{S}(\text{SVEP}(f, C))$ is a subset of $K \cap (m_0 - 1)\, V$, where V is a bounded open balanced neighborhood of $\mathbf{0}$. Assume that the Assumption 9.5 does not hold. Then for every $m > m_0$, there exists $x_m \in K \backslash K \cap W_m$ such that

$$f(x_m, y) \geq_C \mathbf{0}, \quad \text{for all } y \in K \cap W_m. \tag{9.28}$$

where $W_m = \underbrace{V + V + \cdots + V}_{m}$.

We show that $f(x_m, y) \geq_{C_0} \mathbf{0}$ whenever $y \in K \cap W_m$. Indeed, assume that $f(x_m, y) = \mathbf{0}$ for all y. By (9.28), $f(x_m, y) \geq_C$. Let $z \in K \backslash W_m$ such that $f(x_m, z) \not\geq_C \mathbf{0}$. From assumption (iii) of Lemma 9.4, we obtain

$$f(x_m, (1 - t)y + tz)) \not\geq_C \mathbf{0}, \quad \text{for all } t \in]0, 1[.$$

Using $(1-t)y+tz \to y$ if $t \to 0^+$ and $y \in K \cap W_m$, there exists t (small enough) such that $(1 - t)y + tz) \in W_m$, which is a contradiction of (9.28). Therefore, $f(x_m, y) \geq_{C_0}$ for all $y \in K \cap W_m$. Since C is pointed, $f(x_m, y) \not\leq_C \mathbf{0}$ for all $y \in K \cap W_m$. Thus by strong C-quasimonotonicity of f, we have

$$f(y, x_m) \leq_C \mathbf{0}, \quad \text{for all } y \in K \cap W_m. \tag{9.29}$$

Let x_0 be a point in $\mathbb{S}(\text{MSVEP}(f, C))$. It follows from $x_0 \in (m - 1)\, V$ and $x_m \in K \backslash W_m$ (for m sufficiently large) that there exists a positive number $t \in]0, 1[$ such that

$$z_m = (1 - t)x_m + tx_0 \notin (m - 1)\, V, \quad \text{for all } z_m \in W_m \cap K.$$

From (9.28), $x_0 \in \mathbb{S}(\text{MSVEP}(f, C))$ and the convexity of the set $\{y \in K : f(x, y) \leq_C \mathbf{0}\}$, we have

$$f(y, z_m) \leq_C \mathbf{0}, \quad \text{for all } y \in K \cap W_m. \tag{9.30}$$

Hence, $z_m \in \mathbb{S}^M_{K,loc}$ and so by Lemma 9.4, $z_m \in \mathbb{S}(\text{SVEP}(f, C))$. Therefore, the sequence $\{z_m\}$ is unbounded which contradicts the boundedness of $\mathbb{S}(\text{SVEP}(f, C))$.

For the second part, if f is strong properly C-quasimonotone, we get the result arguing as in proof of the converse part of Theorem 9.8. If f is not strong properly C-quasimonotone, from Lemma 9.5, $\mathbb{S}^M_{K,loc} \neq \emptyset$ and so by Lemma 9.4, $\mathbb{S}(\text{MSVEP}(f, C)) \neq \emptyset$. □

The following theorem is a vector version of Theorem 4.2 in [21].

Theorem 9.12 *Let $f : K \times K \to Y$ be a strong C-quasimonotone bifunction. If $\mathbb{S}(\text{MSVEP}(f, C))$ is nonempty, then the Assumption 9.6 holds. Conversely, if f satisfies the conditions of Lemma 9.4 and 9.5, and Assumption 9.6 for a bounded neighborhood V of* $\mathbf{0}$, *and moreover condition* (iii) *of Theorem 9.8 for $f|\text{co}(K \cap W)$ holds for every bounded neighborhood W with* $\text{int}(W) \supseteq \overline{V}$, *then* $\mathbb{S}(\text{SVEP}(f, C))$ *is nonempty.*

Proof We only prove the converse part. For this, if f is strong properly C-quasimonotone, we get $\mathbb{S}(\text{SVEP}(f, C)) \neq \emptyset$, by arguing as in Theorem 9.10. If f is not strong properly C-quasimonotone, then form Lemma 9.5, we have $\mathbb{S}^M_{K,loc} \neq \emptyset$ and so by Lemma 9.4, $\mathbb{S}(\text{SVEP}(f, C)) \neq \emptyset$. □

Farajzadeh et al. [46] described the definition of C-pseudomonotonicity and C-upper sign continuity in terms of $\leq_C$ or $\not\leq_{C_0}$. They used the usual technique of Fan-KKM lemma and derived the existence results for a solution of SVEPs. While, Gong [53] used separation theorem for convex sets and established some existence results for a solution of SVEPs. The arc-wise connectedness and closedness of the solution set of a SVEP are also discussed. They also presented a necessary and sufficient condition for the solution of SVEPs. Fang and Huang [45] discussed the relationship between a SVEP and a minimal element problem. The iterative algorithm for finding a solution of SVEPs is studied by Wang and Li [96].

The existence of solutions of VEP (9.2) has been investigated in [26, 30, 31, 42, 80]. While, Capătă and Kassay [32] studied the existence of solutions of implicit vector variational problems.

9.2.3 Existence Results for Implicit Weak Vector Variational Problems

Throughout this subsection, unless otherwise specified, we assume that K is a nonempty convex subset of a topological vector space X and (Y, C) is an ordered topological vector space with a proper closed convex pointed cone C such that $\text{int}(C) \neq \emptyset$.

Definition 9.7 (C-Diagonally Convex Function) A bifunction $f : K \times K \to Y$ is called *C-diagonally convex* if for any finite set $\{x_1, x_2, \ldots, x_m\} \subset K$ and any $x_0 = \sum_{i=1}^{m} \lambda_i x_i$, where $\lambda_i \geq 0$ for all $i = 1, 2, \ldots, m$ and $\sum_{i=1}^{m} \lambda_i = 1$, we have

$$f(x_0, x_0) \not>_C \sum_{i=1}^{m} \lambda_i f(x_0, x_i).$$

The bifunction f is called *C-diagonally concave* if $-f$ is C-diagonally convex.

Definition 9.8 (Strongly C-Diagonally Convex Function) A bifunction $f : K \times K \to Y$ is called *strongly C-diagonally convex* if for any finite set $\{x_1, x_2, \ldots, x_m\} \subset$

K and any $x_0 = \sum_{i=1}^{m} \lambda_i x_i$, where $\lambda_i \geq 0$ for all $i = 1, 2, \ldots, m$ and $\sum_{i=1}^{m} \lambda_i = 1$, we have

$$f(x_0, x_0) \leq_C \sum_{i=1}^{m} \lambda_i f(x_0, x_i).$$

The bifunction f is called *strongly C-diagonally concave* if $-f$ is strongly C-diagonally convex.

Remark 9.4

(a) If $l : K \times K \to Y$ is C-diagonally convex (concave) and $g : K \times K \to Y$ is strongly C-diagonally convex (concave) , then $l + g$ is also C-diagonally convex (concave).
(b) When $Y = \mathbb{R}$ and $C = \mathbb{R}_+$, the definition of C-diagonal convexity (concavity) reduces to the definition of diagonal convexity (concavity) of a function.

We present the following existence result for solutions of IWVVP.

Theorem 9.13 *Let $l, g : K \times K \to Y$ be vector-valued bifunctions. Assume that the following conditions hold:*

(i) *l is C-diagonally convex;*
(ii) *g is strongly C-diagonally convex;*
(iii) *For each $A \in \mathscr{F}(K)$, l and g are C-continuous on* $\mathrm{co}(A)$*;*
(iv) *For each $A \in \mathscr{F}(K)$ and each $x, y \in \mathrm{co}(A)$ and every net $\{x_\alpha\}_{\alpha \in \Gamma}$ in K converging to x with*

$$l(x_\alpha, x_\alpha) + g(x_\alpha, x_\alpha) \not>_C l(x_\alpha, \lambda y + (1 - \lambda)x) + g(x_\alpha, \lambda y + (1 - \lambda)x),$$

for all $\alpha \in \Gamma$ and all $\lambda \in [0, 1]$, we have

$$l(x, x) + g(x, x) \not>_C l(x, y) + g(x, y);$$

(v) *There exist a nonempty closed compact subset B of K and $\tilde{y} \in B$ such that $l(x, x) + g(x, x) >_C l(x, \tilde{y}) + g(x, \tilde{y})$ for all $x \in K \setminus B$.*

Then the IWVVP (9.11) *has a solution $\bar{x} \in B$.*

Proof For all $y \in K$, define a set-valued map $T : K \to 2^K$ by

$$T(y) = \{x \in K : l(x, x) + g(x, x) \not>_C l(x, y) + g(x, y)\}.$$

Clearly, for all $y \in K$, $T(y)$ is nonempty, since $y \in T(y)$ by the properness of C. By conditions (i) and (ii), T is a KKM-map. We also have

(a) $T(\tilde{y}) \subseteq B$, so that $\mathrm{cl}_K\,(T(\tilde{y})) \subset \mathrm{cl}_K(B) = B$, and hence, $\mathrm{cl}_K\,(T(\tilde{y}))$ is compact in K;

(b) For each $A \in \mathscr{F}(K)$ with $\tilde{y} \in A$ and $y \in \text{co}(A)$,

$$T(y) \cap \text{co}(A) = \{x \in \text{co}(A) : l(x,x) + g(x,x) \not> _C l(x,y) + g(x,y)\}$$

is closed in $\text{co}(A)$ by condition (iii);

(c) For each $A \in \mathscr{F}(K)$ with $\tilde{y} \in A$, let $x \in \left(\text{cl}_K\left(\bigcap_{y\in\text{co}(A)} T(y)\right)\right) \cap \text{co}(A)$, then $x \in \text{co}(A)$ and there is a net $\{x_\alpha\}_{\alpha\in\Gamma}$ in $\bigcap_{y\in\text{co}(A)} T(y)$ such that $x_\alpha \to x$. For each $y \in \text{co}(A)$, since $\lambda y + (1-\lambda)x \in \text{co}A$ for all $\lambda \in [0,1]$, we have $x_\alpha \in T(\lambda y + (1-\lambda)x)$, for all $\alpha \in \Gamma$ and all $\lambda \in [0,1]$. This implies that

$$l(x_\alpha, x_\alpha) + g(x_\alpha, x_\alpha) \not> _C l(x_\alpha, \lambda y + (1-\lambda)x) + g(x_\alpha, \lambda y + (1-\lambda)x)$$

for all $\alpha \in \Gamma$ and all $\lambda \in [0,1]$. By condition (v), we have

$$l(x,x) + g(x,x) \not> _C l(x,y) + g(x,y).$$

It follows that $x \in \left(\bigcap_{y\in\text{co}(A)} T(y)\right) \cap \text{co}(A)$. Hence

$$\left(\text{cl}_K\left(\bigcap_{y\in\text{co}(A)} T(y)\right)\right) \cap \text{co}(A) = \left(\bigcap_{y\in\text{co}(A)} T(y)\right) \cap \text{co}(A).$$

By Lemma 1.18, we have $\bigcap_{y\in K} T(y) \neq \emptyset$. Hence, there exists $\bar{x} \in \bigcap_{y\in K} T(y)$, and therefore,

$$l(\bar{x},\bar{x}) + g(\bar{x},\bar{x}) \not> _C l(\bar{x},y) + g(\bar{x},y), \quad \text{for all } y \in K,$$

from which it follows that $\bar{x}$ is a solution of IWVVP (9.11). □

When $l(x,y) = f(y)$ and $g(x,x) = \mathbf{0}$ for all $x, y \in K$, we have the following result.

Theorem 9.14 *Let $\phi : K \to Y$ be a vector-valued function and $f : K \times K \to Y$ be a vector-valued bifunction. Assume that the following conditions hold.*

(i) *For all $x \in K$, $y \mapsto f(x,y) + \phi(y)$ is C-quasiconvex;*
(ii) *For each $A \in \mathscr{F}(K)$, ϕ is C-lower semicontinuous on* $\text{co}(A)$*;*
(iii) *For each $A \in \mathscr{F}(K)$ and for each $y \in \text{co}(A)$, $x \mapsto f(x,y)$ is C-upper semicontinuous on* $\text{co}(A)$*;*
(iv) *For each $A \in \mathscr{F}(K)$ and each $x, y \in \text{co}(A)$ and every net $\{x_\alpha\}_{\alpha\in\Gamma}$ in K converging to x with*

$$\phi(x_\alpha) - \phi(\lambda y + (1-\lambda)x) \not> _C f(x_\alpha, \lambda y + (1-\lambda)x),$$

for all $\alpha \in \Gamma$ and all $\lambda \in [0,1]$, we have

$$\phi(x) - \phi(y) \not> _C f(x,y);$$

(v) *There exist a nonempty closed and compact subset B of K and $\tilde{y} \in B$ such that $\phi(x) - \phi(\tilde{y}) >_C f(x, \tilde{y})$ for all $x \in K \setminus B$.*

Then there exists $\bar{x} \in B$ such that

$$\varphi(\bar{x}) - \varphi(y) \not>_C f(\bar{x}, y), \quad \textit{for all } y \in K.$$

Proof It is on the lines of the proof of Theorem 9.13. □

9.3 Duality of Implicit Weak Vector Variational Problems

Let $h : X \to Y$ and $\xi \in \mathcal{L}(X, Y)$. The *vector conjugate function* [101] denoted by h^*_{sup}, of h at ξ is defined by

$$h^*_{\text{sup}}(\xi) = \text{Sup}_{\text{int}(C)}\{\langle \xi, y\rangle - h(y) : y \in X\}.$$

Let $y \in X$. The *vector biconjugate function* denoted as h^{**}_{sup}, of h at y is defined by

$$h^{**}_{\text{sup}}(y) = \text{Sup}_{\text{int}(C)} \bigcup \left\{\langle \xi, y\rangle - h^*_{\text{sup}}(l) : \xi \in \mathcal{L}(X, Y)\right\}.$$

Note that both h^*_{sup} and h^{**}_{sup} are set-valued maps and $h^*_{\text{sup}} : \mathcal{L}(X, Y) \to 2^Y$, $h^{**}_{\text{sup}} : X \to 2^Y$. Throughout this section, we assume that $h^*_{\text{sup}}(\xi) \neq \emptyset$ and $h^{**}_{\text{sup}}(y) \neq \emptyset$.

We now define the dual problem of (9.11), denoted by DIWVVP, as follows: Find $\bar{x} \in X$, $-\xi^* \in \partial^w l(\bar{x}, \bar{x})$ and $\bar{y} \in g^*_{\text{sup}}(\bar{x}, \xi^*)$ satisfying $\langle \xi^*, \bar{x}\rangle - g(\bar{x}, \bar{x}) = \bar{y}$ such that

$$\bar{y} - \langle \xi^*, \bar{x}\rangle \not>_C g^*_{\text{sup}}(\bar{x}, \xi) - \langle \xi, \bar{x}\rangle, \quad \text{for all } \xi \in \mathcal{L}(X, Y). \tag{9.31}$$

It is called the *dual implicit weak vector variational problem* (in short, DIWVVP) and $(\bar{x}, \xi^*)$ is called a solution of DIWVVP. The following two results show the relationships between solutions of IWVVP and DIWVVP.

Theorem 9.15 *Assume that $l(y, .)$ and $g(y, .)$ are C-convex for each fixed $y \in X$. If $\bar{x}$ is a solution of* IWVVP (9.11) *and $g(\bar{x}, .) : x \mapsto g(\bar{x}, x)$ is externally stable, then there exists $\xi^* \in \mathcal{L}(X, Y)$ such that $(\bar{x}, \xi^*)$ is a solution of* DIWVVP (9.31).

Proof Let $\bar{x}$ be a solution of IWVVP (9.11). Then

$$g(\bar{x}, \bar{x}) + l(\bar{x}, \bar{x}) \not>_C g(\bar{x}, z) + l(\bar{x}, z), \quad \text{for all } z \in X.$$

Let $\mathbf{0}^*$ be the zero operator from X to Y. Then

$$\langle \mathbf{0}^*, g\rangle \not>_C [g(\bar{x}, z) + l(\bar{x}, z)] - [g(\bar{x}, \bar{x}) + l(\bar{x}, \bar{x})], \quad \text{for all } z \in X.$$

By the definition of $\partial^w l(y, z)$, we have

$$\mathbf{0}^* \in \partial^w (g(\bar{x}, \bar{x}) + l(\bar{x}, \bar{x})).$$

It follows from [100] that

$$\partial^w (g(\bar{x}, \bar{x}) + l(\bar{x}, \bar{x})) \subseteq \partial^w g(\bar{x}, \bar{x}) + \partial^w l(\bar{x}, \bar{x}).$$

Hence,

$$\mathbf{0}^* \in \partial^w g(\bar{x}, \bar{x}) + \partial^w l(\bar{x}, \bar{x}),$$

or equivalently, there exists $\xi^* \in \mathcal{L}(X, Y)$ such that

$$\xi^* \in \partial^w g(\bar{x}, \bar{x}) \cap (-\partial^w l(\bar{x}, \bar{x})).$$

Then from Lemma 2.16, we obtain

$$\langle \xi^*, \bar{x} \rangle - g(\bar{x}, \bar{x}) \in g^*_{\text{sup}}(\bar{x}, \xi^*), \tag{9.32}$$

$$\langle -\xi^*, \bar{x} \rangle - l(\bar{x}, \bar{x}) \in l^*_{\text{sup}}(\bar{x}, -\xi^*).$$

As $g(\bar{x}, .) : x \mapsto g(\bar{x}, x)$ is externally stable,

$$g(\bar{x}, \bar{x}) \in g^{**}_{\text{sup}}(\bar{x}, \bar{x}) = \text{Sup}_{\text{int}(C)}\{\langle \xi, \bar{x} \rangle - g^*_{\text{sup}}(\bar{x}, \xi) \ : \ \xi \in \mathcal{L}(X, Y)\}.$$

Thus,

$$g(\bar{x}, \bar{x}) \not<_C \langle \xi, \bar{x} \rangle - g^*_{\text{sup}}(\bar{x}, \xi), \quad \text{for all } \xi \in \mathcal{L}(X, Y).$$

From (9.32), there exists $\bar{y} \in g^*_{\text{sup}}(\bar{x}, \xi^*)$ such that

$$\langle \xi^*, \bar{x} \rangle - g(\bar{x}, \bar{x}) = \bar{y},$$

that is,

$$g(\bar{x}, \bar{x}) = \langle \xi^*, \bar{x} \rangle - \bar{y}.$$

So,

$$\langle \xi^*, \bar{x} \rangle - \bar{y} \not<_C \langle \xi, \bar{x} \rangle - g^*_{sup}(\bar{x}, \xi), \quad \text{for all } \xi \in \mathcal{L}(X, Y).$$

$$\bar{y} - \langle \xi^*, \bar{x} \rangle \not>_C g^*_{\text{sup}}(\bar{x}, \xi) - \langle \xi, \bar{x} \rangle, \quad \text{for all } \xi \in \mathcal{L}(X, Y).$$

Thus $(\bar{x}, \xi^*)$ is a solution of DIWVVP (9.31). □

Theorem 9.16 *Assume that $l(y,.)$ and $g(y,.)$ are C-convex for each fixed $y \in X$. If $(\bar{x}, \xi^*)$ is a solution of* DIWVVP (9.31) *and*

$$\partial^w(g(\bar{x},\bar{x}) + l(\bar{x},\bar{x})) = \partial^w g(\bar{x},\bar{x}) + \partial^w l(\bar{x},\bar{x}), \tag{9.33}$$

then $\bar{x}$ is a solution of IWVVP (9.11).

Proof This is obtained by inverting the reasoning in the proof of Theorem 9.15 step by step. □

For details regarding condition (9.33), see [100].

Remark 9.5 Theorems 9.16 and 9.16 are extensions of Theorem 1 in [37] for vector-valued bifunctions.

Indeed, let $Y = \mathbb{R}$ and $C = \mathbb{R}_+$. Then DIWVVP reduces to find $\bar{x} \in X$, $\xi^* \in X^*$ such that $-\xi^* \in \partial l(\bar{x},\bar{x})$ and

$$g^*(\bar{x},\xi^*) - \langle \xi^*, \bar{x}\rangle \leq g^*(\bar{x},\xi) - \langle \xi, \bar{x}\rangle, \quad \text{for all } \xi \in X^*,$$

where $\partial l(\bar{x},\bar{x})$ and $g^*(\bar{x},\xi^*)$ are convex subdifferential and convex conjugate functions, respectively. This is a dual problem of IVP which was studied by Dolcetta and Matzeu [37].

9.4 Gap Functions and Variational Principles

Let X and Y be topological vector spaces, and $C \subseteq Y$ be a pointed, proper, closed convex cone with $\text{int}(C) \neq \emptyset$. Let K be a nonempty convex subset of X and $f : K \times K \to Y$ be a vector-valued bifunction such that $f(x,x) = \mathbf{0}$ for all $x \in K$. Let us recall a *vector optimization problem* for a set-valued map $F : X \to 2^Y$, denoted by SVVOP,

$$\min_{x \in K} F(x); \tag{9.34}$$

which consists in finding $\bar{x} \in K$ for which there exists $\bar{y} \in F(\bar{x})$ such that $\bar{y} \in \min_C F(K)$, that is, $F(\bar{x}) \cap \min_C F(K) \neq \emptyset$, where

$$\begin{aligned}\min_C F(K) &= \{x \in F(K) : \text{ there exists no } y \in F(K) \text{ such that } y \leq_{C_0} x\} \\ &= \{x \in F(K) : (F(K) - x) \cap (C_0) = \emptyset\}\end{aligned}$$

and $F(K) = \bigcup_{x \in K} F(x)$. If we replace the acronym min by w − min, we obtain, so called, *weak vector optimization problem* for a set-valued map, denoted by SVWVOP.

By means of a gap function, one can convert vector equilibrium problems into a vector optimization problem for set-valued maps or for single-valued maps. It plays a vital role to design the algorithms for finding the approximate solutions of vector equilibrium problems. The variational principles are more general than the gap functions, in the sense that variational principles provide the characterizations of solutions of vector equilibrium problems by means of solving the zero inclusion problems for set-valued maps. The set-valued maps involved in the formulations of variational principles depend on the data of vector equilibrium problems, but not on their solution sets. Ansari et al. [9, 11] suggested several set-valued gap functions and variational principles for weak vector equilibrium problems (in short, WVEPs). These set-valued gap functions enable one to convert WVEPs into a vector optimization problem for set-valued maps. Konnov [70] modified the approach of [9, 11] and presented a gap function which allows one to reduce WVEPs to a vector optimization problem for single-valued maps. Ceng et al. [34] also studied the gap functions and variational principles for VEPs and WVEPs. Mastroeni et al. [79] used the image space approach [52] to discuss the gap functions for VEPs. Zhang et al. [104] introduced a scalar-valued gap function for WVEPs. Mirazaee and Soleimani-damaneh [80] studied set-valued gap functions for VEPs and investigated their differential properties using Hadamard directional differentials. Altangerel et al. [2] proposed some variational principles for vector equilibrium problems by using so-called Fenchel duality. They gave the characterizations of the solutions of VEPs by means of solving a set-valued vector optimization problem, where set-valued map is defined on the basis of Fenchel duality depending on the data, but not on the solutions set of VEPs. Li et al. [76] established two set-valued gap functions for a vector equilibrium problem by virtue of conjugate dual problems. Sun and Li [90] used weak and strong duality results to suggest gap functions for vector equilibrium problems.

9.4.1 Gap Function for Vector Equilibrium Problems

Let X and Y be topological vector spaces, and $C \subseteq Y$ be a pointed, proper, closed convex cone with $\operatorname{int}(C) \neq \emptyset$. Let K be a nonempty convex subset of X and $f : K \times K \to Y$ be a vector-valued bifunction such that $f(x, x) = \mathbf{0}$ for all $x \in K$.

We define the following gap function associated with a vector equilibrium problem.

Definition 9.9

(a) A set-valued map $G : X \to 2^Y$ is said to be a *gap function* for VEP (9.2) if the following conditions hold:

   (i) $\bar{x} \in \mathbb{S}(\text{VEP}(f, C))$ if and only if $\mathbf{0} \in G(\bar{x})$,
   (ii) $\mathbf{0} \in \max\limits_{x \in K} G(x)$;

(b) A set-valued map $G_{\mathrm{w}} : X \to 2^Y$ is said to be a *gap function* for WVEP (9.1) if the following conditions hold:

(i) $\bar{x} \in \mathbb{S}(\mathrm{WVEP}(f, C))$ if and only if $\mathbf{0} \in G_{\mathrm{w}}(\bar{x})$,

(ii) $\mathbf{0} \in \mathrm{w} - \max_{x \in K} G_{\mathrm{w}}(x)$;

We associate with the VEP (9.2), the following set-valued map $G : K \to 2^Y$ defined by

$$G(x) := \min_{y \in K} f(x, y), \tag{9.35}$$

and to the WVEP (9.1), the set-valued map $G_{\mathrm{w}} : K \to 2^Y$ defined by

$$G_{\mathrm{w}}(x) := \mathrm{w} - \min_{y \in K} f(x, y) \tag{9.36}$$

Theorem 9.17

(a) *The set-valued map* $G(x) := \min_{y \in K} f(x, y)$ *is a gap function for* VEP (9.2).

(b) *The set-valued map* $G_{\mathrm{w}}(x) := \mathrm{w} - \min_{y \in K} f(x, y)$ *is a gap function for* WVEP (9.1).

Proof For the sake of simplicity, we prove only (b). Case (a) is analogous. We prove (i) and (ii) of Definition 9.9 (b).

$$\begin{aligned}
\text{(i)} \quad \bar{x} \in \mathbb{S}(\mathrm{WVEP}(f, C)) \;&\Leftrightarrow\; f(\bar{x}, y) \not<_C \mathbf{0} \quad \text{for all } y \in K \\
&\Leftrightarrow\; \mathbf{0} \in \mathrm{w} - \min_C \{f(\bar{x}, y) : y \in K\} \\
&\qquad \text{(taking into account that } f(\bar{x}, \bar{x}) = \mathbf{0}) \\
&\Leftrightarrow\; \mathbf{0} \in G_{\mathrm{w}}(\bar{x}).
\end{aligned}$$

$$\text{(ii)} \quad \mathbf{0} \in \mathrm{w} - \max_{x \in K} G_{\mathrm{w}}(x) \Leftrightarrow z \not>_C \mathbf{0} \text{ for all } z \in G_{\mathrm{w}}(x) \text{ and all } x \in K. \tag{9.37}$$

Let $x \in K$ and $z \in G_{\mathrm{w}}(x)$, then $f(z, y) - z \not<_C \mathbf{0}$ for all $y \in K$. In particular, for $y := z$, we obtain $-z \not<_C \mathbf{0}$ which leads to (9.37). □

Remark 9.6 [25]

(a) $G(x) \cap (C_0) = \emptyset$ for all $x \in K$.
Indeed, assume that for some $\hat{x} \in K$, $G(\hat{x}) \cap (-(C_0)) \neq \emptyset$. Then there exists $\hat{y} \in C_0$ such that $\hat{y} \in \min_{y \in K} f(\hat{x}, y)$, that is, $(f(\hat{x}, K - \hat{y})) \cap (-(C_0)) = \emptyset$. Since $f(x, x) = \mathbf{0}$ for all $x \in K$, $\mathbf{0} \in f(\hat{x}, K)$, and therefore, $(-\hat{y}) \cap (-(C_0)) = \emptyset$, a contradiction.

(b) $\bar{x} \in \mathbb{S}(\mathrm{VEP}(f, C))$ if and only if $G(\bar{x}) \cap C \neq \emptyset$.
Indeed, since G is a gap function by Theorem 9.17, we have $\bar{x} \in \mathbb{S}(\mathrm{VEP}(f, C))$ if and only if $\mathbf{0} \in G(\bar{x})$. Then by part (a), $G(\bar{x}) \cap C \neq \emptyset$.

9.4.2 Variational Principle for Weak Vector Equilibrium Problems

Extending the terminology of Auchmuty [17] and Blum and Oettli [28], we say that a *variational principle* holds for WVEP (9.1) if there exists a set-valued map $F : K \to 2^Y$, depending on the data of WVEP (9.1) but not on its solution set such that the solution set of WVEP (9.1) coincides with the solution set of the following vector optimization problem for set-valued map F, denoted by SVVOP,

$$\mathrm{w} - \min_{x \in K} F(x), \tag{9.38}$$

that is, to find $\bar{x} \in K$ for which there exists $\bar{y} \in F(\bar{x})$ such that $\bar{y} \in \mathrm{w} - \min_C F(K)$, that is, $F(\bar{x}) \cap \mathrm{w} - \min_C F(K) \neq \emptyset$, where $\mathrm{w} - \min_C F(K) = \{x \in F(K) :$ there is no $y \in F(K)$ such that $y <_C x\}$ and $F(K) = \bigcup_{x \in K} F(x)$.

In order to formulate first variational principle, we define a set-valued map $\Phi : K \to 2^Y$ as follows:

$$\Phi(x) = \mathrm{w} - \min_C f(x, K), \quad \text{for all } x \in K. \tag{9.39}$$

Let dom $(\Phi) := \{x \in K : \Phi(x) \neq \emptyset\}$.

Lemma 9.6 *For each $x \in K$,*

$$z \in \Phi(x) \quad \textit{implies} \quad z \not>_C \mathbf{0}. \tag{9.40}$$

Proof Assume contrary that there exists $z \in \Phi(x)$ such that $z >_C \mathbf{0}$. Then there exists $y \in K$ such that

$$f(x, y) = z >_C \mathbf{0} = f(x, x),$$

which is a contradiction, since $f(x, y) \in \mathrm{w} - \min_C f(x, K)$. □

We associate to WVEP (9.1) with the following set-valued weak vector optimization problem (in short, SVWVOP) for set-valued map Φ:

$$\mathrm{w} - \max_{x \in K} \Phi(x), \tag{9.41}$$

that is, to find $\bar{x} \in K$ for which there exists $\bar{y} \in \Phi(\bar{x})$ such that $\bar{y} \in \mathrm{w} - \max_C \Phi(K)$, that is, $\Phi(\bar{x}) \cap \mathrm{w} - \max_C \Phi(K) \neq \emptyset$, where $\mathrm{w} - \max_C \Phi(K) = \{a \in \Phi(K) :$ there is no $b \in \Phi(K)$ such that $b > a\}$ and $\Phi(K) = \bigcup_{x \in K} \Phi(x)$. We denote by $\mathbb{S}(\mathrm{SVWVOP}_\Phi)$ the solution set of SVWVOP (9.41).

Theorem 9.18

(a) *$\bar{x} \in K$ is a solution of* WVEP (9.1) *if and only if* $\mathbf{0} \in \Phi(\bar{x})$.
(b) $\mathbb{S}(\mathrm{WVEP}(f, C)) \subseteq \mathbb{S}(\mathrm{SVWVOP}_\Phi)$.

Proof

(a) Suppose $\bar{x}$ solves WVEP (9.1). Then

$$f(\bar{x}, y) \not<_C \mathbf{0}, \quad \text{for all } y \in K.$$

Assume that $\mathbf{0} \notin \Phi(\bar{x})$, then there exists $z \in K$ such that

$$f(\bar{x}, z) <_C \mathbf{0},$$

which is a contradiction.

Conversely, let $\mathbf{0} \in \Phi(\bar{x})$. Assume that $\bar{x}$ does not solve WVEP (9.1), then there exists $y \in K$ such that $f(\bar{x}, y) <_C \mathbf{0}$. This implies that $\mathbf{0} \notin \Phi(\bar{x})$, a contradiction. This proves the first part of the theorem.

(b) Let $\bar{x} \in \mathbb{S}(\text{WVEP}(f, C))$. Then, $\mathbf{0} \in \Phi(\bar{x})$ due to part (a). It now follows from (9.41) that $\bar{x} \in \mathbb{S}(\text{SVWVOP}_{\Phi})$, as desired.

□

We remark that the inclusion in Theorem 9.18 (b) can be strict as the following example shows.

Example 9.4 Let $X = Y = \mathbb{R}$ and $C = [0, \infty)$. Let

$$f(x, y) = \sin(x - y), \quad \text{for all } x, y \in \mathbb{R}.$$

Then for all $x \in \mathbb{R}, f(x, \mathbb{R}) = [-1, 1]$. Hence for all $x \in \mathbb{R}$, $\Phi(x) = \{-1\}$. It is clear that $\mathbb{S}(\text{WVEP}(f, C)) = \emptyset$ and $\mathbb{S}(\text{SVWVOP}_{\Phi}) = \mathbb{R}$.

In order to formulate second variational principle, we define set-valued maps $S : K \to 2^K$ and $\Psi : K \to 2^Y$ by

$$S(x) := \{y \in K : \sigma(f(x, y)) \leq \sigma(f(x, z)) \text{ for all } z \in K\},$$
$$\Psi(x) := f(x, S(x)).$$

Lemma 9.7 *For each $x \in K$,*

$$z \in \Psi(x) \quad \textit{implies} \quad z \leq_C \mathbf{0}. \tag{9.42}$$

Proof By definition, if $z \in \Psi(x)$, then there exists $y \in K$ such that $z = f(x, y)$. Since $y \in S(x)$, we have

$$\sigma(f(x, y)) \leq \sigma(f(x, x)) = \mathbf{0}.$$

From (2.31), it follows that $z = f(x, y) \leq_C \mathbf{0}$. □

We now define the following SVVOP for set-valued map Ψ:

$$\max_{x \in K} \Psi(x), \tag{9.43}$$

that is, to find $\bar{x} \in K$ for which there exists $\bar{y} \in \Psi(\bar{x})$ such that $\bar{y} \in \max_C \Psi(K)$, that is, $\Psi(\bar{x}) \cap \max_C \Psi(K) \neq \emptyset$. We denote by $\mathbb{S}(\text{VWVOP}_\Psi)$ the solution set of problem (9.43).

Theorem 9.19

(a) *$\bar{x} \in K$ is a solution of* WVEP (9.1) *if and only if* $\mathbf{0} \in \Psi(\bar{x})$.
(b) *If the solution set of* WVEP (9.1) *nonempty, then* $\mathbb{S}(\text{WVEP}(f, C)) = \mathbb{S}(\text{VWVOP}_\Psi)$.

Proof

(a) Suppose $\bar{x}$ solves WVEP (9.1). Then

$$f(\bar{x}, y) \not<_C \mathbf{0}, \quad \text{for all } y \in K,$$

or, due to (2.31),

$$\sigma(f(\bar{x}, y)) \geq 0, \quad \text{for all } y \in K.$$

Therefore,

$$\sigma(f(\bar{x}, \bar{x})) = 0 \leq \sigma(f(\bar{x}, y)), \quad \text{for all } y \in K,$$

and we conclude that $\bar{x} \in S(\bar{x})$. It follows that

$$\mathbf{0} = f(\bar{x}, \bar{x}) \in \Psi(\bar{x}).$$

Conversely, let $\mathbf{0} \in \Psi(\bar{x})$. Assume that $\bar{x}$ does not solve WVEP (9.1), then there exists $y \in K$ such that $f(\bar{x}, y) <_C \mathbf{0}$. Now, the relation (2.31) gives $\sigma(f(\bar{x}, y)) < 0$ and

$$\sigma(f(\bar{x}, y')) \leq \sigma(f(\bar{x}, y)) < 0, \quad \text{for all } y' \in S(\bar{x}).$$

Again, from (2.31), we have

$$f(\bar{x}, y') <_C \mathbf{0}, \quad \text{for all } y' \in S(\bar{x}),$$

that is, $\mathbf{0} \notin \Psi(\bar{x})$, a contradiction. This proves part (a) of the theorem.

(b) Let $\mathbb{S}(\text{WVEP}(f, C)) \neq \emptyset$. Take any $\bar{x} \in \mathbb{S}(\text{WVEP}(f, C))$, then $\mathbf{0} = f(\bar{x}, \bar{x}) \in \Psi(\bar{x})$ due to (a). From (9.42), we have that for each $x \in K$ and for all $z \in \Psi(x)$,

$$z \leq_C f(\bar{x}, \bar{x}) = \mathbf{0}.$$

If there exists $z' \in \Psi(x)$, $z' \neq \mathbf{0}$ such that $z' \geq_C \mathbf{0}$, then we must have

$$z' \geq_C \mathbf{0} \quad \text{and} \quad z' \leq_C \mathbf{0},$$

that is, $\mathbf{0} \neq z' \in C \cap (-C)$. This contradicts that C is a pointed cone. Therefore, $\bar{x} \in \mathbb{S}(\text{VWVOP}_{\Psi})$.

Conversely, take any $\bar{x} \in \mathbb{S}(\text{VWVOP}_{\Psi})$. By the definition, there exists $\bar{y} \in \Psi(\bar{x})$ such that $\bar{y} \in \max_C \Psi(K)$. From (9.42), it follows that $\bar{y} \leq_C \mathbf{0}$. Assume, for contradiction, that $\mathbf{0} \notin \Psi(\bar{x})$; then $\bar{y} \neq \mathbf{0}$. Since $\mathbb{S}(\text{WVEP}(f, C)) \neq \emptyset$, there exists $x^* \in \mathbb{S}(\text{WVEP}(f, C))$, but $\mathbf{0} \in \Psi(x^*)$ due to (a). Therefore, there exists $\mathbf{0} \in \Psi(K)$, $\mathbf{0} \neq \bar{y}$ and $\bar{y} \leq_C \mathbf{0}$. This contradicts that $\bar{y} \in \max_C \Psi(K)$. Thus, assertion (b) is true.

□

We remark that in Theorem 9.19 (b), it is possible that $\mathbb{S}(\text{WVEP}(f, C))$ is empty and $\mathbb{S}(\text{VWVOP}_{\Psi})$ is not empty by considering Example 9.4.

9.4.3 Variational Principle for Minty Weak Vector Equilibrium Problems

This subsection deals with the variational principle for Minty weak vector equilibrium problem (in short, MWVEP) defined by (9.4).

We define a set-valued map $\Phi^* : K \to 2^Y$ by

$$\Phi^*(x) = \text{w} - \max_C f(K, x), \quad \text{for all } x \in K, \tag{9.44}$$

where $\text{w} - \max_C f(K, x) = \{y \in f(K, x) : \text{ there is no } z \in f(K, x) \text{ such that } z >_C y\}$. We associate with WVEP the following SVVOP:

$$\text{w} - \min_{x \in K} \Phi^*(x), \tag{9.45}$$

that is, to find $\bar{x} \in K$ for which there exists $\bar{y} \in \Phi^*(\bar{x})$ such that $\bar{y} \in \text{w}-\min_C \Phi^*(K)$, that is, $\Phi^*(\bar{x}) \cap \text{w} - \min_C \Phi^*(K) \neq \emptyset$.

We denote by $\mathbb{S}(\text{DSVWVOP}_{\Phi^*})$ the solution set of the problem (9.45). Note that, in the sense of vector optimization for set-valued maps, WVOP (9.41) and (9.45) are dual to each other. Therefore, SVVOP (9.45) is closely related to MWVEP (9.4). Similar to the proof of Theorem 9.18 we can get the following characterization of the solutions for MWVEP (9.4).

Proposition 9.5

(a) $\bar{x} \in K$ *is a solution of* MWVEP (9.4) *if and only if* $\mathbf{0} \in \Phi^*(\bar{x})$.
(b) $\mathbb{S}(\text{MWVEP}(f, C)) \subseteq \mathbb{S}(\text{DSVWVOP}_{\Phi^*})$.

Definition 9.10 A bifunction $f : K \times K \to Y$ is called *C-bi-pseudomonotone* if f and $-f$ both are C-pseudomonotone.

Example 9.5 Let $X = Y = \mathbb{R}^2$, $K = [0,1] \times [0,1]$ and $C = \mathbb{R}^2_+$. Define bifunctions $f_1, f_2 : K \times K \to Y$ by

$$f_1(x,y) = \left(x_1 - y_1, x_2^2 - y_2^2\right)$$

and

$$f_2(x,y) = (x_2(x_1 - y_1), y_2(x_2 - y_2)),$$

where $x = (x_1, x_2)$ and $y = (y_1, y_2)$. Then f_1 and f_2 are C-bi-pseudomonotone. Since $f_1(x,y) + f_1(y,x) = \mathbf{0}$ for all $x, y \in K$, f_1 is also C-monotone. But f_2 is not C-monotone, because for $x = (1,1)$, $y = \left(\frac{1}{2}, \frac{1}{2}\right)$, we have

$$f_2(x,y) + f_2(y,x) = \left(\frac{1}{4}, -\frac{1}{4}\right) \not\leq_C \mathbf{0}.$$

Now, we can derive the following existence results for solutions of WVEP.

Corollary 9.1 *Let K be a nonempty convex subset of a Hausdorff topological vector space X and $f : K \times K \to Y$ be a vector-valued bifunction such that $f(x,x) = \mathbf{0}$ for all $x \in K$.*

(a) *If f is C-pseudomonotone and $\bar{x} \in \mathbb{S}(\text{WVEP}(f,C))$, then $\mathbf{0} \in \Phi^*(\bar{x})$.*
(b) *If $\mathbf{0} \in \Phi^*(\bar{x})$, $\bar{x} \in K$, and either $-f$ is C-pseudomonotone, or $f(x,\cdot)$ is explicitly C-quasiconvex and $f(\cdot,y)$ is C-hemicontinuous for all $x, y \in K$, then $\bar{x} \in \mathbb{S}(WVEP(f,C))$.*
(c) *If f is C-pseudomonotone, then $\mathbb{S}(\text{WVEP}(f,C)) \subseteq \mathbb{S}(\text{DSVWVOP}_{\Phi^*})$.*

Corollary 9.2 *Let K be a nonempty convex subset of a Hausdorff topological vector space X and $f : K \times K \to Y$ be a vector-valued bifunction such that $f(x,x) = \mathbf{0}$ for all $x \in K$. Suppose that at least one of the following conditions hold:*

(i) *f is C-bi-pseudomonotone; or*
(ii) *f is C-pseudomonotone, $f(x,\cdot)$ is explicitly C-quasiconvex and $f(\cdot,y)$ is C-hemicontinuous for all $x, y \in K$.*

Then $\bar{x} \in K$ solves WVEP (9.1) *if and only if $\mathbf{0} \in \Phi^*(\bar{x})$.*

By analogy with the second variational principle from Sect. 9.4.2, we define the following variational principle for MWVEP. We also present another set-valued vector optimization problem, which guarantees for the solution sets of WVEP (9.1) and SVVOP to coincide. From now on, unless otherwise specified, we suppose that Y is a locally convex Hausdorff topological vector space. We define set-valued maps $S^* : K \to 2^K$ and $\Psi^* : K \to 2^Y$ by

$$S^*(x) := \{y \in K : \sigma^*(f(y,x)) \geq \sigma^*(f(z,x)) \text{ for all } z \in K\}$$
$$\Psi^*(x) := f(S^*(x), x),$$

where

$$\sigma^*(y) := \min_{\xi \in B} \langle \xi, y \rangle, \quad \text{for all } y \in Y,$$

and B is a base for C^*.

Define the following SVVOP for set-valued map Ψ^*:

$$\min_{x \in K} \Psi^*(x), \tag{9.46}$$

that is, to find $\bar{x} \in K$ for which there exists $\bar{y} \in \Psi^*(\bar{x})$ such that $\bar{y} \in \min_C \Psi^*(K)$, that is, $\Psi^*(\bar{x}) \cap \min_C \Psi^*(K) \neq \emptyset$. We denote by $\mathbb{S}(\text{DSVWVOP}_{\Psi^*})$ the solution set of the problem (9.46).

Proposition 9.6

(a) $\bar{x} \in K$ *is a solution of* MWVEP (9.4) *if and only if* $\mathbf{0} \in \Psi^*(\bar{x})$.
(b) *If the solution set of* MWVEP (9.4) *is nonempty, then* $\mathbb{S}(\text{MWVEP}(f, C)) = \mathbb{S}(\text{DSVWVOP}_{\Psi^*})$.

In view of the above proposition, we have the following characterization of the solutions for WVEP (9.1).

Corollary 9.3 *Assume that the assumptions of Corollary 9.2 hold.*

(a) $\bar{x} \in K$ *is a solution of* WVEP (9.1) *if and only if* $\mathbf{0} \in \Psi^*(\bar{x})$.
(b) *If the solution set of* WVEP (9.1) *is nonempty, then* $\mathbb{S}(\text{WVEP}(f, C)) = \mathbb{S}(\text{DSVWVOP}_{\Psi^*})$.

9.4.4 Variational Principle for WVEP(f, h)

Let $h : K \times K \to Y$ be a vector-valued bifunction such that

(i) $h(x, x) = \mathbf{0}$ for all $x \in K$,
(ii) $h(x, y) \succ_C \mathbf{0}$ for all $x \neq y$, $x, y \in K$.

Consider the following WVEP associated to the function $f + h$: find $\bar{x} \in K$ such that

$$f(\bar{x}, y) + h(\bar{x}, y) \nleq_C \mathbf{0}, \quad \text{for all } y \in K. \tag{9.47}$$

We denote by $\mathbb{S}(\text{WVEP}(f + h, C))$ the solution set of WVEP (9.47).

To obtain the equivalence of $\mathbb{S}(\text{WVEP}(f, C))$ and $\mathbb{S}(\text{WVEP}(f + h, C))$, we need the following additional condition on h:

(iii) For all $x, y \in K$ and $\alpha \in [0, 1]$, the function $\tau(\alpha) = h(x, \alpha y + (1 - \alpha)x)$ is homogenous with a degree $\sigma > 1$, that is, $\tau(\alpha) = \alpha^\sigma \tau(1)$.

Proposition 9.7 *Suppose that $f(x,\cdot)$ is C-convex for each $x \in K$ and h satisfies* (i)–(iii). *Then* $\mathbb{S}(\text{WVEP}(f, C)) = \mathbb{S}(\text{WVEP}(f + h, C))$,

Proof Suppose $\bar{x}$ solves WVEP (9.1). If $\bar{x} \notin \mathbb{S}(\text{WVEP}(f + h, C))$, then there is $y \in K$ such that

$$f(\bar{x}, y) + h(\bar{x}, y) <_C \mathbf{0},$$

hence that

$$f(\bar{x}, y) <_C -h(\bar{x}, y) <_C \mathbf{0},$$

a contradiction to the supposition that $\bar{x}$ solves WVEP (9.1).

Suppose $\bar{x}$ solves WVEP (9.47). Assume contrary that $\bar{x} \notin \mathbb{S}(\text{WVEP}(f, C))$, then there is $y \in K$ such that $f(\bar{x}, y) <_C \mathbf{0}$. Set $y_\alpha = \alpha y + (1-\alpha)\bar{x}$. For each $\alpha \in]0,1[$, we have

$$\begin{aligned} f(\bar{x}, y_\alpha) + h(\bar{x}, y_\alpha) &<_C \alpha f(\bar{x}, y) + (1-\alpha) f(\bar{x}, \bar{x}) + \alpha^\sigma h(\bar{x}, y) \\ &= \alpha f(\bar{x}, y) + \alpha^\sigma h(\bar{x}, y). \end{aligned}$$

Hence, there exists $\alpha' \in]0,1[$ such that

$$f(\bar{x}, y) + \alpha^{\sigma-1} h(\bar{x}, y) <_C \mathbf{0},$$

when $\alpha \in]0, \alpha'[$. It follows that

$$f(\bar{x}, y_\alpha) + h(\bar{x}, y_\alpha) \leq_{C_0} \alpha(f(\bar{x}, y) + \alpha^{\sigma-1} h(\bar{x}, y)) <_C \mathbf{0},$$

a contradiction, since $y_\alpha \in K$. □

By combining Theorem 9.3 and Proposition 9.7, we have the following existence result for a solution to WVEP (9.47).

Theorem 9.20 *Assume that the assumptions of Theorem 9.3 and Proposition 9.7 hold. Then there exists a solution to* WVEP (9.47).

Remark 9.7 In Banach spaces, it suffices to suppose that $\|\tau(\alpha)\| \leq \alpha^\sigma \|\tau(1)\|$.

In order to formulate variational principle, we define a set-valued map $G : K \to 2^Y$ by

$$G(x) := \text{w} - \min\{f(x, y) + h(x, y) : y \in K\},$$

where h is the same as defined above and can be termed as a perturbation bifunction. We associate to WVEP (9.47) with the following SVWVOP for set-valued map G:

$$\text{w} - \max_{x \in K} G(x). \tag{9.48}$$

We denote by $\mathbb{S}(\text{SVWVOPS}_G)$ the solution set of SVWVOP (9.48).

Using Theorem 9.19 with $f := f + h$, we obtain the following result.

Proposition 9.8

(a) $\bar{x} \in K$ *is a solution of* WVEP (9.47) *if and only if* $\mathbf{0} \in G(\bar{x})$.
(b) *If the solution set of* WVEP (9.47) *is nonempty, then* $\mathbb{S}(\text{WVEP}(f + h, C)) \subseteq \mathbb{S}(\text{SVWVOPS}_G)$.

By combining Propositions 9.7 and 9.8, we obtain the following result.

Theorem 9.21 *Assume that the assumptions of Proposition 9.7 hold.*

(a) $\bar{x} \in K$ *is a solution of* WVEP (9.47) *if and only if* $\mathbf{0} \in G(\bar{x})$.
(b) *If the solution set of* WVEP (9.47) *is nonempty, then* $\mathbb{S}(\text{WVEP}(f + h, C)) \subseteq \mathbb{S}(\text{SVWVOPS}_G)$.

Consider another SVVOP whose solution set coincides with the solution set of WVEP. Let Y be a locally convex space and C^* be the dual cone of C. In order to formulate second variational principle, we define set-valued maps $S : K \to 2^K$ and $\Phi : K \to 2^Y$ by

$$S(x) := \{y \in K : \sigma(f(x, y) + h(x, y)) \le \sigma(f(x, z) + h(x, z)) \text{ for all } z \in K\},$$

$$\Phi(x) := f(x, S(x)) + h(x, S(x)).$$

We now define the following SVVOP for set-valued map Φ:

$$\max_{x \in K} \Phi(x). \tag{9.49}$$

We denote by $\mathbb{S}(\text{SVVOPS}_\Phi)$ the solution set of problem (9.49).

Applying Theorem 9.19 with $f = f + h$, we obtain the following result.

Proposition 9.9

(a) $\bar{x} \in K$ *is a solution of* WVEP (9.47) *if and only if* $\mathbf{0} \in \Phi(\bar{x})$.
(b) *If the solution set of* WVEP (9.47) *is nonempty, then* $\mathbb{S}(\text{WVEP}(f + h, C)) = \mathbb{S}(\text{SVVOPS}_\Phi)$.

By combining Propositions 9.7 and 9.9, we have the following result.

Theorem 9.22 *Assume that the assumptions of Proposition 9.7 hold.*

(a) $\bar{x} \in K$ *is a solution of* WVEP (9.47) *if and only if* $\mathbf{0} \in \Phi(\bar{x})$.
(b) *If the solution set of* WVEP (9.47) *is nonempty, then* $\mathbb{S}(\text{WVEP}(f, C)) = \mathbb{S}(\text{SVVOPS}_\Phi)$.

9.5 Vectorial Form of Ekeland's Variational Principle

An existence result for an approximate minimizer of a lower semicontinuous and bounded below function is given by Ekeland in 1972 [39] (see also, [40, 41]), now known as Ekeland's variational principle (in short, EVP). It is appeared as one of the most useful tools to solve the problems in optimization, optimal control theory, game theory, nonlinear equations, dynamical systems, etc; See for example [14–16, 43, 57, 91] and the references therein. In [29, 83], Blum, Oettli and Théra provided the existence results for a solution of an equilibrium problem in the setting of complete metric spaces, and showed that their existence result is equivalent to Ekeland-type variational principle for bifunctions, Caristi-Kirk fixed point theorem for set-valued maps [33] and a maximal element theorem. For further details on Ekeland's variational principle for bifunctions, we refer [1, 5–7] and the references therein. Ansari [4], Araya et al. [13] and Bianchi et al. [24] extended the EVP for vector-valued bifunctions, see also [47]. By using the vectorial form of EVP, the existence of solutions of WVEP is studied by these authors. Araya [12] considered vectorial form of Ekeland's variational principle, vectorial form of Caristi's fixed point theorem and vectorial form of Takahashi's minimization theorem, and established the equivalence among these results. Some examples to illustrate these results are also presented in [12].

In this section, we first establish Ekeland-type variational principle for vector-valued bifunctions in the setting of complete metric spaces. Then by using this result, we derive the existence of solutions of weak vector equilibrium problems. Some equivalent results to the vectorial form of Ekeland-type variational principle are also presented.

9.5.1 *Vectorial Form of Ekeland-Type Variational Principle*

We present the following vectorial form of Ekeland-type variational principle for vector-valued bifunctions in the setting of complete metric spaces.

Theorem 9.23 *Let (X, d) be a complete metric space, Y be a locally convex Hausdorff topological vector space, C be a pointed, proper, closed and convex cone in Y with $\text{int}(C) \neq \emptyset$, and $e \in Y$ be a fixed vector such that $e \in \text{int}(C)$. Let $f : X \times X \to Y$ be a vector-valued bifunction such that the following conditions hold.*

(i) $f(x,x) = \mathbf{0}$ *for all* $x \in X$*;*
(ii) $f(x,y) + f(y,z) \in f(x,z) + C$ *for all* $x, y, z \in X$*;*
(iii) *For each fixed* $x \in X$*, the function* $f(x,\cdot) : X \mapsto Y$ *is* (e, C)*-lower semicontinuous and* C*-bounded below.*

Then for every $\varepsilon > 0$ *and for every* $\hat{x} \in X$, *there exists* $\bar{x} \in X$ *such that*

(a) $f(\hat{x}, \bar{x}) + \varepsilon d(\hat{x}, \bar{x})e \in -C$
(b) $f(\bar{x}, x) + \varepsilon d(\bar{x}, x)e \notin -C$ *for all* $x \in X$, $x \neq \bar{x}$.

Proof For the sake of convenience, we set $d_\varepsilon(x, y) = (1/\varepsilon)d(x, y)$. For all $x \in X$, define

$$S(x) = \{y \in X : x = y \text{ or } \xi_e\left(f(x, y)\right) + d_\varepsilon(x, y) \leq 0\}$$

and set

$$\mathcal{V}(x) := \inf_{y \in S(x)} \xi_e\left(f(x, y)\right),$$

where ξ_e is same as defined in Definition 2.30. Clearly, $x \in S(x)$, and so $S(x)$ is nonempty for all $x \in X$. Also, $\mathcal{V}(x) \leq 0$ for all $x \in X$. Since $f(x, \cdot)$ is (e, C)-lower semicontinuous and C-bounded below, Lemma 2.14 (a) implies that $\xi_e\left(f(x, \cdot)\right)$ is lower semicontinuous and bounded below. Since $d_\varepsilon(x, \cdot)$ is lower semicontinuous, $S(x)$ is closed for all $x \in X$.

Let $x_0 = \hat{x} \in X$. Since $\xi_e\left(f(x, \cdot)\right)$ is bounded below, we have

$$\mathcal{V}(x_0) = \inf_{y \in X} \xi_e\left(f(x_0, y)\right) > -\infty.$$

Let $m \in \mathbb{N}$ and assume that x_{m-1} has been defined with $\mathcal{V}(x_{m-1}) > -\infty$. Choose $x_m \in S(x_{m-1})$ such that

$$\xi_e\left(f(x_{m-1}, x_m)\right) \leq \mathcal{V}(x_{m-1}) + \frac{1}{m}.$$

Let $y \in S(x_m) \setminus \{x_m\}$, then

$$\xi_e\left(f(x_m, y)\right) + d_\varepsilon(x_m, y) \leq 0. \tag{9.50}$$

Since $x_m \in S(x_{m-1})$, we have

$$\xi_e\left(f(x_{m-1}, x_m)\right) + d_\varepsilon(x_{m-1}, x_m) \leq 0. \tag{9.51}$$

Adding (9.50) and (9.50), we obtain

$$\xi_e\left(f(x_{m-1}, x_m)\right) + \xi_e\left(f(x_m, y)\right) + d_\varepsilon(x_{m-1}, x_m) + d_\varepsilon(x_m, y) \leq 0.$$

Using the triangle inequality for metric d_ε, we obtain from above inequality

$$\xi_e(f(x_{m-1}, x_m)) + \xi_e(f(x_m, y)) + d_\varepsilon(x_{m-1}, y) \leq 0. \tag{9.52}$$

By condition (ii) and using Theorem 2.7 (e) and (f), we have

$$\xi_e\left(f(x_{m-1},y)\right) \leq \xi_e\left(f(x_{m-1},x_m)\right) + \xi_e\left(f(x_m,y)\right). \tag{9.53}$$

Combining (9.52) and (9.53), we obtain

$$\xi_e\left(f(x_{m-1},y)\right) + d_\varepsilon(x_{m-1},y) \leq 0,$$

and so $y \in S(x_{m-1})$ which implies that $S(x_m) \subseteq S(x_{m-1})$. Therefore, we obtain

$$\begin{aligned}\mathcal{V}(x_m) &= \inf_{y\in S(x_m)} \xi_e\left(f(x_m,y)\right) \geq \inf_{y\in S(x_{m-1})} \xi_e\left(f(x_m,y)\right)\\ &\geq \inf_{y\in S(x_{m-1})} \left[\xi_e\left(f(x_{m-1},y)\right) - \xi_e\left(f(x_{m-1},x_m)\right)\right]\\ &= \mathcal{V}(x_{m-1}) - \xi_e\left(f(x_{m-1},x_m)\right) \geq -\frac{1}{m}.\end{aligned}$$

Thus, for $y \in S(x_m) \setminus \{x_m\}$, from (9.50) and by definition of $\mathcal{V}$, we have

$$d_\varepsilon(x_m,y) \leq -\xi_e\left(f(x_m,y)\right) \leq -\mathcal{V}(x_m) \leq \frac{1}{m} \to 0 \text{ as } m \to \infty.$$

This shows that $d_\varepsilon(x_m,y) \to 0$ as $m \to \infty$. Since $x_m \in S(x_m)$, the diameter of $S(x_m)$, $\delta(S(x_m)) \to 0$ as $m \to \infty$. By Cantor's intersection theorem, there exists exactly one point $\bar{x} \in X$ such that $\bigcap_{m=0}^{\infty} S(x_m) = \{\bar{x}\}$.

This implies that $\bar{x} \in S(x_0) = S(\hat{x})$, that is,

$$\xi_e\left(f(\hat{x},\bar{x})\right) + d_\varepsilon(\hat{x},\bar{x}) \leq 0, \quad \text{that is,} \quad \xi_e\left(f(\hat{x},\bar{x})\right) \leq -d_\varepsilon(\hat{x},\bar{x}).$$

From Theorem 2.7 (g), we have

$$f(\hat{x},\bar{x}) \in -d_\varepsilon(\hat{x},\bar{x})e - C, \quad \text{that is,} \quad f(\hat{x},\bar{x}) + d_\varepsilon(\hat{x},\bar{x})e \in -C.$$

and so (a) holds.

Moreover, $\bar{x}$ also belongs to all $S(x_m)$ and, since $S(\bar{x}) \subseteq S(x_m)$ for all m, we have

$$S(\bar{x}) = \{\bar{x}\}.$$

It follows that $x \notin S(\bar{x})$ whenever $x \neq \bar{x}$ implying that

$$\xi_e\left(f(\bar{x},x)\right) + d_\varepsilon(\bar{x},x) > 0 \quad \text{or} \quad \xi_e\left(f(\bar{x},x)\right) > -d_\varepsilon(\bar{x},x).$$

From Theorem 2.7 (g), we have

$$f(\bar{x},x) \notin -d_\varepsilon(\bar{x},x)e - C,$$

that is,

$$f(\bar{x}, x) + d_\varepsilon(\bar{x}, x)e \notin -C, \quad \text{for all } x \in X \text{ and } x \neq \bar{x},$$

that is, (b) holds. □

Remark 9.8

(a) If (X, d) is a metric space (not necessarily, complete), $\preccurlyeq$ is a quasi-order on X defined as

$$x \preccurlyeq y \quad \text{if and only if} \quad x = y \text{ or } \xi_e\left(f(x, y)\right) + d_\varepsilon(x, y) \leq 0,$$

and the set $S(x) = \{y \in X : x \preccurlyeq y\}$ is $\preccurlyeq$-complete, even then the conclusion of Theorem 9.23 holds.

(b) If X is replaced by a nonempty closed subset K of X, even then the conclusion of Theorem 9.23 holds.

(c) The conclusion (b) of Theorem 9.23 implies that

$$f(\bar{x}, x) + \varepsilon d(\bar{x}, x)e \notin -\operatorname{int}(C), \quad \text{for all } x \in X.$$

Indeed, from (b) we have

$$f(\bar{x}, x) + \varepsilon d(\bar{x}, x)e \notin -\operatorname{int}(C), \quad \text{for all } x \in X, \ \ x \neq \bar{x}.$$

Suppose that $f(\bar{x}, \bar{x}) + \varepsilon d(\bar{x}, \bar{x})e \in -\operatorname{int}(C)$. Since $f(\bar{x}, \bar{x}) = \mathbf{0}$ by condition (i) of Theorem 9.23, $d(\bar{x}, \bar{x}) \geq 0$ and $e \in \operatorname{int}(C)$, we obtain $f(\bar{x}, \bar{x}) + \varepsilon d(\bar{x}, \bar{x})e \in \operatorname{int}(C)$, a contradiction of our supposition.

Corollary 9.4 *Let (X, d), Y, C, and e be the same as in Theorem 9.23 and let $\phi : X \to Y$ be a (e, C)-lower semicontinuous and C-bounded below function. For every given $\varepsilon > 0$, there is a $\hat{x} \in X$ such that $\phi(x) - \phi(\hat{x}) \notin \varepsilon e - C$ for all $x \in X$, then there exists $\bar{x} \in X$ such that*

(a) $\phi(x) - \phi(\bar{x}) \notin -\varepsilon e - C$ *for all* $x \in X$
(b) $\phi(x) - \phi(\bar{x}) + \varepsilon d(\bar{x}, x)e \notin -C$ *for all* $x \in X$, $x \neq \bar{x}$.

Proof Set $f(x, y) = \phi(y) - \phi(x)$ for all $x, y \in X$. Then all the conditions of Theorem 9.23 are satisfied and hence there exists $\bar{x} \in X$ such that

$$\phi(\bar{x}) - \phi(\hat{x}) + \varepsilon d(\hat{x}, \bar{x})e \in -C,$$

and

$$\phi(x) - \phi(\bar{x}) + \varepsilon d(\bar{x}, x)e \notin -C, \quad \text{for all } x \in X, \ \ x \neq \bar{x},$$

that is, (b) holds.

Since $d(\hat{x}, \bar{x}) \geq 0$ and $e \in \text{int}(C)$, we have $\varepsilon d(\hat{x}, \bar{x}) \in \text{int}(C)$. Then (9.5.1) implies that

$$\phi(\bar{x}) - \phi(\hat{x}) \in -\text{int}(C) - C \subseteq -\text{int}(C). \tag{9.54}$$

By hypothesis, $\hat{x}$ satisfies

$$\phi(x) - \phi(\hat{x}) \notin -\varepsilon e - C, \quad \text{for all } x \in X. \tag{9.55}$$

We claim that $\phi(x) - \phi(\bar{x}) \notin -\varepsilon e - C$, for all $x \in X$.

Suppose to the contrary that

$$\phi(x) - \phi(\bar{x}) \in -\varepsilon e - C, \quad \text{for some } x \in X. \tag{9.56}$$

Then from (9.54) and (9.56), we have

$$\phi(x) - \phi(\bar{x}) \in -\varepsilon e - \text{int}(C) - C \subseteq -\varepsilon e - C,$$

contradicting (9.55). Hence (a) holds. □

If $C = \mathbb{R}_+$ and $Y = \mathbb{R}$, then from Theorem 9.23, we have the following Ekeland's variational principle for bifunctions.

Corollary 9.5 *Let (X, d) be a complete metric space. Let $f : X \times X \to \mathbb{R}$ be a real-valued function such that the following conditions hold:*

(i) $f(x, x) = 0$ *for all* $x \in X$;
(ii) $f(x, z) \leq f(x, y) + f(y, z)$ *for all* $x, y, z \in X$;
(iii) *For each fixed* $x \in X$, *the function* $f(x, \cdot) : X \mapsto \mathbb{R}$ *is lower semicontinuous and bounded below.*

Then for every $\varepsilon > 0$ and for every $\hat{x} \in X$, there exists $\bar{x} \in X$ such that

(a) $f(\hat{x}, \bar{x}) + \varepsilon d(\hat{x}, \bar{x}) \leq 0$
(b) $f(\bar{x}, x) + \varepsilon d(\bar{x}, x) > 0, \quad$ *for all* $x \in X$, $x \neq \bar{x}$.

9.5.2 Existence of Solutions for Weak Vector Equilibrium Problems Via Vectorial Form of EVP

Throughout this subsection, unless otherwise specified, we assume that Y, C, and e are the same as in the previous subsection and (X, d) is a complete metric space.

Definition 9.11 Let $f : K \times K \to Y$ and $\varepsilon \in Y$ be given. A point $\bar{x} \in K$ is called an *ε-equilibrium point* of f if

$$f(\bar{x}, y) + \varepsilon d(\bar{x}, y) \notin -\text{int}(C), \quad \text{for all } y \in K.$$

Theorem 9.24 *Let K be a nonempty compact (not necessarily convex) subset of X, $f : K \times K \to Y$ satisfy conditions (i)–(iii) of Theorem 9.23 and for each fixed $y \in K$, the map $x \mapsto f(x, y)$ be (e, C)-upper semicontinuous. Then there exists a solution $\bar{x} \in K$ of* WVEP (9.1).

Proof By Theorem 9.23 along with Remark 9.8 (c), for each $m \in \mathbb{N}$, there exists $x_m \in K$ such that

$$f(x_m, y) + \frac{1}{m} d(x_m, y)e \notin -\operatorname{int}(C), \quad \text{for all } y \in K,$$

that is, for each $m \in \mathbb{N}$, $x_m \in K$ is an ε-equilibrium point of f for $\varepsilon = \frac{1}{m}e$. By Theorem 2.7 (j), we have

$$\xi_e\left(f(x_m, y)\right) + \frac{1}{m} d(x_m, y) \geq 0, \quad \text{for all } y \in K \text{ and all } m \in \mathbb{N}.$$

Since K is compact, we can choose a subsequence $\{x_{m_k}\}$ of $\{x_m\}$ such that $x_{m_k} \to \bar{x}$ as $m \to \infty$. Then by (e, C)-upper semicontinuity of $f(\cdot, y)$ on K, we have $\xi_e \circ f(\cdot, y)$ is upper semicontinuous and thus

$$\xi_e\left(f(\bar{x}, y)\right) \geq \limsup_{k \to \infty} \left(\xi_e\left(f(x_{m_k}, y)\right) + \frac{1}{m_k} d(x_{m_k}, y) \right) \geq 0, \quad \text{for all } y \in K.$$

Hence, again by Theorem 2.7 (j),

$$f(\bar{x}, y) \notin -\operatorname{int}(C), \quad \text{for all } y \in K$$

and thus $\bar{x}$ is a solution of WVEP (9.1). □

When K is not necessarily compact, we have the following existence result for a solution of WVEP (9.1).

Theorem 9.25 *Let $(X, \|\cdot\|)$ be a Banach space equipped with the weak topology, K a nonempty closed subset of X, $f : K \times K \to Y$ satisfy conditions* (i)–(iii) *of Theorem 9.23 and for each fixed $y \in K$, the map $x \mapsto f(x, y)$ be (e, C)-upper semicontinuous. Let the following coercivity condition hold:*

> *there exists $r > 0$ such that for all $x \in K \setminus K_r$, there exists $y \in K$ with $\|y\| < \|x\|$ satisfying $f(x, y) \in -C$, where $K_r = \{x \in K : \|x\| \leq r\}$.*

Then there exists a solution $\bar{x} \in K$ of WVEP (9.1).

Proof For all $x \in K$, define

$$S(x) = \{y \in K : \|y\| \leq \|x\| \text{ and } \xi_e\left(f(x, y)\right) \leq 0\}.$$

Then for all $x \in K$, $S(x) \neq \emptyset$, and for each $x, y \in K$, $y \in S(x)$ implies that $S(y) \subseteq S(x)$. Indeed, for $z \in S(y)$, we have $\|z\| \leq \|y\| \leq \|x\|$. Condition (iii) in

Theorem 9.23 implies that

$$\xi_e\left(f(x,z)\right) \le \xi_e\left(f(x,y)\right) + \xi_e\left(f(y,z)\right) \le 0.$$

Since $\xi_e \circ f(x,\cdot)$ is lower semicontinuous on K, $S(x)$ is closed for all $x \in K$. Also, since $K_{\|x\|}$ is weakly compact, $S(x)$ is weakly compact subset of $K_{\|x\|}$ for all $x \in K$. Then by Theorem 9.24, there exists $\bar{x}_r \in K_r$ such that

$$f(\bar{x}_r, y) \notin -\operatorname{int}(C), \quad \text{for all } y \in K_r. \tag{9.57}$$

Assume that there exists $x \in K$ such that $f(\bar{x}_r, x) \in -\operatorname{int}(C)$. Set $a = \min\limits_{y \in S(x)} \|y\|$ (the minimum is achieved because $S(x)$ is nonempty and weakly compact and the norm is continuous). We consider the following two cases:

CASE 1: When $(a \le r)$. Assume that $y_0 \in S(x)$ such that $\|y_0\| = a$. Then $\|y_0\| = a \le r$ and $\xi_e\left(f(x,y_0)\right) \le 0$. Since $f(\bar{x}_r, x) \in -\operatorname{int}(C)$, we have $\xi_e\left((\bar{x}_r, x)\right) < 0$ by Theorem 2.7 (j) and thus

$$\xi_e\left(f(\bar{x}_r, x)\right) + \xi_e\left(f(x, y_0)\right) < 0. \tag{9.58}$$

By condition (iii), we obtain

$$\xi_e\left(f(\bar{x}_r, y_0)\right) \le \xi_e\left(f(\bar{x}_r, x)\right) + \xi_e\left(f(x, y_0)\right). \tag{9.59}$$

Combining (9.58) and (9.59), we get

$$\xi_e\left(f(\bar{x}_r, y_0)\right) < 0 \quad \Rightarrow \quad f(\bar{x}_r, y_0) \in -\operatorname{int}(C)$$

contradicting (9.57).

CASE 2: When $(a > r)$. Assume that $y_0 \in S(x)$ such that $\|y_0\| = a$. Then $\|y_0\| = a > r$ and by coercivity condition we can choose an element $y_1 \in K$ such that $\|y_1\| < \|y_0\| = a$ and satisfying $f(y_0, y_1) \in -C$, that is, $\xi_e\left(f(y_0, y_1)\right) \le 0$. Therefore, $y_1 \in S(y_0) \subseteq S(x)$ contradicting $\|y_1\| < a = \min\limits_{y \in S(x)} \|y\|$. Thus, there is no $x \in K$ such that $F(\bar{x}_r, x) \in -\operatorname{int}(C)$, that is, $\bar{x}_r$ is a solution of WVEP (9.1). □

The following results can be easily derived from Theorems 9.24 and 9.25, respectively, by taking $f(x,y) = \phi(y) - \phi(x)$ for all $x, y \in K$, where $\phi : K \to Y$ be a vector-valued function. These results are the vectorial form of the Weierstrass existence theorem.

Corollary 9.6 *Let K be a nonempty compact subset of X and $\phi : K \to Y$ be (e,C)-lower semicontinuous and C-bounded below. Then there exists $\bar{x} \in K$ such that $\phi(y) - \phi(\bar{x}) \notin -\operatorname{int}(C)$ for all $y \in K$.*

Corollary 9.7 *Let $(X, \|\cdot\|)$ be a Banach space equipped with the weak topology, K a nonempty closed subset of X and $\phi : K \to Y$ be (e, C)-lower semicontinuous and C-bounded below. Assume that the following coercivity condition holds:*

there exists $r > 0$ such that for all $x \in K \setminus K_r$, there exists $y \in K$ with $\|y\| < \|x\|$ satisfying $\phi(y) - \phi(x) \in -C$, where $K_r = \{x \in K : \|x\| \leq r\}$.

Then there exists $\bar{x} \in K$ such that $\phi(y) - \phi(\bar{x}) \notin -\operatorname{int}(C)$ for all $y \in K$.

Definition 9.12 We say that $x_0 \in X$ satisfies *Condition (A)* if and only if every sequence $\{x_m\} \subseteq X$ satisfying $f(x_0, x_m) \in -C$ for all $m \in \mathbb{N}$ and $f(x_m, x) + \frac{1}{m}d(x_m, x)e \notin -\operatorname{int}(C)$ for all $x \in X$ and all $m \in \mathbb{N}$, has a convergent subsequence.

Theorem 9.26 *Let (X, d) be a complete metric space, and $f : X \times X \to Y$ satisfy condition* (i)–(iii) *of Theorem 9.23 and (e, C)-upper semicontinuous in the first argument. If some $x_0 \in X$ satisfies Condition (A), then there exists $\bar{x} \in X$ such that $f(\bar{x}, x) \notin -\operatorname{int}(C)$ for all $x \in X$.*

Proof From Theorem 9.23 along with Remark 9.8 (c), for each $m \in \mathbb{N}$, there exists $x_m \in X$ such that

$$f(x_m, x) + \frac{1}{m}d(x_m, x)e \notin -\operatorname{int}(C), \quad \text{for all } x \in X, \tag{9.60}$$

and

$$f(\hat{x}, x_m) + \frac{1}{m}d(\hat{x}, x_m)e \in -C. \tag{9.61}$$

In view of Theorem 2.7 (g), (9.60) and (9.60) can be rewritten, respectively, as

$$\xi_e\left(f(x_m, x)\right) \geq -\frac{1}{m}d(x_m, x), \quad \text{for all } x \in X, \tag{9.62}$$

and

$$\xi_e\left(f(\hat{x}, x_m)\right) \leq -\frac{1}{m}d(\hat{x}, x_m). \tag{9.63}$$

Since $d(\hat{x}, x_m) \geq 0$, we have

$$\xi_e\left(f(\hat{x}, x_m)\right) \leq 0 \quad \Leftrightarrow \quad f(\hat{x}, x_m) \in -C, \quad \text{for all } m \in \mathbb{N}.$$

From Condition (A), there exists a subsequence of $\{x_m\}$ converges to some $\bar{x} \in X$. Then by using the upper semicontinuity of $\xi_e(f(\cdot, x))$ and (9.62), we obtain

$$\xi_e\left(f(\bar{x}, x)\right) \geq 0, \quad \text{for all } x \in X.$$

Again by applying Theorem 2.7 (j), we get

$$f(\bar{x}, x) \notin -\operatorname{int}(C), \quad \text{for all } x \in X.$$

This completes the proof. □

Remark 9.9 If we replace X by a nonempty closed subset K of X in Definition 9.12 and Theorem 9.26, then the conclusion of Theorem 9.26 also holds and gives the existence of a solution of WVEP.

9.5.3 Some Equivalences

We establish some equivalences among Ekeland-type variational principle for vector-valued bifunctions, existence of solutions for WVEP, Caristi-Kirk type fixed point theorem, and Oettli and Théra type theorem.

Theorem 9.27 *Let (X, d), Y, C, and e be the same as in Theorem 9.23. Let $f : X \times X \to Y$ be a vector-valued bifunction satisfying the conditions* (i)–(iii) *of Theorem 9.23. Then the following statements are equivalent:*

(a) (Vectorial Form of Ekeland-type Variational Principle). *For every $\hat{x} \in X$, there exists $\bar{x} \in X$ such that*

$$\bar{x} \in \hat{S} := \{x \in X : f(\hat{x}, x) + d(\hat{x}, x)e \in -C,\ x \neq \hat{x}\}$$

and

$$f(\bar{x}, x) + d(\bar{x}, x)e \notin -C, \quad \textit{for all } x \in X \textit{ and } x \neq \bar{x}. \tag{9.64}$$

(b) (Existence of Solutions for WVEP). *Assume that*

$$\begin{cases} \textit{for every } \tilde{x} \in \hat{S} \textit{ with } f(\tilde{x}, y) \in -\operatorname{int}(C) \textit{ for all } y \in X, \textit{ there exists } x \in X \\ \textit{such that } x \neq \tilde{x} \textit{ and } f(\tilde{x}, x) + d(\tilde{x}, x)e \in -C. \end{cases} \tag{9.65}$$

Then there exists $\bar{x} \in \hat{S}$ such that $f(\bar{x}, x) \notin -\operatorname{int}(C)$ for all $x \in X$.

(c) (Caristi-Kirk type Fixed Point Theorem). *Let $\Phi : X \to 2^X$ be a set-valued map such that*

$$\begin{cases} \textit{for every } \tilde{x} \in \hat{S}, \textit{ there exists } x \in \Phi(\tilde{x}) \textit{ satisfying} \\ f(\tilde{x}, x) + d(\tilde{x}, x)e \in -C. \end{cases} \tag{9.66}$$

Then there exists $\bar{x} \in \hat{S}$ such that $\bar{x} \in \Phi(\bar{x})$.

(d) (OETTLI AND THÉRA TYPE THEOREM) *Let D be a subset of X such that*

$$\begin{cases} \text{for every } \tilde{x} \in \hat{S} \setminus D, \text{ there exists } x \in X \\ \text{such that } x \neq \tilde{x} \text{ and } F(\tilde{x}, x) + d(\hat{x}, x)e \in -C. \end{cases} \tag{9.67}$$

Then there exists $\bar{x} \in \hat{S} \cap D$.

Proof (a) $\Rightarrow$ (d): Let (a) and the hypothesis of (d) hold. Then (a) gives $\bar{x} \in \hat{S}$ such that

$$f(\bar{x}, x) + d(\bar{x}, x)e \notin -C, \quad \text{for all } x \in X \text{ and } x \neq \bar{x}.$$

From (9.67), we have $\bar{x} \in D$. Hence $\bar{x} \in \hat{S} \cap D$, and (d) holds.

(d) $\Rightarrow$ (a): Let (d) hold. For all $\hat{x} \in X$, define

$$\Gamma(\hat{x}) = \{x \in X : f(\hat{x}, x) + d(\hat{x}, x)e \in -C, \ x \neq \hat{x}\}.$$

Choose $D := \{\hat{x} \in X : \Gamma(\hat{x}) = \emptyset\}$. If $\hat{x} \notin D$, then from the definition of D, there exists $x \in \Gamma(\hat{x})$. That is, for $\hat{x} \notin D$, there exists $x \in X$ such that

$$x \neq \hat{x} \text{ and } f(\hat{x}, x) + d(\hat{x}, x)e \in -C.$$

Hence (9.67) is satisfied, and by (d), there exists $\bar{x} \in \hat{S} \cap D$. Then $\Gamma(\bar{x}) = \emptyset$, that is, $F(\bar{x}, x) + d(\bar{x}, x)e \notin -C$ for all $x \neq \bar{x}$. Hence (a) holds.

(b) $\Rightarrow$ (d): Suppose that both (b) and the hypothesis of (d) hold. Assume, for contradiction, that $\tilde{x} \notin D$ for all $\tilde{x} \in \hat{S}$ satisfying $f(\tilde{x}, y) \in -\operatorname{int}(C)$ for all $y \in K$. Then by (9.67), for all $\tilde{x} \in \hat{S}$

$$\text{there exists } x \in X \text{ such that } x \neq \tilde{x} \text{ and } f(\tilde{x}, x) + d(\tilde{x}, x)e \in -C. \tag{9.68}$$

Hence (9.65) is satisfied and by (ii), there exists $\bar{x} \in \hat{S}$ such that

$$f(\bar{x}, x) \notin -\operatorname{int}(C), \quad \text{for all } x \in X. \tag{9.69}$$

We claim that $f(\bar{x}, x) + d(\bar{x}, x)e \notin -C$ for all $x \in X$, $x \neq \bar{x}$ which leads to a contradiction of (9.68). Assume, contrary that, there exists $x \in X$ such that $x \neq \bar{x}$ and

$$f(\bar{x}, x) + d(\bar{x}, x)e \in -C, \quad \text{that is,} \quad f(\bar{x}, x) \in -d(\bar{x}, x)e - C \tag{9.70}$$

Since $e \in \operatorname{int}(C)$ and $d(\bar{x}, x) \geq 0$, we have

$$d(\bar{x}, x)e \in \operatorname{int}(C) \tag{9.71}$$

Combining (9.70) and (9.71), we obtain

$$F(\bar{x}, x) \in -\operatorname{int}(C) - C \subseteq -\operatorname{int}(C)$$

a contradiction of (9.69).

(d) $\Rightarrow$ (b): Suppose that both (d) and the hypothesis of (b) hold. Choose $D := \{\tilde{x} \in X : f(\tilde{x}, y) \notin -\operatorname{int}(C)$ for all $y \in X\}$. Then by hypothesis (9.65), for every $\tilde{x} \in \hat{S}$ with $f(\tilde{x}, y) \notin -\operatorname{int}(C)$ for all $y \in X$, there exists $x \in X$ such that $x \neq \tilde{x}$ and $f(\tilde{x}, x) + d(\tilde{x}, x)e \in -C$, that is, for every $\tilde{x} \in \hat{S} \setminus D$, there exists $x \in X$ such that $x \neq \tilde{x}$ and $f(\tilde{x}, x) + d(\hat{x}, x)e \in -C$. Then by (d), there exists $\bar{x} \in \hat{S} \cap D$. This implies that $\bar{x} \in \hat{S}$ and $f(\bar{x}, y) \notin -\operatorname{int}(C)$ for all $y \in X$. Hence (b) holds.

(c) $\Rightarrow$ (d): Let (c) and the hypothesis of (d) hold. Define a set-valued map $\Phi : X \to 2^X$ by

$$\Phi(\tilde{x}) = \{x \in X : x \neq \tilde{x}\}.$$

Assume, for contradiction, that $\tilde{x} \notin D$ for all $\tilde{x} \in \hat{S}$. By (9.67), for every $\tilde{x} \in \hat{S} \setminus D$, there exists $x \in X$ such that $x \neq \tilde{x}$ and $f(\tilde{x}, x) + d(\hat{x}, x)e \in -C$, that is, for every $\tilde{x} \in \hat{S}$, there exists $x \in \Phi(\tilde{x})$ satisfying $f(\tilde{x}, x) + d(\hat{x}, x)e \in -C$. Then (c) implies that there exists $\bar{x} \in \hat{S}$ such that $\bar{x} \in \Phi(\bar{x})$. But this is clearly impossible from the definition of Φ. Hence $\tilde{x} \in D$ for some $\tilde{x} \in \hat{S}$, and (d) holds.

(d) $\Rightarrow$ (c): Suppose that both (d) and the hypothesis of (c) hold. Choose $D := \{\tilde{x} \in X : \tilde{x} \in \Phi(\tilde{x})\}$. By (9.66), for every $\tilde{x} \in \hat{S}$, there exists $x \in \Phi(\tilde{x})$ satisfying $f(\tilde{x}, x) + d(\tilde{x}, x)e \in -C$, that is, for every $\tilde{x} \in \hat{S} \setminus D$, there exists $x \in X$ such that $x \neq \tilde{x}$ satisfying $f(\tilde{x}, x) + d(\tilde{x}, x)e \in -C$. Then by (d) furnishes some $\bar{x} \in \hat{S} \cap D$. From the definition of D, we have $\bar{x} \in \Phi(\bar{x})$. Hence (c) holds. □

9.6 Sensitivity Analysis of Vector Equilibrium Problems

In this section, we present some results on parametric optimization for equilibrium problems. Parametric optimization investigates how changes in the data change the optimal solution and the value of the optimal solution. This is especially important as most data of optimization problems are indeed uncertain (for instance due to measurement errors, numerical rounding errors, etc). Then one would like to know how small changes in the data affect the obtained optimal solution. Two main questions arising in this regard are about continuity properties of certain maps, which give insight into the problem's stability, and differentiability properties, which can describe the sensitivity of the solution. As an example of parametric scalar optimization problems, consider the weighted sum scalarization with parameters w in Chap. 4.

Let Λ, X be topological vector spaces. Moreover, let a set-valued map $A : \Lambda \to 2^X$ and a real-valued function $f : \Lambda \times X \to \mathbb{R}$ be given. A parametric scalar optimization is given as

$$\begin{aligned} &\text{minimize } f(t,x), \\ &\text{subject to } x \in A(t). \end{aligned} \tag{$P_1(t)$}$$

Then $\varphi(t) := \inf_{x \in A(t)} f(t,x)$ is called *optimal value function* and the set $S(t) := \{x \in A(t) : f(t,x) = \varphi(t,x)\}$ is the *optimal set mapping*. A stability analysis aims at the question whether the solution of the original problem ($P_1(t)$) for some fixed $t \in \Lambda$ depends on the input data. Sensitivity analysis, on the other hand, tries to answer the question how small changes in the input data affect the optimal solution. Although the optimal value function $\varphi(t)$ is often not differentiable at all $t \in \Lambda$, sometimes directional differentiability can be shown. For a detailed analysis to parametric optimization, we refer to [58].

For a set-valued setting, let Λ, X be topological vector spaces and Y be a Hausdorff topological vector space. Furthermore, let the set-valued maps $A : \Lambda \to 2^X$ and $F : \Lambda \times X \to 2^Y$ be given. A parametric set optimization problem is given by

$$\begin{aligned} &\text{minimize } F(t,x), \\ &\text{subject to } x \in A(t). \end{aligned} \tag{$P_2(t)$}$$

For an overview on sensitivity analysis for set-valued optimization problems, see, for instance, [65, Chap. 13].

9.6.1 ε-Weak Vector Equilibrium Problems

The concept of ε-solution is very adaptable in the cases where feasible regions are nonconvex or nonclosed sets. In fact, the original problems are the special cases of ε-approximate problems such as the famous Ekeland's variational principle, which is an ε-solution rule for optimization problem. The concept of ε-solution is also the basis of numerical computing, for example, stability, well-posedness and so on. The interesting example of the ε-equilibrium problem is the generalized game for ε-strategy in economics.

The notion of approximate solutions adapted in this subsection follows from the concept of ε-efficiency originally introduced in multiple objective programming by Loridan [77]. Later, White [98] introduced six alternative definitions of ε-efficient solutions and established the relationships between these concepts. ε-efficiency for more general vector optimization problems are considered in [77, 82]. Tammer [92, 93] studied the existence of ε-solution for (vector) variational inequality problem, and generalized Ekeland's variational principle.

Kimura et al. [68] considered an ε-weak vector equilibrium problem (in short, ε-WVEP) and studied several existence results for its solution. They also investigated continuity properties (upper semicontinuity and lower semicontinuity) of the solution map for ε-WVEP. Qiu and Yang [85] studied a more general form of ε-WVEP and proved the equivalence between the solution sets of WVEP (9.1) and their ε-WVEP. Wei-zhong et al. [97] considered ε-WVEP with constraints and obtained some necessary and sufficient conditions for its solution.

Since the vector equilibrium problem is a very general mathematical model covering vector optimization, vector variational inequalities and so on as special cases, the main motivation of this subsection is to consider ε-WVEP and present the existence results for its solution. We also study the behavior (upper semicontinuity and lower semicontinuity) of the solution map of the ε-WVEP.

We observe that the results in this subsection can be employed to study the behavior of solution maps of ε-vector optimization problems, ε-vector variational inequalities, ε-generalized games and so on.

Throughout this subsection, unless otherwise specified, let X be a Hausdorff topological vector space and Y be a topological vector space. Let K be a nonempty subset of X and $C \subset Y$ be a solid, pointed convex cone. Let $f : X \times X \to Y$ be a vector-valued bifunction. For fixed $\varepsilon \in \text{int}(C)$, the *$\varepsilon$-weak vector equilibrium problem* (in short, ε-WVEP) is to find $\bar{x} \in K$ such that

$$f(\bar{x}, y) + \varepsilon \notin -\text{int}(C), \quad \text{for all } y \in K. \tag{9.72}$$

The set of solutions of the ε-WVEP (9.72) is denoted by $\mathbb{S}(\text{WVEP}(K, f, \varepsilon))$.

Let $\Omega : \text{int}(C) \to 2^X$ be a set-valued map such that $\Omega(\varepsilon)$ is the solutions set of ε-WVEP for $\varepsilon \in \text{int}(C)$, that is,

$$\Omega(\varepsilon) = \{x \in K : f(x, y) + \varepsilon \notin -\text{int}(C) \text{ for all } y \in K\}.$$

Qiu and Yang [85] considered the following *ε-weak vector equilibrium problem:* For each $\varepsilon \in \text{int}(C)$ and $e \geq 0$, find $\bar{x} \in K$ such that

$$f(\bar{x}, y) + \varepsilon e \notin -\text{int}(C), \quad \text{for all } y \in K. \tag{9.73}$$

We denote by $\mathbb{S}(\text{WVEP}(K, f, \varepsilon, e))$ the set of solutions of the ε-WVEP (9.73).

If $e = 1$, then ε-WVEP (9.73) reduces to ε-WVEP (9.72). When $e = 0$, then ε-WVEP (9.73) becomes WVEP (9.1).

Proposition 9.10 *[85] Let $\varepsilon \in \text{int}(C)$ be a fixed vector. Then*

$$\bigcap_{e>0} \mathbb{S}(\text{WVEP}(K, f, \varepsilon, e)) = \mathbb{S}(\text{WVEP}(K, f)).$$

Proof Let $\bar{x} \in \mathbb{S}(\text{WVEP}(K, f))$, then

$$f(\bar{x}, y) \notin -\text{int}(C), \quad \text{for all } y \in K. \tag{9.74}$$

If there exists $e_0 > 0$ such that $\bar{x} \notin \mathbb{S}(\text{WVEP}(K,f,\varepsilon,e_0))$, then there is $y_0 \in K$ such that $f(\bar{x},y_0) + \varepsilon e_0 \in -\operatorname{int}(C)$, that is, $f(\bar{x},y_0) \in -\operatorname{int}(C) - \varepsilon e_0 \subseteq -\operatorname{int}(C)$ which contradicts (9.74). Hence, $\bar{x} \in \bigcap_{e>0} \mathbb{S}(\text{WVEP}(K,f,\varepsilon,e))$.

Now we prove that $\bigcap_{e>0} \mathbb{S}(\text{WVEP}(K,f,\varepsilon,e)) \subseteq \mathbb{S}(\text{WVEP}(K,f))$. Suppose that $\bar{x} \in \bigcap_{e>0} \mathbb{S}(\text{WVEP}(K,f,\varepsilon,e))$, but $\bar{x} \notin \mathbb{S}(\text{WVEP}(K,f))$. Then there is $\hat{y} \in K$ such that $f(\bar{x},\hat{y}) \in -\operatorname{int}(C)$. Consequently, there exists a neighborhood U of $\mathbf{0}$ such that $f(\bar{x},\hat{y}) + U \subseteq -\operatorname{int}(C)$. Since $\varepsilon \in \operatorname{int}(C)$, there is $e_0 > 0$ such that $\varepsilon\, e_0 \in U$. Then we have

$$f(\bar{x},\hat{y}) + \varepsilon\, e_0 \in -\operatorname{int}(C),$$

that is, $\bar{x} \notin \mathbb{S}(\text{WVEP}(K,f,\varepsilon,e_0))$ which contradicts our supposition. This completes the proof. □

We remark that ε-WVEP is closely related to the following weak vector equilibrium problem (in short, WVEP) which is to find $\bar{x} \in \operatorname{cl}(K)$ such that

$$f(x,y) \notin -\operatorname{int}(C), \quad \text{for all } y \in K. \tag{9.75}$$

Let $\mathbb{S}(\text{WVEP}(\operatorname{cl}(K),f))$ denote the solution set of WVEP (9.75), that is,

$$\mathbb{S}(\text{WVEP}(\operatorname{cl}(K),f)) = \{x \in \operatorname{cl}(K) : f(x,y) \notin -\operatorname{int}(C) \text{ for all } y \in K\}.$$

If K is closed, then WVEP (9.75) reduces to WVEP (9.1) studied in Sect. 9.1. We may regard solutions of ε-WVEP as approximate solutions of WVEP (9.75). We remark that $\mathbb{S}(\text{WVEP}(\operatorname{cl}(K),f)) \neq \emptyset$ does not imply $\Omega(\varepsilon) \neq \emptyset$ for all $\varepsilon \in \operatorname{int}(C)$.

Example 9.6 Let $X = \mathbb{R}$, $Y = \mathbb{R}^2$, $C = \mathbb{R}^2_+$, and $K = \left(0, \frac{\pi}{2}\right)$. Let $f : X \times X \to Y$ be defined by

$$f(x,y) = \left(-|x \cdot \tan y|, -|x^2 \cdot \tan y|\right).$$

Then $0 \in \mathbb{S}(\text{WVEP}(\operatorname{cl}(K),f))$ but $\Omega(\varepsilon) = \emptyset$ for each $\varepsilon > 0$.

Kimura et al. [68] established several existence results for solutions of ε-weak vector equilibrium problems (in short, ε-WVEP). First we derive that $\Omega(\varepsilon)$ is nonempty for $\varepsilon \in \operatorname{int}(C)$ under suitable conditions.

Theorem 9.28 *Let K be a nonempty subset of X such that $\operatorname{cl}(K)$ is compact, and $f : X \times X \to Y$ be C-lower semicontinuous on $X \times X$. Assume that $\mathbb{S}(\text{WVEP}(\operatorname{cl}(K),f))^* := \{x \in \operatorname{cl}(K) : f(x,y) \notin -\operatorname{int}(C)$ for all $y \in \operatorname{cl}(K)\} \neq \emptyset$. Then ε-WVEP has at least one solution for each $\varepsilon \in \operatorname{int}(C)$.*

Proof Let $\varepsilon \in \operatorname{int}(C)$ and $x \in \mathbb{S}(\text{WVEP}(\operatorname{cl}(K),f))^*$. Then by C-lower semicontinuity of f, for each $y \in \operatorname{cl}(K)$, there are neighborhoods $\mathcal{U}_y$ of x and $\mathcal{V}_y$ of y such that

$$f(u,v) \in (f(x,y) - \varepsilon) + \operatorname{int}(C), \quad \text{for all } (u,v) \in \mathcal{U}_y \times \mathcal{V}_y.$$

Since $\bigcup_{y \in \mathrm{cl}(K)} \mathcal{V}_y \supset \mathrm{cl}(K)$ and $\mathrm{cl}(K)$ is compact, we can choose $y_i \in \mathrm{cl}(K)$, $i = 1, 2, \ldots, m$, such that $\bigcup_{i=1}^{m} \mathcal{V}_{y_i} \supset \mathrm{cl}(K)$. Then for $\mathcal{U} := \bigcap_{i=1}^{m} \mathcal{U}_{y_i}$, we have

$$f(u, y) \in \bigcup_{i=1}^{m} ((f(x, y_i) - \varepsilon) + \mathrm{int}(C)), \quad \text{for all } u \in \mathcal{U} \text{ and } y \in \mathrm{cl}(K).$$

Hence,

$$f(u, y) + \varepsilon \in \bigcup_{i=1}^{m} (f(x, y_i) + \mathrm{int}(C)), \quad \text{for all } u \in \mathcal{U} \text{ and } y \in \mathrm{cl}(K).$$

Since $x \in \mathbb{S}(\mathrm{WVEP}(\mathrm{cl}(K), f))^*$ and $y_1, y_2, \ldots, y_m \in \mathrm{cl}(K)$, we have

$$(f(x, y_i) + \mathrm{int}(C)) \cap (-\mathrm{int}(C)) = \emptyset, \quad \text{for all } i = 1, 2, \ldots, m,$$

from which it follows that

$$\left(\bigcup_{i=1}^{m} (f(x, y_i) + \mathrm{int}(C)) \right) \cap (-\mathrm{int}(C)) = \emptyset.$$

Consequently,

$$f(u, y) + \varepsilon \notin -\mathrm{int}(C), \quad \text{for all } u \in \mathcal{U} \text{ and } y \in \mathrm{cl}(K).$$

Moreover, $K \cap \mathcal{U} \neq \emptyset$ because $x \in \mathrm{cl}(K)$. Let $\bar{x} \in K \cap \mathcal{U}$. Then $f(\bar{x}, y) + \varepsilon \notin -\mathrm{int}(C)$ for all $y \in \mathrm{cl}(K)$. In particular, $\bar{x} \in \Omega(\varepsilon)$. Therefore, ε-WVEP has at least one solution. □

Remark 9.10 Theorem 9.28 says that if K is a nonempty compact subset of X and $f : K \times K \to Y$ is C-lower semicontinuous, then every solution of WVEP (9.1) is a solution of ε-WVEP (9.72) for each $\varepsilon \in \mathrm{int}(C)$.

Example 9.7 Let $f : \mathbb{R} \times \mathbb{R} \to \mathbb{R}^2$ be defined by

$$f(x, y) = \begin{cases} \left(\dfrac{y}{x}, -1 \right), & \text{if } x \geq y > 0, \\ \left(\dfrac{x}{y}, -2 \right), & \text{if } y > x > 0, \\ (x + y, -3), & \text{if } 0 > x + y, \\ (0, -4), & \text{otherwise}, \end{cases}$$

$C = \mathbb{R}^2_+$ and $K = (-1, 1)$. Then $1 \in \mathbb{S}(\mathrm{WVEP}(\mathrm{cl}(K), f))^*$, that is, $\mathbb{S}(\mathrm{WVEP}(\mathrm{cl}(K), f))^* \neq \emptyset$, $\mathrm{cl}(K)$ is compact, and f is C-lower semicontinuous on $\mathbb{R} \times \mathbb{R}$. Thus, ε-WVEP has at least one solution for each $\varepsilon > 0$ by Theorem 9.28. Actually, $\bar{x} = 1 - \frac{\varepsilon}{2}$ is a solution of ε-WVEP.

The following example shows that the assumption 'cl (K) is compact' is essential along with other assumptions in Theorem 9.28.

Example 9.8 Let $f : \mathbb{R} \times \mathbb{R} \to \mathbb{R}^2$ be defined by $f(x, y) = (-1, -|x - y|)$, $K = [-1, 1]$ and $C = \mathbb{R}^2_+$. Then $\mathbb{S}(\text{WVEP}(\text{cl}\,(K), f))^* = \emptyset$ and also for each $\varepsilon \in (0, 1)$, $\Omega(\varepsilon) = \emptyset$.

Corollary 9.8 *Let K be a nonempty subset of X such that* cl (K) *is compact, and $f : X \times X \to Y$ be C-lower semicontinuous on $X \times X$ such that for some $x \in \mathbb{S}(\text{WVEP}(\text{cl}(K), f))$, $f(x, \cdot)$ is C-upper semicontinuous on* bd (K), *where* $\mathbb{S}(\text{WVEP}(\text{cl}\,(K), f)) := \{x \in \text{cl}\,(K) : f(x, y) \notin -\text{int}(C) \text{ for all } y \in K\} \neq \emptyset$. *Then* ε-WVEP *has at least one solution for each* $\varepsilon \in \text{int}(C)$.

Proof Let $\bar{x} \in \mathbb{S}(\text{WVEP}(\text{cl}(K), f))$ such that $f(\bar{x}, \cdot)$ is C-upper semicontinuous on bd (K). Then $\bar{x} \in \text{cl}\,(K)$ and

$$f(\bar{x}, y) \notin -\text{int}(C), \quad \text{for all } y \in K. \tag{9.76}$$

We want to show that $\bar{x} \in \mathbb{S}(\text{WVEP}(\text{cl}\,(K), f))^*$. Suppose to the contrary that there exists $\hat{y} \in \text{bd}\,(K)$ such that

$$f(\bar{x}, \hat{y}) \in -\text{int}(C).$$

By C-upper semicontinuity of $f(\bar{x}, \cdot)$ at $\hat{y} \in \text{bd}\,(K)$, there exists a neighborhood $\mathcal{V}$ of $\hat{y}$ such that

$$f(\bar{x}, v) \in \left(\frac{f(\bar{x}, \hat{y})}{2} - \text{int}(C)\right), \quad \text{for all } v \in \mathcal{V}.$$

Since $\hat{y} \in \text{bd}\,(K)$, $\mathcal{V} \cap K \neq \emptyset$. Therefore, there exists $y' \in \mathcal{V} \cap K \neq \emptyset$ such that

$$f(\bar{x}, y') \in \left(\frac{f(\bar{x}, \hat{y})}{2} - \text{int}(C)\right) \subset -\text{int}(C).$$

This contradicts to (9.76). Hence,

$$f(\bar{x}, y) \notin -\text{int}(C), \quad \text{for all } y \in \text{bd}\,(K),$$

that is,

$$f(\bar{x}, y) \notin -\text{int}(C), \quad \text{for all } y \in \text{cl}\,(K).$$

Therefore, $\bar{x} \in \mathbb{S}(\text{WVEP}(\text{cl}(K), f))^*$, that is, $\mathbb{S}(\text{WVEP}(\text{cl}(K), f))^* \neq \emptyset$. The result follows from Theorem 9.28. □

Theorem 9.29 *Let K be a nonempty subset of X such that* $\mathrm{cl}\,(K)$ *is compact convex, and $f : X \times X \to Y$ be C-lower semicontinuous on $X \times X$ such that $f(x,x) \notin -\mathrm{int}(C)$ for all $x \in X$. Assume that the following conditions hold:*

(i) *$f(x,\cdot)$ is C-quasiconvex on X for each $x \in X$;*
(ii) *$f(\cdot,y)$ is C-upper semicontinuous on X for each $y \in X$.*

Then ε-WVEP *has at least one solution, that is, $\Omega(\varepsilon)$ is nonempty for each $\varepsilon \in$* $\mathrm{int}(C)$.

Proof For each $y \in \mathrm{cl}\,(K)$, let

$$G(y) := \{x \in \mathrm{cl}\,(K) : f(x,y) \notin -\mathrm{int}(C)\}.$$

First, we show that $G(y)$ is a KKM-map. Suppose to the contrary that there exist $\lambda_i \in [0,1]$, $x_i \in \mathrm{cl}\,(K)$, for $i = 1,2,\ldots,m$, such that

$$\sum_{i=1}^{m} \lambda_i x_i =: x \notin \bigcup_{i=1}^{m} G(x_i).$$

Then,

$$f(x,x_i) \in -\mathrm{int}(C), \quad \text{for all } i = 1,2,\ldots,m.$$

Moreover, $x \in \mathrm{cl}\,(K)$ because of the convexity of $\mathrm{cl}\,(K)$. Hence, by C-quasiconvexity of $f(x,\cdot)$, we have

$$f(x,x) = f\left(x, \sum_{i=1}^{m} \lambda_i x_i\right) \in -\mathrm{int}(C),$$

which contradicts to the fact that $f(x,x) \notin -\mathrm{int}(C)$ for all $x \in X$.

By C-upper semicontinuity of f on X and Proposition 2.22, $A := \{y \in X : f(x,y) \in -\mathrm{int}(C)\}$ is an open subset of X. Then $G(y) = \mathrm{cl}\,(K) \cap (A^{c})$ is a closed subset of X. Therefore, $G(y)$ is closed for each $y \in K$. Since $\mathrm{cl}\,(K)$ is compact, $G(y)$ is compact for each $y \in K$. Thus, by Fan-KKM Lemma 1.14, we have

$$\mathbb{S}(\mathrm{WVEP}(\mathrm{cl}(K),f))^{*} = \bigcap_{y \in K} G(y) \neq \emptyset.$$

Therefore, by Theorem 9.28, ε-WVEP has at least one solution. □

Remark 9.11 Theorem 9.29 is the only one of the variations of Theorem 9.28. Using various existence results for VEP, we may obtain conditions of nonemptiness of $\mathbb{S}(\mathrm{VEP}(\mathrm{cl}(K),f))^{*}$. Then we can easily derive some existence results for ε-WVEP. If we assume closedness of C, we may utilize existence results for WVEP presented in Sect. 9.1.

Next result shows that the nonemptiness of the set of solutions of ε-WVEP implies the nonemptiness of the set of solutions of WVEP under mild conditions.

Theorem 9.30 *Let K be a nonempty subset of X such that $\mathrm{cl}(K)$ is compact. Assume that the following conditions hold.*

(i) $f(\cdot, y)$ *is C-upper semicontinuous on X for all $y \in X$;*
(ii) $\Omega(\varepsilon) \neq \emptyset$ *for all $\varepsilon \in \mathrm{int}(C)$.*

Then the solution set $\mathbb{S}(\mathrm{WVEP}(K,f))$ *of* WVEP *is nonempty.*

Proof Let $\{\varepsilon_\lambda\} \subset \mathrm{int}(C)$, $\varepsilon_\lambda \to \mathbf{0}$ and $x_\lambda \in \Omega(\varepsilon_\lambda)$. Then by compactness of $\mathrm{cl}(K)$, we assume, without loss of generality, that $x_\lambda \to x$ and $x \in \mathrm{cl}(K)$. Suppose to the contrary that $f(x, y) \in -\mathrm{int}(C)$ for some $y \in K$. Then by condition (i) (see Proposition 2.25), there is a λ_0 such that for all $\lambda \geq \lambda_0$

$$f(x_\lambda, y) \in f(x, y) - \varepsilon - \mathrm{int}(C),$$

that is, $f(x_\lambda, y) + \varepsilon \in -\mathrm{int}(C)$. This contradicts to the fact that $x_\lambda \in \Omega(\varepsilon_\lambda)$. Hence, $f(x, y) \notin -\mathrm{int}(C)$ for all $y \in K$ and thus $x \in \mathbb{S}(\mathrm{WVEP}(K,f))$. □

We now show that the solution mapping Ω of ε-WVEP is upper semicontinuous on $\mathrm{int}(C)$ under some suitable conditions.

Theorem 9.31 *Let K be a nonempty compact subset of X and $f : K \times K \to Y$ be a vector-valued function such that $f(\cdot, y)$ is C-upper semicontinuous on K for all $y \in K$. If $\Omega(\varepsilon)$ is nonempty for each $\varepsilon \in \mathrm{int}(C)$, then Ω is upper semicontinuous on* $\mathrm{int}(C)$.

Proof Let $\varepsilon_\lambda \to \varepsilon$ and $x_\lambda \in \Omega(\varepsilon_\lambda)$. Since K is compact, we may assume, without loss of generality, that $x_\lambda \to x \in K$. Suppose to the contrary that $x \notin \Omega(\varepsilon)$. Then there exists $y \in K$ such that $f(x, y) + \varepsilon \in -\mathrm{int}(C)$. Since Y is a topological vector space, there exists a neighborhood U of $\mathbf{0}$ such that

$$f(x, y) + \varepsilon + U + U \subset -\mathrm{int}(C).$$

Then

$$f(x, y) + \varepsilon + U + U - \mathrm{int}(C) \subset (-\mathrm{int}(C) - \mathrm{int}(C)) \subset -\mathrm{int}(C).$$

Since $\varepsilon_\lambda \to \varepsilon$, $x_\lambda \to x$ and $\Omega(\varepsilon)$ is nonempty for each $\varepsilon \in \mathrm{int}(C)$, there exists a $\hat{\lambda}$ such that for every $\lambda \geq \hat{\lambda}$,

$$f(x_\lambda, y) + \varepsilon_\lambda \in -\mathrm{int}(C).$$

This contradicts to the fact that $x_\lambda \in \Omega(\varepsilon_\lambda)$. Hence, $x \in \Omega(\varepsilon)$. Therefore, by Lemma 1.9, $\Omega(\varepsilon)$ is upper semicontinuous on $\mathrm{int}(C)$. □

From Theorems 9.29 and 9.31, we can easily obtain the following corollary.

Corollary 9.9 *Let K be nonempty compact convex subset of X and $f : K \times K \to Y$ be C-lower semicontinuous such that $f(x, x) \notin -\operatorname{int}(C)$ for all $x \in K$. Assume that the following conditions hold:*

(i) *$f(x, \cdot)$ is C-quasiconvex for each $x \in K$;*
(ii) *$f(\cdot, y)$ is C-upper semicontinuous for each $y \in K$;*

Then Ω is upper semicontinuous on $\operatorname{int}(C) \cup \{\mathbf{0}\}$.

We now establish that the solution mapping Ω of ε-WVEP is lower semicontinuous on $\operatorname{int}(C)$ under suitable assumptions.

Theorem 9.32 *Let K be a nonempty compact convex subset of X and $f : K \times K \to Y$ satisfy the following conditions:*

(i) *$f(x, \cdot)$ is C-lower semicontinuous for each $x \in K$;*
(ii) *$f(\cdot, y)$ is strictly C-quasiconcave for each $y \in K$;*
(iii) *$\Omega(\varepsilon)$ is nonempty for each $\varepsilon \in \operatorname{int}(C)$.*

Then Ω is lower semicontinuous on $\operatorname{int}(C)$.

Proof Let $\varepsilon \in \operatorname{int}(C)$. Let $\mathcal{V}$ be an open subset of X with $\mathcal{V} \cap \Omega(\varepsilon) \neq \emptyset$. Suppose that $x \in \mathcal{V} \cap \Omega(\varepsilon)$ and that $\hat{x} \in \Omega(\alpha \cdot \varepsilon)$, where $\alpha \in]0, 1[$. We choose $x' \in]x, \hat{x}[\cap \mathcal{V}$. Obviously, $\hat{x} \in \Omega(\varepsilon)$. From condition (ii), we have

$$f(x', v) \notin -\varepsilon - \operatorname{cl}(C), \quad \text{for all } v \in X.$$

Since $-\varepsilon - \operatorname{cl}(C)$ is a closed set, for each $v \in X$, there exists a positive number $t_v > 0$ such that

$$f(x', v) - t_v \cdot \varepsilon \notin -\varepsilon - \operatorname{cl}(C).$$

Since K is compact, from condition (i) and by Lemma 2.12, we obtain the C-compactness of $\bigcup_{v \in X} f(x', v)$. Clearly, $f(x', v) - t_v \cdot \varepsilon + \operatorname{int}(C)$ is a neighborhood of $f(x', v)$ and

$$\bigcup_{v \in X} \{f(x', v) - t_v \cdot \varepsilon + \operatorname{int}(C)\} \supset \bigcup_{v \in X} f(x', v).$$

Hence, there exist $v_1, v_2, \ldots, v_m \in K$ such that

$$\bigcup_{i=1}^{m} \{f(x', v_i) - t_{v_i} \cdot \varepsilon + \operatorname{int}(C)\} \supset \bigcup_{v \in X} f(x', v). \tag{9.77}$$

Since $f(x', v_i) - t_{v_i} \cdot \varepsilon \notin -\varepsilon - \operatorname{cl}(C)$, $i = 1, 2, \ldots, m$, there exist corresponding numbers $t^1, t^2, \ldots, t^m \in]0, 1[$ such that

$$f(x', v_i) - (t_{v_i} + t^i) \cdot \varepsilon \notin -\varepsilon - \operatorname{cl}(C), \quad \text{for all } i = 1, 2, \ldots, m.$$

Let $\tau = \min\{t_1, t_1, \ldots, t_m\}$. Then by Proposition 2.9,

$$\left(\bigcup_{i=1}^{m} f(x', v_i) - (t_{v_i} + \tau) \cdot \varepsilon \right) \bigcap \Big(- \varepsilon - \mathrm{cl}\,(C) \Big) = \emptyset.$$

From (9.77), we have

$$f(x', v) - \tau \cdot \varepsilon \in \left(\bigcup_{i=1}^{m} f(x', v_i) - (t_{v_i} + \tau) \cdot \varepsilon \right), \quad \text{for all } v \in X.$$

Accordingly,

$$f(x', v) - \tau \cdot \varepsilon \notin -\varepsilon - \mathrm{int}(C), \quad \text{for all } v \in X,$$

that is,

$$x' \in \Omega((1 - \tau) \cdot \varepsilon).$$

Therefore, $x' \in \Omega(\varepsilon')$ for all $\varepsilon' \in (1-\tau)\varepsilon + \mathrm{int}(C)$. Thus, Ω is lower semicontinuous on $\mathrm{int}(C)$. □

Theorem 9.33 *Let K be a nonempty compact convex subset of X and $f : K \times K \to Y$ satisfy the following conditions:*

(i) *$f(x, \cdot)$ is C-lower semicontinuous for all $x \in K$;*
(ii) *$f(\cdot, y)$ is strictly $(-C)$-properly quasiconvex for all $y \in K$;*
(iii) *$\Omega(\varepsilon)$ is nonempty for each $\varepsilon \in \mathrm{int}(C)$.*

Then Ω is lower semicontinuous on $\mathrm{int}(C)$.

Proof Let $\hat{\varepsilon} \in \mathrm{int}(C)$ be arbitrary but fixed and $\mathcal{V}$ be an open set with $\mathcal{V} \cap \Omega(\hat{\varepsilon}) \neq \emptyset$. Let $\hat{x} \in \mathcal{V} \cap \Omega(\hat{\varepsilon})$. Then we show that there exist $\bar{x} \in \mathcal{V}$ and $\mu > 0$ such that for all $\varepsilon \in (1 - \mu)\hat{\varepsilon} + \mathrm{int}(C)$, we have

$$f(\bar{x}, y) + \varepsilon \notin - \mathrm{int}(C), \quad \text{for all } y \in K.$$

We note that $(1 - \mu)\hat{\varepsilon} + \mathrm{int}(C)$ is a neighborhood of $\hat{\varepsilon}$.

First we select $\bar{x} \in \mathcal{V}$ in the following way. Let $\alpha \in]0, 1[$, $x_0 \in \Omega(\alpha\hat{\varepsilon})$ and

$$\bar{x} \in \mathcal{V} \cap \{x \in K : x = \lambda\hat{x} + (1 - \lambda)x_0, \ 0 < \lambda < 1\}.$$

Next we find corresponding $\mu \in]0, 1 - \alpha[$. Because of the way in selecting $\bar{x}$, we have

$$f(\bar{x}, y) \in f(\hat{x}, y) + \mathrm{int}(C),$$

or

$$f(\bar{x}, y) \in f(x_0, y) + \mathrm{int}(C).$$

Let $K' := \{y \in K : f(\bar{x}, y) \notin f(x_0, y) + \mathrm{int}(C)\}$. By condition (i) and Proposition 2.22, $A := \{y \in K : f(\bar{x}, y) \in f(x_0, y) + \mathrm{int}(C)\}$ is an open subset of X. Then $A^c = \{y \in K : f(\bar{x}, y) \notin f(x_0, y) + \mathrm{int}(C)\}$ is a closed subset of K. Hence $K' = (K \cap A^c)$ is a closed set, that is, compact set. From Proposition 2.11, we have $f(\bar{x}, v) \in f(\hat{x}, v) + \mathrm{int}(C)$ for all $v \in K'$. Thus, for each $v \in K'$, there exists $\mu_v \in]0, 1-\alpha[$ such that

$$f(\bar{x}, v) \in f(\hat{x}, v) + \mu_v \cdot \hat{\varepsilon} + \mathrm{int}(C).$$

Hence,

$$\mathfrak{M}^{\bar{x}} \subset \mathfrak{M}^{\hat{x}} + \bigcup_{v \in K'} (\mu_v \cdot \hat{\varepsilon} + \mathrm{int}(C)),$$

where $\mathfrak{M}^{\bar{x}}$ and $\mathfrak{M}^{\hat{x}}$ denote $\bigcup_{v \in K'} \{f(\bar{x}, v)\}$ and $\bigcup_{v \in K'} \{f(\hat{x}, v)\}$, respectively. From compactness of K' and condition (i), $\mathfrak{M}^{\bar{x}}$ is C-compact by Lemma 2.12. In addition,

$$\bigcup_{v \in K'} \{\mu_v \cdot \hat{\varepsilon} + \mathrm{int}(C)\} = \bigcup_{v \in K'} \{\mu_v \cdot \hat{\varepsilon}\} + \mathrm{int}(C) = \bigcup_{v \in K'} \{\mu_v \cdot \hat{\varepsilon}\} + \mathrm{int}(C) + \mathrm{int}(C),$$

and $\mathfrak{M}^{\hat{x}} + \bigcup_{v \in K'} (\mu_v \cdot \hat{\varepsilon} + \mathrm{int}(C))$ is an open covering of $\mathfrak{M}^{\bar{x}}$. So we can choose a finite subset $\{\mu_{v_1}, \mu_{v_2}, \ldots, \mu_{v_m}\} \subset \{\mu_v : v \in K'\}$ such that

$$\mathfrak{M}^{\bar{x}} \subset \mathfrak{M}^{\hat{x}} + \bigcup_{i=1}^{m} (\mu_{v_i} \cdot \hat{\varepsilon} + \mathrm{int}(C)).$$

Putting $\mu = \min\{\mu_{v_1}, \mu_{v_2}, \ldots, \mu_{v_m}\}$, we have

$$\mathfrak{M}^{\bar{x}} \subset \mathfrak{M}^{\hat{x}} + \mu \cdot \hat{\varepsilon} + \mathrm{int}(C).$$

Hence,

$$\mathfrak{M}^{\bar{x}} - \mu \cdot \hat{\varepsilon} \subset \mathfrak{M}^{\hat{x}} + \mathrm{int}(C). \tag{9.78}$$

Since $\hat{x} \in \Omega(\hat{\varepsilon})$, we have

$$\left(\mathfrak{M}^{\hat{x}} + \hat{\varepsilon}\right) \cap (-\mathrm{int}(C)) = \emptyset.$$

Hence, by Proposition 2.9,

$$\left(\mathfrak{M}^{\hat{x}} + \hat{\varepsilon} + \operatorname{int}(C)\right) \cap (-\operatorname{int}(C)) = \emptyset.$$

Therefore, by (9.78), we obtain

$$\left(\mathfrak{M}^{\bar{x}} + (1-\mu)\hat{\varepsilon}\right) \cap (-\operatorname{int}(C)) = \emptyset.$$

On the other hand for each $v \in (K \setminus K')$, we have

$$f(\bar{x}, v) \in f(x_0, v) + \operatorname{int}(C).$$

Since $x_0 \in \Omega(\alpha\hat{\varepsilon})$, we get

$$f(\bar{x}, v) + \alpha\hat{\varepsilon} \notin -\operatorname{int}(C).$$

Also, since $\alpha < (1-\mu)$, we have

$$\left(\bigcup_{v \in (K \setminus K')} \{f(\bar{x}, v) + (1-\mu)\hat{\varepsilon}\}\right) \cap -\operatorname{int}(C) = \emptyset,$$

from which it follows that

$$\left(\bigcup_{v \in K} \{f(\bar{x}, v) + (1-\mu)\hat{\varepsilon}\}\right) \cap -\operatorname{int}(C) = \emptyset.$$

Let $\mathcal{U} = (1-\mu)\hat{\varepsilon} + \operatorname{int}(C)$. Then $\mathcal{U}$ is an open set containing $\hat{\varepsilon}$. For every $\varepsilon \in \mathcal{U}$,

$$\bigcup_{v \in K} \{f(\bar{x}, v) + (1-\mu)\hat{\varepsilon}\} + \operatorname{int}(C) \supset \bigcup_{v \in K} \{f(\bar{x}, v) + \varepsilon\}.$$

Therefore by Proposition 2.9, we obtain

$$\left(\bigcup_{v \in K} \{f(\bar{x}, v) + \varepsilon\}\right) \cap -\operatorname{int}(C) = \emptyset,$$

from which it follows that $f(\bar{x}, v) + \varepsilon \notin -\operatorname{int}(C)$ for all $v \in K$, that is, $\bar{x} \in \Omega(\varepsilon)$ for all $\varepsilon \in \mathcal{U}$. Hence, Ω is lower semicontinuous at $\hat{\varepsilon}$. Since $\hat{\varepsilon}$ was arbitrary, Ω is lower semicontinuous on $\operatorname{int}(C)$. □

We remark that from the proof of Theorem 9.33, one can see that the condition (ii) in Theorem 9.33 can be replaced by the condition that $\Omega(\varepsilon) \neq \emptyset$ for some net $\{\varepsilon_\lambda\} \subset \operatorname{int}(C)$ such that $\varepsilon_\lambda \to \mathbf{0}$.

From Theorems 9.29 and 9.33, we obtain the following result.

Corollary 9.10 *Let K be a nonempty compact convex subset of X and $f : K \times K \to Y$ be C-lower semicontinuous such that $f(x,x) \notin -\text{int}(C)$ for all $x \in K$. Assume that the following conditions hold:*

(i) *$f(x,\cdot)$ is C-quasiconvex for all $x \in K$;*
(ii) *$f(\cdot,y)$ is strictly $(-C)$-properly quasiconvex for all $y \in K$;*
(iii) *$f(\cdot,y)$ is C-upper semicontinuous for all $y \in K$*

Then Ω is continuous on $\text{int}(C)$.

From Theorems 9.31 and 9.33, we have the following corollary.

Corollary 9.11 *Let K be a nonempty compact convex subset of X and $f : K \times K \to Y$ be C-lower semicontinuous such that the following conditions hold:*

(i) *$f(\cdot,y)$ is C-upper semicontinuous for all $y \in K$;*
(ii) *$f(\cdot,y)$ is strictly $(-C)$-properly quasiconvex for all $y \in K$;*
(iii) *$\Omega(\varepsilon)$ is nonempty for each $\varepsilon \in \text{int}(C)$.*

Then Ω is continuous on $\text{int}(C)$.

9.6.2 Parametric Weak Vector Equilibrium Problems

In this subsection, we consider the parametric weak vector equilibrium problems (in short, PWVEPs) which are more general than ε-weak vector equilibrium problems studied in the previous subsection. We focus on the stability properties, such as upper semicontinuity and lower semicontinuity of the solution mappings for PWVEPs. Bianchi and Pini [22] studied upper hemicontinuity of the solution mapping for PWVEPs. They also studied Hölder continuity of the solution mapping for PWVEPs when the underlying bifunction satisfies some strong monotonicity assumption. Kimura and Yao [67] established some existence results for solutions of PWVEPs by using well-known Fan-KKM lemma. They also studied the upper semicontinuity and lower semicontinuity of the solution map of the PWVEPs. Gong [54] also studied sufficient conditions for the upper semicontinuity and lower semicontinuity of the solution map of the PWVEPs. Xu and Li [99] used scalarization method to study the lower semicontinuity of the solution map for parametric vector equilibrium problems (in short, PVEPs) However, Salamon and Bogdan [86, 87, 89] gave some sufficient conditions for closedness of the solution map of the PWVEPs and the solution map of the PVEPs. Zhang et al. [103] further studied lower semicontinuity of the solution map for PVEPs. Huang et al. [62] considered a more general form of PWVEP, called parametric implicit weak vector equilibrium problem, and established some sufficient conditions for the upper semicontinuity and lower semicontinuity of the solution map of their problem. Li and Li [75] obtained an explicit expression of the contingent derivative of the solution map for the PWVEPs. By using a nonlinear scalarization function, Chen and Li [35] studied the Hölder continuity of the solution map. Fan et al. [44] perturbed bifunction involved in the formulation of PWVEP and gave sufficient conditions for lower / upper semicontinuity and Hölder continuity of the solution map for the PWVEP. Most of the results presented in this subsection are taken from [67].

Let K be a nonempty subset of a Hausdorff topological vector space X and Y be a topological vector space with a solid pointed convex cone $C \subset Y$. Let Γ and Λ be two (index sets) nonempty subsets of two Hausdorff topological spaces, respectively. Let $f : \Gamma \times K \times K \to Y$ be a parameterized vector-valued trifunction and $A : \Lambda \to 2^K \setminus \{\emptyset\}$ be a constraint mapping. The *parametric weak vector equilibrium problem* (in short, PWVEP) is defined as follows: For given $p \in \Gamma$ and $\lambda \in \Lambda$, find $\bar{x} \in A(\lambda)$ such that

$$f(p, \bar{x}, y) \notin -\operatorname{int}(C), \quad \text{for all } y \in A(\lambda). \tag{9.79}$$

The solution mapping Ω of PWVEP is a set-valued map from $\Gamma \times \Lambda$ to 2^K defined by

$$\Omega(p, \lambda) = \{x \in A(\lambda) : f(p, x, y) \notin -\operatorname{int}(C) \text{ for all } y \in A(\lambda)\}. \tag{9.80}$$

When A is a constant mapping, say $A(\lambda) = K$ for all $\lambda \in \Lambda$, then we have the following form of PWVEP, and it is denoted by WVEP$_p$. For given $p \in \Gamma$, find $\bar{x} \in K$ such that

$$f(p, \bar{x}, y) \notin -\operatorname{int}(C), \quad \text{for all } y \in K. \tag{9.81}$$

If A is a constant mapping, say $A(\lambda) = K$ for all $\lambda \in \Lambda$, and $f(p, x, y) = g(x, y)$ for all $p \in \Gamma$, $x, y \in K$, where $g : K \times K \to Y$ is a vector-valued bifunction, then PWVEP reduces WVEP studied in Sect. 9.1. In addition, if $f(p, x, y) = g(x, y) + \varepsilon$ for all $p \in \Gamma$, $x, y \in K$, and for any given $\varepsilon \in \operatorname{int}(C)$, then PWVEP becomes the ε-WVEP which is defined as follows: For a given $\varepsilon \in \operatorname{int}(C)$, find $\bar{x} \in K$ such that

$$g(\bar{x}, y) + \varepsilon \notin -\operatorname{int}(C), \quad \text{for all } y \in K.$$

The solution map for ε-WVEP involving $g : K \times K \to Y$ is defined by

$$S(\varepsilon) = \{x \in K : g(x, y) + \varepsilon \notin -\operatorname{int}(C) \text{ for all } y \in K\}. \tag{9.82}$$

Throughout this subsection, unless otherwise specified, K, X, Y, C, Γ and Λ are the same as above, and further we assume that $A : \Lambda \to 2^K \setminus \{\emptyset\}$ is a constraint mapping and $f : \Gamma \times K \times K \to Y$ be a parameterized vector-valued trifunction such that $f(p, x, x) \notin -\operatorname{int}(C)$ for all $p \in \Gamma$ and $x \in K$.

We establish some existence results for solutions of PWVEP.

Theorem 9.34 *Let*

(i) $A(\lambda)$ *be closed and convex for all* $\lambda \in \Lambda$,

(ii) $\{x \in A(\lambda) : f(p, x, y) \in -\operatorname{int}(C)\}$ *be open for all* $y \in A(\lambda)$,

(iii) $\{y \in A(\lambda) : f(p, x, y) \in -\operatorname{int}(C)\}$ *be convex for all* $y \in A(\lambda)$,
(iv) *for each* $\lambda \in \Lambda$, *there exist a compact set* $\mathcal{B}_\lambda \subset K$ *and* $\hat{y} \in A(\lambda) \cap \mathcal{B}_\lambda$ *such that* $f(p, x, \hat{y}) \in -\operatorname{int}(C)$ *for all* $x \in A(\lambda) \setminus \mathcal{B}_\lambda$.

Then PWVEP *has at least one solution for each* $p \in \Gamma$ *and* $\lambda \in \Lambda$.

Proof Let $G : A(\lambda) \to 2^{A(\lambda)}$ be defined by

$$G(y) := \{x \in A(\lambda) : f(p, x, y) \notin -\operatorname{int}(C)\}, \quad \text{for all } y \in A(\lambda),$$

where $p \in \Gamma$ and $\lambda \in \Lambda$ are fixed. Since $f(p, x, x) \notin -\operatorname{int}(C)$, $G(y)$ is nonempty for all $y \in A(\lambda)$. Condition (ii) implies that $G(y)$ is closed for each $y \in A(\lambda)$.

By condition (iv) and closedness of $G(y)$ for each $y \in A(\lambda)$, for corresponding $\hat{y} \in A(\lambda) \cap \mathcal{B}_\lambda$, $G(\hat{y})$ is compact.

Finally, we show that G is a KKM-map. Suppose to the contrary that there exists $\mu_i \in [0, 1]$, $x_i \in A(\lambda)$, $i = 1, 2, \ldots, m$, such that

$$\sum_{i=1}^{m} \mu_i x_i =: x \notin \bigcup_{i=1}^{m} G(x_i).$$

Since $x_i \in A(\lambda)$, $i = 1, 2, \ldots, m$, by the convexity of $A(\lambda)$, we have $x \in A(\lambda)$. Hence, $f(p, x, x_i) \in -\operatorname{int}(C)$, $i = 1, 2, \ldots, m$. This implies that

$$f\left(p, x, \sum_{i=1}^{m} \mu_i x_i\right) = f(p, x, x) \in -\operatorname{int}(C),$$

because of condition (iii). This contradicts the fact that $f(p, x, x) \notin -\operatorname{int}(C)$ for all $p \in \Gamma$.

By applying Fan-KKM Lemma 1.14, we get

$$\Omega(p, \lambda) = \bigcap_{y \in A(\lambda)} G(y) \neq \emptyset,$$

for each $p \in \Gamma$ and $\lambda \in \Lambda$. □

The following result is an easy consequence of Theorem 9.34.

Corollary 9.12 *Let* K *be a nonempty compact subset of a Hausdorff topological vector space* X, *and let*

(i) $A(\lambda)$ *be compact convex for all* $\lambda \in \Lambda$;
(ii) $f(p, \cdot, y)$ *be* C*-upper semicontinuous on* $A(\lambda)$ *for all* $p \in \Gamma, y \in A(\lambda)$;
(iii) $f(p, x, \cdot)$ *be* C*-quasiconvex on* $A(\lambda)$ *for all* $p \in \Gamma, x \in A(\lambda)$.

Then PWVEP *has at least one solution for each* $p \in \Gamma$ *and* $\lambda \in \Lambda$.

Theorem 9.35 *Let*

(i) $\mathrm{cl}\,(A(\lambda))$ *be compact and convex for all* $\lambda \in \Lambda$,
(ii) $f(p, \cdot, y)$ *be* C*-upper semicontinuous on* $\mathrm{cl}\,(A(\lambda))$ *for all* $p \in \Gamma$ *and* $y \in \mathrm{cl}\,(A(\lambda))$,
(iii) $f(p, x, \cdot)$ *be* C*-quasiconvex on* $\mathrm{cl}\,(A(\lambda))$ *for all* $p \in \Gamma$ *and* $x \in \mathrm{cl}\,(A(\lambda))$,
(iv) $f(p, \cdot, \cdot)$ *be* C*-lower semicontinuous on* $X \times X$ *for all* $p \in \Gamma$,
(v) *for some* $\hat{p} \in \Gamma$ *and* $\Gamma' \subset \Gamma$, $f(p, x, y) \in f(\hat{p}, x, y) + \mathrm{int}(C)$ *for all* $p \in \Gamma'$ *and* $(x, y) \in K \times K$.

Then PWVEP (9.79) *has at least one solution for each* $p \in \Gamma'$ *and* $\lambda \in \Lambda$.

Proof For $\hat{p} \in \Gamma$ of condition (v) and each fixed $\lambda \in \Lambda$, let $G : A(\lambda) \to 2^{A(\lambda)}$ be defined by

$$G(y) := \{x \in \mathrm{cl}\,(A(\lambda)) : f(\hat{p}, x, y) \notin -\mathrm{int}(C)\}, \quad \text{for each } y \in A(\lambda).$$

Then by Corollary 9.12,

$$\mathcal{S} = \bigcap_{y \in \mathrm{cl}\,(A(\lambda))} G(y) \neq \emptyset, \quad \text{for each } \lambda \in \Lambda.$$

Let $p \in \Gamma'$, $\lambda \in \Lambda$ and $x \in \mathcal{S}$. Then by condition (v), $f(p, x, y) \notin -\mathrm{cl}\,(C)$ for all $y \in \mathrm{cl}\,(A(\lambda))$. Since $(-\mathrm{cl}\,(C))^{\mathrm{c}}$ is a neighborhood of $f(p, x, y)$ for all $y \in \mathrm{cl}\,(A(\lambda))$ and $(-\mathrm{cl}\,(C))^{\mathrm{c}} + C = (-\mathrm{cl}\,(C))^{\mathrm{c}}$, by condition (iv), for each $y \in \mathrm{cl}\,(A(\lambda))$, there is a neighborhood $\mathcal{U}_{(x,y)}$ of (x, y) such that

$$f(p, u, v) \in (-\mathrm{cl}\,(C))^{\mathrm{c}}, \quad \text{for all } (u, v) \in \mathcal{U}_{(x,y)}.$$

Let $\mathcal{U}_{(x,y)} = \mathcal{W}_y \times \mathcal{V}_y$, where $\mathcal{W}_y$ and $\mathcal{V}_y$ denote neighborhoods of x and y, respectively. Since $\mathrm{cl}\,(A(\lambda))$ is compact, we can choose a finite subset $\{y_1, y_2, \ldots, y_m\}$ of $\mathrm{cl}\,(A(\lambda))$ such that

$$\bigcup_{i=1}^{m} \mathcal{V}_{y_i} \supset \mathrm{cl}\,(A(\lambda)).$$

Let $\mathcal{W} = \bigcap_{i=1}^{m} \mathcal{W}_{y_i}$. Then for any $w \in \mathcal{W}$, we have

$$f(p, w, y) \in (-\mathrm{cl}\,(C))^{\mathrm{c}}, \quad \text{for all } y \in \mathrm{cl}\,(A(\lambda)).$$

Hence,

$$f(p, w, y) \notin -\mathrm{int}(C), \quad \text{for all } y \in \mathrm{cl}\,(A(\lambda)).$$

Since $\mathcal{W}$ is a neighborhood of x and $x \in \mathrm{cl}\,(A(\lambda))$, $\mathcal{W} \cap A(\lambda) \neq \emptyset$. Thus, for all $x' \in \mathcal{W} \cap A(\lambda)$

$$f(p, x', y) \notin -\mathrm{int}(C), \quad \text{for all } y \in \mathrm{cl}\,(A(\lambda)).$$

Therefore,

$$f(p, x', y) \notin -\mathrm{int}(C), \quad \text{for all } y \in A(\lambda).$$

Consequently, $x' \in \Omega(p, \lambda)$, that is, $\Omega(p, \lambda) \neq \emptyset$. □

Example 9.9 Let $f : \Gamma \times K \times K \to Y$, where Γ, K, Y are $\mathbb{R}$. Let Γ' and Λ be $\mathbb{R}_{++} = \{t \in \mathbb{R} : t > 0\}$, $\hat{p} = 0$ and $A(\lambda) = \{x \in \mathbb{Q} : 0 < x < 2 + \lambda\}$ and $C = \mathbb{R}_+$. Let

$$f(p, x, y) = (y^2 - 2)^2 - (x^2 - 2)^2 + p.$$

Then PWVEP (9.79) has at least one solution for each $p \in \Gamma'$ and $\lambda \in \Lambda$.

Corollary 9.13 *Let K be a nonempty subset of a Hausdorff topological vector space X such that* $\mathrm{cl}\,(K)$ *is compact and convex. Let $g : X \times X \to Y$ be C-lower semicontinuous on $X \times X$ such that the following conditions hold.*

(i) $g(x, x) = \mathbf{0}$ *for all $x \in X$.;*
(ii) $g(x, \cdot)$ *is C-quasiconvex on X for all $x \in X$;*
(iii) $g(\cdot, y)$ *is C-upper semicontinuous on X for all $y \in X$.*

Then ε-WVEP (9.72) *has at least one solution for each* $\varepsilon \in \mathrm{int}(C)$.

Proof Putting, $\Gamma = \{\mathbf{0}\} \cup \mathrm{int}(C)$, $\Gamma' = \mathrm{int}(C)$, $\Lambda = \{\mathbf{0}\}$, $A(\lambda) = K, f(p, x, y) = g(x, y) + p$, $p \in \mathrm{int}(C)$, and $\hat{p} = \mathbf{0}$ in Theorem 9.35, we get the conclusion. □

We now investigate conditions under which the solution mapping of PWVEP is upper semicontinuous. From Lemma 1.10, we establish the following upper semicontinuity property of the solution mapping Ω for PWVEP.

Theorem 9.36 *Let Γ and Λ be nonempty closed subsets of two Hausdorff spaces, respectively. Assume that*

(i) *A is continuous and compact-valued,*
(ii) *f is C-upper semicontinuous on $\Gamma \times K \times K$,*
(iii) *$\Omega(p, \lambda) \neq \emptyset$ for each $p \in \Gamma$ and $\lambda \in \Lambda$.*

Then Ω is upper semicontinuous on $\Gamma \times \Lambda$.

Proof Let $p \in \Gamma$ and $\lambda \in \Lambda$. By condition (ii), $\Omega(p, \lambda)$ is closed for all $(p, \lambda) \in \Gamma \times \Lambda$. By condition (i), $A(\lambda)$ is compact for each $\lambda \in \Lambda$. Hence, $\Omega(p, \lambda)$ is compact. We can apply Lemma 1.10. Let $(p_\lambda, \lambda_\alpha) \to (p, \lambda)$ and $x_\alpha \in \Omega(p_\lambda, \lambda_\alpha)$, then for each α

$$f(p_\alpha, x_\alpha, y_\alpha) \notin -\mathrm{int}(C), \quad \text{for all } y_\alpha \in A(\lambda_\alpha). \tag{9.83}$$

Since A is upper semicontinuous, without loss of generality , we can assume $x_\alpha \to x$ for some $x \in A(\lambda)$.

Suppose to the contrary that $x \notin \Omega(p, \lambda)$. Then there exists $y \in A(\lambda)$ such that

$$f(p, x, y) \in -\operatorname{int}(C).$$

By condition (ii), there exists a neighborhood $\mathfrak{U}$ of (p, x, y) such that

$$f(p', x', y') \in -\operatorname{int}(C), \quad \text{for all } (p', x', y') \in \mathfrak{U}.$$

Let $\mathfrak{U} := \mathcal{P} \times X \times Y$. Since A is continuous and so lower semicontinuous, and $y \in Y \cap A(\lambda)$, that is, $Y \cap A(\lambda) \neq \emptyset$, there exists $\mathcal{L}$ which is a neighborhood of λ such that

$$Y \cap A(\lambda') \neq \emptyset, \quad \text{for all } \lambda' \in \mathcal{L}.$$

Since $(p_\alpha, \lambda_\alpha, x_\alpha) \to (p, \lambda, x)$, there exists α' such that

$$(p_\alpha, \lambda_\alpha, x_\alpha) \in \mathcal{P} \times \mathcal{L} \times X, \quad \text{for all } \alpha \geq \alpha'.$$

Hence, for each $\alpha \geq \alpha'$, there exists $y_\alpha \in Y \cap A(\lambda_\alpha)$ such that

$$f(p_\alpha, x_\alpha, y_\alpha) \in -\operatorname{int}(C).$$

This contradicts to (9.83). Hence $x \in \Omega(p, \lambda)$. Thus, Ω is upper semicontinuous at (p, λ). Since $(p, \lambda) \in \Gamma \times \Lambda$ was arbitrary, Ω is upper semicontinuous on $\Gamma \times \Lambda$. □

For fixed $\lambda \in \Lambda$, we have the following result.

Corollary 9.14 *Assume that*

(i) *$A(\lambda)$ be compact for all $\lambda \in \Lambda$,*
(ii) *$f(\cdot, \cdot, y)$ be C-upper semicontinuous on $\Gamma \times K$ for all $y \in A(\lambda)$;*
(iii) *$\Omega(p, \lambda) \neq \emptyset$ for all $p \in \Gamma$ and $\lambda \in \Lambda$.*

Then $\Omega(\cdot, \lambda)$ is upper semicontinuous on Γ for each fixed $\lambda \in \Lambda$.

Corollary 9.15 *Let K be a nonempty compact subset of a Hausdorff topological vector space X, Γ be a topological space, $f : \Gamma \times K \times K \to Y$ be a vector-valued bifunction such that $f(p, x, x) \notin -\operatorname{int}(C)$ for all $p \in \Gamma$ and $x \in K$, and $\Omega : \Gamma \to 2^K$ be a set-valued map defined by*

$$\Omega(p) = \{x \in K : f(p, x, y) \notin -\operatorname{int}(C), \quad \textit{for all } y \in K\}.$$

Assume that the following conditions hold.

(i) *$f(\cdot, \cdot, y)$ is C-upper semicontinuous on $\Gamma \times K$ for all $y \in K$;*
(ii) *$\Omega(p) \neq \emptyset$ for each $p \in \Gamma$.*

Then Ω is upper semicontinuous on Γ.

Now we investigate conditions under which the solution map of PWVEP (9.79) is lower semicontinuous.

Theorem 9.37 *Assume that*

(i) *A is continuous at λ_0 and $A(\lambda_0)$ is compact convex;*
(ii) *f is C-lower semicontinuous on $P \times K \times K$ for a certain neighborhood P of p_0;*
(iii) *$f(p_0, \cdot, y)$ is strictly $(-C)$-properly quasiconvex on K for fixed $p_0 \in \Gamma$ and each $y \in X$;*
(iv) *for fixed $p_0 \in \Gamma$ and $\lambda_0 \in \Lambda$, $\Omega(p_0, \lambda_0)$ contains at least two points.*

Then Ω is lower semicontinuous at (p_0, λ_0).

Proof Let $\mathcal{V} \subset X$ such that $\mathcal{V} \cap \Omega(p_0, \lambda_0) \neq \emptyset$. Then there exists $x \in \mathcal{V} \cap \Omega(p_0, \lambda_0)$. Since $\Omega(p_0, \lambda_0)$ has at least two elements, we can choose $\bar{x} \in \Omega(p_0, \lambda_0) \setminus \{x\}$. Because of the convexity of $A(\lambda_0)$ and condition (iii), $\Omega(p_0, \lambda_0)$ is convex. Then there is an $\mu \in]0, 1[$ such that $x' := (1 - \mu)x + \mu\bar{x}$ belongs to $\mathcal{V}$. From condition (iii), for each $y \in A(\lambda_0)$,

$$f(p_0, x', y) \in f(p_0, x, y) + \text{int}(C),$$

or

$$f(p_0, x', y) \in f(p_0, \bar{x}, y) + \text{int}(C).$$

Thus by Proposition 2.10, we have

$$f(p_0, x', y) \notin -\text{cl}\,(C), \quad \text{for all } y \in A(\lambda),$$

because $\{x, \bar{x}\} \subset \Omega(p_0, \lambda_0)$. Hence for each $y \in A(\lambda_0)$, there exists a neighborhood $\mathcal{W}^y$ of $f(p_0, x', y)$ such that $\mathcal{W}^y \cap (-\text{cl}\,(C)) = \emptyset$. By condition (ii), for each $y \in A(\lambda_0)$, there exists a neighborhood $\mathfrak{U}$ of (p_0, x', y) such that

$$f(\tilde{p}, \tilde{x}', \tilde{y}) \in \mathcal{W}^y + C, \quad \text{for all } (\tilde{p}, \tilde{x}', \tilde{y}) \in \mathfrak{U}.$$

Let $\mathfrak{U} := \mathcal{P}_y \times \mathcal{X}_y \times \mathcal{Y}_y$ where $\mathcal{P}_y$, $\mathcal{X}_y$ and $\mathcal{Y}_y$ are neighborhoods of p_0, x' and y, respectively. Since $\bigcup\limits_{y \in A(\lambda_0)} \mathcal{Y}_y \supset A(\lambda_0)$ and $A(\lambda_0)$ is compact, then there is a finite subset $\{y_1, y_2, \ldots, y_m\} \subset A(\lambda_0)$ such that

$$\bigcup_{i=1}^{m} \mathcal{Y}_{y_i} \supset A(\lambda_0).$$

Let $\mathfrak{P} := \bigcap\limits_{i=1}^{m} \mathcal{P}_{y_i}$, $\mathfrak{Q} := \bigcap\limits_{i=1}^{m} \mathcal{X}_{y_i}$ and $\mathfrak{Y} := \bigcup_{i=1}^{m} \mathcal{Y}_{y_i}$. Then for each $\tilde{p} \in \mathfrak{P}$ and $\tilde{x}' \in \mathfrak{Q}$

$$f(\tilde{p}, \tilde{x}', \tilde{y}) \in \bigcup_{i=1}^{m} (\mathcal{W}^{y_i} + C), \quad \text{for all } \tilde{y} \in \mathfrak{Y}.$$

Now $\bigcup_{i=1}^{m} \mathcal{W}^{y_i} \cap (-\mathrm{cl}\,(C)) = \emptyset$, that is, $\bigcup_{i=1}^{m} \mathcal{W}^{y_i} \subset (-\mathrm{cl}\,(C))^{\mathrm{c}}$. Therefore, by Proposition 2.10, we have

$$\bigcup_{i=1}^{m}(\mathcal{W}^{y_i} + C) = \bigcup_{i=1}^{m}\mathcal{W}^{y_i} + C \subset (-\mathrm{cl}\,(C))^{\mathrm{c}}.$$

Hence, for each $\tilde{p} \in \mathfrak{P}$ and $\tilde{x}' \in \mathfrak{Q}$

$$f(\tilde{p}, \tilde{x}', \tilde{y}) \notin -\mathrm{int}(C), \quad \text{for all } \tilde{y} \in \mathfrak{Y}. \tag{9.84}$$

Since $\mathfrak{Y} \supset A(\lambda_0)$ is an open neighborhood of $A(\lambda_0)$ and A is upper semicontinuous at λ_0, there is a neighborhood $\mathcal{L}_1$ of λ_0 such that

$$A(\lambda') \subset \mathfrak{Y} \text{ for all } \lambda' \in \mathcal{L}_1.$$

Since both of $\mathcal{V}$ and $\mathfrak{Q}$ are neighborhoods of x', $\mathcal{V} \cap \mathfrak{Q} \neq \emptyset$. Let $\bar{\mathcal{V}} := \mathcal{V} \cap \mathfrak{Q}$. Since $x' \in \bar{\mathcal{V}} \cap A(\lambda_0)$, $\bar{\mathcal{V}} \cap A(\lambda_0) \neq \emptyset$. As A is lower semicontinuous at λ_0, there is a neighborhood $\mathcal{L}_2$ of λ such that

$$A(\lambda'') \cap \bar{\mathcal{V}} \neq \emptyset, \quad \text{for all } \lambda'' \in \mathcal{L}_2.$$

Let $\mathfrak{L} := \mathcal{L}_1 \cap \mathcal{L}_2$. Then $A(\tilde{\lambda}) \subset \mathfrak{Y}$ for all $\tilde{\lambda} \in \mathcal{L}$ and $\mathfrak{Q} \supset \bar{\mathcal{V}} \cap A(\tilde{\lambda}) \neq \emptyset$. Therefore, from (9.84), we have that for each $(\tilde{p}, \tilde{\lambda}) \in \mathfrak{P} \times \mathfrak{L}$ which is an open set containing (p_0, λ_0), there is $x^* \in A(\tilde{\lambda}) \cap \bar{\mathcal{V}}$ such that

$$f(\tilde{p}, x^*, u) \notin -\mathrm{int}(C), \quad \text{for all } u \in A(\tilde{\lambda}),$$

that is, $x^* \in \Omega(\tilde{p}, \tilde{\lambda}) \cap \mathcal{V} \neq \emptyset$. Hence, Ω is lower semicontinuous at (p_0, λ_0). □

Corollary 9.16 *Assume that*

(i) *A is continuous and convex-valued;*
(ii) *f is C-lower semicontinuous on $\Gamma \times K \times K$;*
(iii) *$f(p, \cdot, y)$ is strictly $(-C)$-properly quasiconvex on K for all $p \in \Gamma$ and $y \in K$;*
(iv) *$\Omega(p, \lambda)$ contains at least two points for each $p \in \Gamma$ and $\lambda \in \Lambda$.*

Then Ω is lower semicontinuous on $\Gamma \times \Lambda$.

Remark 9.12 The lower semicontinuity of the solution map Ω cannot be guaranteed if f is real-valued continuous and A is constant (see [50, p.151]). Stronger condition is needed, for example, Ω is single-valued or conditions (iii) and (iv) hold.

Example 9.10 Let $X = \mathbb{R}^2$, $\Gamma = [-1, 2]$, $\Lambda = \{0\}$, $A(0) = K = \{(x_1, x_2) : x_1 + x_2 = 1, x_1 \geq 0, x_2 \geq 0\}$, $C = \mathbb{R}_+$. Let $f : \Gamma \times X \times X \to \mathbb{R}$ be defined by

$$f(p, x, y) = (-(1 + p)y_1 - y_2) - (-(1 + p)x_1 - x_2).$$

Then A is constant, f is continuous on $\Gamma \times X \times X$ and

$$\Omega(p,0) = \begin{cases} \{(0,1)\}, & \text{if } p \in [-1,0), \\ K, & \text{if } p = 0, \\ \{(1,0)\}, & \text{if } p \in (0,2]. \end{cases}$$

It is easy to see that Ω is not lower semicontinuous at $(0,0)$.

In the following result we do not assume that $\Omega(p,\lambda)$ contains at least two points.

Theorem 9.38 *Let*

(i) *A be continuous and $A(\lambda)$ be convex for all $\lambda \in \Lambda$,*
(ii) *f be C-lower semicontinuous on $\Gamma \times K \times K$,*
(iii) *$f(p,\cdot,y)$ be strictly $(-C)$-properly quasiconvex on K for all $p \in \Gamma$ and $y \in K$;*
(iv) *there exist $\Gamma' \subset \Gamma$ and $\hat{p} \in \Gamma$ such that for any $p \in \Gamma'$,*

$$f(p,x,y) \in f(\hat{p},x,y) + \text{int}(C), \quad \textit{for all } x,y \in K,$$

(v) *$\Omega(p,\lambda) \neq \emptyset$ for each $p \in \Gamma$ and $\lambda \in \Lambda$.*

Then Ω is lower semicontinuous on $\Gamma' \times \Lambda$.

Proof Let $p \in \Gamma'$, $\lambda \in \Lambda$, and $\mathcal{V}$ be an open set satisfying $\mathcal{V} \cap \Omega(p,\lambda) \neq \emptyset$. Let $x \in \mathcal{V} \cap \Omega(p,\lambda)$. Then by condition (iv), there exists $\bar{p} \in \Gamma$ such that

$$f(p,x,y) \in f(\hat{p},x,y) + \text{int}(C), \quad \text{for all } x,y \in K.$$

Suppose that $\bar{x} \in \Omega(\hat{p},\lambda)$ which is nonempty by condition (iv) and that $x' \in \mathcal{V} \cap \{u \in A(\lambda) : u = \mu x + (1-\mu)\bar{x},\ 0 < \alpha < 1\}$. By condition (iv), $\bar{x} \in \Omega(p,\lambda)$. Because of the way in selecting x', by condition (iii), we have

$$f(p,x',y) \in f(p,x,y) + \text{int}(C),$$

or

$$f(p,x',y) \in f(\hat{p},x,y) + \text{int}(C).$$

Therefore, $f(p,x',y) \notin -\text{cl}(C)$ for all $y \in A(\lambda)$. Then for each $y \in A(\lambda)$, there exists a neighborhood $\mathcal{W}_y$ of $f(p,x',y)$ such that $\mathcal{W}_y \cap (-\text{cl}(C)) = \emptyset$. By condition (ii), for each $y \in A(\lambda)$, there exists a neighborhood $\mathfrak{U}_y$ of (p,x',y) such that

$$f(\tilde{p},\tilde{x}',\tilde{y}) \in \mathcal{W}_y + C, \quad \text{for all } (\tilde{p},\tilde{x}',\tilde{y}) \in \mathfrak{U}_y.$$

Let $\mathfrak{U}_y := \mathcal{P}_y \times \mathcal{X}_y \times \mathcal{Y}_y$, where $\mathcal{P}_y, \mathcal{X}_y$ and $\mathcal{Y}_y$ are neighborhoods of p, x' and y, respectively. Since $A(\lambda)$ is compact, there exists a finite subset $\{y_1, y_2, \ldots, y_m\}$ of

$A(\lambda)$ such that $\bigcup_{i=1}^{m} \mathcal{Y}_{y_i} \supset A(\lambda)$. Let $\mathfrak{P} = \bigcap_{i=1}^{m} \mathcal{P}_{y_i}$, $\mathfrak{Q} = \left(\bigcap_{i=1}^{m} \mathcal{X}_{y_i}\right) \cap \mathcal{V}$ and $\mathfrak{Y} = \bigcup_{i=1}^{m} \mathcal{Y}_{y_i}$. Then for each $(\tilde{p}, \tilde{x}, \tilde{y}) \in \mathfrak{P} \times \mathfrak{Q} \times \mathfrak{Y}$

$$f(\tilde{p}, \tilde{x}, \tilde{y}) \notin -\operatorname{int}(C).$$

In addition, $\mathfrak{Y}$ is a neighborhood of $A(\lambda)$ and $\mathfrak{Q}$ is an open set with $\mathfrak{Q} \cap A(\lambda) \neq \emptyset$. Since A is continuous, by the same argument as that in the proof of Theorem 9.37 we can show that there is a neighborhood $\mathcal{L}$ of λ such that

$$A(\tilde{\lambda}) \subset \mathfrak{Y} \text{ and } A(\tilde{\lambda}) \cap \mathfrak{Q} \neq \emptyset, \quad \text{for all } \tilde{\lambda} \in \mathcal{L}.$$

Hence, for each $(\tilde{p}, \tilde{\lambda}) \in \mathfrak{P} \times \mathcal{L}$ which is an open set containing (p, λ), there exists $\bar{x} \in \mathfrak{Q} \cap A(\tilde{\lambda})$ such that

$$f(\tilde{p}, \bar{x}, \tilde{y}) \notin -\operatorname{int}(C), \quad \text{for all } \tilde{y} \in A(\tilde{\lambda}),$$

that is, $\tilde{x} \in \Omega(\tilde{p}, \tilde{\lambda}) \cap \mathcal{V} \neq \emptyset$. Hence, Ω is lower semicontinuous at (p, λ). Since (p, λ) was arbitrary, Ω is lower semicontinuous on $\Gamma' \times \Lambda$. □

By using similar argument as in the proof of Theorem 9.37, we can prove the following result.

Theorem 9.39 *Assume that*

(i) *$A(\lambda)$ is compact and convex for all $\lambda \in \Lambda$;*
(ii) *$f(\cdot, x, \cdot)$ is C-lower semicontinuous on $\Gamma \times K$ for all $x \in K$;*
(iii) *$f(p, \cdot, y)$ is strictly $(-C)$-properly quasiconvex on K for all $p \in \Gamma$ and $y \in K$;*
(iv) *$\Omega(p, \lambda)$ contains at least two points for each $p \in \Gamma$ and $\lambda \in \Lambda$.*

Then $\Omega(\cdot, \lambda)$ is lower semicontinuous on Γ for each $\lambda \in \Lambda$.

By the same argument as that in the proof of Theorem 9.38, we can prove the following result.

Theorem 9.40 *Assume that*

(i) *$A(\lambda)$ is compact and convex for all $\lambda \in \Lambda$;*
(ii) *$f(\cdot, x, \cdot)$ is C-lower semicontinuous on $\Gamma \times K$ for all $x \in K$;*
(iii) *$f(p, \cdot, y)$ is strictly $(-C)$-properly quasiconvex on K for all $p \in \Gamma$ and $y \in K$;*
(iv) *there exist $\Gamma' \subset \Gamma$ and $\hat{p} \in \Gamma$ such that for any $p \in \Gamma$,*

$$f(p, x, y) \in f(\hat{p}, x, y) + \operatorname{int}(C), \quad \textit{for all } x, y \in K;$$

(v) *$\Omega(p, \lambda) \neq \emptyset$ for each $p \in \Gamma$ and $\lambda \in \Lambda$.*

Then $\Omega(\cdot, \lambda)$ is lower semicontinuous on Γ' for each $\lambda \in \Lambda$.

By combining results for upper semicontinuous and lower semicontinuous cases, we can derive the following results concerning the continuity of the solution map of PWVEP.

Theorem 9.41 *Assume that*

(i) *A is continuous and $A(\lambda)$ is compact and convex for all λ;*
(ii) *f is C-lower semicontinuous and C-upper semicontinuous on $\Gamma \times K \times K$;*
(iii) *$f(p, \cdot, y)$ is strictly $(-C)$-properly quasiconvex on K for all $p \in \Gamma$ and $y \in K$;*
(iv) *$\Omega(p, \lambda) \neq \emptyset$ for each $p \in \Gamma$ and $\lambda \in \Lambda$.*

Then Ω is continuous on $\Gamma \times \Lambda$.

Proof By Theorem 9.36, Ω is upper continuous on $\Gamma \times \Lambda$. Let $p \in \Gamma$ and $\lambda \in \Lambda$. If $\Omega(p, \lambda)$ is a singleton, then Ω is continuous at (p, λ) because Ω is upper semicontinuous on $\Gamma \times \Lambda$. If $\Omega(p, \lambda)$ has at least two points, then by Theorem 9.37, Ω is continuous at (p, λ). Hence, Ω is continuous on $\Gamma \times \Lambda$. □

Theorem 9.42 *Let K be a nonempty compact subset of a Hausdorff topological vector space X. Assume that*

(i) *$A(\lambda)$ is compact and convex for all $\lambda \in \Lambda$;*
(ii) *f is C-upper semicontinuous on $\Gamma \times K$ for all $y \in A(\lambda)$;*
(iii) *$f(\cdot, x, \cdot)$ is C-lower semicontinuous on $\Gamma \times K$ for all $x \in K$;*
(iv) *$f(p, \cdot, y)$ is strictly $(-C)$-properly quasiconvex on K for each $p \in \Gamma$ and $y \in K$;*
(v) *$\Omega(p, \lambda) \neq \emptyset$ for each $p \in \Gamma$ and $\lambda \in \Lambda$.*

Then $\Omega(\cdot, \lambda)$ is continuous on Γ for each $\lambda \in \Lambda$.

Proof By Corollary 9.14, $\Omega(\cdot, \lambda)$ is upper semicontinuous on Γ for each $\lambda \in \Lambda$. Let $p \in \Gamma$ and $\lambda \in \Lambda$. If $\Omega(p, \lambda)$ is a singleton, then Ω is continuous at (p, λ) because Ω is upper semicontinuous on $\Gamma(\cdot, \lambda)$ for each $\lambda \in \Lambda$. If $\Omega(p, \lambda)$ has at least two points, then by Theorem 9.39, $\Omega(\cdot, \lambda)$ is continuous at (p, λ) for each $\lambda \in \Lambda$. Hence, $\Omega(\cdot, \lambda)$ is continuous on $\Gamma \times \Lambda$. □

Theorem 9.43 *Assume that*

(i) *A is continuous and convex-valued;*
(ii) *f is C-lower semicontinuous and C-upper semicontinuous on $\Gamma \times K \times K$;*
(iii) *$f(p, x, \cdot)$ is C-quasiconvex on $A(\lambda)$ for each $p \in \Gamma, x \in A(\lambda)$;*
(iv) *$f(p, \cdot, y)$ is strictly $(-C)$-properly quasiconvex on K for each $p \in \Gamma$ and $y \in K$.*

Then Ω is continuous on $\Gamma \times \Lambda$.

Proof By Theorem 9.36, $\Omega(p, \lambda)$ is nonempty for each $p \in \Gamma$ and $\lambda \in \Lambda$. Hence, by Theorem 9.41, Ω is continuous on $\Gamma \times \Lambda$. □

Theorem 9.44 *Let K be a nonempty compact subset of a Hausdorff topological vector space X. Let $g : K \times K \to Y$ be a vector-valued bifunction such that the following conditions hold.*

(i) $g(x,x) = \mathbf{0}$ *for all* $x \in K$
(ii) $g(\cdot, y)$ *is C-upper semicontinuous for all* $y \in X$;
(iii) $S(\varepsilon)$ *is nonempty for all* $\varepsilon \in \text{int}(C)$.

Then S is upper semicontinuous on $\text{int}(C)$.

Proof Let $\Gamma = \text{int}(C)$, $\Lambda = \{\mathbf{0}\}$, $A(\mathbf{0}) = K$ and $f(\varepsilon, x, y) = g(x,y) + \varepsilon$. Then $f(\cdot,\cdot,y)$ is C-upper semicontinuous on $\Gamma \times K$. The result then follows from Corollary 9.14. □

Theorem 9.45 *Let K be a nonempty compact convex subset of a Hausdorff topological vector space X. Let $g : K \times K \to Y$ be a vector-valued bifunction such that the following conditions hold.*

(i) $g(x,x) = \mathbf{0}$ *for all* $x \in K$;
(ii) $g(x,\cdot)$ *is C-lower semicontinuous for all* $x \in K$;
(iii) $g(\cdot, y)$ *is strictly $(-C)$-properly quasiconvex for all* $y \in K$;
(iv) $S(\varepsilon)$ *is nonempty for each* $\varepsilon \in \text{int}(C)$.

Then, S is lower semicontinuous on $\text{int}(C)$.

Proof Let $\Gamma' = \text{int}(C)$, $\Lambda = \{\mathbf{0}\}$, $A(\mathbf{0}) = K$, $f(\varepsilon, x, y) = g(x,y) + \varepsilon$, $\Gamma = \{\mathbf{0}\} \cup \text{int}(C)$ and $\mathbf{0} = \hat{p}$. The conclusion follows from Theorem 9.40. □

9.6.3 Parametric Strong Vector Equilibrium Problems

Gong et al. [53, 55, 56] considered parametric strong vector equilibrium problems and studied the upper semicontinuity and lower semicontinuity of the solution map for such class of problems.

Let K be a nonempty subset of a Hausdorff topological vector space X. Let Y be a topological vector space with a closed pointed convex cone $C \subset Y$. Let Γ and Λ be two (index sets) nonempty subsets of two Hausdorff topological spaces, respectively. Let $f : \Gamma \times K \times K \to Y$ be a parameterized vector-valued trifunction, and $A : \Lambda \to 2^K \setminus \{\emptyset\}$ be a constraint mapping. The *parametric strong vector equilibrium problem* (in short, PSVEP) is defined as follows: For given $p \in \Gamma$ and $\lambda \in \Lambda$,

$$\begin{aligned} &\text{find } x \in A(\lambda) \text{ such that} \\ &f(p,x,y) \in C, \quad \text{for all } y \in A(\lambda). \end{aligned} \tag{9.85}$$

The solution map S of PSVEP is a set-valued map from $\Gamma \times \Lambda$ to 2^K defined by

$$S(p,\lambda) = \{x \in A(\lambda) : f(p,x,y) \in C \text{ for all } y \in A(\lambda)\}. \tag{9.86}$$

We observe that the cone C may have empty interior, see Example 9.11. If A is a constant mapping, say $A(\lambda) = K$ for all $\lambda \in \Lambda$, and $f(p,x,y) = g(x,y)$ for all $p \in \Gamma$, $x,y \in K$, where $g : K \times K \to Y$ is a vector-valued bifunction, then PSVEP reduces to the following strong vector equilibrium problem (SVEP, in short) studied in Sect. 9.1 (see also [53]):

$$\begin{aligned} &\text{find } x \in K \text{ such that} \\ &g(x,y) \in C, \quad \text{for all } y \in K. \end{aligned} \tag{9.87}$$

Throughout this subsection, unless otherwise specified, we assume that K is a nonempty subset of a Hausdorff topological vector space, Y is a locally convex space with a closed pointed convex cone $C \subset Y$, and Γ and Λ are two nonempty sets in two Hausdorff spaces, respectively. Let $A : \Lambda \to 2^K \setminus \emptyset$ be a constraint mapping and $f : \Gamma \times K \times K \to Y$ be a vector-valued trifunction with $f(p,x,x) \in C$ for all $p \in \Gamma$ and $x \in K$.

We present some existence results for solutions of PSVEP (9.85).

Theorem 9.46 *Assume that*

(i) $A(\lambda)$ *is a compact and convex set for each* $\lambda \in \Lambda$;
(ii) $f(p,\cdot,y)$ *is* C*-upper semicontinuous on* $A(\lambda)$ *for each* $p \in \Gamma, y \in A(\lambda)$;
(iii) $f(p,x,\cdot)$ *is* C*-properly quasiconvex on* $A(\lambda)$ *for each* $p \in \Gamma, x \in A(\lambda)$.

Then PSVEP *has at least one solution for each* $p \in \Gamma$ *and* $\lambda \in \Lambda$.

Proof By condition (i) and (ii), we can see that for each $p \in \Gamma, y \in A(\lambda)$, the set $\{x \in A(\lambda) : f(p,x,y) \in C\}$ is closed. In view of Corollary 3 in [49], PSVEP has at least one solution for each $p \in \Gamma$ and $\lambda \in \Lambda$. □

Example 9.11 Let $\Gamma, \Lambda = \mathbb{N}$, $X = \mathbb{R}$, $A(m) = [0,m]$, and $Y = \ell^2$. Let C be the nonnegative cone of Y. Let

$$f(p,x,y) = (g_1(x,y),\ldots,g_m(x,y),\ldots) + (\delta_p(1),\ldots,\delta_p(m),\ldots),$$

where

$$g_m(x,y) = (x-y)y^m\frac{1}{m}, \quad \text{for each } m \in \mathbb{N},$$

and

$$\delta_p(m) = \begin{cases} 1, & \text{if } p = m, \\ 0, & \text{if } p \neq m. \end{cases}$$

Hence from Theorem 9.46, we see that PSVEP has at least one solution. For each $p \in \Gamma$ and $\lambda \in \Lambda$, $\lambda \in S(p,\lambda)$

We now investigate conditions under which the solution map of PSVEP is upper semicontinuous.

Theorem 9.47 *Assume that*

(i) *A is continuous with compact values;*
(ii) *f is C-upper semicontinuous on $\Gamma \times K \times K$;*
(iii) *$S(p, \lambda) \neq \emptyset$ for each $p \in \Gamma$ and $\lambda \in \Lambda$.*

Then S is upper semicontinuous on $\Gamma \times \Lambda$.

Proof Let $p \in \Gamma$ and $\lambda \in \Lambda$. By condition (ii), $S(p, \lambda)$ is closed for each $(p, \lambda) \in \Gamma \times \Lambda$. By condition (i), $A(\lambda)$ is compact for each $\lambda \in \Lambda$. Hence $S(p, \lambda)$ is compact. Therefore, we can apply Lemma 1.10. Let $(p_\alpha, \lambda_\alpha) \to (p, \lambda)$ and $x_\alpha \in S(p_\alpha, \lambda_\alpha)$. Then for each α, we have

$$f(p_\alpha, x_\alpha, y) \in C, \quad \text{for all } y \in A(\lambda_\alpha). \tag{9.88}$$

Since A is upper semicontinuous, without loss of generality, we may assume that $x_\alpha \to x$ for some $x \in A(\lambda)$.

Assume to the contrary that $x \notin S(p, \lambda)$. Then there exists $y \in A(\lambda)$ such that

$$f(p, x, y) \notin C.$$

By condition (ii), there exists a neighborhood $\mathfrak{U} = \mathcal{P} \times \mathcal{X} \times \mathcal{Y}$ of (p, x, y) such that

$$f(p', x', y') \notin C, \quad \text{for all } (p', x', y') \in \mathfrak{U}, \tag{9.89}$$

where $\mathcal{P}$, $\mathcal{X}$, and $\mathcal{Y}$ are open neighborhoods of p, x and y, respectively.

Since A is lower semicontinuous and $y \in \mathcal{Y} \cap A(\lambda)$, i.e., $\mathcal{Y} \cap A(\lambda) \neq \emptyset$, there exists $\mathcal{L}$ which is an open neighborhood of λ such that

$$\mathcal{Y} \cap A(\lambda') \neq \emptyset, \quad \text{for all } \lambda' \in \mathcal{L}.$$

Since $(p_\alpha, \lambda_\alpha, x_\alpha) \to (p, \lambda, x)$, there exists α' such that

$$(p_\alpha, \lambda_\alpha, x_\alpha) \in \mathcal{P} \times \mathcal{L} \times \mathcal{X}, \quad \text{for all } \alpha \geq \alpha'.$$

Hence for each $\alpha \geq \alpha'$, there exists $y_\alpha \in \mathcal{Y} \cap A(\lambda_\alpha)$. Thus, for each $\alpha \geq \alpha'$, we have

$$(p_\alpha, x_\alpha, y_\alpha) \in \mathfrak{U}.$$

By (9.89), we get

$$f(p_\alpha, x_\alpha, y_\alpha) \notin C.$$

This contradicts to (9.88). Hence $x \in S(p, \lambda)$. Thus S is upper semicontinuous at (p, λ). Since $(p, \lambda) \in \Gamma \times \Lambda$ is arbitrary, S is upper semicontinuous on $\Gamma \times \Lambda$. □

The following result is a consequence of Theorems 9.46 and 9.47.

Theorem 9.48 *Let K be a nonempty compact subset of a Hausdorff topological vector space X. Assume that*

(i) *A is continuous;*
(ii) *$A(\lambda)$ is a compact and convex set for each $\lambda \in \Lambda$;*
(iii) *f is C-upper semicontinuous on $\Gamma \times K \times K$;*
(iv) *$f(p, x, \cdot)$ is C-properly quasiconvex on $A(\lambda)$ for each $p \in \Gamma$, $x \in A(\lambda)$.*

Then PSVEP *has at least one solution for each $p \in \Gamma$ and $\lambda \in \Lambda$, and S is upper semicontinuous on $\Gamma \times \Lambda$.*

Next, we investigate conditions under which the solution map of PSVEP (9.85) is lower semicontinuous.

Theorem 9.49 *Let K be a nonempty convex subset of a Hausdorff topological vector space X. Assume that*

(i) *A is continuous on Λ;*
(ii) *$A(\lambda)$ is a compact and convex set for each $\lambda \in \Lambda$;*
(iii) *for each $p \in \Gamma$, $\lambda \in \Lambda$, and $y \in A(\lambda)$, $f(p, \cdot, y)$ is strictly $(-C)$-quasiconvex on K;*
(iv) *f is C-pseudocontinuous on $\Gamma \times K \times K$;*
(v) *$S(p, \lambda)$ has at least two points for each $p \in \Gamma$ and $\lambda \in \Lambda$.*

Then S is lower semicontinuous on $\Gamma \times \Lambda$.

Proof Let $p \in \Gamma$, $\lambda \in \Lambda$. Let $\mathfrak{V}$ be an open set of K with $S(p, \lambda) \cap \mathfrak{V} \neq \emptyset$. Let $x_1, x_2 \in S(p, \lambda)$ with $x_1 \neq x_2$ and $x_1 \in \mathfrak{V}$. We choose $x' \in (x_1, x_2) = \{tx_1 + (1-t)x_2 : 0 < t < 1\}$ with $x' \in \mathfrak{V}$. By condition (iii),

$$f(p, x', y) \in C \setminus \{\mathbf{0}\}, \quad \text{for all } y \in A(\lambda).$$

By condition (iv), for each $y \in A(\lambda)$, there exist open neighborhoods $\mathcal{P}_y$ of p, $\mathcal{U}_y$ of x', and $\mathcal{V}_y$ of y such that

$$f(q, u, v) \in C, \quad \text{for all } (q, u, v) \in \mathcal{P}_y \times \mathcal{U}_y \times \mathcal{V}_y. \tag{9.90}$$

Since $A(\lambda)$ is compact, there exist $y_1, y_2, \ldots, y_m \in A(\lambda)$ such that

$$\bigcup_{i=1}^{m} \mathcal{V}_{y_i} \supset A(\lambda). \tag{9.91}$$

Let $\mathcal{P} = \bigcup_{i=1}^{m} \mathcal{P}_{y_i}$, $\mathcal{U} = \bigcup_{i=1}^{m} \mathcal{U}_{y_i}$. Then (9.90) and (9.91), we have

$$f(q, u, v) \in C, \quad \text{for all } (q, u, v) \in \mathcal{P} \times \mathcal{U} \times \left(\bigcup_{i=1}^{m} \mathcal{V}_{y_i} \right). \tag{9.92}$$

Since $x' \in A(\lambda) \cap \mathfrak{V} \cap \mathcal{U}$ and A is continuous, there exists an open neighborhood $\mathcal{M}_l$ of λ such that

$$(\mathfrak{V} \cap \mathcal{U}) \cap A(\mu) \neq \emptyset, \quad \text{for all } \mu \in \mathcal{M}_l. \tag{9.93}$$

Since A is continuous, there exists an open neighborhood $\mathcal{M}_u$ of λ such that

$$\bigcup_{i=1}^{m} \mathcal{V}_{y_i} \supset A(\mu), \quad \text{for all } \mu \in \mathcal{M}_u.$$

Hence by (9.92), for each $\mu \in \mathcal{M}_u$, we have

$$f(q, u, v) \in C, \quad \text{for all } q \in \mathcal{P}, u \in \mathcal{U}, v \in A(\mu). \tag{9.94}$$

Let $\mathcal{M} = \mathcal{M}_u \cap \mathcal{M}_l$. Then by (9.93) and (9.94), for each $(q, \mu) \in \mathcal{P} \times \mathcal{M}$, we obtain

$$\mathfrak{V} \cap S(q, \mu) \neq \emptyset.$$

Therefore, S is lower semicontinuous on $\Gamma \times \Lambda$. □

Theorem 9.50 *Let K be a nonempty convex subset of a Hausdorff topological vector space X. Assume that*

(i) *A is continuous on Λ;*
(ii) *$A(\lambda)$ is a compact and convex set for each $\lambda \in \Lambda$;*
(iii) *For each $p \in \Gamma, \lambda \in \Lambda$, and $y \in A(\lambda)$, $f(p, \cdot, y)$ is strictly $(-C)$-quasiconvex on K;*
(iv) *f is C-pseudocontinuous on $\Gamma \times K \times K$;*
(v) *$S'(p, \lambda) = \{x \in K : f(p, x, y) \in C \setminus \{\mathbf{0}\}$ for all $y \in A(\lambda)\} \neq \emptyset$ for each $p \in \Gamma$ and $\lambda \in \Lambda$.*

Then S is lower semicontinuous on $\Gamma \times \Lambda$.

Proof Let $p \in \Gamma$, $\lambda \in \Lambda$, and $x \in S'(p, \lambda)$. If $S(p, \lambda) \neq \{x\}$, by Theorem 9.49, S is lower semicontinuous at (p, λ). Hence we may assume that $S(p, \lambda) = \{x\}$. Let U be an open set of K with $U \cap S(p, \lambda) \neq \emptyset$, i.e., $x \in U$. Since $x \in S'(p, \lambda)$, we have

$$f(p, x, y) \in C \setminus \{\mathbf{0}\}, \quad \text{for all } y \in A(\lambda).$$

Hence by conditions (ii) and (iv), there exist open neighborhoods $\mathcal{P}, \mathcal{U}$, and $\mathfrak{V}$ of p, x, and $A(\lambda)$, respectively, such that

$$f(q, u, v) \in C, \quad \text{for all } (q, u, v) \in \mathcal{P} \times \mathcal{U} \times \mathfrak{V}.$$

Therefore by condition (i), there exists a neighborhood $\mathcal{M}$ of λ such that for each $\mu \in \mathcal{M}$, $U \cap \mathcal{U} \cap A(\mu) \neq \emptyset$ and

$$f(q, u, v) \in C, \quad \text{for all } q \in \mathcal{P}, u \in \mathcal{U}, v \in A(\mu).$$

Clearly, $\mathcal{U} \cap U \neq \emptyset$. Hence for each $q \in \mathcal{P}$ and $\mu \in \mathcal{M}$, we have

$$U \cap S(q, \mu) \neq \emptyset.$$

Consequently, S is lower semicontinuous at (p, λ). Since (p, λ) is arbitrary, S is lower semicontinuous on $\Gamma \times \Lambda$. □

Corollary 9.17 *Let K be a nonempty convex subset of a Hausdorff topological vector space X. Assume that*

(i) *A is continuous on Λ;*
(ii) *$A(\lambda)$ is a compact and convex set for each $\lambda \in \Lambda$;*
(iii) *for each $p \in \Gamma, \lambda \in \Lambda$, and $y \in A(\lambda)$, $f(p, \cdot, y)$ is strictly $(-C)$-quasiconvex on K;*
(iv) *f is C-pseudocontinuous on $\Gamma \times K \times K$;*
(v) *for each $p \in \Gamma$ and $\lambda \in \Lambda$, either $S'(p, \lambda) \neq \emptyset$ or $S(p, \lambda)$ has at least two elements.*

Then S is lower semicontinuous on $\Gamma \times \Lambda$.

By combining results for upper semicontinuity and lower semicontinuity, we can easily derive the results concerning the continuity of the solution map of PSVEP.

9.6.4 Well-Posedness for Parametric Weak Vector Equilibrium Problems

The concept of well-posedness is one of the approaches to study sensitivity analysis, stability, approximation and numerical analysis of solutions to nonlinear problems. There are several notions of well-posedness related to optimization problems, see, for example, [8, 38, 72, 78, 102, 105, 106] and the references therein. These notions of well-posedness can be divided mainly into three categories, namely, Tykhonov type [95], Levitin-Polyak type [73] and Hadamard type [59]. Generally speaking, in the study of Tykhonov well-posedness of a problem one introduces the notion of "approximate sequence" for the solution and requires some convergence of such sequences to a solution of the problem; Levitin-Polyak well-posedness of a problem means the convergence of the approximating solution sequence to the problem with some constraints; while, Hadamard well-posedness of a problem means the continuous dependence of the solutions on the data or on the parameter of the problem; See, for example, [8, 38, 72, 78, 102, 105, 106] and the references therein.

Kimura et al. [69] studied the parametric well-posedness for parametric weak vector equilibrium problems and showed that under suitable conditions, the well-posedness defined by approximating solutions net is equivalent to the upper semicontinuity of the solution map of perturbed problem. They also studied the relationship between well-posedness and parametric well-posedness. Bianchi et al. [25] studied two kinds of well-posedness for vector equilibrium problems. The first notion is linked to the behaviour of suitable maximizing sequences, while the second one is defined in terms of Hausdorff convergence of the map of approximate solutions. They also compared them, and, in a concave setting, they gave sufficient conditions on the data in order to guarantee well-posedness. Li and Li [74] discussed two types of Levitin-Polyak well-posedness for weak vector equilibrium (in short, WVEP) for moving cone, and investigated criteria and characterizations for them. By using nonlinear scalarization function and a gap function for WVEP, they considered a general optimization problem. Then they proved the equivalence between the Levitin-Polyak well-posedness for the optimization problem and Levitin-Polyak well-posedness for the WVEP. Salamon [88] studied and analyzed the Hadamard well-posedness for parametric strong vector equilibrium problems. Peng et al. [84] studied the generalized Hadamard well-posedness and generic Hadamard well-posedness of WVEPs by considering the perturbation of vector-valued functions and as well as of feasible sets.

Throughout this subsection, we let K be a nonempty subset of a Hausdorff topological vector space X and Y be a topological vector space with a solid pointed convex cone $C \subset Y$. Let Γ be (index set) nonempty subset of a Hausdorff topological space. Let $f : \Gamma \times K \times K \to Y$ be a parameterized vector-valued trifunction. The solution map $S(p)$ of WVEP$_p$ (9.81) is a set-valued map from Γ to 2^K defined by

$$S(p) = \{x \in K : f(p, x, y) \notin -\operatorname{int}(C) \text{ for all } y \in K\}. \tag{9.95}$$

Based on [57, Definition 3.4.2, p.119] or [36, Definition 4.74, p.230], we define the parametric (uniquely) well-posedness for parametric weak vector equilibrium problem (9.81).

Definition 9.13 Let $p \in \Gamma$ and $\{p_\lambda\} \subset \Gamma$ be a net converging to p. A net $\{x_\lambda\} \subseteq K$ is said to be an *approximating net* for WVEP$_p$ (9.81) corresponding to $\{p_\lambda\}$ if there exists a net $\{\varepsilon_\lambda\} \subset \operatorname{int}(C)$ converging to $\mathbf{0}$ such that

$$f(p_\lambda, x_\lambda, y) + \varepsilon_\lambda \notin -\operatorname{int}(C), \quad \text{for all } y \in K.$$

Definition 9.14 The family $\{\text{WVEP}_p : p \in \Gamma\}$ is said to be *parametrically well-posed* if

(a) the solution set $S(p)$ of WVEP$_p$ is nonempty for all $p \in \Gamma$;
(b) for given $p \in \Gamma$ and $\{p_\lambda\} \subset \Gamma$ with $p_\lambda \to p$, every approximating net for WVEP$_p$ corresponding to $\{p_\lambda\}$ has a subnet converging to some point of $S(p)$.

As a special case, we have the following definition.

Definition 9.15 The family $\{\text{WVEP}_p : p \in \Gamma\}$ is said to be *parametrically unique well-posed* if

(a) there exists a unique solution x_p to WVEP_p for all $p \in P$;
(b) for given $p \in \Gamma$ and $\{p_\lambda\} \subset \Gamma$ with $p_\lambda \to p$, every approximating net for WVEP_p corresponding to $\{p_\lambda\}$ converges to x_p.

Let $\Pi : \Gamma \times (\text{int}(C) \cup \{\mathbf{0}\}) \to 2^K$ be a set-valued map defined by

$$\Pi(p, \varepsilon) = \{x \in K : f(p, x, y) + \varepsilon \notin -\text{int}(C) \text{ for all } y \in K\}, \tag{9.96}$$

that is, $\Pi(p, \varepsilon)$ is an ε-solution set of WVEP_p (9.81).

Proposition 9.11 *Let K be a nonempty compact subset of a Hausdorff topological vector space X, and let S and Π be set-valued maps defined by* (9.95) *and* (9.96), *respectively. Assume that the following conditions hold:*

(i) *$f(\cdot, \cdot, y)$ is C-upper semicontinuous on $\Gamma \times K$ for each $y \in K$:*
(ii) *$S(p)$ is nonempty for each $p \in \Gamma$.*

Then S is upper semicontinuous at $p \in \Gamma$ and also Π is upper semicontinuous at $(p, \mathbf{0})$ for each $p \in \Gamma$.

Proof Let $p \in \Gamma$. Assume contrary that for any neighborhood $\mathcal{P}$ of p, there is $p_\alpha \in \mathcal{P}$ such that $S(p_\alpha) \not\subset \mathcal{V}$ for some neighborhood $\mathcal{V}$ of $S(p)$. Then there is $x_\alpha \in S(p_\alpha) \cap \mathcal{V}^c$ for all α. Since K is compact, without loss of generality, we may assume that $x_\alpha \to x$ for some $x \in K \cap \mathcal{V}^c$. Therefore,

$$f(p, x, y) \in -\text{int}(C), \quad \text{for some } y \in K.$$

By the condition (i), we have

$$f(p_\alpha, x_\alpha, y) \in -\text{int}(C), \quad \text{for all } \alpha \succeq \alpha_0 \text{ and for some } \alpha_0. \tag{9.97}$$

On the other hand, from $x_\alpha \in S(p_\alpha)$, we have

$$f(p_\alpha, x_\alpha, y) \notin -\text{int}(C), \quad \text{for all } \alpha.$$

This contradicts (9.97). Hence S is upper semicontinuous on Γ.

Next, suppose that $\Gamma' = \Gamma \times (\text{int}(C) \cup \{\mathbf{0}\})$ and that $f' : \Gamma' \times K \times K \to Y$ is defined by

$$f'(p', x, y) = f(p, x, y) + \varepsilon.$$

Then we have $f'(\cdot, \cdot, y)$ is C-upper semicontinuous on $\Gamma' \times K$ for each $y \in K$ and $\Pi(p, \varepsilon)$ is nonempty for each $p \in \Gamma$ and $\varepsilon \in \text{int}(C) \cup \{\mathbf{0}\}$. By the condition (i) (with S replaced by Π), Π is upper semicontinuous at (p, ε) for each $p \in \Gamma$ and $\varepsilon \in \text{int}(C) \cup \{\mathbf{0}\}$. Hence, Π is upper semicontinuous at $(p, \mathbf{0})$ for each $p \in \Gamma$. □

Bednarczuk [19] first defined the well-posedness as upper semicontinuity of its ε-solution mapping at $\mathbf{0}$ whenever there exists at least one solution. We discuss this concept for WVEP_p (9.81) as follows.

Lemma 9.8 *Assume that $S(p)$ is nonempty compact for each $p \in \Gamma$. Then the family $\{\text{WVEP}_p : p \in \Gamma\}$ is parametrically well-posed if and only if Π is upper semicontinuous at $(p, \mathbf{0})$ for each $p \in \Gamma$.*

Proof Suppose Π is upper semicontinuous at $(p, \mathbf{0})$. Note that $\Pi(p, \mathbf{0}) = S(p)$ is compact. By Lemma 1.10, for each $\{p_\lambda\} \subset \Gamma$ with $p_\lambda \to p$, $\{\varepsilon_\lambda\} \subset \text{int}(C) \cup \{\mathbf{0}\}$ with $\varepsilon_\lambda \to \mathbf{0}$ and $\{x_\lambda\} \subset K$ with $x_\lambda \in \Pi(p_\lambda, \varepsilon_\lambda)$, we have $\{x_\mu\} \subset \{x_\lambda\}$ such that $x_\mu \to x$ for some $x \in S(p)$. Hence for each $\{p_\lambda\} \subset \Gamma$ with $p_\lambda \to p$, every approximating net for WVEP_p (9.81) corresponding to $\{p_\lambda\}$ has a subnet converging to some point of $S(p)$. Therefore, the family $\{\text{WVEP}_p : p \in \Gamma\}$ is parametrically well-posed.

Conversely, suppose that the family $\{\text{WVEP}_p : p \in \Gamma\}$ is parametrically well-posed. Let $\{p_\lambda\} \subset \Gamma$ with $p_\lambda \to p$, $\{\varepsilon_\lambda\} \subset \text{int}(C) \cup \{\mathbf{0}\}$ with $\varepsilon_\lambda \to \mathbf{0}$ and $\{x_\lambda\} \subset K$ with $x_\lambda \in \Pi(p_\lambda, \varepsilon_\lambda)$. Then $\{x_\lambda\}$ is an approximating net for WVEP_p (9.81) corresponding to $\{p_\lambda\}$. Hence $x_\lambda \to x$ for some $x \in S(p)$. Therefore, by Lemma 1.10, Π is upper semicontinuous at $(p, \mathbf{0})$. □

When $A(\lambda) = K$ for all $\lambda \in \Lambda$, Theorem 9.35 reduces to the following result.

Theorem 9.51 *Let K be a nonempty compact convex subset of a Hausdorff topological vector space X and Γ be a nonempty set. Let $f : \Gamma \times K \times K \to Y$ be a vector-valued trifunction such that $f(p, x, x) \notin -\text{int}(C)$ for all $p \in \Gamma$ and $x \in K$. Assume that the following conditions hold.*

(i) *$f(p, \cdot, y)$ is C-upper semicontinuous on K for all $p \in \Gamma$ and $y \in K$;*
(ii) *$f(p, x, \cdot)$ is C-quasiconvex on K for all $p \in \Gamma$ and $x \in K$.*

Then WVEP_p (9.81) *has at least one solution for each $p \in \Gamma$.*

Lemma 9.9 *Let K be a nonempty compact convex subset of X and Γ be a nonempty set. Let $f : \Gamma \times K \times K \to Y$ be a vector-valued trifunction such that $f(p, x, x) \notin -\text{int}(C)$ for all $p \in \Gamma$ and $x \in X$. Assume that*

(i) *for each $p \in \Gamma$, there exists $y_p \in K$ such that*

$$f(p, x, y_p) \in -\text{int}(C), \quad \textit{for all } x \in K \setminus \{y_p\};$$

(ii) *for each $p \in \Gamma$, $S(p) \neq \emptyset$.*

Then WVEP_p (9.81) *has a unique solution for each $p \in \Gamma$.*

Proof By condition (i), every $y \in K \setminus \{y_p\}$ cannot be an element of $S(p)$ for each $p \in \Gamma$. By condition (ii), $S(p) \neq \emptyset$ for each $p \in \Gamma$. Therefore, $y_p \in S(p)$ and $y \notin S(p)$ for all $y \in K \setminus \{y_p\}$. Thus, WVEP_p (9.81) has a unique solution for each $p \in \Gamma$. □

Example 9.12 Let $\Gamma = K = [0,1]$, $Y = \mathbb{R}^2$ and $C = \mathbb{R}^2_+$. Let $f : \Gamma \times K \times K \to Y$ be defined by

$$f(p,x,y) = \left(-(x-p)^2 + (y-p)^2\right)\left(\cos\left(\frac{\pi-2}{4}+p\right), \sin\left(\frac{\pi-2}{4}+p\right)\right).$$

Then f satisfies conditions (i) and (ii) of Lemma 9.9. Hence, for each $p \in \Gamma$, WVEP$_p$ (9.81) has a unique solution. Indeed, for each $p \in \Gamma$, $S(p) = \{p\}$.

Lemma 9.10 [66] *Let K and D be nonempty compact convex sets in two topological vector spaces, respectively. If a vector-valued bifunction $f : K \times D \to Y$ satisfies the following conditions* (i) *and* (ii) *or* (i)′ *and* (ii)′,

(i) $f(\cdot, y)$ *is C-lower semicontinuous and C-quasiconvex on K for every $y \in D$,*
(ii) $f(x, \cdot)$ *is C-upper semicontinuous and $(-C)$-properly quasiconvex on D for all $x \in K$,*
(i)′ $f(\cdot, y)$ *is C-lower semicontinuous and C-properly quasiconvex on K for every $y \in D$,*
(ii)′ $f(x, \cdot)$ *is C-upper semicontinuous and $(-C)$-quasiconvex on D for every $x \in K$,*

then there exists $(x, y) \in K \times D$ such that

$$\begin{aligned} f(x,y) - f(u,y) &\notin \operatorname{int}(C), \quad \textit{for all } u \in K,\\ f(x,v) - f(x,y) &\notin \operatorname{int}(C), \quad \textit{for all } v \in E. \end{aligned} \tag{9.98}$$

A point $(x, y) \in K \times D$ is said to be *C-saddle point* of f on $K \times D$ if it satisfies (9.98).

Example 9.13 Let $g : \mathbb{R}^2 \to \mathbb{R}$ be a real-valued function defined by

$$g(u, v) = -u^2 - v^2,$$

and $C = \{(z_1, z_2, z_3) : z_1 > 0\} \cup \{\mathbf{0}\}$ be an ordering cone of $Y = \mathbb{R}^3$. Let $K = [0, 1]$ and $D = [0, 1]$ and $f : K \times D \to Y$ be defined by

$$f(x,y) = \begin{cases} (x, y, g(x,y)), & \text{if } (x,y) \in (K \setminus \{1\}) \times D,\\ \left(x, y, -\dfrac{4}{3}\right), & \text{if } (x,y) = (1,0),\\ \left(x, y, -\dfrac{3}{2}(y+1)\right), & \text{otherwise.} \end{cases}$$

Then f satisfies the conditions (i) and (ii) of Lemma 9.10. Hence, it has at least one C-saddle point on $K \times D$. In fact, $(1, 0) \in K \times D$ is a C-saddle point of f on $K \times D$.

Lemma 9.11 *Let K be a nonempty compact convex subset of a Hausdorff topological vector space X and Γ be a nonempty set. Let $g : K \to Y$ be strictly $(-C)$-proper* *quasiconvex, C-upper semicontinuous on K, and* $g(\bar{x}) \notin g(\hat{x}) - \mathrm{int}(C)$ *for some* $\bar{x} \neq \hat{x} \in K$. *Then there exists* $x' \in K$ *such that* $g(x') \in g(\hat{x}) + \mathrm{int}(C)$.

Proof Since g is strictly $(-C)$-properly quasiconvex on K, we have

$$g(\bar{x}') \in (g(\bar{x}) + \mathrm{int}(C)) \cup (g(\hat{x}) + \mathrm{int}(C)),$$

where $\bar{x}' = \dfrac{\bar{x} + \hat{x}}{2}$. Hence $g(\bar{x}') \notin -\mathrm{cl}\,(C)$. Let $\varepsilon \in \mathrm{int}(C)$. Then $h\,(g(\bar{x}') - g(\hat{x})) =:$ $\alpha > 0$, where

$$h(z) := \inf\{t \in \mathbb{R} : z \in t \cdot \varepsilon - \mathrm{int}(C)\}, \quad \text{for } z \in Y.$$

Thus $g(\bar{x}') \notin g(\hat{x}) + \frac{1}{2}\alpha\varepsilon - \mathrm{int}(C)$. Clearly, $g(\hat{x}) + \frac{1}{2}\alpha\varepsilon - \mathrm{int}(C)$ is a neighborhood of $g(\hat{x})$ and $\left(g(\hat{x}) + \frac{1}{2}\alpha\varepsilon - \mathrm{int}(C)\right) - \mathrm{int}(C) = g(\hat{x}) + \frac{1}{2}\alpha\varepsilon - \mathrm{int}(C)$. Since g is $(-C)$-convex on K, there is a neighborhood $\mathcal{U}$ of $\hat{x}$ such that

$$g(x) \in g(\hat{x}) + \frac{1}{2}\alpha\varepsilon - \mathrm{int}(C), \quad \text{for all } x \in \mathcal{U}.$$

Therefore, we can choose $x' \in \mathcal{U} \cap \{u \in X : u = \mu\hat{x} + (1-\mu)\bar{x}',\ 0 < \mu < 1\}$. Then $g(x') \in (g(\hat{x}) + \mathrm{int}(C)) \cap \left(\frac{1}{2}\alpha\varepsilon - \mathrm{int}(C)\right)$. Thus, $g(x') \in g(\hat{x}) + \mathrm{int}(C)$. □

Theorem 9.52 *Let K be a nonempty compact convex subset of a Hausdorff topological vector space X,* Γ *be a nonempty set and* $f : \Gamma \times K \times K \to Y$ *be a vector-valued trifunction such that* $f(p, x, x) = \mathbf{0}$ *for all* $p \in \Gamma$ *and* $x \in K$. *Assume that*

(i) $f(p, \cdot, y)$ *is C-upper semicontinuous and strictly* $(-C)$*-properly quasiconvex on K for each* $p \in \Gamma$ *and* $y \in K$;
(ii) $f(p, x, \cdot)$ *is C-lower semicontinuous and C-quasiconvex on K for each* $p \in \Gamma$ *and* $x \in K$.

Then WVEP$_p$ (9.81) *has a unique solution for each* $p \in \Gamma$.

Proof Let $p \in \Gamma$. Then by Lemma 9.10, there exists $(x, y) \in K \times K$ such that

$$f(p, x, y) - f(p, u, y) \notin -\mathrm{int}(C), \quad \text{for all } u \in K, \tag{9.99}$$

$$f(p, x, v) - f(p, x, y) \notin -\mathrm{int}(C), \quad \text{for all } v \in K. \tag{9.100}$$

From (9.99), we have

$$f(p, u, y) \notin f(p, x, y) + \mathrm{int}(C), \quad \text{for all } u \in X. \tag{9.101}$$

Taking $v = x$, (9.100) implies $f(p, x, y) \notin \text{int}(C)$. Suppose $f(p, x, y) \neq \mathbf{0}$. Then

$$f(p, y, y) = \mathbf{0} \notin f(p, x, y) - \text{int}(C).$$

By Lemma 9.11, there exists $x' \in X$ such that

$$f(p, x', y) \in f(p, x, y) + \text{int}(C),$$

which contradicts to (9.101). Thus, $f(p, x, y) = \mathbf{0}$. Suppose that $x \neq y$. Then, $f(p, y, y) = \mathbf{0}$ and by Lemma 9.11, there exists $x' \in X$ such that

$$f(p, x', y) \in f(p, x, y) + \text{int}(C).$$

This contradicts to (9.101). Thus,

$$x = y =: \hat{x}.$$

Then by (9.100), we have

$$f(p, \hat{x}, v) - f(p, \hat{x}, \hat{x}) \notin -\text{int}(C), \quad \text{for all } v \in X,$$

that is,

$$f(p, \hat{x}, v) \notin -\text{int}(C), \quad \text{for all } v \in X.$$

Therefore, $\hat{x} \in S(p)$ and thus $S(p) \neq \emptyset$ for each $p \in \Gamma$. By (9.101), we have

$$f(p, u, \hat{x}) \notin \text{int}(C), \quad \text{for all } u \in X.$$

Therefore, by Lemma 9.11, we have

$$f(p, u, \hat{x}) \in -\text{int}(C), \quad \text{for all } u \in X \setminus \{\hat{x}\}.$$

Thus, by Lemma 9.9, $\{\hat{x}\} = S(p)$, that is, WVEP$_p$ (9.81) has a unique solution. □

Theorem 9.53 *Let K be a nonempty compact convex subset of a Hausdorff topological vector space X and Γ be a topological space. Let $f : \Gamma \times K \times K \to Y$ be a vector-valued trifunction such that $f(p, x, x) = \mathbf{0}$ for all $p \in \Gamma$ and $x \in K$. Assume that the following conditions hold.*

(i) *$f(\cdot, \cdot, y)$ is C-upper semicontinuous on $\Gamma \times X$ for all $y \in K$;*
(ii) *$f(p, \cdot, y)$ is strictly $(-C)$-properly quasiconvex on X for each $p \in \Gamma$ and $y \in K$;*
(iii) *$f(p, x, \cdot)$ is C-lower semicontinuous and C-quasiconvex on K for each $p \in \Gamma$ and $x \in K$.*

Then $\{\text{WVEP}_p : p \in \Gamma\}$ *is parametrically unique well-posed.*

Proof By Theorem 9.52, WVEP_p (9.81) has a unique solution x_p for all $p \in \Gamma$. Note that for each $y \in K$ the map $(p, \varepsilon, x) \mapsto f(p, x, y) + \varepsilon$ is C-upper semicontinuous on $\Gamma \times Y \times K$. Hence by Corollary 9.15, Π is upper semicontinuous on $\Gamma \times (\text{int}(C) \cup \{\mathbf{0}\})$. Clearly, $S(p)$ is compact for each $p \in \Gamma$. Therefore, by Lemma 9.8, $\{\text{WVEP}_p : p \in \Gamma\}$ is parametrically unique well-posed. □

Corollary 9.18 *Let K be a nonempty compact convex subset of a Hausdorff topological vector space X, Γ be a nonempty subset of a Hausdorff space and $f : \Gamma \times K \times K \to Y$ be a vector-valued trifunction such that $f(p, x, x) \notin -\text{int}(C)$ for all $p \in \Gamma$ and $x \in K$. Assume that the following conditions hold.*

(i) *$f(\cdot, \cdot, y)$ is C-upper semicontinuous on $\Gamma \times K$ for each $y \in K$;*
(ii) *$S(p) \neq \emptyset$ for each $p \in \Gamma$.*

Then $\{\text{WVEP}_p : p \in \Gamma\}$ is parametrically well-posed.

Proof By Corollary 9.15, Π is upper semicontinuous on $\Gamma \times (\text{int}(C) \cup \{\mathbf{0}\})$. By condition (i), $S(p)$ is closed for each $p \in \Gamma$. Since K is compact, $S(p)$ is compact for each $p \in \Gamma$. Therefore, by Lemma 9.8, $\{\text{WVEP}_p : p \in \Gamma\}$ is parametrically well-posed. □

References

1. S. Al-Homidan, Q.H. Ansari, J.-C. Yao, Some generalizations of Ekeland-type variational principle with applications to equilibrium problems and fixed point theory. Nonlinear Anal. **69**(1), 126–139 (2008)
2. L. Altangerel, R.I. Bot, G. Wanka, Variational principles for vector equilibrium problems related to conjugate duality. J. Nonlinear Convex Anal. **8**(2), 179–196 (2007)
3. Q.H. Ansari, Vector equilibrium problems and vector variational inequalities, in: *Vector Variational Inequalities and Vector Equilibria. Mathematical Theories*, ed. by F. Giannessi (Kluwer Academic, Dordrecht/Boston/London, 2000), pp. 1–15
4. Q.H. Ansari, Vectorial form of Ekeland-type variational principle with applications to vector equilibrium problems and fixed point theory. J. Math. Anal. Appl. **334**, 561–575 (2007)
5. Q.H. Ansari, *Metric Spaces – Including Fixed Point Theory and Set-Valued Maps* (Narosa Publishing House, New Delhi, 2010). Also Published by Alpha Science International Ltd. Oxford, U.K. (2010)
6. Q.H. Ansari, Ekeland's variational principle and its extensions with applications, in: *Topics in Fixed Point Theory*, ed. by S. Almezel, Q.H. Ansari, M.A. Khamsi (Springer, Cham/Heidelberg/New York/Dordrecht/London, 2014), pp. 65–100
7. Q.H. Ansari, L.-J. Lin, Ekeland type variational principle and equilibrium problems, in: *Topics in Nonconvex Optimization, Theory and Applications*, ed. by S.K. Mishra (Springer, New York/Berlin, 2011), pp. 147–174
8. Q.H. Ansari, J.-C. Yao (eds.), *Recent Developments in Vector Optimization* (Springer, Heidelberg, 2012)
9. Q.H. Ansari, I.V. Konnov, J.C. Yao, On generalized vector equilibrium problems. Nonlinear Anal. **47**, 543–554 (2001)
10. Q.H. Ansari, X.Q. Yang, J.C. Yao, Existence and duality of implicit vector variational problems. Numer. Funct. Anal. Optim. **22**(7–8), 815–829 (2001)

11. Q.H. Ansari, I.V. Konnov, J.C. Yao, Characterizations of solutions for vector equilibrium problems. J. Optim. Theory Appl. **113**, 435–447 (2002)
12. Y. Araya, Ekeland's variational principle and its equivalent theorems in vector optimization. J. Math. Anal. Appl. **346**, 9–16 (2008)
13. Y. Araya, K. Kimura, T. Tanaka, Existence of vector equilibria via Ekeland's variational principle. Taiwan. J. Math. **12**(8), 1991–2000 (2008)
14. J.-P. Aubin, *Mathematical Methods of Game and Economic Theory* (North-Holland, Amsterdam/New York/Oxford, 1979)
15. J.-P. Aubin, I. Ekeland, *Applied Nonlinear Analysis* (Wiley, New York, 1984)
16. J.-P. Aubin, H. Frankowska, *Set-Valued Analysis* (Birkhäuser, Boston/Basel/Berlin, 1990)
17. G. Auchmuty, Variational principles for variational inequalities. Numer. Funct. Anal. Optim. **10**, 863–874 (1989)
18. C. Baiocchi, A. Capelo, *Variational and Quasivariational Inequalities, Applications to Free Boundary Problems* (Wiley, New York, 1984)
19. E. Bednarczuk, An approach to well-posedness in vector optimization: consequence to stability, parametric optimization. Control Cybernet. **23**, 107–122 (1994)
20. M. Bianchi, R. Pini, A note on equilibrium problems for properly quasimonotone bifunctions. J. Global Optim. **20**, 67–76 (2001)
21. M. Bianchi, R. Pini, Coercivity conditions for equilibrium problems. J. Optim. Theory Appl. **124**, 79–92 (2005)
22. M. Bianchi, R. Pini, Sensitivity for parametric vector equilibria. Optimization **55**(3), 221–230 (2006)
23. M. Bianchi, N. Hadjisavvas, S. Schaible, Vector equilibrium problems with generalized monotone bifunctions. J. Optim. Theory Appl. **92**(3), 527–542 (1997)
24. M. Bianchi, G. Kassay, R. Pini, Ekelend's principle for vector equilibrium problems. Nonlinear Anal. **66**, 1454–1464 (2007)
25. M. Bianchi, G. Kassay, R. Pini, Well-posedness for vector equilibrium problems. Math. Methods Oper. Res. **70**, 171–182 (2009)
26. G. Bigi, A. Capătă, G. Kassay, Existence results for strong vector equilibrium problems and their applications. Optimization **61**(5), 567–583 (2012)
27. D. Blackwell, An analog of the minimax theorem for vector payoffs. Pac. J. Math. **6**(1), 1–8 (1956)
28. E. Blum, W. Oettli, Variational principles for equilibrium problems, in: *Parametric Optimization and Related Topics III*, ed. by J. Guddat et al. (Peter Lang, Frankfurt am Main, 1993), pp. 79–88
29. E. Blum, W. Oettli, From optimization and variational inequalities to equilibrium problems. Math. Stud. **63**, 123–145 (1994)
30. A. Capătă, Existence and generalized duality of strong vector equilibrium problems. Acta Univ. Sapientiae Math. **4**(1), 36–52 (2012)
31. A. Capătă, G. Kassay, On vector equilibrium problems and applications. Taiwan. J. Math. **15**(11), 365–380 (2011)
32. A. Capătă, G. Kassay, Existence results for strong vector equilibrium problems given by a sum of two functions with applications. J. Global Optim. **63**(1), 195–211 (2015)
33. J. Caristi, W.A. Kirk, Geometric fixed point theory and inwardness conditions, in *The Geometry of Metric and Linear Spaces*, Lecture Notes in Mathematics, vol. 490 (Springer, Berlin/Heidelberg/New York, 1975), pp. 74–83
34. L.C. Ceng, G. Mastroeni, J.C. Yao, Existence of solutions and variational principles for generalized vector systems. J. Optim. Theory Appl. **137**, 485–495 (2008)
35. C.R. Chen, M.H. Li, Hölder continuity of solutions to parametric vector equilibrium problems with nonlinear scalarization. Numer. Funct. Anal. Optim. **35**, 685–707 (2014)
36. G.Y. Chen, X.X. Huang, X.Q. Yang, *Vector Optimization: Set-Valued and Variational Analysis* (Springer, Berlin/Heidelberg, 2005)
37. I.C. Dolcetta, M. Matzeu, Duality for implicit variational problems and numerical applications. Numer. Funct. Anal. Optim. **2**(4), 231–265 (1980)

38. A.L. Dontchev, T. Zolezzi, *Well-posed Optimization Problems* (Springer, Berlin/Heidelberg, 1993)
39. I. Ekeland, Sur les problému variationnels. C.R. Acad. Sci. Paris **275**, 1057–1059 (1972)
40. I. Ekeland, On the variational principle. J. Math. Anal. Appl. **47**, 324–353 (1974)
41. I. Ekeland, On convex minimization problems. Bull. Am. Math. Soc. **1**(3), 445–474 (1979)
42. C.A. Elisabeta, Existence theorems for vector equilibrium problems via quasi-relative interior. Sci. World J. **2013**, Article ID 150130 (2013)
43. F. Facchinei, J.-S. Pang, *Finite Dimensional Variational Inequalities and Complementarity Problems*, vols. I & II (Springer, New York/Berlin/Heidelberg, 2003)
44. X. Fan, C. Cheng, H. Wang, Sensitivity analysis for vector equilibrium problems under functional perturbations. Numer. Funct. Anal. Optim. **35**(5), 564–575 (2014)
45. M. Fang, N.-J. Huang, KKM type theorems with applications to generalized vector equilibrium problems in FC-spaces. Nonlinear Anal. **67**, 809–817 (2007)
46. A.P. Farajzadeh, A. Amini-Harandi, D. O'Regan, R.P. Agarwal, Strong vector equilibrium problems in topological vector spaces via KKM maps. CUBO Math. J. **12**, 219–230 (2010)
47. C. Finet, L. Quarta, Vector-valued perturbed equilibrium problems. J. Math. Anal. Appl. **343**, 531–545 (2008)
48. F. Flores-Bazán, F. Flores-Bazán, Vector equilibrium problems under recession analysis. J. Global Optim. **26**(2), 141–166 (2003)
49. J.Y. Fu, Generalized vector quasi-equilibrium problems. Math. Methods Oper. Res. **52**, 57–64 (2000)
50. M. Fukushima, *Bases of Nonlinear Analysis* (Asakura, Tokyo, 2001)
51. F. Giannessi, Theorems of alternative, quadratic programs and complementarity problems, in: *Variational Inequalities and Complementarity Problems*, ed. by R.W. Cottle, F. Giannessi, J.L. Lions (Wiley, New York, 1980), pp. 151–186
52. F. Giannessi, *Constrained Optimization and Image Space Analysis* (Springer, New York, 2005)
53. X.H. Gong, Strong vector equilibrium problems. J. Global Optim. **36**, 339–349 (2006)
54. X.H. Gong, Continuity of the solution set to parametric weak vector equilibrium problems. J. Optim. Theory Appl. **139**, 35–46 (2008)
55. X.H. Gong, J.-C. Yao, Lower semicontinuity of the set of efficient solutions for generalized systems. J. Optim. Theory Appl. **138**, 197–205 (2008)
56. X.H. Gong, K. Kimura, J.-C. Yao, Sensitivity analysis of strong vector equilibrium problems. J. Nonlinear Convex Anal. **9**(1), 83–94 (2008)
57. A. Göpfert, C. Tammer, H. Riahi, C. Zălinescu, *Variational Methods in Partially Ordered Spaces* (Springer, Berlin/Heidelberg, 2003)
58. J. Guddat, F. Guerra Vazquez, H.T. Jongen, *Parametric Optimization: Singularities, Pathfollowing and Jumps* (Vieweg+Teubner, Wiesbaden, 1990)
59. J. Hadamard, Sur les problèmes aux dérivées partielles et leur signification physique. Princeton Univ. Bull., **13**, 49–52 (1902)
60. N. Hadjisavvas, S. Schaible, From scalar to vector equilibrium problems in the quasimonotone case. J. Optim. Theory Appl. **96**, 297–305 (1998)
61. N. Hadjisavvas, S. Schaible, Quasimonotonicity and pseudomonotonicity in variational inequalities and equilibrium problems, in: *Generalized Convexity, Generalized Monotonicity: Recent Results*, ed. by J.P. Crouzeix, J.E. Martinez-Legaz, M. Volle (Kluwer Academic, Dordrecht/Boston/London, 1998), pp. 257–275
62. N.-J. Huang, J. Li, H.B. Thompson, Stability for parametric implicit vector equilibrium problems. Math. Comput. Model. **43**, 1267–1274 (2006)
63. A.N. Iusem, W. Sosa, New existence results for equilibrium problems. Nonlinear Anal. **52**, 621–635 (2003)
64. G. Kassay, M. Miholca, Existence results for vector equilibrium problems given by a sum of two functions. J. Global Optim. **63**, 195–211 (2015)
65. A.A. Khan, C. Tammer, C. Zălinescu, *Set-Valued Optimization: An Introduction with Applications* (Springer, Berlin/New York, 2015)

66. K. Kimura, T. Tanaka, Existence theorem of cone saddle-point applying a nonlinear scalarization. Taiwan. J. Math. **10**(2), 563–571 (2006)
67. K. Kimura, J.C. Yao, Sensitivity analysis of vector equilibrium problems. Taiwan. J. Math. **12**(3), 649–669 (2007)
68. K. Kimura, Y.C. Liou, J.C. Yao, Semicontinuity of the solution mapping of ε-vector equilibrium problem. Pac. J. Optim. **3**(2), 345–359 (2007)
69. K. Kimura, Y.C. Liou, S.-Y. Wu, J.-C. Yao, Well-posedness for parametric vector equilibrium problems with applications. J. Ind. Manag. Optim. **4**, 313–327 (2008)
70. I.V. Konnov, On vector equilibrium and vector variational inequality problems, in: *Generalized Convexity and Generalized Monotonicity*, ed. by N. Hadjisavvas, S. Komlósi, S. Schaible (Springer, Berlin/Heidelberg, 2001), pp. 247–263
71. G.M. Lee, D.S. Kim, B.S. Lee, On noncooperative vector equilibrium. Indian J. Pure Appl. Math. **27**(8), 735–739 (1996)
72. B. Lemaire, C.O.A. Salem, J.P. Revalski, Well-posedness by peturbations of variational problems. J. Optim. Theory Appl. **115**, 345–368 (2002)
73. E.S. Levitin, B.T. Polyak, Convergence of minimizing sequences in conditional extremum problems. Sov. Math. Dokl. **7**, 764–767 (1966)
74. S.J. Li, H.M. Li, Levitin-Polyak well-posedness of vector equilibrium problems. Math. Methods Oper. Res. **69**, 125–140 (2009)
75. S.J. Li, H.M. Li, Sensitivity analysis of parametric weak vector equilibrium problems. J. Math. Anal. Appl. **380**, 354–362 (2011)
76. S.J. Li, S.F. Yao, C.R. Chen, Saddle points and gap functions for vector equilibrium problems via conjugate duality in vector optimization. J. Math. Anal. Appl. **343**, 853–865 (2008)
77. P. Loridan, Epsilon-solutions in vector minimization problems. J. Optim. Theory Appl. **43**(2), 265–276 (1984)
78. R. Lucchetti, J.P. Revalski (eds.), *Recent Developments in Well-Posed Variational Problems* (Kluwer Academic, Dordrecht, 1995)
79. G. Mastroeni, B. Panicucci, M. Passacantando, J.-C. Yao, A separation approach to vector quasi-equilibrium problems: Saddle point and gap function. Taiwan. J. Math. **13**(2B), 657–673 (2009)
80. H. Mirazaee, M. Soleimani-damaneh, Derivatives of set-valued maps and gap functions for vector equilibrium problems. Set-Valued Var. Anal. **22**, 673–689 (2014)
81. U. Mosco, Implicit variational problems and quasi-variational inequalities, in: *Lecture Notes in Mathematics*, vol. 543 (Springer, Berlin, 1976)
82. A.B. Németh, Between pareto efficiency and pareto epsilon-efficiency. Optimization **20**(5), 615–637 (1989)
83. W. Oettli, M. Théra, Equivalents of Ekeland's principle. Bull. Aust. Math. Soc. **48**, 385–392 (1993)
84. D.-T. Peng, J. Yu, N.-H. Xiu, The uniqueness and well-posedness of vector equilibrium problems with a representation for the solution set. Fixed Point Theory Appl. **2014**, Article ID 115 (2014)
85. Q. Qiu, X. Yang, Scalarization of approximate solution for vector equilibrium problems. J. Ind. Manag. Optim. **9**(1), 143–151 (2013)
86. J. Salamon, Closedness of the solution map for parametric vector equilibrium problems. Stud. Univ. Babes-Bolyai Math. **LIV**(3), 137–147 (2009)
87. J. Salamon, Closedness of the solution mapping to parametric vector equilibrium problems. Acta Univ. Sapientiae Math. **1**(2), 193–200 (2009)
88. J. Salamon, Closedness and Hadamard well-posedness of the solution map for parametric vector equilibrium problems. J. Global Optim. **47**, 173–183 (2010)
89. J. Salamon, M. Bogdan, Closedness of the solution map for parametric weak vector equilibrium problems. J. Math. Anal. Appl. **364**, 483–491 (2010)
90. X.-K. Sun, S.-J. Li, Gap functions for generalized vector equilibrium problems via conjugate duality and applications. Appl. Anal. **92**(10), 2182–2199 (2013)

91. W. Takahashi, *Nonlinear Functional Analysis – Fixed Point Theory and Its Applications* (Yokohama Publishers, Yokohama, 2000)
92. C. Tammer, A generalization of ekellandzs variational priciple. Optimization **25**(2–3), 129–141 (1992)
93. C. Tammer, Existence results and necessary conditions for epsilon-efficient elements, in *Multicriteria Decision, Proceedings of the 14th Meeting of the German Working Group "Mehrkriterielle Entscheidung"*, ed. by B. Brosowski et al. (Peter Lang, Frankfurt am Main, 1992), pp.97–110
94. N.X. Tan, P.N. Tinh, On the existence of equilibrium points of vector functions. Numer. Funct. Anal. Optim. **19**, 141–156 (1998)
95. A.N. Tykhonov, On the stability of the functional optimization problem. USSR Comput. Math. Math. Phys. **6**, 28–33 (1966)
96. S.-H. Wang, Q.-Y. Li, A projection iterative algorithm for strong vector equilibrium problem. Optimization **64**(10), 2049–2063 (2015)
97. L. Wei-zhong, H. Shou-jun, Y. Jun, Optimality conditions for ε-vector equilibrium problems. Abstr. Appl. Anal. **2014**, Article ID 509404 (2014)
98. D.J. White, Epsilon efficiency. J. Optim. Theory Appl. **49**(2), 319–337 (1986)
99. S. Xu, S.J. Li, A new proof approach to lower semicontinuity for parametric vector equilibrium problems. Optim. Lett. **3**, 453–459 (2009)
100. X.Q. Yang, A Hahn-Banach theorem in ordered linear spaces and its applications. Optimization **25**, 1–9 (1992)
101. X.Q. Yang, Vector variational inequality and its duality. Nonlinear Anal. Theory Methods Appl. **21**, 869–877 (1993)
102. H. Yang, J. Yu, Unified approaches to well-posedness with some applications. J. Global Optim. **31**, 371–381 (2005)
103. W.Y. Zhang, Z.M. Fang, Y. Zhang, A note on the lower semicontinuity of efficient solutions for parametric vector equilibrium problems. Appl. Math. Lett. **26**, 469–472 (2013)
104. W. Zhang, J. Chen, S. Xu, W. Dong, Scalar gap functions and error bounds for generalized mixed vector equilibrium problems with applications. Fixed Point Theory Appl. **2015**, Article ID 169 (2015)
105. T. Zolezzi, Well-posedness criteia in optimization with application to the calculus of variations. Nonlinear Anal. **25**, 437–453 (1995)
106. T. Zolezzi, Extended well-posedness of optimization problems. J. Optim. Theory Appl. **91**, 257–266 (1996)

Chapter 10
Generalized Vector Equilibrium Problems

We have seen in Chaps. 8 and 9 that the vector optimization problems can be solved by using generalized vector variational inequality problems (in short, GVVIPs), and that the vector equilibrium problems (in short, VEPs) contain vector variational inequality problems and vector optimization problems as special cases. The vector equilibrium problems with set-valued maps are called generalized vector equilibrium problems (in short, GVEPs). There are several possible ways to extend vector equilibrium problems for set-valued maps in such a way that they contain GVVIPs as special cases; See, for example [5, 6, 11, 19–23, 29, 30, 37, 47, 48, 51, 52] and the references therein. During the last two decades, several existence results for solutions of GVEPs and their special cases, namely, weak form of generalized vector variational inequality problems, weak form of implicit vector variational inequality problems, etc., have been discussed. The basic tools to prove the existence results for solutions of GVEPs are Browder type fixed point theorems, Kakutani fixed point theorem, coincidence theorems, KKM type theorems, intersection theorems and maximal element theorems; See, for example, [4–11, 19–25, 27, 29, 41, 47–49, 51, 60, 63] and the references therein. The connectedness of the solution set of generalized vector equilibrium problems is studied by Chen et al. [17], Han and Huang [35] and Liu et al. [50]; See also references therein. The duality for generalized vector equilibrium problems is defined by using a rule that says that "dual of the dual is the primal", and it has been discussed in [7].

The generalized vector equilibrium problems related to a set-valued bifunction provide only the weak form of the corresponding GVVIPs. Therefore, several authors considered the generalized vector equilibrium problems for a vector-valued trifunction. Such problems include the strong form of corresponding GVVIPs. See, for example, [12, 13, 18, 28, 32, 36, 64] and the references therein.

Most of the results on the existence of solutions for GVEPs require either compactness or some kind of coercivity condition in the setting of topological vector spaces. However, the boundedness is assumed in the setting of reflexive Banach spaces. Recession method that was initiated by Flores-Bazán and Flores-Bazán [26]

Q.H. Ansari et al., *Vector Variational Inequalities and Vector Optimization*,
Vector Optimization, DOI 10.1007/978-3-319-63049-6_10

for vector equilibrium problems, does not require boundedness, compactness or any kind of coercivity condition to prove the necessary and/or sufficient conditions for the solution set of GVEPs to be nonempty and bounded/compact. This topic is mainly studied in [3, 42, 57, 59] and the references therein.

Sometime an exact solution of GVEPs may not exist if the data of the problem is not sufficient. In such cases, we study the approximate solutions or ε-solutions of GVEPs. Kimura and Yao [38, 39] considered ε-generalized vector (weak/strong) equilibrium problems and studied the existence of solutions of these problems. They also studied the behavior of the solution maps of these problems. Namely, they studied the upper semicontinuity and lower semicontinuity of solution maps of these problems. The study of behavior of solution maps plays an important role in stability and sensitivity analysis for generalized vector equilibrium problems. Gong [31] and Qu and Cheng [56] used vectorial form of Ekeland's variational principle for set-valued maps to study the existence of solutions of ε-generalized vector weak/strong equilibrium problems. The parametric generalized vector equilibrium problems are more general than those of ε-generalized vector equilibrium problems. Several authors considered parametric generalized vector equilibrium problems (in short, PGVEPs) and studied the behavior of their solution maps; See, for example, [1, 2, 14–16, 33–35, 43–46, 54, 55, 58, 59, 61, 62] and the references therein.

The main motivation of this chapter is to study the mathematical theory of generalized vector equilibrium problems, namely, existence results for solutions with or without monotonicity, duality and sensitivity analysis.

10.1 Introduction

Let K be a nonempty subset of a topological vector space X, Y be a topological vector space with an ordering cone C such that $\text{int}(C) \neq \emptyset$ and $F : K \times K \to 2^Y$ be a set-valued map with nonempty values. The *generalized weak vector equilibrium problems* (in short, GWVEPs) are defined as follows:

$$\text{find } \bar{x} \in K \text{ such that } F(\bar{x}, y) \nsubseteq -\text{int}(C), \quad \text{for all } y \in K \tag{10.1}$$

and

$$\text{find } \bar{x} \in K \text{ such that } F(\bar{x}, y) \cap (-\text{int}(C)) = \emptyset, \quad \text{for all } y \in K; \tag{10.2}$$

the latter problem can also be written in the following form:

$$\text{find } \bar{x} \in K \text{ such that } F(\bar{x}, y) \subseteq Y \setminus (-\text{int}(C)), \quad \text{for all } y \in K.$$

The set of solutions of generalized weak vector equilibrium problems (10.1) and (10.2) are denoted by Sol(GWVEP) and Sol(GWVEP)*, respectively. Clearly, Sol(GWVEP)* $\subseteq$ Sol(GWVEP).

The *generalized vector equilibrium problem* (in short, GVEP) is to find $\bar{x} \in K$ such that

$$F(\bar{x}, y) \nsubseteq -C \setminus \{\mathbf{0}\}, \quad \text{for all } y \in K. \tag{10.3}$$

We denote by Sol(GVEP) the set of solutions of GVEP (10.3).

The *generalized strong vector equilibrium problems* (in short, GSVEPs) are defined in the following ways:

$$\text{find } \bar{x} \in K \text{ such that } F(\bar{x}, y) \subseteq C, \quad \text{for all } y \in K; \tag{10.4}$$

and

$$\text{find } \bar{x} \in K \text{ such that } F(\bar{x}, y) \cap C \neq \emptyset, \quad \text{for all } y \in K. \tag{10.5}$$

The set of solutions of problems (10.4) and (10.5) are denoted by Sol(GSVEP) and Sol(GSVEP)*, respectively. Clearly, a solution of problem (10.4) is a solution of problem (10.5).

These problems contain corresponding generalized vector variational inequality problems as special cases, and can be used to study vector optimization problems for nondifferentiable (in some sense) or/and nonconvex vector optimization problems. Several other formulations of GVEPs are studied in [6, 19, 23, 52, 53].

We denote by $\langle u, x \rangle$ the evaluation of $u \in \mathcal{L}(X, Y)$ at $x \in X$. Let $T : K \to 2^{\mathcal{L}(X,Y)}$ be a set-valued map with nonempty values. Let $F(x, y) = \langle T(x), y - x \rangle = \{\langle u, y - x \rangle : u \in T(x)\}$. Then the GVEPs reduce to the weak form of the corresponding generalized vector variational inequality problems, namely, the generalized weak vector equilibrium problems reduce to the following weak form of generalized weak vector variational inequality problem:

$(\text{GVVIP})_w$: Find $\bar{x} \in K$ such that for all $y \in K$, there exists $\bar{\zeta} \in T(\bar{x})$ satisfying

$$\langle \bar{\zeta}, y - \bar{x} \rangle \notin -\operatorname{int}(C). \tag{10.6}$$

For other special cases of GVEPs, see [5, 8] and the references therein.

10.2 Generalized Abstract Vector Equilibrium Problems

Let X be a locally convex Hausdorff topological vector space, Y and Z be topological vector spaces, $A \subseteq X$ be a nonempty convex compact set, $B \subseteq Z$ be a nonempty convex set, and $P \subseteq Y$. Let $F : A \times B \to 2^Y$ be a set-valued map with nonempty values. We consider the following *generalized abstract vector equilibrium problems* (in short, GAVEPs):

$$\text{find } \bar{x} \in A \text{ such that } F(\bar{x}, y) \cap P \neq \emptyset, \quad \text{for all } y \in B, \tag{10.7}$$

and

$$\text{find } \bar{x} \in A \text{ such that } F(\bar{x}, y) \subseteq P, \quad \text{for all } y \in B. \tag{10.8}$$

With $P := Y \setminus (-\operatorname{int}(C))$ it is obvious that (10.7) contains (10.1), and with $P := C$ it is obvious that (10.8) contains (10.4). Using the lower inverse $F^{-}(P)$ and the upper inverse $F^{+}(P)$ of the set-valued map F, which are defined as

$$F^{-}(P) := \{(x, y) \in A \times B : F(x, y) \cap P \neq \emptyset\},$$

$$F^{+}(P) := \{(x, y) \in A \times B : F(x, y) \subseteq P\},$$

both GAVEP (10.7) and (10.8) can be written as

$$\text{find } \bar{x} \in A \text{ such that } (\bar{x}, y) \in F^{-1}(P), \quad \text{for all } y \in B, \tag{10.9}$$

with $F^{-1} := F^{-}$ for (10.7) and $F^{-1} := F^{+}$ for (10.8).

In order to obtain existence results for problem (10.9), we introduce another set-valued map $G : A \times B \to 2^{Y}$, and a set $Q \subseteq Y$. Again, we write G^{-1} for G^{-} or G^{+}. Let $T : B \to 2^{A}$ be an upper semicontinuous mapping with nonempty, closed and convex values.

Theorem 10.1 *Let $F^{-1}(P)$ and $G^{-1}(Q)$ have the following properties:*

(i) *$(x, y) \in F^{-1}(P)$ for all $y \in B$ and $x \in T(y)$;*
(ii) *$\{x \in A : (x, y) \notin G^{-1}(Q)\}$ is open in A for all $y \in B$;*
(iii) *$\{y \in B : (x, y) \notin F^{-1}(P)\}$ is convex for all $x \in A$;*
(iv) *$F^{-1}(P) \subseteq G^{-1}(Q)$;*
(v) *for every $x \in A$, $(x, y) \in G^{-1}(Q)$ for all $y \in B$ implies $(x, y) \in F^{-1}(P)$ for all $y \in B$.*

Then there exists $\bar{x} \in A$ such that $(\bar{x}, y) \in F^{-1}(P)$ for all $y \in B$.

Proof In view of (v) it suffices to find $\bar{x} \in A$ such that $(\bar{x}, y) \in G^{-1}(Q)$ for all $y \in B$. If such an $\bar{x}$ does not exist, then the compact set A is covered by the sets

$$V(y) := \{x \in A : (x, y) \notin G^{-1}(Q)\}, \quad \text{for all } y \in B,$$

which are open by (ii). Let $V(y_1), V(y_2), \ldots, V(y_m)$ be a finite subcover of A, and let $\beta_1(\cdot), \beta_2(\cdot), \ldots, \beta_m(\cdot)$ be a continuous partition of unity subordinate to this open cover. The functions $\beta_i : A \to \mathbb{R}$ are continuous, nonnegative, add up to unity, and from $\beta_i(x) > 0$ follows $x \in V(y_i)$. Then $p(x) := \sum_{i=1}^{m} \beta_i(x) y_i$ defines a continuous function $p : A \to B$. The set-valued map $T(p(\cdot)) : A \to 2^{A}$ is upper semicontinuous with nonempty, closed and convex values. Hence, by Kakutani Fixed Point Theorem 1.38, $T(p(\cdot))$ has a fixed point $\hat{x} \in T(p(\hat{x}))$. Let $I := \{i : \beta_i(\hat{x}) > 0\}$. Then for all $i \in I$, $\hat{x} \in V(y_i)$, hence $(\hat{x}, y_i) \notin G^{-1}(Q)$ and thus

$(\hat{x}, y_i) \notin F^{-1}(P)$ by (iv). Then from $p(\hat{x}) \in \text{co}(\{y_i : i \in I\})$ and (iii) it follows that $(\hat{x}, p(\hat{x})) \notin F^{-1}(P)$. Since $\hat{x} \in T(p(\hat{x}))$, this contradicts (i). □

Remark 10.1 Note that the condition (iv) may be considered as an abstract monotonicity requirement.

Let us single out two prototypical cases of Theorem 10.1. First we consider the case $G(x, y) := F(x, y)$, $Q := P$. Then condition (iv) and (v) of Theorem 10.1 are automatically satisfied, and we obtain

Theorem 10.2 *Let $F^{-1}(P)$ have the following properties:*

(i) $(x, y) \in F^{-1}(P)$ *for all* $y \in B$ *and* $x \in T(y)$;
(ii) $\{x \in A : (x, y) \notin F^{-1}(P)\}$ *is open in* A *for all* $y \in B$;
(iii) $\{y \in B : (x, y) \notin F^{-1}(P)\}$ *is convex for all* $x \in A$.

Then there exists $\bar{x} \in A$ *such that* $(\bar{x}, y) \in F^{-1}(P)$ *for all* $y \in B$.

Now we turn to the case $Z := X$, $B := A$, $T := I$ the identity map, $G(x, y) := F(y, x)$. Then we obtain

Theorem 10.3 *Let for all* $x, y \in A$ *the following properties hold:*

(i) $(y, y) \in F^{-1}(P)$;
(ii) $\{z \in A : (x, z) \notin F^{-1}(Q)\}$ *is open in* A;
(iii) $\{z \in A : (x, z) \notin F^{-1}(P)\}$ *is convex;*
(iv) $(x, y) \in F^{-1}(P)$ *implies* $(y, x) \in F^{-1}(Q)$;
(v) *for every* $u \in]x, y[$, $(u, x) \in F^{-1}(Q)$ *and* $(u, y) \notin F^{-1}(P)$ *imply* $(u, z) \in F^{-1}(\text{int}(Q))$ *for all* $z \in]x, y[$;
(vi) $\{z \in [x, y] : (z, y) \notin F^{-1}(P)\}$ *is open in* $[x, y]$;
(vii) $(y, y) \notin F^{-1}(\text{int}(Q))$.

Then there exists $\bar{x} \in A$ *such that* $(\bar{x}, y) \in F^{-1}(P)$ *for all* $y \in A$.

Proof The result follows from Theorem 10.1 upon choosing $G(x, y) := F(y, x)$ and $T := I$ the identity map. It only remains to verify condition (v) of Theorem 10.1. To this end, let $x \in A$ with $(y, x) \in F^{-1}(Q)$ for all $y \in A$. Assume contrary that $(x, y) \notin F^{-1}(P)$ for some $y \in A$. By (i), $y \neq x$. From (vi), there exists $u \in]x, y[$ such that $(u, y) \notin F^{-1}(P)$. Since $(u, x) \in F^{-1}(Q)$, we obtain from (v) that $(u, u) \in F^{-1}(\text{int}(Q))$, contradicting (vii). □

Remark 10.2 Observe that we may use two different inverses for $F^{-1}(P)$ on one hand and $F^{-1}(Q)$, $F^{-1}(\text{int}(Q))$ on the other hand.

Wang et al. [60] used Brouwer fixed point theorem to prove the existence results for solutions of GSVEP (10.4). They also discussed the closedness and the convexity of the solution set.

The structure of Theorems 10.2 and 10.3 is perhaps better understood by considering a special case. Let $C \subseteq Y$ be a proper closed convex cone (not necessarily pointed) with nonempty interior.

We turn first to Theorem 10.2. We choose $P := Y \setminus (-\operatorname{int}(C))$, and $F^{-1} := F^{-}$. Then

$$(x, y) \notin F^{-1}(P) \quad \Leftrightarrow \quad F(x, y) \subseteq -\operatorname{int}(C).$$

Thus we obtain from Theorem 10.2, the following result.

Corollary 10.1 *Let $F : A \times B \to 2^Y$ be a set-valued map such that the following conditions hold.*

(i) $F(x, y) \not\subseteq -\operatorname{int}(C)$ *for all* $y \in B$ *and* $x \in T(y)$;
(ii) $\{x \in A : F(x, y) \subseteq -\operatorname{int}(C)\}$ *is open in* A *for all* $y \in B$;
(iii) $\{y \in B : F(x, y) \subseteq -\operatorname{int}(C)\}$ *is convex for all* $x \in A$.

Then there exists $\bar{x} \in A$ *such that* $F(\bar{x}, y) \not\subseteq -\operatorname{int}(C)$ *for all* $y \in B$.

Now we turn to Theorem 10.3. We choose $P := Y \setminus (-\operatorname{int}(C))$, $Q := Y \setminus \operatorname{int}(C)$ and $F^{-1} := F^{-}$. Then

$$\begin{aligned} (x, y) \notin F^{-1}(P) \quad &\Leftrightarrow \quad F(x, y) \subseteq -\operatorname{int}(C), \\ (x, y) \notin F^{-1}(Q) \quad &\Leftrightarrow \quad F(x, y) \subseteq \operatorname{int}(C), \\ (x, y) \notin F^{-1}(\operatorname{int}(Q)) \quad &\Leftrightarrow \quad F(x, y) \subseteq C, \end{aligned}$$

and $F(y, y) \subseteq C$ implies $F(y, y) \not\subseteq -\operatorname{int}(C)$, provided $F(y, y) \neq \emptyset$. Thus, from Theorem 10.3, we obtain the following result.

Corollary 10.2 *Let $F : A \times A \to 2^Y \setminus \{\emptyset\}$ be such that, for all $x, y \in A$, the following properties hold:*

(i) $\emptyset \neq F(y, y) \subseteq C$;
(ii) $\{z \in A : F(x, z) \subseteq \operatorname{int}(C)\}$ *is open in* A;
(iii) $\{z \in A : F(x, z) \subseteq -\operatorname{int}(C)\}$ *is convex;*
(iv) $F(x, y) \not\subseteq -\operatorname{int}(C)$ *implies* $F(y, x) \not\subseteq \operatorname{int}(C)$;
(v) *for every* $u \in]x, y[$, $F(u, x) \not\subseteq \operatorname{int}(C)$ *and* $F(u, y) \not\subseteq -\operatorname{int}(C)$ *imply* $F(u, z) \not\subseteq \operatorname{int}(C)$ *for all* $z \in]x, y[$;
(vi) $\{z \in [x, y] : F(z, y) \subseteq \operatorname{int}(C)\}$ *is open in* $[x, y]$.

Then there exists $\bar{x} \in A$ *such that* $F(\bar{x}, y) \not\subseteq -\operatorname{int}(C)$ *for all* $y \in A$.

Concerning the assumption of Corollary 10.2, we observe the following:

- The condition (ii) is satisfied if the set-valued map $F(x, \cdot)$ is upper semicontinuous.
- The condition (vi) is satisfied if the mapping $F(\cdot, y)$ is upper semicontinuous along line segments in A.
- The condition (iii) is satisfied if for every $u \in A$, the mapping $t(\cdot) := F(u, \cdot)$ has the property that

$$t(\lambda x + (1 - \lambda y) \subseteq \lambda t(x) + (1 - \lambda)t(y) - C, \quad \text{for all } x, y \in A \text{ and } \lambda \in [0, 1].$$

- The condition (v) is satisfied if for every $u \in A$ the mapping $t(\cdot) := F(u, \cdot)$ has the property that $t(x) \neq \emptyset$ for all $x \in A$ and

$$t(\lambda x + (1-\lambda y) + C \supseteq \lambda t(x) + (1-\lambda)t(y), \text{ for all } x, y \in A \text{ and } \lambda \in [0, 1].$$

Indeed, let $a \in t(x)$, $a \notin \text{int}(C)$, and $b \in t(y)$, $b \in -\text{int}(C)$. Let $z := \lambda x + (1-\lambda)y$ with $0 < \lambda < 1$. Setting $c := \lambda a + (1-\lambda)b$ it follows from $(1-\lambda)b \in -\text{int}(C)$ and $\lambda a \notin \text{int}(C)$ that $c \notin C$. Since $t(z) \neq \emptyset$, from (10.2), there exists $d \in t(z)$ such that $c - d \in C$, hence $d \notin C$. Thus $t(z) \nsubseteq C$, and (v) is true.

For further details, we refer to [6].

Theorem 10.4 *Let K be a nonempty compact convex subset of a Hausdorff topological vector space X and W be a nonempty subset of a topological vector space Y. Let $F : K \times K \to 2^Y$ be a set-valued map with nonempty values such that the following conditions:*

(A0) *$F(x, x) \subseteq W$ for all $x \in K$;*
(A1) *For all $x, y \in K$, $F(x, y) \subseteq W$ implies $F(y, x) \subseteq -W$;*
(A2) *For all $x \in K$, the set $\{\xi \in K : F(x, \xi) \subseteq -W\}$ is closed;*
(A3) *For all $x \in K$, the set $\{\xi \in K : F(x, \xi) \nsubseteq W\}$ is convex;*
(A4) *For all $x, y \in K$, $F(\xi, x) \subseteq -W$ for all $\xi \in]x, y]$ implies $F(x, y) \subseteq W$, where $]x, y]$ denotes the line segment joining x and y but not containing x.*

Then the solution set to the problem

$$\text{find } \bar{x} \in K \text{ such that } F(\bar{x}, y) \subseteq W, \quad \text{for all } y \in K \tag{10.10}$$

and that of the problem

$$\text{find } \bar{x} \in K \text{ such that } F(y, \bar{x}) \subseteq -W, \quad \forall y \in K, \tag{10.11}$$

are nonempty, closed and both coincide.

Proof We first find $\bar{x} \in K$ such that

$$\bar{x} \in \bigcap_{y \in K} \{x \in K : F(y, x) \subseteq -W\}.$$

To that end, we shall use the famous Fan-KKM Lemma 1.14. Set

$$G(y) = \{x \in K : F(y, x) \subseteq -W\}.$$

Assumption (A2) implies that for each $y \in K$, $G(y)$ is a closed subset of a compact set K and hence $G(y)$ is compact. In order to apply the Fan-KKM Lemma 1.14, we need to prove that for any finite subset $\{y_1, y_2, \ldots, y_m\}$ of K, $\text{co}(\{y_1, y_2, \ldots, y_m\}) \subseteq \bigcup_{i=1}^m G(y_i)$. If $y = \sum_{i=1}^m \alpha_i y_i \notin \bigcup_{i=1}^m G(y_i)$ for some $\alpha_i \geq 0$, $i = 1, 2, \ldots, m$,

$\sum_{i=1}^{m} \alpha_i = 1$, then $y \notin G(y_i)$ for all $i = 1, 2, \ldots, m$. Thus for each $i = 1, 2, \ldots, m$, $F(y_i, y) \nsubseteq -W$ which implies $F(y, y_i,) \nsubseteq W$ by assumption (A1). Thus $F(y, y) \nsubseteq W$ because of assumption (A3), which contradicts assumption (A0). This proves that for any finite subset $\{y_1, y_2, \ldots, y_k\}$ of K, $\text{co}(\{y_1, y_2, \ldots, y_k\}) \subseteq \bigcup_{i=1}^{k} G(y_i)$. Hence by Fan-KKM Lemma 1.14, there exists $\bar{x} \in K$ such that $\bar{x} \in \bigcap_{y \in K} G(y)$, that is, $F(y, \bar{x}) \subseteq -W$ for all $y \in K$, in other words, the second problem has a solution.

For any $y \in K$, $]\bar{x}, y] \subseteq K$. Therefore, $F(z, \bar{x}) \subseteq -W$ for all $z \in]\bar{x}, y]$. By assumption (A4), $F(\bar{x}, y) \subseteq W$. Hence every solution of the second problem is a solution of the first problem. By assumption (A1), every solution of the first problem is a solution of the second problem. Hence, we deduce that the solution sets of both the problems coincide. The closedness is a consequence of (A2). □

Since a nonempty closed convex bounded subset of a reflexive Banach space is weakly compact, we have the following corollary.

Corollary 10.3 *Let K be a nonempty closed, convex and bounded subset of a reflexive Banach space X and W be a nonempty subset of a normed space Y. Let $F : K \times K \to 2^Y$ be a set-valued map with nonempty values such that the following conditions:*

(B0) *$F(x, x) \subseteq W$ for all $x \in K$;*
(B1) *For all $x, y \in K$, $F(x, y) \subseteq W$ implies $F(y, x) \subseteq -W$;*
(B2) *For all $x \in K$, the set $\{\xi \in K : F(x, \xi) \subseteq -W\}$ is (sequentially) weakly closed;*
(B3) *For all $x \in K$, the set $\{\xi \in K : F(x, \xi) \nsubseteq W\}$ is convex;*
(B4) *For all $x, y \in K$, $F(\xi, x) \subseteq -W$ for all $\xi \in]x, y]$ implies $F(x, y) \subseteq W$, where $]x, y]$ denotes the line segment joining x and y but not containing x.*

Then the solution set to the problem (10.10) *and* (10.11) *are nonempty, weakly closed and both coincide.*

10.3 Existence Results for Generalized Vector Equilibrium Problems

In this section, we present several existence results for solutions of GWVEPs with or without monotonicity condition. We first use Corollary 10.3 to derive the existence results for solutions of GWVEP (10.2) and then existence results for solutions of GSVEP (10.4).

Definition 10.1 Let K be a nonempty convex subset of a Banach space X and C be a proper closed convex cone in Y. A set-valued map $S : K \to 2^Y \setminus \{\emptyset\}$ is said to be *weakly lower semicontinuous at* $x \in K$ if for any $y \in S(x)$ and for any sequence $\{x_m\}$ in K converges weakly to x, there exists a sequence $\{y_m\}$ in $S(x_m)$ converges strongly to y.

S is *weakly lower semicontinuous on* K if it is weakly lower semicontinuous at each point of K.

We now adapt the abstract Corollary 10.3 to problem (10.2) and we give simpler verifiable conditions on F ensuring the validity of all assumptions imposed in Corollary 10.3.

Let K be a nonempty closed convex bounded subset of a reflexive Banach space X and C be a proper closed convex cone in a normed space Y such that $\text{int}(C) \neq \emptyset$. Let $F : K \times K \to 2^Y \setminus \{\emptyset\}$ have the following properties:

(f_0) For all $x \in K$, $F(x,x) \subseteq l(C) := C \cap (-C)$;
(f_1) For all $x, y \in K$, $F(x,y) \cap (-\text{int}(C)) = \emptyset$ implies $F(y,x) \cap \text{int}(C) = \emptyset$;
(f_2) For all $x \in K$, the mapping $F(x,\cdot) : K \to 2^Y \setminus \{\emptyset\}$ is C-convex;
(f_3) For all $x, y \in K$, the set $\{\xi \in [x,y] : F(\xi, y) \cap (-\text{int}(C)) = \emptyset\}$ is closed;
(f_4) For all $x \in K$, $F(x,\cdot)$ is weakly lower semicontinuous.

Remark 10.3

(a) One can check immediately that the C-convexity of $F(x,\cdot)$ implies that for all $x \in K$, the set

$$\{\xi \in K : F(x,\xi) \not\subseteq Y \setminus (-\text{int}(C))\}$$

is convex. Hence condition (B3) (with $W = Y \setminus (-\text{int}(C))$) of Corollary 10.3 holds.

(b) It can be easily seen that the weakly lower semicontinuity of $F(x,\cdot)$ asserts the (sequential) weak closedness of

$$\{\xi \in K : F(x,\xi) \subseteq Y \setminus (-\text{int}(C))\}$$

for all $x \in K$. Thus condition (B2) (with $W = Y \setminus (-\text{int}(C))$) of Corollary 10.3 is satisfied.

(c) Assumptions (f_0), (f_2) and (f_3) imply that, given any $x \in K$,

$$\mathbf{0} \in F(y,x) + (Y \setminus (-\text{int}(C)) \text{ for all } y \in K \text{ implies}$$
$$F(x,y) \cap (-\text{int}(C)) = \emptyset, \quad \text{for all } y \in K. \tag{10.12}$$

Hence condition (B4) (with $W = Y \setminus (-\text{int}(C))$) of Corollary 10.3 holds.

Indeed, for every $y \in K$ consider $x_t = x + t(y - x)$ for $t \in]0,1[$. Clearly $x_t \in K$. The C-convexity of $F(x_t,\cdot)$ implies

$$tF(x_t,y) + (1-t)F(x_t,x) \subseteq F(x_t,x_t) + C \subseteq C.$$

Since $\mathbf{0} \in F(x_t,x) + (Y \setminus (-\text{int}(C))$, there exists $\xi(x_t,x) \in F(x_t,x)$ such that $\xi(x_t,x) \notin \text{int}(C)$. From a previous inclusion one has

$$tF(x_t,y) \subseteq -(1-t)\xi(x_t,x) + C \subseteq (Y \setminus (-\text{int}(C)) + C \subseteq Y \setminus (-\text{int}(C)).$$

It turns out that $F(x_t, y) \subseteq Y \backslash (-\operatorname{int}(C))$ or, equivalently, $F(x_t, y) \cap (-\operatorname{int}(C)) = \emptyset$. Letting $t \downarrow 0$, we obtain by assumption (f_3), $F(x, y) \cap (-\operatorname{int}(C)) = \emptyset$. Since y was arbitrary, the desired result is proved.

The following result shows that (10.12) also holds if we replace the C-convexity of $F(x, \cdot)$ by the explicitly δ-quasiconvexity for each $x \in K$.

Proposition 10.1 *Assume that the set-valued map $F : K \times K \to 2^Y \setminus \{\emptyset\}$ satisfies assumption (f_0) and (f_3) such that $F(x, \cdot)$ is explicitly δ-C-quasiconvex for each $x \in K$. Then* (10.12) *holds.*

Proof For a given $x \in K$, let

$$\mathbf{0} \in F(y, x) + (Y \setminus (-\operatorname{int}(C))), \quad \text{for all } y \in K. \tag{10.13}$$

Suppose there exists $y \in K$ such that $F(x, y) \cap (-\operatorname{int}(C)) \neq \emptyset$. (f_3) can be written, in an equivalent way, as
(f_3) : for all $x, y \in K$, the set $\{\xi \in [x, y] : F(\xi, y) \cap (-\operatorname{int}(C)) \neq \emptyset\}$ is open (in $[x, y]$).

Since $x \in \{\xi \in [x, y] : F(\xi, y) \cap (-\operatorname{int}(C)) \neq \emptyset\} := M$, there exists $\alpha \in]0, 1[$ such that

$$z := x + \alpha(y - x) = \alpha y + (1 - \alpha)x \in M,$$

that is,

$$z \in [x, y] \text{ and } F(z, y) \cap (-\operatorname{int}(C)) \neq \emptyset. \tag{10.14}$$

Now by explicitly δ-C-quasiconvexity of $F(z, \cdot)$, we have

$$F(z, y) \subseteq F(z, z) + C \subseteq C \subseteq W := Y \setminus (-\operatorname{int}(C))$$

which implies that

$$F(z, y) \cap (-\operatorname{int}(C)) = \emptyset,$$

a contradiction of (10.14). Therefore, we can only have

$$F(z, x) \subseteq F(z, z) + C \subseteq C \subseteq W,$$

that is,

$$F(z, x) \cap (-\operatorname{int}(C)) = \emptyset. \tag{10.15}$$

Relations (10.14) and (10.15) imply

$$[F(z, y) - F(z, x)] \cap (-\operatorname{int}(C)) \neq \emptyset.$$

By explicitly δ-C-quasiconvexity of $F(z,\cdot)$, we have

$$F(z,x) \subseteq F(z,z) + \text{int}(C) \subseteq C + \text{int}(C) \subseteq \text{int}(C)$$

which contradicts (10.13). Hence $F(x,y) \subseteq Y \setminus (-\text{int}(C))$ for all $y \in K$. □

Theorem 10.5 *Let K be a nonempty closed, convex and bounded subset of a reflexive Banach space X and C be a proper closed convex cone in a normed space Y such that* $\text{int}(C) \neq \emptyset$. *Let $F : K \times K \to 2^Y$ be a set-valued map with nonempty values such that the conditions $(f_0) - (f_4)$ hold. Then the solution set to GWVEP* (10.2) *is nonempty and weakly closed.*

In view of Proposition 10.1, condition (f_2) in Theorem 10.5 can be replaced by following condition:

(f_{22}) $F(x,\cdot)$ is explicitly δ-C-quasiconvex for each $x \in K$.

For the existence of a solution of strong GVEP, the basic assumptions on the set-valued map $F : K \times K \to 2^Y \setminus \{\emptyset\}$ are listed below:

(f_1') For all $x, y \in K$, $F(x,y) \subseteq C$ implies $F(y,x) \subseteq -C$;
(f_2') For all $x \in K$, the mapping $F(x,\cdot) : K \to 2^Y \setminus \{\emptyset\}$ is properly C-quasiconvex;
(f_3') For all $x, y \in K$, the set $\{\xi \in [x,y] : F(y,\xi) \subseteq C\}$ is closed;
(f_4') For all $x \in K$, $F(x,\cdot)$ is weakly lower semicontinuous on K.

Remark 10.4

(a) One can check that the proper C-quasiconvexity of $F(x,\cdot)$ implies that the set

$$\{\xi \in K : F(x,\xi) \not\subseteq C\}$$

is convex for all $x \in K$. Hence condition (B3) (with $W = C$) of Corollary 10.3 is satisfied.

(b) It can be easily seen that the weakly lower semicontinuity of $F(x,\cdot)$ asserts the weak closedness of

$$\{\xi \in K : F(x,\xi) \subseteq -C\}$$

for all $x \in K$. Thus condition (B3) (with $W = C$) of Corollary 10.3 is satisfied.

(c) Similar to Part (c) of Remark 10.3, one can prove, under assumptions (f_0), (f_3') and C-convexity of $F(x,\cdot)$, that given any $x \in K$,

$$\mathbf{0} \in F(y,x) + C \text{ for all } y \in K \text{ implies } F(x,y) \cap (Y \setminus C) = \emptyset \text{ for all } y \in K. \tag{10.16}$$

Hence condition (B4) (with $W = C$) of Corollary 10.3 holds.

We obtain the following result from Corollary 10.3 by specializing $W = C$.

Theorem 10.6 *Let $K \subseteq X$ be a nonempty weakly compact convex set and $F : K \times K \to 2^Y \setminus \{\emptyset\}$ be a set-valued map satisfying conditions $(f_0) - (f_4')$ such* *that for all $x \in K$, $F(x,\cdot)$ is C-convex. Then the solution set* Sol(GSVEP) *of the problem* (10.4) *is nonempty and weakly compact.*

Proof The weak compactness is obtained as usual. To prove the nonemptiness of Sol(GSVEP), we show that the assumption of Corollary 10.3 are satisfied when specialized to $W = C$. This was proved in Remark 10.4. □

To study the existence results for solutions of GWVEP (10.1) under some kind of monotonicity, we consider the following generalized weak vector equilibrium problem, which is closely related to the GWVEP (10.1) and can be termed as the dual of GWVEP (10.1) (in short, DGWVEP):

$$\text{Find } \bar{x} \in K \text{ such that } F(y,\bar{x}) \nsubseteq \text{int}(C), \quad \text{for all } y \in K. \tag{10.17}$$

The solution set of DGWVEP is denoted by Sol(DGWVEP).

Definition 10.2 Let C be a closed convex solid cone in Y. A set-valued map $F : K \times K \to 2^Y \setminus \{\emptyset\}$ is said to be

(a) *C-pseudomonotone* if for all $x, y \in K$, we have

$$F(x,y) \nsubseteq -\text{int}(C) \quad \text{implies} \quad F(y,x) \nsubseteq \text{int}(C);$$

(b) *maximal C-pseudomonotone* if it is C-pseudomonotone and for all $x, y \in K$,

$$F(z,x) \nsubseteq \text{int}(C) \text{ for all } z \in]x,y] \quad \text{implies} \quad F(x,y) \nsubseteq -\text{int}(C).$$

We mention the following lemma which provides the relationship between Sol(GWVEP) and Sol(DGWVEP).

Lemma 10.1 *If F is maximal C-pseudomonotone, then* Sol(GWVEP) = Sol(DGWVEP).

Proof By C-pseudomonotonicity of F, we have the inclusion Sol(GWVEP) $\subseteq$ Sol(DGWVEP).

Let $\bar{x} \in$ Sol(DGWVEP). Then $F(y,\bar{x}) \nsubseteq \text{int}(C)$ for all $y \in K$. For any $y \in K$, $]\bar{x},y] \subseteq K$. Therefore, $F(z,\bar{x}) \nsubseteq \text{int}(C)$ for all $z \in]x,y]$. Since F is maximal C-pseudomonotone, we have $F(\bar{x},y) \nsubseteq -\text{int}(C)$. Hence $\bar{x} \in$ Sol(GWVEP). □

Lemma 10.2 *Let K be a nonempty convex subset of a topological vector space X, C be a proper closed convex solid cone in a topological vector space Y and $F : K \times K \to 2^Y \setminus \{\emptyset\}$ be a set-valued map. Then*

(a) Sol(GWVEP) $\subseteq$ Sol(DGWVEP) *if F is C-pseudomonotone;*
(b) Sol(DGWVEP) $\subseteq$ Sol(GWVEP) *if $F(x,x) \subseteq C$, the set-valued map $y \mapsto F(x,y)$ is explicitly δ-C-quasiconvex, and the set-valued map $x \mapsto F(x,y)$ is u-hemicontinuous for all $x, y \in K$.*

Proof

(a) It directly follows from the C-pseudomonotonicity of F.
(b) Let $\bar{x} \in \mathrm{Sol(DGWVEP)}$. Assume contrary that $\bar{x} \notin \mathrm{Sol(GWVEP)}$. Then there exists $y \in K$ such that

$$F(\bar{x}, y) \subseteq -\operatorname{int}(C).$$

By u-hemicontinuity of $F(\cdot, y)$, it follows that, for some $\alpha \in]0, 1[$,

$$F(x_\alpha, y) \subseteq -\operatorname{int}(C), \tag{10.18}$$

where $x_\alpha = \alpha y + (1-\alpha)\bar{x}$. By explicit δ-C-quasiconvexity of F, we now have either

$$F(x_\alpha, y) \subseteq F(x_\alpha, x_\alpha) + C \subseteq C$$

or

$$F(x_\alpha, \bar{x}) \subseteq F(x_\alpha, x_\alpha) + C \subseteq C.$$

The first relation contradicts (10.18). Thus, we must have $F(x_\alpha, \bar{x}) \subseteq C$, hence,

$$F(x_\alpha, \bar{x}) - F(x_\alpha, y) \subseteq \operatorname{int}(C).$$

Again by explicit δ-C-quasiconvexity of F, we have

$$F(x_\alpha, \bar{x}) \subseteq F(x_\alpha, x_\alpha) + \operatorname{int}(C) \subseteq \operatorname{int}(C),$$

which contradicts our assumption that $\bar{x}$ is a solution of DGWVEP (10.17).

□

Theorem 10.7 *Let X and Y be Hausdorff topological vector spaces, K be a nonempty compact convex subset of X and C be a proper closed convex and solid cone in Y. Let $F : K \times K \to 2^Y \setminus \{\emptyset\}$ be a set-valued map such that following conditions hold.*

(i) *For each $x \in K$, $F(x, x) \subseteq C$;*
(ii) *F is C-pseudomonotone and $F(x, \cdot)$ is explicitly δ-C-quasiconvex for each $x \in K$;*
(iii) *For each $y \in K$, $F(\cdot, y)$ is u-hemicontinuous;*
(v) *For each $y \in K$, $U(y) := \{x \in K : F(y, x) \subseteq \operatorname{int}(C)\}$ is open.*

Then there exists $\bar{x} \in$ Sol(GWVEP).

Proof For all $y \in K$, define set-valued maps $S_1, S_2 : K \to 2^K$ by

$$S_1(y) = \{x \in K : F(x, y) \nsubseteq -\operatorname{int}(C)\}$$

and

$$S_2(y) = K \setminus U(y).$$

Then S_1 is a KKM map. Indeed, let $u \in \text{co}(\{y_1, y_2, \ldots, y_m\})$ for any finite subset $\{y_1, y_2, \ldots, y_m\}$ of K. Assume, for contradiction, that

$$u \notin \bigcup_{i=1}^{m} S_1(y_i).$$

Then $u \in K$ and

$$F(u, y_i) \subseteq -\text{int}(C), \quad \text{for all } i = 1, 2, \ldots, m.$$

Since $F(x, \cdot)$ is explicit δ-C-quasiconvex, we have

$$F(u, y_i) \subseteq F(u, u) + C \subseteq C, \quad \text{for some } i,$$

which is a contradiction.

By C-pseudomonotonicity of F, we have $S_1(y) \subseteq S_2(y)$ for all $y \in K$. Since S_1 is a KKM-map, so is S_2.

Since for all $y \in K$, $U(y)$ is open and thus $S_2(y)$ is closed subset of a compact set K and hence compact. Thus, by Fan-KKM Lemma 1.14, we have

$$\text{Sol(DGWVEP)} = \bigcap_{y \in K} S_2(y) \neq \emptyset.$$

From Lemma 10.2, we have Sol(DGWVEP) $\subseteq$ Sol(GWVEP). Consequently, there exists $\bar{x} \in$ Sol(GWVEP), as required. □

By adopting the technique of Theorem 10.7, Farajzadeh et al. [25] proved the existence of solutions of GVEP (10.3) under the upper sign continuity but used a different form of C-pseudomonotonicity.

Definition 10.3 (*υ*-**Coercivity**) A set-valued map $S : K \to 2^Y$ is said to be *υ-coercive* on K if there exist a compact subset B of X and $\tilde{y} \in B \cap K$ such that $K \setminus B \subseteq U(\tilde{y})$.

We now have the following result for existence of solutions of GWVEP (10.1) in the unbounded case.

Theorem 10.8 *Let X, Y, C and F be the same as in Theorem 10.7. Let K be a nonempty closed convex subset of X. In addition, suppose that F is υ-coercive on K. Then there exists $\bar{x} \in$* Sol(GWVEP).

Proof Let S_1 and S_2 be the same as in the proof of Theorem 10.7. Choose compact subset B of X and $\tilde{y} \in B \cap K$ such that $K \setminus B \subseteq U(\tilde{y})$.

Since $K \setminus B \subseteq U(\bar{y})$, we have $S_2(\bar{y}) \subseteq K \cap B$. Hence, $S_2(\bar{y})$ is a compact subset of K. Thus, as in the proof of Theorem 10.7, we have

$$\mathrm{Sol(GWVEP)} = \bigcap_{y \in K} S_2(y) \neq \emptyset.$$

From Lemma 10.2, we have Sol(DGWVEP) $\subseteq$ Sol(GWVEP). Consequently, there exists $\bar{x} \in$ Sol(GWVEP), as required. □

Remark 10.5

(a) The assertion of Theorems 10.7 and 10.8 remains valid if we replace the condition of explicit δ-C-quasiconvexity of F with the following condition:

$$\text{The set } \{y \in K : F(x, y) \subseteq -\operatorname{int}(C)\} \text{ is convex for each } x \in K.$$

(b) The topologies on X and Y need not be equivalent. For instance, if X and Y are normed spaces, we can use the weak topology on X and the norm topology on Y.

Konnov and Yao [41] proved above mentioned results in the setting of moving cones.

When $F(x, \cdot)$ is not necessarily explicitly δ-C-quasiconvex but only C-quasiconvex, we have the following existence result for solutions of GWVEP in the unbounded case.

Theorem 10.9 *Let K be a nonempty convex subset of a Hausdorff topological vector space X and C be a proper closed convex cone in a topological vector space Y such that* $\operatorname{int}(C) \neq \emptyset$. *Let $F : K \to 2^Y$ be set-valued map with nonempty values such that the following conditions hold.*

(i) *$F(x, x) \subseteq C$ for all $x \in K$;*
(ii) *F is C-quasiconvex and maximal C-pseudomonotone;*
(iii) *For each $y \in K$, the set-valued map $x \mapsto F(x, y)$ is upper semicontinuous with compact values on K;*
(iv) *There exist a nonempty compact convex subset D of K and an element $\tilde{y} \in D$ such that $F(z, \tilde{y}) \subseteq -\operatorname{int}(C)$ for all $z \in K \setminus D$.*

Then there exists $\bar{x} \in$ Sol(GWVEP).

Proof For each $y \in K$, we define set-valued maps $S, T : K \to 2^K$ by

$$S(y) = \{x \in K : F(y, x) \nsubseteq \operatorname{int}(C)\}$$

and

$$T(y) = \{x \in K : F(x, y) \nsubseteq -\operatorname{int}(C)\}.$$

By condition (i), $T(y)$ is nonempty for all $y \in K$. Also, $T(y)$ is closed for each $y \in K$. Indeed, let $\{x_\lambda\}_{\lambda \in \Lambda}$ be a net in $T(y)$ such that $\{x_\lambda\}$ converges to x. Then we have $F(x_\lambda, y) \nsubseteq -\operatorname{int}(C)$ for each $y \in K$, that is, there exists $z_\lambda \in F(x_\lambda, y)$ such that $z_\lambda \notin -\operatorname{int}(C)$ for all $\lambda \in \Lambda$. Let $A = \{x_\lambda\} \bigcup \{x\}$. Then A is compact and $z_\lambda \in F(A, y)$ which is compact by condition (iii). Therefore, $\{z_\lambda\}$ has a convergent subnet with limit z. Without loss of generality, we may assume that $\{z_\lambda\}$ converges to z. Then by the upper semicontinuity of $F(\cdot, y)$, we have $z \in F(x, y)$. Consequently, $z \in F(x, y)$ and $z \notin -\operatorname{int}(C)$, i.e., $F(x, y) \nsubseteq -\operatorname{int}(C)$. Hence $x \in T(y)$ and so $T(y)$ is closed as claimed.

Since F is C-quasiconvex, it is easy to see that T is a KKM-map (see for example, the proof of Theorem 10.7). Then T is a KKM-map with closed values and $T(\tilde{y})$ is contained in a compact set D by condition (iv) and hence $T(\tilde{y})$ is compact. It follows from Fan-KKM Lemma 1.14 that there exists $\bar{x} \in D$ such that $\bar{x} \in T(y)$ for all $y \in K$. By C-pseudomonotonicity of F, we have $T(y) \subseteq S(y)$ for all $y \in K$. Therefore, we obtain that $\bar{x} \in S(y)$ for all $y \in K$, that is, $F(y, \bar{x}) \nsubseteq \operatorname{int}(C)$ for all $y \in K$. By Proposition 10.3, $\bar{x} \in D$ is a solution of GWVEP (10.1). □

Theorem 10.9 is studied by Ansari et al. [7] for the moving cone.

Now we establish existence results for a solution of GWVEP (10.1) in the setting of topological vector spaces but by using a Browder type fixed point theorem for set-valued maps.

Theorem 10.10 *Let K be a nonempty convex subset of a Hausdorff topological vector space X and C be a proper closed convex cone in a topological vector space Y such that* $\operatorname{int}(C) \neq \emptyset$. *Let $F : K \times K \to 2^Y \setminus \{\emptyset\}$ be a C-pseudomonotone set-valued map such that the following conditions hold:*

(i) *For each $x \in K$, $F(x, x) \subseteq C$;*
(ii) *For each $x \in K$, the set-valued map $y \mapsto F(x, y)$ is C-quasiconvex-like and explicitly δ-C-quasiconvex;*
(iii) *For each $x \in K$, the set-valued map $y \mapsto F(x, y)$ is upper semicontinuous with compact values;*
(iv) *For each $y \in K$, the set-valued map $x \mapsto F(x, y)$ is u-hemicontinuous;*
(v) *There exist a nonempty compact convex subset B of K and a nonempty compact subset D of K such that for each $x \in K \setminus D$, there exists $\tilde{y} \in B$ such that $F(x, \tilde{y}) \subseteq -\operatorname{int}(C)$.*

Then the solution set Sol(GWVEP) *of* GWVEP (10.1) *is nonempty and compact subset of B.*

Proof Define set-valued maps $S, T : K \to 2^K$ by

$$S(x) = \{y \in K : F(y, x) \subseteq \operatorname{int}(C)\}$$

and

$$T(x) = \{y \in K : F(x, y) \subseteq -\operatorname{int}(C)\},$$

for all $x \in K$. Then by C-pseudomonotonicity of F, $S(x) \subseteq T(x)$ for all $x \in K$.

Also, for each $x \in K$, $T(x)$ is convex. To see this, let $y_1, y_2 \in T(x)$. Then for each $x \in K$, $F(x, y_1) \subseteq -\operatorname{int}(C)$ and $F(x, y_2) \subseteq -\operatorname{int}(C)$. Since the set-valued map $y \mapsto F(x, y)$ is C-quasiconvex-like, for all $\alpha \in [0, 1]$, we have either

$$F(x, \alpha y_1 + (1-\alpha)y_2) \subseteq F(x, y_1) - C \subseteq -\operatorname{int}(C) - C \subseteq -\operatorname{int}(C),$$

or

$$F(x, \alpha y_1 + (1-\alpha)y_2) \subseteq F(x, y_2) - C \subseteq -\operatorname{int}(C) - C \subseteq -\operatorname{int}(C).$$

In both the cases, we get $F(x, \alpha y_1 + (1-\alpha)y_2) \subseteq -\operatorname{int}(C)$. Hence, $\alpha y_1 + (1-\alpha)y_2 \in S(x)$, and therefore, $T(x)$ is convex.

For each $y \in K$, $S^{-1}(y) = \{x \in K : F(y, x) \subseteq \operatorname{int}(C)\}$ is open in K, equivalently, the complement $[S^{-1}(y)]^c = \{x \in K : F(y, x) \nsubseteq \operatorname{int}(C)\}$ is closed in K. Indeed, let $\{x_\lambda\}_{\lambda \in \Lambda}$ be a net in $[S^{-1}(y)]^c$ such that $\{x_\lambda\}$ converges to x. Then we have $F(y, x_\lambda) \nsubseteq \operatorname{int}(C)$ for each $y \in K$, that is, there exists $z_\lambda \in F(y, x_\lambda)$ such that $z_\lambda \notin \operatorname{int}(C)$ for all $\lambda \in \Lambda$. Let $A = \{x_\lambda\} \bigcup \{x\}$. Then A is compact and $z_\lambda \in F(y, A)$ which is compact by condition (iii). Therefore, $\{z_\lambda\}$ has a convergent subnet with limit z. Without loss of generality, we may assume that $\{z_\lambda\}$ converges to z. Then by the upper semicontinuity of $F(y, \cdot)$, we have $z \in F(y, x)$. Consequently, $z \in F(y, x)$ and $z \notin \operatorname{int}(C)$, i.e., $F(y, x) \nsubseteq \operatorname{int}(C)$. Hence $x \in [S^{-1}(y)]^c$ and so $[S^{-1}(y)]^c$ is closed.

We verify that T has no fixed point. Assume contrary that T has a fixed point $x \in K$. Then $x \in T(x)$, that is, $F(x, x) \subseteq -\operatorname{int}(C)$. By condition (i), $F(x, x) \subseteq C$. Therefore, $F(x, x) \subseteq -\operatorname{int}(C) \cap C = \emptyset$, a contradiction. Indeed, if there were a $v \in -\operatorname{int}(C) \cap C$, then $\mathbf{0} = -v + v \in \operatorname{int}(C) + C \subseteq \operatorname{int}(C)$. This implies that $C = Y$ because $\operatorname{int}(C) \ni \mathbf{0}$ is an absorbing set in Y, which contradicts the assumption that C is proper. Therefore, T has no fixed point.

Since T has no fixed point, we reach to a conclusion that either S or T would not satisfy at least one of the hypotheses of Corollary 1.3. But, as we have seen above that S and T satisfy all the hypotheses of Corollary 1.3 except $S(x)$ is nonempty for all $x \in K$. Hence, there must be a $\bar{x} \in K$ such that $S(\bar{x}) = \emptyset$, that is,

$$F(y, \bar{x}) \nsubseteq \operatorname{int}(C), \quad \text{for all } y \subset K.$$

By Lemma 10.2, we have

$$F(\bar{x}, y) \nsubseteq -\operatorname{int}(C), \quad \text{for all } y \in K,$$

as desired. □

Definition 10.4 Let C be a closed convex solid cone in Y. A set-valued map $F : K \times K \to 2^Y \setminus \{\emptyset\}$ is said to be C-*quasimonotone* if for all $x, y \in K$, we have

$$F(x, y) \nsubseteq -C \quad \text{implies} \quad F(y, x) \nsubseteq \operatorname{int}(C).$$

We now provide existence results for a solution of GWVEP under C-quasimonotonicity, which strictly contains the class of C-pseudomonotone set-valued maps.

Lemma 10.3 *Let K be a nonempty convex subset of a topological vector space X and C be a proper, closed and convex cone in a topological vector space Y such that* $\text{int}(C) \neq \emptyset$. *Let $F : K \times K \to 2^Y \setminus \{\emptyset\}$ be a C-quasimonotone, explicitly C-quasiconvex-like set-valued map such that $F(\cdot, z)$ is u-hemicontinuous for any $z \in K$. Then for each pair of points $x \in K$, $y \in \text{ri}\,(K)$, at least one of the following must hold:*

(a) $F(y,x) \subseteq \text{int}(C)$ *implies* $F(x,y) \subseteq -\text{int}(C)$,

or

(b) $F(x,z) \nsubseteq -\text{int}(C)$ *for all* $z \in K$.

Proof Let $F(y,x) \subseteq \text{int}(C)$ for some $x \in K$, $y \in \text{ri}\,(K)$, and let there exist $z \in K$ such that $F(x,z) \subseteq -\text{int}(C)$. Since $y + (z-y) := z \in K$, it follows from the definition of relative algebraic interior point that there is $\varepsilon > 0$ such that $]y-\varepsilon(z-y), y+\varepsilon(z-y)[\subseteq K$. Then for any β with $0 < \beta < 1$, we have

$$\gamma_\beta = \beta(y - \varepsilon(z-y)) + (1-\beta)(y + \varepsilon(z-y)) = y + (2\beta - 1)\varepsilon(y-z) \in K.$$

Therefore by choosing β with $\frac{1}{2} < \beta < 1$ and letting ε to be sufficient small if necessary, we conclude that there exists $\gamma_0 \in]0,1[$ such that $y + \gamma(y-z) \in K$ for all $\gamma \in]0, \gamma_0[$.

By u-hemicontinuity of $F(\cdot, x)$, it follows that, for some $\alpha \in]0,1[$,

$$F(z_\alpha, x) \subset \text{int}(C),$$

where $z_\alpha := y + \alpha(y-z) \in K$. Since F is C-quasimonotone, we have

$$F(x, z_\alpha) \subseteq -C.$$

Since $z_\alpha = (1+\alpha)y - \alpha z$, we have $y = \beta z + (1-\beta)z_\alpha \in]z_\alpha, z[$ with $\beta = \frac{\alpha}{1+\alpha}$. Besides, $y \in]z_\alpha, z[$ and F is explicitly C-quasiconvex-like. In case $F(x, z_\alpha) \subseteq -\text{int}(C)$, it follows that either

$$F(x,y) \subseteq F(x,z) - C \subseteq -\text{int}(C),$$

or

$$F(x,y) \subseteq F(x,z_\alpha) - C \subseteq -\text{int}(C).$$

Otherwise, if $F(x,z_\alpha) \nsubseteq -\text{int}(C)$, then $F(x,z_\alpha) - F(x,z) \nsubseteq -\text{int}(C)$, and we must have

$$F(x,y) \subseteq F(x,z_\alpha) - \text{int}(C) \subseteq -\text{int}(C).$$

In all the cases, we obtain the desired result. □

Theorem 10.11 *Let K be a nonempty convex subset of a Hausdorff topological vector space X such that* $\text{ri}\,(K) \neq \emptyset$, *and C be a proper closed convex cone in a topological vector space Y such that* $\text{int}(C) \neq \emptyset$. *Let $F : K \times K \to 2^Y \setminus \{\emptyset\}$ be a C-quasimonotone set-valued map satisfying the following conditions:*

(i) *For each $x \in K$, $F(x,x) \subseteq C$;*
(ii) *For each $x \in K$, the set-valued map $y \mapsto F(x,y)$ is explicitly C-quasiconvex-like and explicitly δ-C-quasiconvex;*
(iii) *For each $x \in K$, the set-valued map $y \mapsto F(x,y)$ is upper semicontinuous with compact values;*
(iv) *For each $y \in K$, the set-valued map $x \mapsto F(x,y)$ is u-hemicontinuous;*
(v) *There exist a nonempty compact convex subset B of K and a nonempty compact subset D of K such that for each $x \in K \setminus D$, there exists $\tilde{y} \in B \cap \text{ri}\,(K)$ such that $F(x,\tilde{y}) \subseteq -\text{int}(C)$.*

Then there exists a solution $\bar{x} \in K$ to GWVEP (10.1).

Proof Define $S, T : K \to 2^K \setminus \{\emptyset\}$ by

$$S(x) = \{y \in \text{ri}\,(K) : F(y,x) \subseteq \text{int}(C)\}$$

and

$$T(x) = \{y \in \text{ri}\,(K) : F(x,y) \subseteq -\text{int}(C)\},$$

for all $x \in K$.

From Lemma 10.3 it follows that, for each $x \in K$, either $S(x) \subseteq T(x)$, or

$$F(x,y) \not\subseteq -\text{int}(C), \quad \text{for all } y \in K.$$

From the proof of Theorem 10.10, we see that

- for each $x \in K$, $T(x)$ is convex
- for each $y \in K$, $S^{-1}(y)$ is open in K
- T has no fixed point

It follows that there exists $\bar{x} \in K$ such that $S(\bar{x}) = \emptyset$, i.e.,

$$F(y,\bar{x}) \not\subseteq \text{int}(C), \quad \text{for all } y \in \text{ri}\,(K).$$

Take any $z \in K$ and $y' \in \text{ri}\,(K)$ and suppose that

$$F(z,\bar{x}) \subseteq \text{int}(C).$$

Then by u-hemicontinuity of $F(\cdot, \bar{x})$, there is an $y \in]z, y'[\subseteq \text{ri}\,(K)$ such that

$$F(y, \bar{x}) \subseteq \text{int}(C),$$

a contradiction. Hence,

$$F(y, \bar{x}) \nsubseteq \text{int}(C), \quad \text{for all } y \in K.$$

By Lemma 10.2 (b), we have

$$F(\bar{x}, y) \nsubseteq -\text{int}(C), \quad \text{for all } y \in K,$$

as desired. □

Ansari et al. [8] considered the GWVEP for moving cone and derived the existence results for solutions under C_x-quasimonotonicity.

10.3.1 Existence Results Without Monotonicities

Now we provide an existence result of a solution to GWVEP without any kind of monotonicity assumption.

Theorem 10.12 *Let K be a nonempty convex subset of a Hausdorff topological vector space X and C be a proper closed convex cone in a topological vector space Y such that* $\text{int}(C) \neq \emptyset$. *Let $F : K \to 2^Y$ be set-valued map with nonempty values such that the following conditions hold.*

(i) $F(x, x) \nsubseteq -\text{int}(C)$ *for all $x \in K$;*
(ii) *F is C-quasiconvex-like;*
(iii) *For each $y \in K$, the set-valued map $x \mapsto G(x, y)$ is upper semicontinuous with compact values on K;*
(iv) *There exist a nonempty compact convex subset B of K and a nonempty compact subset D of K such that for each $x \in K \setminus D$, there exists $\tilde{y} \in B$ such that* $F(x, \tilde{y}) \subseteq -\text{int}(C)$.

Then the solution set Sol(GWVEP) *of* GWVEP (10.1) *is nonempty and compact subset of B.*

Proof For all $y \in K$, define $Q(y) = \{x \in K : F(x, y) \nsubseteq -\text{int}(C)\}$. Then the solution set Sol(GWVEP) $= \bigcap_{y \in K} Q(y)$. As in the proof of Theorem 10.9, $Q(y)$ is closed for each $y \in K$.

Now we prove that the solution set Sol(GWVEP) is nonempty. Assume contrary that Sol(GWVEP) $= \emptyset$, if possible. Then, for each $x \in K$, the set

$$S(x) = \{y \in K : x \notin Q(y)\} = \{y \in K : F(x, y) \subseteq -\text{int}(C)\} \neq \emptyset.$$

Also, from the proof of Theorem 10.10, $S(x)$ is convex for each $x \in K$.

Thus $S : K \to 2^K$ defines a set-valued map such that for each $x \in K$, $S(x)$ is nonempty and convex. Now for each $x \in K$, the set

$$\begin{aligned} S^{-1}(y) &= \{x \in K : y \in S(x)\} \\ &= \{x \in K : F(x, y) \subseteq -\operatorname{int}(C)\} \\ &= \{x \in K : F(x, y) \nsubseteq -\operatorname{int}(C)\}^c \\ &= [Q(y)]^c \end{aligned}$$

is open in K. From Assumption (iv), for each $x \in K \setminus D$, there exists $\tilde{y} \in B$ such that $F(x, \tilde{y}) \subseteq -\operatorname{int}(C)$, that is, $x \notin Q(\tilde{y})$. By Corollary 1.3, there exists a point $\bar{x} \in S(\bar{x})$, that is, $F(\bar{x}, \bar{x}) \subseteq -\operatorname{int}(C)$, which contradicts to assumption (i). Hence the solution set Sol(GWVEP) is nonempty. We conclude the proof by noting that Sol(GWVEP) $= \bigcap_{y \in K} Q(y)$ being a closed subset of the compact set D is compact and this completes the proof. □

10.4 Duality

We adopt the following rule to obtain the dual formulations of GWVEP (10.1):

> Interchange arguments of the underlying set-valued map and change the sign on the right-hand side of the inclusion.

Then such a dual will satisfy the following fundamental duality property:

> the dual of the dual is the primal.

However in addition to this property, it also satisfy some other properties similar to those in optimization, for example, the solution set of dual problem coincide with the solution set of the primal problem under certain monotonicity and convexity conditions.

Throughout this section, unless otherwise specified, we assume that X and Y are topological vector spaces and K is a nonempty convex subset of X. We denote by $\mathbb{F}(K, Y)$ the family of all set-valued maps from $K \times K$ to 2^Y. Let C be a proper closed convex cone with $\operatorname{int}(C) \neq \emptyset$. For a given set-valued map $F \in \mathbb{F}(K)$, by using above rule, we define the dual form of GWVEP (10.1) in the following way and it is called *dual generalized weak vector equilibrium problem* (in short, DGWVEP):

$$\text{Find } \bar{x} \in K \text{ such that } F(y, \bar{x}) \nsubseteq \operatorname{int}(C), \quad \text{for all } y \in K. \tag{10.19}$$

The solution set of DGWVEP (10.19) is denoted by Sol(DGWVEP). Then clearly the dual of DGWVEP is GWVEP.

Konnov and Schaible [40] introduced this kind of duality concept for equilibrium problems. By extending the terminology of Konnov and Schaible [40], Ansari et al.

[7] (see also [9]) gave some dual formulations of the GWVEP and proved that the solution set of the dual problem coincide with the solution set of the primal GWVEP under certain pseudomonotonicity assumption. By using the dual formulations of GWVEP, they established some existence results for solutions of GWVEP (10.1). Ansari et al. [7] considered the moving cone in the formulation of GWVEP (10.1). However, in this section in particular and in this book in general, we consider the fixed proper closed convex cone.

10.4.1 Generalized Duality

With the help of an operator Φ from $\mathbb{F}(K, Y)$ into itself, we propose the following *dual generalized weak vector equilibrium problem*, denoted by DGWVEP_Φ:

$$\text{Find } \bar{x} \in K \text{ such that } \Phi(F(\bar{x}, y)) \nsubseteq -\operatorname{int}(C), \quad \text{for all } y \in K,$$

and Φ is called *duality operator*. In fact, the operator Φ is nothing but a set of fixed rules applied to GWVEP, rather than a set-valued map. We see that under certain conditions, the dual of DGWVEP_Φ is the primal GWVEP and the solution set $\mathrm{Sol(DGWVEP)}_\Phi$ of DGWVEP_Φ coincides with the solution set Sol(GWVEP) of the primal GWVEP.

For simplicity, we set

$$G(y, x) = -\Phi(F(x, y)),$$

then DGWVEP_Φ can be written as the following problem:

$$\text{Find } \bar{x} \in K \text{ such that } G(y, \bar{x}) \nsubseteq \operatorname{int}(C), \quad \text{for all } y \in K. \tag{10.20}$$

The set of all solutions of problem (10.20) is denoted by $\mathrm{Sol(DGWVEP)}_G$.

Proposition 10.2 *If*

$$\Phi \circ \Phi(F(x, y)) = F(x, y), \quad \textit{for all } x, y \in K,$$

or equivalently,

$$\Phi(-G(y, x)) = F(x, y), \quad \textit{for all } x, y \in K,$$

then the dual of DGWVEP (10.19) *is* GWVEP (10.1).

Proof It is straightforward. □

Definition 10.5 Let K be a nonempty convex subset of a topological vector space X, C be a proper closed convex cone in a topological vector space Y such that $\text{int}(C) \neq \emptyset$, and $F, G : K \times K \to 2^Y$ be set-valued maps. Then F is said to be

(a) *G-C-pseudomonotone* if for all $x, y \in K$,

$$F(x, y) \nsubseteq -\text{int}(C) \quad \text{implies} \quad G(y, x) \nsubseteq \text{int}(C);$$

(b) *maximal G-C-pseudomonotone* if it is G-C-pseudomonotone and for all $x, y \in K$,

$$G(z, x) \nsubseteq \text{int}(C) \text{ for all } z \in]x, y] \quad \text{implies} \quad F(x, y) \nsubseteq -\text{int}(C).$$

The following results give the equivalence between Sol(GWVEP) and Sol(DGWVEP)_G.

Proposition 10.3 *If F is maximal G-C-pseudomonotone, then* Sol(GWVEP) = Sol(DGWVEP)_G.

Proof It lies on the lines of the proof of Lemma 10.1. □

Proposition 10.4 *Let $F, G \in \mathbb{F}(K, Y)$ such that*

(i) *for all $x \in K$, $F(x, x) \subseteq C$,*
(ii) *F is explicitly δ-C-quasiconvex and G-C-pseudomonotone,*
(iii) *for all $x, y \in K$, $F(x, y) \subseteq \text{int}(C)$ implies $G(x, y) \subseteq \text{int}(C)$,*
(iv) *for all $y \in K$, the set-valued map $x \mapsto F(x, y)$ is u-hemicontinuous.*

Then Sol(GWVEP) = Sol(DGWVEP)_G.

Proof It lies on the lines of the proof of Lemma 10.2. □

Theorem 10.13 *Let K be a nonempty and convex subset of a Hausdorff topological vector space X and Y be a topological vector space. Let $F, G \in \mathbb{F}(K, Y)$ such that the following conditions hold.*

(i) *F is G-C-pseudomonotone;*
(ii) *For each $x \in K$, $F(x, x) \nsubseteq -\text{int}(C)$;*
(iii) *For each $x \in K$, there exists $y \in K$ such that $G(y, x) \subseteq \text{int}(C)$;*
(iv) *For each $x \in K$, the set $\{y \in K : F(x, y) \subseteq -\text{int}(C)\}$ is convex;*
(v) *For each $y \in K$, $Q(y) = \{x \in K : G(y, x) \nsubseteq \text{int}(C)\}$ is closed in K;*
(vi) *There exist a nonempty compact convex subset B of K and a nonempty compact subset D of K such that for each $x \in K \setminus D$, there exists $\tilde{y} \in B$ such that $x \in \text{int}_K\{x \in K : F(x, \tilde{y}) \subseteq -\text{int}(C)\}$.*

Then GWVEP (10.1) *has a solution.*

Proof Suppose that the conclusion of this theorem is not true. Then for each $x \in K$, the set

$$\{y \in K : F(x, y) \subseteq -\text{int}(C)\} \neq \emptyset.$$

We define a set-valued map $S : K \to 2^K$ by

$$S(x) = \{y \in K : F(x, y) \subseteq -\operatorname{int}(C)\} \quad \text{for all } x \in K.$$

By (v), $S(x)$ is convex for all $x \in K$. By G-C-pseudomonotonicity of F, we have

$$\begin{aligned}[S^{-1}(y)]^c &= \{x \in K : F(x, y) \nsubseteq -\operatorname{int}(C)\} \\ &\subseteq \{x \in K : G(y, x) \nsubseteq \operatorname{int}(C)\} \\ &= Q(y).\end{aligned}$$

It follows that the complement of $Q(y)$ in K, $[Q(y)]^c \subseteq S^{-1}(y)$ for all $y \in K$. By (v), $Q(y)$ is closed in K for all $y \in K$. Hence $[Q(y)]^c$ is open in K and therefore $[Q(y)]^c \subseteq \operatorname{int}_K S^{-1}(y)$ for all $y \in K$. By (iii), for each $x \in K$, there exists $y \in K$ such that $G(y, x) \subseteq \operatorname{int}(C)$. Hence it follows that

$$K = \bigcup_{y \in K} [Q(y)]^c = \bigcup_{y \in K} \operatorname{int}_K S^{-1}(y).$$

Finally by (vi), for each $x \in K \setminus D$ there exists $\tilde{y} \in B$ such that $x \in \operatorname{int}_K S^{-1}(\tilde{y})$. Hence S satisfies all the conditions of Theorem 1.35. Therefore there exists $x^* \in K$ such that $x^* \in S(x^*)$, that is, $F(x^*, x^*) \subseteq -\operatorname{int}(C)$. However this contradicts to assumption (ii). Hence the result is proven. □

Remark 10.6

(a) If F is C-quasiconvex-like, then condition (iv) of Theorem 10.13 is satisfied (see the proof of Theorem 10.10).
(b) If X is a metrizable locally convex space, for each $y \in K$, $G(y, \cdot)$ is upper semicontinuous with compact values, then condition (v) of Theorem 10.13 is satisfied (see the proof of Theorem 10.10).
(c) The conclusion of Theorem 10.13 still holds if we replace condition (iii) by the maximal G-C-pseudomonotonicity of F.

    Indeed, if condition (iii) does not hold, then there exists $\bar{x} \in K$ such that $G(y, x) \nsubseteq \operatorname{int}(C)$ for all $y \in K$. The maximal G-C-pseudomonotonicity of F implies that $\bar{x} \in K$ is a solution of the GWVEP (10.1).

Next we discuss condition (iii).

Example 10.1 Let $K = [0, 1]$, $C = [0, \infty[$ and $F, G : K \times K \to \mathbb{R}$ be defined by

$$F(x, y) = \begin{cases} 1 - (y/x), & \text{if } y \in [0, 1],\ 0 < x < 1, \\ y - 1, & \text{if } y \in [0, 1],\ x = 1, \\ -y, & \text{if } y \in [0, 1],\ x = 0, \end{cases}$$

and

$$G(x, y) = x - y, \quad \text{for all } x, y \in [0, 1].$$

Then all the conditions of Theorem 10.13 except (iii) are satisfied. To see that (iii) is not satisfied, note that $G(y, 1) = y - 1 \leq 0$ for all $y \in [0, 1]$. In this example GWVEP (10.1) has no solution.

Hence Theorem 10.13 is not always true without condition (iii). If condition (iii) is not satisfied, then an assumption stronger than G-C-pseudomonotonicity of F will guarantee the existence of a solution of GWVEP (10.1).

10.4.2 Additive Duality

In this subsection, we consider the case where

$$\Phi(F(x, y)) = -G(y, x) = -(F(y, x) + H(y, x)),$$

for some $H \in \mathbb{F}(K, Y)$ and specialize the results of previous section. In other words, we study the following *additive dual problem*, denoted by DGWVEP$_H$:

$$\text{Find } \bar{x} \in K \text{ such that } F(y, \bar{x}) + H(y, \bar{x}) \nsubseteq \text{int}(C), \quad \text{for all } y \in K. \tag{10.21}$$

In other words, to define the additive dual problem, we add $H(y, x)$ on the left-hand side of the dual of primal problem, that is, DGWVEP. We denote by Sol(DGWVEP)$_H$ the set of solutions of DGWVEP$_H$.

Proposition 10.5 *If*

$$H(x, y) + H(y, x) \subseteq -\text{int}(C), \quad \textit{for all } x, y \in K,$$

then the additive dual problem of DGWVEP$_H$ *is* GWVEP (10.1).

Proof Since $G(y, x) = F(y, x) + H(y, x)$, DGWVEP$_H$ can be written as to find $\bar{x} \in K$ such that

$$G(y, \bar{x}) \nsubseteq \text{int}(C), \quad \text{for all } y \in K. \tag{10.22}$$

By the rule of additive duality, the additive dual of (10.22) is

$$G(\bar{x}, y) + H(y, \bar{x}) \nsubseteq -\text{int}(C), \quad \text{for all } y \in K,$$

and therefore,

$$F(\bar{x}, y) + H(\bar{x}, y) + H(y, \bar{x}) \nsubseteq -\text{int}(C).$$

Since $H(x, y) + H(y, x) \subseteq -\operatorname{int}(C)$ for all $x, y \in K$, we have

$$F(\bar{x}, y) \nsubseteq -\operatorname{int}(C).$$

This completes the proof. □

Definition 10.6 Let $F, H \in \mathbb{F}(K, Y)$. The set-valued map F is called

(a) *(H)-C-pseudomonotone* if for all $x, y \in K$,

$$F(x, y) \nsubseteq -\operatorname{int}(C) \quad \text{implies} \quad F(y, x) + H(y, x) \nsubseteq \operatorname{int}(C);$$

(b) *maximal (H)-C-pseudomonotone* if it is (H)-pseudomonotone and for all $x, y \in K$,

$$F(z, x) + H(z, x) \nsubseteq \operatorname{int}(C) \text{ for all } z \in]x, y] \quad \text{implies} \quad F(x, y) \nsubseteq -\operatorname{int}(C).$$

Proposition 10.6 *If F is maximal (H)-C-pseudomonotone, then* Sol(GWVEP) = Sol(DGWVEP)$_H$.

Proposition 10.7 *Let $F, H \in \mathbb{F}(K, Y)$ such that*

(i) *for all $x \in K$, $F(x, x) \subseteq C$,*
(ii) *F is explicitly δ-C-quasiconvex and (H)-C-pseudomonotone,*
(iii) *for all $x, y \in K$, $F(x, y) \subseteq \operatorname{int}(C)$ implies $F(x, y) + H(x, y) \subseteq \operatorname{int}(C)$,*
(iv) *for all $y \in K$, the set-valued map $x \mapsto F(x, y)$ is u-hemicontinuous.*

Then Sol(GWVEP) = Sol(DGWVEP)$_H$.

10.4.3 Multiplicative Duality

In order to define the multiplicative dual problem, we multiply by $H(y, x)$ for some $H \in \mathbb{F}(K, Y)$, on the left-hand side and by $\operatorname{int}(C)$ on the right-hand side of the dual problem of primal problem, that is, DGWVEP. In other words, we consider the following *multiplicative dual problem*, denoted by DGWVEP$_{m(H)}$:

$$\text{Find } \bar{x} \in K \text{ such that } F(y, \bar{x}) \times H(y, \bar{x}) \nsubseteq \operatorname{int}(C) \times \operatorname{int}(C), \quad \text{for all } y \in K. \tag{10.23}$$

The set of all solutions of DGWVEP$_{m(H)}$ is denoted by Sol(DGWVEP)$_{m(H)}$. "m" in $m(H)$ refers to "multiplicative".

Proposition 10.8 *If*

$$H(x, y) \times H(y, x) \subseteq \operatorname{int}(C) \times \operatorname{int}(C), \quad \textit{for all } x, y \in K,$$

then the multiplicative dual problem of $\mathrm{DGWVEP}_{m(H)}$ *is* GWVEP (10.1).

Proof Let $G(y,x) = F(y,x) \times H(y,x)$, then $\mathrm{DGWVEP}_{m(H)}$ can be written as to find $\bar{x} \in K$ such that

$$G(y,\bar{x}) \nsubseteq \mathrm{int}(C) \times \mathrm{int}(C), \quad \text{for all } y \in K. \tag{10.24}$$

By the rule of multiplicative duality, the multiplicative dual of (10.24) is

$$G(\bar{x},y) \times H(y,\bar{x}) \nsubseteq -(\mathrm{int}(C) \times \mathrm{int}(C)) \times \mathrm{int}(C), \quad \text{for all } y \in K,$$

and therefore,

$$F(\bar{x},y) \times H(\bar{x},y) \times H(y,\bar{x}) \nsubseteq -\mathrm{int}(C) \times \mathrm{int}(C) \times \mathrm{int}(C).$$

Since $H(x,y) \times H(y,x) \subseteq \mathrm{int}(C) \times \mathrm{int}(C)$ for all $x, y \in K$, we have

$$F(\bar{x},y) \nsubseteq -\mathrm{int}(C).$$

This completes the proof. □

Definition 10.7 Let $F, H \in \mathbb{F}(K,Y)$. The set-valued map F is called

(a) *$m(H)$-C-pseudomonotone* if for all $x, y \in K$,

$$F(x,y) \nsubseteq -\mathrm{int}(C) \quad \text{implies} \quad F(y,x) \times H(y,x) \nsubseteq \mathrm{int}(C) \times \mathrm{int}(C);$$

(b) *maximal $m(H)$-C-pseudomonotone* if it is $m(H)$-C-pseudomonotone and for all $x, y \in K$,

$$F(z,x) \times H(z,x) \nsubseteq \mathrm{int}(C) \times \mathrm{int}(C) \text{ for all } z \in]x,y] \text{ implies}$$
$$F(x,y) \nsubseteq -\mathrm{int}(C).$$

Proposition 10.9 *If F is maximal $m(H)$-pseudomonotone, then* $\mathrm{Sol(GWVEP)} - \mathrm{Sol(DGWVEP)}_{m(H)}$.

Proposition 10.10 *Let $F, H \in \mathbb{F}(K,Y)$ such that*

(i) *for all $x \in K$, $F(x,x) \subseteq C$,*
(ii) *F is explicitly δ-C-quasiconvex and $m(H)$-pseudomonotone,*
(iii) *for all $x, y \in K$, $F(x,y) \subseteq \mathrm{int}(C)$ implies $F(x,y) \times H(x,y) \subseteq \mathrm{int}(C) \times \mathrm{int}(C)$,*
(iv) *for all $y \in K$, the set-valued map $x \mapsto F(x,y)$ is u-hemicontinuous.*

Then $\mathrm{Sol(GWVEP)} = \mathrm{Sol(DGWVEP)}_{m(H)}$.

10.5 Recession Methods for Generalized Vector Equilibrium Problems

As we have seen in the previous sections of this chapter and in Chap. 9 that most of the existence results for solutions of GVEPs and VEPs require the compactness (in topological vector space setting)/boundedness (in reflexive Banach space setting) or some kind of coercivity condition. Flores-Bazán and Flores-Bazán [26] studied the existence of solutions for VEPs under the asymptotic analysis, where neither compactness on K nor any coercivity condition is assumed. They gave some characterizations of nonemptiness of the solution set and also presented several alternative necessary and/or sufficient conditions for the solution set to be nonempty and compact.

Ansari and Flores-Bazán [3] extended the ideas of Flores-Bazán and Flores-Bazán [26] for GVEPs. By using recession method, they provided several alternative necessary and sufficient and/or sufficient conditions for the solution set of GVEPs to be nonempty and bounded. Lee and Bu [42] also considered GWVEP (10.2) in the setting of finite-dimensional Euclidean space but for a moving cone. They used asymptotic cone of the solution set of GWVEP (10.2) to give conditions under which the solution set is nonempty and compact. Sadeqi and Alizadeh [57] improved some results given in [3]. Wang [59] further studied the equivalent characterizations of the solution set of GSVEP (10.5) to be nonempty and bounded based on asymptotic cone theory. In this section, we gather the results that appeared in the above cited references.

Let K be a nonempty closed convex subset of a reflexive Banach space X and Y be a normed space with an ordered cone C, that is, a proper, closed and convex cone such that $\operatorname{int}(C) \neq \emptyset$. Let $F : K \times K \to 2^Y \setminus \{\emptyset\}$ be a set-valued map. The basic assumptions on F are the following:

Assumption 10.8

(f_0) *For all $x \in K$, $F(x,x) \subseteq l(C) := C \cap (-C)$;*
(f_1) *For all $x, y \in K$, $F(x,y) \cap (-\operatorname{int}(C)) = \emptyset$ implies $F(y,x) \cap \operatorname{int}(C) = \emptyset$;*
(f_2) *For all $x \in K$, the set-valued map $F(x,\cdot) : K \to 2^Y \setminus \{\emptyset\}$ is C-convex;*
(f_3) *For all $x, y \in K$, the set $\{\xi \in [x,y] : F(\xi,y) \cap (-\operatorname{int}(C)) = \emptyset\}$ is closed;*
(f_4) *For all $x \in K$, the set-valued map $F(x,\cdot) : K \to 2^Y \setminus \{\emptyset\}$ is weakly lower semicontinuous.*

We introduce the following cones in order to deal with the unbounded case, that is, when K is an unbounded set,

$$R_0 := \bigcap_{y \in K} \{v \in K_\infty : \mathbf{0} \in F(y, z + tv) + W \text{ for all } t > 0 \text{ and } z \in K \\ \text{such that } F(y,z) \subseteq -C\}$$

and

$$R_1 := \bigcap_{y \in K} \{v \in K_\infty : \mathbf{0} \in F(y, y + t\,v) + W \text{ for all } \lambda > 0\},$$

where $W = Y \setminus (-\operatorname{int}(C))$ and K_∞ denotes the recession cone of K.

We note that the sets R_0 and R_1 are nonempty (because of assumption (f_0)) closed cone but not necessarily convex. Clearly, $R_0 \subseteq R_1$.

The proof of the following result is straightforward, and therefore it is omitted.

Proposition 10.11 *Let K be a nonempty closed convex subset of X and let* (10.12) *hold. Then*

$$R_{11} := \bigcap_{y \in K} \{v \in K_\infty : \mathbf{0} \in F(y, y + t\,v) + (Y \setminus (-\operatorname{int}(C))) \textit{ for all } t > 0\}$$
$$\subseteq \bigcap_{y \in K} \{v \in K_\infty : F(y + t\,v, y) \cap (-\operatorname{int}(C)) = \emptyset \textit{ for all } t > 0\}.$$

Remark 10.7 In view of above proposition and Remark 10.3 (c), we have $R_{11} = R_0$ if the conditions (f_0), (f_2) and (f_3) hold. However, Sadeqi and Alizadeh [57] proved that $R_{11} \subseteq R_0$ if (f_0), (f_1), (f_2) and (f_4) hold.

Remark 10.8 If $F : K \times K \to 2^Y \setminus \{\emptyset\}$ is a set-valued map satisfying assumption (f_0) such that for all $x \in K$, $F(x, \cdot) : K \to 2^Y \setminus \{\emptyset\}$ is C-convex, then the conclusion of Proposition 10.11 also holds.

Indeed, let $v \in R_{11}$ and $W = Y \setminus (-\operatorname{int}(C))$. Then $v \in K_\infty$, and for all $y \in K$ and all $t > 0$, there exists $\xi(y, y + t\,v) \in F(y, y + t\,v)$ such that $\xi(y, y + t\,v) \in -W$. On the other hand, for any $y \in K$ and $t > 0$, the C-convexity of $F(y + t\,v, \cdot)$ implies

$$\frac{1}{2}F(y + t\,v, y + t\,v + t\,v) + \frac{1}{2}F(y + t\,v, y) \subseteq F(y + t\,v, y + t\,v) + C \subseteq C.$$

Thus

$$\frac{1}{2}F(y + t\,v, y) \subseteq -\frac{1}{2}\xi(y + t\,v, y + 2t\,v) + C \subseteq W + C \subseteq W.$$

Hence $F(y + t\,v, y) \subseteq W$. Since $y \in K$ and $t > 0$ were arbitrary, we conclude the proof.

Theorem 10.14 *Let K be a nonempty closed convex subset of X and $F : K \times K \to 2^Y \setminus \{\emptyset\}$ be a set-valued map satisfying conditions (f_0), (f_1), (f_2) and (f_4). Then*

$$\left(Sol(GWVEP)^*\right)_\infty \subseteq R_1 \subseteq \bigcap_{y \in K} \{x \in K : F(x, y) \cap (-\operatorname{int}(C)) = \emptyset\}_\infty$$
$$\subseteq \bigcap_{y \in K} \{x \in K : F(y, x) \cap (\operatorname{int}(C)) = \emptyset\}_\infty.$$

If, in addition, there exists $x^* \in K$ such that $F(y, x^*) \subseteq -C$ for all $y \in K$, then $(Sol(GWVEP)^*)_\infty = R_1$.

Proof As before, set $W := Y \setminus (-\operatorname{int}(C))$. Let us prove the first inclusion. Let $v \in (\mathrm{Sol(GWVEP)}^*)^\infty$. Then there exist $\lambda_m \downarrow 0$, $u_m \in \mathrm{Sol(GWVEP)}^*$ such that $\lambda_m u_m \rightharpoonup v$. For $y \in K$ arbitrary, we have $F(u_m, y) \subseteq W$ for all $m \in \mathbb{N}$. By assumption (f_1), we have $F(y, u_m) \subseteq -W$ for all $m \in \mathbb{N}$. Let us fix any $t > 0$. For m sufficiently large, C-convexity of $F(y, \cdot)$ implies

$$(1 - t\lambda_m)F(y, y) + t\lambda_m F(y, u_m) \subseteq F(y, (1 - t\lambda_m)y + t\lambda_m u_m) + C.$$

Hence

$$\mathbf{0} \in F(y, (1 - t\lambda_m)y + t\lambda_m u_m) + W + C \subseteq F(y, (1 - t\lambda_m)y + t\lambda_m u_m) + W.$$

From assumption (f_4), it follows that $\mathbf{0} \in F(y, y + t\,v) + W$. This proves $v \in R_1$.

The proof of the second inclusion is as follows. Let $v \in K_\infty$ such that $\mathbf{0} \in F(y, y+t\,v)+W$ for all $t > 0$ and all $y \in K$. By Proposition 10.11, $F(y+t\,v, y) \subseteq W$ for all $t > 0$ and all $y \in K$. For any fixed $y \in K$, set $x_m := y + m\,v \in K$ for all $m \in \mathbb{N}$. Then $F(x_m, y) \subseteq W$ for all $m \in \mathbb{N}$. By choosing $\lambda_m = \frac{1}{m}$, we have $\lambda_m x_m = \frac{y}{m} + v \to v$ as $m \to +\infty$, that is, $v \in \{x \in K : F(x, y) \subseteq W\}_\infty$. Since y was arbitrary, the proof of the second inclusion is complete.

The last inclusion is a consequence of assumption (f_1).

Let us prove the last part of the theorem. By our hypothesis, there exists $x^* \in K$ such that $F(y, x^*) \subseteq -C$ for all $y \in K$. Let $v \in R_1$. Then for all $y \in K$ and for all $t > 0$, $\mathbf{0} \in F(y, x^* + t\,v) + W$. By Proposition 10.11, $F(x^* + t\,v, y) \subseteq W$. Thus for all $t > 0$, $x^* + t\,v \in \mathrm{Sol(GWVEP)}^*$. Hence $v \in (\mathrm{Sol(GWVEP)}^*)_\infty$ and thus $R_1 \subseteq (\mathrm{Sol(GWVEP)}^*)_\infty$. Consequently, $R_1 = (\mathrm{Sol(GWVEP)}^*)_\infty$. □

Theorem 10.15 *Let K be a nonempty closed convex set in X and $F : K \times K \to 2^Y \setminus \{\emptyset\}$ be a set-valued map satisfying Assumption 10.8. If,*

(C1) *for every sequence $\{x_m\}$ in K, $\|x_m\| \to +\infty$ such that $\frac{x_m}{\|x_m\|} \rightharpoonup v$ with $v \in R_1$ and for all $y \in K$ there exists m_y such that $F(x_m, y) \subseteq Y \setminus (-\operatorname{int}(C))$ for all $m \geq m_y$, there exists $u \in K$ such that $\|u\| < \|x_m\|$ and $F(x_m, u) \subseteq -C$ for $m \in \mathbb{N}$ sufficiently large,*

then GWVEP (10.2) *admits a solution. Indeed,* $\mathrm{Sol(GWVEP)}^*$ *is a nonempty weakly closed set.*

Proof For every $m \in \mathbb{N}$, set $K_m := \{x \in K : \|x\| \leq m\}$. We may assume, without loss of generality, that $K_m \neq \emptyset$ for all $m \in \mathbb{N}$. Let us consider the following problem:

$$\text{Find } \bar{x} \in K_m \text{ such that } F(\bar{x}, y) \subseteq Y \setminus (-\operatorname{int}(C)), \quad \text{for all } y \in K_m. \tag{10.25}$$

Taking into account Remark 10.3, we apply Theorem 10.4 (with $W = Y \setminus (-\operatorname{int}(C))$) to conclude that problem (10.25) admits a solution, say $x_m \in K_m$ for all $m \in \mathbb{N}$.

If $\|x_m\| < m$ for some $m \in \mathbb{N}$, then we claim that x_m is also a solution to problem (10.2). Suppose to the contrary that x_m is not a solution to problem (10.2). Then there exists $y \in K$ with $\|y\| > m$ such that $F(x_m, y) \nsubseteq Y \setminus (-\operatorname{int}(C))$. We choose $z \in K$ with $z \in]x_m, y[$ and $\|z\| < m$. Writing $z = \alpha x_m + (1-\alpha)y$ for some $\alpha \in]0, 1[$, then by C-convexity of $F(x_m, \cdot)$, we have

$$\alpha F(x_m, x_m) + (1-\alpha)F(x_m, y) \subseteq F(x_m, z) + C.$$

This implies that

$$(1-\alpha)F(x_m, y) \subseteq W + C \subseteq W.$$

It follows that $F(x_m, y) \subseteq W = Y \setminus (-\operatorname{int}(C))$, which contradicts to our supposition. Hence x_m is a solution to problem (10.2).

We consider now the case $\|x_m\| = m$ for all $m \in \mathbb{N}$. We may assume, up to a subsequence, that $\frac{x_m}{\|x_m\|} \rightharpoonup v$ for $v \in K$. Then $v \neq \mathbf{0}$ and $v \in K_\infty$. For any fixed $y \in K$ and $t > 0$, we have $F(x_m, y) \subseteq W$ for all $m \in \mathbb{N}$ sufficiently large ($m > \|y\|$). By (f_1), $F(y, x_m) \subseteq Y \setminus (\operatorname{int}(C))$ for all m sufficiently large. For every $t > 0$ and all m sufficiently large, C-convexity of $F(y, \cdot)$ implies

$$\left(1 - \frac{t}{\|x_m\|}\right) F(y, y) + \frac{t}{\|x_m\|} F(y, x_m) \subseteq F\left(y, \left(1 - \frac{t}{\|x_m\|}\right) y + \frac{t}{\|x_m\|} x_m\right) + C.$$

Hence

$$\mathbf{0} \in F\left(y, \left(1 - \frac{t}{\|x_m\|}\right) y + \frac{t}{\|x_m\|} x_m\right) + (Y \setminus (\operatorname{int}(C))) + C.$$

Thus, by assumption (f_4), $\mathbf{0} \in F(y, y + t\,v) + W$. This proves $v \in R_1$. By assumption, there exist $u \in K$ such that $\|u\| < \|x_m\|$ and $F(x_m, u) \subseteq -C$ for m sufficiently large. We claim that x_m is also a solution to GWVEP (10.2). Suppose contrary that x_m is not a solution of GWVEP (10.2). Then there exists $y \in K$, $\|y\| > m$ such that $F(x_m, y) \nsubseteq Y \setminus (-\operatorname{int}(C)) = W$. Since $\|u\| < \|x_m\|$, we can find $z \in]u, y[$ such that $\|z\| < m$. Thus for some $\alpha \in]0, 1[$, by C-convexity of $F(x_m, \cdot)$, we have

$$\alpha F(x_m, u) + (1-\alpha)F(x_m, y) \subseteq F(x_m, z) + C \subseteq W + C.$$

This implies that

$$(1-\alpha)F(x_m, y) \subseteq W,$$

a contradiction to our supposition. Thus, x_m is a solution to GWVEP (10.2). □

Example 10.2 Let $K = \mathbb{R}$ and $C = \mathbb{R}^2_+$. Then the function $F(x, y) = (y - x, x - y)$ does not satisfy condition (C1) while $\mathrm{Sol(GWVEP)}^* = \mathbb{R}$.

We now establish a couple of necessary and sufficient conditions for the nonemptiness of Sol(GWVEP)* for a class of set-valued maps defined on $K \subseteq \mathbb{R}$, which apply to the previous example.

(C2) For every sequence $\{x_m\}$ in K with $|x_m| \to +\infty$, $\frac{x_m}{|x_m|} \to v$, $v \in R_1$, and for all $y \in K$ there exists m_y such that $F(x_m, y) \subseteq Y \setminus (-\operatorname{int}(C))$ for all $m \geq m_y$, there exist $u \in K$ and $\bar{m}$ such that $|u| < |x_{\bar{m}}|$ and $F(x_{\bar{m}}, u) \subseteq Y \setminus (\operatorname{int}(C))$.

(C3) For every sequence $\{x_m\}$ in K with $|x_m| \to +\infty$, there exist m_0 and $u \in K$ such that $F(x_m, u) \subseteq Y \setminus (\operatorname{int}(C))$ for all $m \geq m_0$.

Theorem 10.16 *Let $K \subseteq \mathbb{R}$ be a closed convex set and $F : K \to 2^Y \setminus \{\emptyset\}$ be a set-valued map satisfying Assumption 10.8. Then Sol(GWVEP)* is a closed convex set, and the following three assertions are equivalent.*

(a) Sol(GWVEP)* *is nonempty;*
(b) (C2) *is satisfied;*
(c) (C3) *is satisfied.*

Proof The closedness of Sol(GWVEP)* is obtained as before. We reason as follows to prove the convexity: take $x_1, x_2 \in$ Sol(GWVEP)*, $x_1 < x_2$, and $x \in]x_1, x_2[$. Then if $y \in K$, $y > x$, we write $x := \alpha x_1 + (1 - \alpha)y$ and use the C-convexity of $F(x, \cdot)$ to obtain $F(x, y) \subseteq Y \setminus (-\operatorname{int}(C))$. In case $y \in K$, $y < x$, we write $x = \alpha x_2 + (1-\alpha)y$ and proceed as before to conclude again $F(x, y) \subseteq Y \setminus (-\operatorname{int}(C))$. Thus $x \in$ Sol(GWVEP)*, proving the convexity of Sol(GWVEP)*.

We now prove the equivalences.

(c) $\Rightarrow$ (b): It is obvious.

(a) $\Rightarrow$ (c): It follows by taking as u any element in Sol(GWVEP)*.

(b) $\Rightarrow$ (a): We proceed as in the proof of Theorem 10.15 until being in the case when the sequence $\{x_m\} \subseteq K$ satisfies $m = |x_m| \to +\infty$, $\frac{x_m}{|x_m|} \to v \in R_0$ and for all $y \in K$, m_y exists such that $F(x_m, y) \subseteq Y \setminus (-\operatorname{int}(C))$ for all $m \geq m_y$. By condition (C2), there exist $u \in K$ and $\bar{m}$ such that $|u| < |x_{\bar{m}}|$ and $F(x_{\bar{m}}, u) \subseteq Y \setminus (\operatorname{int}(C))$. We also have $F(x_{\bar{m}}, u) \subseteq Y \setminus (-\operatorname{int}(C))$ because of the choice of $x_{\bar{m}}$. We claim that such $x_{\bar{m}}$ is a solution to GWVEP (10.2). It only remains to check that $F(x_{\bar{m}}, y) \subseteq Y \setminus (-\operatorname{int}(C))$ for all $y \in K$ with $|y| > \bar{m}$. In the case when $x_{\bar{m}} \in [u, y]$ or $x_{\bar{m}} \in [y, u]$, the C-convexity of $F(x_{\bar{m}}, \cdot)$ implies, for some $\alpha \in [0, 1]$,

$$\alpha F(x_{\bar{m}}, u) + (1 - \alpha)F(x_{\bar{m}}, y) \subseteq F(x_{\bar{m}}, x_{\bar{m}}) + C.$$

Then

$$(1 - \alpha)F(x_{\bar{m}}, y) \subseteq C - \alpha F(x_{\bar{m}}, u) \subset Y \setminus (-\operatorname{int}(C)),$$

proving the claim. If on the contrary $u \in [y, x_{\bar{m}}]$ or $u \in [x_{\bar{m}}, y]$, for some $\alpha \in [0, 1]$, we have as before

$$\alpha F(x_{\bar{m}}, y) + (1 - \alpha)F(x_{\bar{m}}, x_{\bar{m}}) \subseteq F(x_{\bar{m}}, u) + C.$$

It follows that

$$(1-\alpha)F(x_{\bar{m}}, y) \subseteq Y \setminus (-\operatorname{int}(C)).$$

This completes the proof of the claim and therefore Sol(GWVEP)* $\neq \emptyset$. □

Now for the strong generalized vector equilibrium problem (10.4), we consider the following cones:

$$R'_0 := \bigcap_{y \in K} \{v \in K_\infty : \mathbf{0} \in F(y, y + t\,v) + C \text{ for all } t > 0 \text{ and } z \in K \quad \text{such that } F(y, z) \subseteq -C\}$$

and

$$R'_1 := \bigcap_{y \in K} \{v \in K_\infty : \mathbf{0} \in F(y, y + t\,v) + C \text{ for all } t > 0\}.$$

Theorem 10.17 *Let $K \subseteq X$ be a nonempty closed convex set and $F : K \times K \to 2^Y \setminus \{\emptyset\}$ be a set-valued map such that the following conditions hold.*

(i) *For all $x \in K$, $F(x, x) \subseteq l(C) := C \cap (-C)$;*
(ii) *For all $x, y \in K$, $F(x, y) \subseteq C$ implies $F(y, x) \subseteq -C$;*
(iii) *For all $x \in K$, the mapping $F(x, \cdot) : K \to 2^Y \setminus \{\emptyset\}$ is C-convex and properly C-quasiconvex;*
(iv) *For all $x, y \in K$, the set $\{\xi \in [x, y] : F(y, \xi) \subseteq C\}$ is closed.*
(v) *For all $x \in K$, $F(x, \cdot)$ is weakly lower semicontinuous on K.*

Then Sol(GSVEP) *is a nonempty and weakly closed set if and only if the following condition holds:*

(C4) *For every sequence $\{x_m\}$ in K, $\|x_m\| \to +\infty$, $\frac{x_m}{\|x_m\|} \rightharpoonup v$ with $v \in R'_1$ and for all $y \in K$, there exists m_y such that $F(x_m, y) \subseteq C$ for all $m \geq m_y$, there exist $u \in K$ and $\bar{m}$ such that $\|u\| < \|x_{\bar{m}}\|$ and $F(x_{\bar{m}}, u) \subseteq -C$.*

Proof To prove (C4) is sufficient for Sol(GSVEP) $\neq \emptyset$, a reasoning similar to the proof of Theorem 10.15 is applied. Instead of considering (10.25), we consider the following problem:

$$\text{Find } \bar{x} \in K_m \text{ such that } F(\bar{x}, y) \subseteq C, \quad \text{for all } y \in K_m. \tag{10.26}$$

Such a problem admits a solution by Lemma 10.6, say $x_m \in K_m$, for all $m \in \mathbb{N}$. If $\|x_m\| < m$ for some $m \in \mathbb{N}$, we show that x_m is also a solution to GSVEP (10.4). In fact, for any fixed $y \in K$ with $\|y\| > m$, we take $z \in K$ with $z \in]x_m, y[$ and $\|z\| < m$. Writing $z := \alpha x_m + (1-\alpha)y$ for some $\alpha \in]0, 1[$, we have

$$\alpha F(x_n, x_m) + (1-\alpha)F(x_m, y) \subseteq F(x_m, z) + C.$$

This implies that

$$(1-\alpha)F(x_m, y) \subseteq C,$$

proving the desired result.

We consider now the case $\|x_m\| = m$ for all $m \in \mathbb{N}$. We may assume, without loss of generality, that $\frac{x_m}{\|x_m\|} \rightharpoonup v$. Then $v \in K_\infty$. For any fixed $y \in K$, $F(x_m, y) \subseteq C$ for all $m \in \mathbb{N}$ sufficiently large ($m > \|y\|$). For every $t > 0$ and all m sufficiently large, C-convexity of $F(y, \cdot)$ implies

$$\left(1 - \frac{t}{\|x_m\|}\right) F(y, y) + \frac{t}{\|x_m\|} F(y, x_m) \subseteq F\left(y, \left(1 - \frac{t}{\|x_m\|}\right) y + \frac{t}{\|x_m\|} x_m\right) + C.$$

Hence

$$\mathbf{0} \in F\left(y, \left(1 - \frac{t}{\|x_m\|}\right) y + \frac{t}{\|x_m\|} x_m\right) + C.$$

Thus, by weakly lower semicontinuity of $F(x, \cdot)$, we have $\mathbf{0} \in F(y, z + tv) + C$. This proves $v \in R_1'$. Now, we can use condition (C4) to ensure the existence of $u \in K$ and $\bar{m}$ such that $\|u\| < \|x_{\bar{m}}\|$ and $F(x_{\bar{m}}, u) \subseteq -C$. Similarly as in the proof of Theorem 10.15, one can check that $x_{\bar{m}}$ is also a solution of GSVEP (10.4). The weakly closedness of Sol(GSVEP) follows as usual.

The "necessity" of (C4) is shown by taking element in Sol(GSVEP) as the point u required in condition (C4). □

For algorithmic purposes it desirable to know a priori when the solution set is bounded, in this case arises the next condition (vi) giving rise to the characterization expressed in Theorem 10.17:

(vi) Any sequence $x_m \in K$ with $\|x_m\| \to +\infty$ such that for all $y \in K$, there exists m_y such that

$$F(x_m, y) \subseteq C \text{ when } m \geq m_y,$$

admits a subsequence $\{x_{m_k}\}$ such that $\left\{\frac{x_{m_k}}{\|x_{m_k}\|}\right\}$ converges strongly.

Remark 10.9 When Y is a finite dimensional space, a condition implying condition (C4) (with $v \in R_1'$) described in the preceding theorem is $R_1' \subseteq -R_1'$ (in particular if $R_1' = \{\mathbf{0}\}$), since in this case, for all $v \in R_1'$,

$$\mathbf{0} \in F(y, y + tv) + C, \quad \text{for all } t \in \mathbb{R} \text{ and all } y \in K.$$

Indeed, (C4) is satisfied by taking $u = x_m - \|x_m\| v$. Notice that $\frac{x_m}{\|x_m\|} \to v$ implies $\|u\| < \|x_m\|$ for all m sufficiently large.

10.6 ε-Generalized Weak Vector Equilibrium Problems

In this section, we extend the ε-weak vector equilibrium problems for set-valued maps and study the behavior of their solution map.

Throughout this section, unless otherwise specified, let X and Y be topological vector spaces and K be a nonempty subset of X. We denote the zero vector of Y by $\mathbf{0}$. Let C be a proper closed convex cone with $\text{int}(C) \neq \emptyset$ and $F : X \times X \to 2^Y \setminus \{\emptyset\}$ be a set-valued map.

For fixed $\varepsilon \in \text{int}(C)$, the *$\varepsilon$-generalized weak vector equilibrium problem* (in short, ε-GWVEP) is to find $\bar{x} \in K$ such that

$$F(\bar{x}, y) + \varepsilon \nsubseteq -\text{int}(C), \quad \text{for all } y \in K. \tag{10.27}$$

Let $\Omega : \text{int}(C) \to 2^X$ be a set-valued map such that $\Omega(\varepsilon)$ is the solution set of ε-GWVEP (10.27) for $\varepsilon \in \text{int}(C)$, that is,

$$\Omega(\varepsilon) = \{x \in K : F(x, y) + \varepsilon \nsubseteq -\text{int}(C) \text{ for all } y \in K\}.$$

Let

$$\text{Sol(GWVEP)}^{\#} = \{x \in \text{cl}\,(K) : F(x, y) \nsubseteq -\text{int}(C) \text{ for all } y \in K\}.$$

For fixed $\varepsilon \in \text{int}(C)$, we also consider the following form of *ε-generalized weak vector equilibrium problem* (in short, ε-GWVEP): Find $\bar{x} \in K$ such that

$$\big(F(\bar{x}, y) + \varepsilon\big) \cap \big(-\text{int}(C)\big) = \emptyset, \quad \text{for all } y \in K. \tag{10.28}$$

Let $\Psi : \text{int}(C) \to 2^X$ be a set-valued map such that $\Psi(\varepsilon)$ is the solution set of ε-GWVEP (10.28) for $\varepsilon \in \text{int}(C)$, that is,

$$\Psi(\varepsilon) = \big\{x \in K : \big(F(x, y) + \varepsilon\big) \cap \big(-\text{int}(C)\big) = \emptyset \text{ for all } y \in K\big\}.$$

Let

$$\text{Sol(GWVEP)}^{*\#} = \big\{x \in \text{cl}\,(K) : F(x, y) \cap \big(-\text{int}(C)\big) = \emptyset \text{ for all } y \in K\big\}.$$

We may regard solutions of ε-GWVEP as approximate solutions of GWVEP. If K is closed, then Sol(GWVEP)$^{\#}$ = Sol(GWVEP) and Sol(GWVEP)$^{*\#}$ = Sol(GWVEP)*.

It is easy to see that Sol(GWVEP)$^{*\#}$ $\subseteq$ Sol(GWVEP)$^{\#}$ and $\Psi(\varepsilon) \subseteq \Omega(\varepsilon)$ for each $\varepsilon \in \text{int}(C)$. The purpose of this section is to establish relationships between $\Omega(\varepsilon)$ and Sol(GWVEP)$^{\#}$ as well as between $\Psi(\varepsilon)$ and Sol(GWVEP)$^{*\#}$ for $\varepsilon \in \text{int}(C)$. We establish some existence results for solutions of ε-generalized weak vector equilibrium problems. We also investigate continuity properties, namely, upper semicontinuity, lower semicontinuity and continuity, of the solution maps $\Omega, \Psi : \mathrm{int}(C) \to 2^X$.

We observe that the results in this section can be employed to study the behavior of solution maps of parametric vector optimization, parametric vector variational inequality problems, parametric vector equilibrium problems and so on.

10.6.1 Existence Results

In this subsection, we establish some existence results for solutions of ε-generalized weak vector equilibrium problems for $\varepsilon \in \mathrm{int}(C)$ under suitable conditions.

Theorem 10.18 *Let K be a nonempty subset of X such that* $\mathrm{cl}\,(K)$ *is compact, and let $F : X\times X \to 2^Y \setminus \{\emptyset\}$ be C-lower semicontinuous at every $(x, y) \in X\times X$. Assume that*

$$Sol(GWVEP)^{\#\#} := \{x \in \mathrm{cl}(K) : F(x, y) \nsubseteq -\mathrm{int}(C) \text{ for all } y \in \mathrm{cl}(K)\} \neq \emptyset.$$

Then ε-GWVEP (10.27) *has at least one solution for each* $\varepsilon \in \mathrm{int}(C)$.

Proof Let $\varepsilon \in \mathrm{int}(C)$, $x \in \mathrm{Sol(GWVEP)}^{\#\#}$, and $\mu := \frac{1}{2}\varepsilon$. Then

$$F(x, y) \nsubseteq -\mathrm{int}(C), \quad \text{for all } y \in \mathrm{cl}\,(K),$$

that is, for each $y \in \mathrm{cl}\,(K)$, there exists $\xi(x, y) \in Y$ such that

$$\xi(x, y) \in F(x, y) \cap \big(-\mathrm{int}(C)\big)^{c}.$$

By C-lower semicontinuity of F, for each $y \in \mathrm{cl}\,(K)$, there exist corresponding neighborhoods $\mathcal{U}^y$ of x and $\mathcal{V}^y$ of y such that

$$F(u, v) \cap \big(\xi(x, y) - \mu + \mathrm{int}(C)\big) \neq \emptyset, \quad \text{for all } (u, v) \in \mathcal{U}^y \times \mathcal{V}^y.$$

Since $\mathrm{cl}\,(K)$ is compact, there exist $y_1, y_2, \ldots, y_m$ such that $\bigcup_{i=1}^m \mathcal{V}^{y_i} \supseteq \mathrm{cl}\,(K)$, where $\mathcal{V}^{y_1}, \mathcal{V}^{y_2}, \ldots, \mathcal{V}^{y_m}$ are corresponding neighborhoods of $y_1, y_2, \ldots, y_m$. Let $\mathcal{U} = \bigcap_{i=1}^m \mathcal{U}^{y_i}$ and

$$\mathcal{A} := \bigcup_{i=1}^{m} \xi(x, y_i) + \varepsilon - \mu + \mathrm{int}(C).$$

Then for each $u \in \mathcal{U}$ and $y \in \mathrm{cl}\,(K)$,

$$\big(F(u, y) + \varepsilon\big) \cap \mathcal{A} \neq \emptyset. \tag{10.29}$$

Note that $\xi(x, y_i) + \varepsilon - \mu = \xi(x, y_i) + \mu \notin -C$ for each $i = 1, 2, \ldots, m$. Then

$$\big(\xi(x, y_i) + \mu + C\big) \cap \big(-C\big) = \emptyset, \quad i = 1, 2, \ldots, m. \tag{10.30}$$

By (10.29) and (10.30), for each $u \in \mathcal{U}$ and $y \in \text{cl}(K)$,

$$F(u, y) + \varepsilon \nsubseteq -C.$$

Since $x \in \text{cl}(K)$, there exists $\hat{x} \in K \cap \mathcal{U}$ such that

$$F(\hat{x}, y) + \varepsilon \nsubseteq -C, \quad \text{for all } y \in K.$$

Thus $\Omega(\varepsilon) \neq \emptyset$. Therefore, ε-GWVEP (10.27) has at least one solution for each $\varepsilon \in \text{int}(C)$. □

Example 10.3 Let $X = \mathbb{R}$, $K = \left]0, \frac{\pi}{2}\right[$ and $Y = \mathbb{R}^2$. Furthermore, let $C = \{(u, v) \in Y : u \geq 0,\ v \geq 0\}$. Then $\text{int}(C) = \{(u, v) \in Y : u > 0,\ v > 0\}$. Let

$$F(x, y) = \text{co}\left((-1, -1 - y), (\cos x - 1, \sin x - 1 - y)\right).$$

Then $\text{Sol(GWVEP)}^{\#\#} = \{0\}$, i.e., $\text{Sol(GWVEP)}^{\#\#} \neq \emptyset$, and $\text{cl}(K) = \left[0, \frac{\pi}{2}\right]$ is compact. F is C-lower semicontinuous at (x, y) for each $x \in X$ and $y \in X$. Hence $\Omega(\varepsilon) \neq \emptyset$ for each $\varepsilon \in \text{int}(C)$.

Next we establish an existence result for a solution of ε-GWVEP (10.28).

Theorem 10.19 *Let K be a nonempty subset of X such that $\text{cl}(K)$ is compact, and let $F : X \times X \to 2^Y \setminus \{\emptyset\}$ be C-upper semicontinuous at every $(x, y) \in X \times X$ and $F(x, y)$ is C-compact for every $(x, y) \in X \times X$. Assume that*

$$\textit{Sol(GWVEP)}^{*\#\#} := \{x \in \text{cl}(K) : F(x, y) \cap (-\text{int}(C)) = \emptyset \quad \textit{for all } y \in \text{cl}(K)\}\} \neq \emptyset.$$

Then ε-GWVEP (10.28) has at least one solution for each $\varepsilon \in \text{int}(C)$.

Proof Let $\varepsilon \in \text{int}(C)$, $x \in \text{Sol(GWVEP)}^{*\#\#}$ and $\mu := \frac{1}{2}\varepsilon$. Then

$$F(x, y) \cap \big(-\text{int}(C)\big) = \emptyset, \quad \text{for all } y \in \text{cl}(K).$$

Let $y \in \text{cl}(K)$. Then $F(x, y) - \mu + \text{int}(C)$ is a neighborhood of $F(x, y)$. By C-compactness of $F(x, y)$, there exist $z_1, z_2, \ldots, z_m \in F(x, y)$ such that

$$G(x, y) := \bigcup_{i=1}^{m} z_i - \mu + \text{int}(C) \supseteq F(x, y).$$

Note that $G(x, y) + \text{int}(C) = G(x, y)$. By the same way, for each $y \in \text{cl}(K)$, we can choose a neighborhood $G(x, y)$ of $F(x, y)$ and corresponding $z_1^y, z_2^y, \ldots, z_{m(y)}^y$ such that

$$F(x, y) - \mu + \text{int}(C) \supseteq G(x, y) = \bigcup_{i=1}^{m(y)} z_i^y - \mu + \text{int}(C),$$

where $m(y)$ is corresponding natural number to y. By C-upper semicontinuity of F, for each $y \in \text{cl}(K)$ there exist neighborhood $\mathcal{U}^y$ of x and $\mathcal{V}^y$ of y such that

$$F(u, v) \subseteq G(x, y), \quad \text{for all } u \in \mathcal{U}^y \text{ for all } y \in \mathcal{V}^y.$$

Clearly, $\bigcup_{y \in \text{cl}(K)} \mathcal{V}^y \supseteq \text{cl}(K)$. Hence by compactness of $\text{cl}(K)$, there exist $y_1, y_2, \ldots, y_k \in \text{cl}(K)$ such that

$$\bigcup_{j=1}^{k} \mathcal{V}^{y_j} \supseteq \text{cl}(K).$$

Let $\mathcal{U} = \bigcap_{j=1}^{k} \mathcal{U}^{y_j}$. Then for each $y \in \text{cl}(K)$

$$F(u, y) \subseteq \bigcup_{j=1}^{k} G(x, y_j), \quad \text{for all } u \in \mathcal{U}. \tag{10.31}$$

Since $z_i^{y_j} \notin -\text{int}(C)$ for each $j = 1, 2, \ldots, k$ and $i = 1, 2, \ldots, m(y_j)$,

$$z_i^{y_j} + \varepsilon - \mu = z_i^{y_j} + \mu \notin -C,$$

we have

$$z_i^{y_j} + \mu \notin -C, \quad \text{for all } j = 1, 2, \ldots, k, \text{ and } i = 1, 2, \ldots, m(y_j). \tag{10.32}$$

Hence,

$$\left(\bigcup_{j=1}^{k} G(x, y_j) + \varepsilon \right) \bigcap (-C) = \emptyset.$$

Furthermore by (10.31) for each $u \in \mathcal{U}$ and $y \in \text{cl}(K)$,

$$F(u, y) + \varepsilon \subseteq \bigcup_{j=1}^{k} G(x, y_j) + \varepsilon.$$

Accordingly, for each $u \in \mathcal{U}$ and $y \in \mathrm{cl}\,(K)$,

$$\big(F(u,y)+\varepsilon\big)\cap\big(-\operatorname{int}(C)\big)=\emptyset.$$

Since $x \in \mathrm{cl}\,(K)$, there exists $\hat{x} \in K \cap \mathcal{U}$. Hence

$$\big(F(\hat{x},y)+\varepsilon\big)\cap\big(-C\big),\quad \text{for all } y \in K.$$

Thus $\Psi(\varepsilon) \neq \emptyset$. Therefore, ε-GWVEP (10.28) has at least one solution for each $\varepsilon \in \operatorname{int}(C)$. □

10.6.2 Upper Semicontinuity of Ω and Ψ

In this subsection, we show that the solution mappings Ω of ε-GWVEP (10.27) and Ψ of ε-GWVEP (10.28) are upper semicontinuous on $\operatorname{int}(C)$.

Theorem 10.20 *Let K be a nonempty compact subset of X and $F : X\times X \to 2^Y\setminus\{\emptyset\}$ be a set-valued map such that $F(x,y)$ is $(-C)$-compact for each $(x,y) \in X \times X$ and for every fixed $y \in X$, $F(\cdot,y)$ is $(-C)$-upper semicontinuous. Assume that $\Omega(\varepsilon)$ is nonempty for each $\varepsilon \in \operatorname{int}(C)$. Then Ω is upper semicontinuous on* $\operatorname{int}(C)$.

Proof Since K is compact, it suffices to show that Ω has closed graph. Let $\varepsilon_\mu \to \varepsilon$ and $x_\mu \in \Omega(\varepsilon_\mu)$. Since K is compact, we can assume, without loss of generality, that $x_\mu \to x \in K$. Suppose to the contrary that $x \notin \Omega(\varepsilon)$. Then there exists $y \in K$ such that

$$\mathcal{A} := F(x,y)+\varepsilon \subseteq -\operatorname{int}(C).$$

Then for each $z \in \mathcal{A}$, there exists a neighborhood V_z of z such that $V_z \subseteq -\operatorname{int}(C)$. Let $e_z \in V_z \cap (z+\operatorname{int}(C))$. Then $e_z - \operatorname{int}(C)$ is a neighborhood of z such that $e_z - \operatorname{int}(C) \subseteq -\operatorname{int}(C)$. Obviously, $\bigcup_{z\in\mathcal{A}}(e_z-\operatorname{int}(C)) \supseteq \mathcal{A}$. Therefore by compactness of $F(x,y)$, there exist $z_1, z_2, \ldots, z_m \in A$ such that

$$\mathcal{B} := \bigcup_{i=1}^{m}(e_{z_i}-\operatorname{int}(C)) \supseteq A.$$

By $(-C)$-upper semicontinuity of $F(\cdot,y)$, there exists a neighborhood $\mathcal{U}$ of x such that

$$F(u,y)+\varepsilon \subseteq \mathcal{B}-\operatorname{int}(C)=\mathcal{B},\quad \text{for all } u \in \mathcal{U}.$$

Let $k \in \operatorname{int}(C)$. Then for each $e_{z_1}, e_{z_2}, \ldots, e_{z_m}$, there exist corresponding positive numbers $t_1, t_2, \ldots, t_m > 0$ such that

$$e_{z_i}+t_ik \in V_{z_i},\quad i=1,2,\ldots,m.$$

Let $t = \min\{t_1, t_2, \ldots, t_m\}$. Then $\mathcal{B} \subseteq \mathcal{B} + tk$ and $\varepsilon + tk - \mathrm{int}(C)$ is a neighborhood of ε. Let $\mathcal{E} := \varepsilon + tk - \mathrm{int}(C)$. We have

$$F(u) + \varepsilon' \subset B + tk - \mathrm{int}(C) = B + tk, \quad \text{for all } u \in \mathcal{U} \text{ and } \varepsilon' \in \mathcal{E}.$$

Then

$$F(u) + \varepsilon' \subseteq -\mathrm{int}(C), \quad \text{for all } u \in \mathcal{U} \text{ and } \varepsilon' \in \mathcal{E}.$$

This contradicts to the facts that $\varepsilon_\mu \to \varepsilon$, $x_\mu \to x$, and $x_\mu \in \Omega(\varepsilon_\mu)$. Hence $x \in \Omega(\varepsilon)$. Therefore Ω is upper semicontinuous on $\mathrm{int}(C)$. □

Theorem 10.21 *Let K be a nonempty compact subset of X and $F : X \times X \to 2^Y \setminus \{\emptyset\}$ be a set-valued map such that for every fixed $y \in X$, $F(\cdot, y)$ is $(-C)$-upper semicontinuous. Assume that $\Psi(\varepsilon)$ is nonempty for each $\varepsilon \in \mathrm{int}(C)$. Then Ψ is upper semicontinuous on* $\mathrm{int}(C)$.

Proof Since K is compact, it suffices to show that Ψ has closed graph. Let $\varepsilon_\mu \to \varepsilon$ and $x_\mu \in \Psi(\varepsilon_\mu)$. Since K is compact, we can assume, without loss of generality, that $x_\mu \to x \in K$. Suppose to the contrary that $x \notin \Psi(\varepsilon)$. Then there exist $y \in K$ and $z \in Y$ such that

$$z \in \big(F(x, y) + \varepsilon\big) \cap \big(-\mathrm{int}(C)\big).$$

Hence there exists a positive number $t > 0$ such that $z + 2t\varepsilon \in -\mathrm{int}(C)$. Since $z + t\varepsilon - \mathrm{int}(C)$ is a neighborhood of z, by $(-C)$-upper semicontinuity of $F(\cdot, y)$, there exists a neighborhood $\mathcal{U}$ of x such that

$$\big(z + t\varepsilon - \mathrm{int}(C) - \mathrm{int}(C)\big) \cap \big(F(u, y) + \varepsilon\big) \neq \emptyset, \quad \text{for all } u \in \mathcal{U}.$$

Hence for each $u \in \mathcal{U}$ and $\varepsilon' \in (1 + t)\varepsilon + \mathrm{int}(C)$, we have

$$\big(F(u, y) + \varepsilon'\big) \cap \big(-\mathrm{int}(C)\big) \neq \emptyset.$$

This contradicts to the facts that $\varepsilon_\mu \to \varepsilon$, $x_\mu \to x$, and $x_\mu \in \Psi(\varepsilon_\mu)$. Hence $x \in \Psi(\varepsilon)$. Therefore Ψ is upper semicontinuous on $\mathrm{int}(C)$. □

10.6.3 Lower Semicontinuity of Ω and Ψ

In this subsection, we establish that the solution mappings Ω of ε-GWVEP (10.27) and Ψ of ε-GEVEP (10.28) are lower semicontinuous on $\mathrm{int}(C)$ under suitable assumptions.

Theorem 10.22 *Let K be a nonempty compact convex subset of X and $F : X \times X \to 2^Y \setminus \{\emptyset\}$ be a set-valued map such that the following conditions hold:*

(i) *$\Omega(\varepsilon)$ is nonempty for each $\varepsilon \in \operatorname{int}(C)$;*
(ii) *$F(\cdot, y)$ is strictly C-quasiconcave on X for each $y \in X$;*
(iii) *$F(x, \cdot)$ is C-lower semicontinuous on X for each fixed $x \in X$.*

Then Ω is lower semicontinuous on $\operatorname{int}(C)$.

Proof Suppose $\varepsilon \in \operatorname{int}(C)$. Let $\mathcal{V}$ be an open set of X with $\mathcal{V} \cap \Omega(\varepsilon) \neq \emptyset$ and $x \in \mathcal{V} \cap \Omega(\varepsilon)$. Let $\hat{x} \in \Omega(\alpha\varepsilon)$ where $\alpha \in]0, 1[$. Then

$$F(\hat{x}, y) + \alpha\varepsilon \notin -\operatorname{int}(C), \quad \text{for all } y \in K.$$

Therefore,

$$F(\hat{x}, y) + \alpha\varepsilon + (1-\alpha)\varepsilon \notin -\operatorname{int}(C), \quad \text{for all } y \in K,$$

that is, $\hat{x} \in \Omega(\varepsilon)$. Now we choose

$$x' \in \mathcal{V} \cap \{x_\mu \in K : x_\mu = \mu x + (1-\mu)\hat{x},\ \mu \in]0, 1[\}.$$

Then by condition (ii),

$$F(x', y) + \varepsilon \nsubseteq -C, \quad \text{for all } y \in K. \tag{10.33}$$

Let $y' \in K$ arbitrary but fixed. Hence by (10.33), we can choose a point $z(y') \in Y$ such that

$$z(y') \in \big(F(x', y') + \varepsilon\big) \cap \big((-C)^{c}\big).$$

Since $z(y') \notin -C$, there exists a positive number $t_{z(y')} > 0$ such that

$$z(y') - t_{z(y')}\varepsilon \notin -C.$$

Because $\big(z(y') - t_{z(y')}\varepsilon + \operatorname{int}(C)\big)$ is a neighborhood of $z(y')$,

$$\big(z(y') - t_{z(y')}\varepsilon + \operatorname{int}(C)\big) \cap \big(F(x', y') + \varepsilon\big) \neq \emptyset.$$

Note that $\big(\operatorname{int}(C) + \operatorname{int}(C)\big) = \operatorname{int}(C)$. Therefore by condition (iii), there exists a neighborhood $\mathcal{U}_{y'}$ of y' such that

$$\big(F(x', v) + \varepsilon\big) \cap \big(z(y') - t_{z(y')}\varepsilon + \operatorname{int}(C)\big) \neq \emptyset, \quad \text{for all } v \in \mathcal{U}_{y'}.$$

Now since $y' \in K$ is arbitrary, for each $y \in K$ there exist corresponding $z(y) \in \big(F(x,y)+\varepsilon\big) \cap \big((-C)^{c}\big)$ and $t_{z(y)} > 0$ such that

$$\big(z(y) - t_{z(y)}\varepsilon + \operatorname{int}(C)\big) \subseteq \big((-\operatorname{int}(C))^{c}\big)$$

and also there exists a corresponding neighborhood $\mathcal{U}_y$ of y such that

$$F(x', v_y) \cap \big(z(y) - t_{z(y)}\varepsilon + \operatorname{int}(C)\big) \neq \emptyset, \quad \text{for all } v_y \in \mathcal{U}_y.$$

Since K is compact, there exist $y_1, y_2, \ldots, y_m \in K$ such that

$$\bigcup_{i=1}^{m} \mathcal{U}_{y_i} \supseteq K.$$

Then for each $y \in K$, we have

$$\bigcup_{i=1}^{m} \big(z(y_i) - t_{z(y_i)}\varepsilon + \operatorname{int}(C)\big) \cap F(x', y) + \varepsilon \neq \emptyset.$$

Since for each $i \in \{1, 2, \ldots, m\}$, $z(y_i) - t_{z(y_i)}\varepsilon \notin -C$, there exist corresponding positive numbers $\tau_1, \tau_2, \ldots, \tau_m > 0$ such that

$$z(y_i) - (t_{z(y_i)} + \tau_i)\varepsilon \notin -C.$$

Let $\tau = \min\{\tau_1, \tau_2, \ldots, \tau_m\}$. Then for each $y \in K$, we have

$$\bigcup_{i=1}^{m} \big(z(y_i) - (t_{z(y_i)} + \tau)\varepsilon + \operatorname{int}(C)\big) \cap \big(F(x', y) + (1-\tau)\varepsilon\big) \neq \emptyset.$$

Since

$$\bigcup_{i=1}^{m} \big(z(y_i) - (t_{z(y_i)} + \tau)\varepsilon + \operatorname{int}(C)\big) \subseteq (-C)^{c},$$

we have $x' \in \Omega\big((1-\tau)\varepsilon\big)$. Note that $\big((1-\tau)\varepsilon + \operatorname{int}(C)\big)$ is a neighborhood of ε and for each $\eta \in \operatorname{int}(C)$

$$F(x', y) + (1-\tau)\varepsilon + \eta \nsubseteq -C.$$

Therefore,

$$\mathcal{V} \cap \Omega(\xi) \neq \emptyset, \quad \text{for all } \xi \in \big((1-\tau)\varepsilon + \operatorname{int}(C)\big).$$

Hence Ω is lower semicontinuous at ε. Since $\varepsilon \in \text{int}(C)$ was arbitrary, Ω is lower semicontinuous on $\text{int}(C)$. □

Theorem 10.23 *Let K be a nonempty compact convex subset of X and $F : X \times X \to 2^Y \setminus \{\emptyset\}$ be a set-valued map such that the following conditions hold:*

(i) *$\Psi(\varepsilon)$ is nonempty for each $\varepsilon \in \text{int}(C)$;*
(ii) *$F(x, y)$ is C-compact for every $(x, y) \in X \times X$;*
(iii) *$F(\cdot, y)$ is strictly C-properly quasiconcave on X for each fixed $y \in X$;*
(iv) *$F(x, \cdot)$ is C-upper semicontinuous on X for each fixed $x \in X$.*

Then Ψ is lower semicontinuous on $\text{int}(C)$.

Proof Suppose $\varepsilon \in \text{int}(C)$. Let $\mathcal{V}$ be an open set in X such that $\mathcal{V} \cap \Psi(\varepsilon) \neq \emptyset$ and let $x \in \mathcal{V} \cap \Psi(\varepsilon)$. Let $\hat{x} \in \Psi(\alpha\varepsilon)$ where $\alpha \in]0, 1[$. Obviously, $\hat{x} \in \Psi(\varepsilon)$. We choose $x' \in \mathcal{V} \cap (x, \hat{x})$. Then by condition (iii), we have

$$\big(F(x', y) + \varepsilon\big) \cap \big(-C\big) = \emptyset, \quad \text{for all } y \in X.$$

Let $y \in X$ arbitrary but fixed. Then for each $z \in F(x', y)$ there exists a corresponding positive number $t_z > 0$ such that

$$z - (1 - 2t_z)\varepsilon \notin -C.$$

Note that $z - (1 - t_z)\varepsilon + \text{int}(C)$ is a neighborhood of $z + \varepsilon$ and that $\bigcup_{z \in F(x',y)} \big(z - (1 - t_z)\varepsilon + \text{int}(C)\big) \supseteq F(x', y)$. By condition (ii), there exist $z_1, z_2, \ldots, z_m \in F(x', y)$ such that

$$A_y := \bigcup_{i=1}^{m} \big(z_i - (1 - t_{z_i})\varepsilon + \text{int}(C)\big) \supseteq F(x', y).$$

Let $\tau_y = \min\{t_{z_1}, t_{z_2}, \ldots, t_{z_m}\}$. Then

$$\big(A_y - \tau_y\varepsilon\big) \cap \big(-C\big) = \emptyset.$$

By condition (iv), there exists a neighborhood V_y of y such that

$$F(x', v) \subseteq A_y, \quad \text{for all } v \in \mathcal{V}_y.$$

Then for each $v \in V_y$,

$$\big(F(x', v) + (1 - \tau_y)\varepsilon\big) \cap \big(-C\big) = \emptyset.$$

Since y is arbitrary, for each $y \in X$ we can choose corresponding τ_y and V_y. Because X is compact, there exist $V_{y_1}, V_{y_2}, \ldots, V_{y_k}$ such that $\bigcup_{j=1}^{k} V_{y_j} \supseteq X$. Let

$\tau = \min\{\tau_{y_1}, \tau_{y_1}, \ldots, \tau_{y_k}\}$. Then for each $y \in X$,

$$\big(F(x', y) + (1 - \tau)\varepsilon\big) \cap \big(-C\big) = \emptyset.$$

Accordingly for each $\eta \in (1 - \tau)\varepsilon + \mathrm{int}(C)$, $x' \in \Psi(\eta)$. Hence Ψ is lower semicontinuous at ε. Since ε was arbitrary, Ψ is lower semicontinuous on $\mathrm{int}(C)$. □

10.6.4 Continuity of Ω and Ψ

By employing results established in Sects. 10.6.2 and 10.6.3, we immediately have the following results for continuity of Ω and Ψ.

Theorem 10.24 *Let K be a nonempty compact convex subset of X and $F : X \times X \to 2^Y \setminus \{\emptyset\}$ be a set-valued map such that the following conditions hold:*

(i) *$\Omega(\varepsilon)$ is nonempty for each $\varepsilon \in \mathrm{int}(C)$;*
(ii) *$F(x, y)$ is $(-C)$-compact for every $(x, y) \in X \times X$;*
(iii) *$F(\cdot, y)$ is strictly C-properly quasiconcave on X for each fixed $y \in X$;*
(iv) *$F(\cdot, y)$ is $(-C)$-upper semicontinuous on X for each fixed $y \in X$;*
(iv) *$F(x, \cdot)$ is C-lower semicontinuous on X for each fixed $x \in X$.*

Then Ω is continuous on $\mathrm{int}(C)$.

Proof The result follows from Theorems 10.20 and 10.22. □

Theorem 10.25 *Let K be a nonempty compact convex subset of X and $F : X \times X \to 2^Y \setminus \{\emptyset\}$ be a set-valued map such that the following conditions hold:*

(i) *$\Psi(\varepsilon)$ is nonempty for each $\varepsilon \in \mathrm{int}(C)$;*
(ii) *$F(x, y)$ is $(-C)$-compact for every $(x, y) \in X \times X$;*
(iii) *$F(\cdot, y)$ is strictly C-properly quasiconcave on X for each fixed $y \in X$;*
(iv) *$F(\cdot, y)$ is $(-C)$-upper semicontinuous on X for each fixed $y \in X$;*
(iv) *$F(x, \cdot)$ is C-upper semicontinuous on X for each fixed $x \in X$.*

Then Ψ is continuous on $\mathrm{int}(C)$.

Proof The result follows from Theorems 10.21 and 10.23. □

The problems and results of this section were considered and established by Kimora and Yao [38] for moving cones.

10.7 ε-Generalized Strong Vector Equilibrium Problems

In this section, we consider the ε-generalized strong vector equilibrium problems and study the behavior of their solution map.

Throughout this section, unless otherwise specified, let X and Y be topological vector spaces and K be a nonempty subset of X. We denote the zero vector of Y by $\mathbf{0}$. Let C be a proper closed convex cone with $\text{int}(C) \neq \emptyset$ and $F : X \times X \to 2^Y \setminus \{\emptyset\}$ be a set-valued map.

For fixed $\varepsilon \in \text{int}(C)$, the *ε-generalized strong vector equilibrium problem* (ε-GSVEP) is to find $x \in K$ such that

$$\big(F(x,y) + \varepsilon\big) \cap C \neq \emptyset, \quad \text{for all } y \in K. \tag{10.34}$$

Let $\Delta : \text{int}(C) \to 2^X$ be the set-valued map such that $\Delta(\varepsilon)$ is the solution set of ε-GSVEP (10.34) for $\varepsilon \in \text{int}(C)$, that is,

$$\Delta(\varepsilon) = \{x \in K : \big(F(x,y) + \varepsilon\big) \cap C \neq \emptyset \text{ for all } y \in K\}.$$

Let

$$\text{Sol(GSVEP)}^{*\#} = \{x \in \text{cl}\,(K) : F(x,y) \cap C \neq \emptyset \text{ for all } y \in K\}.$$

We also consider the following *ε-generalized strong vector equilibrium problem* (ε-GSVEP) for each $\varepsilon \in \text{int}(C)$ which is to find $x \in K$ such that

$$F(x,y) + \varepsilon \subseteq C, \quad \text{for all } y \in K. \tag{10.35}$$

Let $\mho : \text{int}(C) \to 2^X$ be the set-valued map such that $\mho(\varepsilon)$ is the solution set of ε-GSVEP (10.35) for $\varepsilon \in \text{int}(C)$, that is,

$$\mho(\varepsilon) = \{x \in K : F(x,y) + \varepsilon \subseteq C \text{ for all } y \in K\}.$$

Let

$$\text{Sol(GSVEP)}^{\#} = \{x \in \text{cl}\,(K) : F(x,y) \subseteq C \text{ for all } y \in K\}.$$

It is easy to see that $\text{Sol(GSVEP)}^{\#} \subseteq \text{Sol(GSVEP)}^{*\#}$ and $\mho(\varepsilon) \subseteq \Delta(\varepsilon)$ for each $\varepsilon \in \text{int}(C)$. We remark that if the mapping F is single-valued, then $\text{Sol(GSVEP)}^{\#} = \text{Sol(GSVEP)}^{*\#}$ and $\mho(\varepsilon) = \Delta(\varepsilon)$ for each $\varepsilon \in \text{int}(C)$. When K is closed, then $\text{Sol(GSVEP)} = \text{Sol(GSVEP)}^{\#}$ and $\text{Sol(GSVEP)}^{*} = \text{Sol(GSVEP)}^{*\#}$.

The purpose of this section is to establish relationships between $\Delta(\varepsilon)$ and $\text{Sol(GSVEP)}^{*\#}$ as well as between $\mho(\varepsilon)$ and $\text{Sol(GSVEP)}^{\#}$ for $\varepsilon \in \text{int}(C)$. We also investigate continuity properties of the solution mappings $\Delta, \mho : \text{int}(C) \to 2^X$.

10.7.1 Existence Results

We derive that $\Delta(\varepsilon)$ is nonempty for $\varepsilon \in \text{int}(C)$ under suitable conditions.

Theorem 10.26 *Let K be a nonempty subset of X such that* $\mathrm{cl}\,(K)$ *is compact and* $F : X \times X \to 2^Y \setminus \{\emptyset\}$ *be C-lower semicontinuous at every* $(x, y) \in X \times X$. *Assume that*

$$\mathit{Sol(GSVEP)}^{*\#\#} := \{x \in \mathrm{cl}\,(K) : F(x, y) \cap C \neq \emptyset \text{ for all } y \in \mathrm{cl}\,(K)\} \neq \emptyset.$$

Then ε-GSVEP (10.34) *has at least one solution for each* $\varepsilon \in \mathrm{int}(C)$.

Proof Let $\varepsilon \in \mathrm{int}(C)$, $x \in \mathrm{Sol(GSVEP)}^{*\#\#}$ and $\mu := \frac{1}{2}\varepsilon$. Then

$$F(x, y) \cap C \neq \emptyset, \quad \text{for all } y \in \mathrm{cl}\,(K),$$

that is, for each $y \in \mathrm{cl}\,(K)$, there exists $\xi(x, y) \in Y$ such that

$$\xi(x, y) \in F(x, y) \cap C.$$

By C-lower semicontinuity of F, for each $y \in \mathrm{cl}\,(K)$, there exist corresponding neighborhoods $\mathcal{U}^y$ of x and $\mathcal{V}^y$ of y such that

$$F(u, v) \cap \big(\xi(x, y) - \mu + \mathrm{int}(C)\big) \neq \emptyset, \quad \text{for all } (u, v) \in \mathcal{U}^y \times \mathcal{V}^y.$$

Since $\mathrm{cl}\,(K)$ is compact, there exist $y_1, y_2, \ldots, y_m$ such that $\bigcup_{i=1}^m \mathcal{V}^{y_i} \supseteq \mathrm{cl}\,(K)$, where $\mathcal{V}^{y_1}, \mathcal{V}^{y_2}, \ldots, \mathcal{V}^{y_m}$ are corresponding neighborhoods of $y_1, y_2, \ldots, y_m$. Let $\mathcal{U} = \bigcap_{i=1}^m \mathcal{U}^{y_i}$ and

$$\mathcal{A} := \bigcup_{i=1}^{m} (\xi(x, y_i) + \varepsilon - \mu + \mathrm{int}(C)).$$

Then for each $u \in \mathcal{U}$ and $y \in \mathrm{cl}\,(K)$, we have

$$\big(F(u, y) + \varepsilon\big) \cap \mathcal{A} \neq \emptyset. \tag{10.36}$$

Note that $\xi(x, y_i) + \varepsilon - \mu = \xi(x, y_i) + \mu \in \mathrm{int}(C)$ for each $i = 1, 2, \ldots, m$. Then for each $u \in \mathcal{U}$, we obtain

$$\xi(x, y_i) + \mu + \mathrm{int}(C) \subseteq \mathrm{int}(C), \quad i = 1, 2, \ldots, m. \tag{10.37}$$

By (10.36) and (10.37), for each $u \in \mathcal{U}$ and $y \in \mathrm{cl}\,(K)$, we have

$$\big(F(u, y) + \varepsilon\big) \cap \big(\mathrm{int}(C)\big) \neq \emptyset.$$

Since $x \in \mathrm{cl}\,(K)$, there exists $\hat{x} \in K \cap \mathcal{U}$ and hence

$$\big(F(\hat{x}, y) + \varepsilon\big) \cap \big(\mathrm{int}(C)\big) \neq \emptyset, \quad \text{for all } y \in K.$$

Thus $\Delta(\varepsilon) \neq \emptyset$. Therefore ε-GSVEP (10.34) has at least one solution for each $\varepsilon \in \operatorname{int}(C)$. □

Corollary 10.4 *Let K be a nonempty subset of X such that* $\operatorname{cl}(K)$ *is compact and* $F : X \times X \to 2^Y \setminus \{\emptyset\}$ *be C-lower semicontinuous at every $(x, y) \in X \times X$ and $F(x, \cdot)$ is $(-C)$-upper semicontinuous on* $\operatorname{bd}(K)$ *for some $x \in$ Sol(GSVEP)$^{*\#}$, where*

$$Sol(GSVEP)^{*\#} := \{x \in \operatorname{cl}(K) : F(x, y) \cap C \neq \emptyset \text{ for all } y \in K\} \neq \emptyset.$$

Then ε-GSVEP (10.34) *has at least one solution for each* $\varepsilon \in \operatorname{int}(C)$.

Proof Let $\hat{x} \in \mathrm{Sol(GSVEP)}^{*\#}$ for which $F(\hat{x}, \cdot)$ is $(-C)$-upper semicontinuous on $\operatorname{bd}(K)$. Suppose to the contrary that there exists $\hat{y} \in \operatorname{bd}(K)$ such that

$$F(\hat{x}, \hat{y}) \cap C = \emptyset,$$

that is,

$$F(\hat{x}, \hat{y}) \subseteq \big(Y \setminus C\big) - \operatorname{int}(C).$$

Then by $(-C)$-upper semicontinuity of $F(x, \cdot)$ on $\operatorname{bd}(K)$, there exists a neighborhood $\mathcal{V}$ of $\hat{y}$ such that

$$F(\hat{x}, v) \subseteq \big(Y \setminus C\big) - \operatorname{int}(C) \subseteq \big(Y \setminus C\big), \quad \text{for all } v \in \mathcal{V}.$$

However $\hat{y} \in \operatorname{bd}(K)$, that is, $\mathcal{V} \cap K \neq \emptyset$. Let $y' \in \mathcal{V} \cap K$. Then $F(\hat{x}, y') \cap C = \emptyset$. This contradicts to the fact that $\hat{x} \in \mathrm{Sol(GSVEP)}^{*\#}$. Hence $\hat{x} \in \mathrm{Sol(GSVEP)}^{*}$. Therefore by Theorem 10.26, ε-GSVEP (10.34) has at least one solution for each $\varepsilon \in \operatorname{int}(C)$. □

Next we establish an existence result for solutions of ε-GSVEP (10.35).

Theorem 10.27 *Let K be a nonempty subset of X such that* $\operatorname{cl}(K)$ *is compact and* $F : X \times X \to 2^Y \setminus \{\emptyset\}$ *be C-upper semicontinuous at every $(x, y) \in X \times X$ and $F(x, y)$ is C-compact for every $(x, y) \in X \times X$. Assume that*

$$Sol(SGVEP)^{\#\#} := \{x \in \operatorname{cl}(K) : F(x, y) \subset C \text{ for all } y \in \operatorname{cl}(K)\} \neq \emptyset.$$

Then ε-GSVEP (10.35) *has at least one solution for each* $\varepsilon \in \operatorname{int}(C)$.

Proof Let $\varepsilon \in \operatorname{int}(C)$, $x \in \mathrm{Sol(SGVEP)}^{\#\#}$ and $\mu := \frac{1}{2}\varepsilon$. Then $F(x, y) \subseteq C$ for all $y \in \operatorname{cl}(K)$. Let $y \in \operatorname{cl}(K)$. Then $F(x, y) - \mu + \operatorname{int}(C)$ is a neighborhood of $F(x, y)$. By C-compactness of $F(x, y)$, there exist $z_1, z_2, \ldots, z_m \in F(x, y)$ such that

$$G(x, y) := \bigcup_{i=1}^{m} (z_i - \mu + \operatorname{int}(C) \supseteq F(x, y).$$

Note that $G(x, y) + \text{int}(C) = G(x, y)$. By the same way, for each $y \in \text{cl}(K)$ we can choose a neighborhood $G(x, y)$ of $F(x, y)$ and corresponding $z_1^y, z_2^y, \ldots, z_{m(y)}^y$ such that

$$F(x, y) - \mu + \text{int}(C) \supseteq G(x, y) = \bigcup_{i=1}^{m(y)} (z_i^y - \mu + \text{int}(C)),$$

where $m(y)$ is the natural number corresponding to y. By C-upper semicontinuity of F, for each $y \in \text{cl}(K)$, there exist neighborhood $\mathcal{U}^y$ of x and $\mathcal{V}^y$ of y such that

$$F(u, v) \subseteq G(x, y), \quad \text{for all } u \in \mathcal{U}^y \text{ for all } v \in \mathcal{V}^y.$$

Clearly, $\bigcup_{y \in \text{cl}(K)} \mathcal{V}^y \supseteq \text{cl}(K)$. Then by C-compactness of $\text{cl}(K)$, there exist $y_1, y_2, \ldots, y_k \in \text{cl}(K)$ such that

$$\bigcup_{j=1}^{k} \mathcal{V}^{y_j} \supseteq \text{cl}(K).$$

Let $\mathcal{U} = \bigcap_{j=1}^{k} \mathcal{U}^{y_j}$. Then for each $y \in \text{cl}(K)$,

$$F(u, y) \subseteq \bigcup_{j=1}^{k} G(x, y_j), \quad \text{for all } u \in \mathcal{U}.$$

Therefore, for each $u \in \mathcal{U}$ and $y \in \text{cl}(K)$,

$$F(u, y) + \varepsilon \subseteq \bigcup_{j=1}^{k} G(x, y_j) + \varepsilon.$$

Note that

$$z_i^{y_j} + \mu \in \text{int}(C), \quad \text{for all } j = 1, \ldots, k, \text{ and } i = 1, 2, \ldots, m(y_j).$$

Therefore for each x,

$$\bigcup_{j=1}^{k} G(x, y_j) + \varepsilon \subseteq \text{int}(C).$$

Hence for each $u \in \mathcal{U}$ and $y \in \text{cl}(K)$,

$$F(u, y) + \varepsilon \subseteq \text{int}(C).$$

Since $x \in \text{cl}\,(K)$, there exists $\hat{x} \in K \cap \mathcal{U}$. Then

$$F(\hat{x}, y) + \varepsilon \subseteq \text{int}(C), \quad \text{for all } y \in K.$$

Thus $\mho(\varepsilon) \neq \emptyset$. Therefore ε-GSVEP (10.35) has at least one solution for each $\varepsilon \in \text{int}(C)$. □

Corollary 10.5 *Let K be a nonempty subset of X such that* $\text{cl}\,(K)$ *is compact and $F : X \times X \to 2^Y \setminus \{\emptyset\}$ be C-upper semicontinuous at every $(x, y) \in X \times X$ and $F(x, y)$ is C-compact for every $(x, y) \in X \times X$. Assume that* $\text{Sol(SGVEP)}^\# := \{x \in \text{cl}\,(K) : F(x, y) \subset C \text{ for all } y \in K\} \neq \emptyset$ *such that $F(x, \cdot)$ is $(-C)$-lower semicontinuous on* $\text{bd}\,(K)$ *for some $x \in Sol(SGVEP)^\#$. Then ε*-GSVEP (10.35) *has at least one solution for each $\varepsilon \in$* $\text{int}(C)$.

10.7.2 Upper Semicontinuity of Δ and ℧

In this subsection, we show that the solution mappings Δ of ε-GSVEP (10.34) and $\mho$ of ε-GSVEP (10.35) are upper semicontinuous on $\text{int}(C)$ under suitable assumptions.

Theorem 10.28 *Let K be a nonempty compact subset of X and $F : X \times X \to 2^Y \setminus \{\emptyset\}$ be a set-valued map such that the following conditions hold:*

(i) *$\Delta(\varepsilon)$ is nonempty for each $\varepsilon \in$* $\text{int}(C)$*;*
(ii) *$F(x, y)$ is $(-C)$-compact for each $(x, y) \in X \times X$;*
(iii) *For every fixed $y \in X$, $F(\cdot, y)$ is $(-C)$-upper semicontinuous.*

Then Δ is upper semicontinuous on $\text{int}(C)$.

Proof Since K is compact, it suffices to show that Δ has closed graph. Let $\varepsilon_\mu \to \varepsilon$ and $x_\mu \in \Delta(\varepsilon_\mu)$. Since K is compact, we can assume, without loss of generality, that $x_\mu \to x \in K$. Suppose to the contrary that $x \notin \Delta(\varepsilon)$. Then there exists $y \in K$ such that

$$\mathcal{A} := \big(F(x, y) + \varepsilon\big) \cap C = \emptyset.$$

Since C is closed, for each $z \in \mathcal{A}$, there exists a neighborhoods V_z of z such that $V_z \subseteq \big(Y \setminus C\big)$. Let $e_z \in V_z \cap (z + \text{int}(C))$. Then $e_z - \text{int}(C)$ is a neighborhood of z such that $e_z - \text{int}(C) \subseteq \big(Y \setminus C\big)$. Obviously, $\bigcup_{z \in \mathcal{A}} (e_z - \text{int}(C)) \supseteq \mathcal{A}$. Then by $(-C)$-compactness of $F(x, y)$, there exist $z_1, z_2, \ldots, z_m \in A$ such that

$$\mathcal{B} := \bigcup_{i=1}^{m} (e_{z_i} - \text{int}(C)) \supseteq \mathcal{A}.$$

By condition (iii), there exists a neighborhood $\mathcal{U}$ of x such that

$$F(u, y) + \varepsilon \subseteq \mathcal{B} - \text{int}(C) = \mathcal{B}, \quad \text{for all } u \in \mathcal{U}.$$

Let $k \in \text{int}(C)$. Then for each $e_{z_1}, e_{z_2}, \ldots, e_{z_m}$, there exist corresponding positive numbers $t_1, t_2, \ldots, t_m > 0$ such that

$$e_{z_i} + t_i k \in V_{z_i}, \quad i = 1, 2, \ldots, m.$$

Let $t = \min\{t_1, t_2, \ldots, t_m\}$. Then $\mathcal{B} \subseteq \mathcal{B} + tk$ and $\varepsilon + tk - \text{int}(C)$ is a neighborhood of ε. Let $\mathcal{E} := \varepsilon + tk - \text{int}(C)$. We have

$$F(u, y) + \varepsilon' \subseteq \mathcal{B} + tk - \text{int}(C) = \mathcal{B} + tk, \quad \text{for all } u \in \mathcal{U} \text{ and } \varepsilon' \in \mathcal{E}.$$

Let $\mathfrak{U} = \bigcap_{i=1}^{m} U^{z_i} \cap \mathcal{U}$. Then

$$F(u, y) + \varepsilon' \subseteq \big(Y \setminus C\big), \quad \text{for all } u \in \mathfrak{U} \text{ and } \varepsilon' \in \mathcal{E}.$$

This contradicts to the fact that $\varepsilon_\mu \to \varepsilon$, $x_\mu \to x$, and $x_\mu \in \Delta(\varepsilon_\mu)$. Hence $x \in \Delta(\varepsilon)$. Therefore Δ is upper semicontinuous on $\text{int}(C)$. □

Theorem 10.29 *Let K be a nonempty compact subset of X and $F : X \times X \to 2^Y \setminus \{\emptyset\}$ be a set-valued map such that for every fixed $y \in X$, $F(\cdot, y)$ is $(-C)$-lower semicontinuous. Assume that $\mho(\varepsilon)$ is nonempty for each $\varepsilon \in$* int(C). *Then $\mho$ is upper semicontinuous on* int(C).

Proof Since K is compact, it suffices to show that $\mho$ has closed graph. Let $\varepsilon_\mu \to \varepsilon$ and $x_\mu \in \mho(\varepsilon_\mu)$. Since K is compact, we can assume, without loss of generality, that $x_\mu \to x \in K$. Suppose to the contrary that $x \notin \mho(\varepsilon)$. Then there exist $y \in K$ and $z \in Y$ such that

$$z \in \big(F(x, y) + \varepsilon\big) \cap \big(Y \setminus C\big),$$

that is, there exists a positive number $t > 0$ such that

$$z + 2t\varepsilon \in Y \setminus C.$$

Since $z + t\varepsilon - \text{int}(C)$ is a neighborhood of z, by $(-C)$-lower semicontinuity of $x \mapsto F(x, y)$, there exists a neighborhood $\mathcal{U}$ of x such that

$$\big(z + t\varepsilon - \text{int}(C) - \text{int}(C)\big) \cap \big(F(u, y) + \varepsilon\big) \neq \emptyset, \quad \text{for all } u \in \mathcal{U}.$$

Then for each $u \in \mathcal{U}$ and $\varepsilon' \in (1 + t)\varepsilon - \text{int}(C)$, we have

$$\big(F(u, y) + \varepsilon'\big) \cap \big(Y \setminus C\big) \neq \emptyset.$$

This contradicts to the facts that $\varepsilon_\mu \to \varepsilon$, $x_\mu \to x$, and $x_\mu \in \mho(\varepsilon_\mu)$. Hence $x \in \mho(\varepsilon)$. Therefore, $\mho$ is upper semicontinuous on $\operatorname{int}(C)$. □

Remark 10.10 In Theorems 10.28 and 10.29, if the sets Sol(GSVEP)$^{*\#}$ and Sol(GSVEP)$^{\#}$ are nonempty, then the problems are kind of well-posed, i.e., every approximating net has subnet converging to a point of the solution set: For every $\varepsilon_\mu \in \operatorname{int}(C)$ with $\varepsilon_\mu \to \mathbf{0}$ and $x_\mu \in \Delta(\varepsilon_\mu)$ (respectively, $\mho(\varepsilon_\mu)$) has a subnet $\{x_\nu\} \subseteq \{x_\mu\}$ with $x_\nu \to \hat{x}$ for some $\hat{x} \in$ Sol(GSVEP)$^{*\#}$ (respectively, Sol(GSVEP)$^{\#}$). Especially, if the solution set are singleton, every approximating net converges to the solution.

10.7.3 Lower Semicontinuity of Δ and $\mho$

We next establish that the solution mapping Δ of ε-GSVEP (10.34) and $\mho$ of ε-GSVEP (10.35) are lower semicontinuous on $\operatorname{int}(C)$ under suitable assumptions.

Theorem 10.30 *Let K be a nonempty compact convex subset of X and $F : X \times X \to 2^Y \setminus \{\emptyset\}$ be a set-valued map such that the following conditions hold:*

(i) *$\Delta(\varepsilon)$ is nonempty for each $\varepsilon \in \operatorname{int}(C)$;*
(ii) *$F(\cdot, y)$ is strictly $(-C)$-quasiconvex on X for each $y \in X$;*
(iii) *For every fixed $y \in X$, $F(x, \cdot)$ is C-lower semicontinuous.*

Then Δ is lower semicontinuous on $\operatorname{int}(C)$.

Proof Suppose $\varepsilon \in \operatorname{int}(C)$. Let $\mathcal{V}$ be an open set in X with $\mathcal{V} \cap \Delta(\varepsilon) \neq \emptyset$ and $x \in \mathcal{V} \cap \Delta(\varepsilon)$. Let $\hat{x} \in \Delta(\alpha\varepsilon)$ where $\alpha \in]0,1[$. Then

$$\big(F(\hat{x}, y) + \alpha\varepsilon\big) \cap C \neq \emptyset, \quad \text{for all } y \in K.$$

Therefore,

$$\big(F(\hat{x}, y) + \alpha\varepsilon + (1-\alpha)\varepsilon\big) \cap C \neq \emptyset, \quad \text{for all } y \in K,$$

that is, $\hat{x} \in \Delta(\varepsilon)$. Now we choose

$$x' \in \mathcal{V} \cap \big\{x_\mu \in K : x_\mu = \mu x + (1-\mu)\hat{x}, \quad \mu \in]0,1[\big\}.$$

Then by condition (ii), we have

$$\big(F(x', y) + \varepsilon\big) \cap \big(\operatorname{int}(C)\big) \neq \emptyset, \quad \text{for all } y \in K. \tag{10.38}$$

Let $y' \in K$ be arbitrary but fixed. Then by (10.38), we can choose a point $z(y') \in Y$ such that

$$z(y') \in \big(F(x', y') + \varepsilon\big) \cap \big(\operatorname{int}(C)\big).$$

Since $z(y') \in \text{int}(C)$, there exists a positive number $t_{z(y')} > 0$ such that

$$z(y') - t_{z(y')}\varepsilon \in \text{int}(C).$$

Because $\left(z(y') - t_{z(y')}\varepsilon + \text{int}(C)\right)$ is a neighborhood of $z(y')$, we obtain

$$\left(z(y') - t_{z(y')}\varepsilon + \text{int}(C)\right) \cap \left(F(x', y') + \varepsilon\right) \neq \emptyset.$$

Note that $\left(\text{int}(C) + \text{int}(C)\right) = \text{int}(C)$. Therefore by condition (iii), there exists a neighborhood $\mathcal{U}_{y'}$ of y' such that

$$\left(F(x', v) + \varepsilon\right) \cap \left(z(y') - t_{z(y')}\varepsilon + \text{int}(C)\right) \neq \emptyset, \quad \text{for all } v \in \mathcal{U}_{y'}.$$

Since $y' \in K$ is arbitrary, for each $y \in K$ there exist corresponding $z(y) \in \big(F(x, y) + \varepsilon\big) \cap \left(\text{int}(C)\right)$ and $t_{z(y)} > 0$ such that

$$z(y) - t_{z(y)}\varepsilon + \text{int}(C) \subseteq \text{int}(C)$$

and also there exists a corresponding neighborhood $\mathcal{U}_y$ of y such that

$$F(x', v_y) \cap \left(z(y) - t_{z(y)}\varepsilon + \text{int}(C)\right) \neq \emptyset, \quad \text{for all } v_y \in \mathcal{U}_y.$$

Since K is compact, there exist $y_1, y_2, \ldots, y_m \in K$ such that

$$\bigcup_{i=1}^{m} \mathcal{U}_{y_i} \supset K.$$

Hence for each $y \in K$, we have

$$\left(\bigcup_{i=1}^{m} z(y_i) - t_{z(y_i)}\varepsilon + \text{int}(C)\right) \cap \left(F(x', y) + \varepsilon\right) \neq \emptyset.$$

Since for each $i \in \{1, 2, \ldots, m\}$, $z(y_i) - t_{z(y_i)}\varepsilon \in \text{int}(C)$, there exist corresponding positive numbers $\tau_1, \tau_2, \ldots, \tau_m > 0$ such that

$$z(y_i) - (t_{z(y_i)} + \tau_i)\varepsilon \notin -C.$$

Let $\tau = \min\{\tau_1, \tau_2, \ldots, \tau_m\}$. Then for each $y \in K$, we have

$$\bigcup_{i=1}^{m} \left(z(y_i) - (t_{z(y_i)} + \tau)\varepsilon + \text{int}(C)\right) \cap \left(F(x', y) + (1 - \tau)\varepsilon\right) \neq \emptyset.$$

Since

$$\bigcup_{i=1}^{m}\big(z(y_i) - (t_{z(y_i)} + \tau)\varepsilon + \text{int}(C)\big) \subseteq \text{int}(C),$$

we have $x' \in \Delta\big((1-\tau)\varepsilon\big)$. Note that $\big((1-\tau)\varepsilon+\text{int}(C)\big)$ is a neighborhood of ε and for each $\eta \in \text{int}(C)$, $\big(F(x',y)+(1-\tau)\varepsilon+\eta\big)\cap\big(\text{int}(C)\big) \neq \emptyset$. Then $x' \in \Delta((1-\tau)\varepsilon+\eta)$. Therefore,

$$\mathcal{V} \cap \Delta(\xi) \neq \emptyset, \quad \text{for all } \xi \in \big((1-\tau)\varepsilon + \text{int}(C)\big).$$

Hence Δ is lower semicontinuous at ε. Since $\varepsilon \in \text{int}(C)$ is arbitrary, Δ is lower semicontinuous on $\text{int}(C)$. □

Theorem 10.31 *Let K be a nonempty compact convex subset of X and $F : X\times X \to 2^Y \setminus \{\emptyset\}$ be a set-valued map such that the following conditions hold:*

(i) *$\mho(\varepsilon)$ is nonempty for each $\varepsilon \in \text{int}(C)$;*
(ii) *$F(x,y)$ is C-compact for every $(x,y) \in X \times X$;*
(iii) *$F(\cdot,y)$ is strictly C-properly quasiconvex on X for each $y \in X$;*
(iv) *For every fixed $y \in X$, $F(x,\cdot)$ is C-upper semicontinuous.*

Then $\mho$ is lower semicontinuous on $\text{int}(C)$.

Proof Suppose $\varepsilon \in \text{int}(C)$. Let $\mathcal{V}$ be an open set of X with $\mathcal{V} \cap \mho(\varepsilon) \neq \emptyset$ and $x \in \mathcal{V} \cap \mho(\varepsilon)$. Let $\hat{x} \in \mho(\alpha\varepsilon)$ where $\alpha \in]0,1[$. Obviously, $\hat{x} \in \mho(\varepsilon)$. We choose $x' \in \mathcal{V} \cap (x,\hat{x})$. Then by condition (iii), we have

$$F(x',y) + \varepsilon \subseteq \text{int}(C), \quad \text{for all } y \in X.$$

Let $y \in X$ be arbitrary but fixed. Then for each $z \in F(x',y)$, there exists corresponding positive number $t_z > 0$ such that

$$z + (1-2t_z)\varepsilon \in \text{int}(C).$$

Note that $z + (1-t_z)\varepsilon + \text{int}(C)$ is a neighborhood of $z+\varepsilon$ and that $\bigcup_{z\in F(x',y)}\big(z + (1-t_z)\varepsilon+\text{int}(C)\big) \supseteq F(x',y)$. By assumption (ii), there exist $z_1, z_2, \ldots, z_m \in F(x',y)$ such that

$$A_y := \bigcup_{i=1}^{m}\big(z_i + (1-t_{z_i})\varepsilon + \text{int}(C)\big) \supseteq F(x',y) + \varepsilon.$$

Let $\tau_y = \min\{t_{z_1}, t_{z_2}, \ldots, t_{z_m}\}$. Then $A_y - \tau_y\varepsilon \subseteq \text{int}(C)$. By condition (iv), there exists a neighborhood V_y of y such that

$$F(x',v) + \varepsilon \subseteq A_y, \quad \text{for all } v \in \mathcal{V}_y.$$

Therefore, for each $v \in V_y$, we obtain

$$F(x', v) + (1 - \tau_y)\varepsilon \subseteq \text{int}(C).$$

Since y is arbitrary, for each $y \in X$ we can choose corresponding τ_y and V_y. Because K is compact, there exist $V_{y_1}, V_{y_2}, \ldots, V_{y_k}$ such that $\bigcup_{j=1}^k V_{y_j} \supseteq K$. Let $\tau = \min\{\tau_{y_1}, \tau_{y_2}, \ldots, \tau_{y_k}\}$. Hence for each $y \in X$, we have

$$F(x', y) + (1 - \tau)\varepsilon \subseteq \text{int}(C).$$

Accordingly for each $\eta \in (1 - \tau)\varepsilon + \text{int}(C)$, $x' \in \mho(\eta)$. Hence $\mho$ is lower semicontinuous at ε. Since ε is arbitrary, $\mho$ is lower semicontinuous on $\text{int}(C)$. □

10.7.4 Continuity of Δ and ℧

By employing results established in Sects. 10.7.2 and 10.7.3, we immediately have the following results for continuity of Δ and $\mho$.

By combining Theorems 10.28 and 10.30, we obtain the following result.

Theorem 10.32 *Let K be a nonempty compact convex subset of X and $F : X \times X \to 2^Y \setminus \{\emptyset\}$ be a set-valued map such that the following conditions hold:*

(i) *$\Delta(\varepsilon)$ is nonempty for each $\varepsilon \in \text{int}(C)$;*
(ii) *$F(x, y)$ is $(-C)$-compact for every $(x, y) \in X \times X$;*
(iii) *$F(\cdot, y)$ is $(-C)$-upper semicontinuous on X for each $y \in X$;*
(iii) *$F(\cdot, y)$ is strictly C-properly quasiconvex on X for each $y \in X$;*
(iv) *For every fixed $y \in X$, $F(x, \cdot)$ is C-lower semicontinuous.*

Then Δ is continuous on $\text{int}(C)$.

From Theorems 10.29 and 10.31, we have following result.

Theorem 10.33 *Let K be a nonempty compact convex subset of X and $F : X \times X \to 2^Y \setminus \{\emptyset\}$ be a set-valued map such that the following conditions hold:*

(i) *$\mho(\varepsilon)$ is nonempty for each $\varepsilon \in \text{int}(C)$;*
(ii) *$F(x, y)$ is $(-C)$-compact for every $(x, y) \in X \times X$;*
(iii) *$F(\cdot, y)$ is strictly $(-C)$-properly quasiconvex on X for each $y \in X$;*
(iii) *$F(\cdot, y)$ is $(-C)$-lower semicontinuous on X for each $y \in X$;*
(iv) *For every fixed $y \in X$, $F(x, \cdot)$ is C-upper semicontinuous.*

Then $\mho$ is continuous on $\text{int}(C)$.

References

1. L.Q. Anh, P.Q. Khanh, On the Hölder continuity of solutions to parametric multivalued vector equilibrium problems. J. Math. Anal. Appl. **321**, 308–315 (2006)
2. L.Q. Anh, P.Q. Khanh, Uniqueness and Hölder continuity of solution to multivalued equilibrium problems in metric spaces. J. Glob. Optim. **37**, 449–465 (2007)
3. Q.H. Ansari, F. Flores-Bazán, Recession method for generalized vector equilibrium problems. J. Math. Anal. Appl. **321**(1), 132–146 (2006)
4. Q.H. Ansari, J.C. Yao, An existence result for the generalized vector equilibrium problem. Appl. Math. Lett. **12**(8), 53–56 (1999)
5. Q.H. Ansari, J.C. Yao, Generalized vector equilibrium problems. J. Stat. Manag. Syst. **5**, 1–17 (2002)
6. Q.H. Ansari, W. Oettli, D. Schläger, A generalization of vectorial equilibria. Math. Meth. Oper. Res. **46**, 147–152 (1997)
7. Q.H. Ansari, A.H. Siddiqi, S.Y. Wu, Existence and duality of generalized vector equilibrium problems. J. Math. Anal. Appl. **259**, 115–126 (2001)
8. Q.H. Ansari, I.V. Konnov, J.C. Yao, On generalized vector equilibrium problems. Nonlinear Anal. **47**, 543–554 (2001)
9. Q.H. Ansari, S. Schaible, J.C. Yao, Generalized vector equilibrium problems under generalized pseudomonotonicity with applications. J. Nonlinear Convex Anal. **3**(3), 331–344 (2002)
10. M. Balaj, A common fixed point theorem with applications to vector equilibrium problems. Appl. Math. Lett. **23**, 241–245 (2010)
11. M. Balaj, D. O'Regan, An intersection theorem and its applications. Cubo Math. J. **10**(4), 73–83 (2008)
12. L.C. Ceng, S. Schaible, J.C. Yao, A generalization of Minty's lemma and its applications to generalized mixed vector equilibrium problems. Optimization **56**, 565–576 (2007)
13. L.C. Ceng, S.-M. Guu, J.C. Yao, On generalized implicit vector equilibrium problems in Banach spaces. Comput. Math. Appl. **57**, 1682–1691 (2009)
14. B. Chen, X.-H. Gong, Continuity of the solution set to parametric set-valued weak vector equilibrium problems. Pac. J. Optim. **6**, 511–520 (2010)
15. B. Chen, N.-J. Huang, Continuity of the solution mapping to parametric generalized vector equilibrium problems. J. Glob. Optim. **56**, 1515–1528 (2013)
16. G.Y. Chen, S.J. Li, K.L. Teo, Solution semicontinuity of parametric generalized vector equilibrium problems. J. Glob. Optim. **45**, 309–318 (2005)
17. B. Chen, X.-H. Gong, S.-M. Yuan, Connectedness and compactness of weak efficient solutions for set-valued vector equilibrium problems. J. Inequal. Appl. **2008** (2008). Article ID 581849
18. Y. Chiang, O. Chadli, J.C. Yao, Generalized vector equilibrium problems with trifunctions. J. Glob. Optim. **30**, 135–154 (2004)
19. X.P. Ding, T.M. Ding, KKM type theorems and generalized vector equilibrium problems in noncompact FC-spaces. J. Math. Anal. Appl. **331**, 1230–1245 (2007)
20. X.P. Ding, J.Y. Park, Fixed points and generalized vector equilibria in general topological spaces. Acta Math. Acad. Paedagog. Nyházi. (N.S.) **19**, 89–100 (2003)
21. X.P. Ding, J.Y. Park, Generalized vector equilibrium problems in generalized convex spaces. J. Optim. Theory Appl. **120**, 327–353 (2004)
22. M. Fakhar, J. Zafarani, Generalized vector equilibrium problems for pseudomonotone multivalued bifunctions. J. Optim. Theory Appl. **126**, 109–124 (2005)
23. M. Fang, N.-J. Huang, KKM type theorems with applications to generalized vector equilibrium problems in FC-spaces. Nonlinear Anal. **67**, 809–817 (2007)
24. A.P. Farajzadeh, A. Amini-Harandi, On the generalized vector equilibrium problems. J. Math. Anal. Appl. **344**, 999–1004 (2008)
25. A.P. Farajzadeh, A. Amini-Harandi, D. O'Regan, Existence results for generalized vector equilibrium problems with multivalued mappings via KKM theory, Abstr. Appl. Anal. **2008** (2008), Article ID 968478

26. F. Flores-Bazán, F. Flores-Bazán, Vector equilibrium problems under recession analysis. J. Glob. Optim. **26**(2), 141–166 (2003)
27. J.Y. Fu, Vector equilibrium problems: existence theorems and convexity of solution set. J. Glob. Optim. **31**, 109–119 (2005)
28. J.Y. Fu, A.-H. Wan, Generalized vector equilibrium problems with set-valued mappings. Math. Meth. Oper. Res. **56**, 259–268 (2002)
29. P.Gr. Georgiev, T. Tanaka, Vector-valued set valued variants of Ky Fan's inequality. J. Nonlinear Convex Anal. **1**(3), 245–254 (2000)
30. P.Gr. Georgiev, T. Tanaka, Fan's inequality for set-valued maps. Nonlinear Anal. **47**(1), 607–618 (2001)
31. X.H. Gong, Ekeland's principle for set-valued vector equilibrium problems. Acta Math. **34B**(4), 1179–1192 (2014)
32. N.X. Hai, P.Q. Khanh, The solution existence of general variational inclusion problems. J. Math. Anal. Appl. **328**, 1268–1277 (2007)
33. Y. Han, X.H. Gong, Lower semicontinuity of solution mappings to parametric generalized strong vector equilibrium problems. Appl. Math. Lett. **28**, 38–41 (2014)
34. Y. Han, X.H. Gong, Semicontinuity of solution mappings to parametric generalized vector equilibrium problems. Numer. Funct. Anal. Optim. **37**(11), 1420–1437 (2016)
35. Y. Han, N.-J. Huang, Some characterizations of the approximate solutions to generalized vector equilibrium problems. J. Ind. Manag. Optim. **12**, 1135–1151 (2016)
36. S.H. Hou, X.H. Gong, X.M. Yang, Existence and stability of solutions for generalized Ky Fan inequality problems with trifunctions. J. Optim. Theory Appl. **146**, 387–398 (2010)
37. Y.W. Huang, A Farkas-Minkowski type alternative theorem and its applications to set-valued equilibrium problems. J. Nonlinear Convex Anal. **3**, 17–24 (2002)
38. K. Kimura, J.C. Yao, Semicontinuity of solution mappings of parametric generalized vector equilibrium problems. J. Optim. Theory Appl. **138**, 429–443 (2008)
39. K. Kimura, J.C. Yao, Semicontinuity of solution mappings of parametric generalized strong vector equilibrium problems. J. Ind. Manag. Optim. **4**(1), 167–181 (2008)
40. I.V. Konnov, S. Schaible, Duality for equilibrium problems under generalized monotonicity. J. Optim. Theory Appl. **104**, 395–408 (2000)
41. I.V. Konnov, J.-C. Yao, Existence of solutions for generalized vector equilibrium problems. J. Math. Anal. Appl. **233**, 328–335 (1999)
42. G.M. Lee, I.J. Bu, On vector equilibrium problem with multifunctions. Taiwan. J. Math. **10**(2), 399–407 (2006)
43. S.J. Li, Z.M. Fang, Lower semicontinuity of the solution mappings to a parametric generalized Ky Fan inequality. J. Optim. Theory Appl. **147**, 507–515 (2010)
44. S.J. Li, X.B. Li, L.N. Wang, K.L. Teo, The Hölder continuity of solutions to generalized vector equilibrium problems. Eur. J. Oper. Res. **199**, 334–338 (2009)
45. S.J. Li, H.M. Liu, C.-R. Chen, Lower semicontinuity of parametric generalized weak vector equilibrium problems. Bull. Aust. Math. Soc. **81**, 85–95 (2010)
46. S.J. Li, H.M. Liu, Y. Zhang, Z.M. Fang, Continuity of the solution mappings to parametric generalized strong vector equilibrium problems. J. Glob. Optim. **55**, 597–610 (2013)
47. L.J. Lin, Existence results for primal and dual generalized vector equilibrium problems with applications to generalized semi-infinite programming. J. Glob. Optim. **33**, 579–595 (2005)
48. L.J. Lin, Z.T. Yu, G. Kassay, Existence of equilibria for multivalued mappings and its applications to vectorial equilibria. J. Optim. Theory Appl. **114**, 189–208 (2002)
49. L.-J. Lin, Q.H. Ansari, J.Y. Wu, Geometric properties and coincidence theorems with applications to generalized vector equilibrium problems. J. Optim. Theory Appl. **117**, 121–137 (2003)
50. Q.-Y. Liu, X.-J. Long, N.-J. Huang, Connectedness of the sets of weak efficient solutions for generalized vector equilibrium problems. Math. Slovaca **62**, 123–136 (2012)
51. Z. Mitrović, On generalized vector equilibrium problem with bounds. Appl. Math. Lett. **23**, 783–787 (2010)

52. W. Oettli, D. Schläger, Existence of equilibria for monotone multivalued mappings. Math. Meth. Oper. Res. **48**, 219–228 (1998)
53. W. Oettli, D. Schläger, Generalized vectorial equilibria and generalized monotonicity, in *Functional Analysis with current Applications in Science, Technology and Industry*, ed. by M. Brokate, A.H. Siddiqi. Pitman Research Notes in Mathematics Series, vol. 377 (Longman, Essex, 1998), pp. 145–154
54. Z.Y. Peng, X.M. Yang, J.W. Peng, On the lower semicontinuity of the solution mappings to parametric weak generalized Ky Fan inequality. J. Optim. Theory Appl. **152**, 256–264 (2012)
55. Z.Y. Peng, Y. Zhao, X.M. Yang, Semicontinuity of approximate solution mappings to parametric set-valued vector equilibrium problems. Numer. Funct. Anal. Appl. **36**, 481–500 (2015)
56. D.-N. Qu, C.-Z. Cheng, Ekeland variational principle for generalized vector equilibrium problems with equivalences and applications. J. Inequal. Appl. **2013** (2013), Article ID 267585
57. I. Sadeqi, C.G. Alizadeh, Existence of solutions of generalized vector equilibrium problems in reflexive Banach spaces. Nonlinear Anal. **74**, 2226–2234 (2011)
58. I. Sadeqi, M.S. Paydar, Lipschitz continuity of an approximate solution mapping for parametric set-valued vector equilibrium problems. Optimization **65**, 1003–1021 (2016)
59. G. Wang, Existence-stability theorems for strong vector set-valued equilibrium problems in reflexive Banach spaces. J. Inequal. Appl. **2015** (2015), Article ID 239
60. S.-H. Wang, Q.-Y. Li, J.-Y. Fu, Strong vector equilibrium problems on noncompact sets. Bull. Malays. Math. Sci. Soc. **35**(1), 119–132 (2012)
61. R. Wangkeeree, R. Wangkeeree, P. Preechasip, Continuity of the solution mappings to parametric generalized vector equilibrium problems. Appl. Math. Lett. **29**, 42–45 (2014)
62. Y.D. Xu, S.J. Li, On the lower semicontinuity of the solution mappings to a parametric generalized strong vector equilibrium problem. Positivity **17**, 341–353 (2013)
63. M.-G. Yang, N.-J. Huang, Existence results for generalized vector equilibrium problems with applications. Appl. Math. Mech. Engl. Ed. **35**(7), 913–924 (2014)
64. L.C. Zeng, J.C. Yao, Generalized Minty's lemma for generalized vector equilibrium problems. Appl. Math. Lett. **19**, 1320–1326 (2006)

Appendix A
Set-Valued Maps

Let X and Y be two nonempty sets. A *set-valued map* or *multivalued map* or *point-to-set map* or *multifunction* $T : X \to 2^Y$ from X to Y is a map that associates with any $x \in X$ a subset $T(x)$ of Y; the set $T(x)$ is called the *image of* x under T. The set $Dom(T) = \{x \in X : T(x) \neq \emptyset\}$ is called the *domain* of T. Actually, a set-valued map T is characterized by its *graph*, the subset of $X \times Y$ defined by

$$Graph(T) = \{(x, y) : y \in T(x)\}.$$

Indeed, if A is a nonempty subset of the product space $X \times Y$, then the graph of a set-valued map T is defined by

$$y \in T(x) \quad \text{if and only if} \quad (x, y) \in A.$$

The domain of T is the projection of $Graph(F)$ on X. The *image* of T is a subset of Y defined by

$$Im(T) = \bigcup_{x \in X} T(x) = \bigcup_{x \in Dom(T)} T(x).$$

It is the projection of $Graph(T)$ on Y. A set-valued map T from X to Y is called *strict* if $Dom(T) = X$, that is, if the image $T(x)$ is nonempty for all $x \in X$. Let K be a nonempty subset of X and T be a strict set-valued map from K to Y. It may be useful to extend it to the set-valued map T_K from X to Y defined by

$$T_K(x) = \begin{cases} T(x), & \text{when } x \in K, \\ \emptyset, & \text{when } x \notin K, \end{cases}$$

whose domain $Dom(T_K)$ is K.

Q.H. Ansari et al., *Vector Variational Inequalities and Vector Optimization*, Vector Optimization, DOI 10.1007/978-3-319-63049-6

Let T be a set-valued map from X to Y and $K \subseteq X$, then we denote by $T_{|_K}$ the restriction of T to K.

Definition A.1 Let $T : X \to 2^Y$ be a set-valued map. For a nonempty subset A of X, we write

$$T(A) = \bigcup_{x \in A} T(x).$$

If $A = \emptyset$, we write $T(\emptyset) = \emptyset$. The set $T(A)$ is called the *image* of A under the set-valued map T.

Theorem A.1 *Let $\{A_\alpha\}_{\alpha \in \Lambda}$ be a family of nonempty subsets of X and $T : X \to 2^Y$ be a set-valued map.*

(a) *If $A_1 \subseteq A_2$ then $T(A_1) \subseteq T(A_2)$;*

(b) $T\left(\bigcup_{\alpha \in \Lambda} A_\alpha\right) = \bigcup_{\alpha \in \Lambda} T(A_\alpha)$;

(c) $T\left(\bigcap_{\alpha \in \Lambda} A_\alpha\right) \subset \bigcap_{\alpha \in \Lambda} T(A_\alpha)$;

(d) $T(X \setminus A_1) \supseteq T(X) \setminus T(A_1)$.

Definition A.2 Let T_1 and T_2 be two set-valued maps from X to Y.

- The *union* of T_1 and T_2 is a set-valued map $(T_1 \cup T_2)$ from X to Y defined by

$$(T_1 \cup T_2)(x) = T_1(x) \cup T_2(x), \quad \text{for all } x \in X.$$

- The *intersection* of T_1 and T_2 is a set-valued map $(T_1 \cap T_2)$ from X to Y defined by

$$(T_1 \cap T_2)(x) = T_1(x) \cap T_2(x), \quad \text{for all } x \in X.$$

- The *Cartesian product* of T_1 and T_2 is a set-valued map $(T_1 \times T_2)$ from X to $Y \times Y$ defined by

$$(T_1 \times T_2)(x) = T_1(x) \times T_2(x), \quad \text{for all } x \in X.$$

- If T_1 is a set-valued map from X to Y and T_2 is another set-valued map from Y to Z, then the *composition product* of T_2 by T_1 is a set-valued map $(T_2 \circ T_1)$ from X to Z defined by

$$(T_2 \circ T_1)(x) = T_2(T_1(x)), \quad \text{for all } x \in X.$$

Theorem A.2 *Let T_1 and T_2 be set-valued maps from X to Y and A be a nonempty subset of X. Then*

(a) $(T_1 \cup T_2)(A) = T_1(A) \cup T_2(A)$;
(b) $(T_1 \cap T_2)(A) \subseteq T_1(A) \cap T_2(A)$;
(c) $(T_1 \times T_2)(A) \subseteq T_1(A) \times T_2(A)$;
(d) $(T_2 \circ T_1)(A) = T_2(T_1(A))$.

Definition A.3 If T is a set-valued map from X to Y, then the *inverse* T^{-1} of T is defined by

$$T^{-1}(y) = \{x \in X : y \in T(x)\}, \quad \text{for all } y \in Y$$

Further, let B be a subset of Y. The *upper inverse image* $T^{-1}(B)$ and *lower inverse image* $T_{+}^{-1}(B)$ of B under F are defined by

$$T^{-1}(B) = \{x \in X : T(x) \cap B \neq \emptyset\}$$

and

$$T_{+}^{-1}(B) = \{x \in X : T(x) \subseteq B\}.$$

We also write $T^{-1}(\emptyset) = \emptyset$ and $T_{+}^{-1}(\emptyset) = \emptyset$. It is clear from the definition of inverse of T that $(T^{-1})^{-1} = T$ and $y \in T(x)$ if and only if $x \in T^{-1}(y)$.

We have the following relations between domain, graph and image of T and T^{-1}.

$$Dom(T^{-1}) = Im(T), \qquad Im(T^{-1}) = Dom(T) \text{ and}$$

$$Graph(T^{-1}) = \{(y, x) \in Y \times X : (x, y) \in Graph(T)\}.$$

Theorem A.3 *Let $\{B_\alpha\}_{\alpha \in \Lambda}$ be a family of nonempty subsets of Y, $A \subseteq X$ and $B \subseteq Y$. Let $T : X \to 2^Y$ be a set-valued map.*

(a) *If $B_1 \subseteq B_2$, then $T^{-1}(B_1) \subseteq T^{-1}(B_2)$;*
(b) $A \subset T_{+}^{-1}(T(A))$;
(c) $B \supset T(T_{+}^{-1}(B))$;
(d) $T_{+}^{-1}\left(\bigcup_{\alpha \in \Lambda} B_\alpha\right) \supset \bigcup_{\alpha \in \Lambda} T_{+}^{-1}(B_\alpha)$;
(e) $T_{+}^{-1}\left(\bigcap_{\alpha \in \Lambda} B_\alpha\right) = \bigcap_{\alpha \in \Lambda} T_{+}^{-1}(B_\alpha)$;
(f) $T^{-1}(T(A)) \supset A$;

(g) $T^{-1}\left(\bigcap_{\alpha\in\Lambda} B_\alpha\right) \subset \bigcap_{\alpha\in\Lambda} T^{-1}(B_\alpha)$;

(h) $T^{-1}\left(\bigcup_{\alpha\in\Lambda} B_\alpha\right) = \bigcup_{\alpha\in\Lambda} T^{-1}(B_\alpha)$.

Theorem A.4 *Let $T_1, T_2 : X \to 2^Y$ be set-valued maps such that $(T_1 \cap T_2)(x) \neq \emptyset$ for all $x \in X$ and $B \subseteq Y$. Then*

(a) $(T_1 \cup T_2)^{-1}(B) = T_1^{-1}(B) \cup T_2^{-1}(B)$;
(b) $(T_1 \cap T_2)^{-1}(B) \subset T_1^{-1}(B) \cap T_2^{-1}(B)$;
(c) $(T_1 \cup T_2)_+^{-1}(B) = T_{1+}^{-1}(B) \cap T_{2+}^{-1}(B)$;
(d) $(T_1 \cap T_2)_+^{-1}(B) \supset T_{1+}^{-1}(B) \cup T_{2+}^{-1}(B)$;

Theorem A.5 *Let $T_1 : X \to 2^Y$ and $T_2 : Y \to 2^Z$ be set-valued maps. Then for any set $B \subseteq Z$, we have*

(a) $(T_2 \circ T_1)_+^{-1}(B) = T_{1+}^{-1}\left(T_{2+}^{-1}(B)\right)$;
(b) $(T_2 \circ T_1)^{-1}(B) = T_1^{-1}\left(T_2^{-1}(B)\right)$.

Theorem A.6 *Let $T_1 : X \to 2^Y$ and $T_2 : X \to 2^Z$ be set-valued maps. Then for any sets $B \subseteq Y$ and $D \subseteq Z$, we have*

(a) $(T_1 \times T_2)_+^{-1}(B \times D) = T_{1+}^{-1}(B) \cap T_{2+}^{-1}(D)$;
(b) $(T_1 \times T_2)^{-1}(B \times D) = T_1^{-1}(B) \cap T_2^{-1}(D)$.

For further details and applications of set-valued maps, we refer to [1–8] and the references therein.

Appendix B
Some Algebraic Concepts

Let K and D be nonempty subsets of a vector space X. The *algebraic sum* and *algebraic difference* of K and D are, respectively, defined as

$$K + D = \{x + y : x \in K \text{ and } y \in D\}$$

and

$$K - D = \{x - y : x \in K \text{ and } y \in D\}.$$

Let λ be any real number, then λK is defined as

$$\lambda K = \{\lambda x : x \in K\}.$$

We pointed out that $K + K = 2K$ is not true in general for any nonempty subset K of a vector space X.

Definition B.1 Let X be a vector space. The space of all linear mappings from X to $\mathbb{R}$ is called *algebraic dual of* X and it is denoted by X'.

Definition B.2 A subset K of a vector space X is called

- *balanced* if it is nonempty and $\alpha K \subset K$ for all $\alpha \in [-1, 1]$;
- *absolutely convex* if it is convex and balanced;
- *absorbing* or *absorbent* if for each $x \in X$, there exists $\rho > 0$ such that $\lambda x \in K$ for all $|\lambda| \leq \rho$. Note that an absorbing set contains the zero of X.

Theorem B.1 *Let X and Y be vector spaces and $f : X \to Y$ be a linear map.*

(a) *If K is a balanced subset of X, then $f(K)$ is balanced.*
(b) *If K is absorbing and f is onto, then $f(K)$ is absorbing.*
(c) *Inverse image under f of absorbing or balanced subsets of Y are absorbing or balanced, respectively, subset of X.*

Q.H. Ansari et al., *Vector Variational Inequalities and Vector Optimization*, Vector Optimization, DOI 10.1007/978-3-319-63049-6

Definition B.3 Let K be a nonempty subset of a vector space X.

- The set $\text{cor}(K) = \{y \in K : \text{for every } x \in X \text{ there is a } \bar{\lambda} > 0 \text{ with } y + \lambda x \in C \text{ for all } \lambda \in [0, \bar{\lambda}]\}$ is called the *algebraic interior* of K (or the *core* of K) .
- The set K is called *algebraic open* if $K = \text{cor}(K)$.
- The set of all elements of X which neither belong to $\text{cor}(K)$ nor to $\text{cor}(X \setminus K)$ is called the *algebraic boundary* of K.
- An element $x \in X$ is called *linearly accessible* from K if there exists $y \in K$, $y \neq x$, such that $\lambda y + (1 - \lambda)x \in K$ for all $\lambda \in (0, 1]$.
  The union of K and the set of all linearly accessible elements from K is called the *algebraic closure* of K and it is denoted by

$$\text{lin}(K) = K \cup \{x \in X : x \text{ is linearly accessible from } K\}.$$

  The set K is called *algebraic closed* if $K = \text{lin}(K)$.
- The set K is called *algebraic bounded* if for every $y \in K$ and every $x \in K$, there is a $\bar{\lambda} > 0$ such that $y + \lambda x \notin K$ for all $\lambda \geq \bar{\lambda}$.

These algebraic notions have a special geometric meaning. Take the intersections of the set K with each straight line in the vector space X and consider these intersections as subsets of the real line $\mathbb{R}$. Then the set K is algebraic open if these subsets are open; K is algebraic closed if these subsets are closed; and K is algebraic bounded if these subsets are bounded.

Lemma B.1 ([9, Lemma 1.9]) *Let K be a nonempty convex subset of a vector space X.*

(a) $x \in \text{lin}(K)$, $y \in \text{cor}(K) \Rightarrow \{\lambda y + (1 - \lambda)x : \lambda \in [0, 1]\} \subset \text{cor}(K)$;
(b) $\text{cor}(\text{cor}(K)) = \text{cor}(K)$;
(c) $\text{cor}(K)$ *and* $\text{lin}(K)$ *are convex;*
(c) *If* $\text{cor}(K) \neq \emptyset$, *then* $\text{lin}(\text{cor}(K)) = \text{lin}(K)$ *and* $\text{cor}(\text{lin}(K)) = \text{cor}(K)$.

Lemma B.2 ([9, Lemma 1.11]) *Let K be a convex cone in a vector space X with a nonempty algebraic interior. Then*

(a) $\text{cor}(K) \cup \{\mathbf{0}\}$ *is a convex cone,*
(b) $\text{cor}(K) = K + \text{cor}(K)$.

Lemma B.3 *A cone C in a vector space X is reproducing if* $\text{cor}(C) \neq \emptyset$.

Proof Since $\text{cor}(C)$ is nonempty, we let $x \in \text{cor}(C)$ and take any $y \in X$. Then there is a $\bar{\lambda} > 0$ with $x + \bar{\lambda} y \in C$ implying

$$y \in \frac{1}{\bar{\lambda}}C - \left\{\frac{1}{\bar{\lambda}}x\right\} \subset C - C.$$

So, we get $X \subset C - C$ and together with the trivial inclusion $C - C \subset X$, we obtain the assertion. □

Given x and y in a vector space X, we denote by $[x, y]$ and $]x, y[$ the closed and open line segments joining x and y, respectively.

Definition B.4 Let K be a nonempty convex subset of a vector space X. A point $x \in K$ is said to be *relative algebraic interior point of* K if for any $u \in X$ such that $x + u \in K$, there exists $\varepsilon > 0$ such that $]x - \epsilon u, x + \epsilon u[\subseteq K$. The set of all relative algebraic interior points of K is denoted by relint(K). We note that if $0 < \alpha < \epsilon$, then $]x - \alpha u, x + \alpha u[\subseteq]x - \epsilon u, x + \epsilon u[$.

Appendix C
Topological Vector Spaces

Definition C.1 (Directed Set) A set Λ together with a reflexive and transitive ordering relation $\preccurlyeq$ such that every finite set of Λ has an upper bound in Λ (that is, for $\alpha, \beta \in \Lambda$, there is a $\gamma \in \Lambda$ such that $\alpha \preccurlyeq \gamma$ and $\beta \preccurlyeq \gamma$) is called a *directed set*.

Definition C.2 Let X be a topological space and, Λ and Γ be any index sets.

(a) A collection $\mathscr{F} = \{O_\alpha\}_{\alpha \in \Lambda}$ of subsets of X is said to be a *cover* of X if $\bigcup_{\alpha \in \Lambda} O_\alpha = X$; If each member of $\mathscr{F}$ is an open set, then $\mathscr{F}$ is called an *open cover* of X;
(b) A subcollection $\mathscr{C}$ of a cover $\mathscr{F}$ of X is said to be a *subcover* if $\mathscr{C}$ is itself a cover of X; If the number of members of $\mathscr{C}$ is finite, then $\mathscr{C}$ is called a *finite subcover*.
(c) A collection $\mathscr{C} = \{U_\beta\}_{\beta \in \Gamma}$ of subsets of X is said to be a *refinement* of the cover $\mathscr{F} = \{O_\alpha\}_{\alpha \in \Lambda}$ of X if $\mathscr{C}$ is an open cover of X and for each member $U_\beta \in \mathscr{C}$, there is $O_\alpha \in \mathscr{F}$ such that $U_\beta \subseteq O_\alpha$.

Note that an open subcover is a refinement, but a refinement is not necessarily an open cover.

Definition C.3 An open cover $\mathscr{F} = \{O_\alpha\}_{\alpha \in \Lambda}$ of a topological space X is said to be *locally finite* if each point of X has a neighborhood which meets only finitely many O_α.

Definition C.4 A Hausdorff space X is said to be *paracompact* if every open cover of X has a locally finite open refinement.

Definition C.5 Let X be a topological space and $f : X \to \mathbb{R}$. The *support* of f is the set $\operatorname{Supp}(f) := \operatorname{cl}(\{x \in X : f(x) \neq 0\})$.

Since a finite cover is necessarily locally finite, it follows that every open cover of a compact space has locally finite open refinement.

Q.H. Ansari et al., *Vector Variational Inequalities and Vector Optimization*, Vector Optimization, DOI 10.1007/978-3-319-63049-6

Definition C.6 (Partition of Unity) Let X be a topological space. A family $\{\beta\}_{i\in I}$ of continuous functions defined from X into $[0,\infty)$ is called a *partition of unity* associated to an open cover $\{U_i\}_{i\in I}$ of X if

(i) for each $x \in X$, $\beta_i(x) \neq \emptyset$ for only finitely many β_i;
(ii) $\sum_{i\in I} \beta_i(x) = 1$ for all $x \in X$.

Theorem C.1 ([10, pp. 68]) *A Hausdorff space X is paracompact if and only if every open cover of X has a continuous locally finite partition of unity.*

We note that every compact Hausdorff space is paracompact and every metrizable space is paracompact.

Definition C.7 A subset in a topological space is *precompact* (or *relatively compact*) if its closure is compact.

Note that every element in $\mathbb{R}^n$ has a precompact neighborhood.

Definition C.8 ([11, 12]) Let X be a Hausdorff topological vector space and L be a lattice with least one minimal element, denoted by $\mathbf{0}$. A mapping $\Phi : 2^X \to L$ is said to be a *measure of noncompactness* provided that the following conditions hold for all $M, N \in 2^X$:

(i) $\Phi(M) = \mathbf{0}$ if and only if M is precompact;
(ii) $\Phi(\mathrm{cl}(M)) = \Phi(M)$;
(iii) $\Phi(M \cup N) = \max\{\Phi(M), \Phi(N)\}$.

It follows from condition (iii) that if $M \subseteq N$, then $\Phi(M) \leq \Phi(N)$.

Definition C.9 A *net* in a topological space X is a mapping $\alpha \mapsto x_\alpha$ from a directed set Λ into X; we often write $\{x_\alpha\}_{\alpha\in\Lambda}$, $\{x_\alpha : \alpha \in \Lambda\}$, or simply $\{x_\alpha\}$.

We say that a net $\{x_\alpha\}_{\alpha\in\Lambda}$ *converges* to $x \in X$ if for any neighborhood V of x, there exists an α (an index α) such that $x_\beta \in V$ for all $\beta \succcurlyeq \alpha$. The point x is called a limit of the net $\{x_\alpha\}$. When a net $\{x_\alpha\}$ converges to a point x, we denote it by $x_\alpha \to x$.

We say that x is a *cluster point* of the net $\{x_\alpha\}$ if for any neighborhood V of x and for any index α, there exists $\beta \succcurlyeq \alpha$ such that $x_\beta \in V$.

Definition C.10 A vector space X with a topology $\mathcal{T}$ under which the mappings

$$(x, y) \mapsto x + y \text{ from } X \times X \to X$$

and

$$(\alpha, x) \mapsto \alpha x \text{ from } \mathbb{R} \times X \to X$$

are continuous, is called a *topological vector space*.

Theorem C.2 *Let X be a vector space and $\mathcal{B}$ be a family of subsets of X such that the following conditions hold.*

(i) *Each $U \in \mathcal{B}$ is balanced and absorbing;*
(ii) *For any given $U_1, U_2 \in \mathcal{B}$, there exists $U \in \mathcal{B}$ such that $U \subset U_1 \cap U_2$;*
(iii) *For any given $U \in \mathcal{B}$, there exists $V \in \mathcal{B}$ such that $V + V \subset U$.*

Then there is a unique topology $\mathcal{T}$ on X such that $(X, \mathcal{T})$ is a topological vector space and $\mathcal{B}$ is a neighborhood base (or local base) at $\mathbf{0}$.

Theorem C.3 *Let X be a topological vector space.*

(a) *The closure of a balanced set $A \subseteq X$ is balanced.*
(b) *The closure of a convex set $A \subseteq X$ is convex.*
(c) *The interior of a convex set $A \subseteq X$ is convex.*

Theorem C.4 *Let X be a topological vector space and K be a subset of X.*

(a) *If K is balanced and $\mathbf{0} \in \text{int}(K)$, then its interior is also balanced.*
(b) *Any superset of an absorbing set is absorbing. So if K is absorbing then so is its closure. The interior of an absorbing set is not generally absorbing;*
(c) *If K is open, then so is its convex hull.*

Theorem C.5 *Let X be a topological vector space and U be a neighborhood of its zero element* $\mathbf{0}$. *Then λU is a neighborhood of* $\mathbf{0}$ *for all nonzero real* λ.

Definition C.11 A topological vector space X is said to be *locally bounded* if there is a bounded neighborhood of $\mathbf{0}$.

Trivially, every normed space is locally bounded. A well-known example of a locally bounded topological vector space is L^p for $0 < p < 1$ which is not normable (see [13]).

Theorem C.6

(a) *Every neighborhood of zero in a topological vector space X is absorbing.*
(b) *Every neighborhood of zero in a topological vector space X includes a closed balanced neighborhood of zero.*

Theorem C.7 *Let X be a topological vector space with its zero element is denoted by* $\mathbf{0}$ *and K be a nonempty subset of X. Then*

$$\text{cl}(K) = \{K + U : U \text{ is a neighborhood of } \mathbf{0}\}.$$

In particular, $\text{cl}(K) \subset K + U$ *for any neighborhood U of* $\mathbf{0}$.

Proposition C.1 ([14, Proposition 15]) *For each $i = 1, 2, \ldots, m$, let K_i be a compact convex subset of a Hausdorff topological vector space X. Then* $\text{co}\left(\bigcup_{i=1}^{m} A_i\right)$ *is compact.*

Corollary C.1 ([14, Corollary 1]) *In a Hausdorff topological vector space, the convex hull of a finite set is compact.*

Definition C.12 A topological vector space X with its topology $\mathcal{T}$ is said to be *locally convex* if there is a neighborhood base (local base) at zero consisting of convex sets. The topology $\mathcal{T}$ is called locally convex.

Theorem C.8 *Let $\mathcal{T}$ be a locally convex topology on X. Then there exists a local base $\mathcal{B}$ whose members have the following properties:*

(i) *Every member of $\mathcal{B}$ is absolutely convex and absorbing;*
(ii) *If $U \in \mathcal{B}$ and $\lambda > 0$, then $\lambda U \in \mathcal{B}$.*

Conversely, if $\mathcal{B}$ is a filter base on X which satisfies conditions (i) and (ii), then there exists a unique locally convex topology $\mathcal{T}$ on X such that $\mathcal{B}$ is a local base at zero for $\mathcal{T}$.

Definition C.13 ([15, pp. 188]) Let X and Y be vector spaces. A *bilinear functional* or *bilinear form* $B : X \times Y \to \mathbb{R}$, $(x, y) \mapsto B(x, y)$ is a map which is linear in either argument when the other is held fixed.

A *pairing* or *pair* is an ordered pair (X, Y) of linear spaces together with a fixed bilinear functional B. Usually, $B(x, y)$ will be denoted by $\langle x, y\rangle$.

If X is a linear space and X' its algebraic dual then the *natural pairing* of X and X' is that arising from the (natural or canonical) bilinear functional on $X \times X'$, which sends (x, x') into $x'(x)$, that is, $\langle x, x'\rangle = x'(x)$.

If X and Y are any two paired vector spaces, then we also have the natural pairing in the following sense: If y is any fixed element of Y, then the map $y' : X \to \mathbb{R}$, $x \mapsto \langle x, y\rangle$ is obviously a linear functional on X, that is, $y' \in X'$.

It is clear that

$$\langle x, y\rangle = 0, \quad \text{for all } x \in X \quad \text{implies} \quad y = \mathbf{0}$$

equivalently,

$$y \neq \mathbf{0} \quad \text{implies that there is some } x \in X \text{ such that } \langle x, y\rangle \neq 0.$$

Definition C.14 Let X be a vector space. A semi-norm on X is a function $p : X \to \mathbb{R}$ such that the following conditions hold:

(i) $p(x) \geq 0$ for all $x \in X$;
(ii) $p(\lambda x) = |\lambda| p(x)$ for all $x \in X$ and $\lambda \in \mathbb{R}$;
(iii) $p(x + y) \leq p(x) + p(y)$ for all $x, y \in X$.

Theorem C.9 ([14, Corollary, p. TVS II.24]) *Let X be a topological vector space with its topology $\mathcal{T}$. Then $\mathcal{T}$ is defined by a set of semi-norms if and only if $\mathcal{T}$ is locally convex.*

Definition C.15 ([15, pp. 189]) Let X and Y be vector spaces. The map $x \mapsto |\langle x, y\rangle| = p_y(x)$ determines a semi-norm on X for all $y \in Y$. The topology generated by the family of semi-norms $\{p_y : y \in Y\}$ is the weakest topology on X and it is called *weak topology* on X determined by the pair (X, Y), and it is denoted by $\sigma(X, Y)$.

Remark C.1 Clearly, $\sigma(X, Y)$ is a locally convex topology on X and also it is a Hausdorff topology.

Let X and Y be Hausdorff topological vector spaces and $\mathcal{L}(X, Y)$ denote the family of continuous linear functions from X to Y. Let σ be the family of bounded subsets of X whose union is total in X, that is, the linear hull of $\bigcup\{U : U \in \sigma\}$ is dense in X. Let $\mathcal{B}$ be a neighborhood base of $\mathbf{0}$ in Y, where $\mathbf{0}$ is the zero element of Y. When U runs through σ, V through $\mathcal{B}$, the family

$$M(U, V) = \left\{ \xi \in \mathcal{L}(X, Y) : \bigcup_{x \in U} \langle \xi, x \rangle \subseteq V \right\}$$

is a neighborhood base of $\mathbf{0}$ in $\mathcal{L}(X, Y)$ for a unique translation-invariant topology, called the *topology of uniform convergence* on the sets $U \in \sigma$, or, briefly, the σ-topology (see [16, pp. 79–80]).

Lemma C.1 ([16]) *Let X and Y be Hausdorff topological vector spaces and $\mathcal{L}(X, Y)$ be the topological vector space under the σ-topology. Then the bilinear mapping $\langle ., . \rangle : \mathcal{L}(X, Y) \times X \to Y$ is continuous on $\mathcal{L}(X, Y) \times X$.*

We now present an open mapping theorem due to Brezis [17].

Theorem C.10 ([17, Théorème II.5]) *Let X and Y be two Banach spaces and $T : X \to Y$ be a continuous linear surjective mapping. Then there exists a constant $c > 0$ such that $B_c[\mathbf{0}] \subset T(B_1[\mathbf{0}])$, where $B_c[\mathbf{0}]$ denotes the closed ball of radius c around $\mathbf{0}$ in Y and $B_1[\mathbf{0}]$ is the closed unit ball in X.*

Remark C.2 From Theorem C.10, we see that $T : X \to Y$ is an open mapping.

Indeed, let U be an open subset of X. We show that $T(U)$ is open. Let $y \in T(U)$, where $y = T(x)$ for some $x \in U$. Let $r > 0$ such that $B_r[x] \subset U$, i.e., $x + B_r[\mathbf{0}] \subset U$. Then we have $y + T(B_r[\mathbf{0}]) \subset T(U)$. Due to Theorem C.10, we obtain $B_{rc}[\mathbf{0}] \subset T(B_r[\mathbf{0}])$, and consequently, $B_{rc}[y] \subset T(U)$.

Theorem C.11 ([18, Chapter 6, Theorem 1.1]) *If X and Y are Banach spaces and $T : X \to Y$ is a linear operator, then T is bounded if and only if it is continuous from the weak topology of X to the weak topology of Y.*

Definition C.16 Let (X, d) be a metric space and A be a nonempty subset of X. The *diameter of A*, denoted by $\delta(A)$, is defined as

$$\delta(A) = \sup\{d(x, y) : x, y \in A\}.$$

We close this subsection by presenting the following Cantor's intersection theorem.

Theorem C.12 (Cantor's Intersection Theorem) *Let (X, d) be a complete metric space and $\{A_m\}$ be a decreasing sequence (that is, $A_{m+1} \subseteq A_m$) of nonempty closed* *subsets of X such that the diameter of A_m $\delta(A_m) \to 0$ as $m \to \infty$. Then the intersection* $\bigcap_{m=1}^{\infty} A_m$ *contains exactly one point.*

References

1. Q.H. Ansari, *Metric Spaces – Including Fixed Point Theory and Set-Valued Maps* (Narosa Publishing House, New Delhi, 2010). Also Published by Alpha Science International Ltd. Oxford, UK (2010)
2. J.-P. Aubin, A. Cellina, *Differential Inclusions* (Springer, Berlin, 1994)
3. J.-P. Aubin, H. Frankowska, *Set-Valued Analysis* (Birkhäuser, Boston, 1990)
4. C. Berge, *Topological Space* (Oliver & Boyd, Edinburgh, 1963)
5. R.S. Burachik, A.N. Iusem, *Set-Valued Mappings and Enlargements of Monotone Operators* (Springer, New York, 2008)
6. L. Górniewicz, *Topological Fixed Point Theory of Multivalued Mappings* (Kluwer Academic Publishers, Dordrecht, 1999)
7. V.I. Istrătescu, *Fixed Point Theory: An Introduction* (D. Reidel Publishing Company, Dordrecht, 1981)
8. E. Klein, A.C. Thompson, *Theory of Correspondences, Including Applications to Mathematical Economics* (Wiley, New York, 1984)
9. J. Jahn, *Vector Optimization: Theory, Applications, and Extensions* (Springer, Berlin, 2004)
10. C.D. Aliprantis, K.C. Border, *Infinite Dimensional Analysis* (Springer, New York, 1994)
11. W.V. Petryshyn, P.M. Fitzpatrick, Fixed point theorems of multivalued noncompact inward maps. J. Math. Anal. Appl. **46**, 756–767 (1974)
12. W.V. Petryshyn, P.M. Fitzpatrick, Fixed point theorems of multivalued noncompact acyclic mappings. Pac. J. Math. **54**, 17–23 (1974)
13. G. Köthe, *Topological Vector Spaces* (Springer, Berlin, 1983)
14. N. Bourbaki, *Topological Vector Spaces* (Springer, Berlin, 1987)
15. L. Narice, E. Beckenstein, *Topological Vector Spaces* (Marcel Dekker, Inc., New York, 1985)
16. X.P. Ding, E. Tarafdar, Generalized vector variational-like inequalities without monotonicity, in [19] (2000), pp. 113–124
17. H. Brézis, *Analyse Fonctionnelle* (Masson, Paris, 1983)
18. J.B. Conway, *A Course in Functional Analysis*, 2nd edn. (Springer, New York, 1990)
19. F. Giannessi (ed.), *Vector Variational Inequality and Vector Equilibria: Mathematical Theories* (Kluwer Academic, Dordrecht, 2000)

Bibliography

1. L. Altangerel, Scalarized gap functions for vector variational inequalities via conjugate duality. Mongolian Math. J. **11**, 46–54 (2007)
2. L. Altangerel, R.I. Bot, G. Wanka, Conjugate duality in vector optimization and some applications to the vector variational inequality. J. Math. Anal. Appl. **329**, 1010–1035 (2007)
3. L. Altangerel, G. Wanka, O. Wilfer, An oriented distance function application to gap functions for vector variational inequalities, in *Optimization, Simulation, and Control*, ed. by A. Chinchuluun et al. (Springer, New York, 2013), pp. 17–34
4. Q.H. Ansari, J.C. Yao, On strong solutions of the generalized implicit vector variational problem. Adv. Nonlinear Var. Inequal. **2**, 1–10 (1999)

5. A.S. Antipin, On convergence of proximal methods to fixed point of extremal mappings and estimates of their rate of convergence. Comput. Math. Math. Phys. **35**, 539–551 (1995)
6. M. Arana, R. Rufián, R. Osuna, Necessary and sufficient conditions for weak minima, in *Minutes of the XXIV National Congress of Statistics and Operations Research*, Almería, Spain (1998)
7. K. Bergstresser and P.L. Yu, Domination structures and multicriteria problem in N-person games. Theor. Decis. **8**, 5–47 (1977)
8. J.M. Borwein, On the existence of Pareto efficient points. Math. Oper. Res. **8**(1), 64–73 (1983)
9. C. Charitha, J. Dutta, Regularized gap functions and error bounds for vector variational inequalities. Pac. J. Optim. **6**, 497–510 (2010)
10. G.Y. Chen, B.D. Craven, A vector variational inequality and optimization over an efficient set. ZOR-Math. Meth. Oper. Res. **34**, 1–12 (1990)
11. G.Y. Chen, C.J. Goh, X.Q. Yang, On gap functions for vector variational inequalities, in [19], pp. 55–72 (2000)
12. G.Y. Chen, C.J. Goh, X.Q. Yang, Existence of a solution for generalized vector variational inequalities. Optimization, **50**(1), 1–15 (2001)
13. G.Y. Chen, X.X. Huang, Stability results for Ekeland's ε variational principle for vector valued functions. Math. Meth. Oper. Res. **48**, 97–103 (1998)
14. G. Cohen, F. Chaplais, Nested monotony for variational inequalities over a product of spaces and convergence of iterative algorithms. J. Optim. Theory Appl. **59**, 360–390 (1988)
15. H.W. Corley, Existence and Lagrange duality for maximization of set-valued functions. J. Optim. Theory Appl. **54**, 489–501 (1987)
16. H.W. Corley, Optimality conditions for maximization of set-valued functions. J. Optim. Theory Appl. **58**, 1–10 (1988)
17. B.D. Craven, Nondifferentiable optimization by smooth approximation. Optimization, **17**, 3–17 (1986)
18. G. Debreu, A social equilibrium existence theorem. Proc. Natl. Acad. Sci. U. S. A. **38**, 886–893 (1952)
19. F. Ferro, A minimax theorem for vector-valued functions, J. Optim. Theory Appl. **60**(1), 19–31 (1989)
20. F. Ferro, Optimization and stability results through cone lower semicontinuity. Set-valued Anal. **5**, 365–375 (1997)
21. J. Fu, Simultaneous vector variational inequalities and vector implicit complementarity problem. J. Optim. Theory Appl. **93**, 141–151 (1997)
22. C.J. Goh, X.Q. Yang, On the solutions of vector variational inequalities, in *Proceedings of Optimization Techniques and Applications*, ed. by L. Caccetta et al. (Curtin University of Technology, Perth, 1998), pp. 1158–1164
23. X.H. Gong, Efficiency and Henig efficiency for vector equilibrium problems. J. Optim. Theory Appl. **108**, 139–154 (2001)
24. A. Göpfert, Chr. Tammer, C. Zălinescu, On the vectorial Ekeland's variational principle and minimal points in product spaces. Nonlinear Anal. **39**, 909–922 (2000)
25. R. Hartley, On cone-efficiency, cone-convexity and cone-compactness. SIAM J. Appl. Math. **34**(2), 211–222 (1978)
26. D.W. Hearn, The gap function of a convex program. Oper. Res. Lett. **1**, 67–71 (1982)
27. S. Helbig, On the connectedness of the set of weakly efficient points of a vector optimization problem in locally convex spaces. J. Optim. Theory Appl. **65**(2), 257–270 (1990)
28. M.I. Henig, Proper efficiency with respect to cones. J. Optim. Theory Appl. **36**, 387–407 (1982)
29. C. Hillermeier, *Nonlinear Multiobjective Optimization: A Generalized Homotopy Approach* (Birkhäuser Verlag, Basel, 2001)
30. N.-J. Huang, J. Li, J.C. Yao, Gap functions and existence of solutions for system of vector equilibrium problems. J. Optim. Theory Appl. **133**, 201–212 (2006)
31. T. Husain, E. Tarafdar, Simultaneous variational inequalities, minimization problems and related results. Math. Japon. **39**, 221–231 (1994)

32. V. Jeyakumar, A generalization of minimax theorem of Fan via a theorem of the alternative. J. Optim. Theory Appl. **48**, 525–533 (1986)
33. K. Kimura, X.H. Gong, J.C. Yao, Sensitivity analysis of strong vector equilibrium problems. J. Nonlinear Convex Anal. **9**(1), 83–94 (2008)
34. K. Kimura, Y.C. Liou, J.C. Yao, Parametric equilibrium problems with applications to optimization problems under equilibrium constraints. J. Nonlinear Convex Anal. **7**(2), 237–243 (2006)
35. S. Komlòsi, On the Stampacchia and Minty variational inequalities, in *Generalized Convexity and Optimization for Economic and Financial Decisions*, ed. by G. Giorgi, F. Rossi (Pitagora Editrice, Bologna, 1998), pp. 231–260
36. I.V. Konnov, Combined relaxation method for solving vector equilibrium problems. Russ. Math. (Izvestiya Vuzov Math.) **39**, 51–59 (1995)
37. I.V. Konnov, A general approach to finding stationary point and the solution of related problems. Comput. Math. Math. Phys. **36**, 585–593 (1996)
38. I.V. Konnov, A class of combined relaxation methods for decomposable variational inequalities. Optimization **51**(1), 109–125 (2002)
39. S. Kum, G.M. Lee, Remarks on implicit vector variational inequalities. Taiwan J. Math. **6**(3), 369–382 (2002)
40. G.M. Lee, D.S. Kim, B.S. Lee, Generalized vector variational inequality. Appl. Math. Lett. **9**(1), 39–42 (1996)
41. G.M. Lee, S. Kum, On implicit vector variational inequalities. J. Optim. Theory Appl. **104**(2), 409–425 (2000)
42. J. Li, G. Mastroeni, Vector variational inequalities involving set-valued mappings via scalarization with applications to error bounds for gap functions. J. Optim. Theory Appl. **145**, 355–372 (2010)
43. S.J. Li, G.Y. Chen, Properties of gap function for vector variational inequalities, in *Variational Analysis and Applications*, ed. by F. Giannessi, A. Maugeri (Springer, Berlin, 2005), pp. 605–631
44. S.J. Li, H. Yan, G.Y. Chen, Differential and sensitivity properties of gap functions for vector variational inequalities. Math. Meth. Oper. Res. **57**, 377–391 (2003)
45. Z.F. Li, G.Y. Chen, Lagrangian multipliers, saddle points, and duality in vector optimization of set-valued maps. J. Math. Anal. Appl. **215**, 297–316 (1997)
46. Z.F. Li, S.Y. Wang, A type of minimax inequality for vector-valued mappings, J. Math. Anal. Appl. **227**, 68–80 (1998)
47. M.B. Lignola, J. Morgan, Well-posedness for optimization problems with constrains defined by variational inequalities having a unique solution. J. Global Optim. **16**(1), 57–67 (2000)
48. D.T. Luc, Connectedness of the efficient point sets in quasiconcave vector maximization. J. Math. Anal. Appl. **122**, 346–354 (1987)
49. D.T. Luc, N.X. Tan, P.N. Tinh, Convex vector functions and their subdifferential. Acta Math. Vietnam. **23**, 107–127 (1998)
50. D.T. Luc and C.A. Vargas, A saddle point theorem for set-valued maps, Nonlinear Anal. Theory Methods Appl. **18**, 1–7 (1992)
51. H.-Q. Ma, N.-J. Huang, M. Wu, D. O'Regan, A new gap function for vector variational inequalities with an application. J. Appl. Math. **2013**, Article ID 423040, 8 pages (2013)
52. G. Mastroeni, Separation methods for vector variational inequalities. Saddle point and gap function, in *Nonlinear Optimization and Applications*, ed. by G. Di Pillo, F. Giannessi (Kluwer Academic, Dordrecht, 2000), pp. 207–217
53. G. Mastroeni, On Minty vector variational inequality, in [19] (2000), pp. 351–361
54. K.W. Meng, S.J. Li, Differential and sensitivity properties of gap functions for Minty vector variational inequalities. J. Math. Anal. Appl. **337**, 386–398 (2008)
55. J. Morgan, Constrained well-posed two-level optimization problem, in *Nonsmooth optimization and Related Topics*, ed. F.H. Clarke, V.F. Demyanov, F. Giannessi (Plenum Press, New York, 1989), pp. 307–325
56. U. Mosco, Dual variational inequalities. J. Math. Anal. Appl. **40**, 202–206 (1972)

57. E. Muselli, Upper and lower semicontinuity for set-valued mappings involving constraints. J. Optim. Theory Appl. **106**, 527–550 (2000)
58. J.F. Nash, Equilibrium point in n-person games. Proc. Natl. Acad. Sci. U. S. A. **36**, 48–49 (1950)
59. W. Oettli, A remark on vector-valued equilibria and generalized monotonicity. Acta Math. Vietnam. **22**(1), 213–221 (1997)
60. R. Reemtsen, J.J. Rückmann (eds.), *Semi-Infinite Progamming* (Kluwer Academic, Dordrecht, 1998)
61. W.D. Rong, Y.N. Wu, Epsilon-weak minimal solution of vector optimization problems with set-valued maps. J. Optim. Theory Appl. **103**(3), 569–579 (2000)
62. H.H. Schaefer, *Topological Vector Spaces* (Springer, Berlin, 1971)
63. W. Song, Vector equilibrium problems with set-valued mappings, in [19] (2000), pp. 403–432
64. F. Szidarovszky, M.E. Gershon, L. Duckstein, *Techniques for Multiobjective Decision Making in System Management* (Elsevier, Amsterdam, 1986)
65. T. Tanaka, Generalized quasiconvexities, cone saddle points, and minimax theorem for vector-valued functions. J. Optim. Theory Appl. **81**, 355–377 (1994)
66. T. Tanaka, Cone-quasiconvexity of vector-valued functions. Sci. Rep. Hirosaki Univ. **42**, 157–163 (1995)
67. T. Tanaka, Generalized semicontinuity and existence theorems for cone saddle points. Appl. Math. Optim. **36**, 313–322 (1997)
68. T. Tanaka, D. Kuroiwa, The convexity of A and B assures $\text{int}A + B = \text{int}(A + B)$. Appl. Math. Lett. **6**, 83–86 (1993)
69. X.Q. Yang, C.J. Goh, On vector variational inequalities: applications to vector equilibria. J. Optim. Theory Appl. **95**, 431–443 (1997)
70. G.X.-Z. Yuan, G. Isac, K.-K. Tan, J. Yu, The study of minimax inequalities, abstract economies and applications to variational inequalities and Nash equilibria. Acta Appl. Math. **54**, 135–166 (1998)
71. E. Zeidler, *Nonlinear Functional Analysis and Its Applications, I: Fixed Point Theorems* (Springer, New York, 1986)
72. M. Zeleny, Games with multiple payoffs. Int. J. Game Theory **4**, 179–191 (1976)
73. L.C. Zeng, J.C. Yao, An existence result for generalized vector equilibrium problems without pseudomonotonicity. Appl. Math. Lett. **19**, 1320–1326 (2006)

Index

Q.H. Ansari et al., *Vector Variational Inequalities and Vector Optimization*,
Vector Optimization, DOI 10.1007/978-3-319-63049-6